# International Conference on Mathematics: Education, Theory and Application 2016



Journal of Physics: Conference Series Volume 855

Surakarta, Indonesia
6 – 7 December 2016

ISBN: 978-1-5108-4245-8
ISSN: 1742-6588

**Printed from e-media with permission by:**

Curran Associates, Inc.
57 Morehouse Lane
Red Hook, NY 12571

**Some format issues inherent in the e-media version may also appear in this print version.**



Printed by Curran Associates, Inc. (2017)

For permission requests, please contact the Institute of Physics
at the address below.

Institute of Physics
Dirac House, Temple Back
Bristol BS1 6BE UK

Phone: 44 1 17 929 7481
Fax: 44 1 17 920 0979

techtracking@iop.org

**Additional copies of this publication are available from:**

Curran Associates, Inc.
57 Morehouse Lane
Red Hook, NY 12571 USA
Phone: 845-758-0400
Fax: 845-758-2633
Email: curran@proceedings.com
Web: www.proceedings.com

# Table of contents

## Volume 855

**International Conference on Mathematics: Education, Theory and Application**

**6–7 December 2016, Surakarta, Indonesia**

**Accepted papers received: 25 May 2017**
**Published online: 12 June 2017**

### Preface

011001
OPEN ACCESS
International Conference on Mathematics: Education, Theory and Application

011002
OPEN ACCESS
Peer review statement

### Papers

012001
OPEN ACCESS
The Outlier Detection for Ordinal Data Using Scalling Technique of Regression Coefficients

Arisman Adnan and Sigit Sugiarto.....1

012002
OPEN ACCESS
Mixed Estimator of Kernel and Fourier Series in Semiparametric Regression

Ngizatul Afifah, I Nyoman Budiantara and I Nyoman Latra.....8

012003
OPEN ACCESS
On Rainbow $k$-Connection Number of Special Graphs and It's Sharp Lower Bound

Ika Hesti Agustin, Dafik, A.W. Gembong and Ridho Alfarisi.....18

012004
OPEN ACCESS
On the total $H$-irregularity strength of graphs: A new notion

Ika Hesti Agustin, Dafik, Marsidi and Ermita Rizki Albirri.....26

012005
OPEN ACCESS
On the star partition dimension of comb product of cycle and complete graph

Ridho Alfarisi, Darmaji and Dafik.....35

012006
OPEN ACCESS
Structural Equation Modelling with Three Schemes Estimation of Score Factors on Partial Least Square (Case Study: The Quality Of Education Level SMA/MA in Sumenep Regency)

Anik Anekawati, Bambang Widjanarko Otok, Purhadi and Sutikno.....42

012007
OPEN ACCESS
Mixed geographically weighted regression (MGWR) model with weighted adaptive bi-square for case of dengue hemorrhagic fever (DHF) in Surakarta

H N Astuti, D R S Saputro and Y Susanti.....52

012008
OPEN ACCESS
Spread of Ebola disease with susceptible exposed infected isolated recovered ($SEII_hR$) model

Afina Azizah, Purnami Widyaningsih and Dewi Retno Sari Saputro.....62

012009
OPEN ACCESS
A Closed-Loop Supply Chain under Retail Price and Quality Dependent Demand with Remanufacturing and Refurbishing

A Y Christy, B N Fauzi, N A Kurdi, W A Jauhari and D R S Saputro.....68

012010
OPEN ACCESS
On super *H*–antimagicness of an edge comb product of graphs with subgraph as a terminal of its amalgamation

Dafik, Ika Hesti Agustin, A. I. Nurvitaningrum and R. M. Prihandini.....75

012011
OPEN ACCESS
Several classes of graphs and their *r*-dynamic chromatic numbers

Dafik, D.E.W. Meganingtyas, K. Dwidja Purnomo, M. Dicky Tarmidzi and Ika Hesti Agustin.....81

012012
OPEN ACCESS
On The Partition Dimension of a Lollipop Graph and a Generalized Jahangir Graph

Maylinda Purna Kartika Dewi and Tri Atmojo Kusmayadi.....91

012013
OPEN ACCESS
A Regular Production-Remanufacturing Inventory Model for a Two-Echelon System with Price-dependent Return Rate and Environmental Effects Investigation

A R Dwicahyani, W A Jauhari and Jonrinaldi.....96

012014
OPEN ACCESS
Bound of Distance Domination Number of Graph and Edge Comb Product Graph

A.W. Gembong, Slamin, Dafik and Ika Hesti Agustin.....107

012015
OPEN ACCESS
Rainfall model on area of rice production in Sragen, Karanganyar and Klaten by using Generalized Space Time Autoregressive (GSTAR)

S S Handajani, H Pratiwi, Y Susanti, S Subanti, Respatiwulan and Hartatik.....114

012016
OPEN ACCESS
Passive Copy-Move Forgery Detection Using Halftoning-based Block Truncation Coding Feature

Bambang Harjito and Heri Prasetyo.....125

012017
OPEN ACCESS
Comparison of B-Spline Model and Iterated Conditional Modes (ICM) For Data With Measurement Error (ME)

Hartatik and Agus Purnomo.....133

012018
OPEN ACCESS
The Construction of Labeling and Total Irregularity Strength of Specified Caterpillar Graph

Diari Indriati, Widodo, Isnaini Rosyida and Kiki A Sugeng.....144

012019
OPEN ACCESS
Experimentation of cooperative learning model Numbered Heads Together (NHT) type by concept maps and Teams Games Tournament (TGT) by concept maps in terms of students logical mathematics intellegences

Adi Irawan, Mardiyana and Dewi Retno Sari Saputro.....151

012020
OPEN ACCESS
Verification of Radiation Isocenter on Linac Beam 6 MV using Computed Radiography

Muhammad Irsal, Eko Hidayanto and Heri Sutanto.....159

012021
OPEN ACCESS
Students' mathematical representations on secondary school in solving trigonometric problems

Istadi, T A Kusmayadi and I Sujadi.....168

012022
OPEN ACCESS
A periodic review integrated inventory model with controllable safety stock and setup cost under service level constraint and distribution-free demand

N A Kurdhi, A Jamaluddin, W A Jauhari and D R S Saputro.....177

012023
OPEN ACCESS
Identification and estimation of state variables on reduced model using balanced truncation method

Trifena Punana Lesnussa, Didik Khusnul Arif, Dieky Adzkiya and Erna Apriliani.....188

012024
OPEN ACCESS
Discrete time Markov chains (DTMC) susceptible infected susceptible (SIS) epidemic model with two pathogens in two patches

Eka Lismawati, Respatiwulan and Purnami Widyaningsih.....197

012025
OPEN ACCESS
Population Analysis of Disabled Children by Departments in France

Diah Meidatuzzahra, Heri Kuswanto, Nicolas Pech and Amélie Etchegaray.....203

012026
OPEN ACCESS
On The Value at Risk Using Bayesian Mixture Laplace Autoregressive Approach for Modelling the Islamic Stock Risk Investment

Brina Miftahurrohmah, Nur Iriawan and Kartika Fithriasari.....210

012027
OPEN ACCESS
Comparison of Grand Median and Cumulative Sum Control Charts on Shuttlecock Weight Variable in CV Marjoko Kompas dan Domas

N Musdalifah, S S Handajani and E Zukhronah.....219

012028
OPEN ACCESS
The $\mathfrak{C}$-integral of sequence-valued functions on the sequence space $\ell^\infty$

Muslich, Sutrima and Supriyadi Wibowo.....226

012029
OPEN ACCESS
Model reduction of unstable systems using balanced truncation method and its application to shallow water equations

Kiki Mustaqim, Didik Khusnul Arif, Erna Apriliani and Dieky Adzkiya.....234

012030
OPEN ACCESS
Conjecturing and Generalization Process on The Structural Development

Khomsatun Ni'mah, Purwanto, Edy Bambang Irawan and Erry Hidayanto.....241

012031
OPEN ACCESS
Software Piracy Detection Model Using Ant Colony Optimization Algorithm

Nor Astiqah Omar, Zeti Zuryani Mohd Zakuan and Rizauddin Saian.....253

012032
OPEN ACCESS
Error-correcting pairs for a public-key cryptosystem

Ruud Pellikaan and Irene Márquez-Corbella.....261

012033
OPEN ACCESS
Self-exciting point process in modeling earthquake occurrences

H. Pratiwi, I. Slamet, D. R. S. Saputro and Respatiwulan.....271

012034
OPEN ACCESS
The Completion of Non-Steady-State Queue Model on The Queue System in Dr. Yap Eye Hospital Yogyakarta

Arum Helmi Manggala Putri, Retno Subekti and Nikenasih Binatari.....278

012035
OPEN ACCESS
On (local) metric dimension of graphs with *m*-pendant points

Rinurwati, Slamin and H Suprajitno.....293

012036
OPEN ACCESS
Survival Bayesian Estimation of Exponential-Gamma Under Linex Loss Function

S. W Rizki, M. N Mara and E Sulistianingsih.....301

012037
OPEN ACCESS
Super (*a, d*)-Cycle-Antimagic Total Labeling on Triangular Ladder Graph and Generalized Jahangir Graph

Mania Roswitha, Anna Amanda, Titin Sri Martini and Bowo Winarno.....309

012038
OPEN ACCESS
The Development of e-tutorial on Implementation National Curriculum 2013 for Mathematics Teacher

Yenita Roza, Gita Satria and Syarifah Nur Siregar.....316

012039
OPEN ACCESS
Censored Hurdle Negative Binomial Regression (Case Study: Neonatorum Tetanus Case in Indonesia)

Riza Yuli Rusdiana, Ismaini Zain and Santi Wulan Purnami.....325

012040
OPEN ACCESS
Predicting Flood in Perlis Using Ant Colony Optimization

Syaidatul Nadia Sabri and Rizauddin Saian.....334

012041
OPEN ACCESS
Conceptual or procedural mathematics for engineering students at University of Samudra

Saiman, Puji Wahyuningsih and Hamdani.....342

012042
OPEN ACCESS
*H*-Supermagic Labeling on Coronation of Some Classes of Graphs with a Path

H Sandariria, M Roswitha and T A Kusmayadi.....352

012043
OPEN ACCESS
Error Analysis Of Students Working About Word Problem Of Linear Program With NEA Procedure

D.A. Santoso, A Farid and B. Ulum.....358

012044
OPEN ACCESS
Rudder-roll stabilization using fgs-pid controller for sigma-e warship

M Y Santoso, I Munadhif, A Wahidin, Ruddianto, Fathulloh and R T Soelistijono.....366

012045
OPEN ACCESS
The Preventive Control of a Dengue Disease Using Pontryagin Minimum Principal

Eminugroho Ratna Sari, Nur Insani and Dwi Lestari.....376

012046
OPEN ACCESS
Inflation data clustering of some cities in Indonesia

Adi Setiawan, Bambang Susanto and Tundjung Mahatma.....386

012047
OPEN ACCESS
Modelling tourists arrival using time varying parameter

P Suciptawati, K G Sukarsa and Eka N Kencana.....394

012048
OPEN ACCESS
Markov Switching Autoregressive Conditional Heteroscedasticity (SWARCH) Model to Detect Financial Crisis in Indonesia Based on Import and Export Indicators

Sugiyanto, Etik Zukhronah, Yuliana Susanti and Sisca Rahma Dwi.....401

012049
OPEN ACCESS
Application of Semiparametric Spline Regression Model in Analyzing Factors that In uence Population Density in Central Java

Y D Sumantari, I Slamet and Sugiyanto.....408

012050
OPEN ACCESS
Self-Intersection Local Times of Generalized Mixed Fractional Brownian Motion as White Noise Distributions

Herry P. Suryawan and Boby Gunarso.....416

012051
OPEN ACCESS
Study on the factors affecting the quality of public bus transportation service in Bali Province using factor analysis

M Susilawati and D P E Nilakusmawati.....426

012052
OPEN ACCESS
The fundamental properties of quasi-semigroups

Sutrima, Ch. Rini Indrati, Lina Aryati and Mardiyana.....433

012053
OPEN ACCESS
Classication of Status of the Region on Java Island using C4.5, CHAID, and CART Methods

R A Syaraswati, I Slamet and B Winarno.....442

012054
OPEN ACCESS
Online Statistical Modeling (Regression Analysis) for Independent Responses

I Made Tirta, Dian Anggraeni and Martinus Pandutama.....450

012055
OPEN ACCESS
*H*-Supermagic Labeling on Shrubs Graph and $L_m$ o $P_n$

Risala Ulfatimah, Mania Roswitha and Tri Atmojo Kusmayadi.....463

012056
OPEN ACCESS
Internet messenger based smart virtual class learning using ubiquitous computing

K Umam, S N S Mardi and M Hariadi.....470

012057
OPEN ACCESS
Satisfiability modulo theory and binary puzzle

Putranto Utomo.....480

012058
OPEN ACCESS
On The Partition Dimension of $C_m + P_n$ Graph

Hidra Vertana and Tri Atmojo Kusmayadi.....488

012059
OPEN ACCESS
Domination Number of Vertex Amalgamation of Graphs

Y Wahyuni, M I Utoyo and Slamin.....496

012060
OPEN ACCESS
Fisher Scoring Method for Parameter Estimation of Geographically Weighted Ordinal Logistic Regression (GWOLR) Model

Purnami Widyaningsih, Dewi Retno Sari Saputro and Aulia Nugrahani Putri.....501

012061
OPEN ACCESS
Bayesian Hierarchical Random Intercept Model Based on Three Parameter Gamma Distribution

Ika Wirawati, Nur Iriawan and and Irhamah.....509

012062
OPEN ACCESS
Median Filter Noise Reduction of Image and Backpropagation Neural Network Model for Cervical Cancer Classification

D. U. Wutsqa and M. Marwah.....518

012063
OPEN ACCESS
Calculation of Lung Cancer Volume of Target Based on Thorax Computed Tomography Images using Active Contour Segmentation Method for Treatment Planning System

Fiet Patra Yosandha, Kusworo Adi and Catur Edi Widodo.....528

012064
OPEN ACCESS
Parameters Estimation of Geographically Weighted Ordinal Logistic Regression (GWOLR) Model

Shaifudin Zuhdi, Dewi Retno Sari Saputro and Purnami Widyaningsih.....537

IOP Conf. Series: Journal of Physics: Conf. Series **855** (2017) 011001 doi:10.1088/1742-6596/855/1/011001

# Preface

The 1st International Conference on Mathematics: Education, Theory, and Application (ICMETA) was held on December 6-7, 2016 at Universitas Sebelas Maret (UNS), Solo, Indonesia. The ICMETA is a conference that was first accomplished by Department of Mathematics, Universitas Sebelas Maret and planned to be held biennially.

The main objective of the conference is to gather world-class researchers, engineers, and educators engaged in the fields of mathematics and its applications to meet and present their latest activities. It provides a platform to disseminate research findings, and hopefully, it also sparks innovative ideas, foster research relations or partnership between the various institutions. As a scientific meeting event we invited experts from six different countries including Australia, the Netherlands, Indonesia, Japan, Malaysia, and France as keynote speakers.

We are grateful to all speakers for their presentations and all delegates who contributed for the success of this conference.

IOP Conf. Series: Journal of Physics: Conf. Series **855** (2017) 011002 doi:10.1088/1742-6596/855/1/011002

# Peer review statement

All papers published in this volume of *Journal of Physics: Conference Series* have been peer reviewed through processes administered by the proceedings Editors. Reviews were conducted by expert referees to the professional and scientific standards expected of a proceedings journal published by IOP Publishing.

# The Outlier Detection for Ordinal Data Using Scalling Technique of Regression Coefficients

**Arisman Adnan and Sigit Sugiarto**
Department of Mathematics, Faculty of Mathematics and Natural Sciences
Universitas Riau, Indonesia

E-mail: arisman.adnan@lecturer.unri.ac.id, sigit.sugiarto@lecturer.unri.ac.id

**ABSTRACT.** The aims of this study is to detect the outliers by using coefficients of Ordinal Logistic Regression (OLR) for the case of $k$ category responses where the score from 1 (the best) to 8 (the worst). We detect them by using the sum of moduli of the ordinal regression coefficients calculated by jackknife technique. This technique is improved by *scalling* the regression coefficients to their means. *R* language has been used on a set of ordinal data from reference distribution. Furthermore, we compare this approach by using studentised residual plots of jackknife technique for ANOVA (Analysis of Variance) and OLR. This study shows that the jackknifing technique along with the proper scaling may lead us to reveal outliers in ordinal regression reasonably well.

## 1. Introduction

On the category scale, the panellists are asked to give a score, for an example 1, 2, …, 8 on his opinion about the taste of a certain food. This scale is a discrete number and arranged as an ordinal scale which usually ranges from 1 (the best) to 8 (the worst), or from 'extremely tender' to 'extremely tough' for tenderness of meat, for example. Joo et al. [8] reviewed the definition of fresh meat quality, meat quality traits and variations of meat quality. They also presented the recent knowledge underlying the relationship between fresh meat quality traits and muscle fiber characteristics. In the research, they proposed several potential factors including breed, genotype, sex, hormone, growth performance, diet, muscle location, exercise and ambient temperature that can be used to manipulate muscle fiber characteristics and subsequently meat quality in animals.

This problem is a very popular topic on sensory evaluation. A comprehensive review of sensory and consumer science has been conducted by Naes et al. [10]. They also used a 8-category scale for scoring quality attributes. Some researchers omit a central, neutral category to produce a more even distribution of responses. A line or an unstructured scale is a judge response on a solid line from 'little' to 'much' or from 'none' to extremely'.

One of the common used response models for ordinal data analysis is the proportional odds model. This model is an extension of logistic model for binary data where their responses are more than two categories [3]. He discussed loglinear models for ordinal variables which have some advantages compared to the model treating all variable as nominal. Whenever the relationship between response and explanatory variables needs to be distinguished then we use the 'logit'model. The model is based on the success probability on response variables, hence, the procedure has to ensure that fitted probabilities are in the range (0,1). Logit model can be found using logit transformation based on logistic function.

In term of the power test, Avery and Adnan [4] have simulated several cases and investigated the predictor effects on response variable. When comparing this approach with the ANOVA procedure we can say that OLR was more powerful then ANOVA procedure except for very smal sample sizes ($n$=6) and very large sample sizes ($n$=600).

Furthermore, Adnan and Avery [1] have discussed various problems with analysing ordinal data using ANOVA and OLR and proposed several possible solutions such that these approaches could be used. We suggest the use of sample averages to avoid the violation of most of the ANOVA assumptions. By applying the proportional odds model, it was found that the estimated effects from these two methods were quite similar [9].

The most commonly used method to assess the fit of a model is by constructing a goodness-of-fit test. This test determines the adequacy of the fitted model in describing the relationship between response and predictor variables. We can say that a large value of the statistic means that tehre is a big discrepancy between observations and fitted value. Consequently, it gives an indication that the model has not fitted very well. In logistic regression, goodness-of-fit test using the residual deviance is very common but care needs to be taken since the number of parameters is increasing at the same rate as the sample size. If the deviance follows a $\chi^2$-distribution then we expect that the ratio of the mean and the degrees of freedom will be approximately one. Unfortunately, the simulation shows that the deviance does not follow a $\chi^2$-distribution [1]. Hence, jackknifing technique for ordinal logistic regression is an alternative goodness-of-fit test. By applying this method for detecting some outliers along with the proper scaling may lead us to reveal outliers in ordinal regression reasonably well.

## 2. Methods

The term `outlier' refers to values of the response variables which are unusually large or small [3, 7]. Since all of the explanatory variables in our case are factors, any abnormal values are likely to be outliers. The method that we propose here is a resampling method, the so called jackknife technique, i.e. repeated analysis leaving out one observation at a time [6]. This technique is usually used for estimating the bias and standard error of an estimate.

One of the most popular models for logistic regression is a proportional odds model [9] which can be stated in response $Y$ with $j$ category ($j$=1,2,...$k$) and predictor $X$ as

$$\Pr[Y \geq j \mid X] = \frac{1}{1+\exp[-(\theta_j + X\beta)]}; \quad j = 1,2,\ldots,k. \tag{1}$$

The equation can also be written as a linear logistic regression:

$$\operatorname{logit}(\gamma_j) = \theta_j + X\beta; \quad j = 1,2,..k \tag{2}$$

Furthermore, we use this model to identify if any outliers on ordinal data set by using goodness-of-fit test. Now consider the deviance:

$$\begin{aligned} D(y;\hat{\gamma}) &= 2l(\tilde{\gamma}; y) - 2l(\hat{\gamma}, y) \\ &= 2\sum y_{ij} \log \tilde{\gamma}_{ij} - 2\sum y_{ij} \log \hat{\gamma}_{ij} \end{aligned} \tag{3}$$

where $l(\tilde{\gamma};\mathbf{y})$ is a log-*likelihood* function under saturated model while $l(\hat{\gamma},\mathbf{y})$ is a log-*likelihood* di under fitted model with estimated parameters $\hat{\gamma}$. Under certain conditions the deviance $D$ is approximately distributed as $\chi^2$ distribution with a particular degreee of freedom. The deviance is also useful for testing the parameters of two models by examining the difference in the two deviances. The overall discrepancy between observed and predicted values can be calculated using a generalised likelihood ratio statistic. This idea comes from the fact that a generalised linear model is defined and fitted using ideas based on likelihood.

Now consider $n$ binomial observations of the forms of proportion $y_i/n_i$,. $i$=1,2,…$n$. Let the fitted value of $y_i$ is $\hat{y}_i = n_i\hat{p}_i$. The counterpart of standardised residual for logistic model can be stated as [5]:

$$r_i = \frac{y_i - n_i\hat{p}_i}{\sqrt{v_i(1-h_i)}} \qquad (4)$$

where $\hat{v}_i = n_i\hat{p}_i(1-\hat{p}_i)$ and $h_i$ is the $i$th diagonal element of $nxn$ matrix $\mathbf{H}^* = \mathbf{W}^{1/2}\mathbf{X}(\mathbf{X}'\mathbf{W}\mathbf{X})\mathbf{X}'\mathbf{W}^{1/2}$ where **W** is $nxn$ diagonal matrix of weight used in fitting the model. In binomial observations we have two categories, hence we need to consider two residuals $y_i - n_i\hat{p}_i$ and $(n_i - y_i) - n_i(1-\hat{p}_i)$. For the ordinal data with ($k$+1) categories we have $k$ residuals for each multinomial observations. Since these residuals are correlated it is not clear how to analyze them using standardized residuals mentioned above.

Unusual observations in taste panel data can occur in various ways, e.g. misreading of scores, erratic and inconsistent scoring by a panelist. However, since we combine the scores over the panellists for a particular sample, then the important effect is when a sample's score are all high or all low relative to the average for that treatment combination.

Since the treatment effects in ANOVA and OLR are compared to the first level as a reference coefficient, i.e. the first coefficient is zero, then it would be interesting to check if changing the reference coefficient has any effect on the analysis. The sum of the modulus of the three coefficients is plotted again the number of the observation removed.

Since the attribute consists of four levels, we can find four sum of moduli when applying the jackknife technique, depending on which is the reference level, but two of these will be the same as each other. This can be illustrated by imagining the four treatment effects, e.g. their coefficients, which are denoted by $b_1$, $b_2$, $b_3$, and $b_4$. Without loss of generality, we can assume $b_1 < b_2 < b_3 < b_4$. We can now scale numbers to put different classes equal to zero giving four possible scaling as listed in Table 1.

Table 1: Table of four possible scaling

| Number | 1 | 2 | 3 | 4 |
|---|---|---|---|---|
| (i) | 0 | $b_1$-$b_2$ | $b_1$-$b_3$ | $b_1$-$b_4$ |
| (ii) | $b_2$-$b_1$ | 0 | $b_2$-$b_3$ | $b_2$-$b_4$ |
| (iii) | $b_3$-$b_1$ | $b_3$-$b_2$ | 0 | $b_3$-$b_4$ |
| (iv) | $b_4$-$b_1$ | $b_4$-$b_2$ | $b_4$-$b_3$ | 0 |

The sum of the moduli for each column are as follows:

| (1) | (2) | (3) | (4) |
|---|---|---|---|
| $3b_1$-$b_2$-$b_3$-$b_4$, | $b_1$+$b_2$-$b_3$-$b_4$, | $b_1$+$b_2$-$b_3$-$b_4$, | $b_1$+$b_2$+$b_3$-$3b_4$ |

We can see that the second scaling (2) and the third scaling (3) have the same sum of moduli, i.e. when we scale to either of the middle numbers. In general, if the numbers of factor levels is even then the number of different plot is the number of level minus 1.

We now propose an alternative procedure by scaling the coefficients to an average. For example let the estimated coefficients $S_{ki}$, $k$=1,2,3,4; $i$=1,2,3,4 by choosing the first level to be zero, be: 0, $b_2$, $b_3$, and $b_4$. The possible scaling using average, putting each leavel to zero, is listed in Table 2.

Table 2: Table of possible scaling using average

| Number | 1 | 2 | 3 | 4 |
|---|---|---|---|---|
| (i) | 0 | $b_2$ | $b_3$ | $b_4$ |
| (ii) | $-b_2$ | 0 | $b_3-b_2$ | $b_4-b_2$ |
| (iii) | $-b_3$ | $b_2-b_3$ | 0 | $b_4-b_3$ |
| (iv) | $-b_4$ | $b_2-b_4$ | $b_3-b_4$ | 0 |

This is because the estimated effect of differences between levels is always the same. If we calculate the average in each case, we obtain $\bar{S}_k$ for each possibilities:

| (1) | (2) | (3) | (4) |
|---|---|---|---|
| $(b_2+b_3+b_4)/4$, | $(b_3+b_4-3b_2)/4$, | $(b_2+b_4-3b_3)/4$, | $(b_2+b_3-3b_4)/4$ |

Then each coefficient is scaled by $S_{ki}^* = S_{ki} - \bar{S}_k$. The sum of moduli of the $S_{ki}^*$ will be the same for each case. This can be easily verified for each case as follows:

- For $k$=1, choose the first level to be zero

$$S_{1i}^* = \sum_{i=1}^{4} \left| S_{1i} - \bar{S}_1 \right|$$
$$= 1/4\left[ \left| -b_2 - b_3 - b_4 \right| + \left| 3b_2 - b_3 - b_4 \right| + \left| 3b_3 - b_2 - b_4 \right| + \left| 3b_4 - b_2 - b_3 \right| \right]$$

- For $k$=2 (choose the second level to be zero), $k$=3, and $k$=4, we have the same scaled values as for $k$=1.

Since all of the cases give the same scaled values then it is sufficient to calculate the scaled sum of moduli from the first coefficient only.

## 3. Main Results

Let us know consider the meat data set that consist of 948 observations [1]. This dataset has four predictors with different numbers of levels: factor1 (fac1) and favtor3 (fac3) consist of 3 levels while factor2 (fac2) and factor4 (fac4) have 2 levels. The response scores are classified into 8 categories start from 1 (very tenderness) to 8 (very untenderness). We use the reference distribution for this stage by scoring the tenderness of meat. The estimated distributions are listed on Table 3.

Tabel 3: Estimated distribution for two levels of predictor

| Level of | Response | | | | | | | |
|---|---|---|---|---|---|---|---|---|
| predictors | 1 | 2 | 3 | 4 | 5 | 6 | 7 | 8 |
| 1 | 0,047 | 0,160 | 0,259 | 0,202 | 0,178 | 0,112 | 0,036 | 0,006 |
| 2 | 0,024 | 0,121 | 0,241 | 0,209 | 0,201 | 0,142 | 0,044 | 0,018 |

Scores are generated using uniform random numbers conditioned on the estimated probility of a reference distribution. We use the inverse probability integral transform method for generating observations for these discrete cases. We then define $X = x_k$ if and only if

$$F_X(x_{k-1}) < u \le F(x_k), \quad k = 0,1,\dots, K-1$$

Since we have $x_k = 0,1,\dots,7$ i.e 8 categorical responses with probabilities $p_0, p_1, \dots, p_7$ then

$$0 < u \le P_0, \Rightarrow X = 0, \quad P_0 \le u < P_1, \Rightarrow X = 1, \dots, P_6 \le u \le 1, \Rightarrow X = 7$$

where $P_0 = p_0$, $P_1 = p_0 + p_1$, ..., $P_6 = p_0 + p_1 + \dots + p_6$. Based on the probability distribution in Table 3 data were simulated for various numbers of observations per factor. Basically, these observations, i.e. the ordinal responses, were attained using uniform random number generation conditioned on estimated probabilities.

Furthermore, we analysed the outliers, if exist, on modelling fac1 and fac2 only. Following the procedure as the previous discussion, the sum of moduli of the difference between each coefficient $\overline{S}_{ki}$ and their average $\overline{S}_k$ will be the same for each case. Studentised residual plots of jackknife technique for ANOVA and OLR for (fac1) are shown in Figure 1.

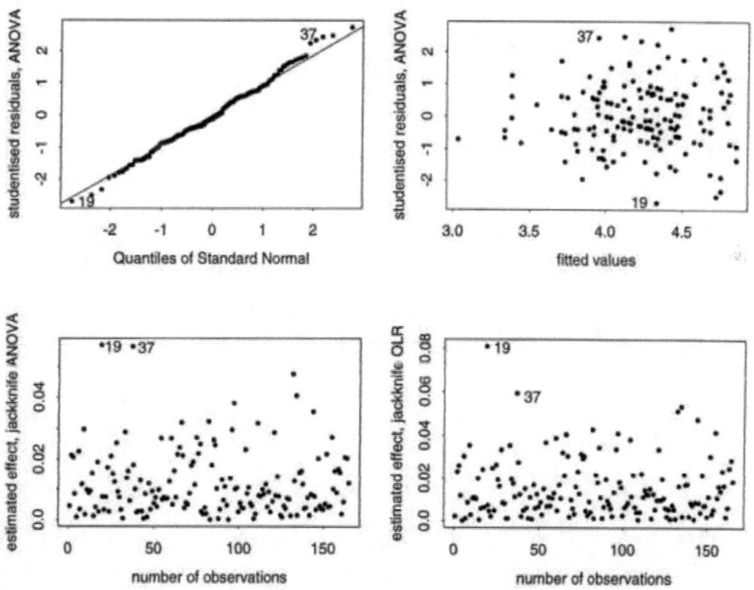


Figure 1: Outliers using plots fitted factor1 effect

This figure shows that observations number 19 and 37 can be considered as probable outliers but they are less clear if you look at the studentised residual plots only. However, the plots of the jackknife technique between ANOVA and OLR reveal similar pattern and indicate more clearly about possibility of these two as probable outliers. Factor2 which consists of two levels indicates one observation only as a probable outlier as shown at the top left in Figure 2, i.e. observation number 19. This observation can be seen clearly for both ANOVA and OLR using the jackknife technique.

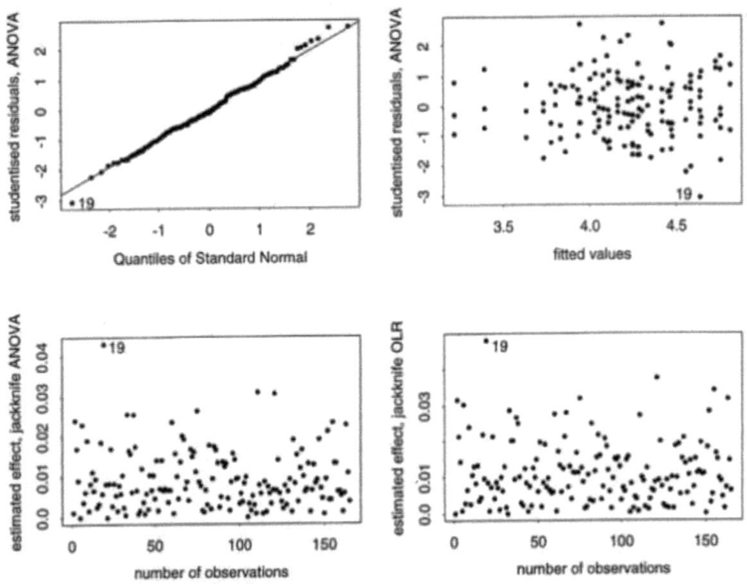


Figure 2: Plots having fitted factor2 effect

It would be interesting if we combine the two factors (i.e. 3 levels of fac1 and 2 levels of fac2 into a single factor consist of 6 levels so called factor12 (fac12). To recode this factor, we simply follow this strategy: fac12=(fac1-1)*level of fac2+factor2.

By recording the new factor based on combination of these two factors then the plot of each coefficient along with the plot of the average for ANOVA are shown in Figure 3.

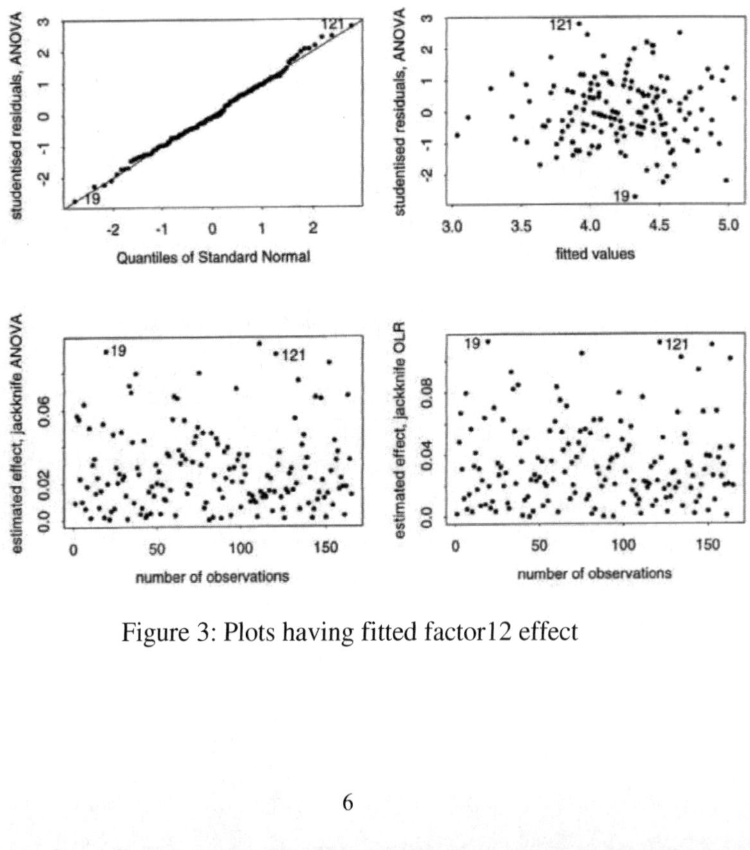


Figure 3: Plots having fitted factor12 effect

Observation number 19 lies at the straight line of the normal probability plot but it has a relatively large studentized residual. Now we consider this observation along with observation number 121 as possible outliers, and we can see from the bottom of Figure 3 that although they have large values so do other points and it is debatable if they are outliers. The jackknife technique support us to show if any outliers in our dataset.

## 4. Conclusions

The residual deviance probably does not have a $\chi^2$-distribution . The main reason for using jackknife technique is because residuals are difficult to define in OLR and we would expect that removing abnormal points to have the largest effect on estimated tretament coefficients. For ANOVA and OLR along with the proper scaling for regression coefficients may lead us to reveal outliers in ordinal regression reasonably well. The choice of the scalling play an important result to find the best method to use.

## References

[1] Adnan, A. and Avery, P.J. (2001). Goodness-of-Fit Tests and Outlier Detection in Taste-panel Experiments. *Proceeding the 6th ISSM*, Manchester, pp.395-398.
[2] Agresti, A. (2010). *Analysis of Ordinal Categorical Data*. 2nd ed., John Wiley & Sons, New York.
[3] Aggarwal, C. C. (2013). *Outlier Analysis*. Springer, New York.
[4] Avery, P.J. and Adnan, A. (2000). Analysis of taste-panel data using ANOVA and OLR. *Proceeding the 5th ISSM*, Paris, pp.381-385.
[5] Collett, D. (2003). *Modelling Binary Data*. Chapman & Hall, London.
[6] Efron, B. &Tibshirani, R.J. (1993), An Introduction to Bootstrap,Chapman &Hall,New York.
[7] Hadi, A.S., Imon, A.H.M.R., and Werner, M. (2009). Detection of outliers. Wiley Interdisciplanary Reviews: Computational Statistics, 1 (1): 57-70.
[8] Joo S.T., Kimb, G.D., Hwanga,Y.H., RyubY.C., (2013). Control of fresh meat quality through manipulation of muscle fiber characteristics. *Meat Science*, 95(4): 828–836.
[9] McCullagh, P. (1980). Regression models for ordinal data. *Journal of Royal Statistical Society B*, 42(2): 109-142.
[10] Naes, T., Brockhoff, P.B., Tomic, D. (2010). Statistics for sensory and consumer science. John Wiley & Sons, UK.

IOP Conf. Series: Journal of Physics: Conf. Series **855** (2017) 012002 doi:10.1088/1742-6596/855/1/012002

# Mixed Estimator of Kernel and Fourier Series in Semiparametric Regression

**[1]Ngizatul Afifah, [2]I Nyoman Budiantara, [3]I Nyoman Latra**

[1]Master's Program Student of Statistics Departement, Institut Teknologi Sepuluh Nopember, Surabaya, Indonesia. [2,3]Lecturer of Statistics Departement, Institut Teknologi Sepuluh Nopember, Surabaya, Indonesia.

[1]ngizatul.afifah@gmail.com, [2] i_nyoman_b@statistika.its.ac.id, [3]i_nyoman_l@statistika.its.ac.id

**Abstract**. Given paired observation $(x_i, v_{1i}, v_{2i}, ..., v_{pi}, t_{1i}, t_{2i}, ..., t_{qi}, y_i), i = 1, 2, ..., n,$ follow the additive semiparametric regression model $y_i = \mu(x_i, \mathbf{v}_i, \mathbf{t}_i) + \varepsilon_i$, where

$$\mu(x_i, \mathbf{v}_i, \mathbf{t}_i) = f(x_i) + \sum_{j=1}^{p} g_j(v_{ji}) + \sum_{s=1}^{q} h_s(t_{si})$$

$\mathbf{v}_i = (v_{1i}, v_{2i}, ..., v_{pi})'$, and $\mathbf{t}_i = (t_{1i}, t_{2i}, ..., t_{qi})'$. Random errors $\varepsilon_i$ is a normal distribution with mean 0 and variance $\sigma^2$. To obtain a mixed estimator $\mu(x_i, \mathbf{v}_i, \mathbf{t}_i)$, the regression curve $f(x_i)$ is approached by linier parametric, $g_j(v_{ji})$ is kernel with bandwidths $\mathbf{\Phi} = (\phi_1, \phi_2, ..., \phi_p)'$ and the regression curve component fourier series $h_s(t_{si})$ is approached by $H_s(t_{si}) = b_s t_{si} + \frac{1}{2}a_{0s} + \sum_{k=1}^{N} a_{ks} \cos kt_{si}$ with oscillation paremeter $N$. The estimator $\sum_{j=1}^{p} g_j(v_{ji})$ is $\sum_{j=1}^{p} \hat{\mathbf{g}}_{j\phi_j}(v_{ji})$ where $\sum_{j=1}^{p} \hat{\mathbf{g}}_{j\phi_j}(v_{ji}) = \mathbf{V}(\mathbf{\Phi})\mathbf{y}$. Penalized Least Squares (PLS) method give

$$\underset{\mathbf{c},\boldsymbol{\beta}}{Min}\left\{L(\mathbf{c}) + L(\boldsymbol{\beta}) + \sum_{s=1}^{q} \theta_s S(H_s(t_{si}))\right\}$$

with smoothing parameter $\boldsymbol{\theta} = (\theta_1, \theta_2, ..., \theta_q)'$, the estimator $\mathbf{f}(x)$ is $\hat{\mathbf{f}}(x)$ and $\sum_{s=1}^{q} h_s(t_{si})$ is $\sum_{s=1}^{q} \hat{\mathbf{h}}_{\theta_s}(t_{si})$, where $\hat{\mathbf{f}}(x) = \mathbf{X}\hat{\boldsymbol{\beta}}$ and $\sum_{s=1}^{q} \hat{\mathbf{h}}_{\theta_s}(t_{si}) = \mathbf{X}\hat{\mathbf{c}}(\theta)$. So that,

$$\hat{\boldsymbol{\mu}}_{\mathbf{\Phi},\theta,N}(\mathbf{v}_i, \mathbf{t}_i) = \mathbf{Z}(\mathbf{\Phi}, \theta, N)\mathbf{y}$$

is the mixed estimator of $\boldsymbol{\mu}(\mathbf{v}_i, \mathbf{t}_i)$ where $\mathbf{Z}(\mathbf{\Phi}, \theta, N) = \mathbf{C}(\mathbf{\Phi}, \theta, N) + \mathbf{V}(\mathbf{\Phi}) + \mathbf{E}(\mathbf{\Phi}, \theta, N)$ Matrix $\mathbf{C}(\mathbf{\Phi}, \theta, N), \mathbf{V}(\mathbf{\Phi})$ and $\mathbf{E}(\mathbf{\Phi}, \theta, N)$ are depended on $\mathbf{\Phi}, \theta$ and $N$. Optimal $\mathbf{\Phi}, \theta$ and $N$ can be obtained by the smallest Generalized Cross Validation (GCV).

IOP Conf. Series: Journal of Physics: Conf. Series **855** (2017) 012002 doi:10.1088/1742-6596/855/1/012002

## 1. Introduction

One of the statistical methods that is often used by researchers in various research fields is regression analysis. It is used to investigate the pattern of the functional relationship between two or more variables. There are three approaches to estimate the regression curve, namely the approach of parametric regression, nonparametric regression and semiparametric regression [1]. If the regression curve shape is known, it is called parametric regression. If there is no shape information, it is called nonparametric regression [2]. Combination of parametric and nonparametric components is called semiparametric regression [3].

The semiparametric model with spline function has been applied by [4], [5] [6] and [2], kernel function is developed by [7], and [8], fourier series is developed by [9], and [10], local polynomial is developed by [11] and [12]. Kernel estimator is a linier estimator in nonparametric regression estimator that is depended on bandwidth. Kernel estimator is able to model data that have no particular pattern [1]. Fourier series estimators are used when there is a trend repeated data patterns [13]. Fourier series is used to describe the curve that show sine or cosine wave [9].

For nonparametric component, it has two heavy and basic assumptions. The first is each predictor pattern in the multivariable nonparametric regression model of the predictor is considered to have the same pattern. The second is researchers insist on using only one shape of model estimator for every predictor variable. In some cases, the two assumptions causing regression model estimation produced are not correct and tend to produce larger errors [1]. To overcome these problems, some researchers have developed the mixed estimator in which each data pattern is approached with appropriate estimator. They are [1] who developed the combination of spline and kernel estimator for nonparametric regression and [14] who developed the combined estimator fourier series and spline truncated in multivariable nonparametric regression. Meanwhile, there has been no research involving mixed estimator in semiparametric regression.

This study will review mixed estimator of kernel and fourier series in semiparametric regression. The mixed estimator assumed a predictor variable follow linier parametric, other variables follow kernel function and the others follow fourier series. It obtained through Penalized Least Squares (PLS) method.

## 2. Main Results

Given paired observation $(x_i, v_{1i}, v_{2i}, ..., v_{pi}, t_{1i}, t_{2i}, ..., t_{qi}, y_i)$, follow thc additive semiparametric regression model

$$y_i = \mu(x_i, \mathbf{v}_i, \mathbf{t}_i) + \varepsilon_i, \tag{1}$$

where $i = 1, 2, ..., n,$ $\mathbf{v}_i = (v_{1i}, v_{2i}, ..., v_{pi})'$, and $\mathbf{t}_i = (t_{1i}, t_{2i}, ..., t_{qi})'$. Random errors $\varepsilon_i$ has normal distribution with $E(\varepsilon_i) = 0$ dan $Var(\varepsilon_i) = \sigma^2$. The regression curve $\mu(x_i, \mathbf{v}_i, \mathbf{t}_i)$ is assumed to be additive, it can be written as:

$$\mu(x_i, \mathbf{v}_i, \mathbf{t}_i) = f(x_i) + \sum_{j=1}^{p} g_j(v_{ji}) + \sum_{s=1}^{q} h_s(t_{si}) \tag{2}$$

with $f(x_i)$, $g_j(v_{ji})$ and $h_s(t_{si})$ being smooth function.

In equation (2), curve $f(x_i)$ is approached by linier parametric regression. Each regression curve $g_j(v_{ji})$ is approached by kernel function. Each regression curve component $h_s(t_{si})$ is approached by fourier series. Linier parametric regression is

$$f(x_i) = \beta_0 + \beta_1 x_{i1} + \beta_2 x_{i2} + ... + \beta_p x_{ip}. \tag{3}$$

**Lemma 2.1.** *If linier parametric curve component in the equation (2) is estimated by estimator component in the equation (3), then*

$$\mathbf{f}(x) = \mathbf{X}\boldsymbol{\beta} \tag{4}$$

*where*

$$\mathbf{f}(x) = \begin{bmatrix} f(x_1) \\ f(x_2) \\ \vdots \\ f(x_n) \end{bmatrix}, \mathbf{X} = \begin{bmatrix} 1 & x_{11} & \cdots & x_{1p} \\ 1 & x_{21} & \cdots & x_{2p} \\ \vdots & \vdots & \ddots & \vdots \\ 1 & x_{n1} & \cdots & x_{np} \end{bmatrix}, \text{ and } \boldsymbol{\beta} = \begin{bmatrix} \beta_0 \\ \beta_1 \\ \vdots \\ \beta_p \end{bmatrix},$$

*vector* $\mathbf{f}(x)$ is $n \times 1$, *vector* $\boldsymbol{\beta}$ is $(p+1)\times 1$ *and matrix* $\mathbf{X}$ is $n \times (1+p)$.

**Proof.** In equation (3), for $i$ = 1 to $i$ = n then $f(x_1), f(x_2), \ldots, f(x_n)$. So that,

$$\begin{bmatrix} f(x_1) \\ f(x_2) \\ \vdots \\ f(x_n) \end{bmatrix} = \begin{bmatrix} \beta_0 + \beta_1 x_{11} + \ldots + \beta_p x_{1p} \\ \beta_0 + \beta_1 x_{21} + \ldots + \beta_p x_{2p} \\ \vdots \\ \beta_0 + \beta_1 x_{n1} + \ldots + \beta_p x_{np} \end{bmatrix}$$

$$= \begin{bmatrix} 1 & x_{11} & \cdots & x_{1p} \\ 1 & x_{21} & \cdots & x_{2p} \\ \vdots & \vdots & \ddots & \vdots \\ 1 & x_{n1} & \cdots & x_{np} \end{bmatrix} \begin{bmatrix} \beta_0 \\ \beta_1 \\ \vdots \\ \beta_p \end{bmatrix}$$

$$= \mathbf{X}\boldsymbol{\beta}$$

Each kernel curve component is

$$\hat{g}_{j\phi_j}(v_j) = n^{-1} \sum_{i=1}^{n} W_{\phi_j i}(v_j) y_i \tag{5}$$

where

$$W_{\phi_j i}(v_j) = \frac{K_{\phi_j}(v_j - v_{ji})}{n^{-1} \sum_{i=1}^{n} K_{\phi_j}(v_j - v_{ji})} \text{ and } K_{\phi_j}(v_j - v_{ji}) = \frac{1}{\phi_j} K\left(\frac{v_j - v_{ji}}{\phi_j}\right).$$

Regression curve estimator (5) is depended on kernel function $K$ and bandwidth parameter $\phi_j$. Kernel function $K$ can be a epanechnikov kernel, uniform kernel, triangle kernel, squares kernel, triweigth kernel, cosine kernel and Gaussian kernel [15].

**Lemma 2.2.** *If kernel curve component in the equation (2) is estimated by estimator component in the equation (5), then*

$$\sum_{j=1}^{p} \hat{\mathbf{g}}_{j\phi_j}(v_j) = \mathbf{V}(\boldsymbol{\Phi})\mathbf{y} \tag{6}$$

*where*

$$\hat{\mathbf{g}}_{j\phi_j}\left(v_j\right)=\begin{bmatrix}\hat{g}_{j\phi_j}\left(v_{j1}\right)\\ \hat{g}_{j\phi_j}\left(v_{j2}\right)\\ \vdots\\ \hat{g}_{j\phi_j}\left(v_{jn}\right)\end{bmatrix},\ \mathbf{y}=\begin{bmatrix}y_1\\ y_2\\ \vdots\\ y_n\end{bmatrix},\ \text{and } \mathbf{V}\left(\mathbf{\Phi}\right)=\begin{bmatrix} n^{-1}\sum_{j=1}^{p}W_{\phi_j1}\left(v_{j1}\right) & n^{-1}\sum_{j=1}^{p}W_{\phi_j2}\left(v_{j1}\right) & \cdots & n^{-1}\sum_{j=1}^{p}W_{\phi_jn}\left(v_{j1}\right)\\ n^{-1}\sum_{j=1}^{p}W_{\phi_j1}\left(v_{j2}\right) & n^{-1}\sum_{j=1}^{p}W_{\phi_j2}\left(v_{j2}\right) & \cdots & n^{-1}\sum_{j=1}^{p}W_{\phi_jn}\left(v_{j2}\right)\\ \vdots & \vdots & \ddots & \vdots\\ n^{-1}\sum_{j=1}^{p}W_{\phi_j1}\left(v_{jn}\right) & n^{-1}\sum_{j=1}^{p}W_{\phi_j2}\left(v_{jn}\right) & \cdots & n^{-1}\sum_{j=1}^{p}W_{\phi_jn}\left(v_{jn}\right)\end{bmatrix} \quad (7)$$

*Vector* $\hat{\mathbf{g}}_{j\phi_j}\left(v_j\right)$ *is* $n\times 1$, *vector* $\mathbf{y}$ *is* $n\times 1$, *and matrix* $\mathbf{V}\left(\mathbf{\Phi}\right)$ *is* $n\times n$

**Proof.** Each component $\hat{g}_{j\phi_j}\left(v_j\right)$ can be written by (5). So that, for each component $v_j = v_{j1}$ to $v_j = v_{jn}$, then

$$\begin{bmatrix}\hat{g}_{j\phi_j}\left(v_{j1}\right)\\ \hat{g}_{j\phi_j}\left(v_{j2}\right)\\ \vdots\\ \hat{g}_{j\phi_j}\left(v_{jn}\right)\end{bmatrix}=\begin{bmatrix} n^{-1}W_{\phi_j1}\left(v_{j1}\right) & n^{-1}W_{\phi_j2}\left(v_{j1}\right) & \cdots & n^{-1}W_{\phi_jn}\left(v_{j1}\right)\\ n^{-1}W_{\phi_j1}\left(v_{j2}\right) & n^{-1}W_{\phi_j2}\left(v_{j2}\right) & \cdots & n^{-1}W_{\phi_jn}\left(v_{j2}\right)\\ \vdots & \vdots & \ddots & \vdots\\ n^{-1}W_{\phi_j1}\left(v_{jn}\right) & n^{-1}W_{\phi_j2}\left(v_{jn}\right) & \cdots & n^{-1}W_{\phi_jn}\left(v_{jn}\right)\end{bmatrix}\begin{bmatrix}y_1\\ y_2\\ \vdots\\ y_n\end{bmatrix}.$$

In matrix notation can be written as:

$$\hat{\mathbf{g}}_{j\phi_j}\left(v_j\right)=\mathbf{V}_{\mathbf{j}}\left(\phi_{\mathbf{j}}\right)\mathbf{y}, \quad (8)$$

where

$$\mathbf{V}_{\mathbf{j}}\left(\phi_{\mathbf{j}}\right)=\begin{bmatrix} n^{-1}W_{\phi_j1}\left(v_{j1}\right) & n^{-1}W_{\phi_j2}\left(v_{j1}\right) & \cdots & n^{-1}W_{\phi_jn}\left(v_{j1}\right)\\ n^{-1}W_{\phi_j1}\left(v_{j2}\right) & n^{-1}W_{\phi_j2}\left(v_{j2}\right) & \cdots & n^{-1}W_{\phi_jn}\left(v_{j2}\right)\\ \vdots & \vdots & \ddots & \vdots\\ n^{-1}W_{\phi_j1}\left(v_{jn}\right) & n^{-1}W_{\phi_j2}\left(v_{jn}\right) & \cdots & n^{-1}W_{\phi_jn}\left(v_{jn}\right)\end{bmatrix}$$

matrix $\mathbf{V}_{\mathbf{j}}\left(\phi_{\mathbf{j}}\right)$ is $n\times n$. The sum of component (8) from $j = 1$ to $j = p$ is

$$\sum_{j=1}^{p}\hat{\mathbf{g}}_{j\phi_j}\left(v_j\right)=\sum_{j=1}^{p}\left\{\begin{bmatrix} n^{-1}W_{\phi_j1}\left(v_{j1}\right) & n^{-1}W_{\phi_j2}\left(v_{j1}\right) & \cdots & n^{-1}W_{\phi_jn}\left(v_{j1}\right)\\ n^{-1}W_{\phi_j1}\left(v_{j2}\right) & n^{-1}W_{\phi_j2}\left(v_{j2}\right) & \cdots & n^{-1}W_{\phi_jn}\left(v_{j2}\right)\\ \vdots & \vdots & \ddots & \vdots\\ n^{-1}W_{\phi_j1}\left(v_{jn}\right) & n^{-1}W_{\phi_j2}\left(v_{jn}\right) & \cdots & n^{-1}W_{\phi_jn}\left(v_{jn}\right)\end{bmatrix}\mathbf{y}\right\}$$

$$=\begin{bmatrix} n^{-1}\sum_{j=1}^{p}W_{\phi_j1}\left(v_{j1}\right) & n^{-1}\sum_{j=1}^{p}W_{\phi_j2}\left(v_{j1}\right) & \cdots & n^{-1}\sum_{j=1}^{p}W_{\phi_jn}\left(v_{j1}\right)\\ n^{-1}\sum_{j=1}^{p}W_{\phi_j1}\left(v_{j2}\right) & n^{-1}\sum_{j=1}^{p}W_{\phi_j2}\left(v_{j2}\right) & \cdots & n^{-1}\sum_{j=1}^{p}W_{\phi_jn}\left(v_{j2}\right)\\ \vdots & \vdots & \ddots & \vdots\\ n^{-1}\sum_{j=1}^{p}W_{\phi_j1}\left(v_{jn}\right) & n^{-1}\sum_{j=1}^{p}W_{\phi_j2}\left(v_{jn}\right) & \cdots & n^{-1}\sum_{j=1}^{p}W_{\phi_jn}\left(v_{jn}\right)\end{bmatrix}\mathbf{y}$$

$$=\mathbf{V}\left(\mathbf{\Phi}\right)\mathbf{y}.$$

Thus proved that $\sum_{j=1}^{p}\hat{\mathbf{g}}_{j\phi_j}\left(v_j\right)=\mathbf{V}(\boldsymbol{\Phi})\mathbf{y}$.

Each fourier series curve component is approached by function $H_s\left(t_{si}\right)$, where

$$H_s\left(t_{si}\right)=b_s t_{si}+\frac{1}{2}a_{0s}+\sum_{k=1}^{N}a_{ks}\cos kt_{si}. \tag{9}$$

**Lemma 2.3** *If fourier series curve component in the equation (2) is estimated by function in the equation (9), then*

$$\sum_{s=1}^{q}\mathbf{H}_s\left(t_s\right)=\mathbf{Uc} \tag{10}$$

*where*

$$\mathbf{U}=\left[\mathbf{U}_1\quad \mathbf{U}_2\quad \cdots\quad \mathbf{U}_p\right],\ \mathbf{c}=\left[\mathbf{c}_1'\quad \mathbf{c}_2'\quad \cdots\quad \mathbf{c}_p'\right]',\ \mathbf{H}_s\left(t_s\right)=\left[H_s\left(t_{s1}\right)\quad H_s\left(t_{s2}\right)\quad \cdots\quad H_s\left(t_{sn}\right)\right]'. \tag{11}$$

*Vector* $\mathbf{H}_s\left(t_s\right)$ *is* $n\times 1$, *vector* $\mathbf{c}$ *is* $p(N+2)\times 1$ *and matrix* $\mathbf{U}$ *is* $n\times\left(p(N+2)\right)$.

**Proof.** In equation (9), for $i = 1$ to $i = n$ then $H_s\left(t_{s1}\right), H_s\left(t_{s2}\right),\ldots,H_s\left(t_{sn}\right)$. So that

$$\begin{bmatrix} H_s\left(t_{s1}\right)\\ H_s\left(t_{s2}\right)\\ \vdots\\ H_s\left(t_{sn}\right)\end{bmatrix}=\begin{bmatrix} b_s t_{s1}+\frac{1}{2}a_{0s}+\sum_{k=1}^{N}a_{ks}\cos kt_{s1}\\ b_s t_{s2}+\frac{1}{2}a_{0s}+\sum_{k=1}^{N}a_{ks}\cos kt_{s2}\\ \vdots\\ b_s t_{sn}+\frac{1}{2}a_{0s}+\sum_{k=1}^{N}a_{ks}\cos kt_{sn}\end{bmatrix}$$

$$=\begin{bmatrix} t_{s1} & 1 & \cos t_{s1} & \cos 2t_{s1} & \ldots & \cos Nt_{s1}\\ t_{s2} & 1 & \cos t_{s2} & \cos 2t_{s2} & \ldots & \cos Nt_{s2}\\ t_{s3} & 1 & \cos t_{s3} & \cos 2t_{s3} & \ldots & \cos Nt_{s3}\\ \vdots & \vdots & \vdots & \vdots & \ldots & \vdots\\ t_{sn} & 1 & \cos t_{sn} & \cos 2t_{sn} & \ldots & \cos Nt_{sn}\end{bmatrix}\begin{pmatrix} b_s\\ \frac{1}{2}a_{0s}\\ a_{1s}\\ \vdots\\ a_{Ns}\end{pmatrix}$$

In matrix notation can be written as:

$$\mathbf{H}_s\left(t_s\right)=\mathbf{U}_s\mathbf{c}_s \tag{12}$$

where

$$\mathbf{U}_s=\begin{bmatrix} t_{s1} & 1 & \cos t_{s1} & \cos 2t_{s1} & \ldots & \cos Nt_{s1}\\ t_{s2} & 1 & \cos t_{s2} & \cos 2t_{s2} & \ldots & \cos Nt_{s2}\\ t_{s3} & 1 & \cos t_{s3} & \cos 2t_{s3} & \ldots & \cos Nt_{s3}\\ \vdots & \vdots & \vdots & \vdots & \ldots & \vdots\\ t_{sn} & 1 & \cos t_{sn} & \cos 2t_{sn} & \ldots & \cos Nt_{sn}\end{bmatrix},\ \text{and }\mathbf{c}_s=\begin{pmatrix} b_s\\ \frac{1}{2}a_{0s}\\ a_{1s}\\ \vdots\\ a_{Ns}\end{pmatrix}$$

Matrix $\mathbf{U}_s$ is $n\times\left(N+2\right)$ and vector $\mathbf{c}_s$ is $\left(N+2\right)\times 1$.

The sum of component (12) from $s=1$ to $s=q$ is

$$\sum_{s=1}^{q}\mathbf{H}_s(t_s)=\sum_{s=1}^{q}\begin{bmatrix} b_s t_{s1}+\frac{1}{2}a_{0s}+\sum_{k=1}^{N}a_{ks}\cos kt_{s1} \\ b_s t_{s2}+\frac{1}{2}a_{0s}+\sum_{k=1}^{N}a_{ks}\cos kt_{s2} \\ \vdots \\ b_s t_{sn}+\frac{1}{2}a_{0s}+\sum_{k=1}^{N}a_{ks}\cos kt_{sn} \end{bmatrix}=\begin{bmatrix} \sum_{s=1}^{q}b_s t_{s1}+\sum_{s=1}^{q}\frac{1}{2}a_{0s}+\sum_{s=1}^{q}\sum_{k=1}^{N}a_{ks}\cos kt_{s1} \\ \sum_{s=1}^{q}b_s t_{s2}+\sum_{s=1}^{q}\frac{1}{2}a_{0s}+\sum_{s=1}^{q}\sum_{k=1}^{N}a_{ks}\cos kt_{s2} \\ \vdots \\ \sum_{s=1}^{q}b_s t_{sn}+\sum_{s=1}^{q}\frac{1}{2}a_{0s}+\sum_{s=1}^{q}\sum_{k=1}^{N}a_{ks}\cos kt_{sn} \end{bmatrix}=\mathbf{Uc}$$

where matrix $\sum_{s=1}^{q}\mathbf{H}_s(t_s)=\mathbf{Uc}$ is given by equation (10).

PLS is optimization methods in statistics. It combines the goodness of fit with the smoothness of curves, which between them controlled by a smoothing parameter [16] dan [17]. PLS method can be defined as

$$\underset{\mathbf{c},\boldsymbol{\beta}}{Min}\left\{\left[L(\mathbf{c})+L(\boldsymbol{\beta})\right]+\sum_{s=1}^{q}\theta_s S(H_s(t_{si}))\right\} \tag{13}$$

where $\left[L(\mathbf{c})+L(\boldsymbol{\beta})\right]$ is the goodness of fit, $S(H_s(t_{si}))$ is penalty function and $\theta_s$ is a smoothing parameter.

**Lemma 2.4.** *If regression model is given by equation* (1), *linier parametric curve is given by Lemma 2.2, kernel curve is given by Lemma 2.2 and fourier series is given by Lemma 2.3, then goodness of fit function is*

$$L(\mathbf{c})+L(\boldsymbol{\beta})=n^{-1}\left\|(\mathbf{I}-\mathbf{V}(\boldsymbol{\Phi}))\mathbf{y}-\mathbf{P}\right\|^2 \tag{14}$$

*where matrix* $\mathbf{I}$ *is* $n\times n$ *identity matrix and vector* $\mathbf{P}=\mathbf{X}\boldsymbol{\beta}+\mathbf{Uc}$ *is* $n\times 1$.

**Proof.** The equation (1), (2), (4), (6) and (10) give

$$\begin{aligned} L(\mathbf{c})+L(\boldsymbol{\beta}) &= n^{-1}\left\|\mathbf{y}-\mathbf{f}(x)-\sum_{j=1}^{p}\hat{\mathbf{g}}_{j\phi_j}(v_j)-\sum_{s=1}^{q}\mathbf{H}_s(t_s)\right\|^2 \\ &= n^{-1}\left\|\mathbf{y}-\mathbf{X}\boldsymbol{\beta}-\mathbf{V}(\boldsymbol{\Phi})\mathbf{y}-\mathbf{Uc}\right\|^2 \\ &= n^{-1}\left\|(\mathbf{I}-\mathbf{V}(\boldsymbol{\Phi}))\mathbf{y}-(\mathbf{X}\boldsymbol{\beta}+\mathbf{Uc})\right\|^2 \\ &= n^{-1}\left\|(\mathbf{I}-\mathbf{V}(\boldsymbol{\Phi}))\mathbf{y}-\mathbf{P}\right\|^2 \end{aligned}$$

**Lemma 2.5.** *If fourier series is given by Lemma 2.3, then penalty function and smoothing parameter for PLS method is*

$$\sum_{s=1}^{q}\theta_s S(H_s(t_{si}))=\mathbf{c}'\mathbf{D}(\boldsymbol{\theta})\mathbf{c} \tag{15}$$

*where* $\mathbf{D}(\boldsymbol{\theta})=diag\begin{pmatrix}\mathbf{D}_1(\boldsymbol{\theta}) & \mathbf{D}_2(\boldsymbol{\theta}) & \cdots & \mathbf{D}_p(\boldsymbol{\theta})\end{pmatrix}$ *and* $\mathbf{D}_s(\boldsymbol{\theta})=diag\begin{pmatrix}0 & 0 & \theta_s 1^4 & \theta_s 2^4 & \cdots & \theta_s N^4\end{pmatrix}$. *Matrix* $\mathbf{D}(\boldsymbol{\theta})$ *is* $(p\times(N+2))\times(p\times(N+2))$ *and* $\mathbf{D}_s(\boldsymbol{\theta})$ *is* $(N+2)\times(N+2)$.

**Proof.** According Bilodeau [7]

$$\int_0^{\pi} \frac{2}{\pi}\left(H_s^{(2)}(t_{si})\right)^2 dt_{si} = \left(\sum_{k=1}^{N} k^4 a^2_{ks}\right) \tag{16}$$

then

$$\sum_{s=1}^{q} \theta_s S\left(H_s(t_{si})\right) = \sum_{s=1}^{q} \theta_s \int_0^{\pi} \frac{2}{\pi}\left(H_s^{(2)}(t_{si})\right)^2 dt_s = \sum_{s=1}^{q} \theta_s \left(\sum_{k=1}^{N} k^4 a^2_{ks}\right) = \mathbf{c}'\mathbf{D}(\boldsymbol{\theta})\mathbf{c}$$

**Theorem 2.1.** *If the goodness of fit function is given by Lemma 2.4 while penalty function and smoothing parameter is given by Lemma 2.5, then the PLS estimator for* $\mathbf{c}$ *parameter is obtained from optimization:*

$$\underset{\mathbf{c},\boldsymbol{\beta}}{Min}\left\{\left[L(\mathbf{c})+L(\boldsymbol{\beta})\right]+\sum_{s=1}^{q}\theta_s S\left(H_s(t_{si})\right)\right\} = \underset{\mathbf{c},\boldsymbol{\beta}}{Min}\left\{\mathbf{Q}(\mathbf{c},\boldsymbol{\beta})\right\} \tag{17}$$

*where* $\mathbf{Q}(\mathbf{c},\boldsymbol{\beta}) = n^{-1}\left\|\left(\mathbf{I}-\mathbf{V}(\boldsymbol{\Phi})\right)\mathbf{y}-\mathbf{P}\right\|^2 + \mathbf{c}'\mathbf{D}(\boldsymbol{\theta})\mathbf{c}$

**Proof.**

$$\underset{\mathbf{c},\boldsymbol{\beta}}{Min}\left\{\left[L(\mathbf{c})+L(\boldsymbol{\beta})\right]+\sum_{s=1}^{q}\theta_s S\left(H_s(t_{si})\right)\right\} = \underset{\mathbf{a},\boldsymbol{\beta}}{Min}\left\{n^{-1}\left\|\left(\mathbf{I}-\mathbf{V}(\boldsymbol{\Phi})\right)\mathbf{y}-\mathbf{P}\right\|^2 + \mathbf{c}'\mathbf{D}(\boldsymbol{\theta})\mathbf{c}\right\} = \underset{\mathbf{c},\boldsymbol{\beta}}{Min}\left\{\mathbf{Q}(\mathbf{c},\boldsymbol{\beta})\right\}$$

**Theorem 2.2.** *If the PLS estimator for parameter* $\mathbf{c}$ *and* $\boldsymbol{\beta}$ *is given by Theorem 2.1, then PLS estimator for the mixed regression curve* $\boldsymbol{\mu}(x_i,\mathbf{v}_i,\mathbf{t}_i)$ *is given by*

$$\hat{\boldsymbol{\mu}}_{\boldsymbol{\Phi},\theta}(x_i,\mathbf{v}_i,\mathbf{t}_i) = \hat{\mathbf{f}}(x) + \sum_{j=1}^{p}\hat{\mathbf{g}}_{j\phi_j}(v_j) + \sum_{s=1}^{q}\hat{\mathbf{h}}_{\theta_s,N}(t_s) \tag{18}$$

*where* $\hat{\mathbf{f}}(x) = \mathbf{X}\hat{\boldsymbol{\beta}}$, $\sum_{s=1}^{q}\hat{\mathbf{h}}_{\theta_s,N}(t_s) = \mathbf{U}\hat{\mathbf{c}}(\boldsymbol{\theta})$, $\hat{\boldsymbol{\beta}} = \mathbf{A}(\boldsymbol{\Phi},\boldsymbol{\theta},N)\mathbf{y}$ *and* $\hat{\mathbf{c}}(\boldsymbol{\theta}) = \mathbf{B}(\boldsymbol{\Phi},\boldsymbol{\theta},N)\mathbf{y}$.

**Proof.** The PLS estimator for parameter $\mathbf{c}$ and $\boldsymbol{\beta}$ is obtained from optimization Theorem 2.1, so that

$$\begin{aligned}
\mathbf{Q}(\mathbf{c},\boldsymbol{\beta}) &= n^{-1}\left\|\left(\mathbf{I}-\mathbf{V}(\boldsymbol{\Phi})\right)\mathbf{y}-\mathbf{P}\right\|^2 + \mathbf{c}'\mathbf{D}(\boldsymbol{\theta})\mathbf{c}\\
&= n^{-1}\mathbf{y}'\left(\mathbf{I}-\mathbf{V}(\boldsymbol{\Phi})\right)'\left(\mathbf{I}-\mathbf{V}(\boldsymbol{\Phi})\right)\mathbf{y} - 2n^{-1}\mathbf{P}'\left(\mathbf{I}-\mathbf{V}(\boldsymbol{\Phi})\right)\mathbf{y} + n^{-1}\mathbf{P}'\mathbf{P} + \mathbf{c}'\mathbf{D}(\boldsymbol{\theta})\mathbf{c}\\
&= n^{-1}\mathbf{y}'\left(\mathbf{I}-\mathbf{V}(\boldsymbol{\Phi})\right)'\left(\mathbf{I}-\mathbf{V}(\boldsymbol{\Phi})\right)\mathbf{y} - 2n^{-1}\left(\boldsymbol{\beta}'\mathbf{X}+\mathbf{c}'\mathbf{U}'\right)\left(\mathbf{I}-\mathbf{V}(\boldsymbol{\Phi})\right)\mathbf{y} + n^{-1}\left(\boldsymbol{\beta}'\mathbf{X}'+\mathbf{c}'\mathbf{U}'\right)\left(\mathbf{X}\boldsymbol{\beta}+\mathbf{U}\mathbf{c}\right) + \mathbf{c}'\mathbf{D}(\boldsymbol{\theta})\mathbf{c}
\end{aligned}$$

Furthermore, the partial derivatives on $\mathbf{c}$ is $\frac{\partial \mathbf{Q}(\mathbf{c},\boldsymbol{\beta})}{\partial \mathbf{c}} = 0$. So that, the estimator for $\mathbf{c}$ is given by:

$$\hat{\mathbf{c}}(\boldsymbol{\theta}) = \left(\mathbf{U}'\mathbf{U} + n\mathbf{D}(\boldsymbol{\theta})\right)^{-1}\left(\mathbf{U}'\left(\mathbf{I}-\mathbf{V}(\boldsymbol{\Phi})\right)\mathbf{y} - \mathbf{U}'\mathbf{X}\hat{\boldsymbol{\beta}}\right) \tag{19}$$

The partial derivatives on $\boldsymbol{\beta}$ is $\frac{\partial \mathbf{Q}(\mathbf{c},\boldsymbol{\beta})}{\partial \boldsymbol{\beta}} = 0$. So that, the estimator for $\boldsymbol{\beta}$ is given by:

$$\hat{\boldsymbol{\beta}} = \left(\mathbf{X}'\mathbf{X}\right)^{-1}\left(\mathbf{X}\left(\mathbf{I}-\mathbf{V}(\boldsymbol{\Phi})\right)\mathbf{y} - \mathbf{X}'\mathbf{U}\hat{\mathbf{c}}(\boldsymbol{\theta})\right) \tag{20}$$

The equation (19) and (20) give

$$\hat{\boldsymbol{\beta}} = \mathbf{A}(\boldsymbol{\Phi},\boldsymbol{\theta},N)\mathbf{y} \tag{21}$$

where

$$\mathbf{A}(\boldsymbol{\Phi},\boldsymbol{\theta},N)=\left[\left((\mathbf{X}'\mathbf{X})-\mathbf{X}'\mathbf{U}\left(\mathbf{U}'\mathbf{U}+n\mathbf{D}(\boldsymbol{\theta})\right)^{-1}\mathbf{U}'\mathbf{X}\right)\right]^{-1}\left(\mathbf{X}-\mathbf{X}'\mathbf{U}\left(\left(\mathbf{U}'\mathbf{U}+n\mathbf{D}(\boldsymbol{\theta})\right)^{-1}\mathbf{U}'\right)\right)\left(\mathbf{I}-\mathbf{V}(\boldsymbol{\Phi})\right)$$

The equation (19) and (21) give

$$\hat{\mathbf{c}}(\boldsymbol{\theta})=\mathbf{B}(\boldsymbol{\Phi},\boldsymbol{\theta},N)\mathbf{y} \tag{22}$$

where

$$\mathbf{B}(\boldsymbol{\Phi},\boldsymbol{\theta},N)=(\mathbf{U}'\mathbf{U}+n\mathbf{D})^{-1}\mathbf{U}'\left(\mathbf{I}-\mathbf{X}\left[\left((\mathbf{X}'\mathbf{X})-\mathbf{X}'\mathbf{U}\left(\mathbf{U}'\mathbf{U}+n\mathbf{D}(\boldsymbol{\theta})\right)^{-1}\mathbf{U}'\mathbf{X}\right)\right]^{-1}\left(\mathbf{X}-\mathbf{X}'\mathbf{U}\left(\left(\mathbf{U}'\mathbf{U}+n\mathbf{D}(\boldsymbol{\theta})\right)^{-1}\mathbf{U}'\right)\right)\right)\left(\mathbf{I}-\mathbf{V}(\boldsymbol{\Phi})\right)$$

Considering the equation (21) then estimator for linier parametric curve is given by:

$$\hat{\mathbf{f}}(x)=\mathbf{X}\hat{\boldsymbol{\beta}} \tag{23}$$

and considering the equation (22) then estimator for fourier series curve is given by:

$$\sum_{s=1}^{q}\hat{\mathbf{h}}_{\theta_s,N}(t_s)=\mathbf{U}\hat{\mathbf{c}}(\boldsymbol{\theta}). \tag{24}$$

So that, based on the equation (23), (24) and Lemma 2.2, then the mixed estimators for semiparametric regression curve (2) is

$$\hat{\boldsymbol{\mu}}_{\boldsymbol{\Phi},\boldsymbol{\theta},N}(x_i,\mathbf{v}_i,\mathbf{t}_i)=\hat{\mathbf{f}}(x)+\sum_{j=1}^{p}\hat{\mathbf{g}}_{j\phi_j}(v_j)+\sum_{s=1}^{q}\hat{\mathbf{h}}_{\theta_s,N}(t_s)$$

**Corollary.** *If the estimator* $\hat{\boldsymbol{\mu}}_{\boldsymbol{\Phi},\boldsymbol{\theta},N}(x_i,\mathbf{v}_i,\mathbf{t}_i),\hat{\mathbf{f}}(x)$, $\sum_{j=1}^{p}\hat{\mathbf{g}}_{j\phi_j}(v_j)$, $\sum_{s=1}^{q}\hat{\mathbf{h}}_{\theta_s,N}(t_s)$ $\hat{\boldsymbol{\beta}}$, *and* $\hat{\mathbf{c}}(\boldsymbol{\theta})$ *are given by Theorem 2.2, then* $\hat{\mathbf{f}}(x)=\mathbf{C}(\boldsymbol{\Phi},\boldsymbol{\theta},N)\mathbf{y}$, $\sum_{s=1}^{q}\hat{\mathbf{h}}_{\theta_s,N}(t_{si})=\mathbf{E}(\boldsymbol{\Phi},\boldsymbol{\theta},N)\mathbf{y}$ *and* $\hat{\boldsymbol{\mu}}_{\boldsymbol{\Phi},\boldsymbol{\theta},N}(x_i,\mathbf{v}_i,\mathbf{t}_i)=\mathbf{Z}(\boldsymbol{\Phi},\boldsymbol{\theta},N)\mathbf{y}$
*where*

$$\mathbf{C}(\boldsymbol{\Phi},\boldsymbol{\theta},N)=\mathbf{X}\left[\left((\mathbf{X}'\mathbf{X})-\mathbf{X}'\mathbf{U}\left(\mathbf{U}'\mathbf{U}+n\mathbf{D}(\boldsymbol{\theta})\right)^{-1}\mathbf{U}'\mathbf{X}\right)\right]^{-1}\left(\mathbf{X}-\mathbf{X}'\mathbf{U}\left(\left(\mathbf{U}'\mathbf{U}+n\mathbf{D}(\boldsymbol{\theta})\right)^{-1}\mathbf{U}'\right)\right)\left(\mathbf{I}-\mathbf{V}(\boldsymbol{\Phi})\right),$$

$$\mathbf{E}(\boldsymbol{\Phi},\boldsymbol{\theta},N)=\mathbf{U}\left(\mathbf{U}'\mathbf{U}+n\mathbf{D}(\boldsymbol{\theta})\right)^{-1}\mathbf{U}'\left(\mathbf{I}-\mathbf{X}\left[\left((\mathbf{X}'\mathbf{X})-\mathbf{X}'\mathbf{U}\left(\mathbf{U}'\mathbf{U}+n\mathbf{D}(\boldsymbol{\theta})\right)^{-1}\mathbf{U}'\mathbf{X}\right)\right]^{-1}\left(\mathbf{X}-\mathbf{X}'\mathbf{U}\left(\left(\mathbf{U}'\mathbf{U}+n\mathbf{D}(\boldsymbol{\theta})\right)^{-1}\mathbf{U}'\right)\right)\right)\left(\mathbf{I}-\mathbf{V}(\boldsymbol{\Phi})\right)$$

and $\mathbf{Z}(\boldsymbol{\Phi},\boldsymbol{\theta},N)=\mathbf{C}(\boldsymbol{\Phi},\boldsymbol{\theta},N)+\mathbf{V}(\boldsymbol{\Phi})+\mathbf{E}(\boldsymbol{\Phi},\boldsymbol{\theta},N)$.

**Proof.** The equation (21) and (23) given an equation

$$\hat{\mathbf{f}}(x)=\mathbf{X}\left[\left((\mathbf{X}'\mathbf{X})-\mathbf{X}'\mathbf{U}\left(\mathbf{U}'\mathbf{U}+n\mathbf{D}(\boldsymbol{\theta})\right)^{-1}\mathbf{U}'\mathbf{X}\right)\right]^{-1}\left(\mathbf{X}-\mathbf{X}'\mathbf{U}\left(\left(\mathbf{U}'\mathbf{U}+n\mathbf{D}(\boldsymbol{\theta})\right)^{-1}\mathbf{U}'\right)\right)\left(\mathbf{I}-\mathbf{V}(\boldsymbol{\Phi})\right)\mathbf{y}$$

or it can written as:

$$\hat{\mathbf{f}}(x)=\mathbf{C}(\boldsymbol{\Phi},\boldsymbol{\theta},N)\mathbf{y} \tag{25}$$

The equation (22) and (24) given an equation

$$\sum_{s=1}^{q}\hat{\mathbf{h}}_{\theta_s,N}(t_{si})=\mathbf{U}\left(\mathbf{U}'\mathbf{U}+n\mathbf{D}(\boldsymbol{\theta})\right)^{-1}\mathbf{U}'\left(\mathbf{I}-\mathbf{X}\left[\left((\mathbf{X}'\mathbf{X})-\mathbf{X}'\mathbf{U}\left(\mathbf{U}'\mathbf{U}+n\mathbf{D}(\boldsymbol{\theta})\right)^{-1}\mathbf{U}'\mathbf{X}\right)\right]^{-1}\left(\mathbf{X}-\mathbf{X}'\mathbf{U}\left(\left(\mathbf{U}'\mathbf{U}+n\mathbf{D}(\boldsymbol{\theta})\right)^{-1}\mathbf{U}'\right)\right)\right)\left(\mathbf{I}-\mathbf{V}(\boldsymbol{\Phi})\right)\mathbf{y}$$

or it can be written as:

$$\sum_{s=1}^{q}\hat{\mathbf{h}}_{\theta_s,N}\left(t_{si}\right)=\mathbf{E}\left(\mathbf{\Phi},\boldsymbol{\theta},N\right)\mathbf{y} \tag{26}$$

Furthermore, the Theorem 2.2, equation (25) and (26) produce:

$$\begin{aligned}\hat{\boldsymbol{\mu}}_{\mathbf{\Phi},\boldsymbol{\theta},N}\left(x_i,\mathbf{v}_i,\mathbf{t}_i\right)&=\hat{\mathbf{f}}\left(x\right)+\sum_{j=1}^{p}\hat{\mathbf{g}}_{j\phi_j}\left(v_j\right)+\sum_{s=1}^{q}\hat{\mathbf{h}}_{\theta_s,N}\left(t_s\right)\\&=\mathbf{C}\left(\mathbf{\Phi},\boldsymbol{\theta},N\right)\mathbf{y}+\mathbf{V}\left(\mathbf{\Phi}\right)\mathbf{y}+\mathbf{E}\left(\mathbf{\Phi},\boldsymbol{\theta},N\right)\mathbf{y}\\&=\left(\mathbf{C}\left(\mathbf{\Phi},\boldsymbol{\theta},N\right)+\mathbf{V}\left(\mathbf{\Phi}\right)+\mathbf{E}\left(\mathbf{\Phi},\boldsymbol{\theta},N\right)\right)\mathbf{y}\\&=\mathbf{Z}\left(\mathbf{\Phi},\boldsymbol{\theta},N\right)\mathbf{y}\end{aligned}$$

where

$$\mathbf{Z}\left(\mathbf{\Phi},\boldsymbol{\theta},N\right)=\mathbf{C}\left(\mathbf{\Phi},\boldsymbol{\theta},N\right)+\mathbf{V}\left(\mathbf{\Phi}\right)+\mathbf{E}\left(\mathbf{\Phi},\boldsymbol{\theta},N\right) \tag{27}$$

Estimator $\hat{\boldsymbol{\mu}}_{\mathbf{\Phi},\boldsymbol{\theta},N}\left(x_i,\mathbf{v}_i,\mathbf{t}_i\right)$ are depended on the bandwidth $\mathbf{\Phi}=\left(\phi_1,\phi_2,\ldots,\phi_p\right)'$, smoothing parameter $\boldsymbol{\theta}=\left(\theta_1,\theta_2,\ldots,\theta_q\right)'$ and oscillation parameter $N$. Generalized Cross Validation (GCV) method can give the best bandwidth, smoothing parameter and oscillation parameter.

$$GCV\left(\mathbf{\Phi},\boldsymbol{\theta},N\right)=\frac{MSE\left(\mathbf{\Phi},\boldsymbol{\theta},N\right)}{\left(n^{-1}\left(\mathbf{I}-\mathbf{Z}\left(\mathbf{\Phi},\boldsymbol{\theta},N\right)\right)\right)^2} \tag{28}$$

where $MSE\left(\mathbf{\Phi},\boldsymbol{\theta},N\right)=n^{-1}\mathbf{y}'\left(\mathbf{I}-\mathbf{Z}\left(\mathbf{\Phi},\boldsymbol{\theta},N\right)\right)'\left(\mathbf{I}-\mathbf{Z}\left(\mathbf{\Phi},\boldsymbol{\theta},N\right)\right)\mathbf{y}$. Optimal bandwidth $\mathbf{\Phi}_{(\mathrm{opt})}=\left(\phi_{1(\mathrm{opt})},\phi_{2(\mathrm{opt})},\ldots,\phi_{p(\mathrm{opt})}\right)'$, smoothing parameter $\boldsymbol{\theta}_{(opt)}=\left(\theta_{1(opt)},\theta_{2(opt)},\ldots,\theta_{q(opt)}\right)'$ and oscillation parameter $N_{(opt)}$ are obtained by optimation

$$GCV\left(\mathbf{\Phi}_{(opt)},\boldsymbol{\theta}_{(opt)},N_{(opt)}\right)=\min_{\mathbf{\Phi},\boldsymbol{\theta},N}\left\{GCV\left(\mathbf{\Phi},\boldsymbol{\theta},N\right)\right\}.$$

## 3. Conclusions

If given additive regression semiparametric model:

$$\mathbf{y}=\boldsymbol{\mu}\left(x_i,\mathbf{v}_i,\mathbf{t}_i\right)+\boldsymbol{\varepsilon}=\mathbf{f}\left(x_i\right)+\sum_{j=1}^{p}\mathbf{g}_j\left(v_{ji}\right)+\sum_{s=1}^{q}\mathbf{h}_s\left(t_{si}\right)+\boldsymbol{\varepsilon}$$

then

1. Mixed estimator of kernel and fourier series in semiparametric regression is given by

$$\hat{\boldsymbol{\mu}}_{\mathbf{\Phi},\boldsymbol{\theta},N}\left(x_i,\mathbf{v}_i,\mathbf{t}_i\right)=\hat{\mathbf{f}}\left(x\right)+\sum_{j=1}^{p}\hat{\mathbf{g}}_{j\phi_j}\left(v_j\right)+\sum_{s=1}^{q}\hat{\mathbf{h}}_{\theta_s,N}\left(t_s\right)$$

where $\hat{\mathbf{f}}\left(x\right)=\mathbf{C}\left(\mathbf{\Phi},\boldsymbol{\theta},N\right)\mathbf{y}$, $\sum_{j=1}^{p}\hat{\mathbf{g}}_{j\phi_j}\left(v_j\right)=\mathbf{V}\left(\mathbf{\Phi}\right)\mathbf{y}$, $\sum_{s=1}^{q}\hat{\mathbf{h}}_{\theta_s,N}\left(t_s\right)=\mathbf{E}\left(\mathbf{\Phi},\boldsymbol{\theta},N\right)\mathbf{y}$ and $\hat{\boldsymbol{\mu}}_{\mathbf{\Phi},\boldsymbol{\theta},N}\left(x_i,\mathbf{v}_i,\mathbf{t}_i\right)=\mathbf{Z}\left(\mathbf{\Phi},\boldsymbol{\theta},N\right)\mathbf{y}$.

2. Mixed estimator kernel and fourier series $\hat{\boldsymbol{\mu}}_{\mathbf{\Phi},\boldsymbol{\theta},N}\left(x_i,\mathbf{v}_i,\mathbf{t}_i\right)$ is depended on $\mathbf{\Phi}_{(opt)},\boldsymbol{\theta}_{(opt)}$, and $N_{(opt)}$ that can be obtained by the smallest GCV value.

**Acknowledgments**
This work is under the supervision of the second and third author.

**References**

[1]. Budiantara IN, Ratnasari V, Ratna M, and Zain I 2015 The Combination of Spline and Kernel Estimator for Nonparametric Regression and its Properties *Applied Mathematical Sciences* **9** 6083-6094
[2]. Eubank RL 1999 *Nonparametric Regression and Spline Smoothing* (New York: Marcel Dekker, Inc)
[3]. Härdle W, Müller M, Sperlich S, and Werwatz A 2004 *Nonparametric and Semiparametric Models: An Introduction* (Berlin: Springer)
[4]. You J, Chen G, and Zhou X, B-Spline Estimation in a Semiparametric Regression Model with Nonlinier Time Series Errors *American Journal on Applied Sciences* **2** 1343-1349
[5]. Budiantara IN 1999 Estimator Spline Terbobot dalam Regresi Semiparametrik, *Majalah Ilmu Pengetahuan dan Teknologi* **10** 103-109
[6]. Budiantara IN 2005 Model Keluarga Spline Polinomial Truncated dalam Regresi Semiparametrik *Berkala Ilmiah MIPA* **15** 55-61
[7]. Hong SY 2002 Normal Approximation Rate and Bias Reduction for Data-Driven Kernel Smoothing Estimator in a Semiparametric Regression Model *Journal of Multivariate Analysis* **80** 1-20
[8]. Manzana S, and Zerom D 2005 Kernel Estimation of a Partially Linier Additive Model *Statistics and Probability Letters* **72** 313-322
[9]. Pane R 2014 Parametric and Nonparametric Estimators in Fourier Series Semiparametric Regression and Their Characteristics *Applied Mathematical Sciences* **8** 5053-5064
[10]. Asrini LJ and Budiantara 2014 IN Fourier Series Semiparametric Regression Models (Case Study: The Production of Lawland Rice Irrigation in Central Java) *ARPN Jurnal of Engineering and Applied Sciences* **9** 1501-1506
[11]. Chamidah N, Budiantara IN, Sunaryo S and Zain I 2012 Disigning of Child Growth Chart Based on Multi-Response Local Polynomial Modeling *Journal of Mathematical and Statistics* **8** 342-347
[12]. Astuti ET, Budiantara IN, Sunaryo S and Dokhi M 2012 Statistical Modeling for Mortality Data Using Local Generalized Poisson Regression Model *International Journal of Applied Mathematics and Statistics* **33** 92-101
[13]. Budiantara IN, Mariati NPAM, Ratnasari V, Zain I, Ratna M, Sudiarsa IW, Mardianto MFF and Hendayanti NPN 2015 Comparison Truncated Spline and Fourier Series in Multivariable Nonparametric Regression Models (Application: Data of Poverty in Papua, Indonesia) *International Journal of Basic and Applied Sciences IJBAS-IJENS* **15** 9-12.
[14]. Sudiarsa IW, Budiantara IN, Suhartono and Purnami SW 2015 Combined Estimator Fourier Series and Spline Truncated in Multivariable Nonparametric Regression *Applied Mathematical Sciences* **9** 4997-5010
[15]. Härdle W *Applied Nonparametric Regression.Cambrige* 1990 (New York: University Press)
[16]. Bilodeau M 1992 Fourier Smoother and Additive Models *The Canadian Journal of Statistics* **20** 257-269
[17]. Wibowo W, Haryatmi S and Budiantara IN 2012 Penalized Least Square for Semiparametric Regression *International Journal of Academic Research Part A* **4** 281-286

# On Rainbow $k$-Connection Number of Special Graphs and It's Sharp Lower Bound

**Ika Hesti Agustin[1,2], Dafik[1,3], Gembong A.W. [1,2], Ridho Alfarisi [1,4]**

[1] CGANT, University of Jember, Indonesia
[2] Mathematics Depart. University of Jember, Indonesia
[3] Mathematics Edu. Depart. University of Jember, Indonesia
[4] Department of Mathematics, Sepuluh Nopember Institute of Technology, Surabaya, Indonesia

E-mail: ikahesti.fmipa@unej.ac.id, d.dafik@unej.ac.id

**Abstract.** Let $G = (V, E)$ be a simple, nontrivial, finite, connected and undirected graph. Let $c$ be a coloring $c : E(G) \to \{1, 2, \dots, s\}, s \in \mathbb{N}$. A path of edge colored graph is said to be a rainbow path if no two edges on the path have the same color. An edge colored graph $G$ is said to be a rainbow connected graph if there exists a rainbow $u - v$ path for every two vertices $u$ and $v$ of $G$. The rainbow connection number of a graph $G$, denoted by $rc(G)$, is the smallest number of $k$ colors required to edge color the graph such that the graph is rainbow connected. Furthermore, for an $l$-connected graph $G$ and an integer $k$ with $1 \leq k \leq l$, the rainbow $k$-connection number $rc_k(G)$ of $G$ is defined to be the minimum number of colors required to color the edges of $G$ such that every two distinct vertices of $G$ are connected by at least $k$ internally disjoint rainbow paths. In this paper, we determine the exact values of rainbow connection number of some special graphs and obtain a sharp lower bound.

**Keywords:** Rainbow $k$-Connection Number, Special Graphs, Sharp Lower Bound

## 1. Introduction

Suppose $G$ is a simple connected graph with a set of vertices $V(G)$ and edges $E(G)$. For a further reference please see Gross, *et. al.* [6]. Let $G$ be a nontrivial connected graph on which it is defined a coloring $c : E(G) \to \{1, 2, \dots, s\}, s \in N$, of the edges of $G$, where adjacent edges may be colored the same. A $u - v$ path $P$ in $G$ is a rainbow path if no two edges of $P$ are colored the same. The graph $G$ is rainbow-connected (with respect to $c$) if $G$ contains a rainbow $u - v$ path for every two vertices $u$ and $v$ of $G$. In this case, the coloring $c$ is called a rainbow coloring of $G$. If $k$ colors are used, then $c$ is a rainbow $k$-coloring. The minimum $k$ for which there exists a *rainbow $k$-coloring* of the edges of $G$ is the rainbow connection number $rc(G)$. The completes concept can be found in Chartrand in [4].

A simple observation can be proposed that if $G$ has $n$ vertices then $rc(G) \leq n-1$ but is not sharp. Since a given spanning tree can be assigned with distinct colors, and color the remaining edges with one of the already used colors then the upper bound of $rc(G) \leq n-1$, see Caro [1] for detail. It is also easy to understand that $rc(G) \geq diam(G)$, where $diam(G)$ denotes the diameter of $G$, Caro in [1]. Thus, it gives the following

$$diam(G) \leq rc(G) \leq n-1$$

There have been some results regarded to rainbow connection numbers. Chandran, *et.al.* in [2] determined rainbow connection number and connected dominating sets, Chakraborty, *et.al.* in [3] considered hardness and algorithms for rainbow connectivity. Furthermore, Li *et.al.* in [7] stated Rainbow connections of graphs - A survey. Also Li *et.al.* in [8] characterized graphs with rainbow connection number and rainbow connection numbers of some graph operations. Schiermeyer in [10] studied rainbow connection in graphs with minimum degree three.

A well-known result shows that in every $l-$connected graph G with $l \geq 1$, there are $k$ internally disjoint $u-v$ paths connecting any two distinct vertices $u$ and $v$ for every integer $k$ with $1 \leq k \leq l$ [9]. Chartrand et al. [5] defined the rainbow $k$-connectivity $rc_k(G)$ of $G$ to be the minimum integer $j$ for which there exists a $j$-edge-coloring of $G$ such that for every two distinct vertices $u$ and $v$ of $G$, there exist at least $k$ internally disjoint $u-v$ rainbow paths.

By the definition of rainbow $k$-connectivity $rc_k(G)$, we realize that it is almost impossible to derive the exact value or a nice bound of the rainbow $k$-connectivity for a general graph $G$ [9]. To answer the problem: given that any connected graph $G$, determine the rainbow connection number $rc_k(G)$ of any graph $G$? It tends to be NP-hard problem. Thus, the study of rainbow $k$-connectivity of some classes of special graphs is still needed. In this paper we will study the rainbow connection number $rc_k(G)$ of Triangular Ladder, Wheel graphs, and edge comb of graph $G = C_n \trianglerighteq TL_m$ and $G = C_n \trianglerighteq K_m$. The edge comb between $L$ and $H$, denoted by $L \trianglerighteq H$, is a graph obtained by taking one copy of $L$ and $|E(L)|$ copies of $H$ and grafting the $i$-th copy of $H$ at the $i$-th edges of $L$. The result show that all the rainbow $k$-connection number $rc_k(G)$ of the graph studied in this paper achieve the minimum value.

## 2. The Results

Before presenting the main results we need to establish the lower bound of $rc_k(G)$ of any graph $G$ such that the graph $G$ is considered to be a $k$-connected graph. Note that the length of the shortest graph cycle (if any) in a given graph is known as a *girth*, and the length of a longest cycle is known as the *graph circumference*.

**Theorem 1.** *Let $d(u,v)$ be a distance between $u$ and $v$, $C(u,v)$ is a shortest cycle that contains the vertices $u$ and $v$. If $G$ is 2-connected graph then $rc_2(G) \geq max\ \{|C(u,v)| - d(u,v), \forall u,v \subset V(G)\}$, where $C(u,v)$ and $d(u,v)$ are in one cycle.*

*Proof.* Let $G$ be a connected cyclical graph. Thus, the length of second alternative internally disjoint rainbow path for any two vertices $u$ and $v$ is $|C(u,v)| - d(u,v)$ where $C(u,v)$ is a girth that contain vertices $u$ and $v$. The greatest lower bound of

IOP Conf. Series: Journal of Physics: Conf. Series **855** (2017) 012003 doi:10.1088/1742-6596/855/1/012003

$rc_2(G) \geq \max\ \{|C(u,v)| - d(u,v)\}$. By contradiction, if we color the edges of $G$ by any value less than $\max\ \{|C(u,v)| - d(u,v)\}$ then there exist two vertices $u$ and $v$ that do not present two internally disjoint paths. □

We can extend the theorem for $l$-connected graph.

**Lemma 1.** *If $G$ is $l$-connected graph, $l \geq 2$, then for every two vertices $u, v \in V(G)$, there exist at least $l-1$ cycles of $G$ containing the vertices $u$ and $v$.*

*Proof.* We can prove this theorem by contradiction. Suppose that there exist two vertices $u, v \in V(G)$ that contain one less than $l-1$ cycles of $G$. Suppose that the number of cycles that contain $u, v \in V(G)$ is $l-k$ where $k \geq 2$. The set $\{C_i | 1 \leq i \leq l-k\}$ is $l-k$ cycles that contain any two vertices in $V(G)$. One cycle is used to make two internally disjoint paths between $u$ and $v$. Two cycles are used to make three internally disjoint paths between $u$ and $v$. Since $u$ and $v$ are on $l-k$ cycles then the number of disjoint paths between $u$ and $v$ is $l-k+1$. Since $k \geq 2$ and we have two vertices with $l-k+1$ disjoint paths connecting $u$ and $v$, then $G$ is $(l-k+1 < l)$-connected graph. It is a contradiction. □

**Theorem 2.** *Let $d(u,v)$ be a distance between $u$ and $v$, $C_i(u,v)$ be a shortest cycles that contain vertices $u$ and $v$. Let $C_i$ be cycles whose their common edge is $uv$. If $G$ is $l$-connected graph then $rc_l(G) \geq max\{max\{|C_i(u,v)| - d(u,v), 1 \leq i \leq l-1\}, \forall u, v \in V(G)\}$, where $C(u,v)$ and $d(u,v)$ are in one cycle.*

*Proof.* If $G$ is $l$-connected graph, then by Lemma 1 every vertex in $V(G)$ lays on at least $l-1$ cycles. Suppose the element of $\{C_i(u,v) | 1 \leq i \leq l-1\, u, v \in V(G)\}$ have $l-1$ cycles containing $u, v \in V(G)$, the $l-1$ cycles that contain $u$ and $v$ has to be minimum of size $|C_i(u,v)|$. The number of $rc_k(G)$ is at least $\max\ \{|C_i(u,v)| - d(u,v), 1 \leq i \leq l-1\}$. Otherwise there exist two vertices $u, v$ that do not give $k$ internally disjoint rainbow path. □

Now we will present some classes of graphs which can be determined their rainbow $k$-connection number.

**Theorem 3.** *Let $G$ be a triangular ladder graph, the rainbow 2-connection number of $G$ is $rc_2(G) = n$.*

*Proof.* Suppose $G = TL_n$. The graph $G$ has vertex set $V(G) = \{x_i, y_i; 1 \leq i \leq n\}$ and edge set $E(G) = \{x_i x_{i+1}, y_i y_{i+1}, x_i y_{i+1}; 1 \leq i \leq n-1\} \cup \{x_i y_i; 1 \leq i \leq n\}$. Define a color $c$ of the edges $c : E(G) \to \{1, 2, \ldots, s\}, s \in N$:

$$c(e) = \begin{cases} n-i & , e \in \{x_i x_{i+1}; 1 \leq i \leq n-1\} \\ i & , e \in \{y_i y_{i+1}; 1 \leq i \leq n-1\} \\ 1 & , e \in \{x_i y_{i+1}; 1 \leq i \leq n-1\} \cup \{x_1 y_1\} \\ n & , e \in \{x_i y_i; 2 \leq i \leq n\} \end{cases}$$

It is easy to see that the color $c(e)$ reach a maximum value when $e = x_i y_i$ and $c(e) = n$. Thus, $rc_2(G) \leq n$. Now we will show that $rc_2(G) \geq n$. Consider the vertex $u = y_1$ and $v = x_n$. The vertex $u$ and $v$ lay on the cycle of size $2n$. Since distance, $d(u,v) = n$, then by Theorem 1, we have $rc_2(G) \geq 2n - n = n$. It concludes that $rc_2(G) = n$. □

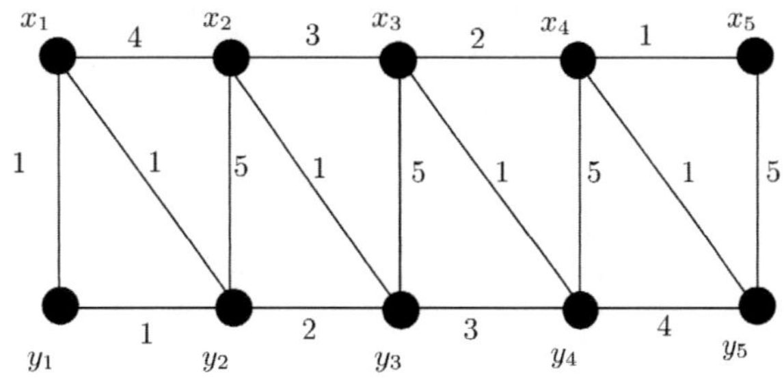


**Figure 1.** Graph $G = TL_5$ with $rc_2(G) = 5$

**Theorem 4.** *Let $G$ be a wheel graph of order $n+1$, the rainbow 3-connection number $G$ is $rc_3(W_n) = n$.*

*Proof.* Given that $G = W_n$. The graph $G$ has vertex set $V(G) = \{x_i; 1 \leq i \leq n\} \cup \{A\}$ and edge set $E(G) = \{Ax_i; 1 \leq i \leq n\} \cup \{x_i x_{i+1}; 1 \leq i \leq n-1\} \cup \{x_1 x_n\}$. Define a color $c$ of the edges $c : E(G) \rightarrow \{1, 2, \ldots, s\}, s \in N$:

$$c(e) = \begin{cases} i, & e \in \{x_i x_{i+1}; 1 \leq i \leq n-1\} \cup \{Ax_i; 1 \leq i \leq n\} \\ n, & e \in \{x_1 x_n\} \end{cases}$$

It is easy to see that the color $c(e)$ reach a maximum value when $e = x_1 x_n$, thus $rc_3(W_n) \leq n$. No we will show that $rc_3(W_n) \geq n$. We will use a contradiction. Suppose that $rc_3(W_n) \leq n-1$, take $rc_3(W_n) = n-1$. Consider edge set $E' = \{x_i x_{i+1} | 1 \leq i \leq n-1\} \cup \{x_1 x_n\}$ and $|E'| = n+1$. If we color $n+1$ edges of $E'$ by $n-1$ colors, then there exist $e_1, e_2 \in E'$ such that $c(e_1) = c(e_2)$, without loss of generality we can choose $e_1 = x_1 x_2$ and $e_2 = x_i x_{i+1}$. Since $W_n$ is 3-connected graph and $rc_3(W_n) = n-1$ then there must exist three disjoint paths between any two vertices. Consider vertex $x_1$ and vertex $x_{i+1}$ which give three disjoint paths between $x_1$ and $x_{i+1}$. The first possible rainbow path is $x_1 A x_{i+1}$, the second is $x_1 x_n x_{n-1} \ldots x_j$, however the third path $x_1 x_2 \ldots x_i x_{i+1}$, for $x_1$ and $x_{i+1}$ is not rainbow path as $c(x_1 x_2) = c(x_i x_{i+1})$. It is a contradiction, thus $rc_3(W_n) \geq n$. It concludes $rc_3(W_n) = n$. □

**Theorem 5.** *If $G = C_n \trianglerighteq TL_m$ then $rc(G) = \frac{n}{2} + 2m - 2$ for $n$ even and $rc_2(G) = 2m+1$ for $n = 4$.*

*Proof.* The graph $G = Cn \trianglerighteq Lt_m$ is a connected graph with vertex set $V(G) = \{x_i | 1 \leq i \leq n\} \cup \{y_{i,j} | 1 \leq i \leq n, 1 \leq j \leq m-1\} \cup \{z_{i,j} | 1 \leq i \leq n, 1 \leq j \leq m-1\}$ and edge set $E(G) = \{x_i x_{i+1} | i \leq i \leq n-1\} \cup \{x_n x_1\} \cup \{x_i y_{i,1} | 1 \leq i \leq n\} \cup \{x_{i+1} z_{i,1} | 1 \leq i \leq n-1\} \cup \{x_1 z_{n,1}\} \cup \{y_{i,j} y_{i,j+1} | 1 \leq i \leq n, 1 \leq j \leq m-2\} \cup \{z_{i,j} z_{i,j+1} | 1 \leq i \leq n, 1 \leq j \leq m-2\} \cup \{y_{i,j} z_{i,j} | 1 \leq i \leq n, 1 \leq j \leq m-1\} \cup \{x_i z_{i,1} | 1 \leq i \leq n\} \cup \{y_{i,j} z_{i,j+1} | 1 \leq i \leq n, 1 \leq j \leq m-2\}$. The value $|V(G)| = n(2m-1)$ and $|E(G)| = 3n + 2n(m-2) + 2n(m-1)$. The diameter of $G$, $diam(G) = \frac{n}{2} + 2(m-1)$. The number $rc(G)$ is given by the following

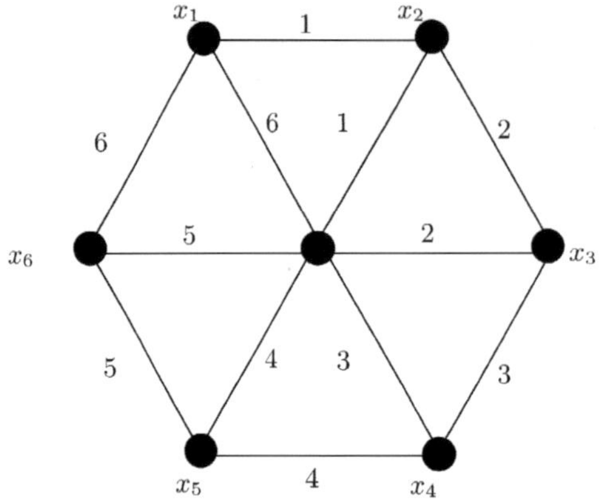


**Figure 2.** Graph $G = W_6$ with $rc_3(W_6) = 6$

coloring function:

$$c(e) = \begin{cases} i \bmod \frac{n}{2} & ,e \in \{x_ix_{i+1} | 1 \le i \le n-1\} \cup \\ & \{y_{i,j}z_{i,j} | 1 \le i \le n, 1 \le j \le m-1\} \\ n \bmod \frac{n}{2}, & e \in \{x_nx_1\} \\ \frac{n}{2}+1 & ,e \in \{x_iy_{i,1} | 1 \le i \le n\} \cup \{x_iz_{i,1} | 1 \le i \le n\} \\ \frac{n}{2}+1+j & ,e \in \{y_{i,j}y_{i,j+1} | 1 \le i \le n, 1 \le j \le m-2\} \\ & \cup \{y_{i,j}z_{i,j+1} | 1 \le i \le n, 1 \le j \le m-2\} \\ \frac{n}{2}+m & ,e \in \{x_{i+1}z_{i,1} | 1 \le i \le n-1\} \cup \{x_1z_{n,1}\} \\ \frac{n}{2}+m+j & ,e \in \{z_{i,j}z_{i,j+1} | 1 \le i \le n, 1 \le j \le m-2\} \end{cases}$$

The maximum value of function is $c(e) = \frac{n}{2} + 2m - 2$ so $rc(G) \le \frac{n}{2} + 2m - 2$. By applying Innequality 1 $rc(G) \ge \frac{n}{2} + 2m - 2$, it implies that $rc(G) = \frac{n}{2} + 2m - 2$.

The number $rc_2(G) \ge 2m+1$ for $n = 4$ and any $m$, is obtained by coloring mapping:

$$c(e) = \begin{cases} i \bmod 2 & ,e \in \{x_ix_{i+1} | 1 \le i \le 3\} \\ 2 & ,e \in \{x_4x_1\} \\ 3 & ,e \in \{x_iy_{i,1} | 1 \le i \le n\} \cup \{x_iz_{i,1} | 1 \le i \le n\} \\ 3+j & ,e \in \{y_{i,j}y_{i,j+1} | 1 \le i \le n, 1 \le j \le m-2\} \\ & \cup \{y_{i,j}z_{i,j+1} | 1 \le i \le n, 1 \le j \le m-2\} \\ m+2 & ,e \in \{x_{i+1}z_{i,1} | 1 \le i \le n-1\} \cup \{x_1z_{n,1}\} \\ m+2+j & ,e \in \{z_{i,j}z_{i,j+1} | 1 \le i \le n, 1 \le j \le m-2\} \\ 2m+1 & ,e \in \{y_{i,j}z_{i,j} | 1 \le i \le n, 1 \le j \le m-1\} \end{cases}$$

To prove $rc_2(G) \le 2m+1$, consider vertex $y_{2,m-1}$ and vertex $z_{1,m-1}$, the vertex $y_{2,m-1}$ and vertex $z_{1,m-1}$ lay on cycle of size of at least $4m-1$. The distance between $y_{2,m-1}$ and $z_{1,m-1}$ is $2(m-1)$ so the lengt of remaining shortest path between $y_{2,m-1}$ and $z_{1,m-1}$ is $2m+1$. This path is the shortest alternative path from $y_{2,m-1}$ to $z_{1,m-1}$ to get the second internally disjoint rainbow path. □

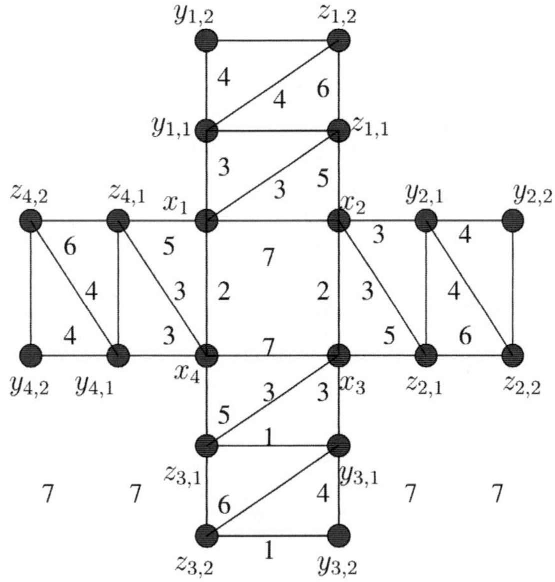


**Figure 3.** Graph edge comb $G = C_4 \trianglerighteq TL_3$ with $rc_2(G) = 7$.

**Theorem 6.** *If* $G = C_n \trianglerighteq K_m$, *then the number* $rc(G) = \frac{n}{2} + 1$ *for* $n$ *even and* $rc_2(G) = 4,$ *for* $n = 4$.

*Proof.* The graph $G = C_n \trianglerighteq K_m$ is a connected graph with vertex set $V(G) = \{x_i | 1 \leq i \leq n\} \cup \{y_{i,j} | 1 \leq i \leq n, 1 \leq j \leq m-2\}$ and edge set $E(G) = \{x_i x_{i+1} | 1 \leq i \leq n-1\} \cup \{x_n x_1\} \cup \{x_i y_{i,j} | 1 \leq i \leq n, 1 \leq j \leq m-2\} \cup \{x_{i+1} y_{i,j} | 1 \leq i \leq n-1, 1 \leq j \leq m-2\} \cup \{x_1 y_{n,j} | 1 \leq j \leq m-2\} \cup (\bigcup_{l=1}^{m-3}(\{y_{i,l} y_{i,j+l} | 1 \leq i \leq n, 1 \leq j \leq m-2-l\})$. The number of vertices and edges of $G$ is $|V(G)| = n + n(m-2)$ and $|E(G)| = n(1 + 2(m-2) + \frac{(m-2)(m-3)}{2})$. The Diameter of $G$, $diam(G) = \frac{n}{2} + 1$

The value $rc(G) = \frac{n}{2} + 1$ obtained by the following edge mapping function:

$$c(e) = \begin{cases} i \bmod \frac{n}{2} & , e \in \{x_i x_{i+1} | 1 \leq i \leq n-1\} \cup \\ & \{x_i y_{i,j} | 1 \leq i \leq n, 1 \leq j \leq m-2\} \cup \\ & (\bigcup_{l=1}^{m-3}(\{y_{i,l} y_{i,j+l} | 1 \leq i \leq n, \\ & 1 \leq j \leq m-2-l\}) \\ n \bmod \frac{n}{2} & , e \in \{x_n x_1\} \\ \frac{n}{2} + 1 & , e \in \{x_1 y_{n,j} | 1 \leq j \leq m-2\} \cup \\ & \{x_{i+1} y_{i,j} | 1 \leq i \leq n-1, 1 \leq j \leq m-2\} \end{cases}$$

The maximum value of $c(e)$ is $\frac{n}{2} + 1$ so $rc(G) \leq \frac{n}{2} + 1$, by applying Innequality 1 $rc(G) \geq \frac{n}{2} + 1$ and finally we get $rc(G) = \frac{n}{2} + 1$.

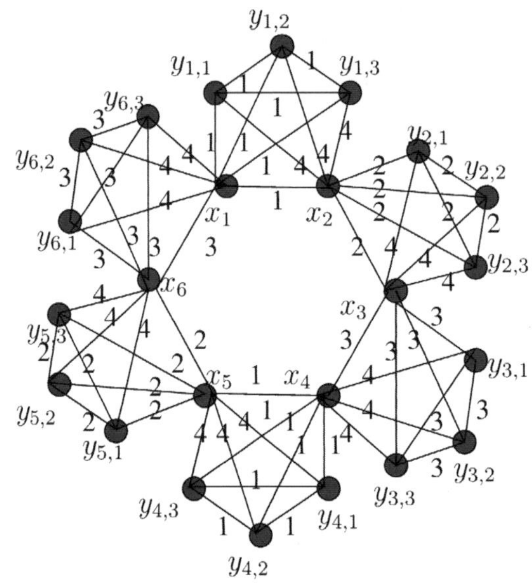

**Figure 4.** Graph edge comb $C_6 \trianglerighteq K_5$ with $rc(G) = 4$.

The value $rc_2(G) \geq 4$ for $n = 4$ and any $m$, is obtained by the following

$$c(e) = \begin{cases} i \bmod 2 & , e \in \{x_i x_{i+1} | 1 \leq i \leq 3\} \cup \{x_i y_{i,j} | 1 \leq i \leq 4, \\ & 1 \leq j \leq m-2\} \cup \{y_{i,j} y_{i,j+1} | 1 \leq i \leq 4, \\ & 1 \leq j \leq m-3\} \\ 4 \bmod 2 & , e \in \{x_4 x_1\} \\ 3 & , e \in \{x_1 y_{4,j} | 1 \leq j \leq m-2\} \cup \\ & \{x_{i+1} y_{i,j} | 1 \leq i \leq 3, 1 \leq j \leq m-2\} \\ 4 & , e \in \{x_i y_{i,j} | 1 \leq i \leq 4, 1 \leq j \leq m-2\} \\ & \cup (\bigcup_{l=1}^{m-3} (\{y_{i,l} y_{i,j+l} | 1 \leq i \leq 4, \\ & 1 \leq j \leq m-2-l\} - \{y_{i,j} y_{i,j+1} | 1 \leq i \leq 4, \\ & 1 \leq j \leq m-3\}) \end{cases}$$

To prove $rc_2(G) \leq 4$ consider vertex $y_{1,j}$ and $y_{2,k}$ for $1 \leq j, k \leq m-2$. This vertices is contained on cycle with size at least 6. The distance between $y_{1,j}$ and $y_{2,k}$ is 2 so the lengt of remaining shortest path between $y_{1,j}$ and $y_{2,k}$ is 4. This path is the shortest alternative path from $y_{1,j}$ to $y_{2,k}$ to make second internally disjoint rainbow path. □

### Concluding Remarks

We have studied the rainbow $k$-connection number of $G$. The result show that all the rainbow $k$-connection number $rc_k(G)$ of the graph studied in this paper achieve the minimum value. We have also characterized any graph to have a minimum $k$-connection number, through the following theorem: If $G$ is $l$-connected graph then

$rc_l(G) \geq \max\ \{|C_i(u,v)| - d(u,v), 1 \leq i \leq l-1\}$, where $|C_i(u,v)|$ is a girth that contains the vertices $u$ and $v$. However, it is just lower bound, we have not found the sharper upper bound of $rc_k(G)$ of any graph. Thus we propose the following open problem.

**Open Problem 1.** *Given that any connected graph $G$, determine a sharp upper bound of the rainbow $k$-connection number $rc_k(G)$ of $G$.*

**Acknowledgement**
We gratefully acknowledge the support from FUNDAMENTAL Ristek Dikti Grant 2017 & CGANT - University of Jember.

**Reference**

[1] Y. Caro, A. Lev, Y. Roditty, Z. Tuza, R. Yuster, On rainbow connection, *Electron. J. Combin.* **15**, R57, 2008
[2] L.S. Chandran, A. Das, D. Rajendraprasad, N.M. Varma, Rainbow connection number and connected dominating sets, Arxiv preprint arXiv:1010.2296v1 [math.CO], 2010
[3] S. Chakraborty, E. Fischer, A. Matsliah, R. Yuster, Hardness and algorithms for rainbow connectivity, 26th International Symposium on Theoretical Aspects of Computer Science STACS 2009, 243-254, 2009
[4] G. Chartrand, G.L. Johns, K.A. McKeon, P. Zhang, Rainbow connection in graphs, *Math. Bohem.* **133**, 85-98, 2008
[5] G. Chartrand, G.L. Johns, K.A. McKeon, P. Zhang, The rainbow connectivity of a graph, *Networks* **54** (2), 75-81, 2009
[6] J.L. Gross, J. Yellen and P. Zhang, *Handbook of Graph Theory*, Second Edition, CRC Press, Taylor and Francis Group, 2014
[7] X. Li, Y. Sun, Rainbow connections of graphs - A survey, arXiv:1101.5747v2 [math.CO], 2011
[8] X. Li, Y. Sun, Characterize graphs with rainbow connection number $m-2$ and rainbow connection numbers of some graph operations, Preprint, 2010
[9] Xueliang Li, Yuefang Sun, On the rainbow $k$-connectivity of complete graphs, Australian Journal Of Combinatorics, 217-226, 2011
[10] I. Schiermeyer, Rainbow connection in graphs with minimum degree three, *IWOCA 2009*, LNCS 5874, 2009

# On the total $H$-irregularity strength of graphs: A new notion

**Ika Hesti Agustin [1,2], Dafik[1,3], Marsidi [1,4], Ermita Rizki Albirri[1]**

[1]CGANT-University of Jember, Jember, Indonesia
[2]Department of Mathematics, University of Jember, Jember, Indonesia
[3]Department of Mathematics Education, University of Jember, Jember, Indonesia
[4]Department of Mathematics Education, IKIP PGRI of Jember, Jember, Indonesia

E-mail: ikahestiagustin@gmail.com, d.dafik@unej.ac.id, marsidiarin@gmail.com, mitalbi@sci.ui.ac.id

**Abstract.** A total edge irregularity strength of $G$ has been already widely studied in many papers. The total $\alpha$-labeling is said to be a total edge irregular $\alpha$-labeling of the graph $G$ if for every two different edges $e_1$ and $e_2$, it holds $w(e_1) \neq w(e_2)$, where $w(uv) = f(u) + f(uv) + f(v)$, for $e = uv$. The minimum $\alpha$ for which the graph $G$ has a total edge irregular $\alpha$-labeling is called the total edge irregularity strength of $G$, denoted by $tes(G)$. A natural extension of this concept is by considering the evaluation of the weight is not only for each edge but we consider the weight on each subgraph $H \subseteq G$. We extend the notion of the total $\alpha$-labeling into a total $H$-irregular $\alpha$-labeling. The total $\alpha$-labeling is said to be a total $H$-irregular $\alpha$-labeling of the graph $G$ if for $H \subseteq G$, the total $H$-weights $W(H) = \sum_{v\in V(H)} f(v) + \sum_{e\in E(H)} f(e)$ are distinct. The minimum $\alpha$ for which the graph $G$ has a total $H$-irregular $\alpha$-labeling is called the total $H$-irregularity strength of $G$, denoted by $tHs(G)$. In this paper we initiate to study the $tHs$ of shackle and amalgamation of any graphs and their bound.

**Keywords:** Total $\alpha$-labeling, Total $H$-irregularity strength, shackle of any graph, amalgamation of any graph.

## 1. Introduction

All graphs in this paper are simple, nontrivial and undirected graphs. A total labeling $f : V(G) \cup E(G) \to \{1, 2, 3, \ldots, \alpha\}$ is called a total $\alpha$-labeling of a graph $G$. The weight of an edge $uv$ of $G$, denoted by $w(uv)$, is the sum of the labels of end vertices $u$ and $v$ and also edge $uv$, i.e. $w(uv) = f(u)+f(uv)+f(v)$. The total $\alpha$-labeling is said to be a total edge irregular $\alpha$-labeling of the graph $G$ if for every two different edges $e_1$ and $e_2$, it holds $w(e_1) \neq w(e_2)$. The minimum $\alpha$ for which the graph $G$ has a total edge irregular $\alpha$-labeling is called the total edge irregularity strength of $G$, denoted by $tes(G)$. A natural extension of this concept is by considering the evaluation of the weight is not only for each edge but we consider the weight on each subgraph $H \subseteq G$. Thus, we extend the notion of the total $\alpha$-labeling into a total $H$-irregular $\alpha$-labeling. The total $\alpha$-labeling is said to be a total $H$-irregular $\alpha$-labeling of the graph $G$ if for $H \subseteq G$, the total $H$-weights $W(H) = \sum_{v\in V(H)} f(v) + \sum_{e\in E(H)} f(e)$ are distinct. The minimum $\alpha$ for which the graph $G$ has a total $H$-irregular $\alpha$-labeling is called the total $H$-irregularity strength of $G$, denoted by $tHs(G)$. The minimum $\alpha$ for which the graph $G$ has a subgraph irregular total $\alpha-$labeling is called *the total $H-$irregularity strength of* $G$, $tHs(G)$.

The beginning of the study of the irregularity strength is introduced by Togni et al. [10] and Frieze et al. [4]. By then, there are some result related to the total $H$-irregularity strength study. Jendrol et al. [6] determined the total edge irregularity strength of complete and bipartite complete graph, Jeyanthi et al. [7] studied about total edge irregularity strength of disjoin union wheel graph, and Baca et al. [2], [3] studied about total edge irregularity strength of generelized of prism graph and any graphs. Furthermore Ahmad et al. [1] found total edge irregularity strength of zigzag graph, as well as Pfender [8] studied about total edge irregularity strength of large graph, and the last Rajasingh et al. [9] also studied total edge irregularity strength of series parallel graph.

In this paper, we study the existence of the total $H$-irregularity $\alpha$-labeling of some graph operations, namely shackle and amalgamation of graph $G$. A shackle of $G_1, G_2, \ldots, G_k$, denoted by $Shack(G_1, G_2, \ldots, G_k)$, is any graph constructed from non-trivial connected and ordered graphs $G_1, G_2, \ldots, G_k$ such that for every $1 \leq i, j \leq k$ with $|i-j| \geq 2, G_i$ and $G_j$ have no common vertex and for every $1 \leq i \leq k-1, G_i$ and $G_i + 1$ share exactly one common vertex, called a linkage vertex, where the $k-1$ linkage vertices are all distinct. Meanwhile, let $\{G_1, G_2, \ldots, G_n\}$ be a finite collection of graphs and each $G_i$ has a fixed vertex $v_{0_i}$ or edge $e_{0_i}$ called a terminal *vertex* or *edge*, respectively [5]. The *vertex-amalgamation* of $G_1, G_2, \ldots, G_n$ denoted by *Amal* $\{G_i, v_{0_i}\}$, is formed by taking all the $G_i$'s and identifying their terminal vertices. Similarly, the *edge-amalgamation* of $G_1, G_2, \ldots, G_n$, denoted by *Amal*$\{G_i, e_{0_i}\}$, is formed by taking all the $G_i$'s and identifying their terminal edges. Furthermore, if $G_i$' are isomorphic graphs then we denote such graphs as *Shack* $\{G, v, n\}$ and *Amal* $\{G, v, n\}$ for vertex, or *Shack* $\{G, e, n\}$ and *Amal* $\{G, e, n\}$ for edge. In this paper we will study the $tHs$ of shackle and amalgamation of any graphs and as well as determine their bound.

## 2. The Results

Prior to show the values of $tHs$ of those graphs, we will show the lower bound of $tHs$ in general graph by the following lemma.

**Lemma 2.1** *Given a graph $H \subset G$. Let $p_H, q_H$ be respectively be number of vertices and edges of $H$ and $|H|$ be the number of subgraphs. The total $H$-irregularity strength satisfies*

$$tHs(G) \geq \lceil \tfrac{p_H + q_H + |H| - 1}{p_H + q_H} \rceil$$

**Proof.** A total $\alpha$-labeling is a labeling $f: V(G) \cup E(G) \rightarrow \{1, 2, 3, \ldots, \alpha\}$. The $H$- irregularity total $\alpha$-labeling of graph $G$ is a total $\alpha$-labeling such that for each subgraph $H \subseteq G$, the weight $W(H) = \sum_{v \in V(K)} f(v) + \sum_{e \in E(K)} f(e)$ are all distinct. Furthermore, since we require the minimum $\alpha$ for which the graph $G$ has a total $H$-irregular $\alpha$-labeling, the set of the total $H$-weight should be consecutive, otherwise it will not give a minimum $tHs$. Thus, the set of total $H$ weight is $W(H) = \{p_H + q_H, p_H + q_H + 1, p_H + q_H + 2, \ldots, p_H + q_H + (|H| - 1)\}$. On the other hand the maximum possible $H$ weight of graph $G$ is at most $tHs(G)(p_H + q_H)$. It implies

$$\begin{aligned} tHs(G)(p_H + q_H) &\geq p_H + q_H + |H| - 1 \\ tHs(G) &\geq \tfrac{p_H + q_H + |H| - 1}{p_H + q_H} \end{aligned}$$

Since $tHs(G)$ should be integer, and we need a sharpest lower bound, it implies

$$tHs(G) \geq \lceil \frac{p_H + q_H + |H| - 1}{p_H + q_H} \rceil.$$

It completes the proof.

Now, we are ready to show our main results.

**Theorem 2.1** *Let $G$ = Shack$(H, v, n)$ be a shackle of any graph $H$. Then the total $H$-irregularity strength satisfies*

$$tHs(\text{Shack}(H, v, n)) = \lceil \frac{m+n+1}{m+2} \rceil$$

*where $p_H$ and $q_H$ are respectively the number of vertices and edges in subgraph $H \subseteq G$ and $m = p_H + q_H - 2$ and $n = |H|$.*

**Proof.** The vertex set and edge set of the graph Shack$(H, v, n)$ can be split into two following sets: $V(\text{Shack}(H, v, n)) = \{v_{ij}; 1 \le i \le p_H - 2, 1 \le j \le n\} \cup \{x_k; 1 \le k \le n+1\}$ and $E(\text{Shack}(H, v, n)) = \{e_{lj}; 1 \le l \le q_H, 1 \le j \le n\}$. Thus, the graph Shack$(H, v, n)$ has $|V(\text{Shack}(H, v, n))| = (n-1)p_H + 1$, $|E(\text{Shack}(H, v, n))| = nq_H$. Since $m = p_H + q_H - 2$, then by Lemma 2.1, we have $tHs(\text{Shack}(H, v, n)) \ge \lceil \frac{p_H+q_H+|H|-1}{p_H+q_H} \rceil = \lceil \frac{m+2+n-1}{m+2} \rceil = \lceil \frac{m+n+1}{m+2} \rceil$. Thus, $tHs(\text{Shack}(H, v, n)) \ge \lceil \frac{m+n+1}{m+2} \rceil$.

Now we will show that $tHs(\text{Shack}(H, v, n)) \le \lceil \frac{m+n+1}{m+2} \rceil$. Define $f$ as a vertex and edge labeling of graph $G$, $f : V(G) \cup E(G) \to \{1, 2, \ldots, \alpha\}$ by the following function.

$$\begin{aligned} f(x_k) &= \lceil \tfrac{k}{m+2} \rceil \\ f(v_{ij}) \cup f(e_{lj}) &= \begin{cases} \lceil \frac{j}{m-(t+1)} \rceil; 1 \le j \le m-t+1, 1 \le t \le m \\ \lceil \frac{j+t-(m+1)}{m+2} \rceil + 1; m-t+2 \le j \le n, 1 \le t \le m \end{cases} \end{aligned}$$

Under the labeling $f$, the total $H$-weight $W(H) = \sum_{v \in V(H)} f(v) + \sum_{e \in E(H)} f(e)$ is $W(H) = \{m+2, m+3, \ldots, m+n+1\}$ forms a consecutive sequence. It implies the set of $H-$weights are distinct. By considering the above label $f$, the minimum $tHs(\text{Shack}(H, v, n))$ can be achieved by the following:

$$\begin{aligned} tHs(\text{Shack}(H, v, n)) &\le \lceil \tfrac{j+t-(m+1)}{m+2} \rceil + 1, \text{ for } j = n, t = m \\ &= \lceil \tfrac{n+m-m-1}{m+2} + \tfrac{m+2}{m+2} \rceil \\ &= \lceil \tfrac{n-1+m+2}{m+2} \rceil \\ &= \lceil \tfrac{m+n+1}{m+2} \rceil \end{aligned}$$

Thus, $tHs(\text{Shack}(H, v, n)) \le \lceil \frac{m+n+1}{m+2} \rceil$. It concludes that $tHs(\text{Shack}(H, v, n)) = \lceil \frac{m+n+1}{m+2} \rceil$. □

**Theorem 2.2** *Let $G = c$Shack$(H, v, n)$ be disjoint union of multiple copies $c$ of shackle of graph $H$. Then*

$$tHs(c\text{Shack}(H, v, n)) = \lceil \frac{m+cn+1}{m+2} \rceil$$

*where $m = p_H + q_H - 2$, $p_H$ and $q_H$ are the number of vertices and edges in $H$ respectively, $n = |H|$ and $c$ is number of copies of $G$.*

**Proof.** The graph $G = c$Shack$(H, v, n)$ is a diconnected graph with vertex set $V(c\text{Shack}(H, v, n)) = \{v^u_{ij}; 1 \le i \le p_H - 2, 1 \le j \le n, 1 \le u \le c\} \cup \{x^u_k; 1 \le k \le n+1, 1 \le u \le c\}$ and edge set $E(c\text{Shack}(H, v, n)) = \{e^u_{lj}; 1 \le l \le q_H, 1 \le j \le n, 1 \le u \le c\}$. Thus, the graph $c$Shack$(H, v, n)$ has $|V(c\text{Shack}(H, v, n))| = c((n-1)p_H + 1)$ and $|E(c\text{Shack}(H, v, n))| = cnq_H$. Since $m = p_H + q_H - 2$, then by Lemma 2.1

$$\begin{aligned} tHs(c\text{Shack}(H, v, n)) &\ge \lceil \tfrac{p_H+q_H+|H|-1}{p_H+q_H} \rceil \\ &= \lceil \tfrac{m+2+cn-1}{m+2} \rceil \\ &= \lceil \tfrac{m+cn+1}{m+2} \rceil \end{aligned}$$

Now we will show that $tHs(c\text{Shack}(H,v,n)) \leq \lceil \frac{m+cn+1}{m+2} \rceil$. The vertex and edge labeling $f$ is a bijective function $f : V(G) \cup E(G) \to \{1, 2, \ldots, \alpha\}$. Let $w = ju$; $1 \leq j \leq n$, $1 \leq u \leq c$ such that $1 \leq w \leq cn$.

$$\begin{aligned} f(x_k^u) &= \lceil \tfrac{u}{m+2} \rceil, 1 \leq u \leq c \\ f(v_{ij}^u) \cup f(e_{lj}^u) &= \begin{cases} \lceil \frac{w}{m-(t+1)} \rceil; 1 \leq w \leq m-t+1, 1 \leq t \leq m \\ \lceil \frac{w+t-(m+1)}{m+2} \rceil + 1; m-t+2 \leq w \leq cn, 1 \leq t \leq m \end{cases} \end{aligned}$$

Under the labeling $f$, the total $H$-weight $W(H) = \sum_{v \in V(H)} f(v) + \sum_{e \in E(H)} f(e)$ is $W(H) = \{m+2, m+3, \ldots, m+cn+1\}$ which form a consecutive sequence. It implies the set of $H-$weights are distinct. Now considering the above label of $f$, the minimum $tHs(c\text{Shack}(H,v,n))$ can be achieved by the following:

$$\begin{aligned} tHs(c\text{Shack}(H,v,n)) &\leq \lceil \tfrac{cn+m-(m+1)}{m+2} \rceil + 1 \\ &= \lceil \tfrac{cn+1}{m+2} + \tfrac{m+2}{m+2} \rceil \\ &= \lceil \tfrac{cn-1}{m+2} + \tfrac{m+2}{m+2} \rceil \\ &= \lceil \tfrac{m+cn-1}{m+2} \rceil \end{aligned}$$

Thus, $tHs(c\text{Shack}(H,v,n)) \leq \lceil \frac{m+cn-1}{m+2} \rceil$. It implies that $tHs(c\text{Shack}(H,v,n)) = \lceil \frac{m+cn-1}{m+2} \rceil$. □

**Theorem 2.3** *Let $G$ be an amalgamation of any connected graph $H$, denoted by $G =$ $\text{Amal}(H,v,n)$. Then the following holds*

$$tHs(\text{Amal}(H,v,n)) = \lceil \frac{r+n-1}{r} \rceil$$

*where $r = p_H + q_H - 1$, $p_H$ and $q_H$ is the number of vertices and edges in $H$ respectively and $n = |H|$.*

**Proof.** The vertex set and edge set of the graph $\text{Amal}(H,v,n)$ can be split into following sets: $V(\text{Amal}(H,v,n)) = \{A\} \cup \{x_{ij}; 1 \leq i \leq p_H - 1, 1 \leq j \leq n\}$ and $E(\text{Amal}(H,v,n)) = \{e_{lj}; 1 \leq l \leq q_H, 1 \leq j \leq n\}$. Thus, the graph $\text{Amal}(H,v,n)$ has $|V(\text{Amal}(H,v,n))| = p_G$, and $|E(\text{Amal}(H,v,n))| = q_G$. Let $n$, $m$ be positive integers with $n \geq 2$ and $m \geq 3$. Thus $|V(\text{Amal}(H,v,n))| = p_G = n(p_H - 1) + 1$ and $|E(\text{Amal}(H,v,n))| = q_G = nq_H$. Then by lemma 2.1,

$$\begin{aligned} tHs(\text{Amal}(H,v,n)) &\geq \lceil \tfrac{p_H+q_H+|H|-1}{p_H+q_H} \rceil \\ &= \lceil \tfrac{r+1+n-1}{r+1} \rceil \\ &= \lceil \tfrac{r+n}{r+1} \rceil \\ &= \lceil \tfrac{r+n-1}{r} \rceil \end{aligned}$$

Thus, the lower bound $tHs(\text{Amal}(H,v,n)) \geq \lceil \frac{r+n-1}{r} \rceil$. Now we will prove that $tHs(\text{Amal}(H,v,n)) \leq \lceil \frac{r+n-1}{r} \rceil$. The vertex and edge labeling $f$ is a bijective function $f : V(G) \cup E(G) \to \{1, 2, \ldots, \alpha\}$.

$$\begin{aligned} f(A) &= 1 \\ f(x_{ij}) \cup f(e_{lj}) &= \begin{cases} \lceil \frac{j}{r-(t-1)} \rceil; 1 \leq j \leq r-t+1, 1 \leq t \leq r \\ \lceil \frac{j+t-(r+1)}{r} \rceil + 1; r-i+2 \leq j \leq n, 1 \leq t \leq r. \end{cases} \end{aligned}$$

Under the labeling $f$, the total $H$-weight $W(H) = \sum_{v \in V(H)} f(v) + \sum_{e \in E(H)} f(e)$ is $W(H) = \{r+1, r+2, \ldots, r+n\}$ form a consecutive sequence. It implies the set of $H-$weights are distinct.

IOP Conf. Series: Journal of Physics: Conf. Series **855** (2017) 012004 doi:10.1088/1742-6596/855/1/012004

Now by considering the above label $f$, the minimum $tHs(\text{Amal}(H,v,n))$ can be achieved by the following:

$$\begin{aligned} tHs(\text{Amal}(H,v,n)) &\leq \lceil \tfrac{n+r-(r+1)}{r} \rceil + 1 \\ &= \lceil \tfrac{n-1}{r} + \tfrac{r}{r} \rceil \\ &= \lceil \tfrac{r+n-1}{r} \rceil \end{aligned}$$

It is clear to concludes that $tHs(\text{Amal}(H,v,n)) = \lceil \frac{r+n-1}{r} \rceil$. □

**Theorem 2.4** *Let $G$ be a disjoint union of multiple copies $c$ of amalgamation of graph $H$, denoted by $G = c\text{Amal}(H,v,n)$. Then*

$$tHs(c\text{Amal}(H,v,n)) = \lceil \frac{r+cn-1}{r} \rceil$$

*where $r = p_H + q_H - 1$, $p_H$ and $q_H$ is the number of vertices and edges in $H$ respectively, $n = |H|$ and $c$ is number of copies of $G$.*

**Proof.** The vertex set and edge set of the graph $G = c\text{Amal}(H,v,n)$ can be split into following sets: $V(G) = \{A^k; 1 \leq k \leq c\} \cup \{x_{ij}{}^k; 1 \leq i \leq p_H - 1, 1 \leq j \leq n, 1 \leq k \leq c\}$ and $E(G) = \{e_{lj}{}^k; 1 \leq j \leq n, 1 \leq l \leq q_H, 1 \leq k \leq c\}$. Thus the graph $c\text{Amal}(H,v,n)$ has with $|V(c\text{Amal}(H,v,n))| = p_G$, and $|E(c\text{Amal}(H,v,n))| = p_G$. Let $n$, $r$, and odd $c$ be positive integers with $n \geq 2$ and $r, c \geq 3$. Thus $|V(G)| = p_G = c(n(p_H - 1) + 1)$ and $|E(G)| = q_G = cnq_H$. Then by lemma 2.1,

$$\begin{aligned} tHs(c\text{Amal}(H,v,n)) &\geq \lceil \tfrac{p_H+q_H+|H|-1}{p_H+q_H} \rceil \\ &= \lceil \tfrac{r+1-cn-1}{r} \rceil \\ &= \lceil \tfrac{r+cn}{r+1} \rceil \\ &= \lceil \tfrac{r+cn-1}{r} \rceil \end{aligned}$$

Thus, the lower bound $tHs(c\text{Amal}(H,v,n)) \geq \lceil \frac{r+cn-1}{r} \rceil$. Now we will show that $tHs(c\text{Amal}(H,v,n)) \leq \lceil \frac{r+cn-1}{r} \rceil$. For any $V$ and $E$, the labeling as follows. Let $w = jk$; $1 \leq j \leq n$, $1 \leq k \leq c$ such that $1 \leq w \leq cn$.

$$\begin{aligned} f(A^k) &= 1, 1 \leq k \leq c \\ f(x_{ij}^k) \cup f(e_{lj}^k) &= \begin{cases} \lceil \frac{w}{r-(t-1)} \rceil; 1 \leq w \leq r-t+1, 1 \leq t \leq r, 1 \leq k \leq c \\ \lceil \frac{w+t-(r+1)}{r} \rceil + 1; r-t+2 \leq w \leq cn, 1 \leq t \leq r, 1 \leq k \leq c. \end{cases} \end{aligned}$$

Under the labeling $f$, the total $H$-weight $W(H) = \sum_{v \in V(H)} f(v) + \sum_{e \in E(H)} f(e)$ is $W(H) = \{r+1, r+2, \ldots, r+cn\}$ form a consecutive sequence. It implies the set of $H-$weights are distinct. Now considering the above label of $f$, the minimum $tHs(c\text{Amal}(H,v,n))$ can be achieved by the following:

$$\begin{aligned} tHs(c\text{Amal}(H,v,n)) &\leq \lceil \tfrac{w+t-(r+1)}{r} \rceil + 1 \\ &= \lceil \tfrac{cn+r-r-1}{r} \rceil + \lceil \tfrac{r}{r} \rceil \\ &= \lceil \tfrac{cn+r-1}{r} \rceil \end{aligned}$$

It concludes the proof. □

**Theorem 2.5** *Let $G$ be a shackle of connected graph $C_m$ graph, denoted by $G = \text{Shack}(C_m, v, n)$. Then*

$$tHs(\text{Shack}(C_m,v,n)) = \lceil \frac{2m+n-1}{2m} \rceil$$

*where $m$ is an order of the cycle graph and $n$ number of $C_m$.*

**Proof.** The graph $\text{Shack}(C_m, v, n)$ is a connected graph with vertex set $V(\text{Shack}(C_m, v, n)) = \{v_{ij}; 1 \le i \le p_{C_m} - 2, 1 \le j \le n\} \cup \{x_k; 1 \le k \le n+1\}$ and edge set $E(\text{Shack}(C_m, v, n)) = \{e_{lj}; 1 \le l \le q_{C_m}, 1 \le j \le n\}$. The cardinalities of the graph $\text{Shack}(C_m, v, n)$ are $|V(\text{Shack}(C_m, v, n))| = (n-1)p_{C_m} + 1$, and $|E(\text{Shack}(C_m, v, n))| = nq_{C_m}$, where $p_{C_m} = |V(C_m)|$, and $q_{C_m} = |E(C_m)|$. Then by Lemma 2.1,

$$\begin{aligned} tHs(\text{Shack}(C_m, v, n)) &\ge \lceil \tfrac{p_{C_m} + q_{C_m} + |C_m| - 1}{p_{C_m} + q_{C_m}} \rceil \\ &= \lceil \tfrac{m+m+n-1}{m+m} \rceil \\ &= \lceil \tfrac{2m+n-1}{2m} \rceil \end{aligned}$$

Now we will show that $tHs(\text{Shack}(C_m, v, n)) \le \lceil \frac{2m+n-1}{2m} \rceil$. Define the vertex and edge labelings $f : V(G) \cup E(G) \to \{1, 2, \ldots, \alpha\}$ as follows

$$\begin{aligned} f(x_k) &= \lceil \tfrac{k}{2m} \rceil \\ f(v_{ij}) \cup f(e_{lj}) &= \begin{cases} \lceil \frac{j}{2m-2-(t+1)} \rceil; 1 \le j \le 2m-t-1, 1 \le t \le 2m-2 \\ \lceil \frac{j+t-(2m-2+1)}{2m} \rceil + 1; 2m-t \le j \le n, 1 \le t \le 2m-2 \end{cases} \end{aligned}$$

Under the labeling $f$, the total $H$-weight $W(H) = \sum_{v \in V(H)} f(v) + \sum_{e \in E(H)} f(e)$ is $W(H) = \{2m, 2m+1, \ldots, 2m+n-1\}$ form a consecutive sequence. It implies the set of $H-$weights are distinct. Now considering the above label of $f$, the minimum $tHs(\text{Shack}(C_m, v, n))$ can be achieved by the following:

$$\begin{aligned} tHs(\text{Shack}(C_m, v, n)) &\le \lceil \tfrac{j+t-(2m-2+1)}{2m} \rceil + 1 \\ &= \lceil \tfrac{n+2m-2-(2m-2+1)}{2m} + \tfrac{2m}{2m} \rceil \\ &= \lceil \tfrac{n-1}{2m} + \tfrac{2m}{2m} \rceil \\ &= \lceil \tfrac{2m+n-1}{2m} \rceil \end{aligned}$$

Thus $tHs(\text{Shack}(C_m, v, n)) \le \lceil \frac{2m+n-1}{2m} \rceil$, it implies that $tHs(\text{Shack}(C_m, v, n)) = \lceil \frac{2m+n-1}{2m} \rceil$. □

**Theorem 2.6** *Let $G$ be a disjoint union of multiple copies $c$ of shackle of graph $C_m$, denoted by $G = c\text{Shack}(C_m, v, n)$. Then*

$$tHs(c\text{Shack}(C_m, v, n)) = \lceil \frac{2m + cn - 1}{2m} \rceil$$

*where $m$ is an order of the cycle graph, $n$ is a number of $C_m$, and $c$ is number of multiple copies of $G$.*

**Proof.** Suppose we denote the vertex and edge sets of the graph $G = c\text{Shack}(C_m, v, n)$ as follows: $V(c\text{Shack}(C_m, v, n)) = \{v^u_{ij}; 1 \le i \le p_{C_m} - 2, 1 \le j \le n, 1 \le u \le c\} \cup \{x^u_k; 1 \le k \le n+1, 1 \le u \le c\}$ and $E(c\text{Shack}(C_m, v, n)) = \{e^u_{lj}; 1 \le l \le q_{C_m}, 1 \le j \le n, 1 \le u \le c\}$. Thus, the graph $c\text{Shack}(C_m, v, n)$ has $|V(c\text{Shack}(C_m, v, n))| = c((n-1)p_{C_m} + 1)$, and $|E(c\text{Shack}(C_m, v, n))| = cnq_{C_m}$, where $p_{C_m} = |V(C_m)|$ and $q_{C_m} = |E(C_m)|$. Then by Lemma 2.1

$$\begin{aligned} tHs(c\text{Shack}(C_m, v, n)) &\ge \lceil \tfrac{p_{C_m} + q_{C_m} + |C_m| - 1}{p_{C_m} + q_{C_m}} \rceil \\ &= \lceil \tfrac{2m+cn-1}{2m} \rceil \end{aligned}$$

Now we will show that $tHs(c\text{Shack}(C_m, v, n)) \le \lceil \frac{2m+cn-1}{2m} \rceil$ by defining the vertex and edge labeling $f : V(G) \cup E(G) \to \{1, 2, \ldots, \alpha\}$ by the following. Let $w = ju$; $1 \le j \le n$, $1 \le u \le c$ such that $1 \le w \le cn$.

$$\begin{aligned} f(x^u_k) &= \lceil \tfrac{u}{2m} \rceil, 1 \le u \le c \\ f(x^u_{ij}) \cup f(e^u_{lj}) &= \begin{cases} \lceil \frac{w}{(2m-2)-(t+1)} \rceil; 1 \le w \le 2m-t-1, 1 \le t \le 2m-2 \\ \lceil \frac{w+t-(2m-2+1)}{2m} \rceil + 1; 2m-t \le w \le cn, 1 \le t \le 2m-2 \end{cases} \end{aligned}$$

Under the labeling $f$, the total $H$-weight $W(H) = \sum_{v\in V(H)} f(v) + \sum_{e\in E(H)} f(e)$ is $W(H) = \{2m, 2m+1, \dots, 2m+cn-1\}$ form a consecutive sequence. It implies the set of $H-$weights are distinct. Now considering the above label of $f$, the minimum $tHs(c\text{Shack}(C_m, v, n))$ can be achieved by the following.

$$\begin{aligned} tHs(\text{Shack}(C_m, v, n)) &\le \lceil \tfrac{w+t-(2m-2+1)}{2m} \rceil + 1 \\ &= \lceil \tfrac{cn+(2m-2)-(2m-2+1)}{2m} + \tfrac{2m}{2m} \rceil \\ &= \lceil \tfrac{cn-1}{2m} + \tfrac{2m}{2m} \rceil \\ &= \lceil \tfrac{2m+cn-1}{2m} \rceil \end{aligned}$$

It is clear to conclude that $tHs(\text{Shack}(C_m, v, n)) = \lceil \frac{2m+cn-1}{2m} \rceil$. □

**Theorem 2.7** *Let $G$ be an amalgamation of connected graph $C_3$, denoted by $G = \text{Amal}(C_3, v, n)$. Then the following holds*

$$tHs(\text{Amal}(C_3, v, n)) = \lceil \frac{n+4}{5} \rceil$$

*where $n$ is a number of $C_3$.*

**Proof.** Let the graph $\text{Amal}(C_3, v, n)$ has with $|V(G)| = p_G$, $|E(G)| = q_G$, $|V(H)| = |V(C_3)| = p_H = p_{C_3}$, and $|E(H)| = |E(C_3)| = q_H = q_{C_3}$. Suppose we denote the vertex and edge sets of the graph $G = \text{Amal}(C_3, v, n)$ as follows: $V(G) = \{A\} \cup \{x_{ij}; 1 \le i \le 2, 1 \le j \le n\}$ and $E(G) = \{Ax_{ij}; 1 \le i \le 2, 1 \le j \le n\} \cup \{x_{1j}x_{2j}; 1 \le j \le n\}$. Thus, the graph $\text{Amal}(C_3, v, n)$ has $|V(\text{Amal}(C_3, v, n))| = 2n+1$, and $|E(\text{Amal}(C_3, v, n))| = 3n$, where $p_{C_3} = |V(C_3)|$ and $q_{C_3} = |E(C_3)|$. Then by Lemma 2.1, we have the following

$$\begin{aligned} tHs(\text{Amal}(C_3, v, n)) &\ge \lceil \tfrac{P_H+q_H+|H|-1}{P_H+q_H} \rceil \\ &= \lceil \tfrac{6+n-1}{6} \rceil \\ &= \lceil \tfrac{n+5}{6} \rceil \\ &= \lceil \tfrac{n+4}{5} \rceil \end{aligned}$$

Thus, the lower bound $tHs\text{Amal}(C_3, v, n)) \ge \lceil \frac{n+4}{5} \rceil$. Now we will show that $tHs(\text{Amal}(C_3, v, n)) \le \lceil \frac{n+4}{5} \rceil$. The vertex and edge labelings $f$ is a bijective function $f : V(G) \cup E(G) \to \{1, 2, \dots, \alpha\}$.

$$\begin{aligned} f(A) &= 1 \\ f(x_{i,j}) \cup f(Ax_{i,j}) \cup f(x_{ij}x2j) &= \begin{cases} \lceil \frac{j}{5-(i-1)} \rceil; 1 \le j \le 6-i, 1 \le i \le 5 \\ \lceil \frac{j+i-(6)}{5} \rceil + 1; 7-i \le j \le n, 1 \le i \le 5. \end{cases} \end{aligned}$$

Under the labeling $f$, the total $H$-weight $W(H) = \sum_{v\in V(H)} f(v) + \sum_{e\in E(H)} f(e)$ is $W(H) = \{6, 7, \dots, 6+(n-1)\}$ which form a consecutive sequence. It implies the set of $H-$weights are distinct. Now considering the above label of $f$, the minimum $tHs\text{Amal}(C_3, v, n))$ can be achieved by the following.

$$\begin{aligned} tHs(\text{Amal}(C_3, v, n)) &\le \lceil \tfrac{j+i-(6)}{5} \rceil + 1 \\ &= \lceil \tfrac{5+n-6}{5} + \tfrac{5}{5} \rceil \\ &= \lceil \tfrac{n+4}{5} \rceil \end{aligned}$$

It concludes the proof. □

**Theorem 2.8** *Let $G$ be a disjoint union of amalgamation of $C_3$ graph, denoted by* $c\text{Amal}(C_3, v, n)$. *Then*

$$tHs(c\text{Amal}(C_3, v, n)) = \lceil \frac{cn+4}{5} \rceil$$

**Proof.** Let the graph $c\text{Amal}(C_3, v, n)$ has with $|V(G)| = p_G$, $|E(G)| = q_G$, $|V(H)| = |V(C_3)| = p_H = p_{C_3}$, and $|E(H)| = |E(C_3)| = q_H = q_{C_3}$. The vertex set and edge set of the graph $G = c\text{Amal}(C_3, v, n)$ can be split into following sets: $V(G) = \{A^k; 1 \le k \le c\} \cup \{x^k_{ij}; 1 \le i \le 2, 1 \le j \le n, 1 \le k \le c\}$ and $E(G) = \{A^k x^k_{ij}; 1 \le i \le 2, 1 \le j \le n, 1 \le k \le c\} \cup \{x^k_{1j} x^k_{2j}; 1 \le j \le n, 1 \le k \le c\}$. Let $n$, $m$, and odd $s$ be positive integers with $n \ge 2$ and $r, c \ge 3$. Thus $|V(G)| = p_G = c(2n+1)$ and $|E(G)| = q_G = 3cn$. Then by lemma 2.1,

$$\begin{aligned} tHs(c\text{Amal}(C_3, v, n)) &\ge \lceil \tfrac{P_H + q_H + |H| - 1}{P_H + q_H} \rceil \\ &= \lceil \tfrac{6+cn-1}{6} \rceil \\ &= \lceil \tfrac{5+cn}{6} \rceil \\ &= \lceil \tfrac{4+cn}{5} \rceil \end{aligned}$$

Thus, the lower bound $tHs(c\text{Amal}(C_3, v, n)) \ge \lceil \frac{cn+4}{5} \rceil$. Now we will prove that $tHs(c\text{Amal}(H, v, n)) \le \lceil \frac{cn+4}{5} \rceil$. Let $l = jk; 1 \le j \le n, 1 \le k \le c$ such that $1 \le l \le cn$. For any $V$ and $E$, the labeling as follows.

$$\begin{aligned} f(A^k) &= 1, 1 \le k \le c \\ f(x^k_{i,j}) \cup f(A^k x^k_{i,j}) \cup f(x^k_{ij} x^k_{2j}) &= \begin{cases} \lceil \frac{l}{5-(i-1)} \rceil; 1 \le l \le 6-i, 1 \le i \le 5, 1 \le k \le c \\ \lceil \frac{l+i-(6)}{5} \rceil + 1; 7-i \le l \le cn, 1 \le i \le 5, 1 \le k \le c. \end{cases} \end{aligned}$$

Under the labeling $f$, the total $H$-weight $W(H) = \sum_{v \in V(H)} f(v) + \sum_{e \in E(H)} f(e)$ is $W(H) = \{6, 7, \ldots, 6 + (cn-1)\}$ which form a consecutive sequence. It implies the set of $H-$weights are distinct. Now considering the above label of $f$, the minimum $tHs(c\text{Amal}(C_3, v, n))$ can be achieved by the following.

$$\begin{aligned} tHs(c\text{Amal}(C_3, v, n)) &\ge \lceil \tfrac{l+i-(6)}{5} \rceil + 1 \\ &= \lceil \tfrac{5+cn-6}{5} + \tfrac{5}{5} \rceil \\ &= \lceil \tfrac{cn+4}{5} \rceil \end{aligned}$$

Thus $tHs(c(\text{Amal}(C_3, v, n))) \le \lceil \frac{cn+4}{5} \rceil$, it implies that $tHs(c(\text{Amal}(C_3, v, n))) = \lceil \frac{cn+4}{5} \rceil$. □

**Concluding Remarks**

We have found the total $H$-irregularity strength of shackle and amalgamation of $G$, namely $tHs(\text{Shack}(H, v, n))$, $tHs(c(\text{Shack}(H, v, n)))$, $tHs(\text{Amal}(H, v, n))$ and $tHs(c(\text{Amal}(H, v, n)))$. Apart from those graphs, the study of the values of $tHs$ are considered to be interesting research topic as it is a new extension of total edge irregularity strength of $G$. Therefore, we propose the following open problem.

**Open Problem 2.1** *Let $G$ be any connected and disconnected graph, apart from the above graphs determine the value of $tHs(G)$.*

**Open Problem 2.2** *Let $tes(G)$ and $tHs(G)$ be total edge irregularity strength and total $H$-irregularity strength of graph $G$. Characterize the connection between $tes(G)$ and $tHs(G)$.*

**Acknowledgement**

We gratefully acknowledge the support from CGANT - University of Jember of year 2017.

IOP Conf. Series: Journal of Physics: Conf. Series **855** (2017) 012004 doi:10.1088/1742-6596/855/1/012004

## References

[1] Ahmad A and Kamran M S. 2012. Total edge irregularity strength of zigzag graph. *Australian Journal of Combinatorics* **54** 141-149.
[2] Baca M and Kamran M S. 2014. Total edge irregularity strength of generalized prism *Applied Mathematics and Computation.* **235** 168-173.
[3] Baca M, Ahmad A, and Saeed O B A. 2014. On edge irregularity strength of graphs *Applied Mathematics and Computation.* **243** 607-610.
[4] Frieze A., Gould R.J., Karonski M., and Pfender F. 2004. On graph irregularity strength. *J, Graph Theory.* **41** 120-137.
[5] Gross J L, Yellen J and Zhang P. 2014 *Handbook of graph Theory.* Second Edition CRC Press Taylor and Francis Group
[6] Jendrol S, Miskuf J, and Sotak R. 2010. Total Edge Irregularity Strength of Complete Graph and Complete Bipartite Graph. *Discrete Math* **310** 400-407.
[7] Jeyanthi P and Sudha A. 2014. Total Edge Irregularity Strength of Disjoint Union of Wheel Graphs. *Electronic Notes in Discrete Mathematics* **48** 175-182.
[8] Pfender F. 2012. Total edge irregularity strength of large graph. *Discrete Math* **312** 229-237.
[9] Rajasingh I and Teresa SA. 2015. Total edge irregularity strength of series parallel graph. *International Journal of Pure and Applied Mathematics* **99** 11-12.
[10] Togni O. 1997. Irregularity strength of the toroidal grid. *Discrete Math.* **165/166** 609-620.

# On the star partition dimension of comb product of cycle and complete graph

**Ridho Alfarisi**[1], **Darmaji**[1] **and Dafik**[2]

[1]Department of Mathematics, Sepuluh Nopember Institute of Technology, Surabaya, Indonesia
[2]CGANT-University of Jember, Indonesia

E-mail: alfarisi38@gmail.com, darmaji@matematika.its.ac.id, d.dafik@unej.ac.id

**Abstract.** Let $G = (V, E)$ be a connected graphs with vertex set $V(G)$, edge set $E(G)$ and $S \subseteq V(G)$. For an ordered partition $\Pi = \{S_1, S_2, S_3, \ldots, S_k\}$ of $V(G)$, the representation of a vertex $v \in V(G)$ with respect to $\Pi$ is the $k$-vectors $r(v|\Pi) = (d(v, S_1), d(v, S_2), ..., d(v, S_k))$, where $d(v, S_k)$ represents the distance between the vertex $v$ and the set $S_k$, defined by $d(v, S_k) = min\{d(v, x)|x \in S_k\}$. The partition $\Pi$ of $V(G)$ is a resolving partition if the $k$-vektors $r(v|\Pi)$, $v \in V(G)$ are distinct. The minimum resolving partition $\Pi$ is a partition dimension of $G$, denoted by $pd(G)$. The resolving partition $\Pi = \{S_1, S_2, S_3, \ldots, S_k\}$ is called a star resolving partition for $G$ if it is a resolving partition and each subgraph induced by $S_i$, $1 \leq i \leq k$, is a star. The minimum $k$ for which there exists a star resolving partition of $V(G)$ is the star partition dimension of $G$, denoted by $spd(G)$. Finding a star partition dimension of $G$ is classified to be a NP-Hard problem. Furthermore, the comb product between $G$ and $H$, denoted by $G \rhd H$, is a graph obtained by taking one copy of $G$ and $|V(G)|$ copies of $H$ and grafting the $i$-th copy of $H$ at the vertex $o$ to the $i$-th vertex of $G$. By definition of comb product, we can say that $V(G \rhd H) = \{(a, u)|a \in V(G), u \in V(H)\}$ and $(a, u)(b, v) \in E(G \rhd H)$ whenever $a = b$ and $uv \in E(H)$, or $ab \in E(G)$ and $u = v = o$. In this paper, we will study the star partition dimension of comb product of cycle and complete graph, namely $C_n \rhd K_m$ and $K_m \rhd C_n$ for $n \geq 3$ and $m \geq 3$.

**Keywords:** Star resolving partition, star partition dimension, comb product, cycle, complete graph.

## 1. Introduction

All graphs in this paper are nontrivial, finite, simple, undirected and connected, for detail definition of graph see [1, 2]. The concept of metric dimension was independently introduced by Slater [3] and Harrary and Melter [4]. In his paper, Slater called this concept as a locating set. Chartrand et.al [5] introduced the concept of partition dimension of $G$. They introduced the same concept of resolving partition with a partition dimension of graphs. For $S \subseteq V(G)$ with vertex $v \in V(G)$, the distance between $v$ and $S$ is $d(v, S) = min\{d(v, x)\}$. Given $k-$partition $\Pi = \{S_1, S_2, S_3, \ldots, S_k\}$ of $V(G)$. The representation $v$ to $\Pi$ is defined by ordered $r(v|\Pi) = (d(v, S_1), d(v, S_2), \ldots, d(v, S_k))$ for every $v \in V(G)$. If $\Pi$ of $V$ is a resolving partition then the different vertices of $G$ have distinct representations, i.e., for every pair of vertices $u, v \in V, r(u|\Pi) \neq r(v|\Pi)$. The minimum of $k$ such that the resolving partition $\Pi$ is partition dimension of $G$ is called the partition dimension number of $G$, denoted by $pd(G)$.

Furthermore, Chartrand *et.al.* [6] determined the bounds of the metric dimensions for any connected graphs and determined the metric dimensions of some well known families of graphs such as tree, path, and complete graph. Furthermore, Saenpholphat *et.al.* in [7] studied a natural extension of partition dimension, they studied a particular case of resolving partition, i.e., connected resolving partition. $\Pi = \{S_1, S_2, \ldots, S_k\}$ is a connected resolving partition if it is a resolving partition and each subgraph induced by $S_i$, $1 \le i \le k$, is connected in $G$. The minimum $k$ for which there is a connected resolving partition of $V(G)$ is said to be connected partition dimension number of $G$, denoted by $cpd(G)$.

In this paper we consider a particular case of resolving partition, namely star resolving partition. We say $\Pi = \{S_1, S_2, ..., S_k\}$ is said to be a star resolving partition if it is a resolving partition and each subgraph induced by $S_i$, $1 \le i \le k$, is star in $G$. The minimum $k$ for which there is a star resolving partition of $V(G)$ is the star partition dimension number of $G$, denoted by $spd(G)$

There are many results related to the star partition dimension study. Saenpholphat *et.al.* in [7] studied the connected partition dimension of graphs, Steiner in [8] on the $k$-path partition of graphs. Ghemeci [9] studied on star partition dimension of Trees. Furthermore, Ghemeci and Tomescu [10] determine on star partition dimension of the generalized of gear graph, Amalia [11] studied star partition dimension of generalized windmill graph.

**Lemma 1.1** [6] *Let $\Pi$ be a resolving partition of $G$ and $u, v \in V(G)$. If $d(u,w) = d(v,w)$ for all $w \in V(G) - \{u,v\}$, then $u$ and $v$ belong to distinct elements of $\Pi$.*

## 2. Main Results

Let $G$ and $H$ be two connected graphs. Let $o$ be a vertex of $H$. The comb product between $G$ and $H$, denoted by $G \triangleright H$, is a graph obtained by taking one copy of $G$ and $|V(G)|$ copies of $H$ and grafting the $i$-th copy of $H$ at the vertex $o$ to the $i$-th vertex of $G$. By the definition of comb product, we can say that $V(G \triangleright H) = \{(a,u) | a \in V(G), u \in V(H)\}$ and $(a,u)(b,v) \in E(G \triangleright H)$ whenever $a = b$ and $uv \in E(H)$, or $ab \in E(G)$ and $u = v = o$. In this section, we will show the results related to star partition dimension of comb product of cycle and complete graph namely $C_m \triangleright K_n$ and $K_n \triangleright C_m$ for $n \ge 3$ and $m \ge 3$.

There are two theorems that we have found in this research. We will prove it in seperate way by introducing an observation in each theorem. The first, we will deal with the star partition dimension of $C_m \triangleright K_n$.

**Observation 2.1** *Let $C_m$ and $K_n$, respectively be a cycle and complete graph. The order and the size of $C_m \triangleright K_n$ are $|V(C_m \triangleright K_n)| = mn$ and $|E(C_m \triangleright K_n)| = \frac{n^2m - mn + 2m}{2}$, respectively.*

**Proof**: The comb product between $C_m$ and $K_n$, denoted by $C_m \triangleright K_n$ is a graph obtained by taking one copy of $C_m$ and $|V(C_m)|$ copies of $K_n$ and grafting the $i-$th copy of $K_n$ at the vertex $o$ to the $i-$th vertex of $C_m$. A graph $C_m \triangleright K_n$ has vertex set $V(C_m \triangleright K_n) = \{y_{j,i} | 1 \le j \le m, 1 \le i \le n\}$ and edge set $E(C_m \triangleright K_n) = \{y_{j,1}y_{j+1,1} | 1 \le j \le m-1\} \cup \{y_{m,1}y_{1,1}\} \cup \{y_{j,i}y_{j,i+k} | 1 \le j \le m, 1 \le i \le n, 1 \le k \le n-i\}$. Thus, $|V(C_m \triangleright K_n)| = mn$ and $|E(C_m \triangleright K_n)| = \frac{n^2m - mn + 2m}{2}$. □

Figure 1 (a) shows an example of $C_m \triangleright K_n$.

**Theorem 2.1** *Let $C_m$ and $K_n$ be two connected graphs of order $m \ge 3$ and $n \ge 3$, respectively. Then star partition dimension of comb product of cycle and complete graph is $spd(C_m \triangleright K_n) = m(n-1)$.*

**Proof:** We will determine the star partition dimension of $C_m \triangleright K_n$. The first step, we will determine the upper bound of star partition dimension obtained by construction star resolving partition. Suppose that $\Pi_S = \{S_1, S_2, S_3, ..., S_{m(n-1)}\}$ where $S_{(n-1)(j-1)+i-1} = \{y_{j,i} | 1 \le j \le m, 2 \le i \le n\}$, and $S_{(n-1)j-n+2} = \{y_{j,1} | 1 \le j \le m\}$. It can be seen that vertices in

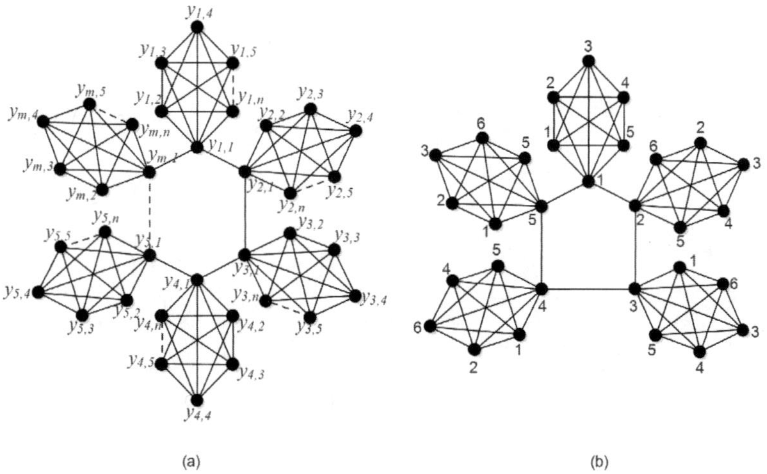


**Figure 1.** (a) A graph of $C_m \triangleright K_n$ , (b) Construction of star resolving partition $\Pi_S$

$S_{(n-1)(j-1)+i-1}$ where $3 \le i \le n$ and $1 \le j \le m$ is a singleton partition class contained trivial graph, it induces a star graph and $S_{(n-1)(j-1)+1} \cup S_{(n-1)j-n+2}$ where $1 \le j \le m$ is partition class that induces star graph $K_{1,1}$. Thus, it can be shown that the representation of every vertex $v \in V(C_m \triangleright P_n)$ is distinct to $\Pi_S$. Thus, $\Pi_S = \{S_1, S_2, S_3, ..., S_{m(n-1)}\}$ is a star resolving partition that consists of $m(n-1)$ partition class. Therefore, the cardinality of $\Pi_S$ is $|\Pi_S| = m(n-1)$. However, $\Pi_S$ is not necessarily to have a minimum cardinality. Thus, the upper bound of star partition dimension $C_m \triangleright K_n$ is $spd(C_m \triangleright K_n) \le m(n-1)$.

Furthermore, we will prove the lower bound of star partition dimension $C_m \triangleright P_n$, namely $spd(C_m \triangleright K_n) \ge m(n-1)$. Now, we consider the subgraph is induced of vertices in one partition class of a star graph. If $\Pi_S$ has cardinality $|\Pi_S| = m(n-1)-1$ then there are at least one partition class that does not induces a star graph. Notice that all vertices in partition class $\Pi_S$ are vertices of $V(C_m \triangleright K_n)$. Without loss of generality, suppose that $\Pi_S = \{S_1, S_2, S_3, \ldots, S_{m(n-1)-1}\}$ then there are two cases as follow:

a) If any vertices $u, v \in V(C_m \triangleright K_n)$ with $u, v$ pass subgraph $K_n$ then, based on Lemma 1.1, it will be obtained a same vertex representation to $\Pi_S$. Given a partition class $S_{nm-m-1} = \{y_{m,n-1}, y_{m,n}\}$, the vertices $y_{m,n-1}$ and $y_{m,n}$ have the same distance to all vertices in $V(C_m \triangleright K_n) - \{y_{m,n-1}, y_{m,n}\}$ or it can be written $d(y_{m,n-1}, w) = d(y_{m,n}, w)$ where $w \in V(C_m \triangleright K_n) - \{y_{m,n-1}, y_{m,n}\}$. It implies that $r(y_{m,n-1}|\Pi_S) = r(y_{m,n}|\Pi_S)$. Thus, the partition class $\Pi_S$ with $|\Pi_S| = m(n-1)-1$ is not star resolving partition even though the partition class $S_{nm-m-1}$ induces star graph.
b) Suppose we take a partition class of $S_{m(n-1)(m-2)+1} = \{y_{m-1,2}\}$, $S_{m(n-1)(m-1)+1} = \{y_{m,2}\}$, $S_{(n-1)(m-1)-n+2} = \{y_{m-1,1}\}$ and $S_{(n-1)m-n+2} = \{y_{m,1}\}$. If we combine it in one partition class of $S_k$ then we can see that the partition class $S_k$ does not induces a star graph. Thus the $\Pi_S$ is not star resolving partition even though the partition class $S_k$ has distinct vertex representation to $\Pi_S$.

Thus, we obtain that $\Pi_S$ with cardinality $|\Pi_S| = m(n-1)-1$ is not star resolving partition. It conludes the lower bound of star partition dimension $C_m \triangleright K_n$ is $spd(C_m \triangleright K_n) \ge m(n-1)$. Therefore the star partition dimension $spd(C_m \triangleright K_n) = m(n-1)$ for $n \equiv 0 (\text{mod } 3)$. It concludes the proof. □

Now, we are dealing with the star partition dimension of $K_n \triangleright C_m$.

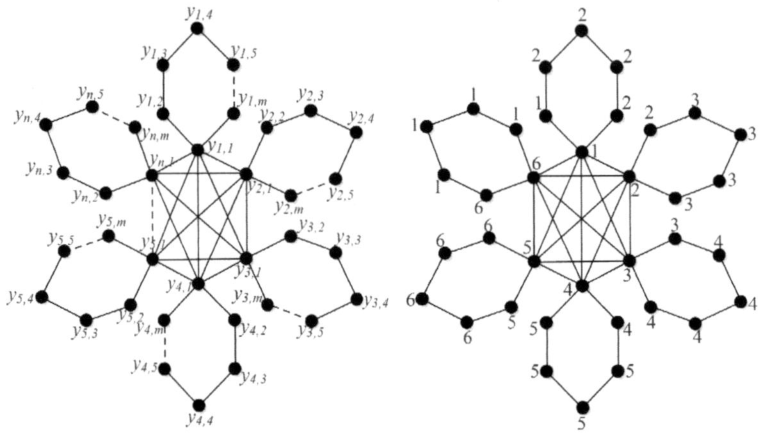

(a) (b)

**Figure 2.** (a) A graph of $K_n \triangleright C_m$ , (b) Construction of star resolving partition $\Pi_S$

**Observation 2.2** *Let $K_n \triangleright C_m$ be a connected graph. The order and the size of $K_n \triangleright C_m$ are $|V(K_n \triangleright C_m)| = mn$ and $|E(K_n \triangleright C_m)| = \frac{n^2+2mn-n}{2}$, respectively.*

**Proof**: The comb product between $K_n$ and $C_m$, denoted by $K_n \triangleright C_m$ is a graph obtained by taking one copy of $K_n$ and $|V(K_n)|$ copies of $C_m$ and grafting the $i-$th copy of $C_m$ at the vertex $o$ to the $i-$th vertex of $K_n$. A graph $K_n \triangleright C_m$ has the vertex set $V(K_n \triangleright C_m) = \{y_{i,j} | 1 \leq j \leq m, 1 \leq i \leq n\}$ and the edge set $E(K_n \triangleright C_m) = \{y_{i,1}y_{i+k,1} | 1 \leq i \leq n, 1 \leq k \leq n-i\} \cup \{y_{i,1}y_{i,m} | 1 \leq i \leq n\} \cup \{y_{i,j}y_{i,j+1} | 1 \leq i \leq n, 1 \leq j \leq m-1\}$. Thus, $|V(K_n \triangleright C_m)| = mn$ and $|E(K_n \triangleright C_m)| = \frac{n^2+2mn-n}{2}$. □

Figure 2 (a) shows an example of comb product of $K_n \triangleright C_m$.

**Theorem 2.2** *Let $K_n$ and $C_m$ be a connected graphs with order $n \geq 3$ and $m \geq 3$, respectively. Then star partition dimension of comb product of complete graph and cycle is*

$$spd(K_n \triangleright C_m) = \begin{cases} n(\lfloor \frac{m}{3} \rfloor + 1), & \textit{if } 3 \leq m \leq 4 \\ n\lceil \frac{m}{3} \rceil, & \textit{if } m \equiv 0 (mod\ 3),\ m \equiv 2 (mod\ 3) \\ n\lfloor \frac{m}{3} \rfloor + 1, & \textit{if } m \equiv 1 (mod\ 3) \end{cases}$$

**Proof:** Now, we will determine the star partition dimension of $K_n \triangleright C_m$. We will divide the proof in three cases.

**Case 1:** *For $m \equiv 0 (mod\ 3)$*

The upper bound of star partition dimension obtained by constructing a star resolving partition. Suppose that $\Pi_S = \{S_1, S_2, S_3, ..., S_{n(\frac{m}{3})}\}$ obtained from the following

$$S_{\frac{m}{3}(i-1)+k} = \{y_{i,j} | 1 \leq k \leq \frac{m}{3}, 3(k-1)+1 \leq j \leq 3k, 1 \leq i \leq n\}$$

It can be seen that vertices in $S_{\frac{m}{3}(i-1)+k}$ induces a star $K_{1,2}$. Thus, it can be shown the representation every vertex $v \in V(K_n \triangleright C_m)$ is distinct to $\Pi_S$. Thus, $\Pi_S = \{S_1, S_2, S_3, ..., S_{n(\frac{m}{3})}\}$ is a star resolving partition that consists of $n(\frac{m}{3})$ partition class. The cardinality of $\Pi_S$ is $|\Pi_S| = n(\frac{m}{3})$. However, $\Pi_S$ is not necessarily to have minimum cardinality. Thus, the upper bound of star partition dimension $K_n \triangleright C_m$ is spd$(K_n \triangleright C_m) \leq n(\frac{m}{3})$.

Furthermore, the lower bound of the star partition dimension $K_n \triangleright C_m$ can be obtained by Lemma 1.1. Now, we consider the subgraph is induced by vertices in one partition class of a star graph. If $\Pi_S$ has cardinality $|\Pi_S| = n(\frac{m}{3}) - 1$, then there are at least one partition class that does not induces a star graph. Notice that the vertices in partition class $\Pi_S$ are vertices of $V(C_m)$. Without loss of generality, suppose that $\Pi_S = \{S_1, S_2, S_3, ..., S_{n(\frac{m}{3})-1}\}$, then there are partition class that does not induces a star graph namely $S_{n(\frac{m}{3})-1} = \{y_{n,j} | \frac{m}{3} - 1 \leq k \leq \frac{m}{3}, 3(k-1)+1 \leq j \leq 3k\}$. Thus, we obtain that $\Pi_S$ of the cardinality $|\Pi_S| = n(\frac{m}{3}) - 1$ is not star resolving partition. It implies that the lower bound of the star partition dimension $K_n \triangleright C_m$ is $\text{spd}(K_n \triangleright C_m) \geq n(\frac{m}{3})$. Once we have the same upper bound and lower bound of star partition dimension $K_n \triangleright C_m$ then star partition dimension $\text{spd}(K_n \triangleright C_m) = n(\frac{m}{3})$ for $m \equiv 0 (\text{mod } 3)$ and $m \geq 5$.

*For* $m = 3$. Suppose lower bound of the star dimension partition is $\text{spd}(K_n \triangleright C_m) \geq n(\frac{m}{3})$. Based on Lemma 1.1, there are some vertex representations in $V(K_n \triangleright C_m)$, namely $r(y_{i,2}|\Pi_S) = r(y_{i,3})$ for $1 \leq i \leq n$ cause the vertices $y_{i,2}$ and $y_{i,3}$ have to be on different partition class. Thus that lower bound of the star partition dimension:

$$spd(K_n \triangleright C_m) \geq \underbrace{(\frac{m}{3}+1) + (\frac{m}{3}+1) + ... + (\frac{m}{3}+1)}_{n\ times} = n(\frac{m}{3}+1)$$

Thus, the lower bound of the star partition dimension $P_n \triangleright C_m$ is $\text{spd}(K_n \triangleright C_m) \geq n(\frac{m}{3}+1)$.

Now, we will determine the upper bound by constructing the star resolving partition namely $\Pi_S = \{S_1, S_2, S_3, ..., S_{n(\frac{m}{3}+1)}\}$ by $S_i = \{y_{i,1}, y_{i,2} | 1 \leq i \leq n\}$ and $S_{n+i} = \{y_{i,3} | 1 \leq i \leq n\}$. It can be seen that vertices in $S_i$ induce a star graph $K_{1,1}$ and vertices in $S_{n+i}$ are singleton partition class that obtain trivial graph of star graph. Thus, it can be shown that the representation of every vertex $v \in V(K_n \triangleright C_m)$ is distinct to $\Pi_S$. The results of observation shows that the coordinate representation of all vertices in $K_n \triangleright C_m$ for $m = 3$ is unique. Thus, $\Pi_S = \{S_1, S_2, S_3, ..., S_{n(\frac{m}{3}+1)}\}$ is star partition dimension which consists of $n(\frac{m}{3}+1)$ partition class. Now the cardinality of $\Pi_S$ is $|\Pi_S| = n(\frac{m}{3}+1)$. However, $\Pi_S$ is not necessarily to have minimum cardinality. Thus, the upper bound of star partition dimension $K_n \triangleright C_m$ is $\text{spd}(K_n \triangleright C_m) \leq n(\frac{m}{3}+1)$. It concludes that the partition dimension $K_n \triangleright C_m$ is $\text{spd}(K_n \triangleright C_m) = n(\frac{m}{3}+1)$ for $m = 3$.

**Case 2:** For $m \equiv 1 (\text{mod } 3)$. The upper bound of star partition dimension is obtained by construction star resolving partition. Suppose $\Pi_S = \{S_1, S_2, S_3, ..., S_{n(\frac{m-1}{3})+1}\}$ where $S_{\frac{m-1}{3}(i-1)+k} = \{y_{i,j} | 1 \leq k \leq \frac{m-1}{3}, 3(k-1)+3 \leq j \leq 3k+2, 1 \leq i \leq n\}$ and $S_{n(\frac{m-1}{3})+1} = \{y_{i,1} | 1 \leq i \leq n\}$. It is clearly to see that for all $v \in V(K_n \triangleright C_m)$, the graph $K_n \triangleright C_m$ has distinct vertex representation to $\Pi_S$. Thus, $\Pi_S = \{S_1, S_2, S_3, \ldots, S_{n(\frac{m-1}{3})+1}\}$ is a star resolving partition consists of $n(\frac{m-1}{3})+1$ partition class. Now, the cardinality of $\Pi_S$ is $|\Pi_S| = n(\frac{m-1}{3})+1$. However, the partition class $\Pi_S$ is not necessarily to have minimum cardinality. Thus, the upper bound of star partition dimension $K_n \triangleright K_m$ is $\text{spd}(K_n \triangleright C_m) \leq n(\frac{m-1}{3}) + 1$.

Furthermore, the lower bound of star partition dimension $K_n \triangleright C_m$ is obtained by Lemma 1.1. Now, we consider the subgraph which is induced by vertices in one partition class of a star graph. Thus, if $\Pi_S$ has cardinality $|\Pi_S| = n(\frac{m-1}{3})$, then there are at least one partition class that does not induces a star graph $K_{1,n}$. Notice that vertices in partition class $\Pi_S$ are vertices of $V(K_n \triangleright C_m)$, without loss of generality, suppose $\Pi_S = \{S_1, S_2, S_3, ..., S_{n(\frac{m-1}{3})}\}$, then there are partition class that does not induces a star graph namely $S_{n(\frac{m-1}{3})} = \{y_{n,j} | m-1 \leq j \leq m\} \cup \{y_{i,1} | 1 \leq i \leq n\}$. Thus, the cardinality $\Pi_S$ is $|\Pi_S| = n(\frac{m-1}{3})$ and $|\Pi_S|$ is not star resolving partition. Therefore, the lower bound of star partition dimension $K_n \triangleright C_m$ is $\text{spd}(K_n \triangleright C_m) \geq$

$n(\frac{m-1}{3})+1$. It concludes that the star partition dimension $spd(K_n \rhd C_m) = n(\frac{m-1}{3})+1$ for $m \equiv 1(\text{mod } 3)$ and $m \geq 5$.

*For* $m = 4$. Suppose the lower bound of the star dimension partition is $spd(K_n \rhd C_m) \geq n(\frac{m-1}{3})+1$. Based on Lemma 1.1, there are vertex representation in $V(K_n \rhd C_m)$ namely $r(y_{i,2}|\Pi_S) = r(y_{i,4})$ for $1 \leq i \leq n$ which cause vertices $y_{i,2}$ and $y_{i,4}$ have to be on different partition class such that the lower bound of the star partition dimension:

$$spd(K_n \rhd C_m) \geq \underbrace{(\frac{m-1}{3}+1)+(\frac{m-1}{3}+1)+\ldots+(\frac{m-1}{3}+1)}_{n\ times} = n(\frac{m-1}{3}+1)$$

Thus, the lower bound of the star partition dimension $K_n \rhd C_m$ is $\text{spd}(K_n \rhd C_m) \geq n(\frac{m-1}{3}+1)$.

Now, we will determine upper bound by constructing a star resolving partition namely $\Pi_S = \{S_1, S_2, S_3, ..., S_{n(\frac{m-1}{3}+1)}\}$ by $S_i = \{y_{i,1}, y_{i,2}, y_{i,3}|1 \leq i \leq n\}$ and $S_{n+i} = \{y_{i,4}|1 \leq i \leq n\}$. It can be seen that vertices in $S_i$ induces a star graph $K_{1,2}$ and vertices in $S_{n+i}$ is singleton partition class that obtain trivial graph of star graph. Thus, it can be shown the representation of all $v \in V(K_n \rhd C_m)$ are distinct to $\Pi_S$. The results of observation shows the representation of all vertices in $K_n \rhd C_m$ for $m = 4$ is unique. Thus, $\Pi_S = \{S_1, S_2, S_3, \ldots, S_{n(\frac{m-1}{3}+1)}\}$ is star partition dimension consists of $n(\frac{m}{3}+1)$ partition class. Now, the cardinality of $\Pi_S$ is $|\Pi_S| = n(\frac{m-1}{3}+1)$. However, the partition class $\Pi_S$ is not necessarily to have a minimum cardinality. Therefore, the upper bound of star partition dimension $K_n \rhd C_m$ is $\text{spd}(P_n \rhd C_m) \leq n(\frac{m-1}{3}+1)$. Once we have got the upper bound and lower bound of star partition dimension $n(\frac{m-1}{3}+1) \leq \text{spd}(K_n \rhd C_m) \leq n(\frac{m-1}{3}+1)$, then the partition dimension $K_n \rhd C_m$ is $\text{spd}(K_n \rhd C_m) = n(\frac{m-1}{3}+1)$ for $m = 4$.

**Case 3:** For $m \equiv 2(\text{mod } 3)$. The upper bound of star partition dimension obtained by constructing star resolving partition. Suppose $\Pi_S = \{S_1, S_2, S_3, ..., S_{n(\frac{m-2}{3})+n}\}$ where$S_{\frac{m-2}{3}(i-1)+k} = \{y_{i,j}|1 \leq k \leq \frac{m-2}{3}, 3(k-1)+3 \leq j \leq 3k+2, 1 \leq i \leq n\}$ and $S_{n(\frac{m-2}{3})+i} = \{y_{i,j}|1 \leq i \leq n, 1 \leq j \leq 2\}$. It can be seen that vertices in $S_{\frac{m-2}{3}(i-1)+k}$ and $S_{n(\frac{m-2}{3})+i}$ induces a star $K_{1,n}$, $1 \leq n \leq 2$. Thus, it can be shown that the representation all vertex $v \in V(P_n \rhd C_m)$ are distinct to $\Pi_S$. Thus, $\Pi_S = \{S_1, S_2, S_3, ..., S_{n(\frac{m-2}{3})+n}\}$ is a star resolving partition that consists of $n(\frac{m-2}{3})+n$ partition class. Therefore, the cardinality of $\Pi_S$ is $|\Pi_S| = n(\frac{m-2}{3})+n = n(\frac{m+1}{3})$. However, the partition class $\Pi_S$ is not necessarily to have minimum cardinality. Thus, the upper bound of star partition dimension $K_n \rhd C_m$ is $\text{spd}(K_n \rhd C_m) \leq n(\frac{m+1}{3})$.

Furthermore, the lower bound of star partition dimension $K_n \rhd C_m$ is obtained by Lemma 1.1. Now, we will consider the subgraph is induced of vertices in one partition class of a star graph. If $\Pi_S$ has cardinality $|\Pi_S| = n(\frac{m+1}{3})-1$, then there are at least one partition class that does not induces a star graph. Notice that all vertices in partition class $\Pi_S$ are vertices of $V(K_n \rhd C_m)$. Without loss of generality, suppose that $\Pi_S = \{S_1, S_2, S_3, ..., S_{n(\frac{m+1}{3})-1}\}$ then there are partition class that does not induces a star graph namely $S_{n(\frac{m+1}{3})-1} = \{y_{i,1}, y_{i,m}|n-1 \leq i \leq n\}$. Thus, we have obtained that $\Pi_S$ with cardinality $|\Pi_S| = n(\frac{m+1}{3})-1$ which is not star resolving partition. Thus, the lower bound of star partition dimension $K_n \rhd C_m$ is $\text{spd}(K_n \rhd C_m) \geq n(\frac{m+1}{3})$. Since we have got the upper bound and lower bound of star partition dimension $K_n \rhd C_m$ is $n(\frac{m+1}{3}) \leq \text{spd}(K_n \rhd C_m) \leq n(\frac{m+1}{3})$, then star partition dimension $\text{spd}(K_n \rhd C_m) = n(\frac{m+1}{3})$ for $n \equiv 2(\text{mod } 3)$.

Based on three cases above, it can be seen that $\text{spd}(K_n \rhd C_m) = n(\frac{m}{3})$ for $m = 3$ and $\text{spd}(K_n \rhd C_m) = n(\frac{m-1}{3}+1)$ for $m = 4$. Those can be rewritten into $\text{spd}(K_n \rhd C_m) = n(\lfloor\frac{m}{3}\rfloor+1)$

for $3 \leq m \leq 4$. Furthermore, $\text{spd}(K_n \triangleright C_m) = n(\frac{m}{3})$ for $m \equiv 0(\text{mod } 3)$ and $m \geq 5$ and $\text{spd}(K_n \triangleright C_m) = n(\frac{m+1}{3})$ for $m \equiv 2(\text{mod } 3)$ and $m \geq 5$, thus for value $m$, it can be combined into $\text{spd}(K_n \triangleright C_m) = n\lceil \frac{m}{3} \rceil$. While $\text{spd}(K_n \triangleright C_m) = n(\frac{m-1}{3}) + 1$ for $m \equiv 1(\text{mod } 3)$, $m \geq 5$, it can be combined into $\text{spd}(K_n \triangleright C_m) = n\lfloor \frac{m}{3} \rfloor + 1$. It concludes the proof. □

## 3. Conclusion

In this paper we have given an aymptotically tight result on the star partition dimension of comb product of cycle and complete graph. The result shows that the star partition dimension of the comb product of cycle and complete graph as follows:

$$spd(C_m \triangleright K_n) = m(n-1)$$

$$spd(K_n \triangleright C_m) = \begin{cases} n(\lfloor \frac{m}{3} \rfloor + 1), & \text{if } 3 \leq m \leq 4 \\ n\lceil \frac{m}{3} \rceil, & \text{if } m \equiv 0(mod\ 3),\ m \equiv 2(mod\ 3) \\ n\lfloor \frac{m}{3} \rfloor + 1, & \text{if } m \equiv 1(mod\ 3) \end{cases}$$

## Acknowledgement

We gratefully acknowledge the support from CGANT - University of Jember of year 2017.

## References

[1] Gross J L, Yellen J and Zhang P 2014 *Handbook of graph Theory* Second Edition CRC Press Taylor and Francis Group
[2] Chartrand G and Lesniak L 2000 *Graphs and digraphs 3rd* ed (London: Chapman and Hall)
[3] Slater P J 1975 Leaves of trees. in:*Proc. 6th Southeast Conf. Comb., Graph Theory, Comput. Boca Rotan* No. 14 549-559
[4] Harary F and Melter R A 1976 On The metric dimension of a graph, *Ars Combin* No.2 191-195
[5] Chartrand G, Salehi E and Zhang P 2000 The Partition dimension of a graphs *Aequationes Math* No.59 45-54
[6] Chartrand G, Eroh L, Johnson M A and Oellermann O R 2000 Resolvability in graphs and the metric dimension of a graph *Discrete Appl. Math* No.105 99-113
[7] Saenpholphat V and Zhang P 2002 Connected partition dimension of a graphs *Discussiones Mathematicae Graph Theory* **22** 305-323
[8] Steiner G 2003 On the $k$-path partition of graphs *Theoretical Computer Science* **290** 2147-2155
[9] Ghemeci R M 2014 On star partition dimension of trees, *Math. Reports* **14(64)** 43-52
[10] Ghemeci R M and Tomescu I 2010 On star partition dimension of the generalized gear graph *Bull. Math. Soc. Sci. Math. Roumanie Tome* **53(101)**, N0.3 261-268.
[11] Amalia R 2013 *Star partition dimension of generalized windmill graph* (Tesis) Institut Teknologi Sepuluh Nopember (ITS)-Surabaya

IOP Conf. Series: Journal of Physics: Conf. Series **855** (2017) 012006 doi:10.1088/1742-6596/855/1/012006

# Structural Equation Modelling with Three Schemes Estimation of Score Factors on Partial Least Square (Case Study: The Quality Of Education Level SMA/MA in Sumenep Regency)

**[1]Anik Anekawati, [2]Bambang Widjanarko Otok, [3]Purhadi, [4]Sutikno**

[1]Student of Doctoral Programme, Departement of Statistics, Institut Teknologi Sepuluh Nopember
[2,3,4]Departement of Statistics, Institut Teknologi Sepuluh Nopember

[1]anekawa10@gmail.com, [2] bambang_wo@statistika.its.ac.id, [3]purhadi@statistika.its.ac.id, [4]sutikno@statistika.its.ac.id

**Abstract.**
Research in education often involves a latent variable. Statistical analysis technique that has the ability to analyze the pattern of relationship among latent variables as well as between latent variables and their indicators is Structural Equation Modeling (SEM). SEM partial least square (PLS) was developed as an alternative if these conditions are met: the theory that underlying the design of the model is weak, does not assume a certain scale measurement, the sample size should not be large and the data does not have the multivariate normal distribution. The purpose of this paper is to compare the results of modeling of the educational quality in high school level (SMA/MA) in Sumenep Regency with structural equation modeling approach partial least square with three schemes estimation of score factors.This paper is a result of explanatory research using secondary data from Sumenep Education Department and Badan Pusat Statistik (BPS) Sumenep which was data of Sumenep in the Figures and the District of Sumenep in the Figures for the year 2015. The unit of observation in this study were districts in Sumenep that consists of 18 districts on the mainland and 9 districts in the islands. There were two endogenous variables and one exogenous variable. Endogenous variables are the quality of education level of SMA/MA (Y1) and school infrastructure (Y2), whereas exogenous variable is socio-economic condition (X1). In this study, There is one improved model which represented by model from path scheme because this model is a consistent , all of its indicators are valid and its the value of R-square increased which is: $Y_1$=0.651$Y_2$. In this model, the quality of education influenced only by the school infrastructure (0.651). The socio-economic condition did not affect neither the school infrastructure nor the quality of education. If the school infrastructure increased 1 point, then the quality of education increased 0.651 point. The quality of education had an $R^2$ of 0.418, which indicates that 41.8 percent of variance in the quality of education is explained by the school infrastructure, the remaining 58.2% is explained by the other factors which were not investigated in this work.

**Key Words**: partial least square, schemes estimation of score factors

IOP Conf. Series: Journal of Physics: Conf. Series **855** (2017) 012006 doi:10.1088/1742-6596/855/1/012006

## 1. Introduction

Research in education often involves latent variables. Statistic analysis technique that can be used to analyse the relationship among the latent variables (LVs) as well as between latent variable (LV) and its indicators is Structural Equation Modeling (SEM). Schumacker and Lomax concluded in his research that SEM is often described using factors that are seen as LV [7]. Variable to be analysed is the latent variable which cannot be observed directly but can be observed using measured indicators.

Several estimation methods have been developed in SEM. The very famoust one is the Maximum Likelihood Estimation (MLE). Since the covariance-based SEM is based on large sample size theory, it must fullfil the requirements of normal multivariate distribution, Bollen therefore suggested an estimation method which accepts an abnormality and is asymptotically efficient [2]. This method is called Weighted Least Square (WLS). WLS is an estimation method adapted from Asymptotically Distribution Free (ADF) method which is the most common method due to its independence on the type of data distribution. Although WLS has advantages compared to MLE, but it requires a larger sample size. If the sample size is too small and not normal multivariate distribution, the researchers will face analysis results which are improper solution [1][3]. As an alternative solution of this problem, variant-based SEM was developed or as known as Partial Least Square (PLS).

SEM PLS is a stastistic technique that generalizes and combines factor analisys, Principal Component Analysis (PCA) and regression analysis by separate estimation procedure between latent variables and its indicators. Additionally, SEM PLS was developed as an alternative method when the theory underlying model design is weak, not based on many assumptions, has a small sample size, and does not have to be a normal multivariate distribution [6].

The shortcoming of SEM PLS is the use of unknown data distribution so that a value of model significance cannot be obtained and therefore a resampling approximation can be used. One of the resampling methods used for a small sample size is bootstrap since a small sample size tends not to distribute normally resulting in an inaccurate parametric statistic estimation. Bootstrap method works using resampling with replacement by creation of pseudodata using information of initial data including its characteristics so that a correct estimation can be achieved [5].

Main stage in PLS algorithm is an iteration process to generate a latent variable score through weight estimation. There are three main schemes in this process, namely centroid, factor and path scheme. In practice, the selection of one weighting scheme has a little relationship in estimation process [9]. Tenenhaus observed and mentioned that the three schemes do not influence the results significantly [8]. However, in the theoretical level, the schemes are very important to understand how PLS can be applied for a different technique from multiple table analysis.

The case of education quality improvement in Sumenep regency has a unique characteristic since it is divided in mainland and islands with a high disparity in education quality. Within the frame of equalization of education quality, a study is required which includes latent variable, namely the quality of the education in high school level, the school infrastructure and the socio-economic condition in Sumenep. Since the sample units are 27 districts, which falls under a small sample size category, it requires analysis technique that includes latent variables with a small sample size, namely SEM PLS.

Based on the above explanation, a research was conducted with the aim to compare modelling results of education quality of high school level in Sumenep regency using approximation method Structural Equation Modeling Partial Least Square with three schemes estimation of score factor.

## 2. Literature Review

### Structural Equation Modeling with Partial Least Square (PLS) Approximation

SEM based on variance or Partial Least Square (PLS) consists generally of two sub-models which are measurement model or outer model and structural model or inner model. Outer model shows how observed variables represent latent variables to be measured, whereas inner model shows the estimation strength among latent variables.

Inner model in PLS describes the relationship among latent variables. The model equation can be written as the following [9]: $\boldsymbol{\eta} = \boldsymbol{\beta}_0 + \boldsymbol{\beta\eta} + \boldsymbol{\Gamma\xi} + \boldsymbol{\zeta}$

IOP Conf. Series: Journal of Physics: Conf. Series **855** (2017) 012006 doi:10.1088/1742-6596/855/1/012006

Since PLS is designed for recursive model, in the latent variable relationship, each dependent latent variable $\boldsymbol{\eta}$ is called as causal chain system from latent variable and can be specified as:

$$\eta_j = \sum_i \beta_{ji}\eta_i + \sum_b \gamma_{jb}\xi_b + \zeta_j$$

Where $\beta_{ji}$ and $\gamma_{jb}$ are the path coefficient that connect LV endogenous ($\eta$) with LV endogenous and LV endogenous with LV exogenous ($\xi$) in the index range i and b, and where $\zeta_j$ is residual from the inner model with the value $E(\zeta_j) = E(\xi_i\zeta_j) = 0$.

Outer model is defined how each indicator block relates to its latent variable. Trujillo reported the outer model equation for block indicators of reflective model as an equation of simple regression [9]: $x_{jk} = \lambda_{jk}\xi_j + \varepsilon_{jk}$
where $\lambda_{jk}$ is coefficient of loading factor from the relationship between the j-th LV ($\xi_j$) with k-th its indicator ($x_{jk}$), and $\varepsilon_{jk}$ is *error* of each measurement variable.

Parameter estimation on SEM PLS model has three categories: weight estimate, path estimate and also mean and location parameter. Weight estimate is used to generate score of LV. Path estimate reflects the relationship between LVs and loading estimate between LV and its indicators. The last category relates to mean and location parameter for indicator and LV. In order to obtain these three estimations, PLS uses iteration proccess.

**Weight Estimate**

In weight estimation, there are three iteration procedures of simple regression or multiple regression considering the relationship of the inner model, the outer model and weight relation. The results of iteration procedures, that are a set of stabil weights, is used to calcualte the variable laten score which is a linear combination of their indicators [8].

Weight calculation is used to estimate LV score with the following steps:

1. Outer model estimation (outside approximation)

In the outside approximation, iterative process begins with an early initialization of each LV as a linear combination (or weighted aggregate) of indicators and written in equation: $l_j = \sum_k \tilde{w}_{jk}x_{jk}$
where $\tilde{w}_{jk}$ is weighted in outer model (outer weights). The basic idea of the outside approximation is to obtain a set of weights to estimate a latent variable accounting for as much variance as possible for the indicators and the constructs [9]. Chin suggested to set initial weights by equal value to present the first approximation of the latent variable as a simple sum of its indicators [9].

2. Inner model estimation (inside approximation)

In the inside approximation, iterative process takes into account the connections among LVs in the inner model to obtain initialization of each LV calculated as a weighted aggregate of its adjacent LVs. Definition of inside approximation is as follows $Z_j = \sum_{i:\beta_{ij\neq 0},\beta_{ji\neq 0}} e_{ij} l_i$
where $e_{ij}$ is inner weights which can be choosen from three scheme which are centroid, factor or path. Each scheme is defined by Trujillo as follows [9] :

i. Inner weight ($e_{ij}$) in centroid scheme is equal to the sign correlation between $l_i$ and $l_j$, where centroid scheme is defined as:

$$e_{ij} = \begin{cases} sign\{cor(l_j, l_i)\} & \xi_j\xi_i \text{ adjacent} \\ 0 & \text{otherwise} \end{cases} \tag{1}$$

ii. Inner weight ($e_{ij}$) in factor scheme is the correlation between $l_i$ and $l_j$ which considers not only the sign direction but also the strength of the paths in inner model. Factor scheme is defined as:

$$e_{ij} = \begin{cases} cor(l_j, l_i) & \xi_j \xi_i \text{ adjacent} \\ 0 & \text{otherwise} \end{cases} \quad (2)$$

iii. Inner weight ( $e_{ij}$ ) in path scheme is the weighting of neighboring LVs depending whether of not the neighboring variables are antecedent or consecuent from LV which is to be estimated. Path scheme is defined as:

$$\left. \begin{array}{l} e_{ij} = cor(l_j, l_i); \text{if } \xi_j \text{is explained by } \xi_i \\ l_j = \sum_i e_{ji} l_i; \text{ e}_{ij} \text{ is coefficient in the regression of } l_i \text{ on } l_j \end{array} \right\} \quad (3)$$

3. Updating outer weight

Once the inside approximation is done, the internal estimation $Z_j$ is considered by observing their indicators. This can be done by updating the weight in outer model (outer weight). The weight outer $w_{jk}$ in the reflective model is expressed in the following equation: $w_{jk} = \left(Z_j^{'} Z_j\right) Z_j{}' x_{jk}$

4. Checking convergence

In each iterative procedure such as S=1,2,3,..., convergence is checked by comparing outer weight in the S step over outer weight in the S-1 step. Wold suggested $\left|\tilde{w}_{jk}^{S-1} - \tilde{w}_{jk}^{S}\right| < 10^{-5}$ as a convergence criterion [9].

**Path Estimate**

Next step is the calculation of the path coefficient estimation and loading, $\hat{\beta}_{ji}$ and $\hat{\lambda}_{jk}$, in the inner and outer model. In the inner model, path coefficient is estimated using Ordinary Least Square (OLS) as the multiple linier regression analysis from relation between $l_j$ and $l_i$, that is $l_j = \sum_i \hat{\beta}_{ji} l_i$ and $\hat{\beta}_{ji} = (l_i{}' l_i)^{-1} l_i{}' l_j$.

In the outer reflective model, loading coefficient is estimated as a simple linier regression from relation between $x_{jk}$ and $l_j$, that is: $x_{jk} = \hat{\lambda}_{jk} l_j$ and $\hat{\lambda}_{jk} = \left(l_j{}' l_j\right)^{-1} l_j{}' x_{jk}$

**Mean and Location Parameter Estimate**

There are two locations parameters to be estimated, namely $\beta_{0j}$ (constants in the inner model) and $\lambda_{0jk}$ (constants in the outer reflective model). The equation with predictor specification above can be written as follows [9]: $E\left(\xi_j \middle| \xi_i\right) = \beta_{0j} + \sum_i \beta_{ji} \xi_i$ for inner model and $E\left(x_{jk} \middle| \xi_j\right) = \lambda_{0jk} + \lambda_{jk} \xi_j$ for outer reflective model.

These location parameter takes the mean of indicator and latent variables into account. Prior to the calculation of the location parameter, the mean to estimate the latent variables is defined as follows [8] : $\hat{m}_j = \sum_k \tilde{w}_{jk} \bar{x}_{jk}$ and $\hat{\xi}_j = l_j + \hat{m}_j$, so that the estimation of the location parameter $\beta_{0j}$ (constant in the inner model) and $\lambda_{0jk}$ (constant in the outer reflective model) can be interpreted as [9]:

$\hat{\beta}_{0j} = b_{0j} = \hat{m}_j - \sum_i b_{ji} \hat{m}_i$, $\hat{\lambda}_{0jk} = \bar{x}_{jk} - \hat{\lambda}_{jk} \hat{m}_j$ and $\hat{\pi}_{0j} = \hat{m}_j - \sum_k \hat{\pi}_{jk} \bar{x}_{jk}$

**Evaluation of SEM PLS Model**

The outer models with reflective indicators can be evaluated through convergent validity, discriminant validity and composite reliability. Inner model is evaluated by observing the percentage of explained variance with the value of $R^2$ for the dependent latent constructs using Stone-Geisser Q test [6]. Here is a more detailed description related to the evaluation of outer and inner model.

1. Outer model

Evaluation of outer models is an assessment of the reliability and validity of the variables or is defined how each block of indicator associated with LVs. There are three criteria for evaluating the measurement model, namely: convergent validity, discriminant validity, and composite realibility.

a. Convergent validity

Convergent validity of the measurement model relates to the principle that the indicators of a construct should be highly correlated and can be seen from the value of the loading factor. Assessment is based on the correlation between the item score/component score and construct score, which is calculated by PLS. The model is declared good if the correlation coefficient is 0.70 more than the construct to be measured. However, for the early stage of a research, scale measurement development with loading value of 0.5 to 0.6 is considered to be sufficient.

b. Discriminant validity

Discriminant validity of the measurement model relates to the principle that indicators of the different variable should be highly correlated and can be seen from value of cross loading. If the correlation between construct and the measurement item is larger than the other constructs, then it indicates that the latent construct predicts that the size of their block is better than the size of the other blocks.

Another method for seeing the discriminant validity is to compare the value of square root of extracted (AVE) of each construct with a correlation among the other constructs in the model. It has a good discriminant validity if the value of the square root of AVE of each construct is greater than the value of correlation between the construct and other constructs in the model [6]. Ghozali suggested value of AVE greater than 0.5[6].

c. Composite realibility

Composite realibility of indicator block that measures a construct can be evaluated by internal consistency and Cronbach's alpha. By using the outputs of PLS, the composite realibility can be calculated with the following formula: $\rho c = \left(\sum_{i=1}^{k}\lambda_i^2\right)^2 \rho c \Big/ \left(\sum_{i=1}^{k}\lambda_i^2\right)^2 + \sum_{i=1}^{k}\mathrm{var}(\varepsilon_i)$

where $\lambda_i$ is component loading of indicator and $\mathrm{var}(\varepsilon_i) = 1-\lambda_i^2$.

2. Inner model (structural model)

Structural model is evaluated using $R^2$ to see the relationship among the LVs of the research model. Stone-Geisser Q-square test for predictive relevance and stability of these estimation results are evaluated by t-test statistics through bootstrapping procedure. The assessment criteria through PLS model as follows: $f^2 = \dfrac{R^2_{include} - R^2_{exclude}}{1 - R^2_{include}}$

where $R^2_{include}$ and $R^2_{exclude}$ is R-square from dependen LV when LV predictor is used or is taken out in model.

## 3. Research Methodology

This work is an explanatory research since the author did not possess initial data or an description of the education quality model. The data used is a secondary data from Sumenep Education Department which is a summary of data on education in SMA/MA in year 2015, Badan Pusat Statistik (BPS) Sumenep that is data of district's sumenep in figure for 2015 and Sumenep in figure for 2015.

The observation units in this research were the districts in Sumenep which consists of 18 districts on mainland and 9 districts on islands so that the total observation units were 27 districts. Additionally, there were 2 endogenous variables, namely the quality of education in SMA/MA (Y1) and the school infrastructure (Y2) and 1 exogenous variable, namely the socio-economic condition in Sumenep.

Indicators of the quality of education SMA/MA (Y1) are a ratio of crude participant figure SMA/MA ($Y_{11}$), the ratio of SMA/MA accredited at least B ($Y_{12}$), a ratio average of national exam score SMA/MA more than 7 ($Y_{13}$), a ratio SMA/MA that have ISO 9001:2008 certificate ($Y_{14}$).

Indicators of the school infrastructure (Y2) are ratio of student number in a class ($Y_{21}$), the ratio of minimum class size ($Y_{22}$), the ratio of laboratorium possession ($Y_{23}$), and the ratio of library possession ($Y_{24}$).

Indicators of socio-economic condition in Sumenep ($X_1$) are the ratio of the number of prosperous families ($X_{11}$), the ratio of household industries ($X_{12}$), the ratio of households using clean water ($X_{13}$), the ratio of households using basic sanitation ($X_{14}$), the ratio of the number of medical personnel ($X_{15}$), and the ratio of healthy house ($X_{16}$).

The analysis step in this research is as follows:

1. Conduct descriptive statistics as an initial overview on the education quality of high school in Sumenep regency in 2015 and all factors that are expected to affect them.
2. Build a conceptual model based on theory and construct path diagram.
   Determination of education quality model in high school level related to many dimensions such as infrastructure and facilities, and socio-economic condition in that region. From that conceptual model, a structural model is built with a hyphothesis that will be tested as follows:
   $H_A$ : Socio-economic condition ($X_1$) and the school infrastructure ($Y_2$) having an influence on the quality of education in SMA/MA ($Y_1$).
   $H_B$ : Socio-economic condition ($X_1$) having an influence on school infrastructure ($Y_2$).
3. Carry out validitation test on outer model
   Evaluation of each indicator variables can be done by observing the value of convergent validity, discriminant validity and the value of composite reliability by means of software Smart PLS.
4. Obtain score of latent variable from each of latent variable
   Estimation of latent variable score SEM model with using the PLS approximation. The score of an exogenous variable is $\hat{\xi}_j = l_j$, and the score of an endogenous variable is $\hat{\eta}_j = k_j$. Since the process to obtain the score of latent variable for both endogenous and exogenous is the same, the scores of latent variable are represented by $l_j$, as illustrated in Figure 1:

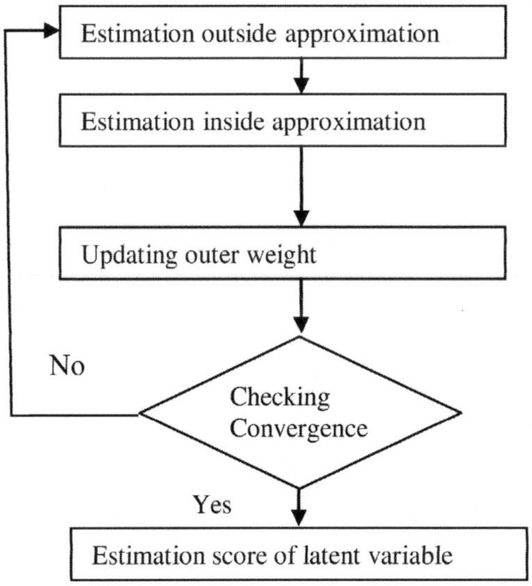


$$l_j = \sum_k \tilde{w}_{jk}^{old} x_{jk}$$

$$Z_j = \sum_{i:\beta_{ii\neq 0},\beta_{ii\neq 0}} e_{ij} l \qquad e_{ij} = \begin{cases} centroid\ scheme \\ factor\ scheme \\ path\ scheme \end{cases}$$

$$w_{jk}^{new} = \left(Z_j' Z_j\right) Z' x_{jk}$$

$$\left| \tilde{w}_{jk}^{old} - \tilde{w}_{jk}^{new} \right| < 10^{-5}$$

$$\hat{\xi}_j = l_j = \sum_k \tilde{w}_{jk} x_{jk}$$
$$\hat{\eta}_j = k_j = \sum_k \tilde{w}_{jk} y_{jk}$$

Figure 1. Procedure to estimate the score of latent variable

5. Select the best model, interprete and conclude results.

## 4. Main Results

### A. Evaluation outer model (measurement model)

a. Convergent validity

The loading factor of each indicator based on three weighted schemes, namely the centroid, factor and path is summarized in the following description. There are four indicators that have a loading factor less than 0.5 for all three schemes, namely $X_{12}$, $X_{14}$, $X_{15}$ and $X_{16}$. It can be concluded that the ratio of household industries, the ratio of households using basic sanitation, the ratio of the number of medical personnel and the ratio of healthy house are invalid or could not be used as an indicator to build construct of socio-economic condition.

The following is a summary of loading factor value of the school infrastructure for 3 schemes. All of the indicators have a loading factor more than 0.5 for three schemes. It can be concluded that the ratio of student number in a class, the ratio of minimum class size, the ratio of laboratorium possession, and the ratio of library possession are valid or could be used as an indicator to build construct of the school infrastructure.

The following is a summary of loading factor value of the quality of education in SMA/MA for 3 schemes. There is one indicator that has a value of loading factor less than 0.5, namely $Y_{13}$. It can be concluded that a ratio average of national exam score SMA/MA more than 7 is invalid or could not be used as an indicator to build construct of the quality of education in SMA/MA.

b. Discriminant validity

The following is a cross loading of each indicator based on three weighted schemes. A value of cross loading and $\sqrt{AVE}$ based on output software Smart PLS. The correlation between socio-economic condition and its indicators is higher than the correlation between socio-economic condition and indicator of other variables, except $X_{14}$. It shows that the socio-economic condition predicts indicator within a group is better than the indicator in the other groups, except the ratio of households using basic sanitation.

The following is the comparison result of the value $\sqrt{AVE}$ of socio-economic condition and the correlation between socio-economic condition and the other variables through three weighted schemes. The value of $\sqrt{AVE}$ of socio-economic condition is smaller than the correlation between socio-economic condition and the other variables. It indicates that socio-economic condition has a low discriminant validity.

Table 1. The $\sqrt{AVE}$ Value and Correlation Between Socio-Economic Condition and The Other Variables

| **Centroid** | | | **Factor** | | | **Path** | | |
|---|---|---|---|---|---|---|---|---|
| **Corellation $X_1$** | | $\sqrt{AVE}$ | **Corellation $X_1$** | | $\sqrt{AVE}$ | **Corellation $X_1$** | | $\sqrt{AVE}$ |
| **$Y_2$** | **$Y_1$** | | **$Y_2$** | **$Y_1$** | | **$Y_2$** | **$Y_1$** | |
| 0.59 | 0.478 | 0.441 | 0.598 | 0.466 | 0.441 | 0.6 | 0.459 | 0.436 |

The correlation between the quality of education and its indicators is smaller than the correlation between the quality of education and indicators of other variables. It shows that the quality of education can not predict inidicators within a group is better than the indicators in the other groups.

Table 2. The $\sqrt{AVE}$ Value and Correlation Between the Quality of Education and the Other Variables

| **Centroid** | | | **Factor** | | | **Path** | | |
|---|---|---|---|---|---|---|---|---|
| **Corellation $Y_1$** | | $\sqrt{AVE}$ | **Corellation $Y_1$** | | $\sqrt{AVE}$ | **Corellation $Y_1$** | | $\sqrt{AVE}$ |
| **$Y_2$** | **$X_1$** | | **$Y_2$** | **$X_1$** | | **$Y_2$** | **$X_1$** | |
| 0.610 | 0.478 | 0.688 | 0.615 | 0.466 | 0.688 | 0.634 | 0.459 | 0.684 |

However, the value of $\sqrt{AVE}$ of the quality of education is greater than the value of the correlation between the quality of education and the others latent variables, as shown in Table 2. It concludes that the quality of education has a high discriminant validity.

The correlation between the school infrastructure and its indicators is smaller than the correlation between the school infrastructure and indicators of other variables. It shows that the school infrastructure cannot predict inidicators within group is better than the indicators in the other groups.

IOP Conf. Series: Journal of Physics: Conf. Series **855** (2017) 012006 doi:10.1088/1742-6596/855/1/012006

However, the value of $\sqrt{AVE}$ of the school infrastructure is greater than the value of the correlation between the school infrastructure and the others latent variables, can be seen in Table 3. It concludes the school infrastructure has high discriminant validity.

Table 3. The $\sqrt{AVE}$ Value and Correlation Between the School Infrastructure and the Other Variables

| **Centroid** | | | **Factor** | | | **Path** | | |
|---|---|---|---|---|---|---|---|---|
| **Corellation $Y_2$** | | $\sqrt{AVE}$ | **Corellation $Y_2$** | | $\sqrt{AVE}$ | **Corellation $Y_2$** | | $\sqrt{AVE}$ |
| **$Y_1$** | **$X_1$** | | **$Y_1$** | **$X_1$** | | **$Y_1$** | **$X_1$** | |
| 0.610 | 0.590 | 0.734 | 0.615 | 0.598 | 0.734 | 0.634 | 0.600 | 0.739 |

c. Composite realibility

The school infrastructure and the quality of education have the value of composite reliability more than 0.7, but the socio-economic condition has less than 0.7. These results can be concluded that the school infrastructure and the quality of education are reliable or indicators of their variables are consistently measuring their constructs. However, the socio-economic condition is not reliable or their indicators are not consistently measuring the socio-economic condition.

**B. Evaluation of Inner Model (Structural Model)**

The evaluation of the inner model can be conducted by viewing the value of R-Square, in which the changes of the value of R-Square can be used to describe whether or not exogenous latent variables significantly effect on the endogenous latent variable. The following table is a summary of the value of R-Square from the three schemes.

Table 4. The Value of R-Square

| **Latent Variable** | **R-square** | | |
|---|---|---|---|
| | **Centroid** | **Factor** | **Path** |
| The school infrastructure | 0.349 | 0.349 | 0.36 |
| The quality of education | 0.393 | 0.393 | 0.411 |

Based on Table 4, it can be concluded that the variance of the school infrastructure can be explained by the socio-economic condition amounted to 34.9% (using the centroid and factors scheme) and the remaining 65.1% is explained by other factors outside this research. The variance of the quality of education can be explained by socio-economic condition and the school infrastructure amounted to 39.3% (using the centroid and factors scheme) and the remaining 60.7% is explained by other factors outside this research. Both the school infrastructure model and the quality of education model are in the category of moderate models.

**C. Significance Test**

The significance test of the model cannot be done using In SEM PLS method, since the data distribution is unknown. Hence, the significance test was carried out using bootstrapping resampling method. The following are the results of significance tests both outer and inner models using $\alpha = 10\%$ with the value of the T-table of 1.65.

a. Ouer Model

The following is a result summary of the validity significance test of latent variables.There are two valid indicators that form the socio-economic condition, namely the ratio of the number of prosperous families and the ratio of households using clean water in the centroid and factor schemes. In contrast, there is only one valid indicator in the path schema, namely the ratio of the number of prosperous families. All of indicators are valid forming of the school infrastructure, except $Y_{24}$ on the centroid scheme, i.e. the ratio of library possession. There is one indicator invalid forming of the quality of education for three schemes, namely a ratio average of national exam score SMA/MA more than 7.

b. Inner Model

The centroid and path scheme gave the same result, namely, the school infrastructure significantly affects the quality of education, but the socio-economic condition does not affect the school infrastructure and the quality of education. In the factor scheme, the socio-economic condition affects the school infrastructure, but does not affect the quality of education. While, the school infrastructure significantly affects the quality of education

**D. Model Improvements and Discussion**

Model improvement was done by eliminating the invalid indicators either based on the results of convergent validity test and significance tests. Indicators that remain to be used in the socio-economic condition are $X_{11}$, $X_{13}$ (for centroid and factor) and $X_{11}$ (for path). Indicators that remain to be used in the quality of education are $Y_{11}$, $Y_{12}$, $Y_{14}$ (for all of schemes). Indicators that remain to be used in the school infrastructure are $Y_{21}$, $Y_{22}$, $Y_{23}$ (for centroid) and $Y_{21}$, $Y_{22}$, $Y_{23}$, $Y_{24}$ (for factor and path).

After the model improvements, the evaluation is done both outer and inner models. The following are the results of the evaluation of the model has been improved.

The result of convergent validity test from three schemes shows valid, because all indicators of three latent variables have a loading factor above 0.5. The result of discriminant validity test shows that the correlation between latent variables and their indicators are higher than the correlation between latent variable and indicators of other variables for three latent variables and for the three schemes. Likewise, the $\sqrt{AVE}$ value of all latent variables is greater than the correlation between the latent variable and the others latent variables on three schemes. The value of composite reliability for three latent variables of the three schemes are more than 0.7. It can be concluded that the indicators on each of the latent variables are valid and reliable. However, the result of significance test of outer model shows only indicator $X_{11}$ on centroid scheme significanty invalid.

The following table shows the result of significance test of inner model for three schemes

Table 5. The Result of Significance Test of Improvement Model

| **Relation** | **Centroid** | | **Factor** | | **Path** | |
|---|---|---|---|---|---|---|
| | **T-value** | **Result** | **T-value** | **Result** | **T-value** | **Result** |
| $X_1$-> $Y_2$ | 3.859 | affected | 2.444 | affected | 1.258 | not affected |
| $Y_2$ -> $Y_1$ | 3.407 | affected | 3.017 | affected | 6.848 | affected |
| $X_1$ -> $Y_1$ | 0.270 | not affected | 0.287 | not affected | 0.153 | not affected |

The following is the evaluation of an improved model of the three schemes. Model of the centroid scheme is inconsistent and has one invalid indicator. Model of the factor scheme is a consistent , all of its indicators are valid but its the value of R-square decreased. Model of the path scheme is a consistent , all of its indicators are valid and its the value of R-square increased. So there is one improved model which represented by model from path scheme which is: $l_1 = 0.651 l_2$ and the full model as illustrated in Figure 2.

The model in the path scheme, the quality of education influenced only by the school infrastructure (0.648). The socio-economic condition did not affect neither the school infrastructure nor the quality of education. If the school infrastructure increased 1 point, then the quality of education increased 0.648 point. The quality of education had an $R^2$ of 0.418, which indicates that 41.8 percent of variance in the quality of education is explained by the school infrastructure, the remaining 58.2% is explained by the other factors which were not investigated in this work, such as support from local or central government, sponsors, cooperation with the stakeholders of the school, the quality of teachers, the quality of the new students etc.

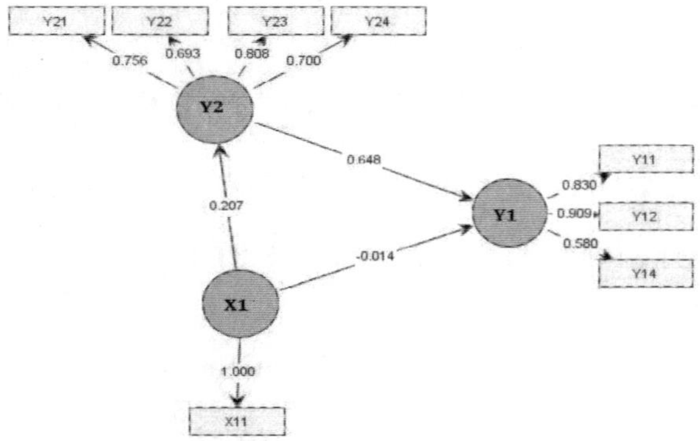


Figure 2. improved model from path scheme

The model in the path scheme shows that the socio-economic condition did not affect the quality of education. This condition contradicts the condition in the district sumenep since there are 185 schools only 13 of which have the status of public schools. It means that almost 93% of schools in the district Sumenep are private schools that were established through local community's initiative and the majority of them are religious schools. It reflects that the motivation of schools establishment tend to lead to idealism motivation, and less attention to the quality of education.

Trujillo stated that the selection of a certain weighted scheme than others has a few associated with the estimation process, whereas R-square of both models obtained in this work are not far different [9]. This agrees with observations made by Tenenhaus on the difference in the weighting results in the three schemes, in which no significant influence was shown [8]. The improved model which obtained in this was categorized as moderate models, which agrees with Chin's opinion [4].

## 5. Recommendations

It is necessary to test the effect of the spatial and necessary to add the variable of teachers' quality

## References

[1]. Anderson JC and Gerbing DW 1984 The Effec of Sampling Error on Convergent, Improper Solution, and Goodnes of fit Indices for Maximum Likelihood Factor Analysis *Psychometrica* **49**.
[2]. Bollen K 1989 *Structural Equations With Latent Variables* (John Wiley & Sons, Inc, New York).
[3]. Boomsma A 1982 The *Robustnes of LISREL Against Small Sample Sizes in Facgtor Analysis Model*, In H. Wold and K. Joreskog (Eds), *System Under Inderec Observation* (New York Elsevier North Holand)
[4]. Chin WW 1998 *The partial least squares approach for structural equation modeling* Eds. Modern Methods for Business Research (Marcoulides GA)
[5]. Efron B 1979 Bootstrap Methods: another Look at the Jackknife *Annuls of Statistics* **7** 1-26.
[6]. Ghozali I 2011 *Structural Equation Modelling Metode Alternatif dengan Partial Least Square PLS* (Undip Semarang)
[7]. Schumacker and Lomax 2004 *A Beginner's Guide To Structural Equation Modeling* (Lawrence Erlbaum Associates Inc New Jersey).
[8]. Tenenhaus M, Vinzi, Chatelin, and Lauro 2005 PLS Path-Modeling *Computational Statistics & Data Analysis* **48** 159–205.
[9]. Trujillo GS 2009 *PATHMOX Approach : Segmentation Trees in Partial Least Squares Path Modeling* (Universitat Politècnica de Catalunya)

# Mixed geographically weighted regression (MGWR) model with weighted adaptive bi-square for case of dengue hemorrhagic fever (DHF) in Surakarta

**H N Astuti**[1, a)], **D R S Saputro**[2], **and Y Susanti**[3]
[1,2,3] Department of mathematics, FMIPA, Universitas Sebelas Maret

E-mail: [a)] h4rdanti@gmail.com

**ABSTRACT.** MGWR model is combination of linear regression model and geographically weighted regression (GWR) model, therefore, MGWR model could produce parameter estimation that had global parameter estimation, and other parameter that had local parameter in accordance with its observation location. The linkage between locations of the observations expressed in specific weighting that is adaptive bi-square. In this research, we applied MGWR model with weighted adaptive bi-square for case of DHF in Surakarta based on 10 factors (variables) that is supposed to influence the number of people with DHF. The observation unit in the research is 51 urban villages and the variables are number of inhabitants, number of houses, house index, many public places, number of healthy homes, number of Posyandu, area width, level population density, welfare of the family, and high-region. Based on this research, we obtained 51 MGWR models. The MGWR model were divided into 4 groups with significant variable is house index as a global variable, an area width as a local variable and the remaining variables vary in each. Global variables are variables that significantly affect all locations, while local variables are variables that significantly affect a specific location.

**Keywords:** local parameters, global parameters, MGWR model, GWR model, DHF, adaptive bi-square

## 1. Introduction

Tobler's first law of Geography (Tobler in Anselin [1]), "Everything is related to everything else, but near things are more related than distant things", has laid foundation for regional science, specifically on spatial dimension. Spatial data refer to geographically-oriented data and possess certain coordinate system as a reference framework. This implies that spatial data contain information about geographic location of an area width. Spatial analysis, in general, requires location-based data containing the location's characteristics. It includes modeling which indicates a concept of cause and effect relationship by using a method sourced from spatial and non-spatial data sources to predict the presence of a spatial pattern.

One of spatial models is geographically weighted regression (GWR). It is a spatial regression model of which geographical location influences dependent variables (Fotheringham, et al. 2002 [9]). In other words, GWR enables to resolve spatial heterogeneity due to heterogeneous conditions of each spatial unit, whereas in fact not all of the independent variables in the GWR exert a spatial influence on dependent ones. For further references on the GWR, see Schabenberger and Gotway (2005, pp.

316–317) [19] and Waller and Gotway (2004, p. 434) [20], and on examples of its application, see Lloyd (2007, pp. 79–86) [18].

A Mixed Geographically Weighted Regression model (MGWR) is a combination of linear regression and the GWR. It is a regression model of which some independent variable coefficients are constant, while some other spatially vary (Brunsdon, et al.[3]). The combination is obtained after testing for spatial variability has been carried out. In the GWR, resulting parameters are only significant for certain spatial units. In several cases, the resulting parameters are significant for both certain spatial units and all of the spatial units.

The MGWR is appropriate for data which are influenced by local and global variables. The GWR and the MGWR differ on their spatial variability. All contributing factors in the GWR have spatial variability, and therefore result in local parameters which are influential for certain spatial units (Wheeler [10], Fotheringham [9], Brunsdon [8]). Meanwhile, some contributing factors in the MGWR have spatial variability, but some other do not. The factors having no spatial variability will generate a global parameter—a parameter which is influential to spatial units, while those having spatial variability will produce a local parameter. In order to explain the influence of the spatial units, weight is utilized. The present research employed bi-square weight which is determined based on the distance of the spatial units.

According to Sefuddin et al. [6], the implementation of the MGWR model with bi-square weight on poverty rates contributes to adequately high goodness of fit. Several other applied researches on the MGWR and GWR models were carried out by Pecci, F and M Sassi [7], Bai et al. [13], Huang, et al. [14], and Wrenn and Sam [15]. Both models yield dissimilar parameter estimation in every spatial unit. The GWR is considered the best model to explain the influence of low-wealth status in districts of North Sulawesi province having the lowest values of MSE and AIC and the highest value of $R^2$ in global regression and MGWR (Pongoh 2015 [11]).

Dengue Hemorrhagic Fever (DHF) is included as one of infectious diseases, which frequently occurs in Indonesia. Data of the number of its sufferers can be modeled using MGWR. The present research seeks to model the number of DHF sufferers in Surakarta by utilizing MGWR model with 'adaptive' (bi-square) weight.

## 2. Mixed Geographically Weighted Regression Model

GWR model. According to Brundson et al. [4], GWR model is influenced by such an aspect as spatial unit, in this case point in the spatial unit, and therefore the model yields estimates for every point. The model is formulated below:

$$y_i = \beta_0 + \sum_{k=1}^{m} \beta_{ki} x_{ki} + \varepsilon_i$$

where $y_i$, $i = 1,2,\dots,n$ represents dependent variable of the i$^{\text{th}}$ datum, $\beta_{ki}$, $k = 1,2,\dots,m$ is regression coefficient for the i$^{\text{th}}$ location and the k$^{\text{th}}$ variable, $x_{ki}$ is the ith value of $x_k$, and $\varepsilon_i$ is a residual which is normally distributed and independent.

### *2.1 MGWR Model.*

The GWR was then developed to MGWR, a model which does not only consider local variable, but also the global one. It is formulated:

$$y_i = \sum_{g=1}^{q} a_g x_{ig}(a) + \sum_{l=q+1}^{p} b_l(u_i, v_i) x_{il}(b) + \varepsilon_i,\ i = 1,2,\dots,n \qquad (1)$$

where $a_g$ is a global parameter, while $b_l$ is a local parameter. Variable $x_{ig}$ is an independent variable which is related to the global parameter, while $x_{il}$ represents an independent variable which is related to local parameter, and $\varepsilon_i$ is a residual in the MGWR model. In the model, an assumption which should be met is heterogeneity. The test of heterogeneity was conducted using Breusch-pagan (BP) test with test statistic, as formulated below:

$$BP = \frac{1}{2}\left[\boldsymbol{r}^T\boldsymbol{Z}(\boldsymbol{Z}^T\boldsymbol{Z})^{-1}\boldsymbol{Z}^T\boldsymbol{r} - n\right] \quad (2)$$

where $\boldsymbol{r}$ is 51x1 matrix and is formulated $\boldsymbol{r} = \begin{pmatrix} r_1 \\ r_2 \\ \vdots \\ r_i \end{pmatrix}$, $r_i = \left(\frac{\varepsilon_i^2}{\sigma^2}\right)$, $\sigma^2 = \frac{1}{n}\sum \varepsilon_i{}^2$, while $\boldsymbol{Z}$ is 51x4 matrix. It is decided that $H_0$ is rejected if $BP \in$ DK with critical area of DK $= \{BP|BP > \chi^2{}_{(k-l)}\}$.

2.2 *Parameter Estimation of the MGWR Model.*

Brundson et al [3] state that since MGWR model includes both local and global parameters, the parameter estimation needs to be conducted using weight least square (WLS) for the former and ordinary least square (OLS) for the latter.

2.3 *'Adaptative' (Bi-square) Weight Matrix*

Adaptive kernel function was utilized in the research since it enables to adapt bandwidths to in size to variations in data density. Therefore, when bandwidth is too large, only few data are involved, and vice versa. Fotheringham et al. [7] propose two weight functions, comprising gaussian and bi-square functions. The latter, as formulated below, was employed in the present research.

$$w_{ij} = \begin{cases} [1 - \left(\frac{d_{ij}}{h_i}\right)^2]^2, & d_{ij} < h_i \\ 0, & d_{ij} \geq h_i \end{cases} \quad (3)$$

where $d_{ij} = ((x_i - x_j)^2 + (y_i - y_j)^2)^{1/2}$ represents Euclidean distance from spatial unit $(x_i, y_i)$ to $(x_i, y_i)$, while $h_i$ is bandwidth.

## 3. Research Method

The present study presents an applied research which models the number of dengue hemorrhagic fever sufferers in Surakarta by applying the MGWR model and bi-square weight. Research data and procedures are described below.

3.1 *Data.*

Data include those on the number of DHF sufferers in 2014 which were sourced from Surakarta Department of Health and used as dependent variable $(Y)$. Factors involved in the research comprise such independent variables $(X_i)$ as the number of inhabitants, $(X_1)$, the number of houses $(X_2)$, *House Index* $(X_3)$, many public places $(X_4)$, the number of healthy homes $(X_5)$, the number of Posyandu $(X_6)$, area width $(X_7)$, level population density $(X_8)$, welfare of the family $(X_9)$ and high region $(X_{10})$ sourced from a public health center (*puskesmas*) in Surakarta.

3.2 *Research Procedures.*

Research procedures include: 1) checking heterogeneity of data on the number of DHF sufferers by using BP test based on equation (2), 2) computing weight using bi-square in each spatial unit based on equation (3), 3) determining significant independent variable $(X_i)$ and GWR model in each spatial unit. During this procedure, both global and local variables for MGWR model were obtained, 4) estimating parameters in MGWR model and finally MGWR model which fits to model (1) for each location was obtained.

## 4. Results and Discussion

Results and discussion cover data description, test of heterogeneity, 'adaptive' (bi-square) weight, GWR model, and MGWR model, each of which is explained further below.

### *4.1 Data Description*

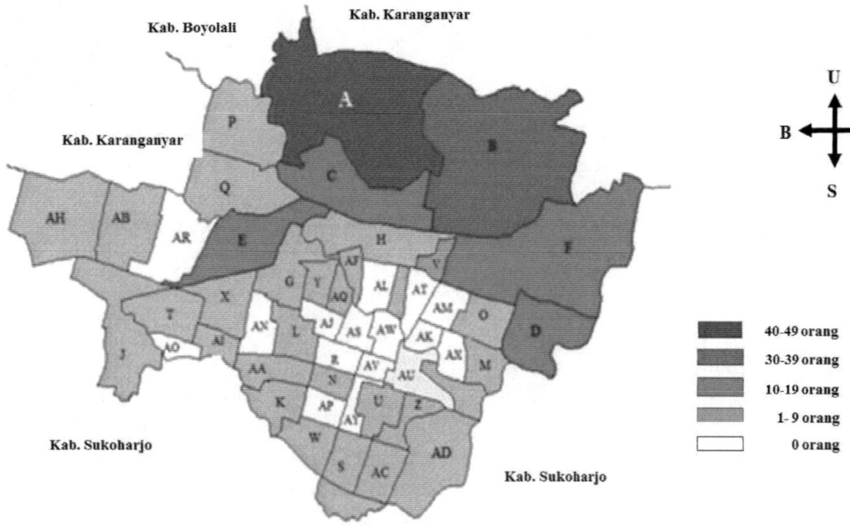


**Figure 1.** Distribution of DHF cases in 51 sub-districts of Surakarta

**Table 1**. Sub-districs in Surakarta

| Code | Sub-district | Code | Sub-district | Code | Sub-district |
|---|---|---|---|---|---|
| A | Kadipiro | R | Kemlayan | AI | Bumi |
| B | Mojosongo | S | Danukusuman | AJ | Timuran |
| C | Nusukan | T | Sondakan | AK | Sudiroprajan |
| D | Pucang Sawit | U | Baluwarti | AL | Setabelan |
| E | Manahan | V | Tegal Harjo | AM | Purwodiningratan |
| F | Jebres | W | Serengan | AN | Penumping |
| G | Mangkubumen | X | Purwosari | AO | Laweyan |
| H | Gilingan | Y | Punggawan | AP | Kratonan |
| I | Joyontakan | Z | Pasar Kliwon | AQ | Ketelan |
| J | Jajar | AA | Panularan | AR | Kerten |
| K | Tipes | AB | Pajang | AS | Keprabon |
| L | Sriwedari | AC | Joyosuran | AT | Kepatihan Wetan |
| M | Kampung Sewu | AD | Semanggi | AU | Kedung Lumbu |
| N | Jayengan | AE | Sangkrah | AV | Kauman |
| O | Jagalan | AF | Kestalan | AW | Kampung Baru |
| P | Banyuanyar | AG | Kepatihan Kulon | AX | Gandekan |
| Q | Sumber | AH | Karang Asem | AY | Gajahan |

Figure 1 demonstrates that Kadipiro sub-district (A) has the most numerous DHF sufferers with total number of 40 persons. Its surrounding sub-districts include Mojosongo (B) with total number of 33 DHF sufferers, Nusukan (C) with total number of 19 DHF sufferers, and Banyuanyar (P)

with total number of 5 DHF sufferers. Other sub-districts have an average number of 1-5 DHF sufferers. This reveals the presence of a spatial indication of distribution of DHF sufferers. Furthermore, there exists a significant data inequality which signifies the presence of heterogeneity in the data. In order to find it out, the Breusch-Pagan (BP) test was carried out.

*4.2 Test of Heterogeneity*

The BP test with null hypothesis ($H_0$) reveals that there is no heterogeneity in data or all spatial units having dissimilar characteristics. It results in BP value of 12.40682 with tabulated value $\chi^2_{(3;0,05)}$ of 7.815 and therefore $H_0$ is rejected. This implies that the DHF sufferers in Surakarta are heterogeneous and have different characteristics.

*4.3 Bi-square Weight Matrix*

A weight matrix quantifies spatial relationship between spatial units. The computation of the weight was conducted in each spatial unit and therefore it may differ. Table 2 presents an example of the computation in which Pasar Kliwon sub-district serves as the center of distribution.

**Table 2.** Euclidean distance and weight with Pasar Kliwon sub-district as the center of distribution

| Sub-district | $d_{ij}$ | Weight | Sub-district | $d_{ij}$ | Weight |
|---|---|---|---|---|---|
| Pajang | 0.053933 | 0.999995251 | Semanggi | 0.005012 | 0.999999959 |
| Karang Asem | 1.293993 | 0.997268126 | Kepatihan Kulon | 0.016671 | 0.999999546 |
| Sondakan | 0.046209 | 0.999996514 | Tegal Harjo | 0.020071 | 0.999999342 |
| Bumi | 0.034495 | 0.999998057 | Jebres | 0.029650 | 0.999998565 |
| Sriwedari | 0.022023 | 0.999999208 | Mojosongo | 0.036018 | 0.999997882 |
| Panularan | 0.026753 | 0.999998831 | Pucang Sawit | 0.024350 | 0.999999032 |
| Purwosari | 0.022023 | 0.999999208 | Sewu | 0.010586 | 0.999999817 |
| Jajar | 0.052368 | 0.999995523 | Jagalan | 0.017328 | 0.999999510 |
| Kemalayan | 0.015652 | 0.999999600 | Nusukan | 0.032242 | 0.999998303 |
| Jayengan | 0.014616 | 0.999999651 | Manahan | 0.037618 | 0.999997690 |
| Tipes | 0.021536 | 0.999999243 | Mangkubumen | 0.026679 | 0.999998838 |
| Serengan | 0.019668 | 0.999999368 | Gilingan | 0.027824 | 0.999998736 |
| Danukusuman | 0.012795 | 0.999999733 | Punggawan | 0.023964 | 0.999999062 |
| Joyontakan | 0.019647 | 0.999999370 | Kestalan | 0.024102 | 0.999999052 |
| Joyosuran | 0.012628 | 0.999999740 | Banyuanyar | 0.047896 | 0.999996255 |
| Pasar Kliwon | 0.000000 | 1.000000000 | Sumber | 0.039800 | 0.999997414 |
| Baluwarti | 0.006392 | 0.999999933 | Kadipiro | 0.047822 | 0.999996266 |
| Sangkrah | 0.006956 | 0.999999921 | | | |

In reference to the above weights of 51 sub-districts in which Pasar Kliwon sub-district presents as the center of distribution, weight matrix ***W*** was determined using the following formula:

$$\boldsymbol{W} = \begin{bmatrix} w_{11} & w_{12} & \cdots & w_{1n} \\ w_{21} & w_{22} & \cdots & w_{2n} \\ \vdots & \vdots & \vdots & \vdots \\ w_{n1} & w_{n2} & \cdots & w_{nn} \end{bmatrix}$$

$$= \begin{bmatrix} 1 & 0.999999619 & \cdots & 0.999996693 \\ 0.99999619 & 1 & \cdots & 0.999996953 \\ \vdots & \vdots & \vdots & \vdots \\ 0.999996693 & 0.999996953 & \cdots & 1 \end{bmatrix}$$

### *4.4 GWR Model*

GWR model refers to a model developed from multiple linear regression by utilizing weight as a basis for spatial analysis. Bi-square function employs bandwidth to form weight matrix for each spatial unit. The optimal bandwidth for the bi-square weight is 33 km. The GWR results in 51 models, one of which is that of Pasar Kliwon sub-district which is formulated:

$$\widehat{y_{Pk}} = 3.829165 + 0.643972X_3 + 3.579917X_7 - 0.000345X_8.$$

### *4.5 MGWR Model*

MGWR model is yielded from global and local variables having significant influence on the number DHF sufferers. Such global variables as house index $X_3$ and level population density $X_8$ were obtained in the model. The result of the analysis of each sub-district from *GWR* model was classified in Table 3.

**Table 3.** The classification of sub-districts according to significant variables

| Group | Sub-district(s) | Significant variable(s) | The number of the sub-districts | Percentage |
|---|---|---|---|---|
| 1 | Kadipiro, Mojosongo, Nusukan, Manahan, Jebres, Mangkubumen, Gilingan, Jajar, Tipes, Sriwedari, Jayengan, Banyuanyar, Sumber, Kemlayan, Sondakan, Baluwarti, Serengan, Purwosari, Punggawan, Panularan, Pajang, Kestalan, Kepatihan Kulon, Karang Asem, Bumi | $X_7$ | 25 | 49.02% |
| 2 | Pucang sawit, Jagalan, Tegal Harjo | $X_7, X_8$ | 3 | 5.88% |
| 3 | Pasar Kliwon, Semanggi, Sangkrah, Kampung Sewu | $X_3, X_7, X_8$ | 4 | 7.84% |
| 4 | Joyontakan, Danukusuman, Joyosuran, Kauman, Timuran, Sudiroprajan, Setabelan, Purwodiningratan, Penumping, Laweyan, Kratonan, Ketelan, Kerten, Keprabon, Kepatihan Wetan, Kedung Lumbu, Gajahan, Gandekan | No significant $X_i$ | 19 | 37.26% |

Considering the GWR model and the classification shown by Table 3, local variable for the MGWR model includes area width ($X_7$), while global variable covers House Index ($X_3$) and level population density ($X_8$). After significant test for global variable, House Index was the significant global variable. The House Index, which refers to the number of houses positively infected with larvae and/or pupae, exerts a significant influence on the number of DHF sufferers in Group 3. Meanwhile, area width gives a significant influence on the number of DHF sufferers in all of the groups. Figure 2 illustrates the mapping of significant variables demonstrated in Table 3.

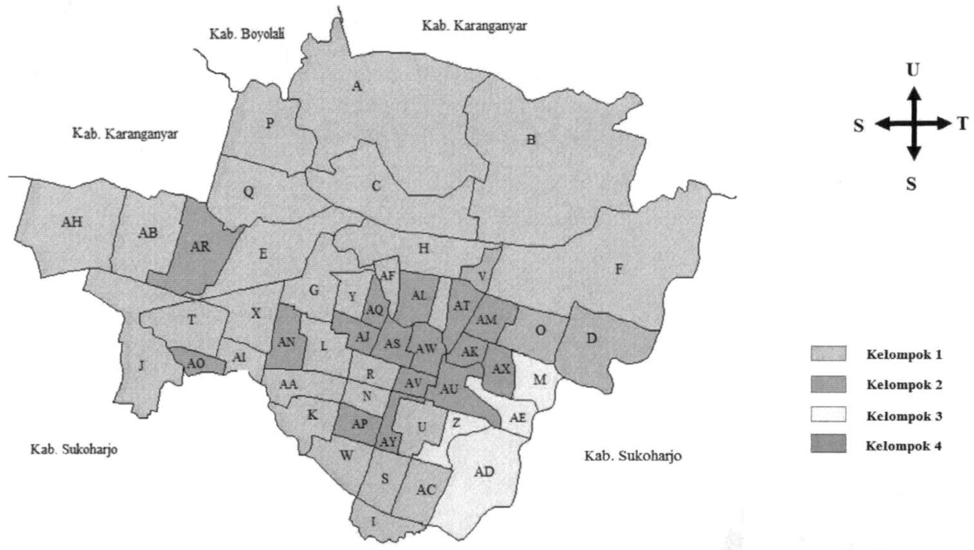


**Figure 2.** The classification of sub-districts based on significant variables

Figure 2 presents that 1) numerous DHF sufferers in Group 1 influenced by area width ($X_7$) are found in 25 sub-districts (49.02%) and spread over Surakarta, except areas bordered with Sukoharjo, 2) numerous DHF sufferers in Group 2 influenced by area width ($X_7$) and level population density ($X_8$) are found in 3 sub-districts (5.88%), 3) numerous DHF sufferers in Group 3 (7.84%) influenced by area width ($X_7$), level population density ($X_8$) and House Index ($X_3$) are found in 4 sub-districts, and 4) numerous DHF sufferers in Group 4 (37.26%) are not influenced by 10 variables predicted to give influence on the number of the DHF sufferers. The parameter estimation is demonstrated by Table 3.

**Table 4.** Parameter estimation of local variables $\hat{\beta}_0$, $\hat{\beta}_7$, and $R^2$ for 11 sub-districts

| Sub-district | ($\hat{\beta}_0$) | calculated t value | ($\hat{\beta}_7$) | calculate d t value | Local $R^2$ |
|---|---|---|---|---|---|
| Pajang | 4.623182 | 6.206654 | 6.113438 | 6.286946 | 0.643961 |
| Laweyan | 4.611724 | 6.048076 | 4.777195 | 2.726309 | 0.363126 |
| Karang Asem | -1.453233 | -0.462694 | 6.939781 | 8.599443 | 0.884828 |
| Sondakan | 4.706330 | 6.481070 | 5.853682 | 5.282221 | 0.581010 |
| Bumi | 4.892310 | 6.574185 | 5.631906 | 3.421980 | 0.429166 |
| Sriwedari | 5.142577 | 6.702331 | 6.289033 | 3.749680 | 0.456800 |
| Panularan | 5.602855 | 6.082733 | 7.522728 | 3.718553 | 0.460204 |
| Purwosari | 5.963288 | 6.117586 | 8.430034 | 4.128316 | 0.520631 |
| Serengan | 3.967844 | 3.731161 | 3.950396 | 1.726267 | 0.347405 |
| Kerten | 4.872237 | 7.177317 | 7.020797 | 9.456836 | 0.775779 |
| Pasar Kliwon | 2.901746 | 3.412516 | 2.365120 | 1.404927 | 0.347395 |

Parameter estimation for the MGWR model was carried out using global parameter estimation, $\widehat{\beta_3}$, of 1.153763 and local parameters, $\widehat{\beta_0}$ and $\widehat{\beta_7}$, as figured out by Table 4 for 11 sub-districts.

**Table 5.** Significant local parameter estimation of $\widehat{\beta_0}$ and $\widehat{\beta_7}$ for the MGWR model

| Sub-district | $\widehat{\beta_0}$ | $\widehat{\beta_7}$ |
|---|---|---|
| Pajang | 4.623182 | 6.113438 |
| Laweyan | 4.611724 | 4.777195 |
| Karang Asem | Not Significant | 6.939781 |
| Sondakan | 4.706330 | 5.853682 |
| Bumi | 4.892310 | 5.631906 |
| Sriwedari | 5.142577 | 6.289033 |
| Serengan | 3.967844 | Not Significant |
| Kerten | 4.872237 | 7.020797 |
| Pasar Kliwon | 2.901746 | Not Significant |

By referring to Table 4, 51 MGWR models for each sub-district are gained. The general formula of MGWR model for predicting the number of DHF sufferers is constructed below:

$$\hat{Y}_i = \hat{\beta}_{i0} + 1.361575X_{i3} + \hat{\beta}_{i7}X_{i7}$$

The model results in $R^2$ value of 0.78, indicating that 78% of total variability is explained by such variables of House Index and area width. An example of MGWR model for Pasar Kliwon sub-district is formulated:

$$\widehat{y_{Pk}} = 2.901746 + 1.153763X_3.$$

It means that an increase in a unit of House Index leads to an increase in a number of one or two DHF sufferers. Provided that the House Index is constant, a number of one or two sufferer(s) exist(s). The model has R-squared value ($R^2$) of 34.7%, indicating that 34.7% of total variability is explained by House Index, while 65.3% by other variables excluded in the observation. The other example of MGWR model was MGWR model for Pajang sub-district is formulated.

$$\widehat{y_{Pj}} = 4{,}623182 + 1{,}153763X_3 + 6{,}113438X_7.$$

It means that an increase in a unit of House Index lead to an increase in a number one or two DHF sufferers. Provides that the House Index is constant, a number of one or two sufferer(s) exist(s). An increase in one km$^2$ of area width leads to an increase in number six or seven DHF sufferers. Provided that area width is constant, a number of six or seven sufferers exists. The model has R-squared value ($R^2$) of 64.4%, indicating that 64.4% of total variability is explained by House Index and area width, while 35.6% by other variables excluded in the observation.

In reference to the two aforementioned MGWR models with global variable of House Index, it is clear that an increase in a unit of House Index causes an increase in a number of one or two DHF sufferer(s) in each sub-district in Surakarta.

## 5. Conclusions

The research concludes that House Index ($X_3$) serves as an independent variable which has global influence, while area width ($X_7$) exerts local influence. The MGWR model results in R-squared value ($R^2$) of 0.78. The general formula of MGWR model for predicting the number of DHF sufferers using bi-square weight is shown below:

$$\hat{Y}_i = \hat{\beta}_{i0} + 1.361575X_{i3} + \hat{\beta}_{i7}X_{i7}.$$

The formula indicates that when other variables are constant, an increase in one unit of House Index leads to an increase in a number of one or two DHF sufferer(s) in each sub-district in Surakarta.

## References

[1] Tobler, W.R. (1970). A computer movie simulating urban growth in the Detroit region. *Economic geography* (46, 234-240).

[2] Anselin, L. (1998). *Spatial econometrics : methods and models.* Dordrecht: Kluwer Academic Publishers.

[3] Brunsdon, C., S. A. Fotheringham, and M. Charlton. (1996). Geographically weighted regression: a method for exploring spatial nonstationary. *Geographical Analysis* (Vol. 28 No. 4, 281-298).

[4] Brunsdon C., S. A. Fotheringham, and M. Charlton. (1999). Some notes on parametric significance tests for geographically weighted regression. *Journal of Regional Science* (39, 497-524).

[5] *Profil kesehatan provinsi Jawa Tengah.* (2013). Semarang: Dinas Kesehatan Provinsi Jawa Tengah (Health Department of Central Java province).

[6] Saefuddin, A., N. A. Setiabudi, and N. A. Achsani. (2011). On comparisson between ordinary linear regression and Geographically Weighted Regression: with application to Indonesian poverty data. *European Journal of Scientific Research* (Vol. 57 No. 2, 275-285).

[7] Pecci, F and M Sassi. (2008). A mixed Geographically Weighted approach to decoupling and rural development in the EU-15. Paper prepared for presentation at the 107th EAAE Seminar "Modelling of Agricultural and Rural Development Policies". Sevilla, Spain, January 29th - February 1st.

[8] Brunsdon, C., Fotheringham, A.S. and Charlton M.E. (1999). Some notes on parametric significance tests for geographically weighted regression. *Journal of Regional Science* (39, 497-524).

[9] Fotheringham, A.S., Brunsdon, C. and Charlton, M. (2002). *Geographically weighted regression: The analysis of spatially varying relationships.* West Sussex: Wiley.

[10] Wheeler, D.C. and Tiefelsdorf, M. (2005). Multicollinearity and correlation among local regression coefficients in geographically weighted regression. *Journal of Geograpical Systems* (7, 161-187).

[11] Pongoh, F., I M Sumertajaya, M N Aidi. (2015). Geographichal Weighted Regression and Mix Geographichal Weighted Regression. *International Journal of Statistics and Applications*, 5(1), 1-4. doi: 10.5923/j.statistics.20150501.01.

[12] Wei, C.H.; Qi, F. (2012). On the estimation and testing of mixed geographically weighted regression models. *Econ. Modell* (29, 2615–2620).

[13] Bai, Y.; Wu, L.; Qin, K.; Zhang, Y.; Shen, Y.; Zhou, Y. (2016). A geographically and temporally weighted regression model for ground-level PM2.5 estimation from satellite-derived 500 m resolution AOD. *Remote Sens* (8, 262).

[14] Huang, B.; Wu, B.; Barry, M. (2010). Geographically and temporally weighted regression for modeling spatio-temporal variation in house prices. *Int. J. Geogr. Inf. Sci* (24, 383–401).

[15] Wrenn, D.H.; Sam, A.G. (2014). Geographically and temporally weighted likelihood regression: Exploring the spatiotemporal determinants of land use change. *Reg. Sci. Urban Econ* ( 44, 60–74).

[16] Wheeler, D.; Tiefelsdorf, M. (2005). Multicollinearity and correlation among local regression coefficients in geographically weighted regression. *J. Geogr. Sys.* (7, 161–187).

[17] Cho, S.; Lambert, D.M.; Kim, S.G.; Jung, S. (2009). Extreme coefficients in geographically weighted regression and their effects on mapping. GIsci. *Remote Sens.* (46, 273–288).

[18] Lloyd, C. D. (2007). *Local models for spatial analysis.* Boca Raton: CRC.

[19] Schabenberger, O. and Gotway, C. A. (2005). *Statistical methods for spatial data analysis.* London: Chapman & Hall.

[20] Waller, L. A. and Gotway, C. A. (2004). *Applied spatial statistics for public health data*. NJ: Wiley, Hoboken.

IOP Conf. Series: Journal of Physics: Conf. Series **855** (2017) 012008 doi:10.1088/1742-6596/855/1/012008

# Spread of Ebola disease with susceptible exposed infected isolated recovered (SEII$_h$R) model

**Afina Azizah, Purnami Widyaningsih, and Dewi Retno Sari Saputro**

Department of Mathematics, Sebelas Maret University, Surakarta, Indonesia

E-mail: afina.azizah@student.uns.ac.id, poer@uns.ac.id, dewi.rss@gmail.com

**Abstract.** Ebola is a deadly infectious disease and has caused an epidemic on several countries in West Africa. Mathematical modeling to study the spread of Ebola disease has been developed, including through models susceptible infected removed (SIR) and susceptible exposed infected removed (SEIR). Furthermore, susceptible exposed infected isolated recovered (SEII$_h$R) model has been derived. The aims of this research are to derive SEII$_h$R model for Ebola disease, to determine the patterns of its spread, to determine the equilibrium point and stability of the equilibrium point using phase plane analysis, and also to apply the SEII$_h$R model on Ebola epidemic in Sierra Leone in 2014. The SEII$_h$R model is a differential equation system. Pattern of ebola disease spread with SEII$_h$R model is solution of the differential equation system. The equilibrium point of SEII$_h$R model is unique and it is a disease-free equilibrium point that stable. Application of the model is based on the data Ebola epidemic in Sierra Leone. The free-disease equilibrium point $(S_e, E_e, I_e, I_{he}, R_e)$=(5743865, 0, 0, 0, 0) is stable.

## 1. Introduction

Ebola disease is a deadly infectious disease. The disease previously called as *Ebola Haemorrhagic Fever* was caused by *Ebola Virus Disease* (*EVD*). Ebola virus firstly was founded in 1976 in Zaire (Kongo) West Africa and it caused an epidemic on the year. After that Ebola caused epidemic again in 1995 in Kongo, in 2000 in Uganda, and in 2014 (the greatest epidemic) take place on several countries in West Africa (CDC [1]).

According to Astacio et al. [2] a mathematical modeling serves a tool to study infectious disease spread. Ebola is one of infectious diseases, the spread of Ebola could be learned using the mathematical model. Astacio et al. [2] in 1996 introduced susceptible infected removed (SIR) model to understand the dynamics of Ebola disease spread. Furthermore in 2004 Chowell et al. [3] added an exposed compartment, that is individual in incubation periods into SIR model and obtained susceptible exposed infected removed (SEIR) model.

Until now, drugs and vaccine to treat Ebola disease have not been found. This requires health care to put infected individuals in an isolation room in order they can not transmit the virus to healthy individuals. In 2015 Koya and Mamo [4] developed SEIR model with isolation treatment and wrote as susceptible exposed infected isolated removed (SEII$_h$R) model. In the model was assumed that the length of an individual who recovered and who died of Ebola was same. The difference between the length of an individual of Ebola until recovered with the deceased will affect to the patterns of Ebola spread. By observing these differences, research of the spread of Ebola was conducted. The aims of this research are to derive SEII$_h$R model for Ebola disease, to determine the patterns of its spread, to determine the equilibrium point and stability at the equilibrium point using phase plane analysis. The model then was applied on Ebola epidemic in Sierra Leone in 2014.

## 2. Susceptible infected removed (SIR) model

In 1927, SIR model for the first time was introduced by Kermack and McKendrik. Then Astacio et al. [1] used the model to explain Ebola outbreak in Kikwit, Zaire. In the SIR model was assumed that constant population of $N$, natural birth and death were ignored, and transmission can only occur through a direct contact between susceptible individual to the infected individual. If we let $\beta$ as an Ebola contact rate, $S(t)$ was denote the number of susceptible at time $t$, and $I(t)$ was denote the number of infected at time $t$, then as many $\beta SI/N$ of susceptible will be infected. Individuals who have been infected with Ebola have possibilities to recover or die (removed). We assumed that duration of the individual on infected compartment to be cured equal to duration of the individual on the compartment to die. If we let $\gamma$ as recovery and death rate, then $\gamma I$ individuals will recover or die of Ebola. Recovered individual was assumed invulnerable and could not transmit Ebola. Therefore $SIR$ model for Ebola epidemic can be written as

$$\begin{aligned}\frac{dS}{dt} &= -\frac{\beta SI}{N}\\ \frac{dI}{dt} &= \frac{\beta SI}{N} - \gamma I\\ \frac{dR}{dt} &= \gamma I,\end{aligned} \tag{1}$$

with $\beta, \gamma > 0$. The model (1) is first order of a nonlinear differential equations system.

## 3. Susceptible exposed infected isolated recovered ($\mathrm{SEII_hR}$) model

The SIR model (1) that was introduced by Astacio et al. [2] regardless the incubation period, further was developed by Chowell et al. [3] by considering the incubation period. The incubation period occurs since susceptible individual began exposed to the virus until showed symptoms of infection. At the period, individuals have not been able to transmit virus to others. Exposed compartment was denoted by E. The SIR model with incubation period later was called as susceptible exposed infected removed (SEIR) model. In the SEIR model, $\beta SI/N$ infected individuals were categorized in exposed compartment. If we let the average of an individual through the incubation period was in $\frac{1}{\sigma}$ days, then $\sigma E$ exposed individuals would be able to transmit Ebola to other people.

Koya and Mamo [4] then evolved SEIR model with isolation treatment. This treatment was given to individuals in exposed and infected compartments by placing them in a particular room in the hospital. Individuals who were given this treatment were categorized in an isolated compartment, $\mathrm{I_h}$. The SEIR model with isolation treatment can be written as susceptible exposed infected isolated removed ($\mathrm{SEII_hR}$) model. If we let $\frac{1}{\lambda}$ and $\frac{1}{\alpha}$ respectively as the duration (in days) of individual in $E$ and $I$ compartment until they were given isolation treatment, then as many $\lambda E + \alpha I$ individuals get isolation treatment. The duration of an isolated Ebola patients survive until cured or died was assumed to be equal, that is $\frac{1}{\omega}$ days. In other words as many $\omega I_h$ individuals were categorized in removed (R) compartment. In addition, $\mathrm{SEII_hR}$ model regarded to natural birth and death, and letting these rate as $\mu$.

In $\mathrm{SEII_hR}$ model, the duration of infected individuals (in infected dan isolated compartment) survive until recover was assumed equal to the duration of individual that eventually died. According to Rivers et al. [5], the average duration of an individual survives from Ebola until cured was different to that eventually died. If the difference was consider, then removed compartment that earlier used to classify recovered and died individuals, converted into recovered (R), a compartment for recovered individuals from Ebola. The parameter $\gamma$ and $\omega$ respectively

IOP Conf. Series: Journal of Physics: Conf. Series **855** (2017) 012008 doi:10.1088/1742-6596/855/1/012008

explain as recovery rate in I dan $I_h$ compartments. Meanwhile, to explain death rate that was caused by Ebola in I and $I_h$ compartments we use parameter $\epsilon$, or as many $\epsilon I + \epsilon I_h$ individuals died due to Ebola and they were removed from population. Individuals who died of Ebola were assumed can not transmit the Ebola virus.

The alterations of assumptions, variables, and parameters in the model would affect on instantaneous change in each compartment. In I compartment, the addition of individual was caused of their having passing through the incubation period by as many as $\sigma E$ individuals. Infected individuals can be cured with cure rate of $\gamma$ or they were given isolation treatment after being infected for $\frac{1}{\alpha}$ days. If we let $\epsilon$ and $\mu$ as death rate respectively due to Ebola and natural death, then as many $\epsilon I + \mu I$ individuals infected would be removed from the population. The instantaneous rate of change on the number of individuals infected can be expressed as

$$\frac{dI}{dt} = \sigma E - \gamma I - \alpha I - \epsilon I - \mu I.$$

The same ways was obtained to determine the instantaneous rate of change in the other compartments. Thus, susceptible exposed infected isolated recovered ($SEII_hR$) model can be written as

$$\begin{aligned}
\frac{dS}{dt} &= \mu N - \frac{\beta SI}{N} - \mu S \\
\frac{dE}{dt} &= \frac{\beta SI}{N} - \sigma E - \lambda E - \mu E \\
\frac{dI}{dt} &= \sigma E - \gamma I - \alpha I - \epsilon I - \mu I \\
\frac{dI_h}{dt} &= \lambda E + \alpha I - \omega I_h - \epsilon I_h - \mu I_h \\
\frac{dR}{dt} &= \gamma I + \omega I_h - \mu R,
\end{aligned} \tag{2}$$

with $\mu, \beta, \gamma, \sigma, \lambda, \alpha, \omega, \epsilon$ are respectively natural birth and death rate, contact rate, recovery rate of I compartment, exposed rate, isolation rate of E compartment, isolation rate of I compartment, recovery rate of $I_hR$ compartment, death rate due to Ebola, and positives. The model of (2) was first order of a nonlinear differential equations system. The spread patterns of Ebola can be determined from solutions of system (2).

### *3.1. Equilibrium*

Equilibrium of differential equations system was achieved when the solution was not change in time, and referred as equilibrium point. The equilibrium points of system (2) were achieved if

$$\begin{aligned}
\mu N - \frac{\beta SI}{N} - \mu S &= 0 \\
\frac{\beta SI}{N} - \sigma E - \lambda E - \mu E &= 0 \\
\sigma E - \gamma I - \alpha I - \epsilon I - \mu I &= 0 \\
\lambda E + \alpha I - \omega I_h - \epsilon I_h - \mu I_h &= 0 \\
\gamma I + \omega I_h - \mu R &= 0.
\end{aligned} \tag{3}$$

The system (3) is a nonlinear equations. Solution of the system (3) was unique, that was $(S, E, I, I, R)$=$(N, 0, 0, 0, 0)$. Thus, the equilibrium point of $SEII_hR$ model (2) can be written as

$$(S_e, E_e, I_e, I_{he}, R_e) = (N, 0, 0, 0, 0). \tag{4}$$

The equilibrium point (4) showed that all individuals in the population were susceptible and the number of infected individuals in the equilibrium point were zero ($I_e = 0$ dan $I_{he} = 0$). The equilibrium point was free-disease equilibrium point.

*3.2. Stability at the equilibrium point*

Stability at the equilibrium point (4) can be determined by observing the direction of trajectories in the phase plane. The equation of $I_h - I$ phase plane at the equilibrium point (4) was

$$\frac{dI}{dI_h} = \frac{-(\gamma + \alpha + \epsilon + \mu)I}{\alpha I - (\omega + \epsilon + \mu)I_h}. \tag{5}$$

The trajectories in the $I_h - I$ phase plane of (5) was shown in Figure 1.

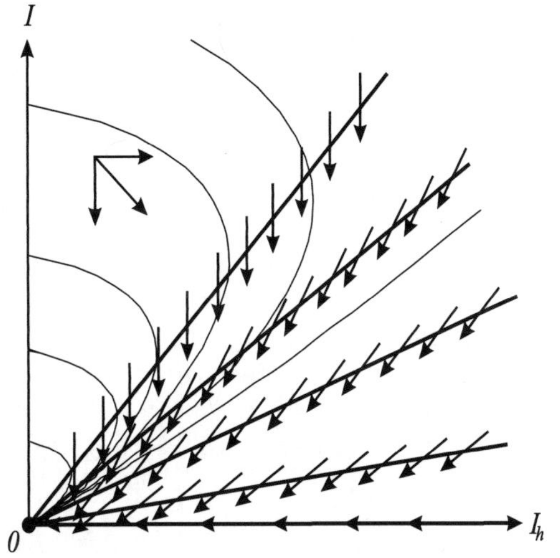


**Figure 1.** Trajectories in $I_h - I$ phase plane.

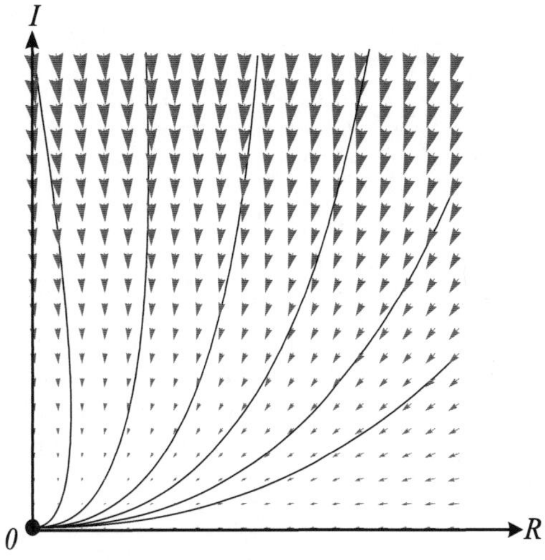


**Figure 2.** Trajectories in $R - I$ phase plane.

Figure 1 shows that for any initial values and parameters given, the value of $I$ decreases from the initial value, while for the value of $I_h$ increases up to a certain point and then decreases to the equilibrium point. Because of the directions of all trajectories toward equilibrium point (0,0), the equilibrium point $(I_{he}, I_e)$ was free-disease equilibrium point that stable.

In the same ways, we determine trajectories in $R - I$ phase plane as shown in Figure 2. From Figure 1 and 2 it is clear that the direction of the trajectories towards equilibrium point (0,0). The equilibrium point (4) was free-disease equilibrium point that stable.

## 4. Discussion Regarding SEII$_h$R Model for Ebola Disease

In this application, the values of parameters were selected based on an epidemic in Sierra Leone (West Africa) in 2014. The following was given values of parameters that refers to Ivorra et al. [6], Rivers et al. [5], and also Koya and Mamo [4]. Given a contact rate ($\beta$) of 0.2605, incubation period ($1/\sigma$) 11.4 days or $\sigma = 0.0877$, the duration of infected individual to be isolated ($1/\alpha$) was 5 days or $\alpha = 0.2$, recovery rate of infected individual ($\gamma$) of 0.05, recovery rate of isolated individual ($\omega$) of 0.17, the length of exposed individual to be isolated ($1/\lambda$) was 3 days or $\lambda = 0.3333$, death rate due to Ebola ($\epsilon$) of 0.185, and natural birth and death rate ($\mu$) of 0.011.

According to data of WHO in Sierra Leone in 2014, the number of health individuals was 5743725 and the number of infected individuals was 14122. The information could be written as $(S_0, E_0, I_0, I_{h0}, R_0)$=(5743725, $0, 14122, 0, 0$). Solution of SEII$_h$R (2) model with these parameters and initial value $(S_0, E_0, I_0, I_{h0}, R_0)$ was shown in Figure 3 and 4.

IOP Conf. Series: Journal of Physics: Conf. Series **855** (2017) 012008 doi:10.1088/1742-6596/855/1/012008

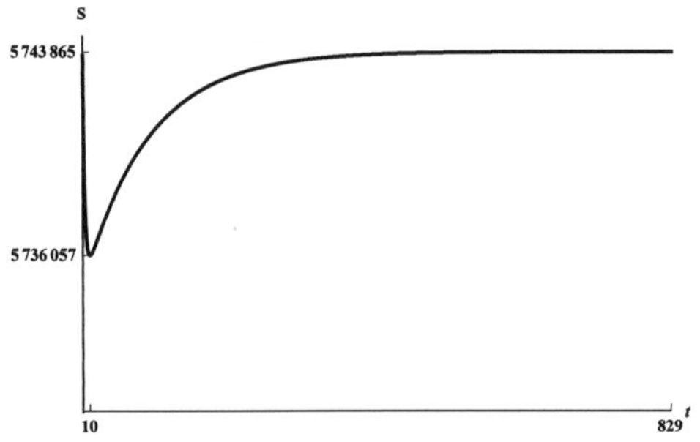


**Figure 3.** The number of individual in S from the application of $SEII_hR$ model in Sierra Leone in the first 829 days.

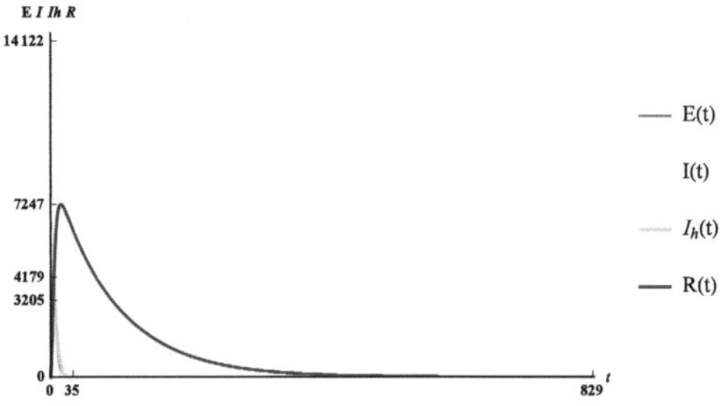


**Figure 4.** The number of individual in E, I, $I_h$, R from the application of $SEII_hR$ model in Sierra Leone in the first 829 days.

From Figure 3 it appears that the number of susceptible individuals (S) decreases from its initial value up to the $10^{th}$ day. This was caused by susceptible start to be infected of Ebola virus. In the next days, the number of susceptible increases because of isolation treatment. Furthermore, the number of susceptible was not vary in time start from the $829^{th}$ day of 5743865. Figure 4 shows that at the $829^{th}$ day, individuals in E, I, $I_h$ and R compartment start to equilibrate. We observe the patterns of its spread in the first 100 days as shown in Figure 5.

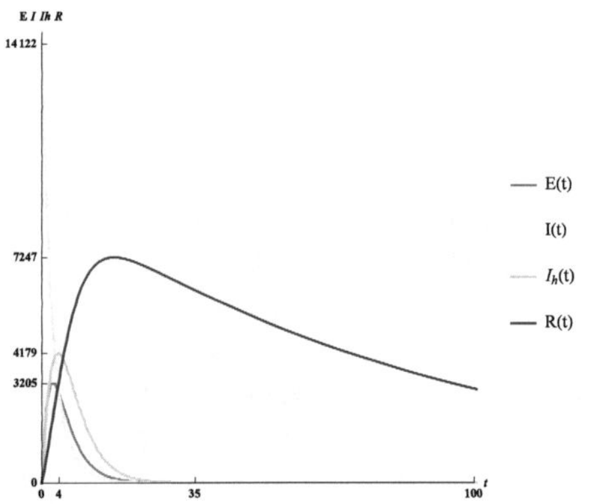


**Figure 5.** The number of individuals E, I, $I_h$, R from the application of $SEII_hR$ model in Sierra Leone in the first 100 days.

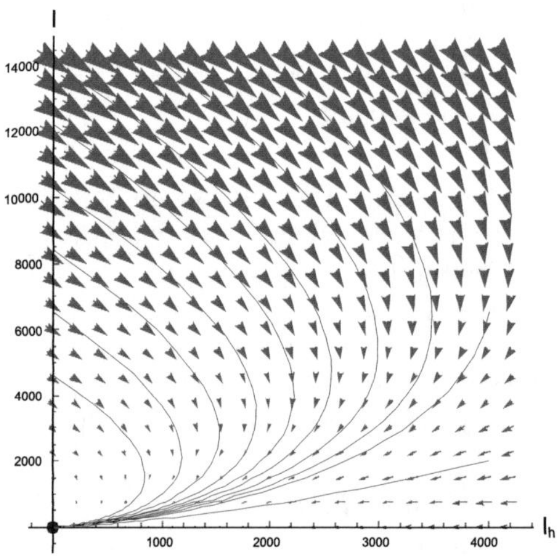


**Figure 6.** Trajectories in $I_h - I$ phase plane in the application of its model

From Figure 5 it appears that the number of infected always decreases from the initial value. Individuals in this compartment potentially to transmit the virus to others, the behaviour of decreases monotonous on the number of infected was in a good condition. This behaviour then begins unchanged at the $33^{rd}$ day. Meanwhile, the number of isolated individuals increases and achieve its peak at the $4^{th}$ day, i.e. as many 4179 isolated. Therefore, health professionals needed to prepare sufficient rooms, equipments, and personnels to face the highest number of isolated individuals. The number of isolated individuals then decreases because the patient to be recover or die, and unchanged start at the $41^{st}$ day.

By observing the patterns of spread in S, E, I, $I_h$, dan R compartments, every compartment begin to equilibrate at the $829^{th}$ day with its equilibrium point was $(S_e, E_e, I_e, I_{he}, R_e)$=(5743865, 0, 0, 0, 0). The equilibrium point corresponds to equilibrium point (4). The number of infected in the equilibrium point was zero, it means that the equilibrium point that achieved was a free-disease equilibrium point.

The trajectories in $I_h - I$ phase plane were shown in Figure 6. The figure shows that for certain parameters and every given initial value, the number of infected individuals is always decreases. Meanwhile, the number of isolated individuals increases up to a certain point, then decrease. Both the number of infected and isolated individuals towards to the equilibrium point. Thus, the equilibrium point was free-disease equilibrium point that stable.

## 5. Conclusions

The $SEII_hR$ model for Ebola disease spread is written in system (2). The pattern of Ebola disease spread is solution of $SEII_hR$ model. The equilibrium point of the model is free-disease equilibrium point that stable. The application of $SEII_hR$ model in Sierra Loene according to Ebola epidemic in 2014 shows that the system tends to equilibrium at the $829^{th}$ day with the equilibrium point is $(S_e, E_e, I_e, I_{he}, R_e)$=(5743865, 0, 0, 0, 0). The equilibrium point is free-disease equilibrium point that stable.

## 6. References

[1] Centers for Disease Control and Prevention (CDC) 2016 *Outbreak Chronology: Ebola Virus Disease* http://www.cdc.gov/vhf/ebola/outbreaks/history/chronology.html accessed on February 29
[2] Astacio J, Briere D, Guillen M, Martinez J, Rodriguez F, and Valenzuela-Campos N 1996 Mathematical models to study the outbreaks of Ebola *Mathematical and Theoretical Biology Report*
[3] Chowell G, Hengartner N W, Castillo-Chavez C, Fenimore P W, and Hyman J M 2004 The basic reproductive number of Ebola and the effects of public health measures : the case of Congo and Uganda *Journal of Theoritical Biology* **229** pp 119-126
[4] Koya P R and Mamo D K 2015 Ebola epidemic disease: modelling, stability analysis, spread control technique, simulation study and data fitting *Journal of Multidisciplinary Engineering Science and Technology* **2** pp 476-484
[5] Rivers C M, Lofgren E T, Marathe M, Eubank S, and Lewis B L 2014 Modeling the impact of interventions on an epidemic of Ebola in Sierra Leone and Liberia *PLOS Current Outbreaks* **1**
[6] Ivorra B, Ngom D, and Ramos A M 2015 A mathematical model to predict the risk of human diseases spread between countries: validation and application to the 2014-15 Ebola virus disease epidemic *BioMed Research International*

# A Closed-Loop Supply Chain under Retail Price and Quality Dependent Demand with Remanufacturing and Refurbishing

**A Y Christy, B N Fauzi, N A Kurdi, W A Jauhari and D R S Saputro**
Mathematics Department and Industrial Engineering Department of Sebelas Maret University, Indonesia

E-mail: arfa@mipa.uns.ac.id, alexlandak@gmail.com

**ABSTRACT.** The demand of a product is linearly dependent on the retail price and quality of the product. We address a closed-loop supply chain where the manufacturer manufactures products according to the demand and sells them through a retailer in the market. A third party collects the used products from costumers and sends to the manufacturer to increase the quality. If the products can retrieve the original quality, thus the process is called remanufacturing. Not every products can retrieve the original quality, thus manufacturer refurbish this products with lower price. We construct four different scenarios – centralized and decentralized led by manufacturer, retailer, and third party. From the comparison of the result obtained in the numerical example, we conclude that the joint profit obtained under centralized, manufacturer-led, and retailer-led policies is higher than third party-led policy.

## 1. Introduction

Reverse logistic management is process to recover the quality of used products by manufacturer. If the used products can retrieve the original quality as a new product, then the recovery process is called remanufacturing. If the used product cannot retrieve their original quality, the manufacturer can sell it with lower quality to the secondary market. This recovery method is called refurbishing. The usual forward supply chain associates with this reverse logistic management, creates a closed-loop supply chain (CLSC).

We address a CLSC where the manufacturer manufactures products according to the demand and sells it through a retailer in the market. A third party collects the used products from costumers and sends to the manufacturer. We use price and quality dependent demand [2,6,13].

Kampstra and Ashayeri [3] reviewed that supply chain model is invented around 1900 and developed until the middle of 1960. Then the research on CLSC [1,4-15] is widely expanding until now. There are plenty of available literatures on CLSC.

Until 2015, many in the literature assumed that all used items returned to the manufacturer can retrieved their original quality as "new product" [1,6]. In the reality, sometimes third party returned non-remanufacturable products, thus they cannot retrieve their original quality as "new product". If manufacturer disposes them as mentioned in Chung et al [1], thus there will be additional cost for disposal management. To avoid this disposal cost, the process of refurbishing mentioned in Konstantaras et al [5] can be added to the model created by Maiti and Giri [6].

The next problem is how to decide the retail price and the product quality maintained by the manufacturer [1-2,4-14]. If we take the price too low, thus the obtained profit will be low. In contrary, if we set the price too high, thus the rate of the demand will drop drastically. Deciding the optimum value of the price is the main idea to obtain the maximum profit. The same problem occurs in deciding the product quality. The highest quality increases the rate of the demand along with the cost of quality improvement.

Our purpose is to construct a of CLSC model by Maiti and Giri [6] modificated by adding the refurbishing process in the Konstantaras [5]. The notations and assumptions are listed in Section 2. Section 3 presents the formulation of the model. The numerical experiment and the analysis under four scenarios is in the Section 4 and this paper is concluded in Section 5.

## 2. Notations and Assumptions

The following notations are used for developing the proposed model:

| | |
|---|---|
| $w$ | Unit wholesale price |
| $v$ | Unit wholesale price for the secondary market |
| $D$ | Demand rate of the product in the market |
| $p$ | Unit selling price |
| $q$ | Product quality maintained by the manufacturer |
| $q_2$ | Product quality to be sold to the secondary market |
| $q_r$ | Product quality of remanufacturable products |
| $q_f$ | Product quality of refurbishable products |
| $\tau$ | Collection rate of the third party channel |
| $\theta$ | Proportion of remanufacturable items of returned products |
| $c_m$ | Unit cost of manufacturing end product from raw materials |
| $c_r$ | Unit cost of manufacturing end product from return products |
| $c_f$ | Unit cost of refurbishing end product from return products |
| $c_q$ | Quality improvement cost at the manufacturer |
| $c_i$ | Unit cost of inspection process from return products |
| $d$ | Basic market demand |
| $\alpha$ | Sensitivity factor of selling price in the demand |
| $\beta$ | Sensitivity factor of product quality in the demand |
| $A_t$ | Average recycling cost of used products collected by the third party |
| $c_t$ | Average recycling price of used products paid by the manufacturer to the third party |
| $\gamma$ | Effective investment by the third party to collect return products |
| $g$ | Inspection cost of unit product at the manufacturer |

The following assumptions are made to develop the proposed model:

(1) The market demand $D$ is linearly dependent on the retail price and quality of the product. We take the function of the demand as $D(p,q) = d - \alpha p + \beta q$.

(2) The quality of the manufactured product is $q$. It lies between 0 and 1, where 1 means the best quality manufacturer can afford. While the quality of returned product $q_r$ ($<q$). That is, the returned products are acceptable for remanufacturing or refurbishing if those are attained minimum quality, i.e., $q_r$. However, after remanufacturing the quality of remanufactured product becomes $q$, and refurbished product becomes $q_2 > q$.

(3) The return rate ($\tau$) of the third party is a fraction of the total demand, but only $\theta$ of returned products are remanufacturable, the rest is refurbhised to the secondary market. All the return products which do not qualify for remanufacturing is refurbished and sold to the secondary market with unit price $v$. A series of processes is performed and only a part of the collected used products are remanufactured and resold.

(4) Unit remanufacturing cost is lower than unit manufacturing cost, i.e., $c_m > c_r$.

(5) We assume that $p > w > 0$, $c_m - c_r > c_t > A_t > 0$.

(6) There is not defective or expired product.

(7) There is not learning process in the manufacturing system. The manufacturer keeps the amount of product as the rate of the demand to avoid unsold product in the retailer.

(8) The amount of the product always satisfy the demand and there is not shortage in the system.

The difference between the model presented in this paper, Maiti and Giri [6], and Konstantaras [5] is shown in the **Table 1**.

IOP Conf. Series: Journal of Physics: Conf. Series **855** (2017) 012009 doi:10.1088/1742-6596/855/1/012009

**Table 1**. The comparison of the CLSC models

| Model | Maiti and Giri | Konstantaras | This paper |
|---|---|---|---|
| There are 3 parties in the system (manufacturer, retailer, and third party as collector) | Yes | No (Manufacturer collects the used product) | Yes |
| All of the used products are collected | No | Yes | Yes |
| There is inspection of the returned products | No | Yes | Yes |
| The system used remanufacturing process | Yes | Yes | Yes |
| The system used refurishing process | No | Yes | Yes |
| The quality is used as decision ariable and can be calculated | Yes | No | Yes |
| The indirect cost (set-up cost, transportation cost, etc) is used in the system | No | Yes | No |

## 3. Model Formulation and Analysis

The diagram consisting forward and reversed logistic of modified model of Maiti and Giri [6] added by refurbishing process is presented on the Fig. 1.

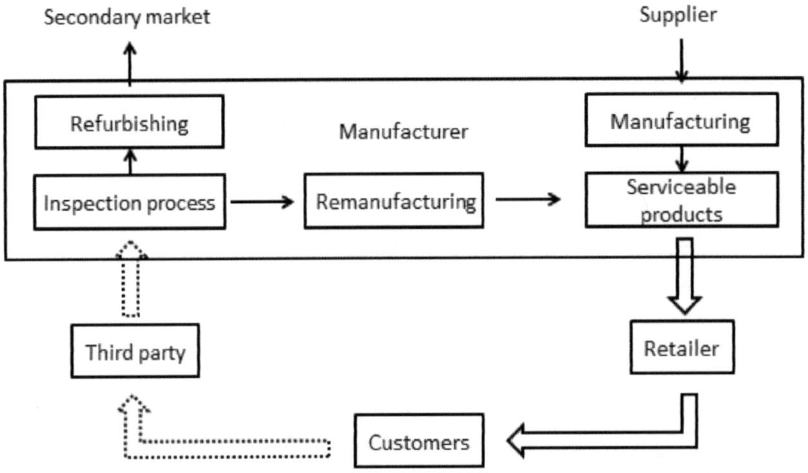


**Fig.1.** Graphical representation of CLSC

The third party returned $\tau D$ products to manufacturer, only $\theta\tau D$ products are remanufacturable, the rest $(1-\theta)\tau D$ products are sold to the secondary market with the lower price. Manufacturer must manufacture $(1-\theta\tau)D$ products to satisfy the demand rate. The serviceable products are sold through a retailer. Thus the profit of the manufacturer, retailer, and third party is

$$\Pi_M = Dw + (1-\theta)\tau D\left(v - c_f - c_q(q_2 - q_f)\right) - (1-\theta\tau)D(c_m + c_q q)$$
$$-\theta\tau D\left(c_r + c_q(q - q_r)\right) - \tau D(c_t + g),$$

$$\Pi_R = Dp - Dw, \text{ and}$$

$$\Pi_T = Dc_t - \tau D\, A_t - \gamma\tau^2 \text{ respectively.}$$

There are four available scenarios, centralized policy, manufacturer-led policy, retailer-led policy and third party-led policy. In centralized policy all players cooperatively decide the value of $p$ and $q$ to maximize the whole joint profit of the system. In the led policies, the leading player decides the value of the specific variable to maximize the profit, then the rest of the players follow the variable stated by the leading player.

### *3.1 Centralized Policy*

The whole joint profit of the system is

$$\Pi = Dp + (1-\theta)\tau D\left(v - c_f - c_q(q_2 - q_f)\right) - (1-\theta\tau)D(c_m + c_q q)$$
$$-\theta\tau D\left(c_r + c_q(q - q_r)\right) - \tau D(A_t - g - \gamma\tau^2). \quad (1)$$

For the optimal solution, the first order necessary conditions are $\frac{\partial \Pi}{\partial p} = 0$ and $\frac{\partial \Pi}{\partial q} = 0$, which given

$$p = \frac{1}{2\alpha}\left(d + \alpha c_r + q\beta + (c_q c_f + A_t + g - v)\alpha\tau + \left(v - c_r - c_q(c_f + q_r)\right)\alpha\theta\tau - \alpha\gamma\tau^3\right), \quad (2)$$

and

$$q = \frac{1}{2c_q\beta}(-c_q d + c_q p\alpha - c_r\beta + p\beta - c_f c_q\beta\tau - A_t\beta\tau - g\beta\tau + v\beta\tau + c_f c_q\beta\theta\tau + c_r\beta\theta\tau +$$
$$c_q q_r\beta\theta\tau - v\beta\theta\tau + \beta\gamma\tau^3). \quad (3)$$

We checked that $\frac{\partial^2 \Pi}{\partial p^2} = -2\alpha < 0$ and $\frac{\partial^2 \Pi}{\partial q^2} = -2c_q\beta < 0$, thus the Eqs (2) and (3) is the optimum solution for Eqs (1).

### *3.2 Manufacturer-led Policy*

This policy is usually applied in the industry which creates high quality of branded product. Manufacturer decides the optimum value of *q* thus they obtain the maximum profit.

$$q = \frac{1}{2c_q\beta}(-c_q d + c_q p\alpha - c_m\beta + w\beta + c_m\beta\theta\tau - c_r\beta\theta\tau + c_q q_r\beta\theta\tau - c_f\beta\tau - c_t\beta\tau - g\beta\tau -$$
$$c_q q_2\beta\tau + c_q q_f\beta\tau + v\beta\tau + c_f\beta\theta\tau + c_q q_2\beta\theta\tau - c_q q_f\beta\theta\tau - v\beta\theta\tau), q \in [0,1]$$

is obtained as the solution of $\frac{d\Pi_M}{dq} = 0$.

We checked that $\frac{\partial^2 \Pi_M}{\partial q^2} = -2c_q\beta < 0$, thus the solution is unique. The retailer and the third party follow the value of $q$ stated by the manufacturer. If the obtained $q > 1$, thus we take $q = 1$. On the other hand, if the obtained $q < 0$, thus we take $q = q_r$.

### *3.3 Retailer-led Policy*

This policy is usually applied in the industry with high demand for unbranded products. Retailer decides the optimum value of *p* thus they obtain the maximum profit.

$$p = \frac{d - w\alpha + q\beta}{2\alpha}$$

is obtained as the solution of $\frac{d\Pi_R}{dp} = 0$.

We checked that $\frac{\partial^2\Pi_R}{\partial p^2} = -2\alpha < 0$, thus the solution is unique. The manufacturer and the third party follow the value of $p$ stated by the retailer.

### *3.4 Third party-led Policy*

This policy is usually applied in the recycling or crafting industry where the value of $A_T$ is high. Third party decides the optimum value of τ thus they obtains the maximum profit.

$$\tau = \frac{(A_t - c_t)(d - \alpha p + \beta q)}{2\gamma}, \tau \in [0,1]$$

is obtained as the solution of $\frac{d\Pi_T}{d\tau} = 0$.

We checked that $\frac{\partial^2\Pi_M}{\partial \tau^2} = -2\gamma < 0$, thus the solution is unique. The manufacturer and the retailer follow the value of τ stated by the third party.

## 4. Numerical Experiment and Analysis

We construct a simulation with the costs $c_m = 2{,}000$, $c_q = 100$, $c_r = 1{,}500$, $c_f = 800$, $g = 1$, , $c_t = 350$ and $A_t = 150$. The used basic demand $d = 100$ with $v = 900$ and $w = 300{,}000$. We used $\tau = 0.75$, except for the third party-led policy. The rest of variable is $q_r = 0.3$, $q_f = 0.1$, $q_2 = 0.5\,q$, $\alpha = 0.0015$, $\beta = 200$, $\gamma = 1$, and $\theta = 0.7$. The obtained results in the maximum of 6 significant digits are shown in **Table 2**.

**Table 2**. The numerical result of the optimum values under difference policies

| Optimal decisions | Policies | | | |
|---|---|---|---|---|
| | Centralized | Manufacturer-led | Retailer-led | Third party-led |
| $p$ | 408,229 | 408,229 | 408,229 | 364,100 |
| $q$ | 1 | 1 | 1 | 0.3 |
| $\tau$ | 0.75 | 0.75 | 0.75 | 1 |
| $\Pi_M$ | $1.38547 \times 10^8$ | $1.38547 \times 10^8$ | $1.38547 \times 10^8$ | $1.53147 \times 10^8$ |
| $\Pi_R$ | $8.835 \times 10^7$ | $8.835 \times 10^7$ | $8.835 \times 10^7$ | $3.2978 \times 10^7$ |
| $\Pi_T$ | 69,749.4 | 69,749.4 | 69,749.4 | 102,769 |
| $\Pi$ | $2.30227 \times 10^8$ | $2.30227 \times 10^8$ | $2.30227 \times 10^8$ | $1.76128 \times 10^8$ |

We obtained $q > 1$ under centralized, manufacturer-led and retailer-led policy. It means that the industry gain the maximum profit if the best quality of product is sold. In contrary, we obtained $q < 0$ under the third party-led policy. That means the manufacturer sold the product with the lowest quality as if it is returned by third party and resold to the retailer directly.

As expected, manufacturer receives the highest profit, i.e. more than 50% of the joint profit in the CLSC system. This is the most common result to be considered as manufacturer plays the greatest role in the system. Third party receives the lowest profit due to the least role in the system.

The sensitivity analysis of $p$ from the obtained numerical result with maximum of 3 significant digits is shown in the **Table 3** and **Fig 2.** While the variable $q$ does not require sensitivity analysis since it is in the maximum value under centralized, manufacturer-led and retailer-led policy, and in the minimum value under third party-led policy.

**Table 3**. The sensitivity analysis of $p$ from the obtained numerical result of $\Pi$ under difference policies

| $p$ | $\Pi$ | |
|---|---|---|
| | Centralized, Manufacturer-led and Retailer-led policy | Third party-led policy |
| 250,000 | $1.92 \times 10^8$ | $1.57 \times 10^8$ |
| 300,000 | $2.13 \times 10^8$ | $1.70 \times 10^8$ |
| 350,000 | $2.25 \times 10^8$ | $1.76 \times 10^8$ |
| 400,000 | $2.30 \times 10^8$ | $1.74 \times 10^8$ |
| 450,000 | $2.28 \times 10^8$ | $1.65 \times 10^8$ |
| 500,000 | $2.18 \times 10^8$ | $1.48 \times 10^8$ |
| 550,000 | $2.00 \times 10^8$ | $1.24 \times 10^8$ |
| 600,000 | $1.75 \times 10^8$ | $9.26 \times 10^7$ |

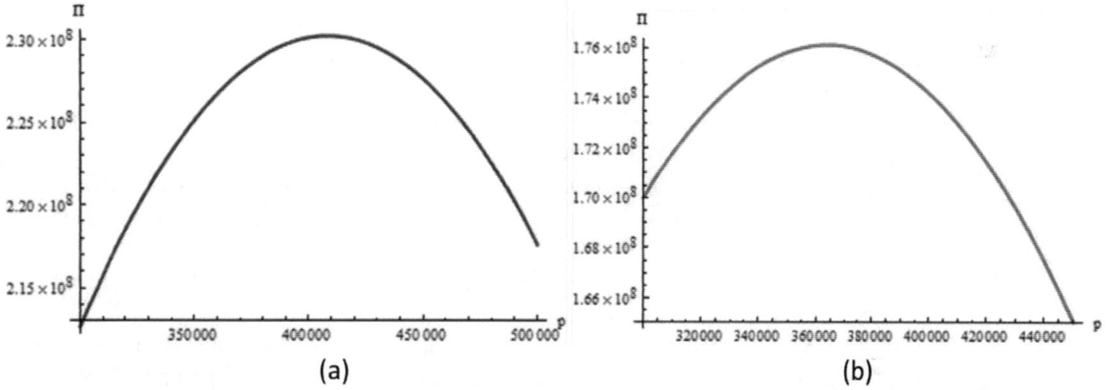


**Fig 2.** Graphical representation of sensitivity analysis of $p$ from the obtained numerical result of $\Pi$ under (a) centralized, manufacturer-led, retailer led policy and (b) third party-led policy

**Table 2** and **Fig 2** show that the results in the **Table 1** are definitely the optimum solution of $\Pi$ under the four scenarios. We find out that centralized, manufacturer-led and retailer-led policy obtain the same result for this numerical simulation. The profit obtained by third party as the leading player in the third party-led policy is definitely the highest compared to the other scenarios, consequently it has the lowest joint profit.

## 5. Conclusions

In this paper we have constructed a CLSC model with manufacturer, retailer and third party. Manufacturer manufactures products and sell it to the costumer through a retailer. The third party collects all the used product and send it to the manufacturer to be remanufactured or refurbished. We construct four scenarios, they are centralized policy, manufacturer-led policy, retailer-led policy and third party-led policy. The numerical experiment is conducted to find out the best scenarios. From the numerical simulation, we obtained the same results under centralized, manufacturer-led and retailer policy. The joint profit obtained under third party-led policy is lower than the other policies.

Several studies can be undertaken for future research. Profit sharing under stochastic demand and including more than one retailer who sells the product. would be more realistic to consider multiple retailers in the system. Further, one may considerthe retailer to act as a collector of return product or recast the current problem as a multi - period decision making problem from more realistic point of view.

IOP Conf. Series: Journal of Physics: Conf. Series **855** (2017) 012009 doi:10.1088/1742-6596/855/1/012009

**References**

[1] Chung, S., H. Wee, and P. Yang, Optimal Policy for a Closed-Loop Supply Chain Inventory System with Remanufacturing, SIAM Journal Math & Comp Model 8 (2008), 867-881.
[2] Harjanto, E, Manajemen Produksi dan Operasi, 2nd ed., Grasindo, Jakarta, 1999.
[3] Kampstra, R. and J. Ashayeri, *Realities of Supply Chain Management*, Journal Log Management 17 (2006), 312-320.
[4] Koh, S., H. Hwang, K. Sohn, and C. Ko, *An Optimal Ordering and Recovery Policy for Reusable Items*, Journal of Computers & Industrial Engineering 43 (2002), 59-73.
[5] Konstantaras, I., K. Skouri, and Y. Jaber, Lot Sizing for a Recoverable Product with Inspection and Sorting, Journal of Computers & Industrial Engineering 58 (2010), 452-462.
[6] Maiti, T. and B. Giri, A Closed-Loop Supply Chain under Retail Price and Quality Dependent Demand, Journal of Manufacturing 37 (2015), 624-637.
[7] Mankiw, G., Teori Makro Ekonomi Terjemahan Imam Nurmawan, 4th ed, Erlangga, Jakarta, 2003.
[8] Nahmias, N. and H. Rivera, *A Deterministic Model for Repairable Item Inventory System with a Finite Repair Rate*. International Journal of Production Research 17 (1979), 215-221.
[9] Nicholson, W., Mikroekonomi Indermediate dan Penerapannya Terjemahan Hani Hutabarat, 3rd ed, Erlangga, Jakarta, 1992.
[10] Rubio, S. and B. Parra, Reverse Logistics: Overview and Challenges for Supply Chain Management, Extremadurs, Spain, 2014.
[11] Saadany, A. and Y. Jaber. *A Production/Remanufacturing Inventory Model with Price and Quality Dependent Return Rate*. Journal of Computers & Industrial Engineering 58 (2010), 52-362.
[12] Schrady, A., A *Deterministic Model for Repairable Items*. Naval Research Logistics Quarterly 14 (1967), 391-398.
[13] Seifbarghy, M., N. Khashayar, and M. Amin, Contract Design in a Supply Chain Considering Price and Quality Dependent Demand with Customer Segmentation, International Journal of & Economics 167 (2015), 108-118.
[14] Tanyal, S., S. Singh, and R. Sharma, A Multi Item Inventory Model for Deteriorating Items with Expiration Date and Allowable Shortages, Indian Journal of Science and Technology 7 (2014), 114-135.
[15] Thierry, M. M. Salomon, V. Nunen, and V. Wassenhove, Strategic Issue in Product Recovery Management, California Managemen Review 7 (1995), 114-135.

# On super $\mathcal{H}-$antimagicness of an edge comb product of graphs with subgraph as a terminal of its amalgamation

**Dafik[1,3], Ika Hesti Agustin [1,2],**
**A. I. Nurvitaningrum[1,2], R. M. Prihandini[1,2]**
[1]CGANT University of Jember, Indonesia
[2]Mathematics Depart. University of Jember, Indonesia
[3]Mathematics Edu. Depart. University of Jember, Indonesia

E-mail: d.dafik@unej.ac.id; ikahesti.fmipa@unej.ac.id

**Abstract.** All graphs in this paper are simple, finite, and undirected graph. Let $r$ be a edges of $H$. The edge comb product between $L$ and $H$, denoted by $L \trianglerighteq H$, is a graph obtained by taking one copy of $L$ and $|E(L)|$ copies of $H$ and grafting the $i$-th copy of $H$ at the edges $r$ to the $i$-th edges of $L$, we call such a graph as an edge comb product of graph with subgraph as a terminal of its amalgamation, denoted by $G = K \trianglerighteq Amal(H, L \subset H, n)$. The graph $G$ is said to admits an $(a,d)$-$H$-antimagic total labeling if there exist a bijection $f : V(G) \cup E(G) \to \{1, 2, \dots, |V(G)| + |E(G)|\}$ such that for all subgraphs isomorphic to $H$, the total $H$-weights $W(H) = \sum_{v \in V(H)} f(v) + \sum_{e \in E(H)} f(e)$ form an arithmetic sequence $\{a, a+d, a+2d, ..., a+(t-1)d\}$, where $a$ and $d$ are positive integers and $t$ is the number of all subgraphs isomorphic to $H$. An $(a,d)$-$H$-antimagic total labeling $f$ is called super if the smallest labels appear in the vertices. In this paper, we will study the super $\mathcal{H}-$antimagicness of disjoint union of edge comb product of graphs with subgraph as a terminal of its amalgamation.

**Keywords:** Graph amalgamation, edge comb product, $\mathcal{H}$-Antimagic total labeling.

## 1. Introduction

Let $G = (V(G), E(G))$ be a simple, finite, and undirected graph with vertex set $V(G)$ and edge set $E(G)$. The graph $G$ is said to admit an $(a,d)$-$H$-antimagic total labeling if there exist a bijection $f : V(G) \cup E(G) \to \{1, 2, \dots, |V(G)| + |E(G)|\}$ such that for all subgraphs isomorphic to $H$, the total $H$-weights $W(H) = \sum_{v \in V(H)} f(v) + \sum_{e \in E(H)} f(e)$ form an arithmetic sequence $\{a, a+d, a+2d, ..., a+(t-1)d\}$, where $a$ and $d$ are positive integers and $t$ is the number of all subgraphs isomorphic to $H$. An $(a,d)$-$H$-antimagic total labeling $f$ is called super if $f : V(G) \to \{1, 2, \dots, |V(G)|\}$.

Some results related to the existence of an $(a,d)$-$H$-antimagic total labeling can be cited in [2, 3, 6, 7] and [8, 9, 10, 11, 12]. Inayah *et al.* in [6] proved that for any $H$ and any integer $k \geq 2$, $shack(H, v, k)$ which contains exactly $k$ subgraphs isomorphic to $H$ admits $H$-super antimagic. But they only covered a connected version of shackle of graph when a vertex as a connector, and their paper did not cover all feasible $d$. Our paper attempt to solve a super $(a,d)$-$H$ antimagic total labeling of disjoint union of edge comb product of graphs with subgraph as a terminal of its amalgamation, denoted by $G = c(K \trianglerighteq Amal(H, L \subset H, n))$, where $c$ is number of copies of $K \trianglerighteq Amal(H, L \subset H, n)$. In this study, we aim to achieve for all feasible $d$.

To show those existence, we will use *an integer set partition technique* introduced by [1, 3]. This technique is used in determining the feasible difference $d$. Let $n, m, d$ and $k$ be positive integers. We consider the partition $\mathcal{P}^n_{m,d}(i,j)$ of the set $\{1, 2, \ldots, mn\}$ into $n$ columns, $n \geq 2$, $m$-rows such that the difference between the sum of the numbers in the $(j+1)$th $m$-rows and the sum of the numbers in the $j$th $m$-rows is always equal to the constant $d$, where $j = 1, 2, \ldots, n-1$. Thus, these sums form an arithmetic sequence with the difference $d$. By the symbol $\mathcal{P}^n_{m,d}(i,j)$, we denote the $j$th $m$-rows in the partition with the difference $d$, where $j = 1, 2, \ldots, n$. Let $\sum \mathcal{P}^n_{m,d}(i,j)$ be the sum of the numbers in $\mathcal{P}^n_{m,d}(i,j)$, thus $d = \sum \mathcal{P}^n_{m,d}(j+1) - \sum \mathcal{P}^n_{m,d}(j)$.

## 2. Some Useful Lemma and Corollary

Let $G$ be a disjoint union of comb product of graphs with subgraph as a terminal of its amalgamation, denoted by $G = c(K \trianglerighteq Amal(H, L \subset H, n))$. The graph $G$ is a simple and disconnected graph with $|V(G)| = p_G$, $|E(G)| = q_G$, $|V(H)| = p_H$, and $|E(H)| = q_H$. The vertex set and edge set of the graph $G = c(K \trianglerighteq Amal(H, L \subset H, n))$ can be split into following sets: $V(G) = \{x_{i,t}; 1 \leq i \leq p_K; 1 \leq t \leq c\} \cup \{x_{i,k,t}; 1 \leq i \leq p_L - 2, 1 \leq k \leq q_K, 1 \leq t \leq c\} \cup \{x_{i,j,k,t}; 1 \leq i \leq p_H - p_L, 1 \leq k \leq q_K, 1 \leq j \leq n; 1 \leq t \leq c\}$ and $E(G) = \{e_{l,k,t}; 1 \leq l \leq q_L, 1 \leq k \leq q_K, 1 \leq t \leq c\} \cup \{e_{l,j,k,t}; 1 \leq l \leq q_H - q_L; 1 \leq j \leq n; 1 \leq k \leq q_K; 1 \leq t \leq c\}$. Thus, the cardinalities of $|V(G)| = p_G = (p_H - p_L)ncq_K + (p_L - 2)cq_K + cp_K$ and $|E(G)| = q_G = (q_H - q_L)ncq_K + q_Lcq_K$.

The upper bound of feasible $d$ such that $G = c(K \trianglerighteq Amal(H, L \subset H, n))$ admits a super $(a,d)$-$H$-antimagic total labeling can be obtained in the following lemma.

**Lemma 1.** [2] *Let $G$ be a simple graph of order $p$ and size $q$. If $G$ is super $(a,d)$-$H$-antimagic total labeling then $d \leq \frac{(p_G - p_H)p_H + (q_G - q_H)q_H}{n-1}$, for $p_G = |V(G)|$, $q_G = |E(G)|$, $p_H = |V(H)|$, $q_H = |E(H)|$, and $n = |H_k|$, $|H_k|$ is number of subgraph which is isomorphic to the graph $H$.*

Thus for $p_G = (p_H - p_L)ncq_K + (p_L - 2)cq_K + cp_K$ and $q_G = (q_H - q_L)ncq_K + q_Lcq_K$, we have the following corollary.

**Corrollary 1.** *For $n \geq 2$, if the graph $G = c(K \trianglerighteq Amal(H, L \subset H, n))$ admits super $(a,d)$-$H$-antimagic total labeling then $d \leq {p_H}^2 + {q_H}^2 - \frac{ncq_K(p_Lp_H + q_Lq_H + (p_L-2)cq_Hp_H + cp_Kp_H + cq_Lq_Kq_H}{ncq_K - 1}$*

**Theorem 1.** [5] *If $G$ is connected graph with $p$ vertices and $q$ edges, then $p \leq q+1$*

We recall a partition $\mathcal{P}^n_{m,d}(i,k)$ introduced in [4]. We will use this partition for a linear combination to develop a bijection of vertex and edge label of the main theorem.

**Lemma 2.** *Let $n$ , $m$ and $s$ be positive integers $1 \leq j \leq n; 1 \leq k \leq s$, the $\sum_{i=1}^m \mathcal{P}^{n,s}_{m,d_1}(i,j,k) = \{a(ns-1)\lceil \frac{i}{a} \rceil - a(ns-1) + i + (k-1)a + (j-1)as;\ 1 \leq i \leq m\}$ forms an arithmetic sequences of difference $d_1 = am$.*

**Proof.** For $j = 1, 2, ..., n;\ k = 1, 2, ..., s$, it gives that $\sum_{i=1}^m \mathcal{P}^{n,s}_{m,d_1}(i,j,k) = \mathcal{P}^{n,s}_{m,d_1}(j,k) = \{\frac{m^2ns}{2} + \frac{m}{2}(1 - a - asn) - mas + am(js + k)\} \longleftrightarrow \mathcal{P}^{n,s}_{m,d_1}(j,k) = \{\frac{m^2ns}{2} + \frac{m}{2}(1 - a - asn) - mas + am(s+1), \frac{m^2ns}{2} + \frac{m}{2}(1 - a - asn) - mas + am(s+2), \ldots, \frac{m^2ns}{2} + \frac{m}{2}(1 - a - asn) - mas + am(2s), \frac{m^2ns}{2} + \frac{m}{2}(1 - a - asn) - mas + am(2s+1), \ldots, \frac{m^2ns}{2} + \frac{m}{2}(1 - a - asn) - mas + am(ns+s)\}$ forms an arithmetic sequences of difference $d_1 = am$. It concludes the proof. □

**Lemma 3.** *Let $n$ , $m$ and $s$ be positive integers $1 \leq j \leq n; 1 \leq k \leq s$; and $1 \leq t \leq c$, the $\sum_{i=1}^{m} \mathcal{P}_{m,d_2}^{n,s,c}(i,j,k,t) = \{(t-1)a + (k-1)ac + (j-1)acs + i + (\lceil \frac{i}{a} \rceil - 1)(acsn-1)\}$ forms an arithmetic sequences of difference $d_2 = am$.*

**Proof.** For $j = 1,2,\dots,n$; $k = 1,2,\dots,s$, and $t = 1,2,\dots,c$, it gives that $\sum_{i=1}^{m} \mathcal{P}_{m,d_2}^{n,s,c}(i,j,k,t) = \mathcal{P}_{m,d_2}^{n,s,c}(j,k,t) = \{m[(t-1)a+(k-1)ac+(j-1)acs-acsn+a]+(\frac{m+m^2}{2})+(acsn-a)(\frac{m^2+am}{2a})\} \longleftrightarrow \mathcal{P}_{m,d_2}^{n,s}(j,k) = \{m[a-acsn]+(\frac{m+m^2}{2})+(acsn-a)(\frac{m^2+am}{2a}), m[a+(k-1)ac+(j-1)acs-acsn+a]+(\frac{m+m^2}{2})+(acsn-a)(\frac{m^2+am}{2a}),\dots, m[(c-1)a+(s-1)ac+(n-1)acs-acsn+a]+(\frac{m+m^2}{2})+(acsn-a)(\frac{m^2+am}{2a}), \{m[acsn]+(\frac{m+m^2}{2})+(acsn-a)(\frac{m^2+am}{2a}), m[a+acs-acsn+a]+(\frac{m+m^2}{2})+(acsn-a)(\frac{m^2+am}{2a}),\dots, m[(c-1)a+(s-1)ac+(n-1)acs-acsn+a]+(\frac{m+m^2}{2})+(acsn-a)(\frac{m^2+am}{2a})$ forms an arithmetic sequences of difference $d_2 = am$. It concludes the proof. □

**Lemma 4.** *Let $m \geq 2$* even *and $s$ be positive integers $1 \leq t \leq c; 1 \leq k \leq s$, the*

$$\sum_{i=1}^{m} \mathcal{P}_{m,d_3}^{c,s}(i,t,k) = \begin{cases} \{csi + t + (k-1)c - cs; i \equiv 1 \pmod 2\} \\ \{csi - t - (k-1)c + 1; i \equiv 0 \pmod 2\} \end{cases}$$

*form an arithmetic sequences of difference $d_3 = 0$.*

**Proof.** for $j = 1,2,...,n$; $k = 1,2,...,s$ it gives $\sum_{i=1}^{m} \mathcal{P}_{m,d_3}^{n,s}(i,t,k) = \mathcal{P}_{m,d_3}^{n,s}(t,k) = \{\frac{m^2cs+m}{2}, \frac{m^2cs+m}{2},\dots,\frac{m^2cs+m}{2}\}$. It concludes the proof. □

With those lemmas in hand, we ready to show our main result in the following section.

## 3. Main Results

In this section we will present our main theorem related to the existence of super $(a,d)-\mathcal{H}$ antimagic total labeling of disjoint union of edge comb product of graphs with subgraph as a terminal of its amalgamation. We will describe a construction of how to obtain the $\mathcal{H}$-antimagic total labeling from a smallest weight of edge-antimagic vertex labeling of graph $G$. We note that if $cK$ is an $(a,d)-EAVL$ graph and $H$ is any graph then $c(K \trianglerighteq Amal(H, L \subset H, n)) \cong G$.

**Lemma 5.** *Let $K$ be a simple, nontrivial, and connected graph. If $K$ admits an $(a,d)$-EAVL then $d \leq \frac{2p_K-4}{q_K-1}$ or $d \in \{1,2\}$.*

**Proof.** Suppose $K$ is a connected graph of order $p_K$ and size $q_K$. If $K$ admits $(a,d)$-*edge antimagic vertex labeling* then the bijection $f(V) = \{1,2,3,...,p_K\}$. The set of *edge weights under vertex labeling* $f$ is $w(uv) = f(u) + f(v)$, where $uv \in E(K)$. The weights $w(uv) = \{a, a+d, a+2d, ..., a+(q_K-1)d\}$ where $a$ is the smallest edge weight. The minimum possible edge weight under labeling $f$ is at least: $1+2$, thus $a \geq 3$. The largest label is $p_K + (p_K - 1)$ Hence $a + (q_K - 1)d \leq 2p_K - 1$. Combining the two inequalities, and also based on Theorem 1 we obtain the upper bound of feasible $d$ for the graph $K$ is said to be $(a,d)$-edge antimagic vertex labeling, namely $d \leq \frac{2(p_K-1)-2}{q_K-1}$. Since the minimum size of graph $K$ is $pK-1$ then $d \leq \frac{2(p_K-1)-2}{(p_K-1)-1} = \frac{2p_K-4}{p_K-2}$ and thus $d \leq 2$. Furthermore for the upper bound of disjoint union, we suppose $p_K = cp_K$ and $q_K = cq_K$ then $d \leq \frac{2(cp_K-1)-2}{(cp_K-1)-1} = \frac{2cp_K-4}{cp_K-2}$. Now, we will show that $d \neq 0$. By contradiction, suppose If $d = 0 \rightarrow 0 \leq \frac{2cp_K-4}{cp_K-2}$ and thus $p_K = 2$. Since we study for a graph of order larger than two then $p_K = 2$ is too trivial, thus $d = 0$ is not our concern. It concludes that the feasible $d \in \{1,2\}$. □

Now we are ready to show our main theorem.

**Theorem 2.** *Let $K, H$ be a two simple, nontrivial, and connected graphs. If the graph $c_K$ admits an $(a, d^*)$-EAVL, then $G = c(K \trianglerighteq Amal(H, L \subset H, n))$ admits a super $(a, d) - \mathcal{H}$-antimagic total labeling with $d = d_v + d_e$, where $d_v$ and $d_e$ are respectively the feasible difference of partitions of integer set of vertex and edge labels.*

**Proof.** Suppose $G$ has a super $(a, d) - \mathcal{H}$-antimagic total labeling $f$, we have a map $f : V(G) \cup E(G) \rightarrow \{1, 2, \dots, (p_H - p_L)ncq_K + (p_L - 2)cq_K + cp_K + (q_H - q_L)ncq_K + q_Lcq_K\}$, Since the graph $cK$ admits an $(a, d^*)$-edge antimagic vertex labeling, the edge weight of $cK$ is $\{a, a + d, a + 2d, \dots, a + (r - 1)d\}$ where $r = 1, 2, \dots, cqK$. According to lemma 5 the feasible difference of graph $cK$ to be $(a, d^*)$-antimagic vertex graph is $d^* \in \{1, 2\}$. Thus, to prove the theorem we will prove it into two cases, namely for $d = 1$ and $d = 2$.

**Case 1.** For $d = 1$, we have $w_1(e_r) = a + (r - 1) \cdot 1$ and the number of $p_L + q_L \in L$ is odd. By using Lemma 2 and Lemma 3, define the vertex and edge labelings $f_1$ in the following way:

$$\begin{aligned}
f_1(V_{(p_H-p_L)ncq_K}) &= P^{ncq_K}_{p_H-p_L,d_v}(i,j,k,t) \oplus cp_K \\
f_1(V_{(p_L-2)cq_K} \cup E_{(q_L-1)cq_K}) &= P^{cq_K}_{p_L-2+(q_L-1),d_v+d_e}(i,j,k) \oplus cp_K + (p_H - p_L)ncq_K \\
f_1(e_r) &= cq_K + 1 - r + cp_K + (p_H - p_L)ncq_K + (p_L - 2 + q_L - 1)cq_K \\
f_1(E_{q_H-q_L)ncq_K}) &= P^{ncq_K}_{q_H-q_L,d_e}(i,j,k,t) \oplus cp_K + (p_H - p_L)ncq_K + (p_L - 2 + q_L - 1)cq_K \\
&\quad +cq_K
\end{aligned}$$

The labeling $f_1$ is a bijective function $f_1 : V(G) \cup E(G) \rightarrow \{1, 2, \dots, (p_H - p_L)ncq_K + (p_L - 2)cq_K + cp_K + (q_H - q_L)ncq_K + q_Lcq_K\}$. The $\mathcal{H}$ weight under the labeling $f_1$ constitute

$$\begin{aligned}
W(H_{j,k,t}) &= w_{(e_r)} + \sum_{u \in V(H_{j,k,t})} f(u) + \sum_{e \in V(E_{j,k,t})} f(e) \\
&= (a + (r-1)) + \sum (P^{ncq_K}_{p_H-p_L,d_v}(i,j,k,t) \oplus cp_K) + \sum (P^{cq_K}_{p_L-2+(q_L-1)}(i,j,k) \oplus \\
&\quad cp_K + (p_H - p_L)ncq_K) + cq_K + 1 - r + cp_K + (p_H - p_L)ncq_K + (p_L - 2)cq_K + \\
&\quad \sum (P^{ncq_K}_{q_H-q_L,d_e}(i,j,k,t) \oplus cp_K + (p_H - p_L)ncq_K + (p_L - 2 + q_L - 1)cq_K + cq_K) \\
&= (a + (r-1)) + C^{ncq_K}_{p_H-p_L,d_v} + d_v(j,k,t) + cp_K(p_H - p_L)ncq_K + C^{cq_K}_{p_L-2+(q_L-1),d_v+d_e} \\
&\quad +(p_L - 2 + q_L - 1)(cp_K + (p_H - p_L)ncq_K) + cq_K + 1 - r + cp_K + (p_H - p_L)ncq_K \\
&\quad +(p_L - 2 + q_L - 1)cq_K + C^{ncq_K}_{q_H-q_L,d_e} + d_e(j,k,t) + (q_H - q_L)(cp_K + (p_H - p_L)ncq_K \\
&\quad +(p_L - 2 + q_L - 1)cq_K + cq_K) \\
&= a + C^{ncq_K}_{p_H-p_L,d_v} + cp_K(p_H - p_L)ncq_K + C^{cq_K}_{p_L-2+(q_L-1),d_v+d_e} \\
&\quad +(p_L - 2 + q_L - 1)(cp_K + (p_H - p_L)ncq_K) + cq_K + cp_K + (p_H - p_L)ncq_K \\
&\quad +(p_L - 2 + q_L - 1)cq_K + C^{ncq_K}_{q_H-q_L,d_e} + (q_H - q_L)(cp_K + (p_H - p_L)ncq_K \\
&\quad +(p_L - 2 + q_L - 1)cq_K + cq_K) + (d_e + d_v)(j,k,t)
\end{aligned}$$

for subgraph $H_{j,k,t}, 1 \le j \le n; 1 \le k \le cq_K; 1 \le t \le c$.

**Case 2.** For $d = 2$, we have $w_1(e_r) = a + (r - 1) \cdot 2$ and the number of $p_L + q_L$ is even. By using Lemma 2, Lemma 3 and Lemma 4, define the vertex and edge labelings $f_2$ in the following way:

$$
\begin{aligned}
f_2(V_{(p_H-p_L)ncq_K}) &= P^{ncq_K}_{p_H-p_L,d_v}(i,j,k,t) \oplus cp_K \\
f_2(V_{(p_L-2)cq_K} \cup E_{(q_L-2)cq_K}) &= P^{cq_K}_{p_L-2+(q_L-2),d_v+d_e}(i,j,k) \oplus cp_K + (p_H-p_L)ncq_K \\
f_2(e_r^1) &= cq_K + 1 - r + cp_K + (p_H-p_L)ncq_K + (p_L-2+q_L-2)cq_K \\
f_2(E_{e^2\in q_L}) &= cq_K + 1 - r + cp_K + (p_H-p_L)ncq_K + (p_L-2+q_L-2)cq_K + cq_K \\
f_2(E_{q_H-q_L)ncq_K}) &= P^{ncq_K}_{q_H-q_L,d_e}(i,j,k,t) \oplus cp_K + (p_H-p_L)ncq_K + (p_L-2+q_L-1)cq_K \\
&\quad +2cq_K
\end{aligned}
$$

The labeling $f_2$ is a bijective function $f_2 : V(G) \cup E(G) \to \{1, 2, \ldots, (p_H - p_L)ncq_K + (p_L - 2)cq_K + cp_K + (q_H - q_L)ncq_K + q_Lcq_K\}$. The $\mathcal{H}$ weight under the labeling $f_2$ constitute

$$
\begin{aligned}
W(H_{j,k,t}) &= w_{(e_r)} + \sum_{u\in V(H_{j,k,t})} f(u) + \sum_{e\in V(E_{j,k,t})} f(e) \\
&= (a+2(r-1)) + \sum(P^{ncq_K}_{p_H-p_L,d_v}(i,j,k,t) \oplus cp_K) + \sum(P^{cq_K}_{p_L-2+(q_L-2),d_v+d_e}(i,j,k) \oplus \\
&\quad cp_K + (p_H-p_L)ncq_K) + cq_K + 1 - r + cp_K + (p_H-p_L)ncq_K + (p_L-2)cq_K \\
&\quad + cq_K + 1 - r + cp_K + (p_H-p_L)ncq_K + (p_L-2+q_L-2)cq_K + cq_K + \\
&\quad \sum(P^{ncq_K}_{q_H-q_L,d_e}(i,j,k,t) \oplus cp_K + (p_H-p_L)ncq_K + (p_L-2+q_L-2)cq_K + 2cq_K) \\
&= (a+2(r-1)) + C^{ncq_K}_{p_H-p_L,d_v} + d_v(j,k,t) + cp_K(p_H-p_L)ncq_K + C^{cq_K}_{p_L-2+(q_L-2),d_v+d_e} \\
&\quad + cq_K + 1 - r + cp_K + (p_H-p_L)ncq_K + (p_L-2)cq_K + cq_K + 1 - r + cp_K + \\
&\quad (p_H-p_L)ncq_K + (p_L-2+q_L-2)cq_K + cq_K + C^{ncq_K}_{q_H-q_L,d_e} + d_e(j,k,t) \\
&\quad + (q_H-q_L)(cp_K + (p_H-p_L)ncq_K + (p_L-2+q_L-2)cq_K + 2cq_K)ncq_K \\
&= (a + C^{ncq_K}_{p_H-p_L,d_v} + cp_K(p_H-p_L)ncq_K + C^{cq_K}_{p_L-2+(q_L-2),d_v+d_e} + cq_K + cp_K + \\
&\quad (p_H-p_L)ncq_K + (p_L-2)cq_K + cq_K + cp_K + (p_H-p_L)ncq_K + (p_L-2+q_L-2) \\
&\quad cq_K + cq_K + C^{ncq_K}_{q_H-q_L,d_e} + (q_H-q_L)(cp_K + (p_H-p_L)ncq_K + (p_L-2+q_L-2)cq_K \\
&\quad + 2cq_K)ncq_K + (d_e+d_v)(j,k,t)
\end{aligned}
$$

for subgraph $H_{j,k,t}$, $1 \le j \le n$; $1 \le k \le cq_K; 1 \le t \le c$.

From the two cases above, it is easy to see if the graph $cK$ admits an $(a, d^*)$-edge antimagic vertex labeling then $G = c(K \trianglerighteq Amal(H, L \subset H, n))$ admits a super $(a, d) - \mathcal{H}$-antimagic total labeling with $d = d_v + d_e$, where $d_v$ and $d_e$ are respectively the feasible difference of partitions of integer set of vertex and edge labels. □

## Concluding Remarks

We have shown the existence of super $(a, d) - H$ antimagic total labeling of disjoint union of edge comb product of graphs with subgraph as a terminal of its amalgamation, denoted by $G = c(K \trianglerighteq Amal(H, L \subset H, n))$. We have proved that $G = c(K \trianglerighteq Amal(H, L \subset H, n))$ admits a super$(a, d)$-$H$ antimagic total labeling for almost feasible difference $d$, namely $d = d_v + d_e$, where $d_v$ and $d_e$ are respectively the feasible difference of partitions of integer set of vertex and edge labels.

## 4. Acknowledgments

Acknowledgement We gratefully acknowledge the support from HIKOM Ristek Dikti Grant 2017 & CGANT - University of Jember.

## 5. References

[1] M. Bača, L. Brankovic, M. Lascsáková, O., Phanalasy, A. Semaničová-Feňovčíková, On $d$-antimagic labelings of plane graphs, *Electr. J. Graph Theory Appli.*, **1**(1), 28-39, (2013)

[2] Dafik, A.K. Purnapraja, R Hidayat, Cycle-Super Antimagicness of Connected and Disconnected Tensor Product of Graphs, *Procedia Computer Science*, **74**, (2015), 9399

[3] Dafik, Slamin, Dushyant Tanna, Andrea Semaničová-Feňovčíková, Martin Bača, Constructions of $H$-antimagic graphs using smaller edge-antimagic graphs, *Ars Combinatoria*, **100** (2017), In Press

[4] Dafik, Moh. Hasan, Y.N. Azizah, Ika Hesti Agustin, A Generalized Shackle of Any Graph $H$ Admits a Super $H$-Antimagic Total Labeling, *Mathematics in Computer Science Journal*, (2016). Submitted

[5] N.Hartsfield, and G.Ringel. 1994. *Pearls in graph theory* . London : Accademic Press Limited.

[6] N. Inayah, R. Simanjuntak, A. N. M. Salman, Super $(a, d) - H$-antimagic total labelings for shackles of a connected graph $H$, *The Australasian Journal of Combinatorics*, **57** (2013), 127138.

[7] P. Jeyanthi, P. Selvagopal, More classes of $H$-supermagic Graphs, *Intern. J. of Algorithms, Computing and Mathematics* **3(1)** (2010), 93-108.

[8] A. Lladó and J. Moragas, Cycle-magic graphs, *Discrete Math.* **307** (2007), 2925 2933.

[9] T.K. Maryati, A. N. M. Salman, E.T. Baskoro, J. Ryan, M. Miller, On $H$- supermagic labelings for certain shackles and amalgamations of a connected graph, *Utilitas Mathematica*, **83** (2010), 333-342.

[10] A. A. G. Ngurah, A. N. M. Salman, L. Susilowati, $H$-supermagic labeling of graphs, *Discrete Math.*, **310** (2010), 1293-1300.

[11] S.T.R. Rizvi, K. Ali, M. Hussain, Cycle-supermagic labelings of the disjoint union of graphs, *Utilitas Mathematica*, (2014), in press.

[12] Roswitha, M. and Baskoro, E. T., $H$-magic covering on some classes of graphs, *American Institute of Physics Conference Proceedings* **1450** (2012), 135-138.

# Several classes of graphs and their $r$-dynamic chromatic numbers

**Dafik[1,2], D.E.W. Meganingtyas[3], K. Dwidja Purnomo[3], M. Dicky Tarmidzi[3], Ika Hesti Agustin[1,3]**

[1] CGANT, University of Jember, Indonesia
[2] Mathematics Edu. Depart. University of Jember, Indonesia
[3] Mathematics Depart. University of Jember, Indonesia

E-mail: `d.dafik@unej.ac.id, ikahesti.fmipa@unej.ac.id`

**Abstract.** Let $G$ be a simple, connected and undirected graph. Let $r, k$ be natural numbers. By a proper $k$-coloring of a graph $G$, we mean a map $c : V(G) \to S$, where $|S| = k$, such that any two adjacent vertices receive different colors. An $r$-dynamic $k$-coloring is a proper $k$-coloring $c$ of $G$ such that $|c(N(v))| \geq min\{r, d(v)\}$ for each vertex $v$ in $V(G)$, where $N(v)$ is the neighborhood of $v$ and $c(S) = \{c(v) : v \in S\}$ for a vertex subset $S$. The $r$-dynamic chromatic number, written as $\chi_r(G)$, is the minimum $k$ such that $G$ has an $r$-dynamic $k$-coloring. By simple observation it is easy to see that $\chi_r(G) \leq \chi_{r+1}(G)$, however $\chi_{r+1}(G) - \chi_r(G)$ does not always show a small difference for any $r$. Thus, finding an exact value of $\chi_r(G)$ is significantly useful. In this paper, we will study some of them especially when $G$ are prism graph, three-cyclical ladder graph, joint graph and circulant graph.

**Keywords:** $r$-dynamic chromatic number, graph coloring, special graphs.

## 1. Introduction

The $r$-dynamic chromatic number, introduced by Montgomery [8] and written as $\chi_r(G)$, is the least $k$ such that $G$ has an $r$-dynamic $k$-coloring. Note that the 1-dynamic chromatic number of graph is equal to its chromatic number, denoted by $\chi(G)$, and the 2-dynamic chromatic number of graph has been studied under the name a dynamic chromatic number, denoted by $\chi_d(G)$. In [8], he conjectured $\chi_2(G) \leq \chi(G) + 2$ when $G$ is regular, which remains open. Akbari *et.al.* [4] proved Montgomery's conjecture for bipartite regular graphs, as well as Lai, *et.al.* [9] proved $\chi_2(G) \leq \Delta(G)+1$ for $\Delta(G) \leq 3$ when no component is the 5-cycle. Some other results can be site in [1, 2, 3, 14] .

By a greedy coloring algorithm, Jahanbekama [7] proved that $\chi_r(G) \leq r\Delta(G) + 1$, and equality holds for $\Delta(G) > 2$ if and only if $G$ is $r$-regular with diameter 2 and girth 5. They improved the bound to $\chi_r(G) \leq \Delta(G) + 2r - 2$ when $\delta(G) > 2r \ln n$ and

$\chi_r(G) \leq \Delta(G) + r$ when $\delta(G) > r^2 \ln n$. For further results of $r$-dynamic chromatic number can be seen in [6, 10, 11, 12, 5] .

The following observation is useful to find the exact values of $r$-dynamic chromatic number.

**Observation 1.** *Let $\delta(G)$ and $\Delta(G)$ be a minimum and maximum degree of a graph $G$, respectively. Then the followings hold*

- $\chi_r(G) \geq \min\{\Delta(G), r\} + 1$,
- $\chi(G) \leq \chi_2(G) \leq \chi_3(G) \leq \cdots \leq \chi_{\Delta(G)}(G)$,
- $\chi_{r+1}(G) \geq \chi_r(G)$ *and if* $r \geq \Delta(G)$ *then* $\chi_r(G) = \chi_{\Delta(G)}(G)$.

Taherkhani in [13], proved the following theorem

**Theorem 1.** *[13] Let $G$ be a d-regular graph and $r$ be a positive integer with $2 \leq r \leq \frac{\delta}{\log(2er(\Delta^2+1))}$. Then the r-dynamic chromatic number of $G$ is $\chi_r(G) \leq \chi(G) + (r-1)\lceil e\frac{\Delta}{\delta}\log(2er(\Delta^2+1))\rceil$, where e euler's number.*

## 2. The Results

We are ready to show our main theorems. There are four theorems found in this study. These deals with prism graph, three-cyclical ladder graph, joint graph and circulant graph.

**Theorem 2.** *Let $\mathbf{P}_{n,2}$ be a prism graph, the r-dynamic chromatic number is:*

$$\chi(\mathbf{P}_{n,2}) = \begin{cases} 2, & n \text{ even} \\ 3, & n \text{ odd} \end{cases} \qquad \chi_d(\mathbf{P}_{n,2}) = \begin{cases} 3, & n = 3k, k \in N \\ 4, & n \text{ otherwise} \end{cases}$$

*For $r \geq 3$, we have*

$$\chi_r(\mathbf{P}_{n,2}) = \begin{cases} 4, & n = 4k, k \in N \\ 6, & n = 3, 7, 11 \\ 5, & n \text{ otherwise} \end{cases}$$

**Proof.** A prism graph, denoted by $\mathbf{P}_{n,2}, n \geq 3$, is a connected graph with vertex set $V(\mathbf{P}_{n,2}) = \{x_i, y_i, 1 \leq i \leq n\}$, and edge set $E(\mathbf{P}_{n,2}) = \{x_i x_{i+1}, y_i y_{i+1}; 1 \leq i \leq n-1\} \cup \{x_n x_1\} \cup \{y_n y_1\} \cup \{x_i y_i; 1 \leq i \leq n\}$. The order and size of $\mathbf{P}_{n,2}, n \geq 3$ are $|V(\mathbf{P}_{n,2})| = 2n$ and $|E(\mathbf{P}_{n,2})| = 3n$. A prism graph is regular graph of degree 3, thus $\mathbf{P}_{n,2}$, $\delta(\mathbf{P}_{n,2}) = \Delta(\mathbf{P}_{n,2}) = 3$. By Observation 1, $\chi_r(\mathbf{P}_{n,2}) \geq \min\{\Delta(\mathbf{P}_{n,2}), r\} + 1 = \min\{3, r\} + 1$. To find the exact value of $r$-dynamic chromatic number of $\mathbf{P}_{n,2}$, we define three cases, namely $\chi(\mathbf{P}_{n,2}), \chi_2(\mathbf{P}_{n,2})$ and $\chi_{r\geq 3}(\mathbf{P}_{n,2})$. For $\chi(\mathbf{P}_{n,2})$, the lower bound $\chi(\mathbf{P}_{n,2}) \geq \min\{3, 1\} + 1 = 2$. We will prove that $\chi(\mathbf{P}_{n,2}) \leq 2$ by defining a map $c_1 : V(\mathbf{P}_{n,2}) \to \{1, 2, \ldots, k\}$ for $n \geq 3$, by the following:

$$c_1(x_1, x_2, \ldots, x_n) = \begin{cases} 21 \ \ldots \ 21, & n \text{ even} \\ 12 \ \ldots \ 12\ 3, & n \text{ odd} \end{cases}$$

$$c_1(y_1, y_2, \ldots, y_n) = \begin{cases} 12 \ \ldots \ 12, & n \text{ even} \\ 3\ 12 \ldots \ 12, & n \text{ odd} \end{cases}$$

It is easy to see that $c_1$ gives $\chi(\mathbf{P}_{n,2}) \leq 2$ for $n$ even, but for $n$ odd, we could not avoid to have $\chi(\mathbf{P}_{n,2}) \leq 3$, otherwise there are at least two adjacent vertices assigned the same colors. Thus $\chi(\mathbf{P}_{n,2}) = 2$ for $n$ even and $\chi(\mathbf{P}_{n,2}) = 3$, for $n$ odd.

For $\chi_2(\mathbf{P}_{n,2})$, the lower bound $\chi_2(\mathbf{P}_{n,2}) \geq \min\{3,2\}+1 = 3$. We will prove that $\chi_2(\mathbf{P}_{n,2}) \leq 3$ by defining a map $c_2 : V(\mathbf{P}_{n,2}) \rightarrow \{1,2,\ldots,k\}$ where $n \geq 3$, by the following

$$c_2(x_1,x_2,\ldots,x_{n-1}) = \begin{cases} 12123, & n=5 \\ 123\ \ldots\ 123, & n \equiv 0 (\text{mod } 3), \\ 123\ \ldots\ 123\ 4, & n \equiv 1 (\text{mod } 3), \\ 123\ \ldots\ 123\ 41234, & n \equiv 2 (\text{mod } 3). \end{cases}$$

$$c_2(y_1,y_2,\ldots,y_{n-1}) = \begin{cases} 23434, & n=5 \\ 312\ \ldots\ 312, & n \equiv 0 (\text{mod } 3), \\ 4\ 123\ldots\ 123, & n \equiv 1 (\text{mod } 3), \\ 4\ 123\ldots\ 123\ 4123, & n \equiv 2 (\text{mod } 3). \end{cases}$$

It is easy to see that $c_2$ gives $\chi_2(\mathbf{P}_{n,2}) \leq 3$, for $n = 3k, k \in N$, but apart $n = 3k$ we could not avoid to have $\chi_2(\mathbf{P}_{n,2}) \leq 4$ otherwise there are at least two adjacent vertices assigned the same colors. Thus $\chi_2(\mathbf{P}_{n,2}) = 3$ for $n = 3k$ and $\chi_2(\mathbf{P}_{n,2}) = 4$ for otherwise.

For $\chi_r(\mathbf{P}_{n,2})$ and $r \geq 3$, the lower bound $\chi_3(\mathbf{P}_{n,2}) \geq \min\{3,3\}+1 = 4$. We will prove that $\chi_3(\mathbf{P}_{n,2}) \leq 4$ by defining a map $c_3 : V(\mathbf{P}_{n,2}) \rightarrow \{1,2,\ldots,k\}$ for $n \geq 3$, by the following.

$$c_3(x_1,x_2,\ldots,x_n) = \begin{cases} 123, & n=3, \\ 1234\ \ldots\ 1234, & n \equiv 0 (\text{mod } 4), \\ 1234\ \ldots\ 1234\ 5, & n \equiv 1 (\text{mod } 4), \\ 123456, & n=6, \\ 1234563, & n=7, \\ 1234\ \ldots\ 1234\ 512345, & n \equiv 2 (\text{mod } 4), n \geq 10, \\ 12345123456, & n=11, \\ 1234\ \ldots\ 1234\ 51234512345, & n \equiv 3 (\text{mod } 4), n \geq 15. \end{cases}$$

$$c_3(y_1,y_2,\ldots,y_{n-1}) = \begin{cases} 456, & n=3, \\ 34\ 1234\ \ldots\ 1234\ 12, & n \equiv 0 (\text{mod } 4), \\ 45\ 1234\ \ldots\ 1234\ 123, & n \equiv 1 (\text{mod } 4), \\ 561234, & n=6, \\ 6412345, & n=7, \\ 4512345\ 1234\ \ldots\ 1234\ 123, & n \equiv 0 (\text{mod } 4), n \geq 10, \\ 56123451234, & n=11, \\ 45\ 1234\ \ldots\ 1234\ 512345123, & n \equiv 0 (\text{mod } 4), n \geq 15. \end{cases}$$

It is easy to see that $c_3$ gives $\chi_3(\mathbf{P}_{n,2}) \leq 4$, for $n = 4k, k \in N$, but for $n = 3,6,7,11$ we are forced to have $\chi_3(\mathbf{P}_{n,2}) \leq 6$ as well as $\chi_3(\mathbf{P}_{n,2}) \leq 5$ for $n$ otherwise. Thus $\chi_3(\mathbf{P}_{n,2}) = 4$, for $n = 4k$, $\chi_3(\mathbf{P}_{n,2}) = 6$ for $n = 3,6,7,11$ and $\chi_3(\mathbf{P}_{n,2}) = 5$ for $n$ otherwise. By Observation 1, since $r \geq \Delta(\mathbf{P}_{n,2}) = 3$, it immediately gives $\chi_3(\mathbf{P}_{n,2})=\chi_r(\mathbf{P}_{n,2})$ for $n \geq 3$. □

IOP Conf. Series: Journal of Physics: Conf. Series **855** (2017) 012011 doi:10.1088/1742-6596/855/1/012011

**Theorem 3.** *Let $G$ be three-cyclical ladder graph ($TCL_n$) for $n \geq 2$, $r$-dynamic chromatic number of $TCL_n$ is*

$$\chi(TCL_n) = \chi_d(TCL_n) = 3, \chi_3(TCL_n) = 4, \chi_4(TCL_n) = 5, \chi_r(TCL_n) = 6, r \geq 5$$

**Proof.** The graph three-cyclical ladder graph, denoted by $TCL_n$, is connected graph with vertex set $V(TCL_n) = \{x_i, y_j, z_j; 1 \leq i \leq n; 1 \leq j \leq n+1\}$ and edge set $E(TCL_n) = \{y_j z_j; 1 \leq j \leq n+1\} \cup \{y_j y_{j+1}; 1 \leq i \leq n\} \cup \{x_i y_i; x_i z_i; x_i y_{i+1}; x_i z_{i+1}; 1 \leq i \leq n\}$. Thus, $p = |V(TCL_n)| = 3n+2, q = |E(TCL_n)| = 6n+1, \Delta(TCL_n) = 5$.

By Observation 1, $\chi_r(TCL_n) \geq \min\{\Delta(TCL_n), r\} + 1 = \min\{5, r\} + 1$. To find the exact value of $r$-dynamic chromatic number of $TCL_n$, we define three cases, namely for $\chi(TCL_n), \chi_d(TCL_n), \chi_3(TCL_n)$ and $\chi_4(TCL_n)$.

For $\chi(TCL_n), \chi_d(TCL_n)$, the lower bound $\chi_1(TCL_n) \geq \min\{5, 2\} + 1 = 3$. We will show that $\chi_1(TCL_n) \leq 3$, by defining a map $c_4 : V(TCL_n) \to \{1, 2, 3, \ldots, k\}$ where $n \geq 2$ by the following

$$c_4(x_i) = 3, 1 \leq i \leq n$$

$$c_4(y_j) = \begin{cases} 1, & j \equiv 1(mod\,2), 1 \leq j \leq n+1, \\ 2, & j \equiv 0(mod\,2), 1 \leq j \leq n+1. \end{cases}$$

$$c_4(z_j) = \begin{cases} 1, & j \equiv 0(mod\,2), 1 \leq j \leq n+1, \\ 2, & j \equiv 1(mod\,2), 1 \leq j \leq n+1. \end{cases}$$

It easy to see that $c_4$ gives $\chi(TCL_n) \leq 3$ and $\chi_d(TCL_n) \leq 3$. Thus $\chi(TCL_n) = 3$ and $\chi_d(TCL_n) = 3$.

For $r = 3$, the lower bound $\chi_3(TCL_n) \geq \min\{5, 3\} + 1 = 4$. We will show that $\chi_3(TCL_n) \leq 4$, by defining a map $c_5 : V(TCL_n) \to \{1, 2, 3, \ldots, k\}$ where $n \geq 2$ by the following

$$c_5(x_i) = \begin{cases} 1, & i \equiv 2(mod\,3), 1 \leq i \leq n, \\ 2, & i \equiv 0(mod\,3), 1 \leq i \leq n, \\ 3, & i \equiv 1(mod\,3), 1 \leq i \leq n. \end{cases}$$

$$c_5(y_j) = \begin{cases} 1, & j \equiv 1(mod\,3), 1 \leq j \leq n+1, \\ 2, & j \equiv 2(mod\,3), 1 \leq j \leq n+1, \\ 3, & j \equiv 0(mod\,3), 1 \leq j \leq n+1. \end{cases}$$

$$c_5(z_j) = 4, \text{for } 1 \leq i \leq n+1$$

It is easy to understand that $c_5$ gives $\chi_3(TCL_n) \leq 4$. Thus $\chi_3(TCL_n) = 4$.

For $r = 4$, the lower bound $\chi_4(TCL_n) \geq \min\{5, 4\} + 1 = 5$. We will show that $\chi_4(TCL_n) \leq 5$, by defining a map $c_6 : V(TCL_n) \to \{1, 2, 3, \ldots, k\}$ where $n \geq 2$ by the following

$$c_6(x_i) = \begin{cases} 3, & i \equiv 1(mod\,3), 1 \leq i \leq n, \\ 4, & i \equiv 2(mod\,3), 1 \leq i \leq n, \\ 5, & i \equiv 0(mod\,3), 1 \leq i \leq n. \end{cases}$$

$$c_6(y_j) = \begin{cases} 1, & j \equiv 1(mod\,2), 1 \leq j \leq n+1, \\ 2, & j \equiv 0(mod\,2), 1 \leq j \leq n+1. \end{cases}$$

$$c_6(z_j) = \begin{cases} 3, & j \equiv 0(mod\,3),\ 1 \le j \le n+1, \\ 4, & j \equiv 1(mod\,3),\ 1 \le j \le n+1, \\ 5, & j \equiv 2(mod\,3),\ 1 \le j \le n+1. \end{cases}$$

It is easy to see that $c_6$ gives $\chi_4(TCL_n) \le 5$. Thus $\chi_4(TCL_n) = 5$.

For $r = 5$, the lower bound $\chi_5(TCL_n) \ge \min\{5, 5\} + 1 = 6$. We will show that $\chi_5(TCL_n) \le 6$, by defining a map $c_7 : V(TCL_n) \to \{1, 2, 3, \ldots, k\}$ where $n \ge 2$ by the following

$$c_7(x_i) = \begin{cases} 4, & i \equiv 1(mod\,3),\ 1 \le i \le n, \\ 5, & i \equiv 2(mod\,3),\ 1 \le i \le n, \\ 6, & i \equiv 0(mod\,3),\ 1 \le i \le n. \end{cases}$$

$$c_7(y_j) = \begin{cases} 1, & j \equiv 1(mod\,3),\ 1 \le j \le n+1, \\ 2, & j \equiv 2(mod\,3),\ 1 \le j \le n+1, \\ 3, & j \equiv 0(mod\,3),\ 1 \le j \le n+1. \end{cases}$$

$$c_7(z_j) = \begin{cases} 4, & j \equiv 0(mod\,3),\ 1 \le j \le n+1, \\ 5, & j \equiv 1(mod\,3),\ 1 \le j \le n+1, \\ 6, & j \equiv 2(mod\,3),\ 1 \le j \le n+1. \end{cases}$$

It clearly shows that $c_7$ gives $\chi_5(TCL_n) \le 6$. Thus $\chi_5(TCL_n) = 6$. Since for $r \ge 5$, we have $r \ge \Delta(TCL_n)$. By Observation 1, $\chi_r(TCL_n) = \chi_5(TCL_n) = 6$. It concludes the proof. □

**Theorem 4.** *Let $P_n + C_m$ be a joint graph of $P_n$ and $C_m$, the $r$-dynamic chromatic number is*

$$\chi_{1 \le r \le 4}(P_n + C_m) = \begin{cases} 5, & m = 3k, k \in N, \\ 6, & m \text{ otherwise.} \end{cases}$$

$$\chi_5(P_n + C_m) = \begin{cases} 6, & m = 3, \\ 8, & m = 5, \\ 7, & m \text{ otherwise.} \end{cases}$$

*For $r \ge 6$, we have*

$$\chi_r(P_n + C_m) = \begin{cases} r + m - 2, & 3 \le m \le r-2, m \ge r-1, n \ge m-1, \\ 2r - 3, & m \text{ lainnya}, n \ge r-1. \end{cases}$$

**Proof.** The graph $P_n + C_m$ is a connected graph with vertex set $V(P_n + C_m) = \{x_i; 1 \le i \le n\} \cup \{y_j; 1 \le j \le m\}$ and edge set $E(P_n + C_m) = \{x_i x_{i+1}; 1 \le i \le n-1\} \cup \{y_j y_{j+1}; 1 \le j \le m\} \cup \{y_m y_1\} \cup \{x_i y_j; 1 \le i \le n; 1 \le j \le m\}$. The order and size of this graph are $p = |V(P_n + C_m)| = m + n, q = |E(P_n + C_m)| = mn + m - 1$. Since all vertices in $P_n$ joint with all vertices in $C_m$, it gives $\Delta(P_n + C_m) = m + 2$

By Observation 1, $\chi_r(P_n + C_m) \ge \min\{\Delta(P_n + C_m), r\} + 1 = \min\{m + 2, r\} + 1$. To find the exact value of $r$-dynamic chromatic number of $P_n + C_m$, we define three cases, namely for $\chi_{1 \le r \le 4}(P_n + C_m), \chi_5(P_n + C_m)$ and $\chi_{r \ge 6}(P_n + C_m)$.

For $\chi_{1\leq r\leq 4}(P_n+C_m)$, define a map $c_8 : V(P_n+C_m) \to \{1,2,\ldots,k\}$ where $n \geq 3$, by the following:

$$c_8(x_0,x_1,x_2,\ldots,x_{n-1}) = \begin{cases} 123\ \ldots\ 123,\ n\equiv 0(\text{mod } 3),\ m\equiv 2(\text{mod } 3), \\ 123\ \ldots\ 123\ 1,\ n\equiv 1(\text{mod } 3),\ m\equiv 2(\text{mod } 3), \\ 123\ \ldots\ 123\ 12,\ n\equiv 2(\text{mod } 3),\ m\equiv 2(\text{mod } 3), \\ 12\ \ldots\ 12,\ n \text{ even},\ m \text{ otherwise}, \\ 12\ \ldots\ 12\ 1,\ n \text{ odd},\ m \text{ otherwise}. \end{cases}$$

$$c_8(y_0,y_1,y_2,\ldots,y_{n-1}) = \begin{cases} 345\ \ldots\ 345,\ m\equiv 0(\text{mod } 3), \\ 345\ \ldots\ 345\ 6,\ m\equiv 1(\text{mod } 3), \\ 45\ \ldots\ 45\ 6,\ m\equiv 2(\text{mod } 3), m \text{ odd}, \\ 45\ \ldots\ 45\ 46,\ m\equiv 2(\text{mod } 3), m \text{ even}. \end{cases}$$

It is easy to see that $c_8$ gives $\chi_{1\leq r\leq 4}(P_n+C_m) = 5$, for $m = 3k, k \in N$ and $\chi_{1\leq r\leq 4}(P_n+C_m) = 6$ for $m$ otherwise.

For $\chi_5(P_n+C_m)$, define a map $c_9 : V(P_n+C_m) \to \{1,2,\ldots,k\}$ where $n \geq 3$, by the following:

$$c_9(x_0,x_1,x_2,\ldots,x_{n-1}) = \begin{cases} 123\ \ldots\ 123,\ n\equiv 0(\text{mod } 3), \\ 123\ \ldots\ 123\ 1,\ n\equiv 1(\text{mod } 3), \\ 123\ \ldots\ 123\ 12,\ n\equiv 2(\text{mod } 3). \end{cases}$$

$$c_9(y_0,y_1,y_2,\ldots,y_{n-1}) = \begin{cases} 456,\ m=3, \\ 45678,\ m=5, \\ 456\ \ldots\ 456\ 457,\ m\equiv 0(\text{mod } 3), m\geq 6, \\ 456\ \ldots\ 456\ 4567,\ m\equiv 1(\text{mod } 3), \\ 456\ \ldots\ 45\ 74567,\ m\equiv 2(\text{mod } 3), m\geq 8. \end{cases}$$

It is easy to see that $c_9$ gives $\chi_5(P_n+C_m) = 6$, for $m = 3$, $\chi_5(P_n+C_m) = 8$, for $m = 5$, and $\chi_5(P_n+C_m) = 7$ for $m$ otherwise.

The last for $\chi_6(P_n+C_m)$, define a map $c_{10} : V(P_n+C_m) \to \{1,2,\ldots,k\}$ where $m \geq 3, n \geq r-2$, by the following

$$c_{10}(x_i) = \begin{cases} 1,\ i\equiv 1(\text{mod } r-2), \\ 2,\ i\equiv 2(\text{mod } r-2), \\ 3,\ i\equiv 3(\text{mod } r-2), \\ \vdots \\ r-3,\ i=n-1, \\ r-2,\ i=n. \end{cases}$$

$$c_{10}(y_j) = \begin{cases} r-1, \; i \equiv 1 (\text{mod } 3), 1 \le i \le n-r+4, \\ r, \; i \equiv 2 (\text{mod } 3), 1 \le i \le n-r+3, \\ r+1, \; i \equiv 3 (\text{mod } 3), 1 \le i \le n-r+2, \\ r+2, \; i = n-r+1, \\ r+3, \; i = n-r, \\ r+4, \; i = n-r-1, \\ \vdots \\ 2n-2, \; i = n-1, \\ 2n-3, \; i = n. \end{cases}$$

It easy to see that $c_{10}$ gives $\chi_6(P_n+C_m) = r+m-2$ for $3 \le m \le r-2, m \ge r-1, n \ge m-1$ and $\chi_6(P_n+C_m) = 2r-3$ for $n \ge r-1$, $m$ otherwise. By Observation 1, since $r \ge \Delta(P_n+C_m) = m+2$, it immediately gives $\chi_6(P_n+C_m)=\chi_r(P_n+C_m)$ for $n \ge 4$. □

**Theorem 5.** *Let $C_n(1, \frac{n}{2})$ be a circulant graph of order 3, the r-dynamic chromatic number is*

$$\chi(C_n(1, \frac{n}{2})) = \begin{cases} 4, \; n = 4, \\ 2, \; n = 4k+2, k \in N, \\ 3, \; n = 4k+4, k \in N. \end{cases} \qquad \chi_d(C_n(1, \frac{n}{2})) = 4$$

*For $r \ge 3$, we have*

$$\chi_r(C_n(1, \frac{n}{2})) = \begin{cases} n, \; n = 4,6,8, \\ 4, \; n = 8k+4, k \in N, \\ 5, \; n = 8k+6, k \in N, \\ 6, \; n \text{ otherwise}. \end{cases}$$

**Proof.** The graph $C_n(1, \frac{n}{2})$ is a connected graph with vertex set $V(C_n(1, \frac{n}{2})) = \{x_i, 0 \le i \le n-1\}$ and edge set $E(C_n(1, \frac{n}{2})) = \{x_i x_{i+1 (\text{mod } n)}, 0 \le i \le n-1\} \cup \{x_i x_{i+\frac{n}{2} (\text{mod } n)}, 0 \le i \le \frac{n}{2}\}$. The order and size of the graph $C_n(1, \frac{n}{2})$ are $p = |V(C_n(1, \frac{n}{2}))| = n, q = |E(C_n(1, \frac{n}{2}))| = \frac{3n}{2}$. Since $C_n(1, \frac{n}{2})$ is a regular graph of degree 3, thus $\delta(C_n(1, \frac{n}{2})) = \Delta(C_n(1, \frac{n}{2})) = 3$.

By Observation 1, $\chi_r(C_n(1, \frac{n}{2})) \ge \min\{\Delta(C_n(1, \frac{n}{2})), r\} + 1 = \min\{3, r\} + 1$. In the same way, to find the exact value of $r$-dynamic chromatic number of $C_n(1, \frac{n}{2})$, we define three cases, namely for $\chi(C_n(1, \frac{n}{2})), \chi_2(C_n(1, \frac{n}{2}))$ and $\chi_{r\ge3}(C_n(1, \frac{n}{2}))$.

For $\chi(C_n(1, \frac{n}{2}))$, define a map $c_{11} : V(C_n(1, \frac{n}{2})) \to \{1, 2, \ldots, k\}$ where $n \ge 3$, by the following:

$$c_{11}(x_0, x_1, \ldots, x_{n-1}) = \begin{cases} 1234, \; n = 4, \\ 12 \ldots 12, \; n = 4k+2, k \in N. \end{cases}$$

$$\begin{aligned} c_{11}(x_0, x_1, \ldots, x_{\frac{n}{2}}) &= 12 \ldots 12\ 13, \; n = 4k+4, k \in N. \\ c_{11}(x_{\frac{n}{2}+1}, x_{\frac{n}{2}+2}, \ldots, x_{n-1}) &= 21 \ldots 21,\ 32, \; n = 4k+4, k \in N. \end{aligned}$$

It easy to see that $c_{11}$ gives $\chi(C_n(1,\frac{n}{2})) = 4$, for $n = 4$, $\chi(C_n(1,\frac{n}{2})) = 2$, for $n = 4k+2, k \in N$, and $\chi(C_n(1,\frac{n}{2})) = 3$, for $n = 4k+4, k \in N$.

For $\chi_2(C_n(1,\frac{n}{2}))$, define a map $c_{12} : V(C_n(1,\frac{n}{2})) \to \{1,2,\ldots,k\}$ where $n \geq 3$, by the following:

$$\begin{aligned} c_{12}(x_0,x_1,\ldots,x_{n-1}) &= 1234, \text{ for } n = 4 \\ c_{12}(x_0,x_1,\ldots,x_{\frac{n}{2}}) &= 12\ldots\ 12, \text{ for } n = 4k+2, k \in N \\ c_{12}(x_{\frac{n}{2}+1},x_{\frac{n}{2}+2},\ldots,x_{n-1}) &= 34\ \ldots\ 34, \text{ for } n = 4k+2, k \in N \end{aligned}$$

It easy to see that $c_{12}$ gives $\chi_2(C_n(1,\frac{n}{2})) = 4$ for any $n$.

For $\chi_r(C_n(1,\frac{n}{2}))$, and $r \geq 3$, define a map $c_{13} : V(C_n(1,\frac{n}{2})) \to \{1,2,\ldots,k\}$ where $n \geq 3$, by the followings

- For $n = 4$, $c_{13}(x_i) = i+1,\ 0 \leq i \leq n-1$
- For $n = 10$

$$c_{13}(x_i) = \begin{cases} 1, & i = 0,7, \\ 2, & i = 5,8, \\ 3, & i = 1,4, \\ 4, & i = 3,6, \\ 5, & i = 2,9. \end{cases}$$

- For $n = 8k+4, k \in N$

$$c_{13}(x_i) = \begin{cases} 1, & i \equiv 0 (\text{mod } 4), 0 \leq i \leq n-4, \\ 2, & i \equiv 1 (\text{mod } 4), 1 \leq i \leq n-3, \\ 3, & i \equiv 2 (\text{mod } 4), 2 \leq i \leq n-2, \\ 4, & i \equiv 3 (\text{mod } 4), 3 \leq i \leq n-1. \end{cases}$$

- For $n = 8k+6, k \in N$

$$c_{13}(x_i) = \begin{cases} 1, & i \equiv 0 (\text{mod } 4), 0 \leq i \leq \frac{n}{2}-7, \\ 2, & i \equiv 1 (\text{mod } 4), 1 \leq i \leq \frac{n}{2}-6, \\ 3, & i \equiv 2 (\text{mod } 4), 2 \leq i \leq \frac{n}{2}-5, \\ 4, & i \equiv 3 (\text{mod } 4), 3 \leq i \leq \frac{n}{2}-4, \\ 5, & i = \frac{n}{2}-12 \text{ atau } i = n-1, \end{cases}$$

$$c_{13}(x_i) = \begin{cases} i, & i \equiv 0 (\text{mod } \frac{n}{2}-2), \frac{n}{2}-2 \leq i \leq n-5, \\ i-1, & i \equiv 1 (\text{mod } \frac{n}{2}-2), \frac{n}{2}-1 \leq i \leq n-4, \\ i-2, & i \equiv 2 (\text{mod } \frac{n}{2}-2), \frac{n}{2} \leq i \leq n-3, \\ i-3, & i \equiv 3 (\text{mod } \frac{n}{2}-2), \frac{n}{2}+1 \leq i \leq n-2. \end{cases}$$

- For $n = 8k + 8, k \in N$

$$c_{13}(x_i) = \begin{cases} 1, \ i \equiv 0(\text{mod } 4), 0 \leq i \leq \frac{n}{2} - 8, \text{ or} \\ \quad i \equiv 0(\text{mod } \frac{n}{2} - 2), \frac{n}{2} - 2 \leq i \leq n - 6, \\ 2, \ i \equiv 1(\text{mod } 4), 1 \leq i \leq \frac{n}{2} - 7, \text{ or} \\ \quad i \equiv 1(\text{mod } \frac{n}{2} - 2), \frac{n}{2} - 1 \leq i \leq n - 5, \\ 3, \ i \equiv 2(\text{mod } 4), 2 \leq i \leq \frac{n}{2} - 6, \text{ or} \\ \quad i \equiv 2(\text{mod } \frac{n}{2} - 2), \frac{n}{2} \leq i \leq n - 4, \\ 4, \ i \equiv 3(\text{mod } 4), 3 \leq i \leq \frac{n}{2} - 5, \text{ or} \\ \quad i \equiv 3(\text{mod } \frac{n}{2} - 2), \frac{n}{2} + 1 \leq i \leq n - 3, \\ 5, \ i = \frac{n-8}{2} \text{or } i = n - 2, \\ 6, \ i = \frac{n-6}{2} \text{or } i = n - 1. \end{cases}$$

- For $n = 8k + 10, k \in N$

$$c_{13}(x_i) = \begin{cases} 1, \ i \equiv 0(\text{mod } 4), 0 \leq i \leq \frac{n}{2} - 6, \text{ or} \\ \quad i \equiv 0(\text{mod } \frac{n}{2} - 2), \frac{n}{2} - 2 \leq i \leq n - 7, \\ 2, \ i \equiv 1(\text{mod } 4), 1 \leq i \leq \frac{n}{2} - 5, \text{ or} \\ \quad i \equiv 1(\text{mod } \frac{n}{2} - 2), \frac{n}{2} - 1 \leq i \leq n - 6, \\ 3, \ i \equiv 2(\text{mod } 4), 2 \leq i \leq \frac{n}{2} - 4, \text{ or} \\ \quad i \equiv 2(\text{mod } \frac{n}{2} - 2), \frac{n}{2} \leq i \leq n - 5, \\ 4, \ i \equiv 3(\text{mod } 4), 3 \leq i \leq \frac{n}{2} - 3, \text{ or} \\ \quad i \equiv 3(\text{mod } \frac{n}{2} - 2), \frac{n}{2} + 1 \leq i \leq n - 4, \text{or } i = n - 1, \\ 5, \ \text{For } i = n - 3, \\ 6, \ \text{For } i = n - 2, \end{cases}$$

It easy to see that $c_{13}$ gives $\chi_3(C_n(1, \frac{n}{2})) = 4, 6, 8$ for $n = 4, 6, 8$, $\chi_3(C_n(1, \frac{n}{2})) = 4$ for $n = 8k + 4$, $\chi_3(C_n(1, \frac{n}{2})) = 5$ for $n = 8k + 6$, and $\chi_3(C_n(1, \frac{n}{2})) = 6$ for $n$ otherwise. By Observation 1, since $r \geq \Delta(C_n(1, \frac{n}{2})) = 4$, it immediately gives $\chi_3(C_n(1, \frac{n}{2}))$=$\chi_r(C_n(1, \frac{n}{2}))$ for $n \geq 4$. □

## Concluding Remarks

We have found some $r-$dynamic chromatic number of several graphs, namely prism graph, three-cyclical ladder graph, joint graph and circulant graph. All numbers attain a best lower bound. For the characterization of the lower bound of for any connected graphs $G$, we have not found any result yet, thus we propose the following open problem.

## Open Problem

Given that any connected graphs $G$, determine the sharp lower bound of $\chi_r(G)$

## Acknowledgement

We gratefully acknowledge the support from DIPA 2017 of CGANT & LEMLIT - University of Jember.

IOP Conf. Series: Journal of Physics: Conf. Series **855** (2017) 012011 doi:10.1088/1742-6596/855/1/012011

## References


[1] M. Alishahi, Dynamic chromatic number of regular graphs, *Discrete Appl. Math.* **160** (2012) 2098–2103.

[2] M. Alishahi, On the dynamic coloring of graphs, *Discrete Appl. Math.* **159** (2011) 152–156.

[3] S. Akbari, M. Ghanbari, S. Jahanbekam, On the list dynamic coloring of graphs, *Discrete Appl. Math.* 157 (2009) 3005–3007.

[4] S. Akbari, M. Ghanbari, S. Jahanbekam, On the dynamic chromatic number of graphs, *in: Combinatorics and Graphs, in: Contemp. Math., Amer. Math. Soc.,* **vol. 531**, 2010, pp. 11–18

[5] S. Akbari, M. Ghanbari, S. Jahanbekam. On The Dynamic Coloring of Cartesian Product Graphs, Ars Combinatoria 114 (2014) 161 – 167

[6] H. Furmanczyk, M. Kubale. Equitable Coloring of Corona Products of Cubic Graphs is Harder Than Ordinary Coloring. Ars Mathematica Contemporanea 10 (2016) 333 – 347

[7] S. Jahanbekam, J. Kim, Suil O, Douglas B. West. On r-Dynamic Coloring of Graph. Discrete Applied Mathematics 206 (2016) 65 – 72

[8] B. Montgomery, Dynamic Coloring of Graphs (Ph.D Dissertation), West Virginia University, 2001.

[9] H.J. Lai, B. Montgomery, H. Poon, Upper bounds of dynamic chromatic number, *Ars Combin.* **68** (2003) 193–201.

[10] H.J. Lai, B. Montgomery. Dynamic Coloring of Graph. Department of Mathematics, West Virginia University, Mongantown WV 26506-6310. 2002

[11] H.J. Lai, B. Montgomery, H. Poon. Upper Bounds of Dynamic Chromatic Number. Ars Combinatoria. 68 (2003) 193 – 201

[12] R. Kang, T. Muller, Douglas B. West. On r-Dynamic Coloring of Grids. Discrete Applied Mathematics 186 (2015) 286 – 290

[13] A. Taherkhani, *r-Dynamic Chromatic Number of Graphs.* **arXiv:1401.6470v1 [math.CO]**, 24 January 2014, Dept of Math, Institutu for Basic Studies and Advance Sciense, Iran.

[14] A. Taherkhani. On r-Dynamic Chromatic Number of Graphs. Discrete Applied Mathematics 201 (2016) 222 – 227

IOP Conf. Series: Journal of Physics: Conf. Series **855** (2017) 012012 doi:10.1088/1742-6596/855/1/012012

# On The Partition Dimension of a Lollipop Graph and a Generalized Jahangir Graph

**Maylinda Purna Kartika Dewi and Tri Atmojo Kusmayadi**

Department of Mathematics, Faculty of Mathematics and Natural Sciences
Sebelas Maret University, Surakarta, Central Java, Indonesia

E-mail: maylindapkd@gmail.com, tri.atmojo.kusmayadi@gmail.com

**Abstract.** Let $G$ be a connected graph with vertex set $V(G)$, such that $V(G)$ can be divided into any partition set $S$. The set $\Pi$ with $S \in \Pi$ is a resolving partition of $G$ if each vertex in $G$ has a distinct representation with respect to $\Pi$, and $\Pi$ is an ordered $k-$partition. The minimum cardinality of resolving $k-$partitions of $V(G)$ is called a partition dimension of $G$, denoted by $pd(G)$. The lollipop graph $L_{m,n}$ is a graph obtained by joining a complete graph $K_m$ to a path $P_n$ with a bridge. A generalized Jahangir graph is a graph consisting of a cycle $C_{mn}$ and one additional vertex which is adjacent to $n$ vertices of $C_{mn}$ at $m$ distance to each other on $C_{mn}$. Many researchers have conducted research in determining the partition dimension for specific graph classes. They are as references to determine some of the graph classes that haven't been studied previously.

In this paper, we determine the partition dimension of a lollipop graph $L_{m,n}$ and a generalized Jahangir graph $J_{m,n}$. The research methods in this paper is a book study.

The results of this paper are as follows. We obtain the partition dimension of a lollipop graph $pd(L_{m,n}) = m$ for $m \geq 3$ and $n \geq 1$. The partition dimension of a generalized Jahangir graph consists of two cases. We showed that $pd(J_{m,n}) = 3$ for $n = 3, 4, 5$ and we prove that $pd(J_{m,n}) = \lfloor \frac{n}{2} \rfloor + 1$ for $n \geq 6$.

## 1. Introduction

The partition dimension of a graph was introduced by Chartrand et al. [1] in 1998. There is another concept of metric dimension form that was introduced by Slater in 1975 and Harary and Melter in 1976 (Darmaji [3]). We assume that $G$ is a graph with a vertex set $V(G)$, such that $V(G)$ can be divided into any partition set $S$. The set $\Pi$ with $S \in \Pi$ is a resolving partition of $G$ if each vertex in $G$ has a distinct representation with respect to $\Pi$, and $\Pi$ is an ordered $k-$partition. The minimum cardinality of resolving $k-$partitions of $V(G)$ is called a partition dimension of $G$, denoted by $pd(G)$.

Many researchers have conducted research in determining the partition dimension for specific graph classes. Tomescu et al. [5] in 2007 found the partition dimension of a wheel graph. In this paper, we determine the partition dimension of a lollipop graph and a generalized Jahangir graph.

## 2. Main Result

### 2.1. Partition Dimension

The following definition and lemma were given by Chartrand et al. [2]. Let $G$ be a connected graph. For a subset $S$ of $V(G)$ and a vertex $v$ of $G$, the distance $d(v, S)$ between $v$ and $S$ is defined as $d(v, S) = min\{d(v, x)|x \in S\}$. For an ordered $k$-partition $\Pi = \{S_1, S_2, \ldots, S_k\}$ of $V(G)$ and a vertex $v$ of $G$, the representation of $v$ with respect to $\Pi$ is defined as the $k$-vector $r(v|\Pi) = (d(v, S_1), d(v, S_2), \ldots, d(v, S_k))$. The partition $\Pi$ is called a resolving partition if the $k$-vectors $r(v|\Pi), v \in V(G)$ are distinct. The minimum $k$ for which there is a resolving $k$-partition of $V(G)$ is called the partition dimension of $G$, denoted by $pd(G)$.

**Lemma 2.1** [2] *Let $G$ be a connected graph, then*

*(i) $pd(G) = 2$ if and only if $G = P_n$ for $n \geq 2$,*
*(ii) $pd(G) = n$ if and only if $G = K_n$, and*
*(iii) $pd(G) = 3$ if $G = n$-cycle for $n \geq 3$.*

**Lemma 2.2** [2] *Let $\Pi$ be a resolving partition of $V(G)$ and $u, v \in V(G)$. If $d(u, w) = d(v, w)$ for all $w \in V(G) - \{u, v\}$, then $u$ and $v$ belong to distinct elements of $\Pi$.*

Let $\Pi = \{S_1, S_2, \ldots, S_k\}$, where $u$ and $v$ belong to the same element, say $S_i$, of $\Pi$. Then $d(u, S_i) = d(v, S_i) = 0$. Since $d(u, w) = d(v, w)$ for all $w \in V(G) - \{u, v\}$, we also have that $d(u, S_j) = d(v, S_j)$ for all $j$, where $1 \leq j \neq i \leq k$. Therefore, $r(u|\Pi) = r(v|\Pi)$ and $\Pi$ is not a resolving partition.

### 2.2. The Partition Dimension of a Lollipop Graph

Weisstein [6] defined the lollipop graph $L_{m,n}$ for $m \geq 3$ as a graph obtained by joining a complete graph $K_m$ to a path $P_n$ with a bridge. See for example, the lollipop graph in Figure 1.

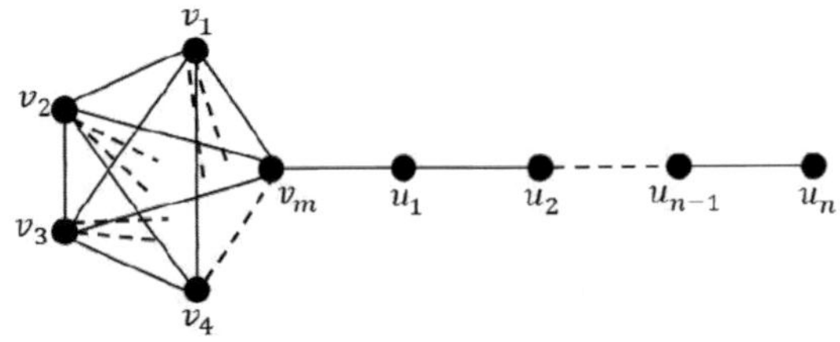


**Figure 1.** The Lollipop Graph

**Theorem 2.1** *Let $L_{m,n}$ be a lollipop graph with $m \geq 3$ and $n \geq 1$, then $pd(L_{m,n}) = m$.*

Let $V(L_{m,n}) = V(K_m) \cup V(P_n)$, where $V(K_m) = \{v_1, v_2, \ldots, v_{m-1}, v_m\}$ and $V(P_n) = \{u_1, u_2, \ldots, u_{n-1}, u_n\}$. Take a vertex $v \in V(K_m)$ such that $v$ is adjacent to $u \in V(P_n)$. Based on Lemma 2.1, the partition dimension of $L_{m,n}$ is $pd(L_{m,n}) \geq m$. Next, we show that the partition dimension of $L_{m,n}$ is $pd(L_{m,n}) = m$. For each $v, u \in V(L_{m,n})$, let $\Pi = \{S_1, S_2, \ldots, S_m\}$ be the resolving partition where $S_i = \{v_i\}$ for $1 \leq i \leq m - 1$ and $S_m = \{v_m, u_1, \ldots, u_n\}$. We obtain the representation between all vertices of $V(L_{m,n})$ with respect to $\Pi$ as follows $r(v_m|\Pi) = (d(v_1, S_i), \ldots, d(v_m, S_i))$ and $r(u_n|\Pi) = (n, n+1, n+1, \ldots, n+1, 0)$ where
$$d(v_m, S_i) = \begin{cases} 0, & \text{for } i = m, \\ 1, & \text{for } i \neq m. \end{cases}$$
Therefore, each vertex $v \in V(L_{m,n})$ has a distinct representation with respect to $\Pi$. Now, we

show that the cardinality of $\Pi$ is $m$. The cardinality of $\Pi$ is the number of $k-$partition in $S_i$ and $S_m$ for $1 \leq i \leq m-1$. The number of $k-$partition in $S_i$ for $1 \leq i \leq m-1$, is $m-1$. The number of $k-$partition in $S_m$ is 1, so we have $\mid \Pi \mid = (m-1)+1 = m$. Furthermore, $\Pi = \{S_1, S_2, \ldots, S_m\}$ is a resolving partition of $L_{m,n}$ with $m$ elements. Hence, the partition dimension of lollipop graph is $pd(L_{m,n}) = m$.

### *2.3. The Partition Dimension of a Generalized Jahangir Graph*

Mojdeh and Ghameshlou [4] defined a generalized Jahangir graph $J_{m,n}$ with $n \geq 3$ as a graph on $nm+1$ vertices consisting of a cycle $C_{mn}$ and one additional vertex which is adjacent to $n$ vertices of $C_{mn}$ at $m$ distance to each other on $C_{mn}$. Let $V(J_{m,n}) = V(C_{mn}) \cup \{u_1\}$ or $V(J_{m,n}) = V(P_m) \cup V(W_n)$ where $V(P_m) = \{v_1, v_2, \ldots, v_{n(m-1)}\}$ and $V(W_n) = \{u_1, u_2, \ldots, u_{n+1}\}$, we have $V(J_{m,n}) = \{u_1, u_2, \ldots, u_{n+1}, v_1, v_2, \ldots, v_{n(m-1)}\}$. Example of a generalized Jahangir graph is in Figure 2.

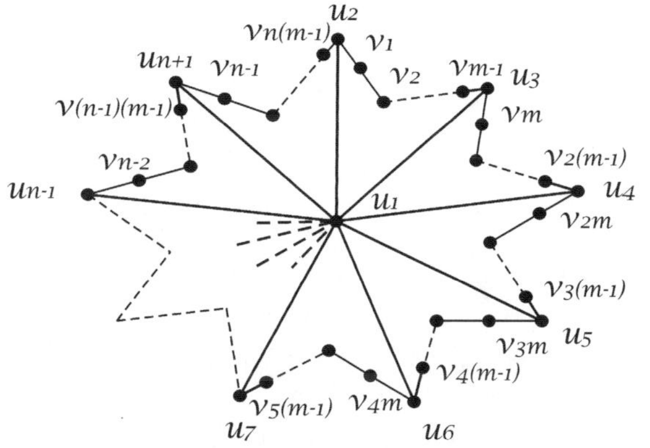


**Figure 2.** The Generalized Jahangir Graph

**Lemma 2.3** *Let $G$ be a subgraph of a generalized Jahangir graph $J_{m,n}$ for $m \geq 3$, $n \geq 3$ and $\Pi$ is a resolving partition of $G$. The distance of vertex $u_1$ to the other vertices are $d(u_1, u_i) = 1$ and $d(u_1, v_k) = 2$ for $1 < i \leq n+1$ and $1 \leq k \leq n(m-1)$. The distance of vertex $u$ to the other vertex $u$ are the same, $d(u_i, u_j) = 2$ for $1 < i \leq n+1$, $1 < j \leq n+1$ and $i \neq j$.*

Let $u_1$ be a central vertex of $V(G) = V(J_{m,n})$. Since each vertex $u_i$ is connected with central vertex, so $d(u_1, u_i) = 1$ and $d(u_i, u_j) = 2$ with $1 < i \leq n+1$, $1 < j \leq n+1$, and $i \neq j$. If vertex $v_k$ is a vertex of path $P_m$ and has a minimum distance to vertex $u_i$, then $d(u_1, v_k) = 2$ for $1 \leq k \leq n(m-1)$.

**Theorem 2.2** *Let $J_{m,n}$ be a generalized Jahangir graph for $m \geq 3$ and $n \geq 3$ then*

$$pd(J_{m,n}) = \begin{cases} 3, & \text{for } n = 3,4,5; \\ \lfloor \frac{n}{2} \rfloor + 1, & \text{for } n \geq 6. \end{cases}$$

Let $J_{m,n}$ be a generalized Jahangir graph for $m \geq 3$, $n \geq 3$ and a vertex set $V(J_{m,n})$. The proof can be divided into two cases according to the values of $n$.

(i) Case $n = 3,4,5$

 (a) For $n = 3$

 Let $a = \{v_1, \ldots, v_m\}$, $b = \{v_1, \ldots, v_{m-1}\}$, $c = \{v_{m+1}, \ldots, v_{n(m-1)}\}$, and $d = \{v_m, \ldots, v_{2(m-1)}\}$, then

$$S_1 = \{u_1, \ldots, u_{n+1}, v_{2m-1}, \ldots, v_{mn-3}\},$$

$$S_2 = \begin{cases} a, & \text{for } m = 3; \\ b, & \text{for } m \geq 4; \end{cases} \qquad S_3 = \begin{cases} c, & \text{for } m = 3; \\ d, & \text{for } m \geq 4. \end{cases}$$

(b) For $n = 4$
Let $a = \{u_1, \ldots, u_{n+2}, v_{(n(m-1))-1}\}$,
$b = \{u_1, \ldots, u_{n+2}, v_{3m-2}, \ldots, v_{n(m-1)}\}$,
$c = \{u_1, \ldots, u_{n+2}, v_{2m-1}, v_{3m-2}, \ldots, v_{n(m-1)}\}$,
$d = \{v_1, \ldots, v_m\}$,
$e = \{v_1, \ldots, v_{m-1}, v_{2(m+1)}, v_{m(n-1)-3}\}$,
$f = \{v_1, \ldots, v_{m-1}, v_{m(n-1)-3}\}$,
$g = \{v_{m+1}, \ldots, v_{n(m-1)}\}$,
$h = \{v_m, \ldots, v_{2(m-1)}, v_{2m-1}, \ldots, v_{(m-1)(n-1)-1}\}$,
$i = \{v_m, \ldots, v_{2(m-1)}, v_{2m}, \ldots, v_{(m-1)(n-1)-1}\}$, and
$j = \{v_m, \ldots, v_{2(m-1)}, v_{2m}, v_{2m+1}, v_{2m+3}, \ldots, v_{(m-1)(n-1)-1}\}$ then

$$S_1 = \begin{cases} a, & \text{for } m = 3; \\ b, & \text{for } m = 5; \\ c, & \text{for } m \geq 4, m \neq 5; \end{cases} \qquad S_2 = \begin{cases} d, & \text{for } m = 3; \\ e, & \text{for } m = 9; \\ f, & \text{for } m \geq 4, m \neq 9; \end{cases}$$

$$S_3 = \begin{cases} g, & \text{for } m = 3; \\ h, & \text{for } m = 5; \\ i, & \text{for } m \geq 4, m \neq 5, 9; \\ j, & \text{for } m = 9. \end{cases}$$

For $n = 5$ is the same as above. The representations of $r(u|\Pi)$ and $r(v|\Pi)$, with $u, v \in V(J_{m,n})$ are

$$\begin{array}{lll} r(u_1|\Pi) = (0,2,2), & r(u_2|\Pi) = (0,1,3), & r(u_3|\Pi) = (0,1,1), \\ r(u_n|\Pi) = (0,3,1), & r(v_1|\Pi) = (1,0,4), & r(v_2|\Pi) = (2,0,5), \\ \ldots, & r(v_{n(m-1)-1}|\Pi) = (0,3,5), & r(v_{n(m-1)}|\Pi) = (0,2,4). \end{array}$$

So the representation of $r(u|\Pi)$ and $r(v|\Pi)$ are distinct and we conclude that $\Pi$ is a resolving partition of $J_{m,n}$ with 3 elements. Furthermore, we prove that there is no resolving partition with 2 elements. Suppose that $J_{m,n}$ has a resolving partition with 2 elements, $\Pi = \{S_1, S_2\}$. According to Lemma 2.3, it is a contradiction. Therefore, $J_{m,n}$ has no a resolving partition with 2 elements. Thus, the partition dimension of $J_{m,n}$ is $pd(J_{m,n}) = 3$ for $n = 3, 4, 5$.

(ii) Case $n \geq 6$.
Let $\Pi = \{S_1, S_i\}$ be a resolving partition of $J_{m,n}$ where $S_1 = \{u_i, v_j\}$, for $1 \leq i \leq n+1$, $1 \leq j \leq n(m-1)$ and $S_i = \{v_k\}$, for $1 < i \leq \lfloor \frac{n}{2} \rfloor + 1, 1 \leq k \leq n(m-1), k \neq j$. Since each vertex of $J_{m,n}$ has a distinct representation with respect to $\Pi$, then $\Pi$ is a resolving partition of $J_{m,n}$ with $\lfloor \frac{n}{2} \rfloor + 1$ elements.
Next, we show that the cardinality of $\Pi$ is $\lfloor \frac{n}{2} \rfloor + 1$. The cardinality of $\Pi$ is the number of $k-$partition in $S_1$ and $S_i$ for $1 < i \leq \lfloor \frac{n}{2} \rfloor$. The number of $k-$partition in $S_1$ is 1. The number of $k-$partition in $S_i$ for $1 < i \leq \lfloor \frac{n}{2} \rfloor$ is $\lfloor \frac{n}{2} \rfloor$. We obtain $\mid \Pi \mid = 1 + \lfloor \frac{n}{2} \rfloor = \lfloor \frac{n}{2} \rfloor + 1$. Therefore, $\Pi = \{S_1, S_2, \ldots, S_{\lfloor \frac{n}{2} \rfloor + 1}\}$ is a resolving partition of $J_{m,n}$.
Now, we show that $J_{m,n}$ has no $pd(J_{m,n}) < \lfloor \frac{n}{2} \rfloor + 1$. We may assume that $\Pi$ is a resolving partition of $J_{m,n}$ where $pd(J_{m,n}) < \lfloor \frac{n}{2} \rfloor + 1$. So, each vertex $v \in V(J_{m,n})$ has a distinct representation $r(v_i|\Pi)$ and $r(v_j|\Pi)$. If we select $\Pi = \{S_1, S_2, S_3, \ldots, S_{\lfloor \frac{n}{2} \rfloor}\}$ as a resolving partition then there is a partition class containing any 2 vertices of $v_j$. By using Lemma 2.3, the distance of vertex $u_1$ to the others have the same distance, and $r(v_i|\Pi) = r(v_j|\Pi)$. It is a contradiction. Hence, the partition dimension of $J_{m,n}$ is $pd(J_{m,n}) = \lfloor \frac{n}{2} \rfloor + 1$ for $n \geq 6$.

## 3. Conclusion

According to the discussion above it can be concluded that the partition dimension of a lollipop graph $L_{m,n}$ and a generalized Jahangir graph $J_{m,n}$ are as stated in Theorem 2.1 and Theorem 2.2.

## References

[1] Chartrand, G., E. Salehi, and P. Zhang, *On the Partition Dimension of a Graph*, Congressus Numerantium **Vol. 131** (1998), 55-66.

[2] Chartrand, G., E. Salehi, and P. Zhang, *The Partition Dimension of a Graph*, Aequation Math. **Vol. 55** (2000), 45-54.

[3] Darmaji, *The Partition Dimension of Multipartite Graphs and Corona Product of Two Connected Graphs*, Disertasi Program Studi Doktor Matematika, ITB, Bandung, 2011.

[4] Mojdeh, D. A., and A. N. Ghameshlou, *Domination in Jahangir Graph* $J_{2,m}$, Int. J. Contemp. Math. Sciences **Vol. 2(24)** (2007), 193199.

[5] Tomescu, I., I. Javaid, and Slamin, *On the Partition Dimension and Connected Partition Dimension of Wheels*, Ars Combinatoria **Vol. 84** (2007), 311-317.

[6] Weisstein, E. W., *CRC Concise Encyclopedia of Mathematics CD-ROM*, $2^{nd}$ ed., CRC Press, Boca Raton, 2003.

# A Regular Production-Remanufacturing Inventory Model for a Two-Echelon System with Price-dependent Return Rate and Environmental Effects Investigation

**A R Dwicahyani[1], W A Jauhari[1], Jonrinaldi[2]**
[1] Department of Industrial Engineering
Sebelas Maret University, Surakarta, Indonesia
[2] Department of Industrial Engineering
Andalas University, Padang, Indonesia

E-mail: anindyard94@gmail.com

**Abstract.** Product take-back recovery has currently became a promising effort for companies in order to create a sustainable supply chain. In addition, some restrictions including government regulations, social-ethical responsibilities, and up to economic factors have contributed to the reasons for the importance of product take-back recovery. This study aims to develop an inventory model in a system of reverse logistic management consisting of a manufacturer and a collector. Recycle dealer collects used products from the market and ships it to manufacturer. Manufacturer then recovers the used products and sell it eventually to the market. Some recovered products that can not be recovered as good as new one will be sold to the secondary market. In this study, we investigate the effects of environmental factors including GHG emissions and energy usage from transportation, regular production, and remanufacturing operations conducted by manufacturer and solve the model to get the maximum annual joint total profit for both parties. The model also considers price-dependent return rate and determine it as a decision variable as well as number of shipments from collector to manufacturer and optimal cycle period. An iterative procedure is proposed to determine the optimal solutions. We present a numerical example to illustrate the application of the model and perform a sensitivity analysis to study the effects of the changes in environmental related costs on the model's decision.

**Keywords:** Inventory model, reverse logistics, collector, regular production, manufacturing remanufacturing, refurbishing, carbon emissions, price-dependent return rate

## 1. Introduction

Nowadays, environmental and sustainability issue have become a promising market-trend among practitioners and researchers, particularly in the sector of industrial and business [1]. In the last two decades, there had been rapidly growth various insights of production systems, logistics, supply chain, up to product design that carry out the concept of 'environmental friendly'. One idea that began to be applied by many global industry companies to maintain their sustainability is reverse logistics. According to [2] reverse logistics is defined as a set of activities that associated with the management of used items to be recovered. They underline that the term recovery itself can simply be either direct reselling a product to the market or enclosed by some particular processes including collection, inspection, cleaning, disassembly, and any other related processes that in the end will lead to remanufacturing or recycling. Thierry et al. [4] categorized five types of recovery processes in accordance with its typical characteristics that are, repair, recycle, remanufacture, refurbish, and cannibalization. A company that considers implementing the system of reverse logistics should differentiate the five processes well to be able to understand which one is the most suitable. Particularly, many companies seek to implement the reverse logistics operations as the related concept actually promises a numerous benefits such as compliance with environmental demands, helps in reducing costs, maintains customer loyalty, shows positive corporate images, improves customer service, increases market share, and provides higher achievement of sustainability goals [4], [5]. In reality, management of reverse logistics has been broadly implemented by various industry including electronic products and automotive parts, such as steel and glass, printers and photocopiers, tires, ships and aircraft engines, automobiles, air conditioners, televisions, and computer parts [6], [7].

In reverse logistics some typical industries usually do an alliance between firms that commonly named by joint ventures collaboration. Many operations including building recycling centres, collaborative transportation and logistics, collection of used item, and joint quality control are performed by the help of various firms collaborate in the joint venture [8]. According to [9], joint venture in reverse supply chain generally accommodates four ways of collaboration including waste disposal, product/part/material in sales, cost sharing, and profit distribution. Vlachos, I [10] emphasizes that "joint ventures can be a better mechanism to manage uncertainty in reverse logistics which can be higher than forward supply chains". Accordingly, many companies have already managed to do the effort as its support to help the system runs more efficiently.

Investigations of reverse logistics management inventory model have attracted the attention of many researchers and practitioners since the year of 1960s. Earlier, the studies about related topics only focused on the single-echelon system and do not accommodate any collaboration among firms in the supply chain ([11]; [12]; [13]; [14]; [15]; [16]; [17]; and [18]). At that time, many researchers presumed that the complexity of the system makes it difficult to develop a multi-echelon model [19]. Those previous studies also have not yet considered the uncertainties that actually happen in the system of reverse logistics, mainly on the return flow of used item that might be dependable on any other variables such as price, quality, interaction between demand and return rate, and so on [7].

Minner, S. [20] initiates the case of reverse logistics model in supply chain by extended his work on forward supply chain strategic safety stock placement to reversed one. The objective of the model was to minimize the investment cost related to the safety stock and determined the optimal service times at different locations. Mitra, S. [19] developed another model that examined the integration between parties among reverse supply chain. He developed two inventory models with returns consisted of a depot and a distributor under deterministic and stochastic demand and returns. The model considered set-up costs, holding costs, as well as shortage costs for the serviceable inventory stock and determined the optimal values of the policy variables that minimize the cost of the system. Afterwards, he extends the model by incorporating a correlation between demand and returns [21].

Chung et al. [22] developed a model considering an integration between four parties in reverse supply chain including supplier, manufacturing, retailer, and third party collector. They intended to obtain the optimal decisions of number of shipments between parties and cycle time that minimized the joint inventory costs of the system. The model was then extended by Yuan and Gao [23] to the more general multiple cycle production and remanufacturing policies with the same objectives. Other studies that analysed multi-echelon reverse logistics inventory system were also conducted by [24], [25], [26], [27], [28] and [29].

In other study, [30] developed a model that investigated a return rate that depends on price and quality factor. They stated that the amount of used item returned to the system were actually relied on purchasing price and quality factor. The higher the purchasing price set by manufacturer, the more amount of used item will be returned to the system because customer might be interested in selling their used products to be exchanged with fresh money. On the contrary, the higher the manufacturer set its acceptable quality level of used item to be collected, the smaller the amount of used item returned to the system. It was due to the quality of the product itself. Many researchers including [25], [27], [29] and [31] then further applied the idea of dependent return rate.

Environmental effects consideration is a fresh direction that surely relevant to the current trend of research. Transforming environmental emission and energy usage into cost components in the system of reverse logistics were done by [32]. In the study, they examined the effects of greenhouse gas (GHG) emission and energy usage on the optimal inventory decisions. They correspondingly found that both costs were proven to have significant influences on the optimal decisions. Therefore, the system should maintain its emission and energy usage hence the cost can be minimized.

In this study, we considers a joint reverse logistics system comprising of two parties, i.e. manufacturer and used item collector, which perform a cooperation in term of inventory. Collector helps the manufacturer to gather the end-of-life products in the market. The model aims to maximize joint total profit per year for both parties and determine the optimal number of shipment, cycle period, and used item purchasing price along with sizing the optimal shipping lot from collector to manufacturer hence the minimum cost and maximum profit can be achieved. We also analyse the environmental effects of green-house-gas emissions and energy usage by considering it as inventory cost components. Hence, the optimal decision will not only minimize the cost related to physical inventory but also environmental cost.

This paper is organized as follows. Section 1 explains about the background of this study and a brief literature review of inventory models that already developed in the related field. Development of the inventory model along with a solution procedure, numerical example and sensitivity analysis are presented in Section 2 and 3, respectively. At the end of the paper, we summarize the results of the study and give directions for future research.

## 2. Model development

The model developed in this study considers a reversed logistics system comprising of two parties depicted in Figure 1. In which, one party stands for used item collector, hereafter referred as collector, and the other is finished product manufacturer.

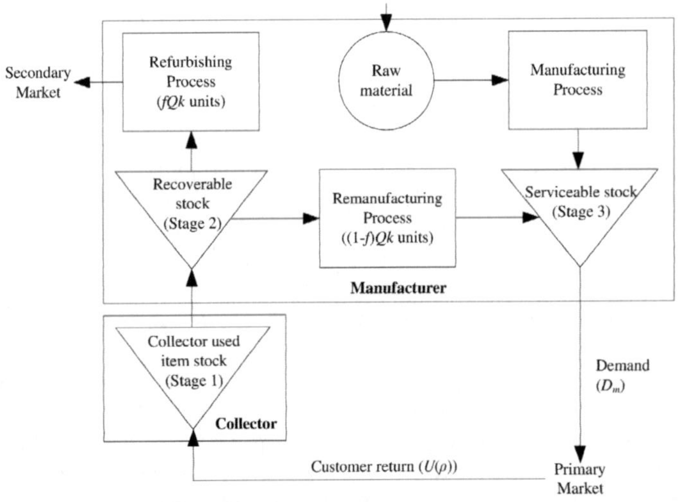


**Figure 1.** Inventory system of reverse logistics within two parties.

A collector is responsible for collecting end-of-life (EoL) items from the market. Collected used items will be delivered lot-by-lot to the manufacturer in $Q$ unit and $k$ deliveries and be hold respectively at a recoverable stock. The manufacturer inspects all of the recoverable items and sorts them subsequently to be remanufacturable, for items with a higher quality level, and otherwise, refurbishable. Refurbishable items that already experienced refurbishing process will be immediately sold to secondary market at a lower price. In order to fulfil demand at the primary market, manufacturer performs two production activities, i.e. remanufacturing and manufacturing (regular production). Afterwards finished-good items will be hold at serviceable stock. As defined by [3], remanufacturing is a process of reconditioning a used product to as-good-as-new state. Whereas manufacturing is a process to transform a completely raw material into finished-good products. Both processes are done sequentially in one period $T$.

We consider a return rate which value depends on purchasing/collection price from the market. El Saadany and Jaber [30] stated that return rate, $U$, is a portion of annual demand, $D_m$, that is returned to the system and formed as a function of price and quality, with $0<U(p, q)/D<1$. In this study, we consider that price factor ($f_p$) is more worthy to be examined instead of quality factor, since the focus of this model is nothing but to maximize the financial profit of both parties. According to El Saadany and Jaber (2010), the function of return rate that depends on price factor is formulated below.

$$f_p=(1-ae^{-\theta\rho}) \quad (1)$$

with, $0<a<1$ and $\theta>1$.

We begin the environmental effects investigation by calculating GHG emissions cost and energy usage cost that generated by some processes, including manufacturing, remanufacturing, as well as transportation related activities. The function of GHG emission cost and energy usage cost related to remanufacturing, manufacturing, and transportation activities were adopted from [32] which originally taken from [33].

The objective function of this model is to maximize the annual joint total profit ($AJTP$) for both parties and subsequently determine the following decision variables.

$k$ : number of delivery from the collector to the manufacturer (delivery/year);
$\rho$ : price factor, ratio of the used item purchasing price to raw material purchasing price (where $0\leq\rho\leq 1$);
$T$ : cycle period (year), consisting of one cycle of remanufacturing and one cycle of regular production in one $T$.

Whereas the parameters used in the model are denoted by these following notations:

$D_m$ : annual demand rate (units / year);
$U(\rho)$ : annual return rate of used items (units / year), $U(\rho) = D_m(1-ae^{-\theta\rho})$;
$A_1$ : set up cost for the collector ($ / batch);
$A_2$ : ordering cost for the manufacturer's recoverable items ($ / order);
$A_m$ : manufacturing set up cost ($ / batch);
$A_r$ : remanufacturing set up cost ($ / batch);
$H_1$ : annual holding cost for the collector ($ / unit / year);
$H_2$ : annual holding cost for the recoverable item stock ($ / unit / year);
$H_3$ : annual holding cost for the serviceable item stock ($ / unit / year);
$K_1$ : fixed cost to process the manufacturer's orders ($);
$P_u$ : used item purchasing price ($/unit), $P_u=\rho P_{rm}$ with $P_u\leq P_{rm};$
$P_c$ : collector selling price of the used items ($ / unit);
$P_{rm}$ : purchasing cost for raw material in the manufacturing activities ($ / unit);
$P_m$ : selling price of the finished product to the primary market ($);
$P_f$ : selling price of the refurbished product to the secondary market ($);
$C_{isc}$ : inspection and sorting cost per used item ($ / unit);
$c_p$ : manufacturing variable cost ($ / unit);
$c_r$ : remanufacturing variable cost ($ / unit);
$c_f$ : refurbishing cost ($ / unit);
$R$ : annual rate of remanufacturing, $R>D_m$ (units/year);
$P$ : annual rate of manufacturing, $P>D_m$ (units/year);
$f$ : proportion of refurbished item to used item ($0<f<1$);
$F_t$ : fixed cost per truck per trip ($ / trucks);
$t_c$ : truck capacity (unit / truck);
$g_t$ : number of gallons per truck per distance travelled (gallons/truck);
$e_t$ : amount of GHG emissions from one gallon of diesel-truck fuel (ton/gallon);
$C_{ec}$ : carbon emissions cost per ton of GHG emissions ($/ton);
$a_p$ : emissions function parameter for the manufacturing (ton year$^2$/unit$^3$);
$b_p$ : emissions function parameter for the manufacturing (ton year/unit$^2$);
$c_p$ : emissions function parameter for the manufacturing (year/unit);

$a_r$ : emissions function parameter for the remanufacturing (ton year$^2$/unit$^3$);
$b_r$ : emissions function parameter for the remanufacturing (ton year/unit$^2$);
$c_r$ : emissions function parameter for the remanufacturing (ton/unit);
$C_{en}$ : cost of energy (\$/KWh);
$c_0p$ : the required energy at the machine to manufacture one unit (KWh/unit);
$c_1p$ : the required energy per year when manufacturing is idle (KWh/year);
$c_0r$ : the required energy at the machine to remanufacture one unit (KWh/unit);
$c_1r$ : the required energy per year when remanufacturing is idle (KWh/year);
$TC_{ri}$ : annual total inventory cost of the recoverable item (\$);
$TC_{si}$ : annual total inventory cost of the serviceable item (\$);
$TC_M$ : annual total inventory cost for the manufacturer (\$);
$TC_c$ : annual total inventory cost for the collector (\$);
$TP_M$ : annual total profit for the manufacturer (\$);
$TP_C$ : annual total profit for the collector (\$);
$AJTP$ : annual joint total profit of the proposed system (\$).

The following assumptions used in the proposed model are listed below:

1. Finite production and remanufacturing rates;
2. Deterministic, constant, and known demand rate;
3. Remanufactured items have the quality as-good-as-new-one;
4. Customer can not distinguish between remanufactured and newly manufactured items;
5. Perfect production processes in the remanufacturing and manufacturing operations;
6. Lead time is zero;
7. Unlimited storage capacity is available;
8. Land transportation is used.

The collector and manufacturer stock levels are shown in Figure 2.

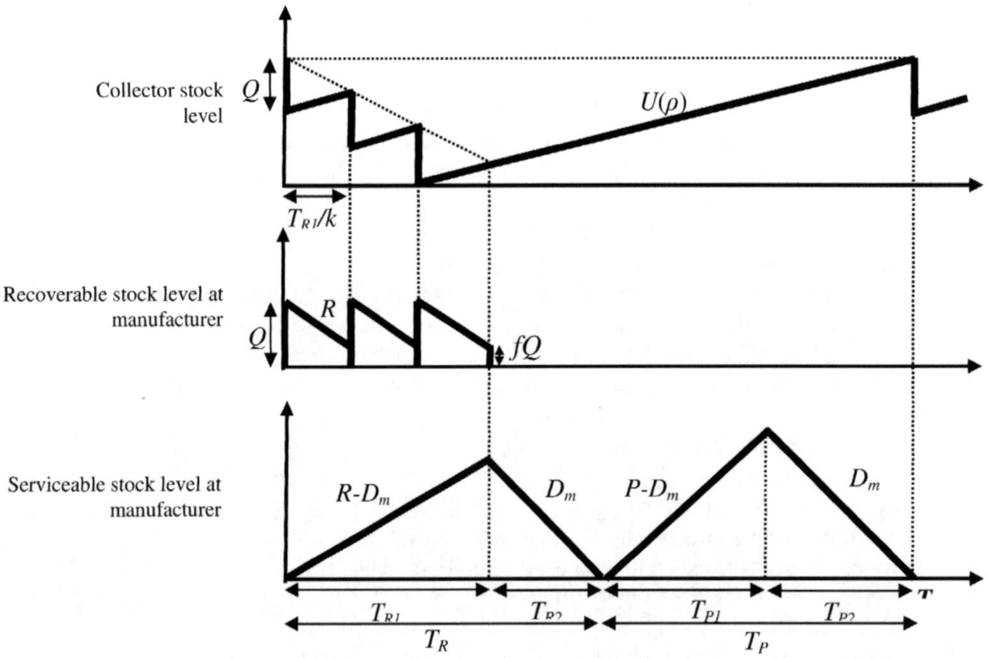


**Figure 2.** The collector stock level and the manufacturer stock level in one period $T$ ($k$=3).

*A. Manufacturer Inventory Model*

The manufacturer hold two types of stock, which are serviceable stock and recoverable stock. Costs related to the serviceable stock ($TC_{si}$) consists of holding cost ($HC_{si}$), production setup cost

IOP Conf. Series: Journal of Physics: Conf. Series **855** (2017) 012013 doi:10.1088/1742-6596/855/1/012013

($SC_{si}$), variable production cost ($CP_{si}$), variable remanufacturing cost ($CR_{si}$), raw material purchasing cost ($C_{rm}$), GHG emission cost ($CGHG_e$), and cost related to energy usage ($C_n$), which is expressed in Equation (2).

$$TC_{si} = HC_{si} + SC_{si} + CP_{si} + CR_{si} + CGHG_e + C_{rm} + C_n$$

$$= \frac{H_3\left(\frac{(1-f)^2(-D_m+R)T^2U^2}{2D_mR} + \frac{D_m(-D_m+P)T^2\left(1-\frac{(1-f)U}{D_m}\right)^2}{2P}\right)}{T} + \frac{A_m+A_r}{T} + C_pD_m\left(1-\frac{(1-f)U}{D_m}\right) + C_r(1-f)U$$

$$\frac{C_{ec}(1-f)(c_r-b_rR+a_rR^2)U}{D_m} + C_{ec}(c_p-b_pP+a_pP^2)\left(1-\frac{(1-f)U}{D_m}\right) + D_mP_{rm}\left(1-\frac{(1-f)U}{D_m}\right)$$

$$+ C_{en}\left(\frac{(1-f)(c_0r+\frac{c_1r}{R})U}{D_m} + \left(c_0p+\frac{c_1p}{P}\right)\left(1-\frac{(1-f)U}{D_m}\right)\right) \quad (2)$$

Costs related to the recoverable stock ($TC_{ri}$) consists of holding cost ($HC_{ri}$), ordering cost ($OC_{ri}$), purchasing cost ($PC_{ri}$), inspection cost ($IC_{ri}$), and refurbishing cost ($FC_{ri}$) as formulated in Equation (3) below.

$$TC_{ri} = HC_{ri} + OC_{ri} + PC_{ri} + IC_{ri} + FC_{ri} = \frac{A_2k}{T} + C_{isc}U + C_ffU + P_cU + \frac{(1-f)H_2U\left(\frac{TU}{k}+\frac{fTU}{k}\right)}{2R} \quad (3)$$

The manufacturer annual profit is obtained from subtracting the manufacturer inventory cost from the manufacturer revenue. Hence, we obtain the following equation.

$$TP_m = D_mP_m + fUP_f - TC_{si} - TC_{ri} \quad (4)$$

*B. Collector Inventory Cost*

The collector inventory cost ($TC_C$) consists of holding cost ($HC_C$), collection cost ($PC_C$), transportation cost ($TC_C$), setup cost ($SC_C$), and GHG emission cost of transportation activities ($CGHG_t$) as formulated in Equation (5) below.

$$TC_C = HC_C + PC_C + SC_C + TC_C + CGHG_t$$

$$= \frac{H_1TU(U-fU+k(R+(-1+f)U))}{2kR} + \rho P_{rm}U + \frac{A_1+kK_1}{T} + \frac{F_tU}{t_c} + \frac{C_{ec}e_tg_tU}{t_c} \quad (5)$$

The collector annual profit is obtained by subtracting the collector inventory cost from the collector revenue as shown by Equation (6) and subsequently, the annual joint profit for manufacturer-collector inventory system is given by Equation (7).

$$TP_C = P_cU - TC_C \quad (6)$$

$$AJTP = TP_m + TP_C \quad (7)$$

## 3. Solution procedure

In this study, we intend to maximize $AJTP(T, k, \rho)$ by determining the optimal value of $T$, $k$, and $\rho$. The optimal solutions of $T$ are found by taking the first partial derivative of $AJTP(T, k, \rho)$ with respect to $T$, then setting the result to zero. Hence, we obtain the optimal values of $T$, denoted by $T^*$, as follows.

$$\frac{dAJTP(T,k,\rho)}{dT} = \frac{A_m+A_r}{T^2} + \frac{A_2k}{T^2} + \frac{A_1+kK_1}{T^2} - \frac{(1-f)H_2U\left(\frac{U}{k}+\frac{fU}{k}\right)}{2R} - \frac{H_3\left(\frac{(1-f)^2(-D_m+R)TU^2}{D_mR} + \frac{D_m(-D_m+P)T\left(1-\frac{(1-f)U}{D_m}\right)^2}{P}\right)}{T} +$$

$$\frac{H_3\left(\frac{(1-f)^2(-D_m+R)T^2U^2}{2D_mR} + \frac{D_m(-D_m+P)T^2\left(1-\frac{(1-f)U}{D_m}\right)^2}{2P}\right)}{T^2} - \frac{H_1U(U-fU+k(R+(-1+f)U))}{2kR} \quad (8)$$

$$T^*(k,\rho)=\frac{\sqrt{2}\sqrt{-D_m k(A_m+A_r+A_2k+A_1+kK_1)PR}}{\sqrt{\begin{array}{c}D_m{}^3H_3kR-2(-1+f)^2H_3kPRU^2-D_m{}^2H_3kR(P-2(-1+f)U)+D_mU(-H_1P(U-fU+k(R+(-1+f)U))\\+(-1+f)((1+f)H_2PU+H_3k((-1+f)RU-P(2R+U-fU))))\end{array}}} \tag{9}$$

The optimal value of $k$ denoted by $k^*$ could be obtained by letting the derivative of $AJTP(T, k, \rho)$ with respect to $k$ equals to zero and then substituting the value of $T^*$ in equation (9) to $k$. Thus, we obtain the following function

$$k^*(\rho)=\frac{\sqrt{(-A_1-A_m-A_r)D_m(-1+f)(H_1+H_2+fH_2)PU^2}}{\sqrt{\begin{array}{c}(-A_2-K_1)(D_m{}^3H_3R-2(-1+f)^2H_3PRU^2-D_m{}^2H_3R(P-2(-1+f)U)+D_mU(-H_1P(R+(-1+f)U)\\+(-1+f)H_3((-1+f)RU-P(2R+U-fU))))\end{array}}} \tag{10}$$

The optimal values of $T$, $k$, and $\rho$ that minimize $AJTP(T, k, \rho)$ for the value of $0\leq\rho\leq 1$ is derived by the following solution procedure.

**Step 1** For each value of $\rho$ for $0<\rho\leq 1$, calculate the value of $T$ and $k$ using equation (9) and (10), respectively. If $k$ is not an integer, then set $k = \lceil k \rceil$ and $k = \lfloor k \rfloor$ then continue to step 2;

**Step 2** Calculate the value of $AJTP(T, k, \rho)$ for $k=\lceil k \rceil$ and $k=\lfloor k \rfloor$;

**Step 3** Find the combination value of $T$, $k$, and $\rho$ that gives the maximum value of $AJTP(T, k, \rho)$, and set as $T^*$, $k^*$, and $\rho^*$.

After obtaining the optimal value of $T$, $k$, and $\rho$, then the optimal delivery batch sizes from the collector to the manufacturer could be determined by the following expression

$$Q=\frac{D_mT_R}{k(1-f)} \tag{11}$$

since $T_R=\frac{(1-f)TU}{D_m}$ and $U=D_m(1-\delta e^{-\theta\rho})$ hence the value of $Q^*$ is generated as

$$Q^*=\frac{T^*D_m(1-\delta e^{-\theta\rho^*})}{k^*} \tag{12}$$

## 4. Numerical example and sensitivity analysis

To illustrate the application of the proposed model, let us consider the inventory system with the following data adopted from [32] and [30]: $D_m$=200 units/year, $P_{rm}$=\$20, $R$=250 units/year, $P$=220 units/year, $H_2$=\$2.5/unit/year, $A_2$=\$100/batch, $C_{isc}$=\$1/unit, $C_f$ =\$5/unit, $H_3$=\$3.5/unit/year, $A_m$=\$150/batch, $A_r$=\$100/batch, $C_p$=\$10/unit, $C_r$=\$5/unit, $P_m$=\$80/unit, $P_f$=\$25/unit, $f$=0.4, $\delta$=1.5, $\theta$=2, $F_t$=100 units, $H_1$=\$3/unit/year, $A_1$=\$50/batch, $K_1$=\$50, $g_t$=375 gallons/truck, $e_t$=0.1008414 tons/gallon, $C_{ec}$=\$10/ton, $t_c$=350 units/truck, $a_p$=0.0000003, $b_p$=0.0012, $c_p$=1.4, $a_r$=0.000000833, $b_r$=0.002, $c_r$=1.4, $c_0p$=25, $c_1p$=500,000, $c_0r$=10, $c_1r$=300,000, and $c_{en}$=\$0.05/KWh. By using the above solution procedure, we generate the results that maximize $AJTP(k, T, \rho)$. The optimization results are shown in Table 1.

**Table 1.** Optimization results based on given numerical data.

| $\rho$ | $U(\rho)$ | $k^*$ | $T^*$ | $T_R$ | $T_P$ | $Q$ | $TP_m$ | $TP_D$ | $AJTP$ |
|---|---|---|---|---|---|---|---|---|---|
| 0.45 | 78 | 1 | 603 | 141 | 462 | 78 | \$8,581.48 | \$1,667.69 | \$10,249.17 |
| 0.46 | 80 | 1 | 595 | 143 | 452 | 80 | \$8,552.10 | \$1,697.73 | \$10,249.80 |
| **0.47*** | **83*** | **1*** | **584*** | **145*** | **439*** | **83*** | **\$8,507.97*** | **\$1,749.68*** | **\$10,257.65*** |

| | | | | | | | | | |
|---|---|---|---|---|---|---|---|---|---|
| 0.48 | 85 | 1 | 577 | 147 | 434 | 85 | \$8,478.50 | \$1,778.06 | \$10,256.56 |
| 0.49 | 87 | 1 | 570 | 149 | 421 | 87 | \$8,449.00 | \$1,805.68 | \$10,254.68 |

As shown in Table 1, it is clear that *AJTP* would be maximized at \$10,257.65 for $\rho^*$=0.47 ($P_c$=$\boldsymbol{\rho}$ $P_{rm}$=0.47 x \$20=\$9.4), $k^*$=1 delivery, $T^*$=584 days, and $Q^*$=83 units. The results suggest that the system would be equally profitable for both parties as long as decision variables *k*, *T*, and $\rho$ are determined at its optimal levels. The manufacturer should perform remanufacturing phase for 145 days, and change to manufacturing operation immediately after remanufacturing phase was finished, for the period of 439 days.

As the value of $\rho$ increases from its initial value, 0.01, *AJTP* is keep increasing and finally reaches the maximum level at 0.47 before it is gradually falling down and hit the rock bottom at $\rho$=1. The results prove that $\rho$ gives concave effects on *AJTP*. Graphical plot of the effects of $\rho$ on *AJTC* is represented in Figure 3.

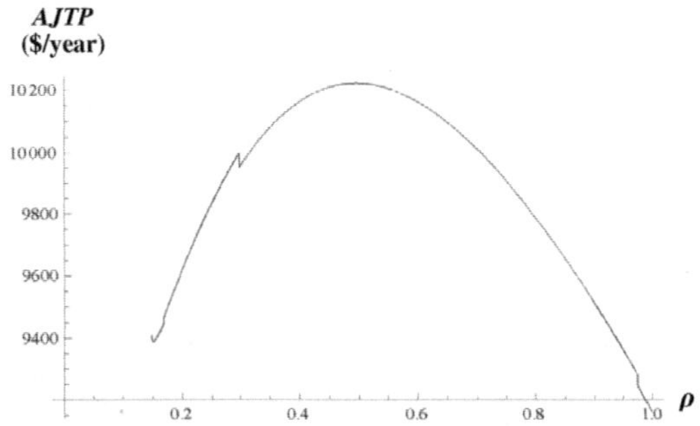


**Figure 3.** Graphical plot of $\rho$ on *AJTP*.

Here, we also conduct a simple sensitivity analysis to study the environmental effects on *AJTP*. The analysis is done by changing the value of $C_{ec}$, denotes the cost related to GHG emission in term of \$/ton, and $C_{en}$, denotes the cost related to energy usage in term of \$/KWh. The results are shown in Table 2 and 3 below.

**Table 2.** Effects of $C_{ec}$ on model's decision.

| $C_{ec}$ | $\rho$ | $U(\rho)$ | $k^*$ | $T^*$ | $T_R$ | $T_P$ | $Q$ | $TP_m$ | $TP_C$ | *AJTP* |
|---|---|---|---|---|---|---|---|---|---|---|
| \$10 | 0.47 | 83 | 1 | 584 | 145 | 439 | 83 | \$8,507.97 | \$1,749.68 | \$10,257.65 |
| \$20 | 0.47 | 83 | 1 | 584 | 145 | 439 | 83 | \$8,496.96 | \$1,660.01 | \$10,156.97 |
| \$30 | 0.44 | 76 | 1 | 610 | 139 | 471 | 76 | \$8,588.72 | \$1,472.69 | \$10,061.40 |
| \$40 | 0.44 | 76 | 1 | 610 | 139 | 471 | 76 | \$8,577.66 | \$1,390.57 | \$9,968.24 |
| \$50 | 0.41 | 68 | 1 | 644 | 131 | 513 | 68 | \$8,683.38 | \$1,198.95 | \$9,882.33 |

The results in Table 2 and 3 indicate that $C_{ec}$ and $C_{en}$ affect almost all of the model's decisions, including $\rho^*$, $T^*$, and $Q^*$, except for $k^*$. The higher the cost related to the environment that should be paid by both parties, the lower the profit that could be achieved by the system. The parameter $C_{ec}$ and $C_{en}$ directly affect the cost related to GHG emission and cost related to energy usage. Unpredictably, the changes in those parameters actually give significant impact on $\rho^*$ and $T^*$ (the changes in *Q* are not considered since the value of *Q* is dependent on $U(\rho)$). Increased value of $C_{ec}$ causes the optimal cycle period, $T^*$ to be longer.

**Table 3.** Effects of $C_{en}$ on model's decision.

| $C_{en}$ | $\rho$ | $U(\rho)$ | $k^*$ | $T^*$ | $T_R$ | $T_P$ | $Q$ | $TP_m$ | $TP_C$ | *AJTP* |
|---|---|---|---|---|---|---|---|---|---|---|
| \$0.05 | 0.47 | 83 | 1 | 584 | 145 | 439 | 83 | \$8,507.97 | \$1,749.68 | \$10,257.65 |
| \$0.15 | 0.47 | 83 | 1 | 584 | 145 | 439 | 83 | \$8,305.28 | \$1,749.68 | \$10,054.97 |

| | | | | | | | | | | |
|---|---|---|---|---|---|---|---|---|---|---|
| \$0.25 | 0.5 | 90 | 1 | 560 | 151 | 409 | 90 | \$8,003.89 | \$1,854.71 | \$9,858.60 |
| \$0.35 | 0.5 | 90 | 1 | 560 | 151 | 409 | 90 | \$7,803.49 | \$1,854.71 | \$9,658.20 |
| \$0.45 | 0.5 | 90 | 1 | 560 | 151 | 409 | 90 | \$7,603.08 | \$1,854.71 | \$9,457.79 |

Whereas, increasing $C_{ec}$ lead to a lower optimal purchasing price of used item ($\rho^*$). On the contrary, increased value of $C_{en}$ cause $T^*$ to be shorter and lead to a higher $\rho^*$. From those results, we can conclude that $C_{ec}$ and $C_{en}$ give different effects on model's decision. In a situation where the uncontrollable costs are increasing, i.e. $C_{en}$ and $C_{ec}$, the system would maintain the maximum level of profit by controlling the optimal value of decision variables, such as $\rho$ and $T$ for this case.

The results and analysis of the proposed model gives some managerial insights to the relevant manufacturing system through its optimal decisions for both manufacturer and recycler dealer so that managers can keep inventories within the supply chain more efficiently at a lower cost. The inventory model developed in this study has considered a variable return rate that is affected by the purchasing price of used items. Accordingly, we may say that the model is suitable to be applied in manufacturing cases which the offered purchasing price influences the behaviour of customer to resell their used item. It is relevant to the real system since customer is more willingly to sell the used product to manufacturer when the bargaining position is higher. It enables the system to control the amount of returned item by determining the correct used item purchasing price.

## 5. Conclusion

This study proposes a collector-manufacturer reverse logistic inventory model with price-dependent return rate and environmental effects investigations. Compared with previous studies, this model has considered the environmental impact of production and the remanufacturing process in the form of emission costs and energy usage costs. In addition, it also provides an optimal solution for the parties involved in the operation of joint ventures. Another consideration is the dependent return rate that rely on the purchasing price of used items. In the proposed model, return rate is assumed to be a demand-like function of used item purchasing price. The optimal cycle period, number of shipments, and used item purchasing price are optimized to achieve maximum annual profit for the system. The optimization of the model is done analytically therefore we guaranteed global optimum decisions.

From the provided numerical example, the results point out that to achieve maximum profit of the system, manufacturer should perform shorter period of remanufacturing and longer period of manufacturing. Price factor is numerically proven to give convex effects on *AJTC*, therefore an optimum value of $\rho$ should be generated. The optimal number of shipments is one shipments per year with delivery lot sizes equal to its return rate per year. This means that it is more beneficial for the system to hold more stock rather than to order/deliver used item too frequent. The results of environmental effects investigation show that costs parameters related to GHG emissions per ton and energy usage per KWh give significant impact on $\rho^*$ and $T^*$ respectively. Hence to maintain its profits, the system should reduce its costs related to GHG emissions and energy usage.

The extension of the proposed study is to consider real stochastic demand and return rate. Accordingly, decisions related to stochastic condition including safety stock and reorder point should be determined. A second extension is to consider more parties in the supply chain, such as supplier and retailer, hence the model will become more comprehensive to be applied in real supply chain. Another direction is to include imperfect manufacturing and remanufacturing processes where defective items are either be reworked, resold, or scrapped.

## References

[1] Schaltegger, S, Lüdeke-Freund F, and Hansen, E G 2012 Business Cases for Sustainability and the Role of Business Model Innovation Developing a Conceptual Framework *Int. J. Innov. Sust. Dev.* **6** 2

[2] De Brito, M P and Dekker, R 2004 *A framework for Reverse Logistics* Reverse Logistics Berlin, Germany: Springer-Verlag

[3] Thierry M C, Salomon M, Van Numen J, and Van Wassenhove L 1995 Strategic issues in product recovery management *California. Man. Rev.* **37**

[4] Chaves, G d L D. and Martins, R S 2005 Diagnóstico da logística reversa na cadeia de suprimentos de alimentos processados no oeste paranaense In *VIII Simpósio de Administração da Produção, Logística e Operações Internacionais (SIMPOI)*

[5] Andrade R, Vieira J M, Lucato W, and Vanalle R 2014 Review of the relationship between reverse logistics and competitiveness *Adv. Mat. Res.* **845**

[6] Xia W –H, Jia D –Y, and He Y –Y 2011 The remanufacturing reverse logistics management based on closed-loop supply chain management processes *Proc. Env. Sci.* **11**

[7] Akçalı, E and Çetinkaya, S 2011 Quantitative models for inventory and production planning in closed-loop supply chains *Int. J. Prod. Res.* **49** 8

[8] Kasper, A C, Berselli, G B T, Freitas B D, Tenório J A S, Bernardes A M, and Veit H M 2011 Printed wiring boards for mobile phones: Characterization and recycling of copper *Waste Man.* **31** 12

[9] Nnorom, I C, Ohakwe J, and Osibanjo O 2009 Survey of willingness of residents to participate in electronic waste recycling in Nigeria – A case study of mobile phone recycling *J. Cleaner Prod.* **17** 18

[10] Vlachos, I 2016 Reverse logistics capabilities and firm performance: the mediating role of business strategy *Int. J. Log.* **19** 5

[11] Shcrady D A 1967 A deterministic inventory model for repairable items *Naval. Res. Log. Quarterly.* **14** 3

[12] Nahmias N and Rivera H 1979 A deterministic model for repairable item inventory system with a finite repair rate *Int. J. Prod. Res.* **17** 3

[13] Mabini M C, Pintelon L M, and Gelders L F 1992 EOQ type formulation for controlling repairable inventories *Int. J. Prod. Eco.* **28** 1

[14] Richter, K 1996a The EOQ repair and waste disposal model with variable setup numbers *Eur. J. Oper. Res.* **95** 2

[15] Richter, K 1996b The extended EOQ repair and waste disposal model *Int. J. Prod. Eco.* **45** 1-3

[16] Fleischmann, M, Bloemhof-Ruwaard, J M, Dekker, R, van der Laan, E, van Nunen, J A E E, and Van Wassenhove, L N 1997 Quantitative models for reverse logistics: A review *Eur. J. Oper. Res.* **103**

[17] Koh S G, Hwang H, Shon K I, and Ko C S 2002 An optimal ordering and recovery policy for reusable items *J. Comp. Ind. Eng.* **43** 1–2

[18] Teunter, R H 2004 Lot-sizing for inventory systems with product recovery *Comp. Ind. Eng.* **46** 3

[19] Mitra, S 2009 Analysis of a Two-chelon Inventory System with Returns *Int. J. Man. Sci.* **37**

[20] Minner, S 2001 Strategic safety stocks in reverse logistics supply chains *Int. J. Prod. Eco.* **71** 1-3

[21] Mitra, S 2012 Inventory Management in a Two-echelon Closed-Loop Supply Chain with Correlated Demands and Returns *Int. J. Comp. Ind. Eng.* **62**

[22] Chung S -L, Wee H -M, and Yang P -C 2008 Optimal policy for a closed-loop supply chain inventory system with remanufacturing *Math. Comp. Modelling.* **48**

[23] Yuan K F and Gao Y 2010 Inventory decision-making models for a closed-loop supply chain system *Int. J. Prod. Res.* **48** 20

[24] Singh, S R and Saxena, N 2013 A Closed Loop Supply Chain System with Flexible Manufacturing and Reverse Logistics Operation under Shortages for Deteriorating Items *Procedia. Tech.* **10**

[25] Jaber, M Y, Zanoni, S, and Zavanella, L E 2014 A consignment stock coordination scheme for the production, remanufacturing, and waste disposal problem *Int. J. Prod. Res.* **52** 1

[26] Gu, Q and Tagaras, G 2014 Optimal collection and remanufacturing decisions in reverse supply chains with collector's imperfect sorting problem *Int. J. Prod. Res.* **52** 17

[27] Giri, B C and Sharma, S 2015 Optimizing a Closed-Loop Supply Chain with Manufacturing Effects and Quality Dependant Return Rate *J. Man. Syst.* **35**

[28] Bazan, E, Jaber, M Y, and Zanoni, S (in press) Carbon emissions and energy effects ona two-level manufacturer-retailer closed-loop supply chain model with remanufacturing subject to different coordination mechanisms *Int. J. Prod. Eco.*

[29] Jauhari, W A, Dwicahyani, A R, and Kurdhi, N A (in press) Lot sizing decisions in a closed-loop supply chain system with remanufacturing *Int. J. Proc. Man.*

[30] El Saadany A M A and Jaber M Y 2010 A production/remanufacturing inventory model with price and quality dependant return rate *Comp. Ind. Eng.* **58**

[31] Dwicahyani, A R, Jauhari W A, and Kurdhi N A 2016 Inventory decision in a closed-loop supply chain with inspection, sorting, and waste disposal *IOP Conf. Ser.: Mat. Sci. Eng.* **114** 1

[32] Bazan, E, Jaber, M Y, and El Saadany, A M A 2015a Carbon emissions and energy effects on manufacturing-remanufacturing inventory model *Comp. Ind. Eng.* **88**

[33] Bogaschewsky, R 1995 *Natürliche Umwelt und Produktion* Wiesbaden: GablerVerlag

# Bound of Distance Domination Number of Graph and Edge Comb Product Graph

**Gembong A.W.** [1,2], **Slamin** [1,3], **Dafik**[1,4], **Ika Hesti Agustin** [1,2]

[1] CGANT Research Group, University of Jember, Indonesia
[2] Mathematics Department, University of Jember, Indonesia
[3] Information System Study Program, University of Jember, Indonesia
[4] Mathematics Education Study Program, University of Jember, Indonesia

E-mail: gembonganggerr@gmail.com, slamin@unej.ac.id, d.dafik@unej.ac.id,ikahesti.fmipa@unej.ac.id

**Abstract.** Let $G = (V, E)$ be a simple, nontrivial, finite, connected and undirected graph. For an integer $1 \leq k \leq diam(G)$, a distance $k$-dominating set of a connected graph $G$ is a set $S$ of vertices of $G$ such that every vertex of $V(G)\backslash S$ is at distance at most $k$ from some vertex of $S$. The $k$-domination number, denoted by $\gamma_k(G)$, of $G$ is the minimum cardinality of a $k$-dominating set of $G$. In this paper, we improve the lower bound on the distance domination number of $G$ regarding to the diameter and minimum degree as well as the upper bound regarding to the order and minimum $k$ distance neighbourhood. In addition, we determine the bound of distance domination number of edge comb product graph.

**Keywords:** distance domination, diameter, edge comb product graph.

## 1. Introduction

Let $G = (V(G), E(G))$ be a simple, finite, connected and undirected graph. Graph terminology related with $k$-dominating number are distance, diameter, neighbourhood of vertex, closed neighbourhood of vertex, $k$-distance neighbourhood of vertex, and $k$-distance closed neighbourhood of vertex. The distance $d(u, v)$ from a vertex $u$ to a vertex $v$ in a connected graph $G$, or simply $d(u, v)$, is the minimum of the lengths of the $u - v$ paths in $G$, for detail definition see [4]. A neighbour of a vertex $v$ in $G$ is a vertex adjacent to $v$. The open neighbourhood of $v$, denoted by $N(v)$, is the set of all neighbours of $v$ in $G$, while the closed neighbourhood of $v$ is the set $N[v] = N(v) \cup \{v\}$. The closed $k$-neighbourhood of $v$, denoted by $N_k[v]$, is defined as the set of all vertices within distance $k$ from $v$ in G; that is, $N_k[v] = \{u|d(u, v) \leq k\}$ [13]. When $k = 1$, $N_k[v] = N[v]$. The degree of a vertex $v$ in $G$, denoted by $d(v)$, is the number of neighbours of $v$ in $G$, that is $|N(v)|$. The minimum and maximum degree among all the vertices of $G$ are denoted by $\delta(G)$ and $\Delta(G)$, respectively.

Domination in graphs is well-known research interest. The study has been done by several authors such as [2], [6], [8], [10], [11], [12], [14],[15], [16]. A dominating set in a graph $G$ is a set $S$ of vertices of $G$ such that every vertex in $V(G) \setminus S$ is adjacent to at least one vertex in S. The domination number of $G$, denoted by $\gamma(G)$, is the minimum cardinality of a dominating set of $G$. For more details about the notion of domination in graphs, see [4] and [5].

In 1975, Meir and Moon [3] introduced the concept of a distance $k$-dominating set (or or $k$-covering) in a graph. A set $S$ is a $k$-dominating set of $G$ if every vertex is within distance $k$ from some vertex of S; that is, for every vertex $v$ of $G$, we have $d(v, S) \leq k$. The $k$-domination number of $G$, denoted by $\gamma_k(G)$, is the minimum cardinality of a $k$-dominating set of $G$. When $k = 1$, the 1-domination number of $G$ is precisely the domination number of $G$; that is, $\gamma_1(G) = \gamma(G)$. The literature on the subject of distance domination in graphs up to the year 1997 can be found in the book chapter [19]. Distance domination is now widely studied, for example [7], [9], [13], [17], [20], and [21].

The $k$-domination number of G is in the class of $NP$-hard problem see [18]. Because of its computational complexity, graph theorists have studied upper and lower bounds on $\gamma_k(G)$ in terms of simple graph parameters like order, size, and degree. Previous results on the bound of $\gamma_k(G)$ is credited to Meir and Moon, Tian and Xu, Henning and Lichiardopol. In 1975, Meir and Moon [3] established an upper bound for the k-domination number of a tree in terms of its order. They proved that for $k \geq 1$, if $T$ is a tree of order $n \geq k+1$, then $\gamma_k(T) \leq \frac{n}{k+1}$. Since $\gamma_k(G) \leq \gamma_k(T)$, then $\gamma_k(G) \leq \frac{n}{k+1}$ see [3]. Tian and Xu [7] improved the Meir-Moon upper bound and showed that for $k \geq 1$, if $G$ is a connected graph of order $n \geq k+1$ with maximum degree $\Delta$, then $\gamma_k(G) \leq \frac{1}{k(n-\Delta+k-1)}$. The Tian-Xu bound was further improved by Henning and Lichiardopol [9] by showing that for $k \geq 2$, if $G$ is a connected graph with minimum degree $\delta \geq 2$ and maximum degree $\Delta$ and of order $n \geq \Delta + k - 1$, then $\gamma_k(G) \leq \frac{n+\delta-\Delta}{\delta+k-1}$. For lower bound of $\gamma_k(G)$, Randy proved that $\gamma_k(G) \geq \frac{d+1}{2k+1}$ for $d$ is diameter of $G$. He also investigated $\gamma_k(G)$ for direct product of graph, where $\gamma_k(G \times H) \leq \gamma_k(G) + \gamma_k(H) - 1$ [13].

In this paper we improve the lower bound on distance domination number of $G$ regarding to the diameter and minimum degree as well as the upper bound regarding to the order and minimum $k$ distance neighborhood. In addition, we determine the bound of distance domination number of graph resulting from operation, namely, edge comb product graph. Edge comb product graph is a varian of comb product graph. For definition of comb product graph operation see [1].

## 2. Results

Prior to show the main result, we first show the upper bound of simple connected graph. We start with the following theorem in [17] explains a propperty of $k$ distance dominating set.

**Theorem 1.** *[17] Let $k \geq 1$ and $S$ be a distance $k$-dominating set of a graph $G$. Then $S$ is a minimal distance $k$-dominating set of G if and only if each $d \in S$ has at least one of the following two properties hold.*

*(i) There exist a vertex $v \in V(G) - S$ such that $N_k(v) \cap S = \{d\}$.*

*(ii) The vertex $d$ is at distance at least $k+1$ from every other vertex $d$ of $S$ in $G$.*

We use the minimum cardinality of $k$ distance neighbourhood of a vertex on $G$, $|X|$, as parameter of our upper bound. As example for $k = 1$, it is easy to obtain that, $|X| = 1 + \delta(G)$. But, for $k \geq 2$ we obtain $|X|$ by evaluating $k$ distance neighbourhood of each vertex in $G$.

**Lemma 1.** *Let $S$ be distance $k$-dominating set and $C = N_k[u] \cap N_k[v], \forall u, v \in S$, then $|C| < n - |S|$.*

*Proof.* Since $S$ is distance $k$ dominating set of $G$ then by Theorem 1, the following two cases is hold.

(i) For all $u \in S$, $u$ have special neighbour at distance at most $k$ that are not dominated by another vertex in $S$. The maximum order of $C$ is $n = |V(G)|$. Since every two vertices in $G$ has its own special neigbour at most $k$ distance and each vertex in $G$ has at least one special neigbour. Then, there exist at least $|S|$ element that not belongs to $C$. So, $|C| < n - |S|$.

(ii) Every two vertices $u, v \in S$, $u \notin N_k[v]$. In other word, any two vertices in $S$ are not adjacent one to another at distance at most $k$. This mean, all vertices in $S$ are not belongs to $C$. So, $|C| < n - |S|$.

□

**Theorem 2.** *Let $|X|$ be minimum cardinality of $k$ distance neighbourhood of a vertex on $G$, the following holds*

$$\gamma_k(G) < \frac{2n}{|X|+1}.$$

*Proof.* Let $S$ be the minimum of distance $k$ dominating set, $|S| = \gamma_k(G)$, and $|V(G)| = n$. We proof the theorem by using the following counting argument.

$$\begin{aligned} \Sigma_{v\in S}(|N_k[v]| - |X|) \geq 0 \\ \Leftrightarrow \Sigma_{v\in S}|N_k[v]| - \Sigma_{v\in S}|X| \geq 0 \\ \Leftrightarrow \Sigma_{v\in S}|N_k[v]| - |S||X| \geq 0 \\ \Leftrightarrow |S||X| \leq \Sigma_{v\in S}|N_k[v]|. \end{aligned}$$

Since, $\Sigma_{v\in S}|N_k[v]| = n + |C|$, where $C = N_k[u] \cap N_k[v], \forall u, v \in S$. We can obtain

$$|S||X| \leq n + |C|. \tag{1}$$

By Lemma 1 we get

$$|C| < n - |S|. \tag{2}$$

By combining Equation 1 and 2 we can get $\gamma_k(G) < \frac{2n}{|X|+1}$. □

This upper bound is tight for some cases, for example, if we take $G$ be regular graph, such as prism, with order $n, \delta = \Delta = 3$ and $k = 2$. By using Henning and Lichiardopol bound we can get $\gamma_k(G) \leq n/4$. By using our bound we get $\gamma_k(G) < 2n/9$ where $|X| = 8$ for $n \geq 5$.

The trivial lower bound for distance domination number of regular graph is $\gamma_k(G) \geq \frac{(d-2)n}{d(d-1)^k-2}$. The lower bound is derived by considering the maximum number of vertices a vertex may dominated. For graph is any simple connected graph with maximum degree $\Delta$, $\gamma_k(G) \geq \frac{(\Delta-2)n}{\Delta(\Delta-1)^k-2}$. We derive distance $k$-domination number of $G$ based on its minimum degree and diameter. This result has similar strategy as in proof of R. Davila *et al.* [13] result, but we add minimum degree parameter as improvement.

**Theorem 3.** *If $G$ is simple connected graph with minimum degree, $\delta \geq 2$, and diameter, $diam(G) = d$, $k \leq d$. Then*

$$\gamma_k(G) \geq \frac{\lceil \frac{d-1}{3} \rceil (\delta - 2) + d + 1}{\lceil \frac{2k-1}{3} \rceil (\delta - 2) + (2k+1)}.$$

*Proof.* Suppose that $P$ is diameter path of $G$ and let $P = \{u_i | i = 1, 2, \ldots, d, d+1\}$. Since $\delta$ is minimum degree of $G$, then $\forall u_i \in P, i = 2, 3, \ldots, d, |N(u_i)| \geq \delta$. Supposed that $P'$ is subgraph of $G$ that is obtained by $P \cup \{w_j | w_j \in N(u_i), u_i \in V(P) \backslash \{u_1, u_{d+1}\}, 1 \leq j \leq \delta - 2\}$. Note that it is not necessary for any two vertices $v_i, v_j \in V(P), N(v_i) \cap N(v_j) = \emptyset$ for $i, j \in \{2, 3, \ldots, d\}$. The maximum number of vertices in $V(P)$ that share its neighbourhood is three. In contrary suppose that there are four vertices in $V(P)$ that share at least one of its neighbourhood. without loss of generality, suppose that for some $w \in V(G), w \in N(v_2) \cap N(v_3) \cap N(v_4) \cap N(v_5)$. Then there is a path $v_1 v_2 w v_5 v_6 \ldots v_d v_{d+1}$ with lenght is equal to $d - 1$. This situation contradiction with our asumption that $d(v_1, d_{d+1}) = d$ where $d$ is diameter. If we include the minimum neighbourhood of every vertices in $P$ a part of $v_1$ and $v_{d+1}$, we can obtain a subgraph $P'$ with $V(P') = V(P) \cup N(v_i), v_i \in V(P), i = 2, 3, \dots d$. Since three vertices of $\{v_2, v_3, \dots, v_d\}$ can share the same neighbour, then the minimum cardinality of $V(P')$ is

$$|V(P')| = \left\lceil \frac{d-1}{3} \right\rceil (\delta - 2) + d + 1.$$

The maximum value of $N_k[s], s \in S$, is obtained by putting $s$ in the middle of diameter path $P$. The maximum number of vertices that can be dominated by $s$ in $P$ is $2k+1$. Suppose that $N_k[s, V(P)] = \{v_1, v_2, \dots v_{2k+1}\}$ is closed $k$ distance neighbourhood of $s$ on $V(P)$, then the maximum number of vertices that can be dominated by $s$ in $P'$ are

$$|N_k[s]| \leq |N_k[s, V(P)]| + |\cup_{i=2}^{2k} N(u), u \in N_k[s, V(P)]|.$$

Since at most three vertices in $N_k[s, V(P)]$ can share the same neoghbour, then $|\cup_{i=2}^{2k} N(u), u \in N_k[s, V(P)]| = \lceil \frac{2k-1}{3} \rceil (\delta - 2)$. Therefore,

$$|N_k[s]| \leq (2k+1) + \left\lceil \frac{2k-1}{3} \right\rceil (\delta - 2).$$

Let $S$ be a minimum $k$-dominating set of G. Thus, $|S| = \gamma_k(G)$. Each vertex of $S$ $k$-dominates at most $(2k+1) + \lceil \frac{2k-1}{3} \rceil (\delta - 2)$ vertices of $P'$, and so $S$ $k$-dominates at most $|S|((2k+1) + \lceil \frac{2k-1}{3} \rceil (\delta - 2))$ vertices of $P'$. However, since $S$ is a $k$-dominating set of $G$, every vertex of $P'$ is $k$ dominated the set $S$, and so $S$ $k$-dominates $|V(P')| = \lceil \frac{d-1}{3} \rceil (\delta - 2) + d + 1$ vertices of $P'$. Therefore, $|S|((2k+1) + \lceil \frac{2k-1}{3} \rceil (\delta - 2)) \geq \lceil \frac{d-1}{3} \rceil (\delta - 2) + d + 1$, or, equivalently, $\gamma_k(G) \geq \frac{\lceil \frac{d-1}{3} \rceil (\delta-2)+d+1}{\lceil \frac{2k-1}{3} \rceil (\delta-2)+(2k+1)}$. □

For $\delta = 2$, then $\gamma_k(G) \geq \frac{d+1}{2k+1}$. Our lower bound is greater than $\frac{d+1}{2k+1}$ when the value $d > 2k$. We can verify thus condition by the following algebraic operation

$$d > 2k \Leftrightarrow 2d > 4k \Leftrightarrow 2kd - 2k + d - 1 > 2kd + 2k - d - 1 \Leftrightarrow (2k+1)(d-1) > (2k-1)(d+1)$$

$$\cdots \Leftrightarrow (2k+1)\frac{(d-1)}{3} > \frac{(2k-1)}{3}(d+1)$$

$$\Rightarrow (2k+1)\left\lceil \frac{(d-1)}{3} \right\rceil > \left\lceil \frac{(2k-1)}{3} \right\rceil (d+1)$$

$$\Leftrightarrow (2k+1)\left\lceil \frac{(d-1)}{3} \right\rceil (\delta - 2) > \left\lceil \frac{(2k-1)}{3} \right\rceil (\delta - 2)(d+1)$$

$$\Leftrightarrow (2k+1)\left\lceil \frac{(d-1)}{3} \right\rceil (\delta - 2) + (d+1)(2k+1) > \left\lceil \frac{(2k-1)}{3} \right\rceil (\delta - 2)(d+1) + (d+1)(2k+1)$$

$$\Leftrightarrow (2k+1)\left(\left\lceil \frac{(d-1)}{3} \right\rceil (\delta - 2) + (d+1)\right) > (d+1)\left(\left\lceil \frac{(2k-1)}{3} \right\rceil (\delta - 2) + (2k+1)\right)$$

$$\Leftrightarrow \frac{\left\lceil \frac{(d-1)}{3} \right\rceil (\delta - 2) + (d+1)}{\left\lceil \frac{(2k-1)}{3} \right\rceil (\delta - 2) + (2k+1)} > \frac{d+1}{2k+1}.$$

We end this section by determining the distance $k$-domination number of an edge comb product graph. The edge comb product between $G$ and $H$, denoted by $G \trianglelefteq H$, is a graph obtained by taking one copy of $G$ and $|E(G)|$ copies of $H$ and grafting the $i$-th copy of $H$ at the $i$-th edges of $G$. The cardinality of edge and vertex of $G \trianglelefteq H$ are defined as follow.

**Definition 1.** *Let $G = (V(G), E(H))$ and $H = (V(H), E(H))$. The edge comb product between $H$ onto $G$ with linking edge $yy' \in E(H)$, denoted by $G \trianglelefteq H$, is graph with vertex set*

$$V(G \trianglelefteq H) = V(G) \cup ((V(H)\backslash\{y, y'\}) \times E(G))$$

*and*

$$\begin{aligned} E(G \trianglelefteq H) =& E(G) \cup \{(y_1e)(y_2e); y_1y_2 \in E(H); y_1, y_2 \notin \{y, y'\}; e \in E(G)\} \\ & \cup \{(x)(y''e); e = ww' \in E(G);\ \textit{if } x = w, \textit{then } yy'' \in E(H)\backslash\{yy'\} \textit{ or} \\ & \textit{if } x = w', \textit{then } y'y'' \in E(H)\backslash\{yy'\}\}. \end{aligned}$$

By simple calculation we can verify that

$$\begin{aligned} |E(G \trianglelefteq H)| &= |E(G)| + |E(G)|(|E(H)| - (|N(y)| + |N(y')| - 1)) + |E(G)|(|N(y)| - 1 + |N(y')| - 1) \\ &= |E(G)||E(H)| \end{aligned}$$

and $|V(G \trianglelefteq H)| = |V(G)| + |E(G)|(|V(H)| - 2)$. we will determine lower and upper bound $\gamma_k(G \trianglelefteq H)$ as a function with $\gamma_k(H)$ as independent variable.

**Theorem 4.** *If $G$ and $H$ be a simple connected graph, then $\gamma_k(G \trianglelefteq H) \leq \gamma_k(H)|E(G)|$.*

*Proof.* Vertex set of graph operation $G \trianglelefteq H$ can be written as $\bigcup_i^{|E(G)|} V(H_i)$, where $H_i$ is graph that isomorphic to $H$ that replace some edge on $G$. If we set

$$\begin{aligned} \gamma_k(G \trianglelefteq H) &= \Sigma_i^{|E(G)|} \gamma_k(H_i) \\ &= \gamma_k(H)|E(G)|, \end{aligned}$$

then $\forall v \in H_i, 1 \leq i \leq |E(G)|$ vertex $v$ is dominated at $k$ distance by some element of $\gamma_k(H_i)$, so $\gamma_k(G \trianglelefteq H) \leq \gamma_k(H)|E(G)|$. □

**Theorem 5.** *For any simple connected graph $G$ and regular graph $H$. If $\gamma_k(H) \geq 2$ and $|V(H)|$ divisible by $|N_k[u]|, u \in V(H)$, then $\gamma_k(G \trianglelefteq H) \geq (\gamma_k(H) - 2)|E(G)| + |V(G)|$.*

*Proof.* Supposed that $|V(G)| = n, |E(G)| = n', |V(H)| = m,$ and $|E(G)| = m'$. First we count how many vertices that can be dominated by $S_k$. We denote $|X|$ as the maximum number of vertices that can be dominated by $S_k$ (the set of distance $k$-dominaiton number).

$$\begin{aligned} |X| \leq& \Sigma_{v \in V(G)}(d(v)(|N_k[u]| - 1)) + |V(G)| + (\gamma_k(G) - 2)|N_k[u]|n' \\ =& 2n'(|N_k[u]| - 1) + n + \left(\left\lceil \frac{m}{|N_k[u]|} \right\rceil - 2\right) n'|N_k[u]| \\ =& 2n'(|N_k[u]| - 1) + n + \left\lceil \frac{m}{|N_k[u]|} \right\rceil |N_k[u]|n' - 2n'|N_k[u]| \\ =& \left\lceil \frac{m}{|N_k[u]|} \right\rceil |N_k[u]|n' + n - 2n'. \end{aligned}$$

Since $V(H)$ divisible by $|N_k[u]|$, then

$$|X| \leq mn' + n - 2n' = |V(G \trianglelefteq H)|.$$

By contradiction we will show that

$$\gamma(G \trianglelefteq H) < (\gamma_k(H) - 2)|E(G)| + |V(G)|$$

is imposible. If $V(G \trianglelefteq H)$ is written as $\bigcup_i^{|E(G)|} V(H_i)$, where $H_i$ is isomorphic graph to $H$ that replace some edge on $G$ and $\gamma(G \trianglelefteq H) < (\gamma_k(H) - 2)|E(G)| + |V(G)|$, then there is exist $v \in S(H_i), v \notin S(G \trianglelefteq H)$. Note that the notation $S(G)$ means $k$ distance dominating set of $G$. We have two cases:
*Case 1*, vertex $v$ is not sharing vertex of any two subgraph $H_i$ and $H_j$. We will count the number of vertices that can be reach by $S$. We denote the maximum number of vertices that can be dominated by $S$ as $|X'|$.

$$\begin{aligned}
|X'| =& \Sigma_{v\in V(G)} d(v)(|N_k[u]| - 1) + |V(G)| + ((\gamma_k(G) - 2) - 1)|N_k[u]| \\
=& \Sigma_{v\in V(G)} (d(v)(|N_k[u]| - 1)) + |V(G)| + (\gamma_k(G) - 2)|N_k[u]| - |N_k[u]| \\
=& |X| - |N_k[u]| \\
\le& mn' + n - 2n' - |N_k[u]| \\
<& |V(G \trianglelefteq H)|
\end{aligned}$$

*Case 2*, vertex $v$ is the element of sharing vertex of some subgraph $H_i$ and $H_j$.

$$\begin{aligned}
|X'| =& (\Sigma_{v\in V(G)} d(v) - \delta)(|N_k[u]| - 1) + |V(G) - 1| + (\gamma_k(G) - 2)|)n'|N_k[u]| \\
=& \Sigma_{v\in V(G)} (d(v)(|N_k[u]| - 1)) + |V(G)| + (\gamma_k(G) - 2)(|N_k[u]|) - (1 + \delta(|N_k[u]| - 1)) \\
\le& |X| - (1 + \delta(|N_k[u]| - 1)) \\
<& |X| - |N_k[u]| \\
<& |V(G \trianglelefteq H)|
\end{aligned}$$

All two cases above shows a contradiction, so $\gamma_k(G) \ge (\gamma_k(H) - 2)|E(H)| + |V(G)|$. It completes the proof. □

## 3. Conclusion

The result of this paper is related to new upper and lower bound for general graph and edge comb product graph operation. Our upper bound for general graph is tight especially if $G$ regular. For lower bound, we find that by including minimum degree and diameter as parameters, we can get better lower bound compared with if we consider only diameter as parameter. For edge comb product graph, we find that the value of its lower and upper bound of distance domination number is depend on the value of distance domination number of second graph in edge comb product operation of graphs.

**Acknowledgement**

This research is supported by DRPM Ditjen Penguatan Risbang Kemenristekdikti under Penelitian Tim Pascasarjana 2016 research grant.

## Bibliography

[1] Akihito Hora and Nobuaki Obata, 2007, *Quantum Probability and Spectral Analysis of Graphs*, Springer Science & Business Media. ISBN 978-3-540-48863-7.
[2] A. Behzad, M. Behzad, and Cheryl E. Praeger, 2008, *Fundamental Dominations in Graphs*, arXiv:0808.4022v1 [math.CO] 29 Aug 2008.
[3] A. Meir and J. W. Moon, 1975, *Relations Between Packing and Covering Number of a Tree*, Pacific J. Math. 61 (1975), 225-233.
[4] G. Chartrand. 2011, *Graph and Digraph*, New York: Taylor and Francis Group.
[5] G. Chartrand, 2011, *Discrete Mathematics*, New York: Taylor and Francis Group.
[6] E. E. DeLaVĩna, R. Pepper, B. Waller, 2010, *Lower Bounds for the Domination Number*, Discuss. Math. Graph Theory 30(3) (2010), 475-487.
[7] F. Tian and J. M. Xu, 2009, *A Note on Distance Domination Numbers of Graphs*, Australasian J. Combin. 43 (2009), 181-190.

[8] Haynes, W., Teresa, 1996, *Fundamental of Dominations in Graphs*, New York: Marcel Dekker, Inc.

[9] M. A. Henning and N. Lichiardopol, *Distance Domination in Graphs with given Minimum and Maximum Degree*, manuscript.

[10] Jose C. Nacher and Tatsuya Akutsu, 2016, *Minimum Dominating Set-Based Methods for Analyzing Biological Networks*, Elseviere Journal: Methods 102 (2016) 57-63.

[11] Peter J. Slater, 1975, *R-Domination in Graphs*, Washingtion: Journal of the Associatlon for Computing Machinery, Vol 23, No 3, July 1976, pp 446-450.

[12] Preeti Gupta, 2013, *Domination in Graph with Application*, Indore: Indian Journal of Research (PAPIREX).

[13] Randy Davila, Caleb Fast, Michael A. Henning dan Franklin Kenter, 2015, *Lower Bounds on the Distance Domination Number of a Graph*, arXiv:1507.08745v1 [math.CO] 31 jul 2015.

[14] B. Reed, 1996, *Paths, Stars and the Number Three*, Combinatorics Probability and Computing 5(3): 277-295 September 1996.

[15] Samantha Dorfling, 1998, *Domination in Graphs With Bounded Degree*, Thesis, Johannesburg: Rand Afrikaans University.

[16] S. Connolly et al, 2016, *Bounds on the Maximum Number of Minimum Dominating Sets*, Elsevier Journal: Discrete Mathematics 339 (2016) 1537-1542.

[17] S K Vaidya, 2014, *Distance k-Domination of Some Path Related Graphs*, India: International Journal of Mathematics and Soft Computing Vol.4, No.1 (2014), 1 - 5.

[18] T. W. Haynes, S. T. Hedetniemi, and P. J. Slater (eds), 1998, *Domination in Graphs: Advanced Topics*, Marcel Dekker, Inc. New York.

[19] M. A. Henning, 1998, *Domination in graphs: advanced topics/edited by Teresa W. Haynes, Stephen T. Hedetniemi, Peter J. Slater* ,Marcel Dekker. Inc, New York, 335-365.

[20] W. Duckworth and B. Mans, *Randomised Greedy Algorithms for finding Small k-Dominating Sets of Regular Graphs* , John Wiley & Sons. Inc, Australia.

[21] Y. Caro and Y. Roditty, 1990, *A Note on the k-Domination Number of a Graph*, Internat. J. Math. & Math. Sci. VOL. 13 NO. (1990) 205-206.

IOP Conf. Series: Journal of Physics: Conf. Series **855** (2017) 012015 doi:10.1088/1742-6596/855/1/012015

# Rainfall model on area of rice production in Sragen, Karanganyar and Klaten by using Generalized Space Time Autoregressive (GSTAR)

**S S Handajani**[1], **H Pratiwi**[2], **Y Susanti**[3], **S Subanti**[4], **Respatiwulan**[5], and **Hartatik**[6]

[1,2,3,4,5]Statistics Department of Mathematics and Natural Sciences Faculty of Sebelas Maret University Indonesia
[6]Informatics of Mathematics and Natural Sciences Faculty of Sebelas Maret University Indonesia

Email: rr_ssh@staff.uns.ac.id, hpratiwi@mipa.uns.ac.id, yulianasusanti@staff.uns.ac.id, srisubanti@staff.uns.ac.id, respatiwulan@staff.uns.ac.id, hartatik@uns.ac.id

**ABSTRACT.** The planning of planting period became non-optimal due to the uncertain rainfall fluctuation condition in the past few years. To anticipate food insecurity caused by such a matter, the development of rainfall prediction models in several rice production areas in Central Java, which represent the center for rice production, is required. This paper provides the application of GSTAR model on the rainfall data in Sragen Regency, Karanganyar Regency, and Klaten Regency with considering time and location conditions. The weighting applied in the model was normalized cross-correlation. By applying the least square method and stepwise procedure, GSTAR model for six, seven in Sragen, seven areas in Karanganyar, and seven areas in Klaten respectively was obtained. The validity of the model was tested by applying RMSE.

## 1. Introduction

One of the important roles of Central Java Province for the regional and national economy is as the food barn. One of staple food resources of the majority of Indonesian communities is rice, making it as the superior food staple commodity in Indonesia and one of the government's main concerns in maintaining the national food security and stability. The availability and access to the commodity and its production continuity highly affect the national food security.

The rice production in Central Java has a surplus, which makes it potential to support the regional food security. Nationally, it was the biggest rice producer with the production of 10.34 million tons of unhulled rice in 2014. By the paddy-into-rice conversion factor assumption with the number of 62.74% and the level of per capita rice consumption of 139.15 kg per year, Central Java is therefore, potential to have the surplus of 1.86 million tons of rice. It is considered significant to support the national rice surplus target which aims for 10 million tons of rice per year. In accordance with the high production capacity, the rice productivity in Central Java is 56.06 quintal per hectare, which is higher than its national average. Considering its important role, the food production in Central Java needs to be maintained by establishing an appropriate cropping pattern.

Rice cropping pattern can be determined through climate and location factors. The global warming has caused climate changes such as temperature increment, sea water level increment and rainfall pattern shifting. *El Niño and La Niña* are the extreme climate conditions with a long duration, and they are allegedly affected by global warming, which seriously affects the agricultural productions [4]. *El Niño* is condition when the *Equatorial* Pacific surface temperatures become warm, especially around the eastern part [3]. *El Niño* causes the eastern Equatorial Pacific surface warm, followed by rainfall increment around Pacific areas and rainfall degradation in the areas of Southeast Asia and Australia. On the contrary, *La Niña* causes western Pacific surface warm and higher rainfalls in the areas of Southeast Asia and Australia. Due to the temperature increment and the frequent *El Niño*, rice production in most of Asian countries experienced degradation in the past few years ([2] , [11]). *La Niña* also causes disadvantages such as floods and the increment of plant pest organism due to above-normal rainfalls and humidity level increment. *El Niño and La Niña* occur irregularly that both of them can shift the time period of rainy season and dry season and also causes changes on food commodity cropping pattern for the farmers.

The planning of planting period became non-optimal due to the uncertain rainfall fluctuation condition in the past few years. To respond food insecurity caused such a condition, the development of the rainfall prediction model in food production areas which represent the centers of food production, especially rice, is required.

Research on rice plants was initiated by [15] and [16] by modeling a robust regression which used M-estimation, S-estimation, and MM-estimation parameter and spasial modelling. In 2015, a new regression model was developed using Geographically Weighted Regression (GWR). This model considered geographical condition and was constructed for corn production in every area in Central Java [14]. A spatial model was also developed in the same year which applied spatial weights (location) in modeling the rice production in Indonesia [15].

The research above will be developed in one of the factors that affect rice production, namely rainfall. Fluctuation in rainfall really affects the result of rice production. The rainfall pattern model, in the form of time series data, needs to be constructed in some rice production centers in Central Java. Some statistical methods used to predict the rainfall were Autoregressive Integrated Moving Average (ARIMA), Wavelet transformation, and Adaptive Neuro-Fuzzy Inference System (ANFIS). Those methods were only based on time in the past and did not consider geographical factor. Thus, there is a need to develop a method to increase the accuracy of rainfall prediction.

Model which considers location and time factors is known as Space Time Autoregressive (STAR) or Generalized Space Time Autoregressive (GSTAR). The STAR model is time and location model, which was firstly introduced by Pfeifer and Deutsch [5]. In this model, the space parameters in all area are considered the same (homogenous). Meanwhile the rainfall data in some centers of rice production area in Central Java are possibly heterogeneous. Ruchjana [6] developed the STAR method into GSTAR, which perfects the STAR through the use of different space parameter for each area. This is in accordance with the heterogeneous state in the field. Therefore, the GSTAR is more suitable method to be used to predict the rainfall in Central Java even for the smallest level of prediction that is rainfall station. According to [6], the model needs stationary data. Some sampled areas in this model considered condition of location and linkage between locations so that the appropriate weighting factor is included in the model.

Suhartono and Subanar [10] suggest various methods to decide the location weighting in the GSTAR model, namely: uniform weighting, binary weighting, distance inversion, and normalized cross correlation weighting. Uniform weighting was not compatible to be used in the GSTAR model because the location characteristic was heterogenic. The binary weighting was also not compatible to be used in GSTAR model. This weighting was subjective because a closer location was scored 1 while a further location was scored 0. Location Correlation was also not compatible based on the location approach. The normalized cross correlation weighting was required to be used in GSTAR model because it considered the correlation of the data which have space and time effects.

The GSTAR model in literatures mentioned above was developed with weighted normalized cross correlation to model the rainfall in the regencies of Sragen, Karanganyar, and Klaten of Central Java which are the centers of rice production.

## 2. Research Method

In Central Java, the regencies of Sragen, Karanganyar, and Klaten are the rice production centers which fulfill the food need of the society of surrounding and outside the areas. For GSTAR rainfall modeling, the researchers took the monthly rainfall data of the years of 2009-2014 from six rice production areas in Sragen Regency: Masaran, Ngrampal, Tanon, Karang Malang, Plupuh, Kedawung; of the years of 1995-2014 from seven rice production areas in Karanganyar Regency: Kebak Keramat, Jaten, Karanganyar, Mojogedang, Karangpandan, Matesih, Tasik Madu; and of the years of 2010-2014 from seven rice production areas in Klaten Regency: Polanharjo, Cawas, Trucuk, Karangdowo, Wonosari, Delanggu, Juwiring.

The steps of the research were checking the data stationarity. Then, autoregressive order (p) in GSTAR model was determined by applying the model of VAR order p. The order determination was done by looking for the least value of AIC. Next, the weighted normalized cross-correlation was calculated based on the rainfall data. After that, a parameter prediction was done with the least square method, and its significance was tested according to autoregressive order (p) and spatial order (λs) for each weight of locations, and in this step, GSTAR model was formed. GSTAR model for each weight was then verified by applying root mean square error (RMSE) and then the normally distributed white noise residual was checked.

## 3. VAR Model

Vector autoregressive model (VAR) which is proposed by [8] in macroeconomics can be stated as

$$Y_t = c + \emptyset_1 Y_{(t-1)} + \emptyset_2 Y_{(t-2)} + \cdots + \emptyset_p Y_{(t-p)+\varepsilon_t}, t = 1, \ldots, T$$

with $Y_t = (y_{1t}, y_{2t}, \ldots, y_{nt})$as the cascading time variable vector in time $t(nx1)$, $\emptyset$ as the coefficient matrix $(nxn)$ and $\varepsilon t$ as the white noise residual vector in $t(nx1)$.

## 4. GSTAR Model

GSTAR model is a special form of VAR model which considers the spatial factor. The general model of GSTAR $(p, \lambda_s)$, (see [1] or [14]) can be stated as:

$$\sum_{k=1}^{p} \sum_{l=0}^{\lambda_s} \sum_{i=1}^{n} \Phi_{kl}^{i} W^{l}(k) Z_{t,k} + e_{i,t}$$

with $\Phi_{kl}^{i}$ as spatial matrix diagonal with spatial lag $l$ and time lag to $k$ in area $i$, $W^{l}(k)$ as dimension weighting matrix $(nxn)$, $e_{i,t}$ as normalized residual $(nx1)$ in area i time $t$, $Z_{i,t}$ as the observation in area $i$ time $t$.

### *4.1. Model Stationarity Assumption*

The forming of GSTAR model began with the data stationarity determination. According to [6], test statistic can be done through augmented Dickey Fuller (ADF) test with the following steps:

1. $H_0$ : The data are not stationary
   $H_1$ : The data are stationary
2. Significance level determination
3. Critical area : $H_0$ is rejected when the value of $ADF\ (t) <$ MacKinnon table $(\alpha; T - p)$
4. Test Statistic :

$$t = \frac{\dfrac{\sum_{t=1}^{n} Z_{t-1} Z_t}{\sum_{t=1}^{n} Z_{t-1}^2} - 1}{\sqrt{\dfrac{\sum_{t=1}^{n} \left(Z_{t-\eta} Z_{t-1}\right)^2}{n-1}}}$$

$n$ as the number of data and $p$ as the number of parameter.

### *4.2. Weighted Normalized Cross-Correlation*

Weighted normalized cross-correlation is the location weightings using the result of normalized cross-correlation between locations in the consistent time lag. According to Wutsqa *et al.* [14] weighted normalized cross-correlation provides any possible correlations between locations. Weighted normalized cross-correlation can be stated as:

$$w_{ij}(k) = \frac{r_{ij}(k)}{\sum_{k=i} |r_{ik}(k)|}$$

with $i \neq j, k = 1,2,\dots,p$, and the value of $r_{ij}(k)$ as the cross-correlation which follows the following equation:

$$r_{ij}(k) = \frac{\sum_{t=k+1}^{n}(Z_i(t) - \bar{Z}_i)(Z_j(t-k) - \bar{Z}_j)}{\sqrt{(\sum_{t=1}^{n}(Z_i(t) - \bar{Z}_i)^2)\left(\sum_{t=1}^{n}(Z_j(t) - \bar{Z}_j)^2\right)}}$$

### *4.3. Model Identification*

GSTAR model has a space and time correlation which is why its model orders are autoregressive order and spatial order. GSTAR model autoregressive order determination can be done with the order from VAR $(p)$ model with the criterion of: the optimal lag length applies the least value of AIC ([14], [13]). According to Tsay [12], *the* value of AIC can be formulaized as:

$$\text{AIC} = \ln\left(\frac{MSE}{n}\right) + \frac{2K^2}{n}.$$

### *4.4. Model Parameter Estimation*

Parameter estimation for GSTAR model can be done by applying the least square method. Based on the above model (2), it can be stated as:

$$\begin{pmatrix} \boldsymbol{Z_1(1)} \\ \vdots \\ \boldsymbol{Z_1(T)} \\ \vdots \\ \boldsymbol{Z_n(1)} \\ \vdots \\ \boldsymbol{Z_n(T)} \end{pmatrix} = \begin{pmatrix} \emptyset_{k0}^{1}\boldsymbol{Z_1(T-k)} \\ \emptyset_{k0}^{2}\boldsymbol{Z_2(T-k)} \\ \emptyset_{k0}^{3}\boldsymbol{Z_3(T-k)} \\ \vdots \\ \emptyset_{k0}^{6}\boldsymbol{Z_6(T-k)} \\ \vdots \\ \emptyset_{k0}^{n}\boldsymbol{Z_n(T-k)} \end{pmatrix} + \begin{pmatrix} \emptyset_{kl}^{1} & \boldsymbol{0} & \boldsymbol{0} & \boldsymbol{0} \\ \boldsymbol{0} & \emptyset_{kl}^{2} & \boldsymbol{0} & \boldsymbol{0} \\ \vdots & \boldsymbol{0} & \ddots & \vdots \\ \boldsymbol{0} & \boldsymbol{0} & \cdots & \emptyset_{kl}^{n} \end{pmatrix}$$

$$\begin{pmatrix} \boldsymbol{V_1(T-k)} \\ \vdots \\ \boldsymbol{V_2(T-k)} \\ \vdots \\ \boldsymbol{V_3(T-k)} \\ \vdots \\ \boldsymbol{V_n(T-k)} \end{pmatrix} + \begin{pmatrix} \boldsymbol{e_1(T)} \\ \vdots \\ \boldsymbol{e_2(T)} \\ \vdots \\ \boldsymbol{e_3(T)} \\ \vdots \\ \boldsymbol{e_n(T)} \end{pmatrix}$$

with $V_i(t) = \sum_{j=1}^{n} w_{ij}(k) Z_i(t)$

The model for $i$ location can be stated as:

$$Z_i = Z_i^* \Phi + \varepsilon$$

so that the $\Phi$ parameter estimation for each location can be calculated separately. First, formulizing the least squares method of $\varepsilon = Z_i - Z_i^* \Phi$ has to be done in order to estimate the $\Phi$ parameter and determining the RSS as follows:

$$\varepsilon'\varepsilon = (Z - Z^*\Phi)'(Z - Z^*\Phi) = Z'Z - Z'Z^*\Phi - \Phi'Z^*Z + \Phi'Z^{*'}Z^*\Phi$$

$$= Z'Z - 2\Phi'Z^*Z + \Phi'Z^{*'}Z^*\Phi.$$

By minimizing, the RSS $\Phi$ parameter estimation was obtained as follows:

$$\frac{\partial \varepsilon'\varepsilon}{\partial \Phi} = -2Z^{*'}Z + 2\Phi Z^{*'}Z^* = 0$$

$$\widehat{\Phi} = (Z^{*\prime}Z^*)^{-1}(Z^{*\prime}Z).$$

### *4.5. Stepwise Regression*

One way to determine a good regression model is by checking its parameter significance. Stepwise regression combines forward and backward method into bidirectional elimination. Stepwise regression method chooses the variable by the biggest partial corelation with the variable that has been input to the model. The variable tha has been inputted to the model can be removed again [7].

### *4.6. GSTAR Model Validation*

Model validation is done to determine the quality of the model and the accuracy of data prediction by finding the value of root mean square error (RMSE):

$$RMSE = \sqrt{\frac{\sum_{t=1}^{n}(Z_t - \widehat{Z_t})^2}{n}}.$$

## 5. Result and Discussion

### *5.1. The Stationarity of Rainfall Data*

First step to the GSTAR modeling was to identify whether not the data were stationary. The results of ADF testing show that the simultaneously the rainfall data from six sub-districts in Sragen regency with zero hypothesis were stationary as indicated by the statistical value of ADF-Fisher-Chi-square = 116.123 with the probability value = 0.000 and with $\alpha = 0.05$; the rainfall data in seven areas of Karanganyar Regency were stationary as shown by the the statistical value of ADF – Fisher Chi-square = 68.5407 with the probability value = 0.000 and with $\alpha = 0.05$; and the rainfall data in seven areas of Klaten Regency were stationary as signified by the statistical value of ADF – Fisher Chi-square = 65.5407 with the probability value = 0.000 and with $\alpha = 0.05$.

### 5.2. *Vector Autoregressive (VAR)*

Following the stationarity of rainfall data, they were then then identified into the model of VAR order $p$. Widarjono [13] claims that identifiying VAR model order is done with optimum length of lag. The criteria to determine the optimum length of lag uses the smallest value of Akaike Information Criterion (AIC). According to Tsay [12], AIC can be formulated as follows:

$$AIC\ (i) = \ln\left(\frac{JKS}{T}\right) + \frac{2k^2 i}{T}$$

where JKS is the value of quadratic residue, $T$ is the number of observation, and $k$ is the number of model parameter. The values of AIC of each lag for rainfall in the areas of Sragen Regency are presented in Table 1.

**Table 1**. Value of *AIC* of each lag on *VAR* model

| Lag | 0 | 1 | 2 | 3 | 4 | 5 | 6 |
|---|---|---|---|---|---|---|---|
| AIC | 69.4 | 69.4 | 69.8 | 70.3 | 70.6 | 70.8 | 71.3 |

The values of AIC of each lag for rainfall of the areas of Karanganyar Regency are presented in Table 2.

**Table 2.** Value of *AIC* of each lag on *VAR* model

| Lag | 0 | 1 | 2 | 3 | 4 | 5 | 6 |
|---|---|---|---|---|---|---|---|
| AIC | 88.97 | 87.1 | 87.4 | 87.6 | 87.9 | 88.2 | 88.4 |

The values of AIC of each lag for rainfall in the areas of Klaten Regency are presented in Table 3.

**Table 3.** Value of *AIC of each lag* on *VAR* model

| Lag | 0 | 1 | 2 | 3 | 4 | 5 |
|---|---|---|---|---|---|---|
| AIC | 80.7 | 80.2 | 80.1 | 80.6 | 80.6 | 80.1 |

According to Table 1 and Table 2, the least value of AIC were 69.4 and 87.305 respectively on lag 1, so that the model order of VAR was 1 on the data of Sragen Regency and Karanganyar Regency. The least value of AIC in Table 3 was 80.079 on lag 5 so that the model order of VAR was 5. Wutsqa *et al.* [14] identify autoregressive order into GSTAR model by using order from VAR model. Thus, GSTAR model autoregressive orders were 1 and 5.

### 5.3. *Generalized Space Time Autoregressive (GSTAR) Model*

The Model of GSTAR as the development of STAR model is as follows:

$$Z_{i,t} = \sum_{k=1}^{p}\sum_{l=0}^{\lambda_s}\sum_{i=1}^{n} \Phi_{kl}^{i} W_k^l Z_{t,k} + e_{i,t}$$

where $Z_{i,t}$ is observation of area $i$ in time of $t$, $\Phi_{kl}^{i}$ is diagonal space time matrix with spatial lag of 1 and $k$ is value of time lag on area of $i$, $W$ is wighting matrix, and $e_{i,t}$ is normal residue. Spatial order 1 is choosen in the establishment of this model for the ease of interpretation. Autoregressive order 1 is for the areas of Sragen Regency and Karanganyar Regency (GSTAR model $(1_1)$ and autoregressive order 5 is for the areas of Klaten Regency Regency (GSTAR model $(5_1)$. The establishment of GSTAR model $(1_1)$ and $(5_1)$ with normalized cross correlation location wighting was then applied on the rainfall data for the areas of Sragen Regency, Karanganyar Regency, and Klaten Regency.

*5.4. GSTAR Model for the Rainfall in the Areas of Sragen Regency*

GSTAR model was established by choosing the weighting result from normalized cross corelation between locations on corresponding lags. The establishment of this weighting factor as claimed by Suhartono and Subanar [10] can be obtained from:

$$W_{ij}(k) = \frac{r_{ij}(k)}{\sum_{k=1}|r_{ik}(k)|}$$

with $i = j, k = 1,2,\dots,p$ and

$$r_{ij}(k) = \sum_{t=k+1}^{n} \frac{[Z_i(t) - \bar{Z}_i][Z_j(t-k) - \bar{Z}_j]}{\sqrt{\sum_{t=1}^{n}[Z_i(t) - \bar{Z}_i]^2 \sum_{t=1}^{n}[Z_j(t) - \bar{Z}_j]^2}}$$

The calculation of weighting factor in this model resulted in:

$$W = \begin{bmatrix} 0 & 0.192 & 0.198 & 0.199 & 0.214 & 0.197 \\ 0.205 & 0 & 0.195 & 0.196 & 0.202 & 0.201 \\ 0.206 & 0.179 & 0 & 0.198 & 0.208 & 0.209 \\ 0.183 & 0.195 & 0.206 & 0 & 0.219 & 0.196 \\ 0.210 & 0.188 & 0.194 & 0.215 & 0 & 0.194 \\ 0.199 & 0.177 & 0.215 & 0.198 & 0.212 & 0 \end{bmatrix}$$

Based on the weighting matrix above, the estimation of model parameter was done by using the least square method and by testing the significance valoue of each model parameter. Then estimation of regression parameter with stepwise procedure was done to achieve a good model. The result of parameter estimation and its significancy test are as follows:

**Table 4.** Value of parameter estimation model, p value and conclusion

| Predictor | Parameter Est. | p-value | Conclusion |
|---|---|---|---|
| $Z_5$(t-1) | 0.80479 | 0.000 | Significant |
| $w_1$(t-1) | 0.77650 | 0.000 | Significant |
| $w_2$(t-1) | 0.66548 | 0.000 | Significant |
| $w_3$(t-1) | 0.73035 | 0.000 | Significant |
| $w_4$(t-1) | 0.85165 | 0.000 | Significant |
| $w_6$(t-1) | 1.22659 | 0.000 | Significant |

$Z_{i,t} = 0.805\, Z_5(t-1) + 0.776\, w_1(t-1)Z_i(t-1) + 0.665\, w_2(t-1)Z_i(t-1) + 0.730\, w_3(t-1)Z_i(t-1) + 0.852\, w_4(t-1)Z_i(t-1) + 1.23\, w_6(t-1)Z_i(t-1)$, with $i = j$, for $i \neq j, Zj = 0$, $w_i$ is the weighting value of the first row.

Then following the the inputting of weighting value, the following was obtained:

$$\begin{pmatrix} Z_1(t) \\ Z_2(t) \\ Z_3(t) \\ Z_4(t) \\ Z_5(t) \\ Z_6(t) \end{pmatrix} = \begin{pmatrix} 0.15Z_2(t-1) + 0.15Z_3(t-1) + 0.15Z_4(t-1) \\ 0.14Z_1(t-1) + 0.13Z_3(t-1) + 0.13Z_4(t-1) \\ 0.15Z_1(t-1) + 0.13Z_2(t-1) + 0.14Z_4(t-1) \\ 0.16Z_1(t-1) + 0.17Z_2(t-1) + 0.18Z_3(t-1) \\ 0.805Z_5(t-1) \\ 0.24Z_1(t-1) + 0.22Z_2(t-1) + 0.26Z_3(t-1) \end{pmatrix} + \begin{pmatrix} 0.17Z_5(t-1) + 0.15Z_6(t-1) \\ 0.13Z_5(t-1) + 0.13Z_6(t-1) \\ 0.15Z_5(t-1) + 0.15Z_6(t-1) \\ 0.19Z_5(t-1) + 0.17Z_6(t-1) \\ 0 \\ 0.24Z_4(t-1) + 0.26Z_5(t-1) \end{pmatrix}$$

Based on the above six prediction models, a conclusion is drawn that the rainfall prediction in Masaran area in month t was slightly affected by the rainfall condition of a month earlier in other areas around it, namely: the areas of Ngrampal, Tanon, Karangmalang, Plupuh, and Kedawung. It also happened in the areas of Ngrampal, Tanon, Karangmalang and Kedawung. However, the rainfall prediction in Plupuh area in month t was only affected by the rainfall condition of a month earlier in its own area.

### 5.5. *GSTAR Model for the Rainfall in the Areas of Karanganyar Regency*

In the forming of GSTAR model $(1_1)$ based on the rainfall data in several areas of Karanganyar Regency, the following weighted normalized cross-correlation was gained:

$$W = \begin{bmatrix} 0 & 0.052 & 0.212 & 0.213 & 0.104 & 0.238 & 0.181 \\ 0.111 & 0 & 0.118 & 0.173 & 0.036 & 0.433 & 0.128 \\ 0.193 & 0.064 & 0 & 0.061 & 0.263 & 0.109 & 0.309 \\ 0.303 & 0.096 & 0.087 & 0 & -0.066 & 0.365 & 0.083 \\ 0.096 & -0.012 & 0.371 & -0.107 & 0 & -0.092 & 0.322 \\ 0.235 & 0.14 & 0.114 & 0.313 & -0.069 & 0 & 0.130 \\ 0.189 & 0.088 & 0.291 & 0.057 & 0.254 & 0.121 & 0 \end{bmatrix}$$

With the least square method and the above stepwise procedure the following result was obtained:

**Table 5.** Value of parameter estimation model, p value and conclusion

| Predictor | Paramete Est. | p-value | Conclusion |
|---|---|---|---|
| $Z_1(t-1)$ | 0.60829 | 0.000 | Significant |
| $Z_2(t-1)$ | 0.4758 | 0.047 | Significant |
| $Z_3(t-1)$ | 0.3517 | 0.048 | Significant |
| $Z_4(t-1)$ | 0.58262 | 0.000 | Significant |
| $Z_5(t-1)$ | 0.61669 | 0.000 | Significant |
| $Z_6(t-1)$ | 0.3595 | 0.049 | Significant |
| $Z_7(t-1)$ | 0.4061 | 0.014 | Significant |
| $w_1(t-1)$ | 0.2757 | 0.017 | Significant |
| $w_3(t-1)$ | 0.4168 | 0.017 | Significant |
| $w_4(t-1)$ | 0.3771 | 0.023 | Significant |
| $w_5(t-1)$ | 0.6441 | 0.000 | Significant |
| $w_6(t-1)$ | 0.3699 | 0.049 | Significant |
| $w_7(t-1)$ | 0.4017 | 0.023 | Significant |

$$Z_{i,t} = 0.608\, Z_1(t-1) + 0.476\, Z_2(t-1) + 0.352\, Z_3(t-1) + 0.583\, Z_4(t-1) + 0.617\, Z_5(t-1) + 0.359\, Z_6(t-1) + 0.406\, Z_7(t-1) + 0.276\, w_1(t-1)Z_i(t-1) + 0.417\, w_3(t-1)Z_i(t-1) + 0.377\, w_4(t-1)Z_i(t-1) + 0.644\, w_5(t-1)Z_i(t-1) + 0.370\, w_6(t-1)Z_i(t-1) + 0.402\, w_7(t-1)Z_i(t-1).$$

Next, following the inputting of the weighting value, the following was obtained:

$$\begin{pmatrix} Z_1(t) \\ Z_2(t) \\ Z_3(t) \\ Z_4(t) \\ Z_5(t) \\ Z_6(t) \\ Z_7(t) \end{pmatrix} = \begin{pmatrix} 0.61Z_1(t-1) + 0.01Z_2(t-1) + 0.06Z_3(t-1) + 0.06Z_4(t-1) \\ 0.48Z_2(t-1) \\ 0.08Z_1(t-1) + 0.03Z_2(t-1) + 0.35Z_3(t-1) + 0.03Z_4(t-1) \\ 0.12Z_1(t-1) + 0.04Z_2(t-1) + 0.03Z_3(t-1) + 0.58Z_4(t-1) \\ 0.06Z_1(t-1) - 0.01Z_2(t-1) + 0.24Z_3(t-1) - 0.07Z_4(t-1) \\ 0.09Z_1(t-1) + 0.05Z_2(t-1) + 0.04Z_3(t-1) + 0.12Z_4(t-1) \\ 0.09Z_1(t-1) + 0.05Z_2(t-1) + 0.04Z_3(t-1) + 0.12Z_4(t-1) \end{pmatrix} + \begin{pmatrix} 0.03Z_5(t-1) + 0.07Z_6(t-1) + 0.05Z_7(t-1) \\ 0 \\ 0.11Z_5(t-1) + 0.05Z_6(t-1) + 0.13Z_7(t-1) \\ -0.03Z_5 + 0.14Z_6(t-1) + 0.03Z_7(t-1) \\ 0.62Z_5 - 0.06Z_6(t-1) + 0.21Z_7(t-1) \\ -0.03Z_5(t-1) + 0.36Z_6(t-1) + 0.05Z_7(t-1) \\ 0.1Z_5(t-1) + 0.05Z_6(t-1) + 0.41Z_7(t-1) \end{pmatrix}$$

The above result shows that rainfall prediction in the areas of Kebak Kramat, Karanganyar, Mojogedang, Karangpandan, Matesih and Tasik Madu in month $t$ was mainly affected by the rainfall condition of a month earlier in their own areas and slightly affected by the other areas. However, for the area of Jaten, the rainfall prediction in month t was only affected by the rainfall condition of a month earlier in its own area.

### 5.6. *GSTAR Model for the Rainfall in the Areas of Klaten Regency*

For captions not placed at the side of the figure, captions should be set to the width of the figure for wider figures, centred across the width of the figure, or, for narrow figures with wide captions, slightly extended beyond the width of the figure. The caption should finish with a full stop (period).
In the forming of GSTAR model ($1_1$) based on the rainfall data in several areas of Klaten Regency with the same method, five weighted normalized cross-correlations were gained. With the least square method and stepwise procedure, the following result was obtained:

**Table 6.** Value of parameter estimation model, p value and conclusion

| Predictor | Parameter Est. | p-value | Concl. |
|---|---|---|---|
| $Z_1(t-1)$ | 0.74698 | 0.000 | Significant |
| $Z_5(t-1)$ | 0.81455 | 0.000 | Significant |
| $Z_6(t-1)$ | 0.5716 | 0.000 | Significant |
| $w_2(t-1)$ | 0.84643 | 0.000 | Significant |
| $w_3(t-1)$ | 0.84914 | 0.000 | Significant |
| $w_4(t-1)$ | 0.95450 | 0.000 | Significant |
| $w_7(t-1)$ | 0.79146 | 0.000 | Significant |
| $w_6(t-2)$ | 0.4992 | 0.001 | Significant |
| $w_6(t-4)$ | 0.2601 | 0.029 | Significant |

$$Z_{i,t} = 0.747\, Z_1(t-1) + 0.815\, Z_5(t-1) + 0.572\, Z_6(t-1) + 0.846\, w_2(t-1)Z_i(t-1)$$

$$+0.849\, w_3(t-1)Z_i(t-1) + 0.954\, w_4(t-1)Z_i(t-1) + 0.791\, w_7(t-1)Z_i(t-1)$$
$$+\, 0.499\, w_6(t-2)Z_i(t-1) + 0.260\, w_6(t-4)Z_i(t-1)$$

Next, by inputting the weightings' value, the following result was obtained:

$$\begin{pmatrix} Z_1(t) \\ Z_2(t) \\ Z_3(t) \\ Z_4(t) \\ Z_5(t) \\ Z_6(t) \\ Z_7(t) \end{pmatrix} = \begin{pmatrix} 0.747Z_1(\mathrm{t}-1) \\ 0.15Z_1(\mathrm{t}-1) + 0.15Z_3(\mathrm{t}-1) + 0.14Z_4(\mathrm{t}-1) + 0.12Z_5(\mathrm{t}-1) \\ 0.17Z_1(\mathrm{t}-1) + 0.13Z_2(\mathrm{t}-1) + 0.14Z_4(\mathrm{t}-1) + 0.14Z_5(\mathrm{t}-1) \\ 0.19Z_1(\mathrm{t}-1) + 0.14Z_2(\mathrm{t}-1) + 0.15Z_3(\mathrm{t}-1) + 0.16Z_5(\mathrm{t}-1) \\ 0.82Z_5(\mathrm{t}-1) \\ 0.57Z_6(\mathrm{t}-1) + 0.1Z_1(\mathrm{t}-2) + 0.09Z_2(\mathrm{t}-2) + 0.08Z_3(t-2) \\ 0.14Z_1(\mathrm{t}-1) + 0.11Z_2(\mathrm{t}-1) + 0.11Z_3(\mathrm{t}-1) + 0.13Z_4(t-1) \end{pmatrix}$$
$$+ \begin{pmatrix} 0 \\ 0.16Z_6(\mathrm{t}-1) + 0.13Z_7(t-1) \\ 0.15Z_6(\mathrm{t}-1) + 0.12Z_7(t-1) \\ 0.17Z_6(\mathrm{t}-1) + 0.14Z_7(t-1) \\ 0 \\ 0.1Z_4(\mathrm{t}-2) + 0.06Z_5(t-2) + 0.07Z_7(t-2) - 0.04Z_1(t-4) \\ 0.14Z_5(\mathrm{t}-1) + 0.15Z_6(t-1) \end{pmatrix}$$
$$+ \begin{pmatrix} 0 \\ 0 \\ 0 \\ 0 \\ 0 \\ -0.04Z_2(t-4) - 0.05Z_3(t-4) - 0.05Z_4(t-4) - 0.04Z_5(t-4) \\ 0 \end{pmatrix} + \begin{pmatrix} 0 \\ 0 \\ 0 \\ 0 \\ 0 \\ -0.04Z_7(t-4) \\ 0 \end{pmatrix}$$

Interpretation of the above models is that the rainfall prediction in Polanharjo area in month t was 75 percent affected by the rainfall condition of a month earlier in its own area but it was'nt affected by the areas of Cawas, Trucuk, Karangdowo, Wonosari, Delanggu and Juwiring. However, for the area of Cawas, it wasn't affected by the rainfall condition of a month earlier in its own area but it was slightly affected by the rainfall condition of a month earlier of a month earlier in other areas around it. Condition in Cawas area had the similarity with th condition of Trucuk and Karangdowo areas. Otherwise the rainfall prediction in Wonosari area in month t was 82 percent affected by the condition in its own area and it wasn't affected by the other around it. For the rainfall prediction in Delanggu area in month t was 57 percent affected by the rainfall condition of a month earlier in its own area but it was slightly affected by the rainfall condition of two month earlier and four month earlier in other areas around it. The last in the Juwiring area in month t was slightly affected by the rainfall condition of a month earlier in other areas around it and it was not affected by the condition in its own area.

### *5.7. Model Validation*

To validate the above GSTAR model, the value of RMSE was calculated $= \sqrt{\frac{\sum_{t=1}^{n}(Z_t - \widehat{Z_t})^2}{n}}$ as one of the model quality measuring tools. GSTAR model $(1_1)$ for the rainfall in the areas of Sragen Regency had the RMSE value of 155.1631. The GSTAR model $(1_1)$ for the rainfall in the area of Karanganyar had the RMSE value of 179.1086. GSTAR model $(5_1)$ for the rainfall in the area of Klaten had the RMSE value of 141.6979.

## 6. Conclusion

The rainfall conditions of the sixth areas of rice production areas of Sragen regency in month t only were affected by the rainfall condition of a month earlier. From the sixth areas in Sragen only Plupuh area was affected by the rainfall condition in its own area which were significant at $\alpha = 5\%$.

Then the rainfall conditions in month t of the seventh areas of rice production areas of Karanganyar regency only were affected by the rainfall condition of a month earlier and only in Jaten area was affected by the rainfall condition in its own area which were significant at $\alpha = 5\%$.

But the rainfall conditions in month t of the seventh areas of rice production areas of Klaten regency were affected by the rainfall condition of a month earlier, two month earlier, and four month earlier. The rainfall conditions in Polanharjo, Wonosari and Delanggu areas were affected by the rainfall condition of a month earlier in its own areas and they were slightly affected by the rainfall condition in other areas around it. While the rainfall conditions in month t in Cawas, Trucuk and Karangdowo weren't affected by the rainfall condition in its own areas but they were slightly affected by the its condition in other areas arount them which were significant at $\alpha = 5\%$.

## Acknowledgments

The author would like to thank to Central Agency of Statistics for supporting the data and LPPM Universitas Sebelas Maret for giving financial support through Grant of Maintenance Research Group 2016.

## References

[1] Borovkova S A, Lopuhaa H P and Ruchjana B N 2002 Generalized STAR Model with Experimental Weight *Proceedings of the 17th Internasional Workshop on Statistical Modeling , paper 139-147*

[2] Fischer G, Shah M and Velthuizen H V 2002 Climate Change and Agriculture Vulnerability *World Summit on Sustainable Development, Johannes-burg, 160 pp*

[3] Fox J J 2000 *The Impact of the1997 – 1998 El Nino on Indonesia. In : R.H. Grove and J. Chappell (ed.) El Nino-History and Crisis Studies from the Asia-Pasific region* (Cambridge UK : The White House Press)

[4] IPCC (Intergovernmental Panel on Climate Change) 2001 *Climate Change 2001 : Impact, Adaptation and Vulnerability* (Cambridge UK : Cambridge University Press)

[5] Preifer P E and Deutsch S J 1980 A Three Stage Iterative Procedure for Space Time Modeling *Technometrics Vol.**22** no. 1 paper 35–47*

[6] Ruchjana B N 2002 Pemodelan Kurva Minyak Bumi Menggunakan Model Generalisasi STAR *Forum Statistika dan Komputasi Institut Pertanian Bogor*

[7] Sembiring R K 1995 *Analisis Regresi* (Bandung : Institut Teknologi Bandung)

[8] Sims C A 1980 Macroeconomics and Reality *Econometrica Vol. **48** no. 1 59–68*

[9] Sri S H, Yuliana S and Hasih P 2015 Geographically Weighted Regression Modeling for Corn Production in Central Java *Global Journal of Pure and Applied Mathematics Volume **11** No. 1 paper 37-46.*

[10] Suhartono and Subanar 2006 The Optimal Determination of Space Weight in GSTAR Model by Using Cross-Correlation Inference *Journal of Quantitative Methods Vol.2 45–53*

[11] Tao F, Yokozawa M, Zhang Z, Hayashi Y, Grassi H and Fu C 2004 Variability in Climatology and Agricultural Production in China in Association with the East Asia Summer Monsoon and El nino South Oscillation *Climate Res. **28** 23-30*

[12] Tsay R S 2005 *Analysis of Financial Time Series* (New Jersey : John Wiley dan Sons)

[13] Widarjono A 2009 *Ekonometrika Pengantar dan Aplikasinya* (Yogyakarta : Ekonisia)

[14] Wutsqa D U, Suhartono and Brodjol S 2010 Generalized Space Time Autoregressive Modelling *Proceedings of the 6th IMT-GT Conference on Mathematics, Statistics and its Applications 752-761*

[15] Yuliana S, Hasih P, Sri S H and Twenty L 2014 M Estimation, S Estimation and MM Estimation in Robust Regression *International Journal of Pure and Applied Mathematics, Volume **91** No. 3 349-360*

[16] Yuliana S, Hasih P, Sri S H and Twenty L 2015 Paddy Availability Modeling in Indonesia Using Spatial Regression *International Journal of Applied Mathematics, Volume **45** No. 4*

# Passive Copy-Move Forgery Detection Using Halftoning-based Block Truncation Coding Feature

**Bambang Harjito and Heri Prasetyo**
Informatics Department of Mathematics and Natural Sciences Faculty of Sebelas Maret University, Indonesia

E-mail: bambang_harjito@staff.uns.ac.id, and heri.prasetyo@staff.uns.ac.id

**ABSTRACT.** This paper presents a new method on passive copy-move forgery detection by exploiting the effectiveness and usability of Halftoning-based Block Truncation Coding (HBTC) image feature. Copy-move forgery detection precisely locates the large size or flat tampered regions of an image. On our method, the tampered input image is firstly divided into several overlapping image blocks to construct the image feature descriptors. Each image block is further divided into several non-overlapping image blocks for processing HBTC. Two image feature descriptors, namely Color Feature (CF) and Bit Pattern Feature (BF) are computed from the HBTC compressed data-stream of each image block. Lexicography sorting rearranges the image feature descriptors in ascending manner for whole image. The similarity between some tampered image regions is measured based on their CF and BF under specific shift frequency threshold. As documented in the experimental results, the proposed method yields a promising result for detecting the tampered or copy-move forgery regions. It has proved that the HBTC is not only suitable for image compression, but it can also be used in the copy-move forgery detection.

## 1. Introduction

HBTC is a simple image compression technique which converts an input image into another image representations, namely two color quantizes and bitmap image. Formerly, HBTC is technique for grayscale image compression. However, some progresses have been made for HBTC to extend its usability in color image compression [1,7], image watermarking, and image retrieval [1-2]. As reported in some literatures, image features derived from HBTC data stream yields a promising result in some applications.

This research extends the usability of the HBTC image feature for passive copy-move forgery detection task. For this task, the tampered or forged regions are determined using similar strategy in [3-6]. The main difference between the proposed method and former schemes [3-8] is on image feature generation. In our proposed method, an image feature is simply derived from HBTC data stream since HBTC offers low computational burden in image feature computation compared to the other schemes.

The rest of this paper is organized as follow: Section II gives a brief introduction about the HBTC image compression for color image. The image feature is subsequently presented in this section.

Section III presents the proposed method for the passive copy-move forgery detection using HBTC image feature. Section IV reports the experimental results on the passive copy-move forgery detection using the proposed method. The conclusion and future directions are given at the end of this paper

## 2. Feature Extraction

This section gives a brief introduction of the HBTC as well as feature extraction technique using HBTC compressed data stream. The HBTC is formerly proposed for grayscale image compression. Some researches extends the usability of the HBTC for color image compression and image retrieval [1-2,8]. Inspired by its successfulness, this paper exploits the effectiveness of HBTC into the copy-move forgery detection domain. The subsequent subsections explain the basic concept of HBTC and the feature extraction.

### 2.1. Halftoning-based Block Truncation Coding (HBTC)

The HBTC compresses an image in effective manner by simply dividing the input image into several non-overlapping (or overlapping) image blocks. The HBTC simply encodes each image block into another representation, i.e. two color quantizes and bitmap image. Let f(x,y) be an image block of size m×n. The HBTC performs the image compression of each block f(x,y) as follow :

$$\mathcal{H}\{f(x,y)\} \Rightarrow \{q_{min}, q_{max}, b(x,y)\} \quad (1)$$

for all pixel position x=1,2,…,m and y=1,2,…,n. The q_min and q_max denote the color quantizers, i.e. min and max quantizer, respectively. These two color quantizers can be simply computed by locating the minimum and maximum values over all pixel intensities. If the input image is in color space, the two color quantizers are also in the same color space. The min and max quantizer can be simply determined using the following strategies:

$$q_{min} = \min_{\forall x,y} f(x,y) \quad (2)$$

$$q_{max} = \max_{\forall x,y} f(x,y) \quad (3)$$

The other data stream can also be obtained after HBTC encoding, namely bitmap image b(x,y). In this research, we employ the Ordered Dither Block Truncation Coding to generate bitmap image. Please refer [1-2] to detail explanation of this bitmap image generation. The two color quantizers and the bitmap image can be utilized to generate image feature descriptor in the copy-move forgery detection.

### 2.2 Color Feature.

The first feature, namely CF, can be easily computed from HBTC color quantizers. Figure 1 shows the schematic diagram of CF computation. The CF computation requires two color codebooks obtained from Vector Quantization (VQ) over a set of training color images. Let $C = \{c_1, c_2, \dots, c_{N_{min}}\}$ and $D = \{d_1, d_2, \dots, d_{N_{max}}\}$ be the min and max color codebook, respectively, to index the min and max color quantizer. The color indexing processes for min and max quantizers are given as follow:

$$i_{min}(\theta) = \text{argmin}_{k=1,2,\dots,N_{\min}} \|q_{min}, c_k\|_2^2, \quad (4)$$

$$i_{max}(\theta) = \text{argmin}_{k=1,2,\dots,N_{\max}} \|q_{max}, c_k\|_2^2, \quad (5)$$

for all $\theta$, where $\theta$ denotes image block position in whole image. Subsequently, the CF for min quantizer can be obtained as:

$$CF_{min}(k) = \Pr\{i_{min}(\theta) = k, \forall\theta\}, \quad (6)$$

for $k = 1,2,\ldots,N_{min}$. Herein, the feature dimensionality of $CF_{min}$ is the same as min color codebook size $N_{min}$. The CF for max quantizer is given as follow:

$$CF_{max}(k) = \Pr\{i_{max}(\theta) = k, \forall\theta\}, \quad (7)$$

for $k = 1,2,\ldots,N_{max}$. The feature dimensionality of $CF_{max}$ is the same as max color codebook size $N_{max}$.

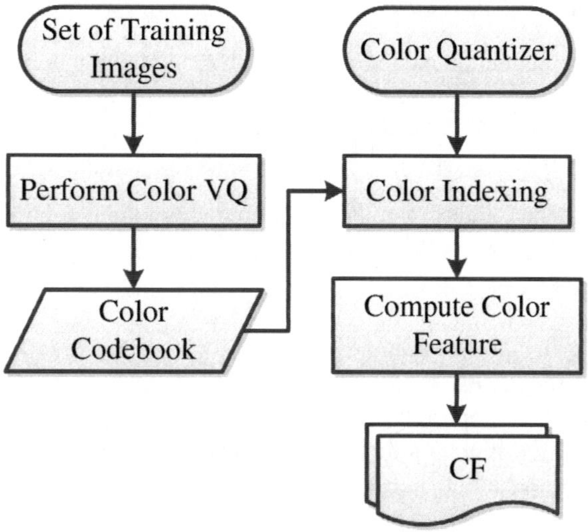


Figure. 1. Schematic Diagram of CF computation.

### 2.3. Bit Pattern Feature ( BF)

Another feature, namely Bit Pattern Feature (BF), can be easily generated from HBTC bitmap image. Figure 2 depicts the schematic diagram of BF computation. Let $B = \{b_1, b_2, \ldots, b_{N_b}\}$ be bit pattern codebook of size $N_b$. This bit pattern codebook can be obtained from binary VQ using soft centroid method under several training images. The bitmap indexing process of bitmap image $b(x,y)$ is formally defined as:

$$i_b(\theta) = \text{argmin}_{k=1,2,\ldots,N_b} \|b(x,y), b_k\|_2^2. \quad (8)$$

The BF can be subsequently calculated using the following formula:

$$BF(k) = \Pr\{i_b(\theta) = k, \forall\theta\}, \quad (9)$$

for $k = 1,2,\ldots,N_b$. The feature dimensionality of BF is identical to the bit pattern codebook size $N_b$.

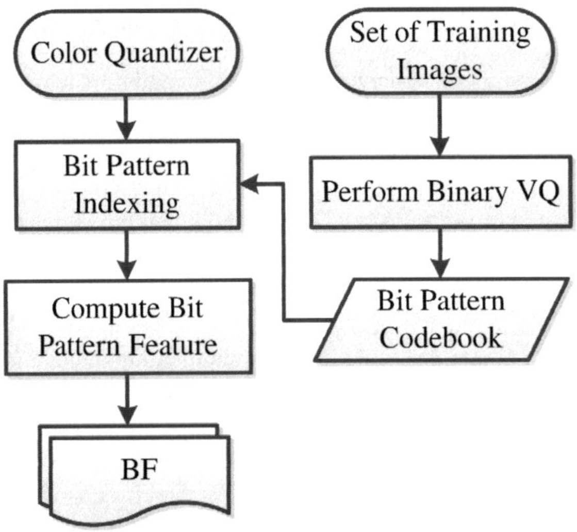


Figure. 2. Schematic Diagram of BF computation.

## 3. Proposed Method

This section presents the proposed passive copy-move forgery detection in detail. Figure 3 shows the schematic diagram of the proposed method. The proposed method receives the forged image containing some tampered or duplication regions on a single image. The proposed method marks the forged/tampered regions of the input image at the end of detection process. The copy-move forgery detection of the proposed method is then given as follow:

(1) Divide the suspicious image into several overlapping image blocks of fixed size.
Firstly, an input image with some tampered regions is divided into several overlapping blocks to yield some image blocks. The size of image block gives great impact at final performance.

(2) Divide an image block into several non-overlapping image blocks for HBTC processing.
Each overlapping image block from the previous step is further processed in this step. In this step, each image block is divided into several non-overlapping image blocks for HBTC processing. Herein, each image block is encoded and compressed using HBTC to yield two color quantizers ($q_{min}$ and $q_{max}$) and bitmap image ($b(x, y)$).

(3) Compute an image feature obtained from the HBTC data stream.
In this step, the proposed method extracts the image features from HBTC data stream (two color quantizers and bitmap image). The CF is derived from two color quantizers , whereas the BF is obtained from bitmap image. In our proposed method, the color and bit pattern codebook sizes are selected as $N_{min} = N_{max} = N_b = 8$ . To achive faster computation, we simply employ the image features with dimensionality 16.

(4) Perform similarity matching for all image blocks.
After obtaining the image features from previous step, this step performs similarity matching between all image regions based on their image features. The proposed method utilizes the lexicographically sorting to detect tampered region as often used in copy-move forgery detection [3-5]. The shift vector is also employed in the proposed method.

(5) Perform post-processing to remove the isolated regions.

To further improve the detection, an additional step is required to remove small regions or isolated image regions. Some morphological operations can be used in this step. This step produces the final copy-move forgery detected image.

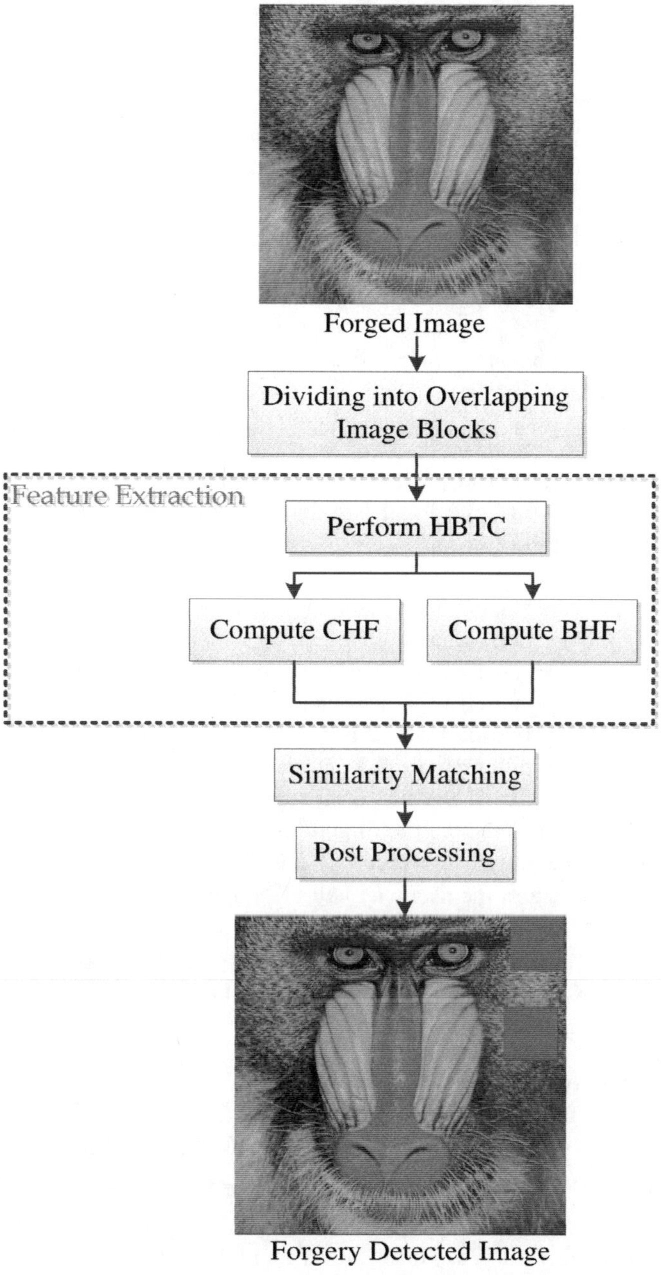


Figure. 3. Schematic Diagram of the Proposed Method

## 4. Experimental Results

Some experiments were carried out to investigate the usability and effectiveness of the proposed method. The performance of the proposed method is examined under three images from SIPI-USC dataset denoted as Baboon, Peppers, and Lake. The tampered regions for each image are manually determined and copied into the other region to create a forged image. The correctness of the proposed method is visually judge when the tampered region can be correctly detected by the algorithm and confirmed by human vision. In addition, the proposed method is objectively measured using two quantitative measurements

### 4.1. Evaluation Metrics

To objectively investigate the performance of proposed method, two quantitative metrics can be used namely accuracy ratio $(p)$ and false negative ratio $(f)$. The value of accuracy ratio decreases when the proposed method detects non-forged region as forged region. The accuracy ratio $(p)$ can be formally defined as:

$$p = \frac{|D_1 \cap R_1| + |D_2 \cap R_2|}{|D_1| + |D_2|}, \quad (10)$$

The other metric, namely false negative ratio, closes to zero while the algorithm detects almost all real forged regions as forged image. The false negative ratio $(f)$ is formally defined as

$$f = \frac{|D_1 \cap R_1| + |D_2 \cap R_2|}{|D_1| + |D_2|} - p, \quad (11)$$

Where $D_1$ and $D_2$ denotes copied and pasted/tampered regions, respectively, in forged/fake image. The symbols $R_1$ and $R_2$ represent copied and pasted/tampered regions, respectively, detected by the copy-move forgery detection algorithm.

### 4.2. Effectiveness of the Proposed Method

This subsection reports the result of proposed copy-move forgery detection over three image sets. For each image, the copied regions are manually determined by user and placed into the other position to yield a forged image. The proposed method tries to detect the copy-move regions for these three images. The successfulness of proposed method is visually examined and quantitatively measured. Fig. 4 shows the detection results of the proposed method. As shown in this figure, the proposed method gives the good result on detecting the forged/tampered regions which can be easily confirmed by human vision. For Baboon image, the accuracy and false negative ratios are $p = 0.98$ and $f = 0.03$. The Pappers image produces the accuracy and false negative ratios at $p = 0.98$ and $f = 0.01$. The proposed method yields the best forgery detection result for Lake image (compared to other two images) with accuracy and false negative ratios $p = 0.99$ and $f = 0.002$. It can be drawn from this experiment that the proposed method offers a promising result in the passive copy-move forgery detection task.

### 4.3. Comparison with Former Existing Schemes

The additional experiments were also conducted to further investigate the performance of the proposed method. Herein, the proposed method is objectively compared to the former scheme [5] over all three images. To make a fair comparison, both methods try to detect the forged regions under an identical feature dimensionality, i.e. 16. The accuracy and false negative ratios are calculated for each image. The average value is subsequently computed over all image tests for both methods. Table I summarizes the performance comparison between the proposed method and former scheme [5]. The proposed method outperforms the former scheme in the passive copy-move forgery detection problem. It proves that the HBTC feature is also suitable for forgery detection, not only for image retrieval and compression.

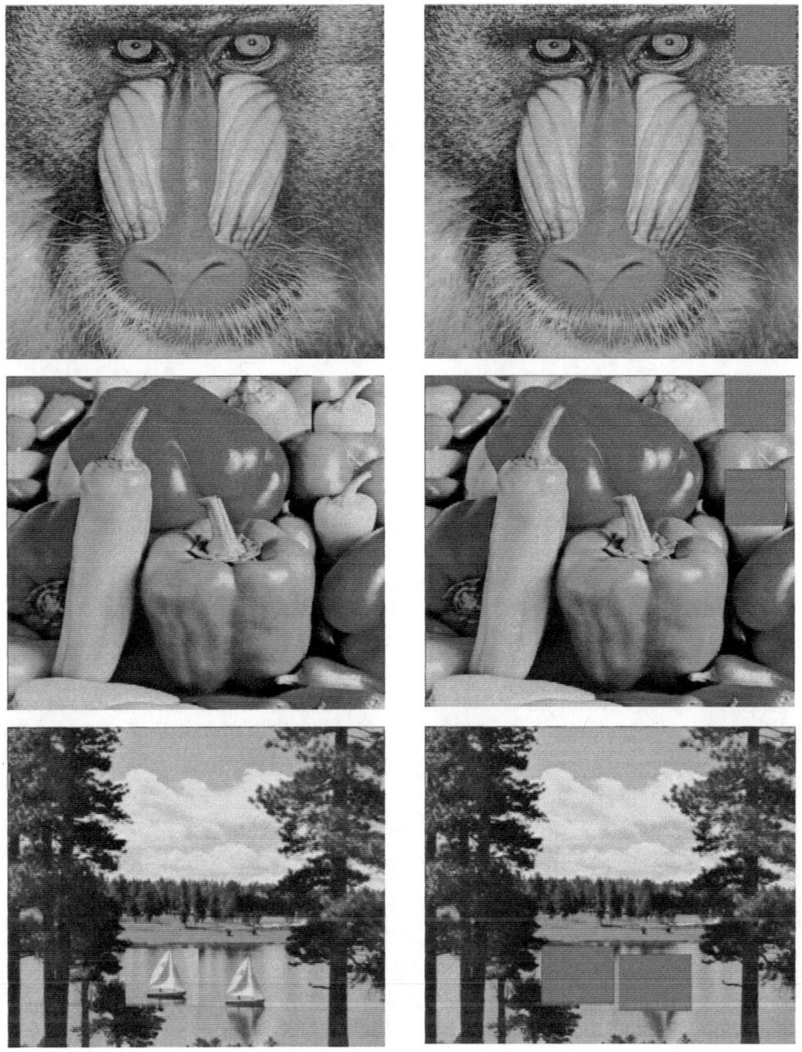

Figure. 4. The results of the copy-move forgery detection using the proposed method for three images. First column shows the forged image, and second column gives the detected forged region using the proposed scheme

Table I. Performance comparison between the proposed method and former scheme [5].

| Test Image | DCT-SVD-Based Scheme | | Proposed Method | |
|---|---|---|---|---|
| | *p* | *f* | *p* | *f* |
| Baboon | 0.97 | 0.03 | 0.98 | 0.03 |
| Peppers | 0.96 | 0.04 | 0.98 | 0.01 |
| Lake | 0.98 | 0.03 | 0.99 | 0.02 |
| Average | 0.97 | 0.03 | **0.98** | **0.02** |

## 5. Conclusions

A simple approach on passive copy-move forgery detection has been presented in this paper. The proposed method exploits the usability of HBTC feature on measuring similarity between the two image regions, i.e. copied and forged regions. The proposed method yields better performance compared to the former scheme under the same feature dimensionality. Another feature generated from HBTC data stream can be investigated to further improve the forgery detection performance. The rotation invariants as well as translation problem and multi resolution condition can also be considered for future works and developments.

**Acknowledgment**

The authors would like to thank the Institute for Research and Community Services of Sebelas Maret University for funding to this research in the academic year of 2016.

**References**

[1] J. M. Guo, and H. Prasetyo, "Content-based image retrieval using features extracted from halftoning-based block truncation coding," IEEE Trans. Image Process., vol. 24, no. 3, 2015.
[2] J. M. Guo, H. Prasetyo, and C. C. Yao, "Image retrieval using indexed histogram of void-and-cluster block truncation coding," Signal Process., vol. 123, pp. 143-156, 2016.
[3] O. M. Al-Qershi, and B. E. Khoo, "Passive detection of copy-move forgery in digital images: State-of-the-art," Forensic Sci. Int., vol 231, pp. 284-295, 2013)
[4] N. B. A. Warif, A. W. A. Wahab, M. Y. I. Idris, R. Ramli, R. Salleh, S. Shamshirband, and K. K. R. Choo, "Copy-move forgery detection: Survey, Challenges, and Future Directions," J. Network Comp. Apps., 2016, http://dx.doi.org/10.1016/j.jnca.2016.09.008
[5] J. Zhao, and J. Guo, "Passive forensics for copy-move image forgery using a method based on DCT and SVD," Forensic Sci. Int., vol. 233, pp. 158-166, 2013.
[6] E. J. Delp and O. R. Mitchell, "Image coding using block truncation coding," IEEE Trans. Commun., vol. 27, pp. 1335–1342, Sept. 1979.
[7] C. S. Huang and Y. Lin, "Hybrid block truncation coding," IEEE Signal Process. Lett., vol. 4, no. 12, pp. 328–330, Dec. 1997.
[8] J. M. Guo, and M. F. Wu, "Improved Block Truncation Coding Based on the Void-and-Cluster Dithering Approach," IEEE Trans. Image Processing, vol. 18, no. 1, pp. 211-213, Jan. 2009

# COMPARISON OF B-SPLINE MODEL AND ITERATED CONDITIONAL MODES (ICM) FOR DATA WITH MEASUREMENT ERROR (ME)

**Hartatik, Agus Purnomo**
Informatics Engineering Departement, Faculty of Mathematics and Natural Sciences , Sebelas Maret University, Surakarta, Indonesia
*E-mail:* hartatik.uns@gmail.com

**Abstract.** Direct observation results are often used to review the estimation model. However, actual data observation findings still need to be re-examined, because of measurement error factors (ME). In the regression modeling if X is a random variable with measurement error then the complicated calculation will not loose from application of Computer and Technology. As is the case for a review of the following model estimation, given data (Xi, Yi), then the regression model is

$$Y_i = g(X_i) + \varepsilon_i$$

where Xi is the element i from the predictor variables X and Yi is the element i of the response variable Y. The variable X is the predictor variables From the findings specific observations usually are constants, but generally found X which is a random variable variable or where Fixed value is not constant. In this case is called the regression model Regression Model with the measurement Errors. Purpose of this research are estimated nonparametric model approach with B-Spline Method to review regression with Measurement Errors are ignored and methods Iterative Conditional Mode (ICM) for review Model regression with measurement error.

Keywords: ICM, Bayesian, B-Spline, regression, measurement error, non-parametric,

## 1. Introduction

Data from the measurement or observation is often taken as a result of the finding of observations used to estimate the model estimates, and spline regression model. The results of the measurement data is generally still to be assessed, because of measurement error (ME) [1].

Many factors influence the measurement result data, including: tools, data retrieval process is correct, human factors, and nature factors. These factors can cause measurement results contain the measurement error (measurement error). But often, and most of researchers take and consider the data as a fixed value and believe that a data observation are correct , without checking how large measurement errors can be ignored [2]. The regression model when significant measurement error would be different if there is no measurement error. In equation (1) as below

$$Y = g(\mathrm{X}) + \varepsilon \tag{1}$$

where **X** is the predictor variables, the response variable **Y**, **g** is a function Certain functions, and $\varepsilon$ is a error random independent with mean zero and variance $\sigma_\varepsilon^2$ [3]. The variable **X** is the predictor variables from the findings of the observation is usually assumed as a variable Fixed (fixed variables). But in reality, often encountered **X** is not fixed but variable random variable **X** measured with errors (*errors in variables*). But this error often overlooked for a Practical Reason and easy of calculation. However, if the error performed repeatedly and variable on the High Risk Level then the consequences would be fatal Against Decision. This cases can be found at epidemiological problems, geology, and survival. Measurement error (ME) is an error appears when a recorded value exactly the same is not true value in the Measurement Process. Thus relates with this definition, there are 3 variables in Measurement Error Model, are

$$W = X + U \tag{2}$$

by $W$ is a variable that states observations called variable replacement (surrogate variable), **X** is the predictor variables for unobserved (latent variables), and **U** is the independent variable measurement errors are assumed normal with mean 0 and variance $\sigma_{\mathbf{u}}^2$.

Based on the regression model (1) and (2), there are two approaches were used to estimate the regression curve is Parametric and nonparametric regression methods. The purpose of this study was to perform steps regression model estimation using S-Plus software, namely the B-Spline method, and ICM [4].

## 2. Research Methods

This research is in the development of science, namely in the field of nonparametric statistics:

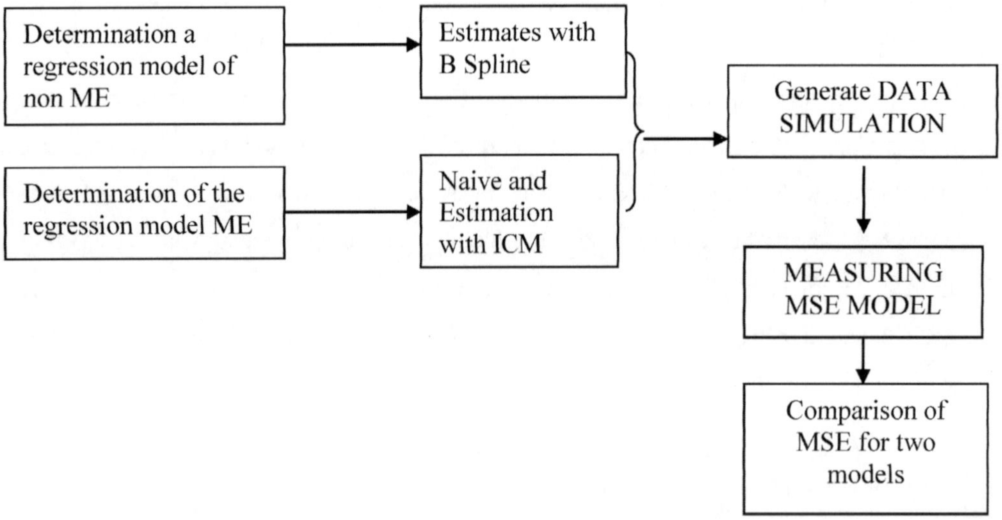

## 3. Comparison of nonparametric Models

### 3.1 B-SPLINE Model

Nonparametric regression model, if there is a spline of order m and a collection of knots that meet $a < \xi_1 < \ldots < \xi_k < b$. Can be defined a number of *2m* knot additional

$\xi_{-(m-1)}, \ldots, \xi_1, \xi_0, \xi_{k+1}, \ldots, \xi_{k+m}$

Where $\xi_{-(m-1)}, \ldots, \xi_0 = a$ dan $\xi_{k+1}, \ldots, \xi_{k+m} = b$.

B-spline order m corresponding to knots $\xi_1, \ldots, \xi_k$ expressed with the following models

$$\mathrm{B}_{j,m}(X) = \frac{X - \xi_j}{\xi_{i+m-1} - \xi_j} \mathrm{B}_{j,m-1}(X) + \frac{\xi_{j+m} - X}{\xi_{j+m} - \xi_{j+1}} \mathrm{B}_{j+1,m-1}(X), \tag{3}$$

with $\mathrm{B}_{j,1}(X) = \begin{cases} 1, X \in [\xi_j, \xi_{j+1}) \\ 0, \textit{the others.} \end{cases}$

As example it is assumed that $\hat{g}(X) = \sum_{j=1}^{p} \mathrm{B}_{j,m}(X)\beta_j$,

where $\mathrm{B}_1(X), \ldots, \mathrm{B}_p(X)$, is a basis for a vector spline of degree m with a point knots $\xi_1, \ldots, \xi_k$ and *p=m+k*. So,

$$S(\mathbf{g}) = \sum_{i=1}^{n} \{Y_i - \hat{g}(X_i)\}^2 + \alpha \int_a^b \{g^{(2)}(x)\}^2 dx$$

Penalized Least Square estimator [5]so that could be written by

$$\mathbf{S}(\mathbf{g}) = (\mathbf{Y} - \mathbf{B}(X)\beta)^T(\mathbf{Y} - \mathbf{B}(X)\beta) + \alpha\beta^T\mathbf{D}\beta, \tag{4}$$

with all elements- *ij* of **D** is

$$\mathbf{D} = \int_a^b \{\mathrm{B}_i''(x)\mathrm{B}_j''(x)\}\, dx.$$

So that the probe is function g

$$\hat{g}(X) = \sum_{j=1}^{p} \mathbf{B}_{j,m}(X)\beta_j, p=m+k \tag{5}$$

with

$$\beta = \left((\mathbf{B}(X))^T \mathbf{B}(X) + \alpha\mathbf{D}\right)^{-1} (\mathbf{B}(X))^T \mathbf{Y}.$$

Nonparametric regression model with B-spline method is applied to the data when the measurement errors are ignored and if the measurement error is not negligible [7]. To estimate the model when the measurement error ignored is by regressing between **W** and **Y**, while for nonparametric regression curve estimation when the measurement error is negligible average regressing between **W** and **Y** called Naive Methods.

**3.2.** ICM **Method for Model ME**

In estimating function **g** in measurement error models with ICM method is required prior information before determining the posterior distribution which then later used for iteration in the ICM method.

**i. Selestion Prior**

Suppose the observation data are assumed normal distribution,

$$Y \sim N\left(g(X_i), \sigma_Y^{\,2}\right)$$

$$W \sim N\left(X_i, \sigma_U^{\,2}\right)$$

Prior to the distribution of **X** and **g** is used for measurement error models (1) are as follows

$$X \sim N\left(\mu_X, \sigma_X^{\,2}\right) \qquad (6)$$

Where $\mu_X$ dan$\sigma_X^2$ is a constant determined.

As for the prior distribution g do approach "partially improper":

$$p(\mathbf{g}) \propto exp-\frac{\alpha}{2}\left(\mathbf{g}^T\mathbf{K}\mathbf{g}\right),$$

where for two matrices Q and R, the matrix ***K*** is defined as $\boldsymbol{K} = QR^{-1}Q^T$

The measures define a metric **K:**

- Define that: $h_i = t_{i+1} - t_i$, i=1,…,n-1
- Determine the matrix A is the matrix size **n x (n-2) with** $q_{ij}$ are as follows

  $q_{j-1,j} = h_{j-1}^{-1}, q_{j,j} = -h_{j-1}^{-1}, q_{j+1,j} = h_j^{-1}$ ,

  **For i=1,…,n, dan *j*=2,…,*n*-1, and** $q_{ij} = 0$ **for** $|i - j| \geq 2$.

**ii. Writing In The Form Of Matrix:**

So that the matrix Q can be written as follows:

$$Q = \begin{bmatrix} q_{12} & q_{13} & q_{14} & \cdots & q_{1,n-1} \\ q_{22} & q_{23} & q_{24} & \cdots & q_{2,n-1} \\ q_{32} & q_{33} & q_{34} & \cdots & q_{3,n-1} \\ \vdots & \vdots & \vdots & \ddots & \vdots \\ q_{n2} & q_{n2} & q_{n3} & \cdots & q_{n,n-1} \end{bmatrix}$$

$$= \begin{bmatrix} \frac{1}{(x_2 - x_1)} & 0 & 0 & 0 & \cdots & 0 & 0 \\ -\frac{1}{(x_2 - x_1)} & \frac{1}{(x_3 - x_2)} & 0 & 0 & \cdots & 0 & 0 \\ \frac{1}{(x_3 - x_2)} & -\frac{1}{(x_3 - x_2)} & 0 & 0 & \cdots & 0 & 0 \\ 0 & -\frac{1}{(x_3 - x_2)} & -\frac{1}{(x_3 - x_2)} & -\frac{1}{(x_3 - x_2)} & \cdots & 0 & 0 \\ \vdots & \vdots & \vdots & & \cdots & & \vdots \\ 0 & 0 & 0 & 0 & \cdots & \frac{1}{(x_n - x_{n-1})} & \frac{1}{(x_n - x_{n-1})} \end{bmatrix}$$

Subsequently determine the matrix R is a matrix of size (n-2) x (n-2) with elements for i and j from 2 to (n-1) is

$$r_{ii} = \frac{1}{3}(h_{i-1} + h_i), \quad i = 2,\dots, n-1$$

$$r_{i,i+1} = r_{i+1,i} = \frac{1}{6}(h_i), \quad i = 2,\dots, n-1$$

**and** $r_{ij} = 0$ **for** $|i - j| \geq 2$.

**iii. Writing in the form of Matrix:**

***R*** can be written in matrix form as follows:

$$R = \begin{bmatrix} r_{22} & r_{23} & r_{24} & \cdots & r_{2,n-2} & r_{2,n-1} \\ r_{32} & r_{33} & r_{34} & \cdots & r_{3,n-2} & r_{3,n-1} \\ r_{42} & r_{43} & r_{44} & \cdots & r_{4,n-2} & r_{4,n-1} \\ \vdots & \vdots & \vdots & & \ddots & \vdots \\ r_{n-1,2} & r_{n-1,2} & r_{n-1,3} & \cdots & r_{n-1,n-2} & r_{n-1,n-1} \end{bmatrix}$$

$$R=\begin{bmatrix} \frac{1}{3}(h_1+h_2) & \frac{1}{6}h_2 & 0 & \cdots\ 0 & 0 \\ \frac{1}{6}h_2 & \frac{1}{3}(h_2+h_3) & \frac{1}{6}(h_3) & \cdots\ 0 & 0 \\ 0 & \frac{1}{6}(h_3) & \frac{1}{3}(h_3+h_4) & \cdots\ 0 & 0 \\ \vdots & \vdots & \vdots & \ddots & \vdots \\ 0 & 0 & 0 & \cdots\ \frac{1}{6}(h_{n-2}) & \frac{1}{3}(h_{n-2}+h_{n-1}) \end{bmatrix}$$

$$R=\begin{bmatrix} \frac{1}{3}(x_3-x_1) & \frac{1}{6}(x_3-x_2) & 0 & \cdots & 0 & 0 \\ \frac{1}{6}(x_3-x_2) & \frac{1}{3}(x_4-x_2) & \frac{1}{6}(x_4-x_3) & \cdots & 0 & 0 \\ 0 & \frac{1}{6}(x_4-x_3) & \frac{1}{3}(x_5-x_3) & \cdots & 0 & 0 \\ \vdots & \vdots & \vdots & & \ddots & \vdots \\ 0 & 0 & 0 & \cdots & \frac{1}{6}(x_{n-1}-x_{n-2}) & \frac{1}{3}(x_n-x_{n-2}) \end{bmatrix},$$

so matrix ***can be write as*** $\boldsymbol{K=QR^{-1}Q^T}$.

Prior to variances in the ICM method is the result of the estimate, is as follows:

$$\hat{\sigma}_U^2 = \frac{\sum_{i=1}^{n}(m_i-1)s_i^2}{\sum_{i=1}^{n}(m_i-1)}, \tag{7}$$

and a estimate of the error variance i *is*

$$\sigma_\varepsilon^2 = \frac{\sum_{i=1}^{n}\{Y_i - g(X)\}^2}{tr\{I - A(\hat{\alpha})\}}, \tag{8}$$

with $\hat{\alpha}$ estimated by the method of GCV.

**iv. Formation Density Posterior with ICM method**

Based on the prior (7), here in after defined posterior distribution of parameters in nonparametric regression. The parameters in the model (1) is $\mathbf{X}, \mathbf{g}, \sigma_{\mathbf{X}}^2 \sigma_\varepsilon^2, \mathrm{dan}\, \sigma_{\mathbf{u}}^2$. Suppose,

$\theta = (X, g, \sigma_\varepsilon^2, \sigma_X^2, \sigma_u^2)$ maka based on the equation (6), the posterior density of $\theta$ is

$$
\begin{aligned}
p(\theta|Y,W) &\propto p(\theta)p(Y|\theta)p(W|\theta) \\
&\propto p(X,g,\sigma_X^2,\sigma_\varepsilon^2,\sigma_U^2)p(Y|g,X,\sigma_\varepsilon^2)p(W|X,\sigma_u^2) \\
&\propto p(Y|g,X,\sigma_\varepsilon^2)p(W|X,\sigma_u^2)p(X)p(g)p(\sigma_X^2)p(\sigma_U^2)p(\sigma_\varepsilon^2)p(\alpha)
\end{aligned}
$$

ICM method approach is used to estimate the parameters $\theta$ . Estimation of parameter $\theta$ with ICM method of determining the posterior distribution of the above equation, which is similar to determining the Bayesian posterior to the method in which all of the parameters have a prior distribution. But in the ICM method, which is also known as Bayesian some, but not all the parameters have distribution, in this case $\sigma_X^2\sigma_\varepsilon^2$, dan $\sigma_u^2$ estimation of data[8].

Thus conditional on $\hat{\sigma}_X^2,\hat{\sigma}_\varepsilon^2$, dan $\hat{\sigma}_u^2$ , the posterior distribution with ICM method is

$$
\begin{aligned}
p(\theta|\mathbf{Y},\mathbf{W}) &\propto p(\mathbf{Y}|\mathbf{g},\mathbf{X},\sigma_\varepsilon^2)p(\mathbf{W}|\mathbf{X},\sigma_u^2)p(\mathbf{X})p(\mathbf{g})p(\sigma_X^2)p(\sigma_U^2)p(\sigma_\varepsilon^2)p(\alpha) \\
&\propto exp\left[-\frac{1}{2\sigma_\varepsilon^2}(Y_i - g(X_i))^2 - \frac{1}{2\sigma_W^2}(W_i - X_i)^2 - \frac{1}{2\sigma_X^2}(X_i - \mu_X)^2 + \right. \\
&\qquad\qquad \left. -\frac{\hat{\alpha}}{2\sigma_\varepsilon^2}\mathbf{g}^T\mathbf{K}\mathbf{g}\right]
\end{aligned} \tag{9}
$$

ICM approach as described ,by finding the posterior mode (9). In the ICM method parameters are updated (update) in each iteration. Unlike the fully Bayesian method, the method of Bayesian posterior of each parameter estimated from the full condition[9].

**Algoritma ICM**

---

1. *Initialize* $\hat{\sigma}_X^2,\hat{\sigma}_\varepsilon^2$, dan $\hat{\sigma}_u^2$, and $g^{(0)}$

   $g^{(0)}$ is estimated using Naïve method, which is regressing between the value of Y with an average of W. The initial value of $\hat{\sigma}_X^2,\hat{\sigma}_\varepsilon^2$, and $\hat{\sigma}_u^2$ derived from the equation (6), (7), (8).

2. Based on one condition, determine the data conditional $g^{(i-1)}$ maximizing (9)

   Value of $X^{(i)}$ which gives the maximum value at (4:17) can not be solved analytically. Therefore, in this case, the method used to maximize the uniform grid (9).

3. Determining vector $g^{(i)}$ conditional on $X^{(i)}$ using Spline.

Based on the equation (9), with conditional on $X^{(i)}$, maximizing the posterior density (9) as well as minimizing

$$\left[ \frac{1}{2\hat{\sigma}_\varepsilon^2} \sum_{i=1}^{n} (Y_i - g(X_i))^2 + \frac{\hat{\alpha}}{2\hat{\sigma}_\varepsilon^2} \mathbf{g}^T \mathbf{K} \mathbf{g} \right].$$

equation (8) is an equation smoothing spline. To that end, estimate **g** can be used B-Spline, with conditional on $X^{(i)}$.

4. iteration $i=i+1$ is repeated until a certain $i$.

---

Simulations conducted using S-Plus for data model that represents the process in statistical data model is exponential and trigonometric[10]. Then performed simulations for each model 1 and model 2, with variations in the value of $n = 25, 50, 100, 250$, $\sigma_\varepsilon^2 = 0.01^2, 0.1^2, 1^2$, , $\sigma_U^2 = 0.01^2, 0.1^2, 0.5^2$ and $1^2$, $\mu_x = 0$, dan $\sigma_x^2 = 1$.

**Model 1**: Exponent Function

$$g(x) = exp(x),$$

where n=250, $\sigma_\varepsilon^2 = 0.01^2$, $\sigma_u^2 = 1^2$, $\mu_x = 0$, and $\sigma_x^2 = 1$.

And from the output S-plus, obtained spline regression curve and the MSE of the above simulation, as shown in Table 1 and Figure 1.

**Table 1**. Values MSE Exponential Function

| Methode | MSE |
|---|---|
| Non ME[1] | 19.90525 |
| Non ME[2] | 25.0184 |
| Naïve | 12.80861 |
| ICM | 4.231545 |

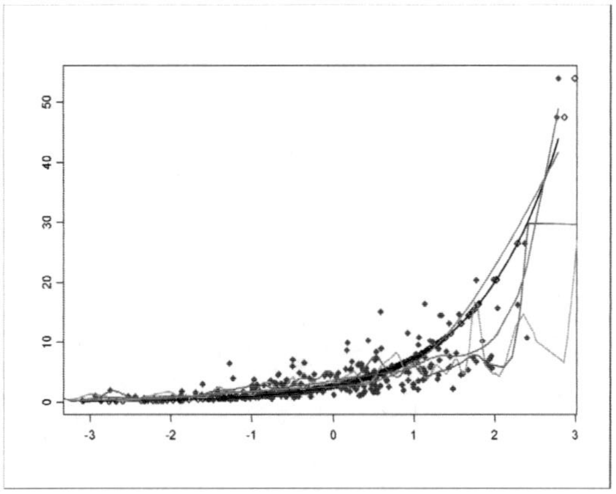

—— Actual curve —— non ME[1]
—— non ME[2] —— Naïve —— ICM

**Figure 1**. Curve Estimation for Exponential Function.

**Model 2** : Trigonometri function

$$g(x) = \frac{sin(\pi x/2)}{1 + 2x^2(sin\, g(x) + 1)},$$

$n$=250, $\sigma_\varepsilon^2 = 0.01^2$, $\sigma_u^2 = 0.1^2$, $\mu_x = 0$, and $\sigma_x^2 = 1$.

Based on the results of the output S-plus, spline regression curve obtained from the above simulation is as follows:

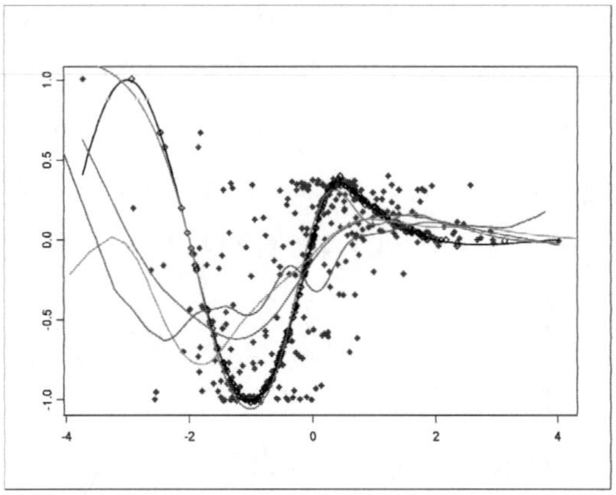

**Figure 2** Estimation Curves for Trigonometry Functions

**Table 2**. Value MSE for Trigonometry Functions

| **methode** | **MSE** |
|---|---|
| Non ME[1] | 0.158436 |
| Non ME[2] | 0.195413 |
| Naïve | 0.147683 |
| ICM | 0.030937 |

Based on simulation results for both models above, Figures 1 and 2 show the regression curves of the three methods, the regression curve of the data by ignoring any measurement error (with B-Spline), and then to estimate the model by not ignoring ME (naïve Method, and also with ICM). Judging from the pictures, the neglect of measurement error (model non-ME) effect on the regression curve. It is seen that the regression curve when the measurement error is ignored much of the actual curve. It appears also from Figure 1 and 2, with the regression curve.

ICM is the closest to the actual curve, g (x). In addition, with Naïve Method, also provide an estimate of the regression curve is closer to the curve g (x) compared with the curve when the measurement error is ignored. It provides information that Measurement Errors That fact should not be overlooked, but the measurement data should be be re-examined before being used to estimate further review.

Likewise, seen from the results of MSE for both the simulation model, the neglect of measurement error (model non-ME) effect on the results of MSE. It is seen that the estimated regression curve when measurement errors are ignored far greater than the value of its MSE ME models with Methode Naive or ICM. It appears also from Tables 1 and 2, the estimated regression curve for the simulated data with ICM method provides the smallest MSE value than the B-Spline and Naive methodes.

## 4. Conclusion

a. Based on the above discussion, it can be concluded that:

The greater $\sigma_U^2$, indicating that the MSE differences between non ME and ICM getting bigger. Based on simulation results, MSE non ME (ignoring measurement error} can achieve four times greater than when there is a measurement error models were estimated by ICM, both for $n$=25, 50,100,and 250. As for Naïve Method also shows the value of MSE is better than if the measurement error is negligible.

b. Based on simulation results also showed that the ICM method provides a better value than the Naïve Method.

## 5. References

[1] Bound, John, Brown,C, and Mathiowetz,N. (2001). "Measurement Error in Survey Data." In Handbook of Econometrics , vol. 5, edited by James J. Heckman, chap. 59. Amsterdam, The Netherlands: Elsevier, 3705–3843.

[2] Blattman, Christopher, Jamison, J.C, Palicz, T.K, Rodrigues,K, and Sheridan,M. (2015). "Measuring the Measurement Error: A Method to Quali-tatively Validate Survey Data." NBER Working Paper Series # 21447.

[3] Fan, J. and Truong, Y. K. (1993), "Nonparametric regression with errors in variables", Annals of Statistics, 21, 1900–25.

[4] Fuller, W. A and Hidiroglou, M. A (1976), " Regression Estimation After Correcting for Attenuation", Journal of the American Statistical Association, 73, 99-169.

[5] Amemiya, Y and Fuller, W. A (1984), "Estimation for Multivariate Errors in Variable Model with Estimated Error Covariance Matrix, " , *Annals of Statistics*, *Annals of Statistics*, 12, 497-509.

[6] Carroll, R. J., Maca, J. D. and Ruppert, D. (1999), "Nonparametric regression with errors in covariates", *Biometrika*, 86, 541–554.

[7] Box, G. E. P. and Tiao, G. (1973), *Bayesian Inference in Statistical Analysis*, Addison–Wesley, London.

[8] Berry, S. A., Carroll, R. J. and Ruppert, D. (2002), "Bayesian smoothing and regression splines for measurement error problems", *Journal of the American Statistical Association*, 9, 160–169.

[9] Besag, J. (1986). " On the Satistical Analysis of Dirty Pictures" (with Discussion), *Journal of the Royal Statistical Society*, Series B, 48, 259-279

[10]Hartatik(2012), Aplikasi S-Plus Dalam Analisa Regresi Guna Estimasi Model Regresi Untuk Data Dengan Kesalahan Pengukuran Sebagai Dasar Dalam Pengambilan Keputusan, Proceeding SNATI 2011,Yogjakarta.

# The Construction of Labeling and Total Irregularity Strength of Specified Caterpillar Graph

**Diari Indriati***, **Widodo, Isnaini Rosyida, Kiki A Sugeng**

Faculty of Mathematics and Natural Sciences, Universitas Sebelas Maret, Surakarta, Indonesia

E-mail: diari_indri@yahoo.co.id, diari_indri@staff.uns.ac.id

**Abstract.** Let $G$ be a simple, connected and undirected graph with vertex set $V$ and edge set $E$. A total $k$-labeling $f : V \cup E \rightarrow \{1, 2, \ldots, k\}$ is defined as totally irregular total $k$-labeling if the weights of any two different both vertices and edges are distinct. The weight of vertex $x$ is defined as $wt(x) = f(x) + \sum_{xy \in E} f(xy)$, while the weight of edge $xy$ is $wt(xy) = f(x) + f(xy) + f(y)$. A minimum $k$ for which $G$ has totally irregular total $k$-labeling is mentioned as total irregularity strength of $G$ and denoted by ts$(G)$. This paper contains investigation of totally irregular total $k$-labeling for caterpillar graphs $S_{n,2,m}$ and determination of their total irregularity strengths. In addition, the total vertex and total edge irregularity strength of this graph also be determined. The results are $tvs(S_{n,2,m}) = \lceil \frac{n+m-1}{2} \rceil$, $tes(S_{n,2,m}) = \lceil \frac{n+m+2}{3} \rceil$, and $ts(S_{n,2,m}) = \lceil \frac{n+m-1}{2} \rceil$ for $n, m \geq 3$.

## 1. Introduction

Let us consider a simple, connected and undirected graph $G$ with a vertex set $(V(G))$ and an edge set $(E(G))$. A labeling of a graph $G$ is a mapping that carries a set of graph elements into a set of integers, called labels (see Wallis [10]). If the domain of mapping is a vertex set, or an edge set, or a union of vertex and edge sets, then the labeling is called *vertex labeling*, *edge labeling*, or *total labeling*, respectively. In his survey, Gallian [2] shows that there are various kinds of labelings on graphs, and one of them is an irregular total labeling.

For a graph $G$, Bača *et al.* [1] defined a labeling $f : V(G) \cup E(G) \rightarrow \{1, 2, \ldots, k\}$ to be a *vertex irregular total $k$-labeling* if for every two different vertices $x$ and $y$ the *vertex-weights* $wt_f(x) \neq wt_f(y)$, where the vertex-weight $wt_f(x) = f(x) + \sum_{xz \in E} f(xz)$. A minimum $k$ for which $G$ has a vertex irregular total $k$-labeling is defined as the *total vertex irregularity strength* of $G$ and denoted by $tvs(G)$. They obtained the exact values of the total vertex irregularity strength for cycle, star, complete graphs and prisms. Moreover, Nurdin *et al.* [7] proved the exact value of the total vertex irregularity strength for any tree $T$ with $n$ pendant vertices and no vertex of degree two, that is

$$tvs(T) = \left\lceil \frac{n+1}{2} \right\rceil. \qquad (1)$$

For a graph $G$, Bača *et al.* [1] also define a labeling $g : V(G) \cup E(G) \rightarrow \{1, 2, \ldots, k\}$ to be an *edge irregular total $k$-labeling* of the graph $G$ if for every two different edges $xy$ and $x'y'$ of $G$ the *edge-weights* $wt_g(xy) = g(x) + g(xy) + g(y)$ and $wt_g(x'y') = g(x') + g(x'y') + g(y')$ are different. The *total edge irregularity strength* denoted by $tes(G)$, is defined as the minimum $k$ for which $G$ has an edge irregular total $k$-labeling. They also obtained the exact values of the *tes* for path, cycle, star, wheel and friendship graphs. The *tes* of generalized web graphs have been determined by Indriati *et al.* [3]. Moreover, Ivančo and Jendrol [5] proved that for any tree $T$, satisfy

$$tes(T) = \max\left\{\lceil(|E(T)|+2)/3\rceil, \lceil(\Delta(T)+1)/2\rceil\right\}. \tag{2}$$

Combining the ideas of vertex irregular total $k$-labeling and edge irregular total $k$-labeling, Marzuki *et al.* [6] introduced another irregular total $k$-labeling called the totally irregular total $k$-labeling. A labeling $h : V(G) \cup E(G) \to \{1, 2, \ldots, k\}$ to be a *totally irregular total $k$-labeling* of the graph $G$ if for every two different vertices $x$ and $y$ the *vertex-weights* $wt_h(x) \neq wt_h(y)$, where the vertex-weight $wt_h(x) = h(x) + \sum_{xz \in E} h(xz)$ and also for every two different edges $xy$ and $x'y'$ of $G$ the *edge-weights* $wt_h(xy) = h(x) + h(xy) + h(y)$ and $wt_h(x'y') = h(x') + h(x'y') + h(y')$ are different. The *total irregularity strength*, $ts(G)$, is defined as the minimum $k$ for which $G$ has a totally irregular total $k$-labeling. For the total irregularity strength of a graph $G$, they observed that

$$ts(G) \geq \max\{tes(G), tvs(G)\}. \tag{3}$$

They determined the total irregularity strength of cycle and path. Ramdani and Salman [8] obtained the total irregularity strength of some cartesian product graphs, namely $K_{1,n}\Box P_2$, $P_n \Box P_2$, $(P_n + P_1)\Box P_2$, and $C_n \Box P_2$. In [9], Ramdani *et al.* determined the total irregularity strength of Gear graphs $G_n, n \geq 3$, fungus graphs $F_{g_n}, n \geq 3$ and disjoint union of stars $mS_n, n, m \geq 2$. In [4], the total irregularity strength of double stars $S_{n,m}$ and caterpillar $S_{n,2,n}$ has been determined. In this paper, we continue to investigate the total irregularity strength of caterpillar $S_{n,2,m}$ for $n, m \geq 3$.

## 2. Caterpillar $S_{n,2,m}$

A caterpillar $S_{n,2,m}$ is a class of graph constructed from the double-star $S_{n,m}$ by inserting one vertex on the bridge connecting of the two centers of two stars. Therefore, this caterpillar contains three stars with the center of the two end-stars have degree $n$ and $m$ respectively, while the center of the middle star has degree two. This graph is a tree with $n + m + 1$ vertices, $n + m$ edges and $n + m - 2$ pendant vertices. Assume $n \leq m$. If $n > m$, then change $n$ by $m$, for example $S_{7,2,5}$ can be written by $S_{5,2,7}$. Maximal degree of the graph is $\Delta = m$.

According to (3), the lower bound of its total irregularity strength is the maximum value between its total edge irregularity strength and its total vertex irregularity strength. The total edge irregularity strength of graph $S_{n,2,m}$ can be found by (2), that is

$$tes(S_{n,2,m}) = \max\left\{\left\lceil\frac{\Delta+1}{2}\right\rceil, \left\lceil\frac{|E|+2}{3}\right\rceil\right\} = \max\left\{\left\lceil\frac{m+1}{2}\right\rceil, \left\lceil\frac{n+m+2}{3}\right\rceil\right\} = \left\lceil\frac{n+m+2}{3}\right\rceil. \tag{4}$$

The next theorem gives the total vertex irregularity strength of $S_{n,2,m}$.

**Theorem 2.1.** *Let consider the caterpillar $S_{n,2,m}$, $n, m \geq 3$. Its total vertex irregularity strength is*

$$tvs(S_{n,2,m}) = \left\lceil\frac{n+m-1}{2}\right\rceil.$$

*Proof.* $S_{n,2,m}$ is a tree with $n + m + 1$ vertices, $n + m$ edges and $n + m - 2$ pendant vertices. There is a vertex (that is a center of middle star) has degree two, therefore (1) can not be used for determining the total vertex irregularity strength of the graph. $S_{n,2,m}$ has $n + m - 2$ pendant vertices, one vertex of degree two and two vertices of degree $n, m \geq 3$ respectively. In order to obtain as small as possible label, start the labeling from the vertex with smallest degree (in this situation, pendant vertices have a smallest degree). After that, we label vertices with greater degree and continue until all vertices are labeled. The smallest vertex-weight is two, then with a consecutive weights, the smallest weight of $n+m-2$ pendant vertices is not smaller than $n+m-2+1 = n+m-1$ which is a sum of two labels, namely the label of pendant vertex and the label of edge incident to this vertex. Then, the greatest label of pendant vertices is not smaller than $\lceil \frac{n+m-1}{2} \rceil$. There is a vertex of degree two, therefore the greatest vertex-label is not smaller than $\lceil \frac{n+m}{3} \rceil$. For vertex of degree $n$, the greatest label is not smaller than $\left\lceil \frac{n+m+1}{n+1} \right\rceil$. The same idea, for vertex of the largest degree, $m$, the greatest label is not smaller than $\left\lceil \frac{n+m+2}{m+1} \right\rceil$. Therefore, the greatest label of all vertices is not smaller than

$$\max\left\{ \left\lceil \frac{n+m-1}{2} \right\rceil, \left\lceil \frac{n+m}{3} \right\rceil, \left\lceil \frac{n+m+1}{n+1} \right\rceil, \left\lceil \frac{n+m+2}{m+1} \right\rceil \right\} = \left\lceil \frac{n+m-1}{2} \right\rceil.$$

Let the vertex set of this graph be $V(S_{n,2,m}) = \{v_i^1 : 1 \le i \le m-1\} \cup \{v_i^3 : 1 \le i \le n-1\} \cup \{v^j : j =$

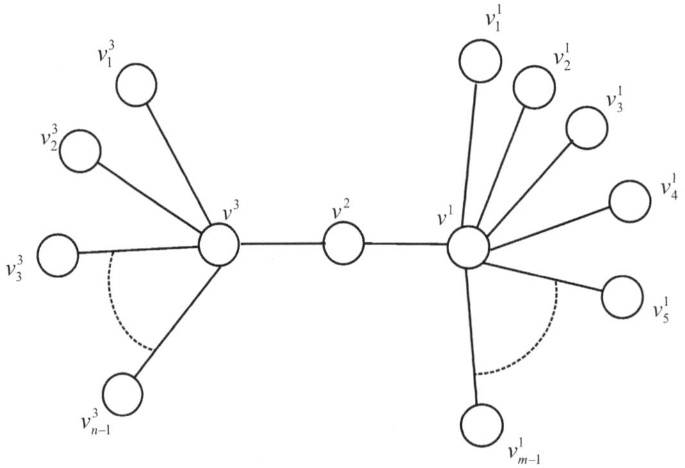


**Figure 1.** *The caterpillar $S_{n,2,m}$*

$1, 2, 3\}$ and the edge set be $E(S_{n,2,m}) = \{v^1v_i^1 : 1 \le i \le m-1\} \cup \{v^3v_i^3 : 1 \le i \le n-1\} \cup \{v^jv^{j+1} : j = 1, 2\}$. Figure 1 shows the illustration of this graph. To determine the exact value of $tvs$, define the vertex irregular total $k$-labeling $h$ as follows.

**Case 1:** For both $n$ and $m$ are odd or even.

Assume $k = \lceil \frac{n+m-1}{2} \rceil$.

$$h(v_i^1) = \begin{cases} 1, & \text{for } 1 \le i \le k, \\ i-k+1, & \text{for } k+1 \le i \le m-1. \end{cases}$$

$$h(v_i^3) = k+i-n, \text{for } 1 \le i \le n-1.$$

$$h(v^j) = \begin{cases} k-1, & \text{for } j = 1, \\ k, & \text{for } j = 2, 3. \end{cases}$$

$$h(v^1v_i^1) = \begin{cases} i, & \text{for } 1 \le i \le k, \\ k, & \text{for } k+1 \le i \le m-1. \end{cases}$$

$$h(v^3v_i^3) = k, \text{for } 1 \le i \le n-1.$$

$$h(v^jv^{j+1}) = \begin{cases} k-n+1, & \text{for } j = 1, \\ k, & \text{for } j = 2. \end{cases}$$

Under the total labeling $h$, it is shown that the greatest label for all vertices is $k = \lceil \frac{n+m-1}{2} \rceil$. It means that $h$ is a total $k$-labeling with $k = \lceil \frac{n+m-1}{2} \rceil$. The weight of vertices are

$$wt_h(v_i^j) = \begin{cases} i+1, & \text{for } 1 \le i \le m-1,\ j=1, \\ m+i, & \text{for } 1 \le i \le n-1,\ j=3. \end{cases}$$

$$wt_h(v^j) = \begin{cases} \frac{k(k+1)}{2} + (m+1-k)k - n, & \text{for } j=1, \\ 3k-n+1, & \text{for } j=2, \\ (n+1)k, & \text{for } j=3. \end{cases}$$

The weight of vertices $v_i^j$ for $j$=1, 3 form a consecutive integers from 2 up to $m$ and for $m+1$ until $m+n-1$, respectively. The weights among vertices $v^1$, $v^2$, and $v^3$ are distinct. Then, it indicates that the weights of every pair of vertices are distinct. Therefore, we conclude that $h$ is a vertex irregular total $k$-labeling and the total vertex irregularity strength is $tvs(S_{n,2,m}) = k = \lceil \frac{n+m-1}{2} \rceil$.

**Case 2:** For $n$ and $m$ have a different parity and odd number less than even number.
Assume $k = \lceil \frac{n+m-1}{2} \rceil$. Define the vertex irregular total $k$-labeling $h$ as follows.

$$h(v_i^1) = \begin{cases} 1, & \text{for } 1 \le i \le k, \\ i-k+1, & \text{for } k+1 \le i \le m-1. \end{cases}$$

$$h(v_i^3) = k+i-n+1, \text{for } 1 \le i \le n-1.$$

$$h(v^j) = \begin{cases} k-2, & \text{for } j=1, \\ k, & \text{for } j=3. \end{cases}$$

$$h(v^2) = \begin{cases} k-1, & \text{for odd number} = 3, \\ 2k-2n+2, & \text{for odd number} \neq 3. \end{cases}$$

$$h(v^1v_i^1) = \begin{cases} i, & \text{for } 1 \le i \le k, \\ k, & \text{for } k+1 \le i \le m-1. \end{cases}$$

$$h(v^3v_i^3) = k, \text{for } 1 \le i \le n-1.$$

$$h(v^1v^2) = \begin{cases} k, & \text{for odd number} = 3, \\ n, & \text{for odd number} \neq 3. \end{cases}$$

$$h(v^2v^3) = \begin{cases} k-1, & \text{for odd number} = 3, \\ n-1, & \text{for odd number} \neq 3. \end{cases}$$

Under total labeling $h$, it is shown that the greatest label for all vertices is $k = \lceil \frac{n+m-1}{2} \rceil$. It means that

$h$ is a total $k$-labeling with $k = \lceil \frac{n+m-1}{2} \rceil$. The weights of vertices are

$$wt_h(v_i^j) = \begin{cases} i+1, & \text{for } 1 \le i \le m-1,\ j=1, \\ 2k+i-n+1, & \text{for } 1 \le i \le n-1,\ j=3. \end{cases}$$

$$wt_h(v^1) = \begin{cases} \frac{k(k+1)}{2} + (m+1-k)k - 2, & \text{for odd number} = 3, \\ \frac{k(k+1)}{2} + (m-k)k + n - 2, & \text{for odd number} \neq 3. \end{cases}$$

$$wt_h(v^2) = \begin{cases} 3k-2, & \text{for odd number} = 3, \\ 2k+1, & \text{for odd number} \neq 3. \end{cases}$$

$$wt_h(v^3) = \begin{cases} 4k-1, & \text{for odd number} = 3, \\ (k+1)n-1 & \text{for odd number} \neq 3. \end{cases}$$

With the same reason as in Case 1, the weights of vertices $v_i^j$ for $j$=1, 3 form a consecutive integers from 2 up to $m$ and for $m+1$ until $m+n-1$, respectively and the weights among vertices $v^1$, $v^2$, and $v^3$ are distinct. Therefore, the weight of every pair of vertices are distinct. We conclude that $h$ is a vertex irregular total $k$-labeling and the total vertex irregularity strength is $tvs(S_{n,2,m}) = k = \lceil \frac{n+m-1}{2} \rceil$.

**Case 3:** For $n$ and $m$ have a different parity and odd number more than even number. Assume $k = \lceil \frac{n+m-1}{2} \rceil$. The vertex irregular total $k$-labeling $h$ is defined as follows.

$$h(v_i^1) = \begin{cases} 1, & \text{for } 1 \le i \le k, \\ i-k+1, & \text{for } k+1 \le i \le m-1. \end{cases}$$

$$h(v_i^3) = k+i-n+1, \text{for } 1 \le i \le n-1.$$

$$h(v^j) = \begin{cases} k-2, & \text{for } j=1, \\ k, & \text{for } j=3. \end{cases}$$

$$h(v^2) = \begin{cases} |m-n|+1, & \text{for } 3n-m \ge 3, \\ k, & \text{for } 3n-m < 3. \end{cases}$$

$$h(v^1v_i^1) = \begin{cases} i, & \text{for } 1 \le i \le k, \\ k, & \text{for } k+1 \le i \le m-1. \end{cases}$$

$$h(v^3v_i^3) = k, \text{for } 1 \le i \le n-1.$$

$$h(v^1v^2) = \begin{cases} n, & \text{for } 3n-m \ge 3, \\ k-n+2, & \text{for } 3n-m < 3. \end{cases}$$

$$h(v^2v^3) = \begin{cases} n-1, & \text{for } 3n-m \ge 3, \\ k-n+1, & \text{for } 3n-m < 3. \end{cases}$$

Under total labeling $h$, it is shown that the greatest label for all vertices is $k = \lceil \frac{n+m-1}{2} \rceil$. It means that

$h$ is a total $k$-labeling with $k = \lceil \frac{n+m-1}{2} \rceil$. The weights of vertices are

$$wt_h(v_i^j) = \begin{cases} i+1, & \text{for } 1 \leq i \leq m-1,\ j=1, \\ 2k+i-n+1, & \text{for } 1 \leq i \leq n-1,\ j=3. \end{cases}$$

$$wt_h(v^1) = \begin{cases} \frac{k(k+1)}{2} + (m-k)k + n - 2, & \text{for } 3n-m \geq 3, \\ \frac{k(k+1)}{2} + (m+1-k)k - n, & \text{for } 3n-m < 3. \end{cases}$$

$$wt_h(v^2) = \begin{cases} 2n + |m-n|, & \text{for } 3n-m \geq 3, \\ 3k - 2n + 3, & \text{for } 3n-m < 3. \end{cases}$$

$$wt_h(v^3) = \begin{cases} nk + n - 1, & \text{for } 3n-m \geq 3, \\ (n+1)k - n + 1, & \text{for } 3n-m < 3. \end{cases}$$

With the same reason as in Case 2, the weight of vertices $v_i^j$ for $j$=1, 3 form a consecutive integers from 2 up to $m$ and for $m+1$ until $m+n-1$, respectively and the weight among vertices $v^1$, $v^2$, and $v^3$ are distinct. It means that the weight of every pair of vertices are distinct. Therefore, $h$ is a vertex irregular total $k$-labeling. The total vertex irregularity strength of this graph is $tvs(S_{n,2,m}) = k = \lceil \frac{n+m-1}{2} \rceil$. From the three cases, it is shown that the total vertex irregularity strength of caterpillar $S_{n,2,m}$ is $\lceil \frac{n+m-1}{2} \rceil$. □

The next theorem proved the total irregularity strength of graph $S_{n,2,m}$ as follows.

**Theorem 2.2.** *Let $S_{n,2,m}$, $n, m \geq 3$ be a caterpillar graph. Then the total irregularity strength of this graph is*

$$ts(S_{n,2,m}) = \left\lceil \frac{n+m-1}{2} \right\rceil.$$

*Proof.* With the same statement as in Theorem 2.1, $S_{n,2,m}$ is a tree with $n+m+1$ vertices, $n+m$ edges and $n+m-2$ pendant vertices. Theorem 2.1 proved that $tvs(S_{n,2,m}) = \lceil \frac{n+m-1}{2} \rceil$. Total edge irregularity strength of this graph is in (4). According to (3), the lower bound of the total irregularity strength is

$$ts(S_{n,2,m}) \geq \max\{tes(S_{n,2,m}), tvs(S_{n,2,m})\} = \max\left\{ \left\lceil \frac{n+m+2}{3} \right\rceil, \left\lceil \frac{n+m-1}{2} \right\rceil \right\} = \left\lceil \frac{n+m-1}{2} \right\rceil.$$

To prove the exact value of the total irregularity strength of this graph, $ts(G)$, we will show the existence of the totally irregular total $k$-labeling as follows. The similar definition of vertex and edge set of $S_{n,2,m}$ is presented in Theorem 2.1. In fact, the vertex irregular total $k$-labeling $h$ which is obtained in Theorem 2.1 also satisfies the condition of totally irregular total $k$-labeling. The vertex weight distinction has been shown in the proof of Theorem 2.1. Now, we only show the distinction of the edge weight of this graph. By assume $k = \lceil \frac{n+m-1}{2} \rceil$, from the vertex irregular total $k$-labeling $h$ in Theorem 2.1, the weight of the edges are as follows.

$$wt_h(v^1 v_i^1) = \begin{cases} k+i, & \text{for } 1 \leq i \leq m-1, \text{ both } n \text{ and } m \text{ are odd or even,} \\ k-1+i, & \text{for } 1 \leq i \leq m-1, \ n \text{ and } m \text{ have a different parity} \end{cases}$$

$$wt_h(v^3 v_i^3) = \begin{cases} 3k-n+i, & \text{for } 1 \leq i \leq n-1, \text{ both } n \text{ and } m \text{ are odd or even,} \\ 3k-n+i+1, & \text{for } 1 \leq i \leq n-1, \ n \text{ and } m \text{ have a different parity} \end{cases}$$

For the case 1: both $n$ and $m$ are odd or even.

$$wt(v^j v^{j+1}) = \begin{cases} 3k - n, & \text{for } j = 1, \\ 3k, & \text{for } j = 2. \end{cases}$$

For the case 2: $n$ and $m$ have a different parity, odd number < even number.

$$wt(v^j v^{j+1}) = \begin{cases} 3k - n, & \text{for } j = 1, \\ 3k - n + 1, & \text{for } j = 2. \end{cases}$$

For the case 3: $n$ and $m$ have a different parity, odd number > even number.

$$wt(v^1 v^2) = \begin{cases} k + n + |m - n| - 1, & \text{for } 3n - m \geq 3, \\ 3k - n, & \text{for } 3n - m < 3. \end{cases}$$

$$wt(v^2 v^3) = \begin{cases} k + n + |m - n|, & \text{for } 3n - m \geq 3, \\ 3k - n + 1, & \text{for } 3n - m < 3. \end{cases}$$

It can be seen that for the case 1, that is both $n$ and $m$ are odd or even, the weights of $v^1 v_i^1$ and $v^3 v_i^3$ form a consecutive integers from $k+1$ up to $k+m-1$ and for $k+m+1$ until $k+m+n-1$, respectively. While for the case 2 and 3, that is $n$ and $m$ have a different parity, the weights of $v^1 v_i^1$ and $v^3 v_i^3$ form a consecutive integers from $k$ up to $k+m-2$ and for $k+m+1$ until $k+m+n-1$, respectively. For the case 1, the weight of $v^1 v^2$ is $k+m$ and the weight of $v^2 v^3$ is $k+m+n$. While for the case 2, the weight of $v^1 v^2$ is $3k-n$ and the weight of $v^2 v^3$ is $3k-n+1$. Same as another cases, the weights of $v^1 v^2$ and $v^2 v^3$ for the case 3 also different. Therefore, for all cases, the weight of each pair of edges are distinct. Then we conclude that $h$ is also totally irregular total $k$-labeling. Therefore, we determine that the total irregularity strength of this graph is $\lceil \frac{n+m-1}{2} \rceil$. □

Furthermore, we conclude this paper with the following open problem for the direction of further research which is still in progress.

**Open problem**: What is the total irregularity strength of caterpillar $S_{n,2,2,\ldots,m}$ for $t$-times of 2's.

## References

[1] Bača, M.,Jendrol, S., Miller, M., Ryan, J. On irregular total labeling. *Discrete Math.* 2007; **307**: 1378-1388.
[2] Gallian, J. A. A Dynamic Survey of Graph Labeling. *The Electronic Journal of Combinatorics* 2014; **17** #DS6.
[3] Indriati, D., Widodo, Wijayanti, I.E., Sugeng, K.A. and Bača, M. On total edge irregularity strength of generalized web graphs and related graphs. *Mathematics in Computer Science* 2015; **9** (2):161-167.
[4] Indriati, D., Widodo, Wijayanti, I.E., Sugeng, K.A. On total irregularity strength of double-stars and related graphs. *Procedia-Computer Science* 2015; **74**: 118-123.
[5] Ivančo, J. and Jendroľ, S. Total edge irregularity strength of trees. *Discussiones Math. Graph Theory* 2006; **26**: 449-456.
[6] Marzuki, C.C., Salman, A.N.M. and Miller, M. On the total irregularity strength of cycles and paths. *Far East Journal of Mathematical Sciences* 2013; **82**: 1-21.
[7] Nurdin, Baskoro, E.T., Salman, A.N.M. and Gaos, N.N. On the total vertex irregularity strength of trees. *Discrete Math.* 2010; **310**: 3043-3048.
[8] Ramdani, R. and Salman, A.N.M. On the total irregularity strength of some cartesian product graphs. *AKCE International J. Graphs Combin.* 2013; **10** (2): 199-209.
[9] Ramdani R., Salman, A.N.M., Assyatun, H., Semaničová - Feňovčíková, A., Bača, M. Total irregularity strength of three families of graphs. *Mathematics in Computer Sci.* 2015; **9**: 229-237 DOI. 10.1007/s11786-015-0229-x.
[10] Wallis, W.D. *Magic graphs*. Boston: Birkhauser; 2001.

IOP Conf. Series: Journal of Physics: Conf. Series **855** (2017) 012019 doi:10.1088/1742-6596/855/1/012019

# Experimentation of cooperative learning model Numbered Heads Together (NHT) type by concept maps and Teams Games Tournament (TGT) by concept maps in terms of students logical mathematics intellegences

Adi Irawan, Mardiyana, and Dewi Retno Sari Saputro.
Study Program of Mathematics Education and Postgraduate of Sebelas Maret University, Indonesia

E-mail: mathadi790@gmail.com, mardiyana@lycos.com,dewi.rss@gmail.com

**ABSTRACT:** This research is aimed to find out the effect of learning model towards learning achievement in terms of students' logical mathematics intelligences. The learning models that were compared were NHT by Concept Maps, TGT by Concept Maps, and Direct Learning model. This research was pseudo experimental by factorial design 3x3. The population of this research was all of the students of class XI Natural Sciences of Senior High School in all regency of Karanganyar in academic year 2016/2017. The conclusions of this research were: 1) the students' achievements with NHT learning model by Concept Maps were better than students' achievements with TGT model by Concept Maps and Direct Learning model. The students' achievements with TGT model by Concept Maps were better than the students' achievements with Direct Learning model. 2) The students' achievements that exposed high logical mathematics intelligences were better than students' medium and low logical mathematics intelligences. The students' achievements that exposed medium logical mathematics intelligences were better than the students` low logical mathematics intelligences. 3) Each of student logical mathematics intelligences with NHT learning model by Concept Maps has better achievement than students with TGT learning model by Concept Maps, students with NHT learning model by Concept Maps have better achievement than students with the direct learning model, and the students with TGT by Concept Maps learning model have better achievement than students with Direct Learning model. 4) Each of learning model, students who have logical mathematics intelligences have better achievement then students who have medium logical mathematics intelligences, and students who have medium logical mathematics intelligences have better achievement than students who have low logical mathematics intelligences.

Keywords: NHT, TGT, Concept Maps, Logical Mathematics intelligences

## 1. Introduction

Learning model is very influential in attempt to increase the achievement of mathematics learning. The use of appropriate learning models can encourage students` sense of happy towards lessons and able to achieve better learning outcomes[1]. The appropriate learning model to make students more active is cooperative. It can solve issues together so that the purpose in learning can be achieved.

The fundamental principle of cooperative learning is the group members are linked together in such a way that they cannot succeed unless everyone succeed, they will actively assist each other to make sure that the assignment is done and the purpose of the group achieved[2]. Cooperative learning model in practice is centered on students so able to active student's cognitive activity. The learner-centered approach makes it possible to activate the cognitive and educational activity of every student and provide them with opportunities to practice oral communication[3]. Expressed that cooperative learning is the pedagogical use of a small group to enable each student to maximize the results of your learning and at the same time learning all others[4].

Cooperative learning involves many activities, so it can enhance the interactions between students. Cooperative learning to involve many activities, such as communication, observation, and support. In addition, students changed their learning behaviors, stimulated their cognitive activity, and improved the relationships among students[5]. Besides that, cooperative learning has far-reaching effects that extend beyond the classroom, into professional participant and personal lives[6]. In mathematics learning, cooperative is able to contribute significant achievements in achievement and students` attitude. Cooperative learning had significant effects on the mathematics achievement and attitudes towards mathematics[2].

Cooperative learning model that can be applied in learning mathematics is a cooperative learning model of Numbered Heads Together (NHT) and Teams Games Tournament (TGT). The learning model Numbered Heads Together (NHT) is a model of learning that can eliminate the gap between students who are intelligent and not intelligent so, each student will play actively in discussion. This learning model demands a responsibility of each student, because in practice, teachers will designate one of the students randomly to answer each question. Effective teaching of mathematics is interactive, two-way process in which students play an active part by answering questions and discussion, explaining and demonstrating their methods to others in the class[7]. It will ensure the involvement of all students and it is a very good attempt to increase individual responsibility in group discussion[8]. Cooperative learning model of Teams Games Tournament (TGT) is a learning model that provides a challenge to students to compete as individuals who represent their respective groups to collect as much as score in game tournaments. This technique provides an opportunity for all members of the group to be able to participate to get value[9].

In addition to the learning models, the other factor that may affect the achievement of learning mathematics is a logical mathematical intelligence possessed by each student. Logical mathematics intelligence is able to complete a wide range of issues related to patterns of figures through thinking capability based on the rules of logic. The logical mathematics intelligence is intelligence that contains a person's ability in inductive and deductive thinking, thinking ability according to the rules of logic, understanding, and analyzing patterns figures and solving the problem using the thinking ability[10]. Logical-mathematical intelligence consisted of the capacity to analyze problems logically, carry out mathematical operations, and investigate issues scientifically. In mathematical education, our aims are solving problem with thinking logically, and reason deductively[11].

Mathematics is a subject which requires lengthy reasoning in solving any problems. Therefore, a good strategy is needed to give well understanding to the students. One of the strategies that are able to give contribution is strategy of Concept Maps. The strategy of Concept Maps is a strategy that is created by connecting the core concepts. The relationship between these concepts is a visual representation of other concepts is interrelated. Concept maps visualized through a graphical representation are commonly defined u.s. two dimen framework diagrams that consist of concepts or nodes (put into an ellipse, circle, rectangle or square) jointed by labelled lines (the words or phrase) to show relationships (propositions) and interrelationships between those selected concepts[12]. Concept maps is an effective strategy for teaching and learning mathematics. The strategy is also capable of improving students ' mastery of content at the higher-order levels of cognition[13].

This research is supported by some research related to cooperative learning model type NHT, TGT, Logical Mathematics Intelligence and the use of Concept Maps strategy. Research regarding to cooperative learning model type NHT performed by [15]*Mahaedy* et al (2006) this research showed that Numbered Heads with incentives was more effective in raising academic achievement than Numbered Heads without incentives and both methods were more advantageous for learning than traditional methods. The difference with this research is located on the strategies used in modifying the learning model of NHT. In this study the researchers modified the NHT with Concept Maps. Research about cooperative learning model of type TGT was conducted by Veloo and Chairhany[16]. This research showed that significant differences in the improved attitude and achievement towards probability. In addition, cooperative learning TGT also creates an active learning environment in solving exercises, and discussions among students and teachers. This study provides evidence that learning probability with TGT benefits students. The difference with this research is TGT modified with Concept Maps. [17]Research conducted by Niroo et al (2012) this research showed that in the case of pre-test, there exists a significant relationship between the mathematical Intelligencer and students ' mathematical functioning in general and in levels of application and reasoning. Based on the results of the research, researchers were taking research review of logical mathematics intelligence. Research that has been done by [18]Kamble and Tembe (2013), the results suggest that the concept mapping strategy improved the students' performance in problem solving than the traditional teaching method. This research aims to find out 1) which produce better learning achievements of mathematics, learning model cooperative of NHT with Concept Maps, TGT with Concept Maps, or Direct learning model, 2) which had better mathematics learning achievements, student with high, medium, or low logical mathematics; 3) on each model of learning, which had better mathematical learning achievements, students with high, medium or low logical mathematics intelligence; 4) on each intelligence, which produce better mathematics learning achievements, learning model cooperative of NHT with Concept Maps, TGT with Concept Maps, or Direct learning model.

## 2. Research Method

This type of research was quasi experimental with 3 x 3 factorial designs. The population in this research was the whole XI students of Science at Senior High School in the regency of Karanganyar in Academic Year 2016/2017. Sampling technique used was Stratified Cluster Random Sampling. This research was conducted in Kebakkramat, Senor High School of Kerjo and Mojogedang. In each of school was taken three classes, with two classes of experimental and one control class. Data collection method used documentation and tests. Before the research conducted, researchers did test normality and homogeneity to the data of student ability, and then tested the balance with variable analysis of one road with the same cells. On the beginning ability data and and learning mathematics achievement normality used Liliefors test, and homogeneity test used Bartlet test. Hypothesis test used the variable analysis test two paths with different cells and test comparison test with double Scheffe.

## 3. Result and Discussion

The prerequisite of analysis have been completed that it distributed normally and homogenous to the group of learning model and logical mathematics intelligence group then preceded with variable analysis test two paths with the different cells. In Table 01, was presented the result summary of variable analysis counting by two paths with different cells.

Table 01. The Summary Of Variable Analysis Two Paths With Different Cells

| Source | S.S | D.F | M.S.S | $F_{obs}$ | $F_{tabel}$ | Decision |
|---|---|---|---|---|---|---|
| A | 9317.32 | 2 | 4658.66 | 30.62 | 3.00 | $H_{0A}$ rejected |
| B | 92377.67 | 2 | 46188.83 | 303.63 | 3.00 | $H_{0B}$ rejected |
| AB | 328.31 | 4 | 82.07 | 0.54 | 2.37 | $H_{0AB}$ accepted |
| Error | 47157.87 | 310 | 152.12 | - | - | - |
| Total | 149181.19 | 318 | - | - | - | - |

Based on Table 01 above can be concluded that: 1) there was a difference in student learning achievement given the learning model of NHT with Concept Maps, TGT with Concept Maps and Direct learning model; 2) there was a difference in student learning achievement which has the logical mathematical intelligence of high, medium and low; 3) there was no interaction between the learning model with logical mathematics intelligence towards the students` learning achievements. Before the double comparison tests conducted in advance were presented the summary of marginal average on each learning model and mathematics intelligence. As shown in the following table 02

Table 02. The Summary of Marginal average

| Learning Models | Logical Mathematics Intellgences | | | Marginal Average |
|---|---|---|---|---|
| | High | Medium | Low | |
| NHT-CM | 93.09 | 69.73 | 49.10 | 71.76 |
| TGT-CM | 85.56 | 62.70 | 45.14 | 64.63 |
| Direct Learning | 77.18 | 57.44 | 36.84 | 56.43 |
| marginal average | 85.35 | 63.19 | 43.18 | |

Comparison test between the lines was done because $H_{0A}$ was declined. In table 03, the summary of testing results of comparison between lines on each learning model was presented.

Table 03. The Summary Of Testing Results Of Double Comparison Between Lines

| $H_0$ | $F_{obs}$ | $2F_{0,05;2;319}$ | Decision Test |
|---|---|---|---|
| $\mu_{1.} = \mu_{2.}$ | 173.59 | 6.00 | Rejected |
| $\mu_{1.} = \mu_{3.}$ | 140.94 | 6.00 | Rejected |
| $\mu_{2.} = \mu_{3.}$ | 599.44 | 6.00 | Rejected |

Based on Table 03 marginal average, it can be concluded that 1) Learning Model of NHT with Concept Maps was better than learning model TGT with Concept Maps. 2) Learning Model of NHT with Concept Maps was better than Direct learning model; 3) Learning Model of TGT with Concept Maps was better than Direct learning model.

The test double comparison between lines was done because $H_{0B}$ was declined. In table 04, the summary of testing results of comparison between columns on each logical mathematics intelligence was presented.

Table 04. The Summary of Testing Results of Double Comparison between Columns

| $H_0$ | $F_{obs}$ | $2F_{0,05;2;319}$ | Decision Test |
|---|---|---|---|
| $\mu_{.1} = \mu_{.2}$ | 70.71 | 6 | Rejected |
| $\mu_{.1} = \mu_{.3}$ | 62.73 | 6 | Rejected |
| $\mu_{.2} = \mu_{.3}$ | 254.80 | 6 | Rejected |

Based on the table 04, it can be concluded that: 1) the students who have high logical mathematics intelligence had better learning mathematics achievement then students who have medium logical mathematics intelligence. 2) The students who have high logical mathematics intelligence had better learning mathematics achievement then students who have low logical mathematics intelligence. 3) The students who have medium logical mathematics intelligence had better learning mathematics achievement then students who have the low logical mathematics intelligence.

Double comparison tests between cells were not done because $H_{0AB}$ accepted, then there was no interaction between the learning model with logical mathematics intelligence towards the students` learning achievement. Comparison test between cells on the same line and between cells on the same column did not need to be done. The conclusion of comparison between cells on the same line was based on the conclusion of comparisons test between lines, namely; 1) learning model of NHT with Concept Maps better than learning model TGT with Concept Maps; 2) cooperative learning model of NHT with Concept Maps are better than Direct learning model; 3) learning model of TGT with Concept Maps are better than Direct learning model.

The comparison conclusion between cells in the same column were based on comparisons between test conclusion of double column namely; 1) students who have high logical mathematics intelligence had better learning mathematics achievement then students who have medium logical mathematics intelligence.; 2) students who have high logical mathematics intelligence had better learning mathematics achievement then students who have low logical mathematics intelligence.; 3) students that have medium logical mathematics intelligence had better learning achievement then students who have low logical mathematics intelligence.

## 4. Conclution and Suggestions

Based on the results of research data analysis and discussion, it can be concluded. 1) Student Learning Achievement given the cooperative learning model of NHT with Concept Maps was better than learning achievement of students who were given the cooperative learning model TGT with Concept Maps and models of learning directly. The students` learning achievements that were given the cooperative learning model of TGT with Concept Maps produced better learning achievement then of students who were given Direct learning model; 2) learning achievements of students who have high logical mathematics intelligence were better than students who have the medium and low logical mathematics intelligent. The learning achievements of students who have medium logical mathematics intelligence was better than students who have low logical mathematics intelligence; 3) on each logical mathematics intelligence, students who were given learning model of NHT with Concept Maps have better learning achievements than students who were given learning model of TGT with Concept Maps, the students who were given the learning model of NHT with Concept Maps have better learning achievement than students who were given direct learning model, and the students` learning achievement who were given the learning model of TGT Concept Maps have better learning achievement than students who were given direct learning model; 4) On each learning model, students who have high logical mathematics intelligence have better learning achievements than students who have medium logical mathematics intelligence, and students who have medium logical mathematics intelligence have better learning achievements than students who have low logical mathematics intelligence.

Based on the results of the study, the researchers gave advice as follows: 1) Teachers should be able to develop creative learning processes, effective and enjoyable so that student learning achievement increased. It can be realized if the teachers used appropriate mathematics learning models. One of learning model that can be applied in learning mathematics in terms of opportunities material was learning models Numbered Head Together; 2) Teacher should use the strategy Concept Maps in modifying the learning model because these strategies were able to make students understanding mathematics concepts quickly and optimally so that learning mathematics achievement will be better; 3) teacher should notice the students` logical mathematics intelligence, so that in cooperative learning, teachers can create heterogeneous groups due to the level of the students` logical mathematics intelligence ; 4) on this study, researchers used logical mathematics intelligence review. Logical mathematical intelligence was selected because it was intelligence that contributed greatly in learning mathematics. For further researcher, it was possible to conduct different views, such as on learning style, creativity, Advertise Question, Reading comprehension and others.

## 5. Reference

[1] Aunurrahman. 2013.*Belajar Dan Pembelajaran*. Bandung: Alfabeta

[2] Hossain, A., Tarmizi, R.,A. 2013. Effects of cooperative learning on students' achievement and attitudes in secondary mathematics. *Procedia - Social and Behavioral Sciences* 93 (2013) 473 – 477

[3] Saltymakov, M.,S., Frantcuzskaia, E., O. 2015. Cooperative Learning Approach to

Delivering Professional Modules to Bachelor and Master Students: TPU Experience. Procedia - Social and Behavioral Sciences.215. 90 – 97

[4]Serafin C 2016 The Re-conceptualization of Cooperative Learning in an Inquiry-oriented Teaching Procedia - Social and Behavioral Sciences 217 201 – 207

[5]Othman H, Asshaari I, Bahaludin H, Tawil N M , Ismail N A 2012 Students' Perceptions on Benefits Gained from Cooperative Learning Experiences in Engineering Mathematics Courses Procedia - Social and Behavioral Sciences 60 500 – 506

[6] Kupczynski, L., Mundy, M. A., Goswami, J., & Meling V. 2012. Cooperative Learning in Distance Learning: A Mixed Methods Study. *International Journal of Intruction*. Vol. 5. No.2. 82.

[7]Sulaiman T Abdurahman A R, Rahim S S A 2010 Teaching Strategies Based on Multiple Intelligences Theory among Science and Mathematics Secondary School Teachers Procedia Social and Behavioral Sciences 8 512–518

[8] Nur, M. 2005. *Pembelajaran Kooperatif.* Surabaya: Pusat Sains dan Matematika Sekolah UNESA.

[9] Huda. M. 2013. *Cooperative Learning*. Yogyakarta: Pustaka Pelajar.

[10] Iskandar. (2012). *Psikologi Pedidikan (Sebuah Orientasi Baru).* Jakarta : Referensi.

[11] Karamikabir N 2012 Gardner's multiple intelligence and mathematics education Procedia - Social and Behavioral Sciences 31 778 – 781

[12] Djanette B, Fouad C 2014 Determination of university students' misconceptions about light using concept maps Procedia-Social and Behavioral Sciences 152 582 – 589

[13] Awofala, A. O. A.2011. Effect of Concept Mapping Strategy on Students' Achievement in Junior Secondary School Mathematics. *International Journal of Mathematics Trends and Technology.*Vol. 2. Issue3. 11.

[14]Maheady, L., Michielli-Pendl,J., Mallette, B.& Harper, G.F. 2006. The Effects of Numbered Heads Together with and Without an Incentive Package on the Science Test Performance of a Diverse Group of Sixth Graders. *Jurnal of Beharvioral Education*, Volume 15, Number 1, page 25-39.

[15] Veloo A Chairhany S (2013) Fostering Studens' Attitudes and Achieveent In probability Using Teas Gaes Tournaents Procedia Social and Behavioral Sciences 93 59-64

[16] Niroo M. Nejhad G. H. H., Haghani. M. 2012 The effect of Gardner theory application

on mathematical/logical intelligence and student's mathematical functioning relationship Procedia - Social and Behavioral Sciences 47 2169 – 2175

[17] Kamble S. K., Tembe B.L.2013The effect of use of concept maps on problem solving performance and attitude in Mechanical engineering course Procedia - Social and Behavioral Sciences 83 748 – 754

IOP Conf. Series: Journal of Physics: Conf. Series **855** (2017) 012020 doi:10.1088/1742-6596/855/1/012020

# Verification of Radiation Isocenter on Linac Beam 6 MV using Computed Radiography

**Muhammad Irsal, Eko Hidayanto, and Heri Sutanto**

Post Graduate Program of Physics Department of Diponegoro University, Indonesia

E-mail : muhammadirsal@st.fisika.undip.ac.id

**Abstract**. Radiation isocenter is more important part of quality assurance for the linear accelerator (Linac) due to radiation isocenter is a main location in irradiation radiotherapy, isocenter can shift when the gantry and collimator rotation. In general, the radiation isocenter verification using a special film. This research was conducted radiation isocenter verification using computed radiography with digital image processing techniques. Image acquisition was done using the modalities of Linac 6 MV with star shot method is star-shaped beam due to rotation of the collimator, gantry and couch. Then do the delineation on each beam to determine the centroid and beam diameter. By the results of verification of radiation isocenter performed on collimator and the couch, it shows that the size diameter for rotational collimator is 0.632 mm and 0.458 mm for the couch. Based on AAPM report 40 about the size of the Linac radiation isocenter diameter used in this study is still in good condition and worth to be operated because the value of the radiation isocenter diameter is below 2 mm.

## 1. Introduction

The use of ionizing radiation for the treatment (radiotherapy) of cancer due to interaction of ionizing radiation with tissue can lead to death of cell, either direct or indirect. This is the base of radiotherapy. Radiotherapy utilize the sensitivity of cancer cells, which half of cancer cells get more sensitiv than normal cells, therefore radiation therapy is a cancer treatment method that is commonly used for this [1]. One of the modality of modern radiotherapy used in the treatment of cancer is a Linear Accelerator (Linac). This radiotherapy modality produces electron beam and X-rays (photons) generated by the generator and have the energy that can be varied [2].

Quality Assurance (QA) plays a very important role in ensuring that the treatment given to patients have been correctly conducted. Modern treatment techniques such as Three-Dimensional Conformal Radiotherapy (3DCRT), Intensity Modulated Radiation Therapy (IMRT) and Streotatic Radio Surgery (SRS) aims to deliver maximum radiation dose to the cancer cells and the minimum radiation dose to the surrounding normal tissue. Therefore, various QA tests should be conducted on Linear Accelerator for ensuring accurate treatment delivery [3].

The success of radiotherapy depends on various factors, one of them is the accuracy of the location of irradiation on the planning to the implementation of performed irradiation. Verification radiation isocenter has a very important role on the quality of radiation therapy administered [4].

IOP Conf. Series: Journal of Physics: Conf. Series **855** (2017) 012020 doi:10.1088/1742-6596/855/1/012020

Radiation isocenter is a very fundamental part of commissioning and QA testing because isocenter Linac is a major role of irradiating radiotherapy [5].

The main reason for the lack of commitment to QA by many hospitals are financial problems. An adequate QA program requires an increase in staff and the latest equipment, both require substantial funds. According to an analysis by Peters, the total cost of radiotherapy QA program about 3% of annual hospital bill [6]. Computed radiography (CR) was first used in the late 80s as a portal imaging with X-rays megavolt energy. CR has several advantages in radiotherapy because it can save operating costs compared to the use of film and Electronic Portal Imaging Device or EPID. In addition, the film storage can affect the outcome of radiotherapy QA. One of QA should be done on Linac in radiotherapy, the radiation isocenter verification. Verification of radiation isocenter generally performed only once when commissioning data collection for the first Linac activated by the Linac vendor. Based Task Group 40 American Association of Physicists in Medicine (AAPM), radiation isocenter verification must be regularly performed once a year so that the research expected of medical physicists can radiation isocenter verification regularly once a year. Verify isocenter radiation aimed to determine linac beam diameter. From the results of research conducted by the Peace of An experimental study on using a diagnostic computed radiography system as a quality assurance tool in radiotherapy showed a maximum variation values between film and CR that is -0.13 mm to verify the radiation isocenter so that CR can be used for QA in radiotherapy [3]. In this research, the authors were done radiation isocenter diameter measurement Linac beam using computed radiography. Radiation isocenter diameter measurement generally used a special film, but it takes a long time in data processing. In this research using CR for easier and faster to display the results of images and data processing.

In the radiotherapy modern treatment, equipment got to be very accurate, that the provision of radiation therapy mostly uses mechanical of collimator, gantry, and an couch that aims to optimize the target range and minimize exposure of surrounding healthy tissue [7]. Based on Task Group 40 American Association of Physicists in Medicine (AAPM) about verification of radiation isocenter conducted once a year [8]. provides recommendations tolerance limit radiation isocenter which is 2 mm, because the radiation isocenter can be shifted during the gantry and collimator rotation [9]. Therefore it is necessary to verify the radiation isocenter on the X-ray beam plane of 6 MV linear accelerator using computed radiography.

## 2. Literature Study

### *2.1. Linear Accelerator*

Linear accelerator uses electromagnetic waves with high frequency to accelerate charged particles like electrons to produce energy that is directed in a tube linear. Produced high energy electrons can be used directly for the treatment of tumors close to the surface, or directed towards a target that generates X-rays with high energy (megavolt) used for the treatment of tumors in high depth.

For generating a photon beam (X-ray), the high-energy electron beam is passed at a target made of a thin heavy metal so that the interaction Bremstrahlung happens. Bremstrahlung is the X-rays emitted when an electron at high speed passes with a nucleus (nuclei), then the strong Coulomb attractive force causes the electrons to sharply deviate from the initial trajectory [10].

The result of the generation of X-rays have a high intensity in the direction of the target axis. To achieve required flatness, flattening filter made of steel is mounted.

### *2.2. Definition and Basic Principles of Computed Radiography (CR)*

Computed radiography is the digitization of images using a sheet or photostimulable plate for the acquisition of image data [13]. Photostimulable plate on computed radiography stores latent image which is processed by using a laser and can be applied to analog cassette-based screens and films [11].

Principles of pictures imaging on CR is not much different from the conventional radiography, fundamental differences in CR that it does not use screens and films, but it use imaging plate. In the imaging plate, Photostimulable Phosphor Plate with a thickness of less than 1 mm. Photostimulable phosphorus will capture signals attenuation of X-rays.

The process of imaging on CR begins when the Imaging Plate (IP) is exposed with X-rays, so it will produce a latent image. Exposed IP is inserted into the slot on Imaging Plate Reader Device. Then, IP is scanned with a helium-neon laser (emission of red light with a wavelength of 633 nM) so that, the crystal on the IP produce blue-violet light (wavelength 390-400 nM). This light is detected by the photo sensor and sent via Analog Digital Converter to a computer for processing. After the image is obtained, the IP is transferred to another section of the Imaging Plate Reader Device to remove the remnants of the image to a reusable IP [12].

Imaging plate that has been exposed is inserted into the imaging plate reader. In the IP Reader, the cassette will automatically open, and the IP is taken out. IP is scanned, removed and returned to the cassette to be used for further examination. The image that has been scanned is inputted into a computer for processing and displayed on a monitor or a film.

The most important components in the acquisition of image data consisting of imaging plates, cassettes, and imaging plate reader. Imaging plate is a sheet that can capture and store latent X-ray shadow which the shadow of the latent is formed on the film when the X-rays passing through the patient and captured by the detector. Image acquisition functions performed by Photostimulable Plate or Imaging Plate (IP) is a radiographic film. This IP receives X-ray beam that has passed through the patient's body [13]. As for the characteristics of imaging plate is shown in Figure 1.

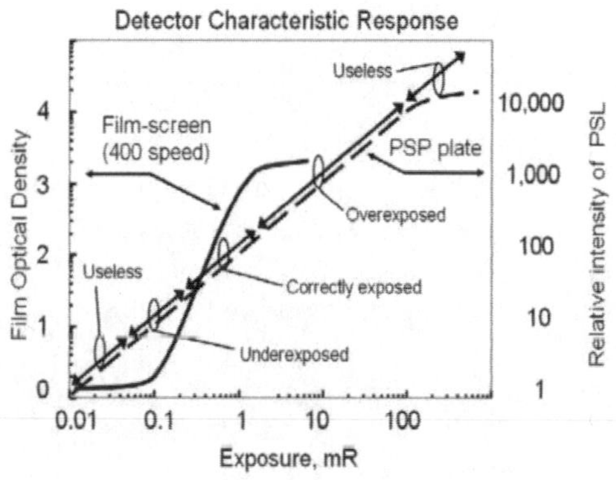


**Figure 1.** Characteristics of imaging plate [14]

Figure 1 Characteristics of imaging plate is a chart of comparing response screen the movie speed rare-earth 400 (S shaped curve) with photostimulable phosphor in computed radiography (curved dotted lines). Doubled arrow line indicates used exposure divided into underexposed, proper exposure and over exposure regions. Useless area is an area where the given response generating the diagnosis information low value dueto quantum noise and saturation PSP. From the chart, it can be seen that the PSP on CR requires higher exposure, but provides dynamic a widerrange. The wider the dynamic range, the wider the given exposure range CR.

## 2.3. *QA for Linear Accelerator*

AAPM report No. 46 discussing the report of the Task Rroup 40 radiation therapy committee: a comprehensive QA for radioation oncology describes the use of linear accelerator that at that time the radiation treatment is the most widely used which would require greater scrutiny and more careful review of the QA test parameters. In addition, a new generation of linear accelerator has been controlled by a computer so that adds to the complexity of tool parameters based on the frequency of measurements, including the parameters described in AAPM Report 13 and some additional parameters that correspond to hardware latest generation [8].

Since the publication of TG-40, several new technologies have been developed and now commonly used in clinical treatment of radiotherapy, named multileaf collimation (MLC), asymmetric jaws, dynamic and virtual wedges, electronic portal imaging devices (EPIDs) and image guidance devices cone-beam CT (CBCT). In addition, TG-40 does not considerate the radiotherapy treatment procedure using stereotactic radiosurgery (SRS), streotactic body radiation therapy (SBRT), total body photon irradiation (TBI), and intensity-modulated radiotherapy treatment (IMRT). So that, AAPM Task Group 142 report: quality assurance of medical accelerator on the latest QA of use Linac is made [15]. This is comparison table of QA TG 40 and TG 142. As for the QA Linear Accelerator is shown in Table 1.

**Table 1.** QA Linear Accelerator

| Frequency | Procedure | Tolerance TG 40 | Tolerance TG142 |
|---|---|---|---|
| Annual | Mechanical checks | | |
| | Collimator rotation isocenter | 2 mm diameter | - |
| | Gantry rotation isocenter | 2 mm diameter | - |
| | Couch rotation isocenter | 2 mm diameter | - |
| | Coincidence of collimetry, gantry, couch axes with isocenter | 2 mm diameter | - |
| | Coincidence of radiation and mechanical isocenter | 2 mm diameter | 2 mm from baseline |

## *2.4. Verification of Radiation Isocenter*

Verification of radiation isocenter is a very fundamental part of the test of commissioning and QA, isocenter is very important in the treatment of radiotherapy using a Linac as a center of irradiation radiotherapy. Isocenter QA radiation in radiotherapy Linac is a star shot method [5]. As for the radiation isocenter star shot method is shown in Figure 2.

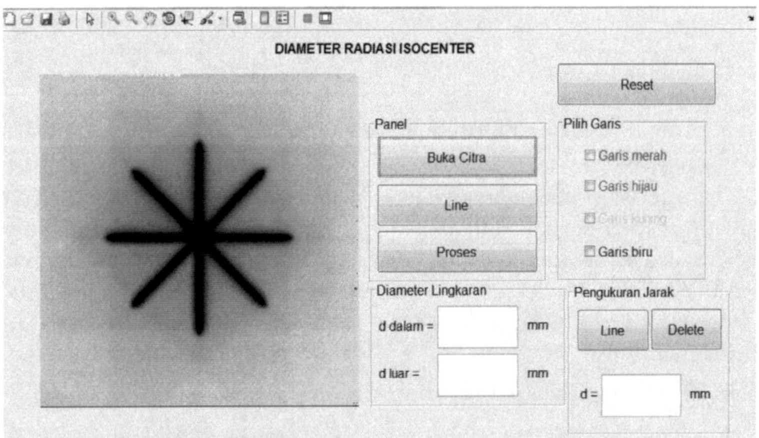


**Figure 2.** Radiation isocenter star shot method

A stellar beam shots is obtained by placing the film radiation in the field of rotational motion of a gantry, collimator, and an couch. Film will be exposed to radiation from different angles of rotation so that the beam intersect to form stars. radiation isocenter Verify is done by determining the smallest circle of the intersection of all the files, with the center of the circle called radiation isocenter and the radius of the circle called the size of the radiation isocenter. To determine shifts radiation isocenter can be seen from the result of the file to the isocenter alignment shifting or used laser Linac for determining the location of irradiation [7]

## 3. Method

### *3.1. Instruments and Materials*

This research use Linac modality with 6 MV photon energy. Then, it uses computed radiography (CR) and the Matlab r2015b. Used Digital image is an image with the TIFF format CR result of the acquisition modalities. This digital image is used as an inputted image that will be processed through the computational programming. The variables in this study is rotational collimator, the angle of rotation and the couch.

### *3.2. Procedure*

Before the study began, firstly research preparation is conducted to ensure QA of Linac output in a constant state during the study by conducting a check output of output daily. Examination results is compared with the results of commissioning according to AAPM recommendations to the deviation of daily output that may not have a deviation of more than 3% compared to the commissioning. In addition, the calibration on the CR is done before being used in research. CR is calibrated by using Linac irradiation at a distance of 100 cm between the Linac tube to the CR plate and the field size of 5 x 5 $cm^2$ by placing phantom above CR plate. Exposure is done by varying the monitor unit (MU) from 0.5 to 2.5. Then the results were analyzed and obtained a calibration graph the relationship between pixel-dose radiation. As for the schematic of CR calibration is shown in Figure 3.

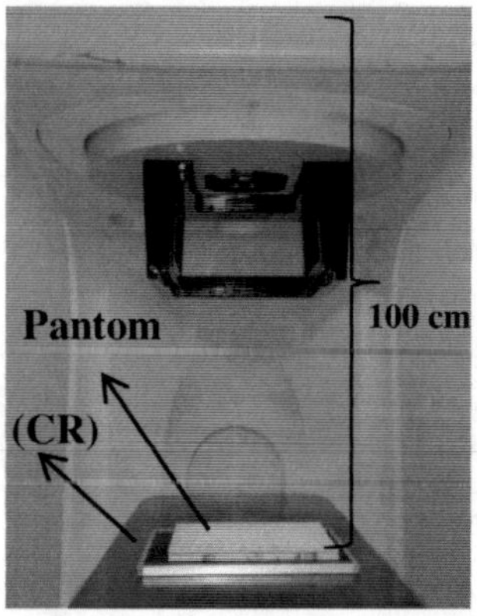


**Figure 3.** Schematic of CR calibration

Measurement radiation field size is determined on in-line and cross-line axis. Measurement of irradiation on CR cassette with the distance of 100 cm between the tubes air against the CR cassette by using 2 MU exposure, field size of 5 x 5 $cm^2$. Analysis of measurement data are treated in a plot on the profile beam and determined Full Width at Half Maximum (FWHM) using matlab. Verification of radiation isocenter is conducted using the method of star pattern image to determine the size of the diameter of the radiation isocenter component parameters that are collimator and couch. CR is placed perpendicular to the central axis of X-ray beam. Irradiation is done with the distance of 100 cm between the device tube to the CR cassette. Jaws is opened with the area of 0.2 x 20 $cm^2$ field for verification of radiation isocenter on used angle collimator were 0, 45, 90 and 135 while on the couch angles used 270, 315, 0 and 45. Then CR is irradiated with a beam energy of 0.5 MU. the results Analysis of the CR image processed by using Matlab with delineation method was done to determine the diameter size of the radiation isocenter.

## 4. Results

### *4.1. Linac Calibration Test*

Calibration on the Linac, according to AAPM TG 142, includes several timing parameters such as daily, monthly and yearly with calibration testing which is dosimeter parts, mechanical and safety. In this research, calibration test of the X-ray output consistency is done. The percentage output of the X-ray photons in July and August 2016 is shown in Figure 4.

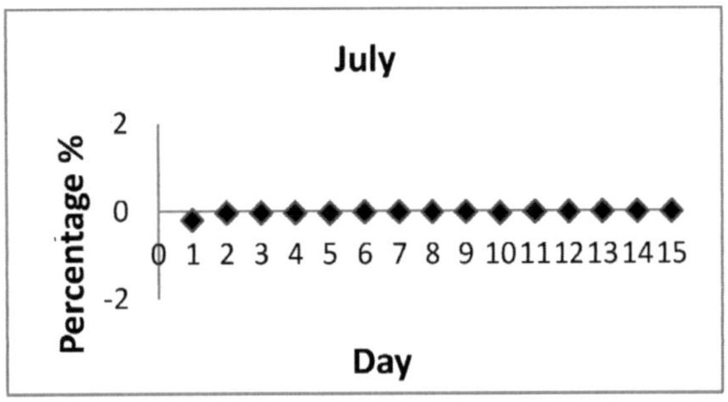


(a)

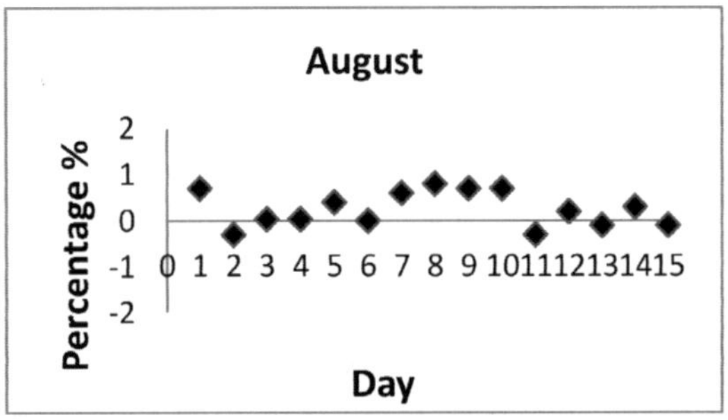


(b)

**Figure 4.** Percentage output of the X-ray photons in:
(a) July (b) August

By the results on the last two months show output variation device is different for each month. By the pictures, we can observe that Linac output is varies in July with a range between -0.02% to -0.2%, range between 0.8% to -0.3% in August. According to this result, the Linac is still in good condition because it came under the range of -3% to 3%.

### *4.2. Computed Radiography Calibration Test*

CR calibration test by making relationship graph between the pixel value and monitor unit value. CR calibration test using photon irradiation with energy variation start from 0.5 up to 2.5 MU (or in a dose of 0.5 cGy up to 2.5 cGy). As for the computed radiography calibration test results shown in Figure 5.

.

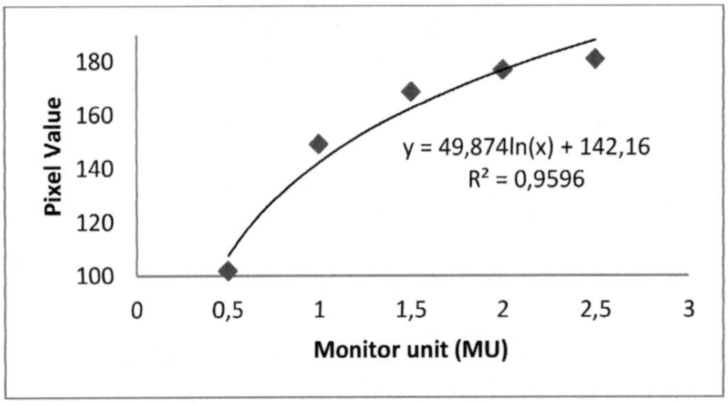


**Figure 5.** Computed radiography calibration test results

From the computed radiography calibration test results shown in Figure 5 looks the same as AAPM report 93 result about computed radiography responses in the Figure 1 increased over radiation exposure, and then it can determine exposure dose using graph equation above.

The increase in pixel value with increasing radiation exposure relationship with a dynamic range which is ratio of largest and smallest intensity from input photon beam that can be imaged. The smallest intensity affected by noise system. To avoid the system noise affected, so dynamic range should be greater than system noise. While the greatest intensity influenced by receptor saturation. In computed radiography response graph (Figure 1), output of phosphor photostimulable light comparable with exposure.

### *4.3. Measurements Linac Area Field*

Measurements carried out with area of the field of 5 x 5 $cm^2$, which is irradiated with an energy 2 MU. The measurement results are determined using the FWHM on cross-plane and in-plane axis of the photon beam. The results of irradiation field size measurement are shown in Table 2.

**Table 2.** Results of irradiation field size measurement

| Irradiation field ($cm^2$) | Irradiation field size measurement | | | |
|---|---|---|---|---|
| | Cross-plane | Relativ error | In-plane | Relative error |
| 5 x 5 | 6.6 | 0.32 | 6.6 | 0.32 |

Table 2 shows the results of irradiation field size measurement by using a CR with results of error value in 0.32 cross-plane and 0.32 in-plane field Also, it looks the response CR is good enough to used field size form.

### 4.4. *Imaging Analysis of LINAC beam using CR*

CR imaging analysis produced by beam of LINAC done to verified radiation isocenter by relating Signal to Noise Ratio with chosen Monitor Unit. Analysis done to evaluate quality of CR image in order to get reference image.

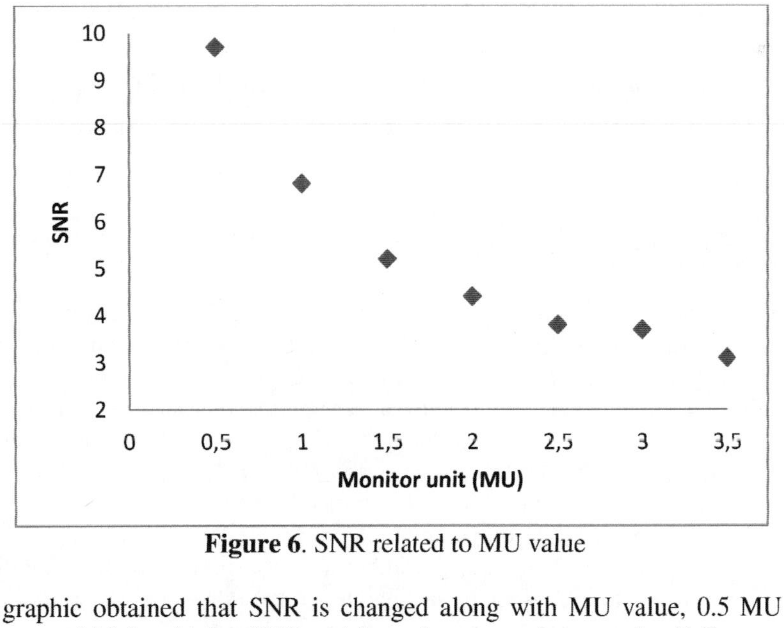


**Figure 6**. SNR related to MU value

Based on graphic obtained that SNR is changed along with MU value, 0.5 MU is chosen as research parameter which has higher SNR which produce lower image noise [14].

### *4.4. Verification of Radiation Isocenter Linac Beam 6 MV*

Verification of radiation isocenter is performed by using *star shot* which the radiation beam shaped a star due to the rotation of one of the Linac component, the collimator and the couch. To determine the diameter of the radiation isocenter, image processing using Matlab 2015b was done. As for the results of verification of radiation isocenter on collimator and couch are shown in Figure 7.

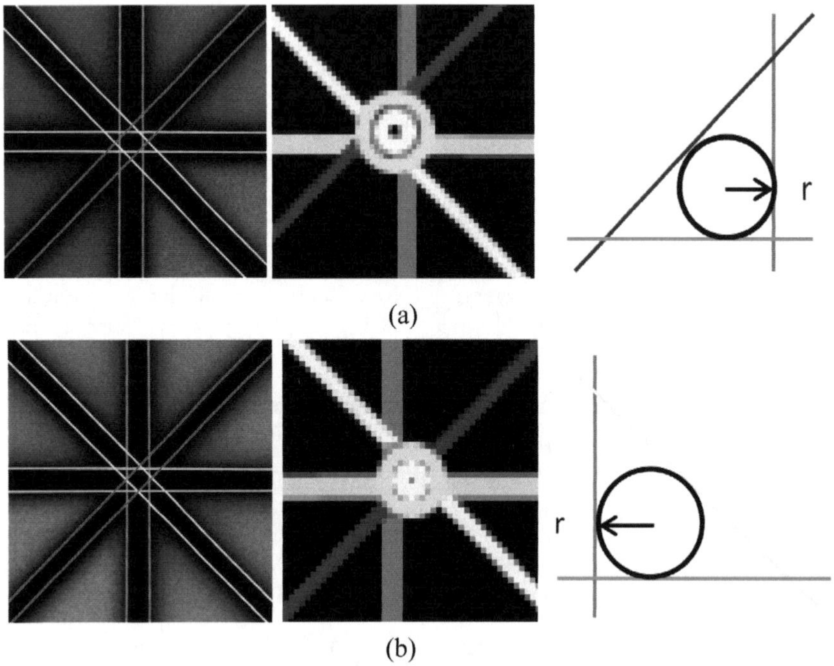


**Figure 7.** Results of verification of radiation isocenter (a) collimator (b) couch

**Table 3.** Verification of radiation isocenter on Linac

| Rotation Angle | Diameter (mm) | Mean | Standar Deviasi |
|---|---|---|---|
| Collimator : 0, 45, 90, 135 | 0.632 | 0.595 | 0.13 |
| Couch : 270, 315, 0, 45 | 0.458 | 0.472 | 0.03 |

Determination of radiation isocenter diameter on Linac is very important and will have an impact on the quality og radiation therapy given to patients. Most of the radiotherapy treatment appropriation using a gantry combination, collimator, and couch. For the Linac, radiation isocenter value measurement is not very significant. The results of verification of radiation isocenter are shown in Table 3. performed on a collimator with rotation angle of 0, 45, 90, 135 obtained radiation isocenter diameter that is 0.632 mm and on the couch with an angle of 270, 315, 0, 45 obtained radiation isocenter diameter that is 0.458 mm. But these results have consequences that shape radiation isocenter diameter cause enlargement of penumbra or increase radiation dose on irradiation field shape, especially small field sizes. This can happen because of asymmetry and instability of the tools used. Based on the use Linac for radiotherapy irradiation, collimator is more often used for irradiation method on every patient is different. AAPM report 40 about radiation isocenter diameter Linac size used in this research is still good condition and worth to be operated because that value of radiation isocenter diameter below 2 mm.

## 5. Conclusions

By the results of verification of radiation isocenter performed on collimator and the couch, it shows that the size of radiation isocenter diameter for collimator is 0.632 mm and 0.458 mm for the couch. Based on AAPM report 40 about the size of the Linac radiation isocenter diameter used in this research is still in good condition and worth to be operated because the value of the radiation isocenter diameter is below 2 mm.

## References

[1] Choirul A 2011 *Monte Carlo Simulation for Electron Contamination On Linac Beam 6 MV Aircraft Production Elekta SL15 Thesis* Mathematic and Natural Science Medical Physics of Post Graduate Programme (Jakarta: University of Indonesia)

[2] Deinar F 2010 Dose Distribution to Asymmetry Field 6 MV X-rays with Depth Variation *Thesis* Mathematic and Natural Science Medical Physics of Post Graduate Programme (Jakarta: University of Indonesia)

[3] Peace T, Subramanian B, and Ravindran P 2008 An Experimental Study on Using a Diagnostic Computed Radiography System as a Quality Assurance Tool in Radiotherapy *Australasian Physical and Engineerig Sciences in Medicine* **31** **3** 216-234

[4] Yin Z, Kai D, Garth C, Erik T, Elwood A, and Ken K W 2015 Alignment of Multiradiation Isocenter for Megavoltage Photon Beam *Journal of Applied Clinical Medical Physics* **16** **6** 314-324

[5] Gonzalez A, Castro I, and Martinez A J 2004 A Procedure to Determine the Radiation Size in a Linear Accelerator *Research Gate Article Medical Physics* **31** **6** 1489-1493

[6] Khan F M 2014 *Khans The Physics of Radiation Therapy* (Philadelphia: Lippincott Williams and Wilkins)

[7] Depuydt T, Penne R, Verelle D, Hrbacek J, Lang S, Leysen K, Vandevondel I, Poels K, Reynders T, Gavaert T, Duchateau M, Tournel K, Boussaer M, Cosentino D, Garibaldi C, Solberg T, De Ridder M 2012 Computer-Aided Analysis of Star Shot Films for High-Accuracy Radiation Therapy Treatment Unit *IOB Publishing Phys. Med. Biol.* **57** 2997–3011

[8] Gerald J K, Lawrence C, Michael G, William F H, Steven L, Robert J M, Jatinder R P, James A P, Lawrance E R, Goran K S, Mona, and Linda W 1994 Comprehensive QA For Radioation Oncology: Report of Taskgroup 40 Radiation Therapy Comitte AAPM *Medical Physics* **21** **4** 581-618

[9] Weiliang, D., Jennifer, L.J., Wei, J., dan Rajat, J.K., 2016, *On the Selection of Gantry and Collimator Angles for Isocenter Localization Using Winston-Lutz Tests*, Journal of Applied Clinical Medical Physics, Vol 17 Number 1.

[10] Juni S P 2010 *Dose X-rays 6 MV Beam Transmission for Irregular Field with Block Variation Thesis* Mathematic and Natural Science Medical Physics of Post Graduate Programme (Jakarta: University of Indonesia)

[11] Krupinski E A, Williams M B, Strauss K J, Breeden W K, Rzeszotarski M S, Applegate K, Wyatt M, Bjork S, and Seibert J A 2007 Digital Radiography Image Quality (Image Acquisition *Journal of The American College of Radiology* **4** 389-400

[12] Papp J 2006 *Quality Management in The Imaging Science Thrid Edition* (Saint Louis: Mosby)

[13] Ballinger P W and Eugene D F 2003 *Merrill's Atlas of Radiographic Positions and Radiologic Prosedures Tenth Edition Volume Three* (Saint Louis: Mosby)

[14] Seibert J A, Terese M B, Ted C, Walter H, Andrew K, John R M, Ehsan S, Shepard J S, Brent K S, Keith J S, Orhan H S, Doug T, Robert A U, John C W, and Charles E W 2006 Acceptance Testing and Quality Control Storage Phospor Imaging System *American Association of Physicists in Medicine*

[15] Klein Eric E 2009 Task Group 142 Report: Quality Assurance of Medical Accelerator *Medical Physics* **36** **9** 4197-4212

# Students' mathematical representations on secondary school in solving trigonometric problems

**Istadi[1], T A Kusmayadi[2,3] and I Sujadi[3]**

[1] Student of Post Graduate Degree of Mathematics Education of Sebelas Maret University, Indonesia
[2] Lecturer of Department of Mathematics, Faculty of Mathematics and Natural Sciences of Sebelas Maret University, Indonesia
[3] Lecturer of Department of Mathematics Education, Faculty of Teacher Training and Education of Sebelas Maret University, Indonesia

E-mail: istadi.47@gmail.com, tri.atmojo.kusmayadi@gmail.com, imamsujadi@ymail.com

**Abstract**. This research aimed to analyse students' mathematical representations on secondary school in solving trigonometric problems. This research used qualitative method. The participants were 4 students who had high competence of knowledge taken from 20 students of 12th natural-science grade SMAN-1 Kota Besi, Central Kalimantan. Data validation was carried out using time triangulation. Data analysis used Huberman and Miles stages. The results showed that their answers were not only based on the given figure, but also used the definition of trigonometric ratio on verbal representations. On the other hand, they were able to determine the object positions to be observed. However, they failed to determine the position of the angle of depression at the sketches made on visual representations. Failure in determining the position of the angle of depression to cause an error in using the mathematical equation. Finally, they were unsuccessful to use the mathematical equation properly on symbolic representations. From this research, we could recommend the importance of translations between mathematical problems and mathematical representations as well as translations among mathematical representaions (verbal, visual, and symbolic) in learning mathematics in the classroom.

## 1. Introduction

Representation is one of the process standard should enable students to know and do from kindergarten to K-12. The representation standard should enable students to create and use representations to organize, record, and communicate mathematical ideas. Representations can be expressed in the form of visual, verbal, and symbolic. Visual representations include illustrate, show, or work with mathematical ideas using diagrams, pictures, number lines, graphs, and other math drawings. Verbal representations include using language (words and phrases) to interpret, discuss, define or describe mathematical ideas, bridging informal and formal mathematical language. Symbolic representations include recording or working with mathematical ideas using numerals, variables, tables, and other symbols [9, 11].

Other representation standard should enable students to select, apply, and translate among mathematical representations to solve problems. For instance, translations between mathematical problems and mathematical representations as well as translations among mathematical

IOP Conf. Series: Journal of Physics: Conf. Series **855** (2017) 012021 doi:10.1088/1742-6596/855/1/012021

representations. Thus, students must be able to move flexibly in between form of representations. [4, 9, 11]. It suggests using representations could help to facilitate when solving problem more efficient. For instance, visual representation could facilitate in solving algebraic problems than using symbolic representation, despite previous students were more likely to prefer symbolic representations over visual representations [10].

Translations between mathematical problems and mathematical representations as well as translations among mathematical intended to get the description of students' matematical representations abilities by themselves without provided (helped) by their teachers [18]. Illustration translations between mathematical problems and mathematical representations could be shown in Figure 1 as well as illustration translations among mathematical representations could be shown in Figure 2 as follows:

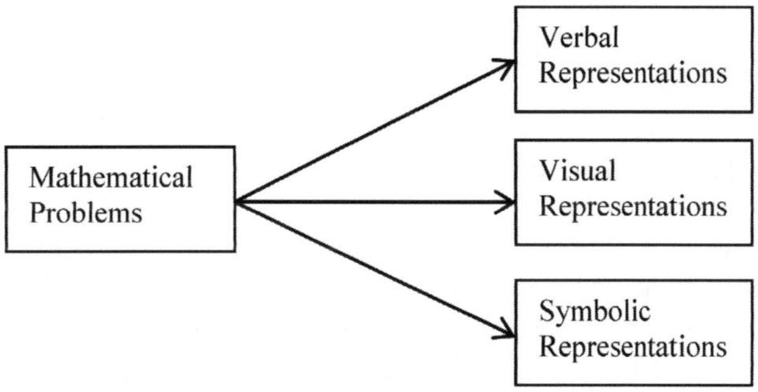


**Figure 1.** Illustration translations between mathematical problems and mathematical representations.

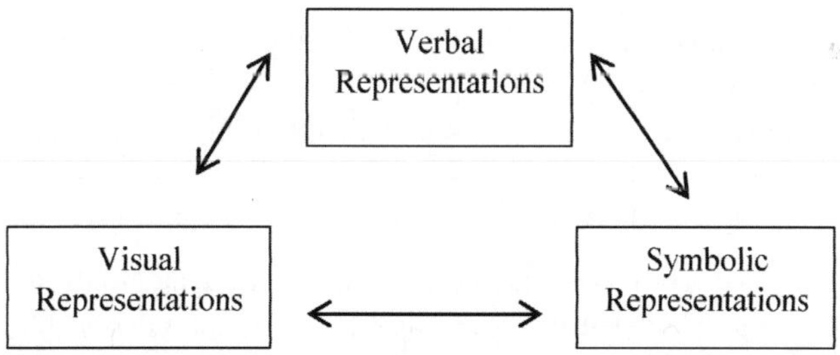


**Figure 2.** Illustration translations among mathematical representations.

Some of the benefits of representations could encourage students' mathematical ideas, especially in problem-solving ability (e.g., [11], [15], [20]). Besides in solving problem, representations useful in understanding the abstract concepts of mathematics. For instance, in transition between arithmetic and algebra by geometric representations as well as in teaching factoring second-degree polynomials (e.g, [3], [6], [12], [13]). Representations are not only to goal curriculum standard, but an important aspect to improving students' educational value as follows: (1) to help students consolidate their knowledge and improve skills, (2) to help teachers and students enrich their concept of mathematics and mathematical teaching, (3) to help students overcome the psychological barriers, (4) to help teachers assess students' learning result, and (5) to help teachers improve their own literacy [23].

Research related to mathematical representations previously showed that relationship between visual representations and problem-solving as well as in solving routin and non-routine problems (e.g., [14], [17]). Other research related to translation among mathematical representations showed students

IOP Conf. Series: Journal of Physics: Conf. Series **855** (2017) 012021 doi:10.1088/1742-6596/855/1/012021

were successful than the pre-service teachers in understanding of functions, likewise in representing the fractions on number lines through other representations (e.g., [1], [2]). From the research, representations were used in analyzing the understanding of prime numbers, making generalizations on algebra material, as well as in representing the law of cosines without using the Pythagorean theorem and trigonometry (e.g., [16], [21], [22]).

This research focused on trigonometry. Trigonometry is one of content standard that students should enable in school mathematics [11], either in national curriculum 2006 (KTSP) or national curriculum 2013 (K-2013). On the other hand, trigonometry is one of difficult materials in the classroom according to the local mathematics teacher experiences. The research related to trigonometry showed that there were difficulties in understanding trigonometric function by standard instruction [19] also in solving routine and no routine problems [5].

Based on the explanations above, this research was conducted to answer the question: How do students' mathematical representations abilities on secondary school in solving trigonometric problems?

## 2. Method

This research used qualitative method. The participants were 4 students, named Anne, Bob, Clara, and Tina who had high competence of knowledge taken from 20 students of 12th natural-science grade in SMAN-1 Kota Besi, Central Kalimantan. They were given 2 problems to determine their abilities of translation between mathematical problems and mathematical representations as well as translation among mathematical representations (verbal, visual and symbolic). The form of the problems are shown as follows:

Problem 1
Use the figure and right-triangle definition of sine to explain why, $\sin 20° < \sin 40° < \sin 60°$ and $\cos 20° > \cos 40° > \cos 60°$.

Problem 2
The pilot of aircraft is flying at an altitude of 1,300 m above the sea level saw a cruise ship sailing in the position of the angle of depression is $40°$.

- Make a sketch regard to the angle of depression, both pilot position and cruise ship position as well.
- How far the distance of cruise ship to the point at sea level which precise under the aircraft.

Meanwhile, data validation was carried out using time triangulation with data collection 2 times (test and interview). Data analysis used Miles and Huberman model which consists of three stages: (1) data reduction, (2) data display, and (3) conclusion drawing/verification.

## 3. Results

This research aimed to analyse the students' mathematical representations abilities, including verbal, visual and symbolic representations. Verbal representations analysis used Problem 1, as well as visual and symbolic representations used Problem 2. In this paper, their responses (i.e., test results and interviews) are not exposed completely. However, some responses in representing their abilities on mathematical representations. The results with respect to each their abilities on mathematical representations are shown the following.

### *3.1. Students' verbal representations abilities*

Student's verbal representations ability who named Anne on the first test, she answered only based on the given figure. However, on the interview she was able to answer based on the definition of trigonometric ratio. On second test, she was able to answer based on the figure as well as used the definition of trigonometric ratio.

Student's verbal representations ability who named Bob on the first test, he answered only based on the value of the sine without used the definition trigonometric ratio. However, on the interview he was able to answer based on the definition of trigonometric ratio. On the second test, he was able to answer based on the given figure as well as used the definition of trigonometric ratio.

Students's verbal representations abilities who named Clara and Tina were almost similar on the first test. They didn't use the definition trigonometric ratio, but only based on the given figure. However, on the interview they were able to answer based on the definition of trigonometric ratio. On the second test, they were able to answer based on the given figure as well as used th definition of trigonometric ratio.

The excerpt of interview with one of them to explain orally on verbal representations as follows:

| | |
|---|---|
| Interviewer: | Could you explain how to answer this question? |
| Student: | Since the concept of sine is invers by the concept of cosine Sir. |
| Interviewer: | Now, if we use the definition of cosine, How? |
| Student: | Based on the definition of cosine is the ratio of adjacent and hypotenuse. |
| Interviewer: | Thus, could you explain it completely? |
| Student: | Since the adjacent length of the angle $20^\circ$ is longer than the adjacent length of the angle $40^\circ$as well as the adjacent length of the angle $40^\circ$is longer than the adjacent length of the angle $60^\circ$. Thus, $\cos 20^\circ > \cos 40^\circ > \cos 60^\circ$. |

### *3.2. Students' visual representations abilities*

Student's visual representations ability who named Anne on the first test, she was not able to determine the object positions (pilot and cruise ship) in the sketch made, as well as in determining the position of the angle of depression. On the other hand, students who named Bob, Clara, and Tina were successful to determine the object positions (pilot and cruise ship), although they failed to determine the position the angle of depression. On the second test, they (i.e., Anne, Bob, Clara, and Tina) were successful to determine the object positions (pilot and cruise ship). However, they also failed to determine the position of the angle of depression.

The following one of students' solution of problem of visual representations is shown in Figure 3 below:

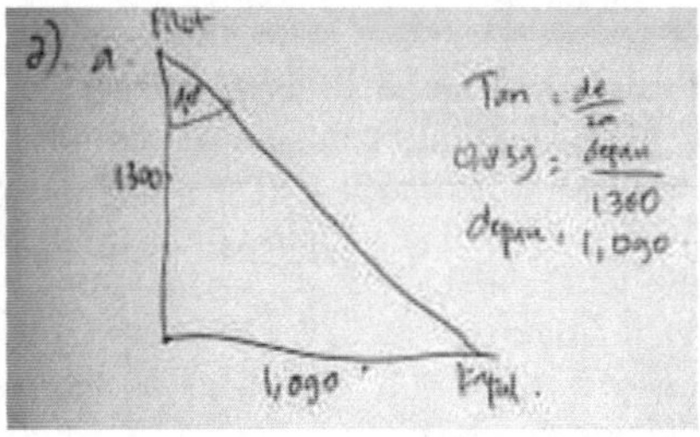

**Figure 3**. Students' solution of problem on visual representations.

### *3.3. Students' symbolic representations abilities*

Student's symbolic representations ability who named Anne on the first test, she was able to determine the mathematical equation based on the sketches made as well as on the second test. However, the result was error because failure in determining the position of the angle of depression on the previous sketch made. On the other hand, students who named Bob, Clara and Tina on the first test, they didn't

IOP Conf. Series: Journal of Physics: Conf. Series **855** (2017) 012021 doi:10.1088/1742-6596/855/1/012021

answer or write the solution of problem. However, they were able to determine the mathematical equation. Meanwhile, on the second test, they were able to determine the mathematical equation. However, they failed to use the mathematical equation properly.

The following one of students' solution of problem of symbolic representations could be shown in Figure 4 below:

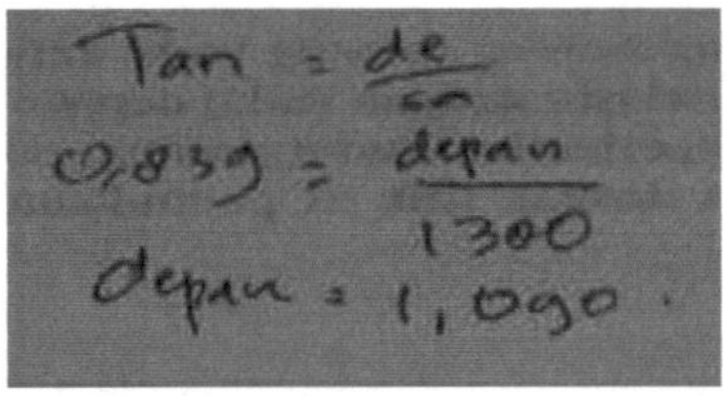


**Figure 4**. Students' solution of problem on symbolic representations.

The summary data analysis of students' mathematical representations abilities could be shown in Table 1, Table 2, Table 3, and Table 4 as follows:

**Table 1**. Summary of Anne's mathematical representations abilities.

| Student | Mathematical Representations | The First Data Collection | The Second Data Collection | Conclusion |
|---|---|---|---|---|
| Anne | Verbal | On the first test, Anne only answer based on the given figure. However, on the interview Anne was able to answer based on the definiton of trigonometric ratio. | On the second test, Anne answer based on the given figure as well as based on the definiton of trigonometric ratio. | Anne was able to answer based on the given figure as well as based on the definition of trigonometric ratio. |
| | Visual | On the first test, Anne failure in determining the object positions and the position of the angle of depression. On the interview, Anne successful in determining the object positions. However, Anne failure in determining position of the angle of depression. | On the second test, Anne successful in determining the object positions. However, Anne failure in determining position of the angle of depression. | Anne successful in determining the object positions. However, Anne failed to determine the position of the angle of depression. |
| | Symbolic | On the first test, Anne Successful in determining the mathematical equation. However, she failed to use the mathematical equation properly. | On the second test, Anne Successful in determining the mathematical equation. However, she failed to use the mathematical equation properly. | Anne successful in determining mathematical equation. However, Anne failed to use mathematical equation properly. |

**Table 2**. Summary of Bob's mathematical representations abilities.

| Student | Mathematical Representations | The First Data Collection | The Second Data Collection | Conclusion |
|---|---|---|---|---|
| Bob | Verbal | On the first test, Bob only answer based on the value of trigonometric ratio. However, on the interview Bob was able to answer based on the definiton of trigonometric ratio. | On the second test, Bob answer based on the given figure as well as based on the definiton of trigonometric ratio. | Bob was able to answer based on the given figure as well as based on the definition of trigonometric ratio. |
| | Visual | On the first test, Bob successful in determining the object positions. However, she failure in determining the position of depression angle. | On the second test, Bob successful in determining the object positions. However, Bob failure in determining the position of depression angle. | Bob successful in determining the object positions. However, Bob failed to determine the position of the angle of depression. |
| | Symbolic | On the first test, Bob failure in determining the mathematical equation. However, on the interview Bob was able to determine the mathematical equation. | On the second test, Bob failure in determining the mathematical equation. However, on the interview Bob was able to determine the mathematical equation. | Bob successful in determining the mathematical equation. However, Bob failed to use the mathematical equation properly. |

**Table 3**. Summary of Clara's mathematical representations abilities.

| Student | Mathematical Representations | The First Data Collection | The Second Data Collection | Conclusion |
|---|---|---|---|---|
| Clara | Verbal | On the first test, Clara only answer based on the given figure. However, on the interview, Clara was able to answer based on the definiton of trigonometric ratio. | On the second test, Clara answer based on the given figure as well as based on the definiton of trigonometric ratio. | Clara was able to answer based on the given figure as well as based on the definition of trigonometric ratio. |
| | Visual | On the first test, Clara failure in determining the object positions and the position of the angle | On the second test, Clara successful in determining the object positions. However, | Clara successful in determining the object positions. However, Clara failed to determine |

IOP Conf. Series: Journal of Physics: Conf. Series **855** (2017) 012021 doi:10.1088/1742-6596/855/1/012021

| | | | | |
|---|---|---|---|---|
| | | of depression. | Anne failure in determining position of the angle of depression. | the position of the angle of depression. |
| | Symbolic | On the first test, Clara failure in determining the mathematical equation. However, on the interview, Clara was able to determine the mathematical equation. | On the second test, Clara also failure in determining the mathematical equation. However, on the interview, Clara was able to determine the mathematical equation. | Clara successful in determining the mathematical equation. However, Clara failed to use the mathematical equation properly. |

**Table 4**. Summary of Tina's mathematical representations abilities.

| Student | Mathematical Representations | The First Data Collection | The Second Data Collection | Conclusion |
|---|---|---|---|---|
| Tina | Verbal | On the first test, Tina only answer based on the given figure. However, on the interview, Tina was able to answer based on the definiton of trigonometric ratio. | On the second test, Tina answer based on the given figure as well as based on the definiton of trigonometric ratio. | Tina was able to answer based on on the given figure as well as based on the definition of trigonometric ratio |
| | Visual | On the first test, Tina failure in determining the object positions and the position of the angle of depression. | On the second test, Tina successful in determining the object positions. However, Tina failure in determining position of the angle of depression. | Tina successful in determining the object positions. However, Tina failed to determine the position of the angle of depression. |
| | Symbolic | On the first test, Tina failure in determining the mathematical equation. However, on the interview, she was able to determine the mathematical equation. | On the second test, Tina failure in determining the mathematical equation. However, on the interview, she was able to determine the mathematical equation. | Tina successful in determining the mathematical equation. However, she failed to use the mathematical equation properly. |

## 4. Discussion

Based on the results above showed students were able to answer based on the given figure, as well as based on the definition of trigonometric ratio. Fathermore, students were able to determine the object positions to be observed. However, they failed to determine the position of the angle of depression. Finally, students were able to determine the mathematical equation. However, they unsuccessful to use the mathematical equation properly. According to Bruner's theory that representation consists of three modes, there are enactive, iconic and symbolic. The first mode of representation is enactive where student should have direct experience. The second mode representation is iconic where student should have been able to memorize of their experiences. The third mode representation is symbolic where student should be able to understand the concept in learning trigonometry. For instance, students successful to prove the problem based on the definition of trigonometric ratio, to determine the object position, as well as to determine the mathematical equation. On the other hand, students unsuccessful to determine the position of the angle of depression as well as to use the mathematical equation properly because they didn't have direct experience, memory and understand concept to solve the problem they may encounter.

According to Goldin's theory, representation is divided into external representations and internal representations. External representations comprise students' representations, such as verbal, visual and symbolic. Meanwhile, internal representations comprise visual imagination, problem-solving strategies, as well as natural language students [7]. Based on Goldin's theory, interaction between the internal representations and the external representations is considered in learning mathematics. For instance, students successful to prove the problem based on the definition of trigonometric ratio, to determine the object position, as well as to determine the mathematical equation. Meanwhile, students unsuccessful to determine the position of the angle of depression. Finally, they failed to use the mathematical equation properly because interaction between the internal representations and the external representations is underdeveloped.

## 5. Conclusions

This research could be concluded students' mathematical representations abilities on secondary school in solving trigonometric problems as follows: (1) students were able to answer based on the given figure, as well as based on the definition of trigonometric ratio, either written or orally on verbal representations, (2) students were able to determine the object positions to be observed. However, they failed to determine the position of the angle of depression on visual representations, and (3) students were able to determine the mathematical equation. However, they unsuccessful to use the mathematical equation properly on symbolic representations. Based on these conclusions, we recommend the importance of translations between mathematical problems and mathematical representations as well as translations among mathematical representations (verbal, visual, and symbolic) in learning mathematics in the classroom. For the future research, we suggest to analyse the students' mathematical representations among secondary students, either on trigonometry or other materials.

## References

[1] Bannister V R P 2014 Flexible conceptions of perspectives and representations: an examination of pre-service mathematics teachers' knowledge *Int. J. of Edu. in Math., Sci. and Technol.* **2** 223–33

[2] Biber A C 2014 Mathematics teacher candidates' skills of using multiple representations for division of fractions *Academic J.* **9** 237–44

[3] Cabahug J A 2012 The Use of bruner's modes of representations in teaching factoring second-degree polynomials *IAMURE: Int. of Edu.* **1** 85–103

[4] Dindyal J 2009 Mathematical problems for secondary classroom *Mathematical Problem Solving Yearbook 2009* ed B Kaur, Y B Har and M Kapur (Singapore: World Scientific Publishing) chapter 11 pp 208–225

[5] Dundar S and Yaman H 2015 How do prospective teachers solve routine and non-routine trigonometry problems? *Int. Outline and Educational Sci.* **7** 41–57

[6] Flores A 2002 Geometric representations in the transition from arithmetic to algebra *Representations and Mathematics Visualizations* ed F Hitt (Mexico: Ciinvestav-IPN) chapter 1 pp 9–30

[7] Godino J D and Font V 2010 The theory of representations as viewed from onto-semiotic approach to mathematics education *Mediterranean J. for Research in Math. Edu.* **9** 189–210

[8] Hegarty M and Kozhevnikov M 1999 Types of visual-spatial representations and mathematical

problem solving *J. of Educational Psy.* **91** 648−89
[9] Huinker D 2015 Representational competence: a renewed focus for classroom practice in mathematics *Wisconsin Teacher of Mathematics* **2015** 4−8
[10] Mielicki M K and Wiley J 2016 Alternative representations for algebraic problem solving: when are graphs better than equations? *J. of Problem Solving* **9** 3−12
[11] NCTM 2000 *Principles and Standards for School Mathematics* (USA: NCTM inc)
[12] Panasuk R M and Beyranevand M L 2011 Preferred representations of middle school algebra students when solving problems *The Math. Educator* **13** 32−52
[13] Pape S J and Tchoshanov M A 2001 The role of representation(s) in developing mathematical Understanding *Theory into Practice* **40** 118−27
[14] Presmeg N C and Balderas-Canas P E 2002 Graduate students' visualizations in two rate of change problems *Representations and Mathematics Visualizations* ed F Hitt (Mexico: Ciinvestav-IPN) chapter 2 pp 47−62
[15] Sajadi M, Parvaneh A and Rostamy-Malkhalifeh M 2013 The examining mathematical word problems solving ability under efficient representation aspect *ISPAC* **2013** 1−11
[16] Scher D and Goldenberg E P 2001 A multirepresentational journey through the law of cosinus *The Roles of Representation in School Mathematics* ed A A Couco (USA: NCTM inc) chapter 10 pp 117−128
[17] Stylianou D A and Pitta-Pantazi D 2002 Visualization and high achievement in mathematics: a critical look at successful visualization strategies *Representations and Mathematics Visualizations* ed F Hitt (Mexico: Ciinvestav-IPN) chapter 3 pp 31−46
[18] Terwel J, Oers B V, Dijk I V and Eeden P V D 2009 Are representations to be provided or generated in primary mathematics education? effects on transfer *Educational Research and Evaluation* **15** 25–44
[19] Weber K 2005 Students' understanding of trigonometric functions *Math. Edu. Research J.* **17** 91−12
[20] Yee S P and Bostic J D 2014 Developing a contextualization of students' mathematical problem solving *J. of Mathematical Behavior* **36** 1−19
[21] Zazkis R and Liljedahl P 2004 Understanding prime: the role of representation *J. for Research in Math. Edu.* **36** 164−86
[22] Zeljic M and Dabic M 2014 Iconic representation as students' success factor in algebraic generlitations *J. Plus Edu.* **10** 173−84
[23] Zhe L 2012 Survey of primary students' mathematical representation status and study on the teaching model of mathematical representation *J. of Math. Edu.* **5** 63−76

# A periodic review integrated inventory model with controllable safety stock and setup cost under service level constraint and distribution-free demand

**N A Kurdhi[1], A Jamaluddin[2], W A Jauhari[3], D R S Saputro[4]**

[1,4]Department of Mathematics,
Faculty of Mathematics and Natural Science, Sebelas Maret University, Indonesia
[2]Department of Physics Education
E-Smart Research Group, Sebelas Maret University, Indonesia
[3]Department of Industrial Engineering
Faculty of Engineering, Sebelas Maret University, Indonesia

E-mail: [1]arfa@mipa.uns.ac.id, [2]anifjamal@gmail.com, [3]wachid_aj@yahoo.com, [4]dewi.rss@gmail.com

**Abstract.** In this study, we consider a stochastic integrated manufacturer-retailer inventory model with service level constraint. The model analyzed in this article considers the situation in which the vendor and the buyer establish a long-term contract and strategic partnership to jointly determine the best strategy. The lead time and setup cost are assumed can be controlled by an additional crashing cost and an investment, respectively. It is assumed that shortages are allowed and partially backlogged on the buyer's side, and that the protection interval (i.e., review period plus lead time) demand distribution is unknown but has given finite first and second moments. The objective is to apply the minmax distribution free approach to simultaneously optimize the review period, the lead time, the setup cost, the safety factor, and the number of deliveries in order to minimize the joint total expected annual cost. The service level constraint guarantees that the service level requirement can be satisfied at the worst case. By constructing Lagrange function, the analysis regarding the solution procedure is conducted, and a solution algorithm is then developed. Moreover, a numerical example and sensitivity analysis are given to illustrate the proposed model and to provide some observations and managerial implications.

## 1. Introduction

In today's global market, the cooperation between the vendor and the buyer is an important way to gain competitive advantages as it lowers supply chain cost. It involves commitment to long-term coordination, information sharing, joint problem solving and shared benefits. This regenerated interest is encouraged by the growing focus on supply chain management. Many firms are recognizing that inventories across the entire supply chain can be more efficiently organized through better integration of the supply chain. Several policies are used to align the activities of the components of a supply chain to establish better performance of supply chain in terms of response time, cost, customer service, and timely supply. Supply chain colaboration is concerned with the development and implementation of such strategies. Several researchers have shown that the vendor and the buyer can increase their mutual benefit or obtain their minimal total cost through strategic colaboration with each other. Goyal [7] was one of the first authors to develop a single-vendor single-buyer integrated inventory model. The joint vendor-buyer optimization was later reinforced by Banerje [3] and Goyal [8]. Pan and Yang [23] then generalized Goyal's [8] model and presented an integrated inventory model by considering lead time as a decision variable, and obtained a shorter lead time and lower joint total expected cost. Other related studies of the integrated inventory system include Lin [16], Shah [26], and Jauhari et al. [10]. A review of related literature on coordination inventory models is given in Glock [6].

In the collaborative inventory management system, the continuous review integrated inventory model continuous to be widely discussed. However, it is noticed that the inventory literature discussing the integration between the vendor and the buyer on the periodic review inventory model is quite sparse. The periodic review inventory model can frequently be found in organizing inventory cases such as smaller retail stores, grocery stores, and drugstores. There are papers related to the periodic review inventory model studies. Pal and Chandra [22] investigated a periodic review inventory model with stock dependent demand and permissible delay in payment. Lin [18] explored a stochastic periodic review inventory model with back-order price discounts and ordering cost dependent on lead time. Soni and Joshi [27] developed a periodic review inventory model by considering lead-time and the backorder rate as control variables in fuzzy stochastic environment. Akyurt [1] studied the single step single-item periodic review inventory models under Poisson demand distribution. Chand et al. [16] examined a periodic review inventory model where the buyer has access to a mix of two delivery modes for the quantity it orders in a period; these include an emergency delivery mode that delivers instantaneously and a regular delivery mode that delivers within the length of the review period. However, all of the above periodic review inventory model focused on determining optimal policy for the buyer only. These models neglect the opportunity that buyer and vendor can cooperate with each other to obtain a better integrated strategy. We also notice that the traditional periodic review inventory models assumed that all products are perfect and protection interval demand is normally distributed. It is common in all industries that a certain percent of produced/ordered products are non-conforming (imperfect) quality, which may be resulted from an imperfect production process, insufficient process control, inadequate work instruction, wrongly planned maintenance, damage in transit, careless handling, or others negative factors. These defective items will impact the on-hand inventory level, the number of shortages and the frequency of orders in the inventory system. Hence, it is important to study the production/order policy in which an arrival lot contains defective items. Recently, some research (see, for instance, [12], [14], [17], [20], [30],) has been done in continuous review multi-stage lot sizing decisions for imperfect production processes. In addition, the information about the probability distribution of the lead time demand is often quite limited. There are many related studies such as Lin [17], Moon et al. [21], Sarkar et al. [25], Annadurai [2], etc.

Most of the integrated inventory models are generally developed without any constraint. Nevertheless, in real situations, this kind of inventory system rarely occurs. Therefore, many companies deal with several constraints such as service level and capital investment. While allowing backlogging, shortages may turn into possible loss of future sales because of an unacceptable service level. Hence, the maximal allowable level of shortages per cycle is usually set as an operating constraint of the business in order to achieve optimal service level while deriving the optimal lot-size decision. In addition, in many cases the shortage cost includes intangible components such as potential delays to other parts of the system and loss of goodwill. A common substitute for a shortage cost is a service level which is measured as proportion of demands satisfied directly from stock. Therefore, several scholars (see, [2] [11] , [12] , [14], [15], [21], [25]) have incorporated a service level constraint into their inventory models. Consequently, the shortage level per cycle is bounded. Furthermore, safety factor (safety stock) is one of the important factors which influences the service level. However, in most periodic review inventory models, safety stock was not taken into account, and they merely focused on the relationship between periodic review and lead time. In other words, they neglected the possible impact of the safety stock on the economic ordering strategy. Such a phenomenon is usually not perfect in a real inventory situation. Hence, we consider the controllable safety stock in the proposed model in this paper.

Furthermore, setup cost is one of the important parts of the inventory organization. Setup cost is related with expenses incurred in surroundings up a machine, assembly line, or work center, to switch from one production career to the subsequently ([29]). The basic integrated inventory model is generally based on the assumption of fixed setup cost. By using investment, the setup cost can be reduced. That is, setup cost is assumed to be variable rather than constant. The initial investment may be high but expected total cost will be reduced in each stage by using the investment. The setup cost reduction is one of the important production activities in an integrated inventory control. The concept of setup cost reduction has been widely applied to the production system. The consequences of set-up cost reduction are significantly increasing production efficiency and decreasing inventory. In practical situations, setup cost can be reduced through worker training, procedural changes and specialized equipment acquisition. Porteus [24] investigated the impact of investing in reducing setup cost by considering the discounted model. Further, Kurdhi et al. [13], Lo [19], Tahami et al. [28] and Vijayashree and Uthayakumar [29], addressed the joint optimization cost of the vendor and the buyer. They showed that by investing in set-up cost reduction, a significant saving in joint total cost can be achieved. Moreover, we set the normal setup cost without any investment as the upper limit of the setup cost level, which implies that if the optimal set-up cost obtained does not satisfy the restriction on the cost, then no set-up cost reduction investment is made. This is to avoid losses caused by the investment.

Based on the above discussions, we present the periodic review integrated vendor-buyer inventory model with controllable lead time and safety stock, imperfect production process, investment for setup cost reduction, partial backordering, and stochastic demand under service level constraint. We assume that the probability density function of demand during protection period (i.e. review period plus lead time) is unknown with given mean and variance. A minimax distribution free procedure and Lagrange multiplier are applied in order to minimize the joint expected total cost of the vendor and the buyer by simultaneously optimizing the review period, safety factor, lead time, setup cost, and the number of shipments in one production cycle. An algorithm procedure is developed to determine the optimal solutions. Later, a numerical example is provided to illustrate the proposed model. Finally, the concluding remark is made. The rest of this paper is organized as follows: The notations and assumptions used in this paper are introduced in Section 2. In Section 3, we formulate the stochastic integrated periodic review inventory model containing single vendor and single buyer subject to a service level constraint in the partial backorder case. The minimax distribution free procedure, the Lagrangian multiplier technique and a detailed solution procedure to solve the proposed model are presented in Section 4. In Section 5, a numerical example and discussion of the results are provided. Finally, some conclusions and suggestions for some future research are given in Section 6.

## 2. Notations and Assumptions

The following notations and assumptions are used in developing mathematical model.

### 2.1. Notations

*Variables:*

$T$ : buyer's review period length
$R$ : buyer's target level
$L$ : buyer's lead time
$S$ : vendor's setup cost per setup
$m$ : number of shipments from vendor to buyer per production run, a positive integer

*Parameters:*

$P$ : vendor's production rate per unit time

$v$ : vendor's production cost per unit
$h_s$ : vendor's average inventory holding cost per unit per unit time
$D$ : buyer's average demand per unit time
$A$ : buyer's ordering cost per order
$\omega$ : vendor's treatment cost of defective items
$\gamma$ : percentage of defective items in each deliver, $0 \leq \gamma < 1$
$x$ : buyer's inspecting rate
$s$ : buyer's unit inspecting cost
$p$ : unit price charged by the vendor to the buyer, $p > v$
$h_{r1}$ : buyer's inventory holding cost per non-defective items per unit per unit time
$h_{r2}$ : buyer's inventory holding cost per defective items per unit per unit time, $h_{r1} \geq h_{r2} > h_s$
$\beta$ : fraction of the shortage that will be backordered, i.e., backorder ratio, $0 \leq \beta < 1$
$\beta_0$ : upper bound of the backorder ratio
$\pi_x$ : backorder price discount offered by the buyer per unit
$\pi_0$ : Marginal profit (i.e. cost of lost demand) per unit
$X$ : The protection interval, $T + L$, demand which has a normal probability density function (p.d.f.) $f_X$ with finite mean $D(T + L)$ and standard deviation $\sigma\sqrt{T + L}$, where $\sigma$ denotes the standard deviation of the demand per unit time
$E(\cdot)$ : Mathematical expectation
$y^+$ : Maximum value of y and 0, i.e., $y^+ = Max\{y, 0\}$.

### 2.2. *Assumptions*

1. There is a single-vendor and a single-buyer for a single product in this model, and the inventory system deals with only one type item
2. When buyer orders quantity $DT$, vendor manufactures $mDT$ at one set-up, with a finite production rate $P$, and $(1-\gamma)P > D$. Each batch is dispatched to the buyer in $m$ equally-sized shipments, where m is a positive integer.
3. The inventory level is reviewed every $T$ units of time. A sufficient quantity is ordered up to the target level $R$, and the ordering quantity is received after $L$ units of time.
4. The lead time consists of $n$ mutually independent components. The $i$-th component has a minimum duration $a_i$ and normal duration $b_i$, and a crashing cost per unit time $c_i$, where $c_1 \leq c_2 \leq \cdots \leq c_n$, $L_i$ the length of lead time with components 1,2,..., $i$ crashed to their minimum duration;
$$L_i = \sum_{j=1}^{n} b_j - \sum_{j=1}^{i}(b_j - a_j), i = 1,2, \dots, n.$$
Also, let $L_0 = \sum_{j=1}^{n} b_j$, $C(L)$ the lead time crashing cost per cycle;
$$C(L) = c_i(L_{i-1} - L) + \sum_{j=1}^{i-1} c_j(b_j - a_j),$$
for a given $L \in [L_i, L_{i-1}]$.
5. The length of the lead time $L$ does not exceed an inventory cycle time $T$ so that there is never more than a single order outstanding in any cycle. That is, $L \leq T$.
6. The target level $R$ = expected demand during the protection interval + safety stock ($SS$), and $SS$ = $k$×(standard deviation of protection interval demand), i.e., $R = D(T + L) + k\sigma\sqrt{T + L}$, where $k$ is the safety factor.
7. The buyer makes decisions in order to obtain profits. Therefore, if the price discount, $\pi_x$, is greater than the marginal profit, $\pi_0$, then the buyer may decide against offering the price discount. The choice of investing in reducing the setup cost for the buyer is available.
8. The investment required to reduce the ordering cost from original level $S_0$ to a target level $S$ is denoted by $I(S)$, where $I(S)$ is a convex and strictly decreasing function.

## 3. Model Formulation

### *3.1. Buyer's expected total cost per unit time*

The unit price charged by the vendor to the buyer is $p$. Hence, the purchase cost per unit time is $pD$. It is assumed that an arriving order items may contain some defective items with defective rate, $\gamma$. Upon arrival of an order, all items are inspected by the buyer with inspected rate, $x$. The defective items in each lot will be returned to the vendor when the next lot is supplied. The unit inspecting cost for items is $s$, so the buyer's inspecting cost per unit time is $sD$. Further, the buyer will have two extra costs: non-defective items holding cost and defective items holding cost. The number of non-defective items is $(1-\gamma)DT$ and the buyer's holding cost of non-defective items per unit time is $h_{r1}$. Hence, the buyer's holding cost of non-defective items per unit time is

$$\frac{h_{r1}}{T}\left[\frac{DT^2}{2}+\frac{\gamma(DT)^2}{2x}\right]+h_{r1}[R-D(T+L)+(1-\beta)E[(X-R)^+]].$$

In addition, the number of defective items in each shipment is $\gamma DT$, inspection period time is $\frac{DT}{x}$, and the buyer's holding cost of defective items per unit per unit time is $h_{r2}$. Hence, the buyer's holding cost of defective items per unit time is $\frac{h_{r2}}{T}\left[\gamma DT^2-\frac{\gamma(DT)^2}{2x}\right]$. Therefore, the buyer's holding cost of non-defective and defective items per unit time is

$$\frac{h_{r1}}{T}\left[\frac{DT^2}{2}+\frac{\gamma(DT)^2}{2x}\right]+h_{r1}[R-D(T+L)+(1-\beta)E[(X-R)^+]]+\frac{h_{r2}}{T}\left[\gamma DT^2-\frac{\gamma(DT)^2}{2x}\right]$$
$$=DT\left[\frac{h_{r1}}{2}+h_{r2}\gamma+\frac{\gamma D}{2x}(h_{r1}-h_{r2})\right]+h_{r1}[R-D(T+L)+(1-\beta)E[(X-R)^+]].$$

Thus, the buyer's expected total cost per unit time is

$TC_b(T,R,L)=$ purchase cost + ordering cost + inspecting cost + total holding cost of non-defective and defective items + stockout cost + lead time crashing cost

$$=(p+s)D+\frac{A+C(L)}{T}+DT\left[\frac{h_{r1}}{2}+h_{r2}\gamma+\frac{\gamma D}{2x}(h_{r1}-h_{r2})\right]$$
$$+h_{r1}[R-D(T+L)+(1-\beta)E[(X-R)^+]]$$
$$+\frac{1}{T}[\pi\beta+\pi_0(1-\beta)]E[(X-R)^+]. \quad (1)$$

### *3.2. Vendor's expected total cost per unit time*

For each production run, the setup cost per unit time is $S/mT$. When the vendor produces the first $DT$ units, he/she will deliver them to the buyer, after that the vendor will make the deliver on the average every $T$ unit of time until the inventory level falls to zero. Hence, the average inventory per unit time can be calculated as follows:

$$\left\{mDT\left[\frac{DT}{P}+(m-1)T\right]-\frac{(mDT)^2}{2P}-\frac{D^2T^2}{D}[1+2+\cdots+(m-1)]\right\}\frac{D}{mDT}$$
$$=\frac{DT}{2}\left[\frac{D}{P}(2-m)+(m-1)\right].$$

Hence, the vendor's average holding cost per unit time is $h_s\frac{DT}{2}\left[\frac{D}{P}(2-m)+(m-1)\right]$, and the number of defective items is $\gamma DT$, the length of inventory cycle is $T$, the treatment cost of defective items is $\omega$. Hence, the treatment cost of defective items per unit time is $\omega\gamma D$. Furthermore, the capital investment, $I(S)$, is a logarithmic function of the setup cost $S$. That is,

$$I(S)=\frac{1}{\tau}\ln\left(\frac{S_0}{S}\right),$$

where $0<S\le S_0$ and $\tau$ is the percentage decrease in $S$ per dollar increase in $I(S)$. The function is consistent with the Japanese experience as informed in Hall [9] and has been used in literature (see, [13], [19], [28], [29]). Moreover, to reduce the ordering cost, the retailer invests capital $I(S)$ to

improve and avoid too highly ordering cost. This capital investment $I(S)$ is the onetime investment cost whose benefits will intend indefinitely into the future. Therefore, the investment cost per unit time is $\delta I(S)$, where $\delta$ is fractional opportunity cost of capital per unit time. Thus, the vendor's expected total cost per unit time is

$$TC_v(T,S,m) = \text{production cost} + \text{setup cost} + \text{holding cost} + \text{inspecting cost} + \text{treatment defective cost} + \text{Investment in setup cost reduction}$$
$$= Dv + \frac{S}{mT} + h_s\frac{DT}{2}\left[\frac{D}{P}(2-m)+(m-1)\right] + \omega\gamma D + \frac{\delta}{\tau}\ln\left(\frac{S_0}{S}\right). \quad (2)$$

*3.3. The joint expected total cost per unit time*

In this subsection, we consider the situation where vendor and buyer coordinate their production and inventory strategies with each other to determine the best policy for their periodic review integrated supply chain system. Therefore, the joint expected total cost per unit time is given by:

$$JTC(T,R,S,L,m) = TC_b(T,R,L) + TC_v(T,S,m)$$
$$= (p+s+v+\omega\gamma)D + \frac{A+C(L)}{T} + DT\left[\frac{h_{r1}}{2} + h_{r2}\gamma + \frac{\gamma D}{2x}(h_{r1}-h_{r2})\right]$$
$$+h_{r1}[R - D(T+L) + (1-\beta)E[(X-R)^+]] + \frac{S}{mT} + \frac{\delta}{\tau}\ln\left(\frac{S_0}{S}\right)$$
$$+\frac{1}{T}[\pi\beta + \pi_0(1-\beta)]E[(X-R)^+] + h_s\frac{DT}{2}\left[\frac{D}{P}(2-m)+(m-1)\right]. \quad (3)$$

For a given safety factor which satisfies the probability that lead time demand at the buyer exceeds buyer's target level, the actual proportion of demands not met from stock should not exceed the desired value of $\alpha$. Hence, the service level constrant can be established as

$$\frac{\text{Expected demand shortages at the end of the cycle}}{\text{Expected quantity avaliable for satisfying the demand per cycle}} \le \alpha.$$

Simbolically, it can be expressed by $\frac{E[(X-r)^+]}{D(T+L)} \le \alpha$. Moreover, we note that the set-up cost level is $S \in (0, S_0]$, which implies that if the optimal set-up cost obtained does not satisfy the restriction on $S$, then no set-up cost reduction investment is made. For this special case, the optimal set-up cost is the initial set-up cost. Therefore, the problem in this research can be represented as

$$\text{Min } JTC(T,R,S,L,m) = (p+s+v+\omega\gamma)D + \frac{A+C(L)}{T} + DT\left[\frac{h_{r1}}{2} + h_{r2}\gamma + \frac{\gamma D}{2x}(h_{r1}-h_{r2})\right]$$
$$+h_{r1}[R - D(T+L) + (1-\beta)E[(X-R)^+]] + \frac{S}{mT} + \frac{\delta}{\tau}\ln\left(\frac{S_0}{S}\right)$$
$$+h_s\frac{DT}{2}\left[\frac{D}{P}(2-m)+(m-1)\right]. \quad (4)$$

subject to $\frac{E[(X-r)^+]}{D(T+L)} \le \alpha,\ 0 < S \le S_0$.

In many practical situations, the distributional information of the protection interval demand is often quite limited. Hence, in this section, we relax the assumption about the normal distribution of the protection interval demand and only assume that the protection interval demand $X$ has given finite first two moments; i.e., the *p.d.f.* $f_X$ belongs to the class **F** of p.d.f.'s with finite mean $D(T+L)$ and standard deviation $\sigma\sqrt{T+L}$. Since the probability distribution of $X$ is unknown, we can not find the exact value of $E[(X-R)^+]$. We propose to apply the minimax distribution free procedure for our problem. The minimax distribution free approach for this problem is to find the "most unfavorable" *p.d.f.* $f_X$ in **F** for each $(T,R,S,L,m)$ and then minimize the total expected annual cost over $(T,R,S,L,m)$; more exactly, we need to solve

$$\text{Min}_{T,R,S,L,m}\ \text{Max}_{f_X\in\mathbf{F}}\ JTC(T,R,S,L,m)$$
$$\text{subject to } \frac{E[(X-r)^+]}{D(T+L)} \le \alpha. \quad (5)$$

For this purpose, we need the following proposition which was asserted by Gallego and Moon [5].

**Proposition 1**. *For any* $f_X \in \mathbf{F}$,

$$E[(X-r)^+] \le \frac{1}{2}\left\{\sqrt{\sigma^2(T+L)+[R-D(T+L)]^2}-[R-D(T+L)]\right\}$$
$$=\tfrac{1}{2}\sigma\sqrt{T+L}\left(\sqrt{1+k^2}-k\right). \quad (6)$$

*Moreover the upper bound (6) is tight.*

Based on the results of (6) and $R = D(T+L)+k\sigma\sqrt{T+L}$, the safety factor $k$ can be viewed as a decision variable instead of $R$, and thus model (5) is reduced to

$$\text{Min } JTC(T,k,S,L,m) = (p+s+v+\omega\gamma)D+\tfrac{A+C(L)}{T}+DT\left[\tfrac{h_{r1}}{2}+h_{r2}\gamma+\tfrac{\gamma D}{2x}(h_{r1}-h_{r2})\right]$$
$$+h_{r1}\sigma\sqrt{T+L}\left[k+\tfrac{1}{2}(1-\beta)\left(\sqrt{1+k^2}-k\right)\right]+\tfrac{S}{mT}+\tfrac{\delta}{\tau}\ln\left(\tfrac{S_0}{S}\right)$$
$$+h_s\tfrac{DT}{2}\left[\tfrac{D}{P}(2-m)+(m-1)\right]. \quad (7)$$

subject to $\sigma\sqrt{T+L}\left(\sqrt{1+k^2}-k\right) \le 2\alpha D(T+L)$, $0 < S \le S_0$.

## 4. Solution Technique

Our problem is to find the optimal values of $T, k, S, L,$ and $m$ such that $JTC(T,k,S,L,m)$ has a minimum value. To solve the nonlinear programming problem in (7), we provisionally ignore the setup cost constraint $0 < S \le S_0$ and add a nonnegative slack variable $S_v^2 \ge 0$ to convert the service level constraint $\sigma\sqrt{T+L}\left(\sqrt{1+k^2}-k\right) \le 2\alpha D(T+L)$ to equality. One has

$$\text{Min } JTC(T,k,S,L,m,S_v) = (p+s+v+\omega\gamma)D+\tfrac{A+C(L)}{T}+DT\left[\tfrac{h_{r1}}{2}+h_{r2}\gamma+\tfrac{\gamma D}{2x}(h_{r1}-h_{r2})\right]$$
$$+h_{r1}\sigma\sqrt{T+L}\left[k+\tfrac{1}{2}(1-\beta)\left(\sqrt{1+k^2}-k\right)\right]+\tfrac{S}{mT}+\tfrac{\delta}{\tau}\ln\left(\tfrac{S_0}{S}\right)$$
$$+h_s\tfrac{DT}{2}\left[\tfrac{D}{P}(2-m)+(m-1)\right]. \quad (8)$$

subject to $\sigma\sqrt{T+L}\left(\sqrt{1+k^2}-k\right)+S_v^2 = 2\alpha D(T+L)$.

The Lagrangian function of (8) is given by

$$JTC(T,k,S,L,m,\lambda,S_v) = (p+s+v+\omega\gamma)D+\tfrac{A+C(L)}{T}+DT\left[\tfrac{h_{r1}}{2}+h_{r2}\gamma+\tfrac{\gamma D}{2x}(h_{r1}-h_{r2})\right]$$
$$+h_{r1}\sigma\sqrt{T+L}\left[k+\tfrac{1}{2}(1-\beta)\left(\sqrt{1+k^2}-k\right)\right]+\tfrac{S}{mT}+\tfrac{\delta}{\tau}\ln\left(\tfrac{S_0}{S}\right)$$
$$+h_s\tfrac{DT}{2}\left[\tfrac{D}{P}(2-m)+(m-1)\right]$$
$$+\lambda\left[\sigma\sqrt{T+L}\left(\sqrt{1+k^2}-k\right)+S_v^2-2\alpha D(T+L)\right], \quad (9)$$

where $\lambda$ is a Lagrange multiplier. And, for any given $(T,k,S,m,\lambda,S_v)$, $JTC(T,k,S,L,m,\lambda,S_v)$ is a concave function in $L \in [L_i, L_{i-1}]$, because

$$\frac{\partial^2 JTC(T,k,S,L,m,\lambda,S_v)}{\partial L^2} = -\frac{\lambda\sigma\left(\sqrt{1+k^2}-k\right)}{4(T+L)^{\frac{3}{2}}}-\frac{\sigma h_{r1}}{4(T+L)^{\frac{3}{2}}}\left[k+\tfrac{1}{2}(1-\beta)\left(\sqrt{1+k^2}-k\right)\right] < 0.$$

Therefore, for fixed $(T,R,S,m,\lambda,S_v)$, the minimum joint expected total cost will occur at the end points of the interval $[L_i, L_{i-1}]$. On the other hand, we can further prove that, for any given $L \in [L_i, L_{i-1}]$ and $m$, by the Kuhn-Tucker necessary conditions for minimization problem, we can obtain the slack variable $S_v^2 = 0$ (hence, the service level constraint is active when the optimal solution is obtained). Therefore for fixed $L \in [L_i, L_{i-1}]$ and $m$, the minimum value of Eq. (9) (in which the variable $S_v = 0$) will occur at the point $(T,k,S,\lambda)$ which satisfies

$$0 = \frac{\partial JTC(T,k,S,L,m,\lambda)}{\partial T} = -\frac{A+C(L)}{T^2}-\frac{S}{mT^2}-2D\alpha\lambda+\frac{\sigma\lambda\left(\sqrt{1+k^2}-k\right)}{2\sqrt{T+L}}+\frac{\sigma h_{r1}\left[k+\frac{1}{2}(1-\beta)\left(\sqrt{1+k^2}-k\right)\right]}{2\sqrt{T+L}}$$
$$+D\left[\tfrac{h_{r1}}{2}+h_{r2}\gamma+\tfrac{\gamma D}{2x}(h_{r1}-h_{r2})\right]+h_s\tfrac{D}{2}\left[\tfrac{D}{P}(2-m)+(m-1)\right],$$

$$0=\frac{\partial JTC(T,k,S,L,m,\lambda)}{\partial k}=\lambda\sigma\sqrt{T+L}\left(\frac{k}{\sqrt{k^2+1}}-1\right)+h_{r1}\sigma\sqrt{T+L}\left[1+\frac{1}{2}(1-\beta)\left(\frac{k}{\sqrt{k^2+1}}-1\right)\right],$$
$$0=\frac{\partial JTC(T,k,S,L,m,\lambda)}{\partial S}=\frac{1}{mT}-\frac{\delta}{\tau S},$$
$$0=\frac{\partial JTC(T,k,S,L,m,\lambda)}{\partial \lambda}=\sigma\sqrt{T+L}\left(\sqrt{1+k^2}-k\right)-2\alpha D(T+L).$$

Simplifying these equations leads to

$$\left(\frac{A+C(L)}{T^2}-D\left[\frac{h_{r1}}{2}+h_{r2}\gamma+\frac{\gamma D}{2x}(h_{r1}-h_{r2})\right]-h_s\frac{D}{2}\left[\frac{D}{P}(2-m)+(m-1)\right]+\frac{S}{mT^2}+2D\alpha\lambda\right)\sqrt{T+L}$$
$$=\frac{h_{r1}\sigma k}{2}+\frac{\sigma}{4}\left(\sqrt{1+k^2}-k\right)[h_{r1}(1-\beta)+2\lambda], \quad (10)$$
$$\lambda=h_{r1}\left[\frac{\sqrt{1+k^2}}{\sqrt{1+k^2}-k}-\frac{1}{2}(1-\beta)\right], \quad (11)$$
$$S=\frac{m\delta T}{\tau}, \quad (12)$$
$$\sqrt{1+k^2}-k=\frac{2\alpha D}{\sigma}\sqrt{T+L}. \quad (13)$$

By substituting (11), (12) and (13) into (10), we have

$$\left(\frac{A+C(L)}{T^2}-D\left[\frac{h_{r1}}{2}+h_{r2}\gamma+\frac{\gamma D}{2x}(h_{r1}-h_{r2})\right]-h_s\frac{D}{2}\left[\frac{D}{P}(2-m)+(m-1)\right]+\frac{\delta}{\tau T}+\right.$$
$$\left.2\alpha Dh_{r1}\left[\frac{4\alpha^2D^2(L+T)+\sigma^2}{8\alpha^2D^2(T+L)}-\frac{1}{2}(1-\beta)\right]\right)\sqrt{T+L}$$
$$=\frac{h_{r1}\sigma[\sigma^2-4D^2(L+T)\alpha^2]}{8\alpha\sigma D\sqrt{L+T}}+\frac{\alpha D}{2}\sqrt{T+L}\left[h_{r1}(1-\beta)+2\left[\frac{4\alpha^2D^2(L+T)+\sigma^2}{8\alpha^2D^2(T+L)}-\frac{1}{2}(1-\beta)\right]\right]. \quad (14)$$

Further, for fixed $L\in[L_i,L_{i-1}]$, $m$ and the active constraint $\sigma\sqrt{T+L}\left(\sqrt{1+k^2}-k\right)=2\alpha D(T+L)$, one can prove that the optimal value $(T^*,k^*,S^*)$ satisfies sufficient condition.

Under the active constraint $\sigma\sqrt{T+L}\left(\sqrt{1+k^2}-k\right)=2\alpha D(T+L)$, if we take the second derivatives of $JTC(T,k,S,L,m)$ with respect to $m$, we have

$$\frac{\partial^2 JTC_U(T,k,S,L,m)}{\partial m^2}=\frac{2S}{m^3T}>0.$$

Therefore, for fixed $(T,k,S,L)$, $JTC(T,k,S,L,m)$ is a convex function of $m$. Since the number of shipments per batch production run, $m$, is an integer, consequently the optimal value of $m$ (denoted by $m^*$) can be obtained when

$$JTC(T_m^*,k_m^*,S_m^*,L_m^*,m^*)\le JTC(T_{m-1}^*,k_{m-1}^*,S_{m-1}^*,L_{m-1}^*,m^*-1)$$

and

$$JTC(T_m^*,k_m^*,S_m^*,L_m^*,m^*)\le JTC(T_{m+1}^*,k_{m+1}^*,S_{m+1}^*,L_{m+1}^*,m^*+1).$$

The following algorithm can be utilized to find the optimal $(T,k,S,L,m)$.

**Algorithm 1.**

*Step 1.* Set $m=1$.

*Step 2.* For each $L_i, i=0,1,\cdots,n$, we use Eq. (14) to evaluate $T_i$, and then use Eq. (12) and (13) to compute $S_i$ and $k_i$, respectively.

*Step 3.* Compare $S_i$ and $S_0$.

(i) If $S_i\le S_0$, then $S_i$ is feasible and we denote the solution found in Step 1 for given $L_i$ and $m$ by $(T_i^*,k_i^*,S_i^*)$.

(ii) If $S_i>S_0$, then $S_i$ is not feasible and for given $L_i$ and $m$, take $S_i^*=S_0$ and the corresponding value of $T_i^*$ and $k_i^*$ can be obtained from (12) and (13).

*Step 4.* Utilize (7) to compute the corresponding joint expected total cost $JTEC(T_i^*,k_i^*,S_i^*,L_i,m)$ for $i=0,1,2,\cdots,n$.

*Step 5.* Find $\text{Min}_{i=0,1,2,\cdots,n}\, JTC(T_i^*,k_i^*,S_i^*,L_i,m)$.
If $JTC(T_m^*,k_m^*,S_m^*,L_m^*,m)=\text{Min}_{i=0,1,2,\cdots,n}JTC(T_i^*,k_i^*,S_i^*,L_i,m)$, then $(T_m^*,k_m^*,S_m^*,L_m^*,m)$ is the optimal solution for the fixed $m$.

*Step 6.* Set $m=m+1$, and repeat Step 2 to Step 5 to get $JTC(T_m^*,k_m^*,S_m^*,L_m^*,m)$.

*Step 7.* If $JTC(T_m^*, k_m^*, S_m^*, L_m^*, m) \leq JTC(T_{m-1}^*, k_{m-1}^*, S_{m-1}^*, L_{m-1}^*, m-1)$ then go to Step 6, otherwise go to Step 8.

*Step 8.* Set $JTC(T^*, k^*, S^*, L^*, m^*) = JTC(T_{m-1}^*, k_{m-1}^*, S_{m-1}^*, L_{m-1}^*, m-1)$, then $JTC(T^*, k^*, S^*, L^*, m^*)$ is the minimum joint expected total cost and $(T^*, k^*, S^*, L^*, m^*)$ is the optimal solution. Hence the optimal target level is $R^* = D(T^* + L^*) + k\sigma\sqrt{T^* + L^*}$.

## 5. Numerical Example

We consider a stochasti periodic review integrated inventory system with the following data: $D = 600$ units per year, $A = \$200$ per order, $\beta = 0.8$, $\sigma = 15$ units per year, $P = 1200$ units per year, $S_0 = \$1000$ per setup, $v = \$60$, $h_s = \$20$ per unit per year, $h_{r1} = \$30$ per unit per year, $h_{r2} = \$25$ per unit per year, $x = 1500$, $s = \$0.5$ per unit, $\omega = \$30$ per unit, and $\gamma = 0.005$, $\delta = 0.1$, $1/\tau = 2800$, where 1 year = 52 weeks. We assume that the worst service levels are $1 - \alpha = 98\%$, $98.5\%, 99\%$ and $99.5\%$, i.e., the maximum allowable proportions of demands which are not met from stock are $\alpha = 2\%, 1.5\%, 1\%$ and $0.5\%$. Besides, it is assumed that the lead time has three components with data shown in Table 1.

Table 1. Lead time and crashing cost data

| Lead time component $i$ | Normal duration $b_i$ (days) | Minimum duration $a_i$ (days) | Unit crashing cost $c_i$ (\$/day) |
|---|---|---|---|
| 1 | 20 | 6 | 0.4 |
| 2 | 20 | 6 | 1.2 |
| 3 | 16 | 9 | 5.0 |

Algorithm 1 procedure is applied to yield the results for various service level $(1 - \alpha)$ values as shown in Table 2. The optimal policy for each $1 - \alpha$ value can be determined by comparing $JTC(T_m^*, k_m^*, S_m^*, L_m^*, m)$, $m = 1,2, \ldots$, and the results are summarized in Table 3. For example, in the case of service level $(1 - \alpha) = 99.5\%$, applying the above procedure yields $T^* = 8.30$ weeks, $L^* = 8$ weeks, $k^* = 2.12$ (i.e. $R^* = 205.96$ units), $S^* = \$319.39$ per setup and $m^* = 2$. The joint expected total cost of integrated model is \$81731. Moreover, as the value of $1 - \alpha$ increases, the integrated policy not only achieves the required service level but also has the higher periode review, safety factor (target level), setup cost and joint expected total cost. Besides, the values of lead time and number of deliveries are not influenced by changes in the value of service level.

Table 2. The results for various service level using the solution procedures ($T$ and $L$ in weeks)

| Service level $(1 - \alpha)$ | $m$ | $L_m^*$ | $T_m^*$ | $k_m^*$ | $S_m^*$ | $JTC(T_m^*, k_m^*, S_m^*, L_m^*, m)$ |
|---|---|---|---|---|---|---|
| 98.0% | 1 | 8 | 9.50 | 0.07 | 182.87 | 81435 |
| | 2 | 8 | 8.18 | 0.11 | 314.71 | 81240* |
| | 3 | 8 | 7.26 | 0.14 | 418.90 | 81272 |
| 98.5% | 1 | 8 | 9.49 | 0.37 | 182.46 | 81506 |
| | 2 | 8 | 8.17 | 0.41 | 314.24 | 81310* |
| | 3 | 8 | 7.25 | 0.44 | 418.46 | 81340 |
| 99.0% | 1 | 8 | 9.50 | 0.85 | 182.75 | 81624 |
| | 2 | 8 | 8.18 | 0.90 | 314.77 | 81426* |
| | 3 | 8 | 7.27 | 0.94 | 419.17 | 81455 |
| 99.5% | 1 | 8 | 9.66 | 2.03 | 185.85 | 81931 |
| | 2 | 8 | 8.30 | 2.12 | 319.39 | 81731* |
| | 3 | 8 | 7.36 | 2.19 | 424.58 | 81759 |

Table 3. Summary of the optimal integrated policy for various service level

| Service level $1 - \alpha$ | $m^*$ | $L^*$ | $T^*$ | $k^*(R^*)$ | $S^*$ | $JTC(T^*, k^*, S^*, L^*, m^*)$ |
|---|---|---|---|---|---|---|
| 98.0% | 2 | 8 | 8.18 | 0.11 (187.67) | 314.71 | 81240 |
| 98.5% | 2 | 8 | 8.17 | 0.41 (190.03) | 314.24 | 81310 |

| | | | | | | |
|---|---|---|---|---|---|---|
| 99.0% | 2 | 8 | 8.18 | 0.90 (194.25) | 314.77 | 81426 |
| 99.5% | 2 | 8 | 8.30 | 2.12 (205.96) | 319.39 | 81731 |

## 6. Conclusions

In this paper, we propose an integrated vendor-buyer inventory model with the consideration of defective items, controllable lead time and safety factor, invesment for setup cost reduction, partial backlogging, stochastic demand with unknown distribution, and there is an upper bound on the actual proportion of demands not met from stock (service level constraint). Under these conditions, we formulated the mathematical problem as a nonlinear programming model with a constraint and recommended minimax distribution free procedure and Lagrange multiplier method to solve it. The proposed algorithm procedure is developed to determine the optimal number of periode review, safety factor, lead time, setup cost and the number of shipments. The results of the numerical example show that the higher service level will produce positive influences on customer loyalty and brand which are crucial to building competitive advantage in market.

Possible extensions for future research could be by considering learning in production. In this case, the vendor experiences learning in the production process while some of the units are defective. The effect of learning in buyer's inspection errors on supply chain cost also can be investigated. Furthermore, the other possibility is considering fuzzy demand, inflation and the time value of money, and delay-in-payments concepts in this model

## References

[1] Akyurt I Z 2016 Comparison of periodic review inventory models under Poisson demand distribution *Social Sciences Research Journal* **5** 150-160

[2] Annadurai K 2016 Minimax distribution-free procedure for mixture inventory model with variable lead time and a service level constraint by reducing order cost *American Journal of Mathematical and Management Sciences*, **35** 1-14

[3] Banerjee A 1986 A joint economic-lot-size model for purchaser and vendor *Decision Sciences* **17** 292–311

[4] Chand S, Li J, and Xu Y 2016 A periodic review inventory model with two delivery modes, fractional lead-times, and age-and-period-dependent backlogging costs *International Journal of Production Economics* **173** 199-206

[5] Gallego G, and Moon I 1993 The distribution free newsboy problem: Review and extensions *Journal of the Operational Research Society* **44** 825–834

[6] Glock C H 2012 The joint economic lot size problem: A review *International Journal of Production Economic* **135** 671-686

[7] Goyal S K 1976 An integrated inventory model for a single supplier single customer problem *International Journal of Production Research* **15** 107–111

[8] Goyal S K 1988 A joint economic-lot-size model for purchaser and vendor: A comment *Decision Sciences* **19** 236–241

[9] Hall R W 1983 *Zero Inventories* (Dow Jones-Irwin, Homewood, Ill, USA)

[10] Jauhari W A, Fitriyani A, and Aisyati A 2016 An integrated inventory model for single-vendor single-buyer system with freight rate discount and stochastic demand *International Journal of Operational Research* **25** 327-350

[11] Jha J K, and Shanker K 2014 An integrated inventory problem with transportation in a divergent supply chain under service level constraint *Journal of Manufacturing Systems* **33** 462-475

[12] Jindal P, and Solanki A 2016 Supply chain coordination under service level constraint and controllable lead time *International Journal of Operations Research and Information Systems* **7** 84-106

[13] Kurdhi N A, Lestari S M P, and Susanti Y 2015 A fuzzy collaborative supply chain inventory model with controllable setup cost and service level constraint for imperfect items *International Journal of Applied Management Science* **7** 93-122

[14] Kurdhi N A, Santoso L P, Handajani S S, and Martini T S 2016 A coordinated supply chain model for imperfect items under power function lead time crashing cost and investment with a service level constraint *International Journal of Information Systems and Supply Chain Management* **9** 65-82

[15] Li G, Kang Y C, and Xu, G 2015 Optimal decision for fashion supply chains with service level constraint and controllable lead time *Mathematical Problems in Engineering* **2015** Article ID 758494 9 pages

[16] Lin Y J 2009 An integrated vendor-buyer inventory model with backorder price discount and effective investment to reduce ordering cost *Computers and Industrial Engineering* **56** 1597–1606

[17] Lin H J 2013 An integrated supply chain inventory model with imperfect-quality items, controllable lead time and distribution-free demand *Yugoslav Journal of Operations Research* **23** 87-109

[18] Lin H J 2015 A stochastic periodic review inventory model with back-order discounts and ordering cost dependent on lead time for the mixtures of distributions *TOP* **23** 386-400

[19] Lo C C 2013 A collaborative business model for imperfect process with setup cost and lead time reductions *Open Journal of Social Sciences* **1** 6-11

[20] Mohabbatdar S, Ahmadi A, and Sajadieh M S 2016 Optimal manufacturer-retailer policies in a supply chain with defective product and price dependent demand *Journal of Optimization in Industrial Engineering* **19** 37-46

[21] Moon I, Shin E, and Sarkar B 2014 Min-max distribution free continuous-review model with a service level constraint and variable lead time *Applied Mathematics and Computation* **229** 310-315

[22] Pal M, and Chandra S 2014 A periodic review inventory model with stock dependent demand, permissible delay in payment and price disount on backorders *Yugoslav Journal of Operations Research* **24** 99-110

[23] Pan C J, and Yang J S 2002 A study of an integrated inventory with controllable lead time *International Journal of Production Research* **40** 1263–1273

[24] Porteus E L 1986 Optimal lot sizing, process quality improvement and setup cost reduction *Operations Research* **34** 137-144

[25] Sarkar B, Chaudhuri K, and Moon I 2015 Manufacturing setup cost reduction and quality improvement for the distribution free continuous-review inventory model with a service level constraint *Journal of Manufacturing Systems* **34** 74-82

[26] Shah N H 2015 Manufacturer-retailer inventory model for deteriorating items with price-sensitive credit-linked demand under two-level trade credit fnancing and proft sharing contract *Cogent Engineering* **2** doi:10.1080/23311916.2015.1012989

[27] Soni H N, and Joshi M 2015 A periodic review inventory model with controllable lead time and backorder rate in Fuzzy-stochastic environment *Fuzzy Information and Engineering* 7 101-114

[28] Tahami H, Mirzazadeh A, Khamseh A A, and Qadikolaei A G 2016 A periodic review inventory model for buyer's unidentified protection interval demand distribution *Cogent Engineering* **3** 1-21

[29] Vijayashree M, and Uthayakumar R 2016 An integrated vendor and buyer inventory model with investment for quality improvement and setup cost reduction *Operations Research and Applications: An International Journal* **3** 1-14

[30] Yang M F, Kuo J Y, Chen W H, and Lin Y 2015 Integrated supply chain cooperative inventory model with payment period being dependent on purchasing price under defective rate condition *Mathematical Problems in Engineering* **2015** Article ID 513435 20 pages.

IOP Conf. Series: Journal of Physics: Conf. Series **855** (2017) 012023 doi:10.1088/1742-6596/855/1/012023

# Identification and estimation of state variables on reduced model using balanced truncation method

**Trifena Punana Lesnussa, Didik Khusnul Arif, Dieky Adzkiya and Erna Apriliani**

Department of Mathematics, Institut Teknologi Sepuluh Nopember, Surabaya 60111, Indonesia

E-mail: trivelesnussa@gmail.com, didik@matematika.its.ac.id, dieky@matematika.its.ac.id, april@matematika.its.ac.id

**Abstract.** In this paper, we study the identification of variables on a model reduction process and estimation of variables on reduced system. We aim to relate variables on reduced and original system, so that we can compare the estimation accuracy of the original system and reduced system. As such, the objective of this paper is to discuss identification and estimation of variables on reduced model. First, model order reduction is done by using balanced truncation method. This process begins with the construction of balanced system. After that, we identify the relationship between variables of the balanced system and the original system. Then, we eliminate variables of the balanced system that have a small influence on the system. Furthermore, we estimate state variables on the original system and reduced system using a Kalman Filter algorithm. Finally, we compare the estimation result of the identified reduced and original system.

## 1. Introduction

Generally, a system in nature has a great order. This is because the system has many state variables. If the number of state variables in a system is increased, the system is closer to the real phenomenon. However, it caused difficulties in analyzing the system. Thus the computational time is increased significantly. Therefore, we need a method to simplify the order of such systems. In this case, we reduce the order of the original system. One way is using model order reduction [1]. There are many methods in model order reduction, such as balanced truncation method, Hankel norm approximation and singular perturbation approximation [1]. Among those methods, balanced truncation method is the frequently used method because of the simplicity of the method [2, 3].

Estimation is a method to compute the state variables of a discrete-time stochactic system based on measurement data [4]. Kalman filter is an optimal estimation method for linear systems. For nonlinear systems, there are many methods that can be applied such as extended Kalman filter, ensemble Kalman filter, fuzzy Kalman filter. We have studied the application of extended Kalman filter and fuzzy Kalman filter on Nonisothermal Continuous Stirred Tank Reactor systems in [5].

We have investigated the application of Kalman filter to the original system and its reduced system in [6, 7]. Then in [8], the authors have developed Kalman filter algorithm on reduced system. However there are some difficulties in comparing the accuracy of estimation between the original system and reduced system because the order is different. More precisely, we want to

IOP Conf. Series: Journal of Physics: Conf. Series **855** (2017) 012023 doi:10.1088/1742-6596/855/1/012023

compare the estimation of state variables. Since the dimension of state variables in the reduced and original models are different, we cannot compare them. Therefore, we need an identification of state variables on reduced model such that we can determine the corresponding state variables on the original model. In this paper, we want to determine a relationship between state variables on the reduced and original models: given a state variable in the reduced model, we can compute the corresponding state variable in the original model.

## 2. Preliminaries

In this section, we describe some definitions that will be used in this paper.

### *2.1. Discrete-Time Linear-Time-Invariant Systems*

A discrete-time linear-time-invariant system $(A, B, C, D)$ is defined as [4]

$$\begin{aligned} x_{k+1} &= Ax_k + Bu_k \\ z_k &= Cx_k + Du_k \end{aligned} \tag{1}$$

where $x_k \in \mathbb{R}^n$, $u_k \in \mathbb{R}^m$, $z_k \in \mathbb{R}^p$ and $A \in \mathbb{R}^{n\times n}$, $B \in \mathbb{R}^{n\times m}$, $C \in \mathbb{R}^{p\times n}$, $D \in \mathbb{R}^{p\times m}$ are constant matrices. There are many properties that a system may have. Stability, controllability and observability are some important properties of a system. As it will be clear later, we focus on the previous three properties.

**Definition 1 (Stability)** *A discrete-time linear-time-invariant system* (1) *is asymptotically stable if and only if* $|\lambda_i(A)| < 1$ *for* $i = 1, 2, \cdots, n$ *with* $\lambda_i(A)$ *denotes eigenvalues of* $A$. *If* $|\lambda_i(A)| \leq 1$ *for* $i = 1, 2, \cdots, n$ *then the system is stable.*

**Definition 2 (Controllability)** *The discrete-time linear-time-invariant system* (1) *is controllable if rank of the controllability matrix* $M_c = [B|AB|\cdots|A^{n-1}B]$ *equals* $n$.

**Definition 3 (Observability)** *The discrete-time linear-time-invariant system* (1) *is observable if rank of the observability matrix* $M_o = [C|CA|\cdots|CA^{n-1}]^T$ *equals* $n$.

**Definition 4 (Controllability and Observability Gramians)** *The controllability gramian of the discrete-time linear-time-invariant system* (1) *is* $W = \sum_{k=0}^{\infty} A^k BB^T (A^T)^k$. *The observability gramian of the discrete-time linear-time-invariant system* (1) *is* $M = \sum_{k=0}^{\infty} (A^T)^k C^T CA^k$.

The controllability and observability gramians are respectively a unique positive definite solution of the following Lyapunov equation:

$$\begin{aligned} AW + WA^T + BB^T &= 0 \\ A^T M + MA + C^T C &= 0 \end{aligned}$$

### *2.2. Model Reduction of Discrete-Time Linear-Time-Invariant Systems*

In this section, we describe the process of model order reduction using balanced truncation method. This process begins with the construction of balanced system (or balanced realization). On balanced system, we identify the relationship between variables of the original and balanced systems. After all variables have been identified, we eliminate the variables that have small influence to the system.

*2.2.1. Balanced Realization* Balanced system is a system such that controllability and observability gramians are equal. Furthermore the gramian is a diagonal matrix. Balanced system is obtained by applying a transformation to the original system $(A, B, C, D)$. Suppose that defined of gramian like $diag(\sigma_1, \sigma_2, \cdots, \sigma_n)$ with $\sigma_1 \geq \sigma_2 \geq \cdots \geq \sigma_n > 0$ and $\sigma_1, \sigma_2, \cdots, \sigma_n$ is singular Hankel value.

From the original system $(A, B, C, D)$, we construct a new system $(\tilde{A}, \tilde{B}, \tilde{C}, \tilde{D})$. The new system can be constructed if and only if realization of the system is asymptotically stable, have controllability gramian $\tilde{W}$ and observability gramian $\tilde{M}$ is equal and it is a diagonal matrix.

Let us describe a procedure to determine transformation matrix $T$. First, we check whether realization $(A, B, C, D)$ is asymptotically stable, controllable and observable. Second, compute the controllability gramian $W$ and observability gramian $M$ by using Definition 4. Third, we use Cholesky factorization on $W$ to determine $\phi$ such that $W = \phi\phi^T$. Next, we use singular value decomposition on $\phi^T M \phi$ to compute $U$ and $\Sigma$ such that $\phi^T M \phi = U\Sigma^2 U^T$. The last step, compute transformation matrix $T = \Sigma^{1/2} U^T \phi^{-1}$.

By applying the transformation matrix to the original realization, $x_k = T\tilde{x}_k$ we obtain an ordered balanced realization $(\tilde{A}, \tilde{B}, \tilde{C}, \tilde{D}) = (T^{-1}AT, T^{-1}B, CT, D)$.

*2.2.2. Identification of Variables* Let a transformation matrix $T$ be given such that it meets the following equation:

$$x_k = T\tilde{x}_k \tag{2}$$

In this work, we focus on the identification of variables in the balanced system and the original system.

*2.2.3. Balanced Truncation Methods* Model reduction by using balanced truncation simply applies the truncation operation to a balanced realization. Let $(\tilde{A}, \tilde{B}, \tilde{C}, \tilde{D})$ be a balanced realization, where the gramian is denoted by $\Sigma$. Remember that the controllability gramian is equal to the observability gramian and the gramian is a diagonal matrix. First we partition the gramian as follows

$$\Sigma = \begin{pmatrix} \Sigma_1 & 0 \\ 0 & \Sigma_2 \end{pmatrix} \tag{3}$$

where $\Sigma_1 = diag(\sigma_1, \sigma_2, \cdots, \sigma_r)$, $\Sigma_2 = diag(\sigma_{r+1}, \sigma_{r+2}, \cdots, \sigma_n)$ and $\sigma_r > \sigma_{r+1}$.

By using the preceding partition, the corresponding partition of balanced system $(\tilde{A}, \tilde{B}, \tilde{C}, D)$ is given by

$$G = \left(\begin{array}{cc|c} \tilde{A}_{11} & \tilde{A}_{12} & \tilde{B}_1 \\ \tilde{A}_{21} & \tilde{A}_{22} & \tilde{B}_2 \\ \hline \tilde{C}_1 & \tilde{C}_2 & D \end{array}\right) \tag{4}$$

where $\tilde{A} = \begin{pmatrix} \tilde{A}_{11} & \tilde{A}_{12} \\ \tilde{A}_{21} & \tilde{A}_{22} \end{pmatrix}$, $\tilde{B} = \begin{pmatrix} \tilde{B}_1 \\ \tilde{B}_2 \end{pmatrix}$, and $\tilde{C} = \begin{pmatrix} \tilde{C}_1 & \tilde{C}_2 \end{pmatrix}$. The reduced model is defined as the subsystem corresponding to $\Sigma_1$. Thus the reduced model can be written as $(\tilde{A}_r, \tilde{B}_r, \tilde{C}_r, \tilde{D}_r)$ where $\tilde{A}_r = \tilde{A}_{11}$, $\tilde{B}_r = \tilde{B}_1$, $\tilde{C}_r = \tilde{C}_1$ and $\tilde{D}_r = D$.

Next, we present a proposition saying that the truncation of a balanced realization is again a balanced realization. Furthermore, the proposition also states that the properties of the original realization are the same with the truncated one.

**Proposition 5 ([1, Lemma 9.4.1])** *Suppose $(\tilde{A}, \tilde{B}, \tilde{C}, \tilde{D})$ is a balanced realization and assumed it is stable, controllable, observable and balanced with gramian*

$$\tilde{M} = \tilde{N} = diag(\sigma_1, \sigma_2, \cdots, \sigma_n), \qquad \sigma_1 \geq \cdots \geq \sigma_r \geq \cdots \geq \sigma_n > 0$$

IOP Conf. Series: Journal of Physics: Conf. Series **855** (2017) 012023 doi:10.1088/1742-6596/855/1/012023

*If $\sigma_r > \sigma_{r+1}$ then reduced system of order $r$ will be stable, controllable, observable, balanced and satisfies $\|G_s - G_{s_r}\|_\infty \leq 2(\sigma_{r+1} + \cdots + \sigma_n)$ with $G_s$ and $G_{s_r}$ is a transfer function of $(\tilde{A}, \tilde{B}, \tilde{C}, \tilde{D})$ and the reduced system $(\tilde{A}_r, \tilde{B}_r, \tilde{C}_r, \tilde{D}_r)$, respectively.*

### 2.3. Kalman Filter on Discrete-Time Linear-Time-Invariant System

In this section, we describe the process of Kalman filter algorithm over the original system and reduced system. Firstly, we analyze an estimation process of state variables in discrete-time linear-time-invariant system (1). Here, we analyze the estimation process and construct the best estimator for discrete-time linear-time-invariant system (1) based on the Kalman filter algorithm [4]. Furthermore, we apply the Kalman filter procedure to original model and reduced model.

In Kalman filter, the model is described as a stochastic discrete-time linear-time-invariant system, as follows

$$\begin{aligned} x_{k+1} &= Ax_k + Bu_k + Gw_k, \\ z_k &= Cx_k + Du_k + v_k, \end{aligned} \tag{5}$$

where $x_k \in \mathbb{R}^n$ is a state vector at time $k$, $u_k \in \mathbb{R}^m$ is an input vector at time $k$, $z_k \in \mathbb{R}^p$ is a measurement vector or output vector at time $k$, $w_k$ is a system noise and $v_k$ is a measurement noise. Variables $w_k$ and $v_k$ are normally distributed random variables, i.e. $w_k \sim N(0, Q)$ and $v_k \sim N(0, R)$. Matrices $A$, $B$, $C$ and $D$ are appropriately dimensioned real constant matrices.

Estimation of state variables in the system (5) can be done by using Kalman filter. First, we estimate the state variables based on a dynamic system (prediction step) and then we improve the estimation results by using the measurement data (correction step). The Kalman filter algorithm for stochastic discrete-time system is as follows [4]:

(i) Initialization
- Define the estimation of initial state $\hat{x}_0$
- Define the covariance of initial state's estimation $P_0$
- Define the variance of system noise $Q$
- Define the variance of measurement noise $R$

(ii) Prediction step
   (a) Error Covariance: $P^-_{k+1} = AP_kA^T + GQG^T$
   (b) Estimation: $\hat{x}^-_{k+1} = A\hat{x}_k + Bu_k$

(iii) Correction step
   (a) Error Covariance: $P_{k+1} = P^-_{k+1} - P^-_{k+1}C^T(CP^-_{k+1}C^T + R)^{-1}CP^-_{k+1}$
   (b) Estimation: $\hat{x}_{k+1} = \hat{x}^-_{k+1} + P_{k+1}C^TR^{-1}(z_{k+1} - C\hat{x}^-_{k+1})$

Initially, we define the estimation of initial state and its covariance. In this step, we also define the variance of system and measurement noises. After that, in each time step, we execute the prediction and correction steps. Notice that in the prediction step, we do not need any measurement data, whereas in the measurement step, we need a measurement data at that time step.

## 3. Main Results

In this section, we discuss the steps in the identification of state variables on the reduced model. Firstly, we construct a discrete-time linear-time-invariant system $(A, B, C, D)$. Then, we examine the properties of the system, such as stability, controllability and observability. If the system is asymptotically stable, controllable and observable, we continue to the next step. If the system is not asymptotically stable or the system is not controllable or the system is not observable then we have to construct another system until the three properties are satisfied.

IOP Conf. Series: Journal of Physics: Conf. Series **855** (2017) 012023 doi:10.1088/1742-6596/855/1/012023

Then, we construct the balanced system. The balanced system is a transformation of the original system using a transformation matrix. On balanced system, we identify the relationship between state variables of the original and balanced systems. After all state variables in balanced system have been identified, we remove state variables of the balanced system that have small influences on the system using balanced truncation method.

The last step is estimation of reduced and original systems by using Kalman filter algorithm. Finally, we compare the estimation result of the reduced and original system. More precisely, first we identify the states corresponding to the estimation results of reduced system, then we compare the identification results against the estimation results over the original system. In this study, we use a reduced system of order 6.

### *3.1. Construction of The Original System*

Given the following discrete-time linear-time-invariant system:

$$\begin{pmatrix} x_1 \\ x_2 \\ x_3 \\ x_4 \\ x_5 \\ x_6 \\ x_7 \\ x_8 \end{pmatrix}_{k+1} = \begin{pmatrix} 0.2 & 0 & 0 & 0 & 0 & 0 & 0 & -1 \\ 0 & -0.5 & 0 & 0 & 0 & 1 & 0 & 0 \\ 0 & 0 & 0.8 & 0 & 0 & 3 & 0 & 0 \\ 0 & -1 & 0 & 0.7 & 0 & 0 & 0 & 0 \\ 0 & 0 & 0 & 0 & -0.5 & 0.4 & 0 & 0 \\ 0 & 0 & 0 & 0 & 0.2 & -0.6 & 0 & 0 \\ 0 & 0 & 0 & 0 & 2 & 0 & -0.81 & 0 \\ 0 & 0 & -2 & 0 & 0 & 0 & 0 & 0.9 \end{pmatrix} \begin{pmatrix} x_1 \\ x_2 \\ x_3 \\ x_4 \\ x_5 \\ x_6 \\ x_7 \\ x_8 \end{pmatrix}_{k} + \begin{pmatrix} 4 \\ 1 \\ 3 \\ 3 \\ -2 \\ 1 \\ 4 \\ 1 \end{pmatrix} u_k$$

$$y_k = \begin{pmatrix} 1 & 5 & 3 & 7 & -3 & 2 & 4 & 6 \end{pmatrix} \begin{pmatrix} x_1 \\ x_2 \\ x_3 \\ x_4 \\ x_5 \\ x_6 \\ x_7 \\ x_8 \end{pmatrix}_{k} + u_k \tag{6}$$

This system (6) is asymptotically stable, controllable and observable. Since the system in (6) satisfies the three properties, we can use the system in the numerical example.

### *3.2. Model Order Reduction*

*3.2.1. Balanced System* We obtain the following balanced system

$$\begin{pmatrix}\tilde{x}_1\\ \tilde{x}_2\\ \tilde{x}_3\\ \tilde{x}_4\\ \tilde{x}_5\\ \tilde{x}_6\\ \tilde{x}_7\\ \tilde{x}_8\end{pmatrix}_{k+1} = \begin{pmatrix} 0.9630 & 0.1108 & -0.0024 & -0.0087 & -0.0148 & -0.0011 & -0.0001 & -0.0000\\ -0.1108 & 0.5633 & -0.4411 & 0.0630 & 0.1315 & 0.0017 & 0.0006 & 0.0003\\ 0.0024 & -0.4411 & -0.8034 & -0.0835 & -0.1420 & -0.0096 & -0.0006 & -0.0003\\ -0.0087 & -0.0630 & 0.0835 & 0.8972 & 0.3007 & -0.0489 & 0.0008 & 0.0006\\ 0.0148 & 0.1315 & -0.1420 & -0.3007 & -0.0347 & 0.3154 & 0.0004 & -0.0007\\ 0.0011 & 0.0017 & -0.0095 & 0.0489 & 0.3154 & -0.0100 & -0.1268 & -0.0702\\ 0.0001 & 0.0006 & -0.0006 & 0.0008 & 0.0004 & -0.1268 & 0.7454 & -0.2847\\ -0.000 & -0.0003 & 0.0003 & 0.0006 & 0.0007 & 0.0702 & 0.2847 & -0.3364 \end{pmatrix}\begin{pmatrix}\tilde{x}_1\\ \tilde{x}_2\\ \tilde{x}_3\\ \tilde{x}_4\\ \tilde{x}_5\\ \tilde{x}_6\\ \tilde{x}_7\\ \tilde{x}_8\end{pmatrix}_k +$$

$$\begin{pmatrix}-9.2733 & -12.2348 & -1.1801 & -1.3131 & 2.3704 & 0.1318 & 0.0109 & -0.0056\end{pmatrix}^T \tilde{u}_k$$

$$\tilde{y}_k = \begin{pmatrix}9.2733 & -12.2348 & -1.1801 & 1.3131 & 2.3704 & 0.1318 & 0.0109 & 0.0056\end{pmatrix}\begin{pmatrix}\tilde{x}_1\\ \tilde{x}_2\\ \tilde{x}_3\\ \tilde{x}_4\\ \tilde{x}_5\\ \tilde{x}_6\\ \tilde{x}_7\\ \tilde{x}_8\end{pmatrix}_k + \tilde{u}_k \tag{7}$$

The transformation matrix used to obtain the balanced system (7) is as follows

$$T = \begin{pmatrix} -2.4836 & 2.7670 & -1.0787 & -0.7136 & 5.0536 & 4.6166 & 0.9975 & -2.4005\\ -0.0078 & -0.0038 & 0.2493 & -0.1043 & 0.4635 & -0.4272 & -0.1333 & 0.2741\\ -0.1688 & -0.1837 & 0.3369 & 0.2326 & -0.0481 & 0.0043 & 0.1288 & -0.3290\\ -0.0529 & -0.1288 & 0.1588 & -0.1614 & 0.4081 & -0.2571 & -0.6292 & 0.4543\\ 0.0083 & 0.0887 & 0.1864 & 0.2261 & -0.1375 & 0.0517 & -0.0523 & 0.2043\\ -0.0030 & -0.0482 & -0.1442 & -0.1395 & 0.0139 & -0.0243 & -0.0338 & 0.1123\\ -0.0010 & -0.4052 & -2.0874 & 0.6429 & -1.1149 & 0.4837 & -0.1793 & 0.8229\\ 2.1179 & -1.9245 & 0.9545 & 0.2277 & -0.6083 & -0.3822 & 0.7209 & -0.6768 \end{pmatrix}$$

*3.2.2. Identification of State Variables in the Balanced System* In the following equation, we describe the relationship between state variables of the original system and the balanced system:

$$\begin{pmatrix}x_1\\ x_2\\ x_3\\ x_4\\ x_5\\ x_6\\ x_7\\ x_8\end{pmatrix} = \begin{pmatrix} -2.4836 & 2.7670 & -1.0787 & -0.7136 & 5.0536 & 4.6166 & 0.9975 & -2.4005\\ -0.0078 & -0.0038 & 0.2493 & -0.1043 & 0.4635 & -0.4272 & -0.1333 & 0.2741\\ -0.1688 & -0.1837 & 0.3369 & 0.2326 & -0.0481 & 0.0043 & 0.1288 & -0.3290\\ -0.0529 & -0.1288 & 0.1588 & -0.1614 & 0.4081 & -0.2571 & -0.6292 & 0.4543\\ 0.0083 & 0.0887 & 0.1864 & 0.2261 & -0.1375 & 0.0517 & -0.0523 & 0.2043\\ -0.0030 & -0.0482 & -0.1442 & -0.1395 & 0.0139 & -0.0243 & -0.0338 & 0.1123\\ -0.0010 & -0.4052 & -2.0874 & 0.6429 & -1.1149 & 0.4837 & -0.1793 & 0.8229\\ 2.1179 & -1.9245 & 0.9545 & 0.2277 & -0.6083 & -0.3822 & 0.7209 & -0.6768 \end{pmatrix}\begin{pmatrix}\tilde{x}_1\\ \tilde{x}_2\\ \tilde{x}_3\\ \tilde{x}_4\\ \tilde{x}_5\\ \tilde{x}_6\\ \tilde{x}_7\\ \tilde{x}_8\end{pmatrix}$$

*3.2.3. Balanced Truncation Method* Since the reduced model is of order 6, the reduced model is as follows. The reduced model is obtained by removing the state variables $\tilde{x}_7$ and $\tilde{x}_8$ from the balanced system.

$$\begin{pmatrix}\tilde{x}_1^r\\ \tilde{x}_2^r\\ \tilde{x}_3^r\\ \tilde{x}_4^r\\ \tilde{x}_5^r\\ \tilde{x}_6^r\end{pmatrix}_{k+1} = \begin{pmatrix} 0.9630 & 0.1108 & -0.0024 & -0.0087 & -0.0148 & -0.0011\\ -0.1108 & 0.5633 & -0.4411 & 0.0630 & 0.1315 & 0.0017\\ 0.0024 & -0.4411 & -0.8034 & -0.0835 & -0.1420 & -0.0096\\ -0.0087 & -0.0630 & 0.0835 & 0.8972 & 0.3007 & -0.0489\\ 0.0148 & 0.1315 & -0.1420 & -0.3007 & -0.0347 & 0.3154\\ 0.0011 & 0.0017 & -0.0095 & 0.0489 & 0.3154 & -0.0100\end{pmatrix}\begin{pmatrix}\tilde{x}_1^r\\ \tilde{x}_2^r\\ \tilde{x}_3^r\\ \tilde{x}_4^r\\ \tilde{x}_5^r\\ \tilde{x}_6^r\end{pmatrix}_k + \begin{pmatrix}-9.2733\\ -12.2348\\ -1.1801\\ -1.3131\\ 2.3704\\ 0.1318\end{pmatrix}\tilde{u}_k^r$$

$$\tilde{y}_k^r = \begin{pmatrix}9.2733 & -12.2348 & -1.1801 & 1.3131 & 2.3704 & 0.1318\end{pmatrix}\begin{pmatrix}\tilde{x}_1^r\\ \tilde{x}_2^r\\ \tilde{x}_3^r\\ \tilde{x}_4^r\\ \tilde{x}_5^r\\ \tilde{x}_6^r\end{pmatrix}_k + \tilde{u}_k^r$$

*3.2.4. Identification of State Variables in the Reduced System* In this section, we describe the relationship between state variables of the original system and the reduced system. This relationship is derived from the identification of state variables in the balanced system. More precisely, we remove the columns of transformation matrix $T$ that corresponds to the removed state variables. The relationship between state variables of the original system and the reduced system is described in the following equation:

$$\begin{pmatrix}x_1\\ x_2\\ x_3\\ x_4\\ x_5\\ x_6\\ x_7\\ x_8\end{pmatrix} = \begin{pmatrix} -2.4836 & 2.7670 & -1.0787 & -0.7136 & 5.0536 & 4.6166\\ -0.0078 & -0.0038 & 0.2493 & -0.1043 & 0.4635 & -0.4272\\ -0.1688 & -0.1837 & 0.3369 & 0.2326 & -0.0481 & 0.0043\\ -0.0529 & -0.1288 & 0.1588 & -0.1614 & 0.4081 & -0.2571\\ 0.0083 & 0.0887 & 0.1864 & 0.2261 & -0.1375 & 0.0517\\ -0.0030 & -0.0482 & -0.1442 & -0.1395 & 0.0139 & -0.0243\\ -0.0010 & -0.4052 & -2.0874 & 0.6429 & -1.1149 & 0.4837\\ 2.1179 & -1.9245 & 0.9545 & 0.2277 & -0.6083 & -0.3822\end{pmatrix}\begin{pmatrix}\tilde{x}_1^r\\ \tilde{x}_2^r\\ \tilde{x}_3^r\\ \tilde{x}_4^r\\ \tilde{x}_5^r\\ \tilde{x}_6^r\end{pmatrix}$$

*3.2.5. Simulation* In this simulation, we compare state variables on the original system and identified state estimation on reduced systems. The estimation of state variables on the reduced system is done by using Kalman filter algorithm. The procedure for estimation of identified state variables on the reduced system is as follows. First, we apply balanced truncation method to the original system. In this step, we obtain a reduced system. Then we apply Kalman filter algorithm on the reduced system. In this step, we obtain estimation of state variables on the reduced system. Finally, we use the identification result. In this step, we obtain the identified state estimation on the reduced system that has the same dimension with the original system. This allows us to compare the identified state estimation on the reduced system and the state on the original system. In this case, we determine the estimation error on the identified state variables. The estimation error is defined as the difference between the estimation results of identified state variables on the reduced system and real states on the original system. The comparison results are shown in Table 1 and the graphics are shown in Figure 1. From Table 1, the estimation error on the identified state estimation is quite close to the real values.

Furthermore we want to compare the estimation result in Table 1 and the estimation error of the original system. In Table 2, we display the estimation error of the original system. The estimation error is defined as the difference between the estimation results of state variables on the original system and real states on the original system. According to Tables 1 and 2,

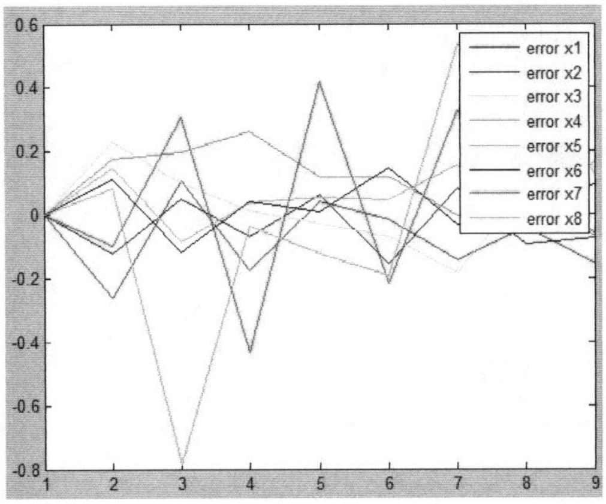


**Figure 1.** Error of estimation results on the identified state estimation of the reduced system and state on the original system. The $x$-axis represents the time index, whereas the $y$-axis represents the error at each time. Each line plot represents a state component. Since dimension of the state space is 8, there are 8 plots in the figure.

**Table 1.** Estimation error of the identified reduced system.

| time | Estimation error of the identified reduced system | | | | | | | |
|---|---|---|---|---|---|---|---|---|
| step | $x_1$ | $x_2$ | $x_3$ | $x_4$ | $x_5$ | $x_6$ | $x_7$ | $x_8$ |
| 1 | -0.508 | -0.029 | -0.041 | -0.002 | -0.069 | 0.0198 | 0.1637 | 0.4973 |
| 2 | 1.1773 | 0.0873 | 0.0631 | 0.0423 | 0.042 | -0.059 | -0.805 | -0.778 |
| 3 | 0.8762 | -0.204 | -0.18 | -0.056 | 0.0319 | 0.0646 | 0.9103 | -0.799 |
| 4 | -0.293 | 0.1614 | 0.1214 | 0.2225 | 0.136 | -0.099 | -0.233 | 0.7497 |
| 5 | -0.46 | -0.039 | 0.1279 | 0.1228 | 0.1459 | 0.0046 | -0.299 | 0.601 |
| 6 | -0.245 | 0.007 | 0.1883 | 0.1205 | -0.078 | 0.0842 | 0.1452 | -0.027 |
| 7 | -0.154 | -0.03 | 0.3448 | 0.1116 | 0.0751 | -0.093 | -0.126 | -0.449 |
| 8 | 0.4889 | -0.014 | 0.1985 | 0.2061 | -0.049 | -0.007 | -0.448 | -1.169 |

**Table 2.** Estimation Error of the Original system of all state variables for the first 8 time steps.

| time | Estimation error of the original system | | | | | | | |
|---|---|---|---|---|---|---|---|---|
| step | $x_1$ | $x_2$ | $x_3$ | $x_4$ | $x_5$ | $x_6$ | $x_7$ | $x_8$ |
| 1 | -0.263 | 0.1048 | -0.174 | 0.0414 | -0.015 | -0.141 | -0.039 | -0.157 |
| 2 | -0.121 | 0.0486 | -0.068 | 0.0603 | -0.156 | 0.0841 | -0.096 | -0.074 |
| 3 | 0.2297 | 0.1002 | 0.0131 | -0.029 | -0.07 | -0.186 | 0.0829 | -0.018 |
| 4 | 0.1753 | 0.1961 | 0.2639 | 0.1207 | 0.1192 | -0.005 | 0.0998 | -0.071 |
| 5 | 0.1488 | -0.088 | 0.0391 | 0.0549 | 0.0446 | 0.1556 | -0.03 | 0.1687 |
| 6 | 0.1108 | -0.118 | 0.0428 | 0.0105 | 0.1477 | -0.033 | -0.025 | 0.1005 |
| 7 | -0.098 | 0.3075 | -0.434 | 0.4199 | -0.216 | 0.3272 | 0.0301 | -0.056 |
| 8 | 0.0831 | -0.787 | -0.036 | -0.123 | -0.191 | 0.5306 | 0.4268 | 0.1269 |

the estimation error of both cases is pretty close. Thus we conclude that the performance of estimation on the identified state variables is acceptable.

In Table 3, we discuss the computational times of both cases. In the first case, we measure the computational time for estimation of state variables on the original system. In the second case, we measure the computational time for estimation of identified state variables. According to Table 3, the computational time for estimation in the reduced system is faster than in the original system. This is because the number of state variables in the reduced system is smaller than the number of state variables in the original system.

**Table 3.** Computational times for the estimation process in the original and reduced systems.

| | Computational time (sec) |
|---|---|
| Original system | 5.21722 |
| Identified reduced system | 3.164235 |

## 4. Conclusions

In this work, we have designed a procedure for identification and estimation of state variables on reduced models of discrete-time linear-time-invariant systems. From the identification process, we obtain the corresponding state variables in the original model from state variables in the reduced model via a linear transformation. Thus we can compare the state variables in the identified reduced model and original model. The estimation error in the original system is quite close to the error in the identified reduced system. However, the computational time for estimation in the reduced system is faster than the computational time for estimation in the original model.

## Acknowledgments

This work has been supported by:

- Penelitian Doktor Baru number 31040/IT2.11/PN.08/2016 with the title of Konstruksi Algoritma Filter Kalman Tereduksi Pada Sistem Tidak Stabil Dan Aplikasinya Pada Aliran Air Sungai;
- Penelitian Unggulan Perguruan Tinggi year 2017 with the title of Estimasi Ketinggian dan Kecepatan Aliran Sungai dengan Filter Kalman Tereduksi Metode Residual.

## References

[1] Green M and Limebeer D 2012 *Linear Robust Control* (Dover Publications)
[2] Sari Y, Arif D and Adzkiya D 2016 *International Conference on Mathematics: Pure, Applied and Computation*
[3] Kartika D, Arif D and Adzkiya D 2016 *International Conference on Mathematics: Pure, Applied and Computation*
[4] Lewis F 1986 *Optimal Estimation* (John Wiley and Sons)
[5] Fitria R and Arif D 2017 *International Journal of Computing Science and Applied Mathematics* **3** 16–20
[6] Arif D, Widodo, Apriliani E and Salmah 2011 *3rd International Conferences and Workshops on Basic and Applied Sciences* p M007
[7] Arif D, Widodo, Salmah and Apriliani E 2012 *Proceedings of the 8th International Conference on Mathematics, Statistics and Its Applications*
[8] Arif D, Widodo, Salmah and Apriliani E 2014 *International Journal of Control and Automation* **7** 257–270

# Discrete time Markov chains (DTMC) susceptible infected susceptible (SIS) epidemic model with two pathogens in two patches

**Eka Lismawati, Respatiwulan, and Purnami Widyaningsih**

Department of Mathematics, Faculty of Mathematics and Natural Sciences, Sebelas Maret University, Surakarta, Indonesia

E-mail: ekalismawati@student.uns.ac.id, respatiwulan@staff.uns.ac.id, poer@uns.ac.id

**Abstract.** The SIS epidemic model describes the pattern of disease spread with characteristics that recovered individuals can be infected more than once. The number of susceptible and infected individuals every time follows the discrete time Markov process. It can be represented by the discrete time Markov chains (DTMC) SIS. The DTMC SIS epidemic model can be developed for two pathogens in two patches. The aims of this paper are to reconstruct and to apply the DTMC SIS epidemic model with two pathogens in two patches. The model was presented as transition probabilities. The application of the model obtain that the number of susceptible individuals decreases while the number of infected individuals increases for each pathogen in each patch.

## 1. Introduction

Health is the important thing in human life. Health can be impaired due to an infectious disease. The spread of infectious diseases that occurs in a long period of time can lead to the occurrence of an epidemic. The epidemic can be represented in a mathematical model (Hethcote [5]).

In 1927, the epidemic model was first introduced by Kermack and McKendrick [6]. In 1994, Allen [2] has been written the susceptible infected susceptible (SIS) epidemic models. The characteristic of the model is a susceptible individual, after a successful contact with an infectious individual, becomes infected and infectious, but it does not develop immunity to the disease. In the SIS epidemic models, the individuals conditions are classified into two categories, there are susceptible (S) and infectious (I). The number of susceptible and infected individuals is random events that depend on time so it follows the stochastic process (Allen [2]). The number of susceptible and infected individuals at time $t+1$ depends only on the individuals at time $t$, so these follow the discrete time Markov process. It can be described as an epidemic model of discrete time Markov chains (DTMC) SIS (Allen [1]).

According to Allen and Kiruparahan [3] in 2005, a susceptible individual can be infected more than one pathogen. In 2007, McCormack and Allen [7] represented that the spread of the disease also can occur in one or more patch, due to their individual displacement occurred from one patch to another. Then Allen et al.[4] developed a model of an epidemic DTMC SIS for two pathogens in the two patches. By referring to Allen et al., the aims of this research are to reconstruct and to apply the DTMC SIS epidemic model with two pathogens in two patches.

## 2. DTMC SIS Epidemic Models with One Pathogen

In 2003, Allen [1] introduced the DTMC SIS epidemic model. There are four assumptions on the model i.e. total population size $N$ is constant, the birth rate that equals to the death rate, homogeneously population, and the individuals born are susceptible individuals. Let $S(t)$ and $I(t)$ denote discrete random variables for the number of susceptible and infected individuals at time $t$. The total population size $N$ is constant so $S(t) + I(t) = N$. Let $S(t) = s$ and $I(t) = i$, $s$ and $i$ are state, the joint probability function for $S(t)$ and $I(t)$ is given by

$$p_{(s,i)}(t) = P\{S(t) = s, I(t)) = i\},$$

where $s, i = 0, 1, 2, \ldots, N$ and $t = 0, \Delta t, 2\Delta t, \ldots$.

A transition is a change in the number of susceptible and infected individuals during the time period $\Delta t$. The time period $\Delta t$ is chosen sufficiently small such that the number of susceptible and infected individuals changes by at most one during the time period $\Delta t$, that is $-1$ if it decreases, 0 if it fixed, or 1 if it increases. During the time period $\Delta t$, if the change in the number of susceptible and infected individuals is $h$ and $j$, where $h$ and $j \in \{-1, 0, 1\}$, so the transitions are from state $s$ to $s+h$ and from state $i$ to $i+j$. According to Allen [1], the DTMC SIS epidemic model satisfy

$$p_{(s+h,i+j),(s,i)}(\Delta t) = \begin{cases} \frac{\beta is}{N}\Delta t, & (h,j) = (-1,1); \\ \gamma i \Delta t, & (h,j) = (1,-1); \\ \delta i \Delta t, & (h,j) = (0,-1); \\ 1 - \left(\frac{\beta is}{N} + \gamma i + \delta i\right)\Delta t, & (h,j) = (0,0); \\ 0, & (h,j) \text{ others,} \end{cases}$$

where $\delta$ denotes the birth rate that equals to the death rate, $\beta$ denotes the contact rate, and $\gamma$ denotes the recovery rate.

## 3. DTMC SIS Epidemic Models with Two Pathogens

According to Allen et al. [4], the individuals can be infected with more than one pathogen through direct contact between an infected individual and a susceptible one. Let $S(t)$ and $I_k(t)$ denote discrete random variables for susceptible and infected individuals with pathogen $k$ at time $t$, where $k = 1$ and 2. The assumptions of the DTMC SIS epidemic model with two pathogens equal to the assumptions of the DTMC SIS epidemic model one pathogen. Let $S(t) = s$, $I_k(t) = i_k$, and the change in the number of susceptible and infected individuals with pathogen $k$ during the time period $\Delta t$ is $j_k$, with $j_k \in \{-1, 0, 1\}$, then the DTMC SIS epidemic model is given by

$$p_{(s+h,i_k+j_k),(s,i_k)}(\Delta t) = \begin{cases} \beta_k \frac{i_k}{N} s \Delta t, & (h,j_k) = (-1,1); \\ \gamma_k i_k \Delta t, & (h,j_k) = (1,-1); \\ \delta_k i_k \Delta t, & (h,j_k) = (0,-1); \\ 1 - a & (h,j_k) = (0,0); \\ 0, & (h,j_k) \text{ others,} \end{cases}$$

where $a = \left(\frac{\beta_1 s}{N} i_1 + \gamma_1 i_1 + \delta_1 i_1 + \frac{\beta_2 s}{N} i_2 + \gamma_2 i_2 + \delta_2 i_2\right)\Delta t$, with $\delta_k, \beta_k, \gamma_k$ denote the birth rate that equals to the death rate, the contact rate, and the recovery rate of pathogen $k$.

## 4. Main Results

### *4.1. The Reconstruction Model*

The reconstruction of the DTMC SIS epidemic model with two pathogens in two patches refers to Allen et al.[4]. The assumptions of the DTMC SIS epidemic model with two pathogens in two patches equal to the assumptions of the DTMC SIS epidemic model with two pathogens. There are two processes in the model i.e. an infection process and an dispersal process. The infection process discusses about the contact of susceptible and infected individuals in the same patch. The dispersal process discusses about an individual movement from one patch to another.

Let $S_d(t)$ and $I_{dk}(t)$ indicate discrete random variables for the number of susceptible individuals at time $t$ in patch $d$ and the number of infected individuals by pathogen $k$ at time $t$ in patch $d$, respectively, where $S_d(t), I_{dk}(t) \in \{0, 1, 2, \ldots, N\}$. Let $N_1$ and $N_2$ indicate the population size in patch $d$, $d = 1, 2$ and the total population size is $N_1 + N_2 = N$, where $N_1$ and $N_2$ are constant. Let $S_d(t) = s_d$ and $I_{dk}(t) = i_{dk}$, the joint probability function for $S_d(t)$ and $I_{dk}(t)$ is given by

$$p_{(s_d,i_{dk})}(t) = P\{S_d(t) = s_d, I_{dk}(t) = i_{dk}\},$$

where $s_d, i_{dk} = 0, 1, 2, \ldots, N$ and $t = 0, \Delta t, 2\Delta t, \ldots$.

If the change in the number of susceptible and infected individuals during the time period $\Delta t$ are $h_d$ and $j_{dk}$, then the transitions are from *state* $s_d$ to state $s_d + h_d$ and from *state* $i_{dk}$ to *state* $i_{dk} + j_{dk}$. The transition probability for $s_d$ to $s_d + h_d$ and $i_{dk}$ to $i_{dk} + j_{dk}$ is

$$\begin{aligned} p_{(s_d+h_d,i_{dk}+j_{dk}),(s_d,i_{dk})}(\Delta t) &= P\{(S_d(t+\Delta t), I_{dk}(t+\Delta t)) = (s_d + h_d, i_{dk} + j_{dk})| \\ &\quad (S_d(t), I_{dk}(t)) = (s_d, i_{dk})\}. \end{aligned}$$

If there is a new infection, so state $(s_d, i_{dk})$ transition to state $(s_d - 1, i_{dk} + 1)$. The transition occurs due to a susceptible individual becomes infected by pathogen $k$. If there are $i_{dk}$ infected individuals in $N_d$ population, then the probability of contact between infected individuals and susceptible individuals is $\frac{i_{dk}}{N_d}$. If $\beta_{dk}$ is the contact rate of infected individuals by pathogen $k$ in patch $d$, then the transition probability from state $(s_d, i_{dk})$ to state $(s_d - 1, i_{dk} + 1)$ satisfy

$$p_{(s_d-1,i_{dk}+1),(s_d,i_{dk})}(\Delta t) = \beta_{dk} \frac{i_{dk}}{N_d} s_d \Delta t.$$

If an infected individual by pathogen $k$ recovers from the infection, then state $(s_d, i_{dk})$ transition to *state* $(s_d + 1, i_{dk} - 1)$. If $\gamma_{dk}$ is the recovery rate of pathogen $k$ in patch $d$, then the transition probability from state $(s_d, i_{dk})$ to state $(s_d + 1, i_{dk} - 1)$ is given by

$$p_{(s_d+1,i_{dk}-1),(s_d,i_{dk})}(\Delta t) = \gamma_{dk} i_{dk} \Delta t.$$

If an infected individual with pathogen $k$ dies from the infection, then state $(s_d, i_{dk})$ transition to state $(s_d, i_{dk} - 1)$. If $\delta_{dk}$ is the death rate of pathogen $k$ in patch $d$, then the transition probability from state $(s_d, i_{dk})$ to state $(s_d, i_{dk} - 1)$ is

$$p_{(s_d,i_{dk}-1),(s_d,i_{dk})}(\Delta t) = \delta_{dk} i_{dk} \Delta t.$$

There is a possibility that during the time period $\Delta t$, no event occurs. The probability no event occurs during the time period $\Delta t$ is given by

$$p_{(s_d,i_{dk}),(s_d,i_{dk})}(\Delta t) = 1 - \left( \frac{\beta_{d1} s_d}{N_d} i_{d1} + \gamma_{d1} i_{d1} + \delta_{d1} i_{d1} + \frac{\beta_{d2} s_d}{N_d} i_{d2} + \gamma_{d2} i_{d2} + \delta_{d2} i_{d2} \right) \Delta t,$$

where $s_d + i_{d1} + i_{d2} \in \{0, 1, 2, \ldots, N_d\}$.

Let $b = (\frac{\beta_{d1} s_d}{N_d} i_{d1} + \gamma_{d1} i_{d1} + \delta_{d1} i_{d1} + \frac{\beta_{d2} s_d}{N_d} i_{d2} + \gamma_{d2} i_{d2} + \delta_{d2} i_{d2})\Delta t$, then the DTMC SIS epidemic model for the infection process in patch $d$ for pathogen $k$ satisfy

$$p_{(s_d+h_d, i_{dk}+j_{dk}),(s_d,i_{dk})}(\Delta t) = \begin{cases} \beta_{dk} \frac{i_{dk}}{N_d} s_d \Delta t, & (h_d, j_{dk}) = (-1, 1); \\ \gamma_{dk} i_{dk} \Delta t, & (h_d, j_{dk}) = (1, -1); \\ \delta_{dk} i_{dk} \Delta t, & (h_d, j_{dk}) = (0, -1); \\ 1 - b, & (h_d, j_{dk}) = (0, 0); \\ 0, & (h_d, j_{dk}) \text{ yang lain}, \end{cases} \quad (1)$$

where $\beta_{dk}$, $\gamma_{dk}$, and $\delta_{dk}$ are positive.

There is a dispersal process after the infection process. The dispersal process refers to the movement of individuals from one patch to another. In this process, the populations size $N_1$ and $N_2$ are constant. Thus, if there is the movement of individuals from patch 1 to patch 2 then also there is the movement of individuals from patch 2 to patch 1. If the susceptible individuals move from one patch to another with probability $p_d$, then the remains of the susceptible individuals which don't move from one patch to another have a probability $1-p_d$. If the infected individuals by pathogen $k$ move from one patch to another with probability $p_{dk}$, then the remains of the infected individuals by pathogen $k$ which don't move from one patch to another have a probability $1 - p_{dk}$. Thus, the DTMC SIS epidemic model for the dispersal process in patch $d$ for pathogen $k$ satisfy

$$p = \begin{cases} p_d, \\ 1 - p_d, \end{cases} \quad \text{and} \quad q = \begin{cases} q_{dk}, \\ 1 - q_{dk}. \end{cases}$$

*4.2. The Model Application*

In this section, the parameters value were selected by referring to Allen et al.[4]. The parameters value are contact rate $\beta_{11} = 0.1$, $\beta_{12} = 0.05$, $\beta_{21} = 0.05$, $\beta_{22} = 0.075$, recovery rate $\gamma_{11} = 0.05$, $\gamma_{12} = 0.025$, $\gamma_{21} = 0.033$, $\gamma_{22} = 0.05$, the birth rate that equals to the death rate $\delta_{dk} = 0$, the population size $N = 200$, $N_1 = 100$ and $N_2 = 100$. By using the parameters into the model (1), it is obtained the transition probabilities. By knowing the transition probabilities, it is also known the transition from state $(s_d, i_{dk})$ to state $(s_d + h_d, i_{dk} + j_{dk})$, then the number of susceptible and infected individuals at time $t$ can be known. By using initial conditions $I_{11}(0) = I_{12}(0) = I_{21}(0) = I_{22}(0) = 1$ and $S_1(0) = S_2(0) = 98$ into the model (1) with the parameters value which refers to Allen et al., the number of susceptible and infected individuals at an interval 600 first period of times is presented in the Figure 1 and 2.

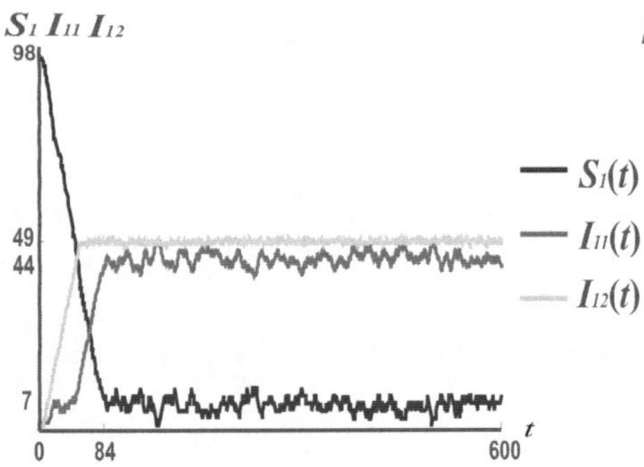


**Figure 1.** The number of $S_1 I_{11} I_{12}$ at an interval 600 first period times.

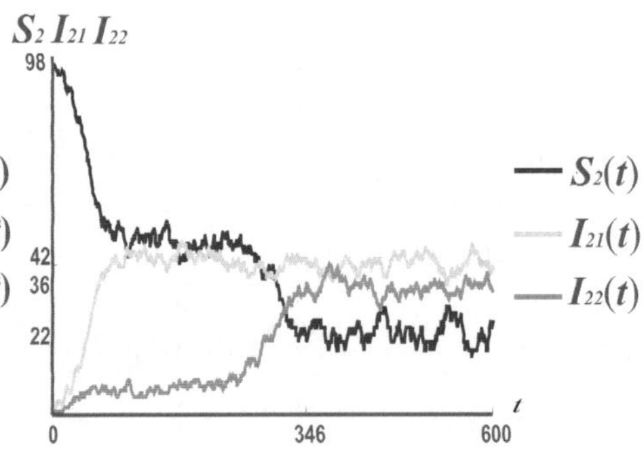


**Figure 2.** The number of $S_2 I_{21} I_{22}$ at an interval 600 first period times.

Figure 1 shows that the number of susceptible individuals at time 84th decreases from 1 to 7. At time 84th to time 600th, the number of susceptible individuals fluctuates in number 5 to 9. The number of infected individuals by the first pathogen increases from 1 to 44 at time 84th, then it fluctuates in number 41 to 47. The number of infected individuals by the second pathogen increases from 1 to 49, then it fluctuates in number 48 to 50. Figure 2 shows that the number of susceptible individuals decreases from 98 to 22 at time 346th, then it fluctuate until time 600th in number 18 to 26. At time 346th, the number of infected individuals by the first pathogen increases from 1 to 42, then it fluctuates until time 600th in number 39 to 45. The number of infected individuals by the second pathogen increases from 1 to 36, then it fluctuates until time 600th in number 31 to 40.

To observe the change in the number of infected individuals at time $t$ and to calculate the transition probabilities, we present the disease spread at an interval 10 first period of times in Figure 3 and 4.

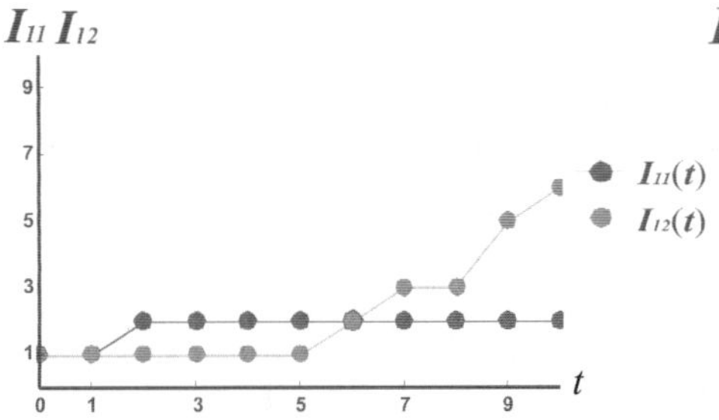


**Figure 3.** The number of $I_{11}I_{12}$ at an interval 10 first period of times.

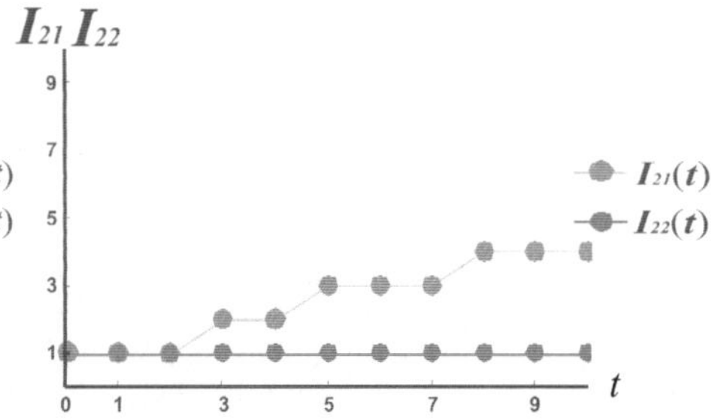


**Figure 4.** The number of $I_{21}I_{22}$ at an interval 10 first period of times.

The transition probabilities of diseases spread in Figure 3 and 4 can be calculated by using model (1). Figure 3 shows the change in the number of infected individuals, $I_{11}(1) = 1$ to $I_{11}(2) = 2$. There is a transition of a susceptible individual that it can be infected by the first pathogen at a time period $\Delta t$. The transition probability of $I_{11}(1) = 1$ to $I_{11}(2) = 2$ by using model (1) is 0.098. Figure 4 shows the change in the number of infected individuals $I_{21}(2) = 1$ to $I_{21}(3) = 2$. The transition probability of $I_{21}(2) = 1$ to $I_{21}(3) = 2$ is 0.049. In the same way, the transition probability at every time can be easily calculated.

The transition probabilities of the movement individuals in the dispersal process were selected by referring to Allen et al.[4]. The transition probability of the movement susceptible individuals is $p_d = 0.01$. Allen et al. assumes that the infected individuals are isolated, then the transition probability of the movement infected individuals is $q_{dk} = 0$. Thus, the transition probability of no occur the movement susceptible individuals is $1 - p_d = 0.99$.

## 5. Conclusions

From this paper, we have two conclusions.

(i) The DTMC SIS epidemic model with two pathogens in two patches is given by

(a) the transition probabilities of the infection process are

$$p_{(s_d+h_d,i_{dk}+j_{dk}),(s_d,i_{dk})}(\Delta t) = \begin{cases} \beta_{dk}\frac{i_{dk}}{N_d}s_d\Delta t, & (h_d, j_{dk}) = (-1, 1); \\ \gamma_{dk}i_{dk}\Delta t, & (h_d, j_{dk}) = (1, -1); \\ \delta_{dk}i_{dk}\Delta t, & (h_d, j_{dk}) = (0, -1); \\ 1 - b, & (h_d, j_{dk}) = (0, 0); \\ 0, & (h_d, j_{dk}) \text{ others,} \end{cases}$$

with $\beta_{dk}$, $\gamma_{dk}$, and $\delta_{dk}$ are positive, where $b = (\frac{\beta_{d1}s_d}{N_d}i_{d1} + \gamma_{d1}i_{d1} + \delta_{d1}i_{d1} + \frac{\beta_{d2}s_d}{N_d}i_{d2} + \gamma_{d2}i_{d2} + \delta_{d2}i_{d2})\Delta t$.

(b) the transition probabilities of the dispersal process are

$$p = \begin{cases} p_d, \\ 1 - p_d, \end{cases} \quad \text{and} \quad q = \begin{cases} q_{dk}, \\ 1 - q_{dk}. \end{cases}$$

(ii) The application of the model which refers to Allen shows the pattern of disease spread DTMC SIS epidemic model with two pathogens in two patches. In patch one, at time 84th, the number of susceptible individuals decreases from 98 to 7, the number of infected individuals by the first pathogen increases from 1 to 44, and the number of infected individuals by the second pathogen increases from 1 to 49. In patch two, at time 346th, the number of susceptible individuals decreases from 98 to 22, the number of infected individuals by the first pathogen increases from 1 to 42, and the number of infected individuals by the second pathogen increases from 1 to 36.

## References

[1] Allen L J S 2008 *An Introduction to Stochastic Epidemic Models* (Texas: Texas Tech University)

[2] Allen L J S 2003 *An Introduction to Stochastic Processes with Applications to Biology* (New Jersey, Upper Saddla River: Prentice Hall)

[3] Allen L J S and Kirupaharan N 2005 Asymptotic Dynamics of Deterministic and Stochastic Epidemic Models with Multiple Pathogens *International Journal of Numerical Analysis and Modeling* **2** 3 pp 329-344

[4] Allen L J S, Kirupaharan N, and Wilson S M 2004 SIS Epidemic Mode ls with Multiple Pathogen Strains *Journal of Difference Equations and Applications* **10** 1 pp 53-75

[5] Hethcote H W 2000 The Mathematics of Infectious Diseases *SIAM Review* **42** 4 pp 599-653

[6] Kermack W O and McKendrick A G 1927 A Contribution to the Mathematical Theory of Epidemics *Proceeding of the Royal Society of London, Series A, Containing Papers of a Mathematical and Physical Character* **115** pp 700-721

[7] McCormack R K and Allen L J S 2007 Multi-patch Deterministic and Stochastic Models for Wildlife Diseases *Journal of Biological Dynamics* **1** 1 pp 63-85

IOP Conf. Series: Journal of Physics: Conf. Series **855** (2017) 012025 doi:10.1088/1742-6596/855/1/012025

# Population Analysis of Disabled Children by Departments in France

**Diah Meidatuzzahra[1], Heri Kuswanto[1], Nicolas Pech[2], Amélie Etchegaray[3]**

[1]Department of Statistics, Faculty of Mathematics and Sciences, Sepuluh Nopember Institute of Technology
[2]Department of Mathematics, Faculty of Sciences, Aix-Marseille University
[3]Centre inter-Regional d'Etudes, d'Action et d'Information (CREAI), Provence-Alpes-Côte d'Azur (PACA) et Corse

E-mail : meyda.azzahra@gmail.com, kuswanto.its@gmail.com, nicolas.pech@univ-amu.fr, amelie.etchegaray@creai-pacacorse.com

**Abstract**. In this study, a statistical analysis is performed by model the variations of the disabled about 0-19 years old population among French departments. The aim is to classify the departments according to their profile determinants (socioeconomic and behavioural profiles). The analysis is focused on two types of methods: principal component analysis (PCA) and multiple correspondences factorial analysis (MCA) to review which one is the best methods for interpretation of the correlation between the determinants of disability (independent variable). The PCA is the best method for interpretation of the correlation between the determinants of disability (independent variable). The PCA reduces 14 determinants of disability to 4 axes, keeps 80% of total information, and classifies them into 7 classes. The MCA reduces the determinants to 3 axes, retains only 30% of information, and classifies them into 4 classes.

## 1. Introduction

The term "disability" is defined as a limitation of a person's ability to interact with their environment, due to a permanent disability or non-permanent that leads to stress and moral disorder, intellectual, physical or social. Disability is the consequence of an impairment that may be physical, cognitive, mental, sensory, emotional, developmental, or some combination of these. A disability may be present from birth, or occur during a person's lifetime. Disability comes in multiple forms and ambiguous definition. It is to be distinguished from the disease or the accident, which can be the disability origin.

In France, the definition disability is governed by French law dated 11 February 2005 on the opportunities and the same rights, and the participation and citizenship of disability people, that "Disability of activity limitations or restrictions on participation in the social life suffered by a person, due to substantial continuing modification of one or more functions"[1].

One of the institutions cared for the disabled is CREAI branch PACA et Corse. CREAI collaborated with Population Environment Development Laboratory (LPED) Aix-Marseille University propounds a project "Geography of Disability". The purpose is to estimate of the population and establishment medical social service. Therefore, statistical analysis is required to estimate the disabled population of multiple databases which were defined as disability determinants.

In this study, statistical analysis performed to model the variations of disabled children (aged 0-19 years) population among French department consisting of 14 determinant variables of children disabilities based on six categories, namely, the professional category of social (CSP) of their parents, the level education of their parents, the premature rates, the tax of revenues, alcohol consumption, facilities and services of medical social for disabled. The aim is to classify departments according to their profiles determinants (socioeconomic and behavioral profiles).

In [2] has used PCA to clustering application of elderly people autonomy-disability). However, in this paper applying two types of methods: principal components analysis (PCA) and multiple correspondence analysis (MCA) to review which one is the best methods for the interpretation of the correlation between the determinants of disability (independent variable). And then, hierarchical clustering can be used to classify the departments according to their profile determinants. The software to support the analysis is SAS (Statistical Analysis System).

## 2. Preliminaries

This section, describe the definitions that will be used in this paper.

### *2.1. Centre inter-Régional d'Etudes, d'Action et d'Information (CREAI)*

CREAI PACA et Corse or Central interregional of studies, action and information branch Provence-Alpes-Côte d'Azur (PACA) and Corse in France was founded in 1965 and belongs to the National Association of CREAI (ANCREAI) for a person with the condition of vulnerability. The CREAI is a private organizations and non-profit status established by statute law in 1901 which is subsidized by the State to optimize information sharing, collaboration and develop synergies of the technical experts to reflection and observation in the sectors of social action and medico-social [3].

### *2.2. The Determinants of Disability*

The hypothesis of the project "Geography of Disability" is the distribution unusual of the disabled population in the region. The distribution of this population related by multiple factors, for example, economic factors, education, environment, lifestyle, etc., called the determinants of disability. [4] [5] [6]. Six groups of determinants identified:

*2.2.1. The professional category of social (CSP).* The CSP based on the data of INSEE. Labor force of 15 years and more having a job by gender, age, and the CSP are divided into 6 positions, is show on table 1.

**Table 1.** The position of CSP

| Number | The professional category of social |
|---|---|
| 1. | Farmer |
| 2. | Artisan, craftsman and trader |
| 3. | Manager and high professions |
| 4. | Intermediate professions |
| 5. | Employee |
| 6. | Labor |

*2.2.2. The education level of their parents.* In [7] identified through HSM survey, 33% of the disabled population aged 20-59 years is not graduation (without diplome). The database is used BTX_TD_FOR2_2012 of INSEE. It is classify 4 of education level, is show on table 2.

**Table 2.** The classify of education level

| Number | The level of education |
|---|---|
| 1. | Without diplome |
| 2. | BEPC / BEP / CAP (Diplome or certificate of professional) |
| 3. | BAC (Baccalaureat) |
| 4. | $BAC^{+}$ |

*2.2.3. The Premature Rates.* The [8] study showed the importance of disabling sequelae preterm infants, before 33 weeks of amenorrhea (WA), and among those born between 33 and 36 WA of age. According [8], if the preterm birth is increased, the risk of disability is also. The 8th day certificate (Cs8) data used as database from 2010 to 2012.

*2.2.4. The Tax of Revenues.* The report of revenues tax is derived from local INSEE Social and Tax File (Philosophy) data. The first quartile of income report is the average wage in the department below which is 25% of wages [3].

*2.2.5. The Consumption of Alcohol.* Consumption of alcohol among women is unknown, therefore the number of premature deaths due to overdose of alcohol (it cause alcoholic psychoses and alcoholic cirrhosis of the liver) in women under the 65 years old were selected.

*2.2.6. The facilities and services of medical social for disabled.* The facilities and services of medical social for 1000 children disabled from age 0-19 years old. According the CREAI data (2016), Paris and PACA region are the least equipped regions mainly located (below 8 places for 1000 children). The Lozère is the most departments equipped in medico-social places with 22 places for 1000 children and also the Orne (18 places for 1000 children) and the Creuse (17 places for 1000 children).

### *2.3. Principal Component Analysis (PCA)*

A principal component analysis (PCA) is concerned with explaining the variance-covariance structure of a set of variables through a few linear combinations of these variables [9]. PCA is a statistical procedure that uses an orthogonal transformation to convert a set of observations of possibly correlated variables into a set of values of linearly uncorrelated variables called principal components. Its general objectives are reduction and interpretation of dimension (axes) data without reducing significantly the characteristics of the data. PCA is also often used to avoid problems of multicollinearity between independent variables in a multiple regression model.

The number of principal components is less than or equal to the number of original variables. This transformation is defined in such a way that the first principal component has the largest possible variance (that is, accounts for as much of the variability in the data as possible), and each succeeding component in turn has the highest variance possible under the constraint that it is orthogonal to the preceding components. The resulting vectors are an uncorrelated orthogonal basis set. PCA is sensitive to the relative scaling of the original variables [10].

### *2.4. Multiple Correspondence Analysis*

Multiple correspondence analysis (MCA) is the factorial method which adapted to tables. It is a set of individuals which described by several qualitative variables. It can be presented in many different ways. In France, following the work of L. Lebart, the most common is to focus on the similarities with correspondence analysis, a method designed to study the relationship between two qualitative variables [11].

In the perspective of simultaneously processing quantitative and qualitative variables for the same individuals, which is one of the strengths of multiple factor analysis (MFA), it is important to focus on the similarities between principal component analysis (PCA) and MCA [12].

MCA can also be viewed as a PCA applied to the complete disjunctive table (CDT). To do this, the CDT must be transformed as follows. Let $y_{ik}$ denote the general term of the CDT. $y_{ik}$ is equal to 1 if individual $i$ possesses the category $k$ and 0 if not. Let's denote $p_k$, the proportion of individuals possessing the category $k$ [13].

### *2.5. Hierarchical Ascendant Classification (HAC)*

A hierarchical ascendant classification or hierarchical clustering method is a procedure which represents the data as a nested sequence of partitions. An example of the corresponding graphical representation, called a dendrogram. It is important to note that the height of a node is proportional to the distance between groups it links. Consequently, the shape of a dendrogram gives information on the number of clusters in a data-set. Thus, cutting a dendrogram horizontally engineers a clustering. The number of clusters is selected by three criteria. The diagram below represents the Cubic Clastering Criterion (CCC), pseudo F and pseudo t2 as a function of the number of clusters.

## 3. Methodology

In this section, explain of the steps in this research.

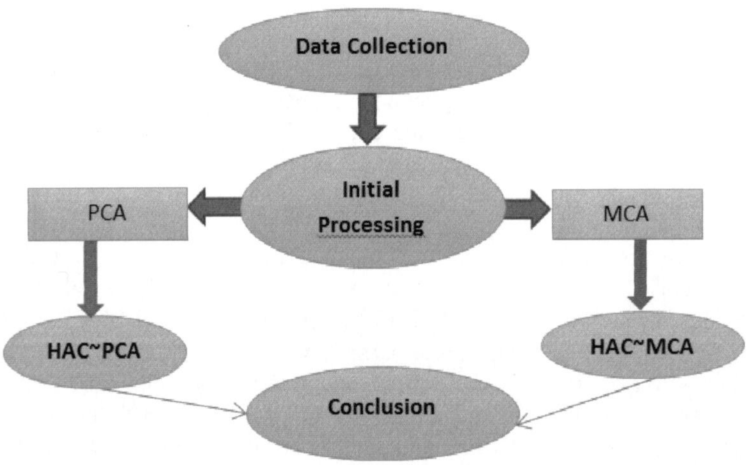


### *3.1. Data Collection*

In the data collection is the first step in a study. The data used in this research is secondary data obtained from [15] [3] and the amount of department in France which is 96 departments as an observation.

### *3.2. Initial Processing*

Initial processing stages to prepare the determinant of disability data. There are 14 determinant variables of children disabilities that will be used to process data, is show on table 3.

**Table 3.** The total of determinants variables

| Number | The determinant variables |
|---|---|
| 1. | Farmer |
| 2. | Artisan, craftsman and trader |
| 3. | Manager and high professions |
| 4. | Intermediate professions |
| 5. | Employee |
| 6. | Labor |
| 7. | Without diplome |
| 8. | BEPC / BEP / CAP (Diplome or certificate of professional) |
| 9. | BAC (Baccalaureat) |
| 10. | $BAC^+$ |
| 11. | The Premature Rates |
| 12. | The Tax of Revenues |
| 13. | The Consumption of Alcohol |
| 14. | The facilities and services of medical social for disabled |

*3.3. Experiment and Data Processing*

In this step will discuss the data processing for the PCA, MCA, and HAC method using software SAS. For each method will be discussed as follows:

For processing the data using PCA method:
1. Obtained the correlation matrix from determinant variables
2. Obtained the eigenvalue of the correlation matrix
3. Consider the number of principal components (axes)
4. Interpretation of the sample principal components

For processing the data using MCA method:
1. Construction of table the data
2. Construction of the complete disjunctive table
3. Consider the contribution of variables to the construction of axes
4. Interpretation the contribution of variables and departments

For processing the data using HAC method:
1. Selection the number of class
2. Characterization and location of class

## 4. Discussion and Result

*4.1. Principal Component Analysis (PCA)*

The PCA reduces 14 determinants of disability to 4 axes, keeps 80% of total information. PCA is also calculating the coordinates of the department and the contribution of the determinants variables per axis. It is not balanced because some departments have too strong contribution. Figure 1 describe the number of axes will be interpretation, and table 4 describe the eigenvalues of the correlation matrix.

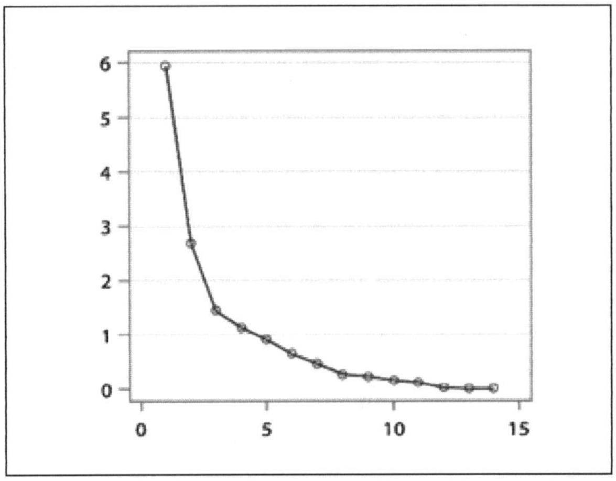


**Figure 1**. The number of axes

**Table 4.** The eigenvalues of the correlation matrix

| | Eigenvalues | Difference | Proportion | Cumulative |
|---|---|---|---|---|
| 1 | 5.94759063 | 3.26410811 | 0.4248 | 0.4248 |
| 2 | 2.68348252 | 1.24735361 | 0.1917 | 0.6165 |
| 3 | 1.43612891 | 0.30611838 | 0.1026 | 0.7179 |
| 4 | 1.13001053 | 0.22447436 | 0.0807 | 0.7998 |

### *4.2. Multiple Correspondence Analysis*
The MCA reduces the determinants to 3 axes, retains only 30% of total information. However, the determinant of the alcohol consumption does not contribute to these axes. The MCA graph is difficult to interpret. The 56 variables supplementary and 96 departments do not identify the group of department.

### *4.3. Hierarchical Ascendant Classification (HAC)*
The Hierarchical Ascendants Classification (HAC) issued by the ACP put the 96 departments into 7 classes. Class 1 is the biggest class because there are 35 departments. In class 2, there are 18 departments. Class 3 includes all the departments of the Mediterranean such as : Hautes Alpes, Alpes-de-Haute-Provence, Alpes-Maritimes, Vars, Bouches-du-Rhône, Vaucluse, Gard, Hérault, Aude, Pyrénées-Orientales Corsica (Corse-du-Sud and Haute-Corse). In class 4 are found 16 departments. Class 5 and Class 6 each only 6 departments. On the contrary, class 7 contains only two departments, Paris and The Hauts de Seine.

The HAC issued by the MCA put the 96 departments into 4 classes. In class 1, there are 23 departments and class 2 is the biggest class because there are 31 departments. Class 3 includes 23 departments and in class 4 only 19 departments.

## 5. Conclusions
Based on the results and the discussion on the methods, the PCA is the best method for interpretation of the correlation between the determinants of disability (independent variable). Although, a more balanced classification of the department is classification on MCA.

The PCA reduces 14 determinants of disability to 4 axes, keeps 80% of total information. PCA is also calculating the coordinates of the department and the contribution of the determinants variables per axis. It is not balanced because some departments have too strong contribution. The Hierarchical Ascending Clusterification (HCA) issued by the PCA made it possible to group the departments into 7 clusters. With 35 departments, cluster 1 is the cluster with the most individuals. On the contrary, cluster 7 contains only two. The MCA reduces the determinants to 3 axes, retains only 30% of total information. However, the determinant of the alcohol consumption does not contribute to these axes. The HCA issued by the MCA put the 96 departments into 4 clusters. In cluster 1, there are 23 departments and cluster 2 is the biggest cluster because there are 31 departments. Cluster 3 includes 23 departments and in cluster 4 only 19 departments.

**Acknowledgment**
The author would like to thank the lectures who have provided advice in the preparation of this research and also thank to other parties who contributed to the completion of this analysis.

**References**

[1] OMS 2011 Organisation Mondiale de la Santé
[2] Combes C, Azema J 2010 Clustering Using Principal Component Analysis: Application of elderly people autonomy-disability
[3] Le Centre interregional s'Etudes, d'Action et d'Informations (CREAI) en faveur des personnes en situation de vulnérabilité
[4] Vigneron E. 2011 Inégalités de santé, inégalités de soins dans les territoires français : Etat des lieux et voies de progrès. Elsevier Masson, France 194p
[5] Salem G., Rican S., Kurzinger M.L. 2000 Atlas de la santé de France, volume 2 : *comportements et maladies*. John Libbey , 222p.
[6] Vigneron E. 2013 Inégalités de santé, inégalités de soins dans les territoires français : Les tribunes de la santé , France p41-53.
[7] Espagnacq M. 2015 *Populations à risque de handicap et restrictions de participation sociale Dossiers solidarité et santé,* Drees. n°68, 2015, 18p.
[8] Etude épidémiologique sur les petits ages gestationnels (EPIPAGE), epidemiologic study on small-for-gestational-age children, France
[9] Johnson R.A, Wichern D.W 2007 *Applied Multivariate Statistical Analysis* United States of America
[10] Baccini A. 2010 *Statistique descriptive multidimensionnelle*, Publications de l'institut de mathématiques de Toulouse
[11] Dagneli P. 1975 *Analyse statistique à plusieurs variables,* Les presses agronomiques de gembloux, 362p
[12] Tenenhaus M. 2010 *Analyse Factorielle des Correspondance Multiple et Classification Ascendante Hierarchique*, France
[13] Bressoux P. 2010 Modélisation statistique appliquée aux sciences sociales, De boeck, 464p.
[14] Ribert A., Enanji A., Lecourtier Y. 1999 *An Incremental Hierarchical Clustering* Canada
[15] INSEE, The National Institute of Statistics and Economic Studies collects, analyses and disseminates information on the French economy and society

IOP Conf. Series: Journal of Physics: Conf. Series **855** (2017) 012026 doi:10.1088/1742-6596/855/1/012026

# On The Value at Risk Using Bayesian Mixture Laplace Autoregressive Approach for Modelling the Islamic Stock Risk Investment

**Brina Miftahurrohmah, Nur Iriawan, and Kartika Fithriasari**
Statistics Department of Mathematics and Natural Sciences Faculty of Institut Teknologi Sepuluh Nopember, Indonesia

E-mail: brina.bm@gmail.com, nur_i@statistika.its.ac.id, kartika_f@statistika.its.ac.id

**Abstract.** Stocks are known as the financial instruments traded in the capital market which have a high level of risk. Their risks are indicated by their uncertainty of their return which have to be accepted by investors in the future. The higher the risk to be faced, the higher the return would be gained. Therefore, the measurements need to be made against the risk. Value at Risk (VaR) as the most popular risk measurement method, is frequently ignore when the pattern of return is not uni-modal Normal. The calculation of the risks using VaR method with the Normal Mixture Autoregressive (MNAR) approach has been considered. This paper proposes VaR method couple with the Mixture Laplace Autoregressive (MLAR) that would be implemented for analysing the first three biggest capitalization Islamic stock return in JII, namely PT. Astra International Tbk (ASII), PT. Telekomunikasi Indonesia Tbk (TLMK), and PT. Unilever Indonesia Tbk (UNVR). Parameter estimation is performed by employing Bayesian Markov Chain Monte Carlo (MCMC) approaches.

## 1. Introduction

Stocks are type of security that signifies ownership in a corporation and represents a claim on part of the corporation's assets and earnings. Stocks are known as the financial instruments traded in the capital market which have a high level of risk. Their risks are indicated by their uncertainty of their return which have to be accepted by investors in the future. The higher the risk to be faced, the higher the return would be gained. Therefore, the measurements need to be made against the risk. By knowing the risk, the investment policy can be made more scalable. Important treatment that should be made in risk management is to identify the cause of the risk. The popular and most commonly method used by investors and commercial banks to determine the extent and occurrence ratio of potential losses in their institutional portfolios is Value at risk (VaR). The accuracy of the calculation of VaR is indispensable when determining the capital will be issued by the company. Thus, the risks which will faced by the company is getting smaller and losses can be handled.

Forecasting the return which will occur in the future is required in determining the likelihood of the risk occurring in the future. Stock return data is data in the form of time series. One method of forecasting time series data is ARIMA Box-Jenkins. Box-Jenkins ARIMA time series is linear methods are developed based on assumptions residual Normal distribution. Box-Jenkins ARIMA method would be appropriate if the observation of time series data is dependent or related to each other. One of the ARIMA models which characterize the condition of the stock return is Autoregressive models. Autoregressive Model is a model that describes the dependent variable is affected by the dependent variable itself in the periods and times previously. This model requires that residuals follow Normal distribution with zero mean and a certain variance. But in reality, many time series data are not stationary in mean and tend to be multimodal. In addition, many time series data are heteroscedasticity which provide marginal patterns and leptokurtic, so the assumption of Normality is violated [1]. Thus, [1] generalize the model of Gaussian Mixture Transition Distribution (GMTD) which was introduced by [2] for the case of non-linear became Mixture Autoregressive (MAR) model. The MAR model consists of a mixture of *K* Gaussian AR components. It is able to model time series with conditional multimodal distributions and heteroscedasticity. Not only that, but also this model can capture features better than other alternative models.

Nguyen and McLachan have been able to prove on the research which has been conducted using linear mixture of experts that by using Laplace distribution, the results obtained will be more robust than using the Normal distribution [3]. Based on the statement, [4] introduced a Mixture Laplace Autoregressive (MLAR) which consists of a mixture of *K* Laplace AR components as an alternative of MAR. The results of the analysis conducted by Nguyen on the data calcium imaging of zebrafish brain showed that the results from the model established by MLAR been able to capture the pattern of actual conditions.

This paper proposes VaR method couple with the Mixture Laplace Autoregressive (MLAR) that would be implemented for analysing the first three biggest capitalization Islamic stock return in *Jakarta Islamic Index* (JII), namely PT. Astra International Tbk (ASII), PT. Telekomunikasi Indonesia Tbk (TLMK), and PT. Unilever Indonesia Tbk (UNVR). Data obtained from the official website of yahoo finance in the form of daily data closing stock price (close price) from January 2010 to July 2016. Stock of the three companies were chosen because the company has the largest market capitalization, which represent the daily market trading, even capable of being a mover index in forming JII and JCI (Joint Stock Price Index) at the di Bursa Efek Indonesia (BEI). In addition, the stocks were the active and liquid stocks which meet the JII criteria during the last 6 years and one of the market leader in the industrial sector, so that it can be used as a benchmark. Parameter estimation is performed by employing Bayesian Markov Chain Monte Carlo (MCMC) approaches.

## 2. Literature

### *2.1 Mixture Laplace Autoregressive*

An Autoregressive model or AR(p) is when a value from a time series is regressed on previous values from that same time series. Equation of AR (*p*) the model can be written

$$\dot{y}_t = \phi_1 \dot{y}_{t-1} + \cdots + \phi_p \dot{y}_{t-p} + a_t; \qquad \dot{y}_t = y_t - \mu, \tag{1}$$

where $\phi_p$ is a parameter autoregressive *p*th-lag and $a_t$ is residual of *t*th-observation.

[1] was introduced Mixture Autoregressive (MAR) methods using a mixture of *K* Gaussian components to equation

$$F\left(y_t \mid \mathcal{F}_{t-1}; \vartheta\right) = \sum_{i=1}^{K} \pi_i \Phi\left( y_t; \phi_{i0} + \sum_{j=1}^{p} \phi_{ij} y_{t-j}, \sigma_i \right), \tag{2}$$

where $\Phi$ is Normal density function, $\pi_i > 0$, $\sum_{i=1}^{K} \pi_i = 1, \phi_i = \left(\phi_{i0}, \phi_{i1}, \ldots, \phi_{ip}\right)^T \in \mathbb{R}^{p+1}$, and $\sigma_i > 0$ for $i = 1, 2, \ldots, K$.

[4] introduced a Mixture Laplace Autoregressive (MLAR) which consists of a mixture of *K* Laplace AR components as an alternative of MAR. $Y_t$ arises from a *K* component MLAR model of order *p* (i.e. an MLAR*(K, p)* model) if $Y_t \mid \mathcal{F}_{t-1}; \vartheta$ has a density of the form

$$F\left(y_t \mid \mathcal{F}_{t-1}; \vartheta\right) = \sum_{i=1}^{K} \pi_i \lambda\left( y_t; \phi_{i0} + \sum_{j=1}^{p} \phi_{ij} y_{t-j}, \sigma_i \right), \tag{3}$$

where

$$\lambda(y;\mu,\sigma)=\left(\sigma\sqrt{2}\right)^{-1}\exp\left(-\sqrt{2}\,|x-\mu|/\sigma\right) \tag{4}$$

is the Laplace density function with mean $\mu$ and variance $\sigma^2$, and

$$\boldsymbol{\vartheta}=\left(\pi_1,\pi_2,\ldots,\pi_{K-1},\boldsymbol{\phi}_1^T,\boldsymbol{\phi}_2^T,\ldots,\boldsymbol{\phi}_K^T,\sigma_1,\sigma_2,\ldots,\sigma_K\right)^T \tag{5}$$

is the model parameter vector. MLAR model residual also has a Laplace Distribution [3].

MLAR model has conditional mean and conditional variance

$$E\left(Y_t\,|\,\mathfrak{F}_{t-1}\right)=\sum_{i=1}^{K}\pi_i\boldsymbol{\phi}_i^T\mathbf{y}_{(t-1)} \tag{6}$$

and

$$Var\left(Y_t\,|\,\mathfrak{F}_{t-1}\right)=\sum_{i=1}^{K}\pi_i\sigma_i^2+\sum_{i=1}^{K}\pi_i\left[\boldsymbol{\phi}_i^T\mathbf{y}_{(t-1)}\right]^2+\left[\sum_{i=1}^{K}\pi_i\boldsymbol{\phi}_i^T\mathbf{y}_{(t-1)}\right]^2. \tag{7}$$

where $\mathbf{y}_{t-1}=\left(1,y_{t-1},\ldots,y_{t-p}\right)^T$. Given a known or estimated parameter vector, Eqs. (6) and (7) can be used to make predictions of $y_t$ based on $\mathfrak{F}_{t-1}$.

### *2.2 Parameter Estimate*

Bayesian models developed from Bayes methods. The basis of this method is the Bayes theorem which the posterior distribution can be written

$$p(\boldsymbol{\theta}\,|\,y)\propto f(y\,|\,\boldsymbol{\theta})\,p(\boldsymbol{\theta}) \tag{8}$$

Prior distribution is an important part of Bayesian inference and represent information about the parameters that are not sure which, combined with the probability distribution of the new data to generate a posterior distribution, then it is used for the conclusion and future decisions involving the prior distribution. The types of prior distributions is informative Prior, Conjugate and Non-conjugate prior, proper and non-proper prior, as well as Pseudo prior.

*Markov Chain Monte Carlo* (MCMC) is a general method which is done by determining the values $\boldsymbol{\theta}$ of distribution approach and then $\boldsymbol{\theta}$ values corrected so it closer to the target posterior distribution $p(\boldsymbol{\theta}\,|\,y)$.

Markov Chain algorithm that has been found and is useful in many multi-dimensional problem is the Gibbs sampler also called alternating conditional sampling, defined in sub vector of $\boldsymbol{\theta}$. Let parameter $\boldsymbol{\theta}$ has been divided into $d$ components or sub vector, $\left(\theta_1,\theta_2,\ldots,\theta_d\right)$. Each iteration of the Gibbs sampler cycle through sub vector of $\theta$, draw each part depends on the value of other $\theta$. At each iteration $t$, a sequence of $d$ sub vector selected and, in turn, each $\theta_j^t$ was given a sample of the conditional distribution of all other components of $\boldsymbol{\theta}$:

$$p\left(\theta_j\,|\,\theta_{-j}^{t-1},y\right), \tag{9}$$

where $\theta_{-j}^{t-1}$ represents all components of $\boldsymbol{\theta}$, except for the actual value of $\theta_j$:

$$\theta_{-j}^{t-1}=\left(\theta_1^t,\theta_2^t,\ldots,\theta_{j+1}^{t-1},\ldots,\theta_d^{t-1}\right). \tag{10}$$

then, each sub vector $\theta_j$ is updated in conditional on a final value of the other components of $\boldsymbol{\theta}$, where iteration value of the components ready to be updated and iterating $t-1$ the values of the other components.

### *2.3 Best Model Selection*

Assume, in general, that the distribution of the data, $y=\left(y_1,y_2,\ldots,y_n\right)$, depends on a *p*-dimensional parameter vector $\theta$. From a frequentist point of view, model assessment is based on the deviance, the difference in the log-likelihoods between the fitted and the saturated model. The saturated model refers to the model with as many parameters as observations, which yields a perfect fit to the data. By analogy, [6] suggested examining the posterior distribution of the classical deviance defined by

$$D(\boldsymbol{\theta}) = -2\ln f(\boldsymbol{y} \mid \boldsymbol{\theta}) + 2\ln g(\boldsymbol{y}) \tag{11}$$

for Bayesian model selection. [6] proposed comparing plots and potential summaries such as the posterior mean of $D(\boldsymbol{\theta})$, and [7] followed these suggestions in the development of DIC as a model choice criterion. By Re-evaluated $DIC = \bar{D} + p_D$ then obtained equation

$$DIC = \bar{D} + 2p_D, \tag{12}$$

*2.4 Value at Risk (VaR)*

Value at Risk (VaR) or also called *Quantile Risk Metrics* describe the estimation of the maximum loss that may occur in the bank's portfolio due to market risk in a certain time period and in a certain degree of statistical confidence. According to [8], VaR is the dominant methodology to predict exactly how much money is at risk every day in the financial markets. Based on the definition of VaR, when talking about the VaR must not be separated in terms of risk. Risk is a combination of probability of occurrence with the consequences or repercussions.

VaR can be calculated analytically, on the basis of Normal linear, historical simulation or monte carlo simulation. VaR can be defined mathematically as the value of the loss at a given confidence level $(1-\alpha)$ and it is equivalent to decreasing the quantile of the probability distribution of the random variable, so

$$P(X < x_\alpha) = \alpha \tag{13}$$

where $x_\alpha$ is -VaR

If distribution function $F(x)$ of $X$ is known then the quantile corresponding to any given value of α can be calculated as

$$x_\alpha = F^{-1}(\alpha) \tag{14}$$

then

$$\text{VaR} = -x_\alpha = -\left(F^{-1}(\alpha)\right). \tag{15}$$

Normal Linear VaR calculation method is a method that assumes that the distribution of returns is Normal, and portfolios must be linear. Therefore

$$P(X < x_\alpha) = P\left(\frac{X-\mu}{\sigma} < \frac{x_\alpha - \mu}{\sigma}\right) = P\left(Z < \frac{x_\alpha - \mu}{\sigma}\right) = \alpha, \tag{16}$$

where $Z$ is a Normal standard variable. So

$$\frac{x_\alpha - \mu}{\sigma} = \Phi^{-1}(\alpha), \tag{17}$$

With symmetry Normal distribution function,

$$\Phi^{-1}(\alpha) = -\Phi^{-1}(1-\alpha). \tag{18}$$

Therefore, with substituting $\text{VaR}_\alpha = -x_\alpha$ to equation (18) then

$$\text{VaR} = \Phi^{-1}(1-\alpha)\sigma - \mu. \tag{19}$$

Thus, the equation used to calculate the VaR on the horizon $h$ is writable

$$\text{VaR}_{h,\alpha} = \Phi^{-1}(1-\alpha)\sigma_h - \mu_h. \tag{20}$$

Hence, we often assume the portfolio is expected to return the risk free rate so that $\mu_h$, the present value of the expected return, is zero. We shall assume this in the following, unless explicitly stated otherwise. Without the drift adjustment, the Normal linear VaR formula is simply

$$\text{VaR}_{h,\alpha} = \Phi^{-1}(1-\alpha)\sigma_h \tag{21}$$

By the symmetry of the Normal distribution function,

$$\mathrm{VaR}_{h,\alpha} = -\left(\Phi^{-1}(1-\alpha)\sigma_h\right) \tag{22}$$

So, under the assumption of i.i.d. returns, it is only when the portfolio is expected to return the discount rate, i.e. $\mu_1 = 0$, that [9]

$$\mathrm{VaR}_{h,\alpha} = \sqrt{h} \times \mathrm{VaR}_{1,\alpha} \tag{23}$$

Thus, VaR to calculate the Laplace distribution mix as follows

$$\mathrm{VaR}_{mix} = \pi_1\sqrt{h} \times \mathrm{VaR}_{1,\alpha} + \pi_2\sqrt{h} \times \mathrm{VaR}_{1,\alpha} + \ldots + \pi_K\sqrt{h} \times \mathrm{VaR}_{1,\alpha} \tag{24}$$

*2.5 Return*

Return is important because it is used as a measure of corporate performance and is also useful as a basis for determining the expected return and risk in the future. The formula used to determine the return is

$$R_t = \frac{d_t - d_{t-1}}{d_{t-1}}, \tag{25}$$

## 3. Previous Research

Based on previous research conducted by [10] with the same data obtained identification results return three stocks as follows

Figure 1 shows that the stock return data PT Astra International Tbk, PT Telekomunikasi Indonesia Tbk and PT Unilever Indonesia Tbk bring nature slope and kurtosis indicating abnormalities in the data return stock. In addition it was shown that the variability of return sufficient high, so that led to stock returns tails at the extreme left end or right. If the forced return using univariate Normal unimodal pattern seen their deviations.

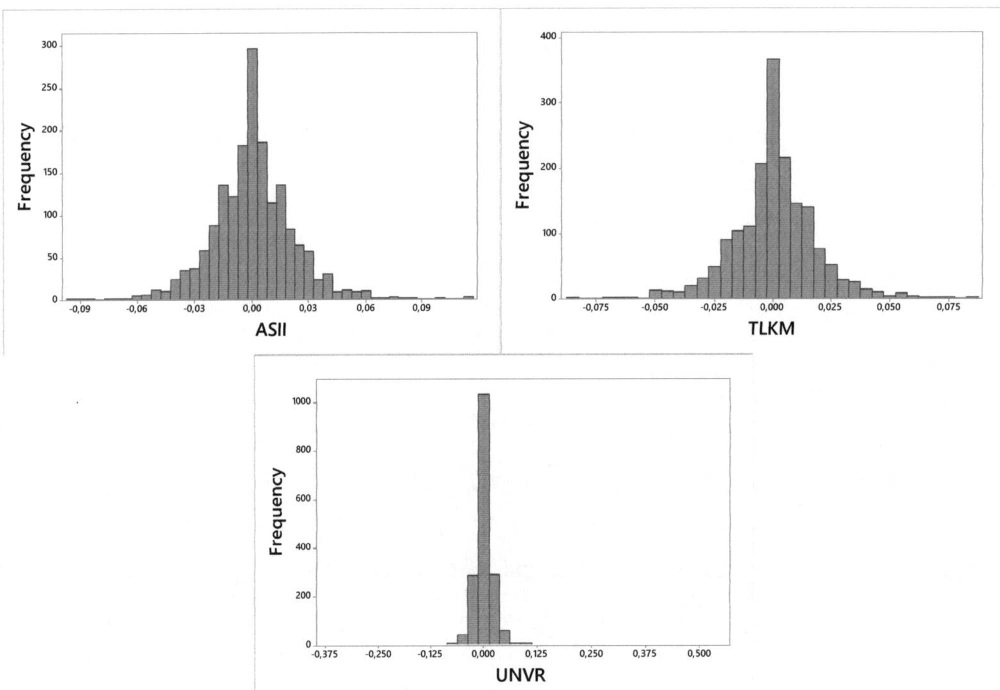


Figure 1. Histogram of stock return data PT Astra International Tbk, PT Telekomunikasi Indonesia Tbk and PT Unilever Indonesia Tbk

Irregularities that occurs is due to the fairly extreme return data or outliers, resulting slope value is greater than zero. The number of returns is worth around zero making patterns return has a high peak is commonly called the leptokurtic. Consequently alleged that on the pattern of stock returns ASII, TLKM and UNVR to have a mixture distribution or a mixture distribution is quite clear. This is supported by results conformance testing distribution (goodness of fit).

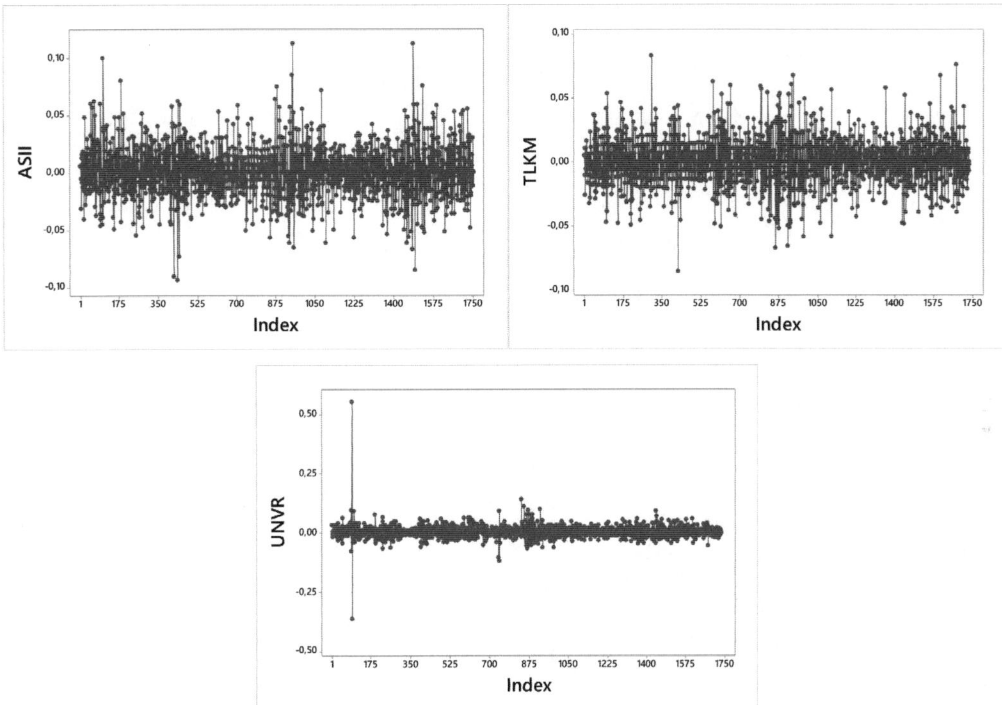


Figure 2. Time series plot of stock return data PT Astra International Tbk, PT Telekomunikasi Indonesia Tbk and PT Unilever Indonesia Tbk

Figure 2 provides information that return data ASII shares, TLKM and UNVR fluctuate around the mean. Thus, visually ASII return data, TLKM and UNVR can be said to have been stationary in the mean. After found that the three stock return data has been stationary in mean

## 4. Result

In this section, will explain the results of the comparison the best model between the MNAR and MLAR methods, estimate of the best model and VaR calculation results based on the parameters obtained from the estimated best model. Bayesian MNAR method performed again as was done by the [10] but with a number of longer period. This was done to compare goodness of fit between Bayesian MNAR and MLAR model based on the value of DIC. Comparison between Bayesian MNAR and MLAR the return data PT. Astra International Tbk (ASII), PT. Telekomunikasi Indonesia Tbk (TLMK), and PT. Unilever Indonesia Tbk (UNVR) shown by Table 1.

Table 1 shows that in PT. Astra International Tbk (ASII), PT. Telekomunikasi Indonesia Tbk (TLMK), and PT. Unilever Indonesia Tbk (UNVR) best model generated using Bayesian MLAR analysis. This is indicated by the value of Bayesian MLAR DIC smaller base than Bayesian MNAR. On the results of the analysis of each share also shows that the same order, Bayesian MLAR also showed better results than the Bayesian MNAR. It shows that in this case, the method MLAR better able to model better than MNAR methods, so that the prediction results generated by Bayesian MLAR also better than MNAR Bayesian methods.

Table 1. Comparison of Bayesian MNAR and MLAR model based on the value of DIC

| Stock | Methods | Models | DIC |
|---|---|---|---|
| ASII | Bayesian MLAR | MLAR(2;[3],[3,6]) | -12572.1 |
| | | MLAR(3;[3],[3,6],0) | -14648.1 |
| | Bayesian MNAR | MNAR(2;[3],[3,6]) | -12266.5 |
| | | MNAR(3;[3],[3,6],0) | -14149.1 |
| TLKM | Bayesian MLAR | MLAR(2;[2,3],[3,4]) | -13725.4 |
| | | MLAR(2;[2],[3]) | -13731.2 |
| | | MLAR(2;[2],[4]) | -13729.9 |
| | | MLAR(2;[3],[4]) | -13725.8 |
| | | MLAR(3;[2,3],[3,4],[2,3,4]) | -16385.7 |
| | | MLAR(3;[2],[3],[4]) | -16383.6 |
| | Bayesian MNAR | MNAR(2;[2,3],[3,4]) | -13423.6 |
| | | MNAR(2;[2],[3]) | -13411.8 |
| | | MNAR(2;[2],[4]) | -13405.3 |
| | | MNAR(2;[3],[4]) | -12151.5 |
| | | MNAR(3;[2,3],[3,4],[2,3,4]) | -14461.9 |
| | | MNAR(3;[2],[3],[4]) | -15897.5 |
| UNVR | Bayesian MLAR | MLAR(2;2,[11]) | -12772.7 |
| | | MLAR(3;1,2,[11],0) | -14934.4 |
| | Bayesian MNAR | MNAR(2;2,[11]) | -10777.8 |
| | | MNAR(3;1,2,[11],0) | -12002.9 |

VaR calculation is done using VaR formula that is based on the return distributed graphic form Laplace. Thus, it is necessary to determine VaR formula based CDF Laplace distribution. The resulting decline in VaR following formula

$$\text{VaR}_{h,\alpha} = -\left(\sigma_h\left(\ln\left(4\left(-\alpha^2+\alpha\right)\right)\right)\right)$$

Similarly, the VaR is based on the Normal curve,

$$\text{VaR}_{h,\alpha} = \sqrt{h} \times \text{VaR}_{1,\alpha}$$

Thus, VaR to calculate the Laplace distribution mix as follows

$$\text{VaR}_{mix} = \pi_1\sqrt{h} \times \text{VaR}_{1,\alpha} + \pi_2\sqrt{h} \times \text{VaR}_{1,\alpha} + \ldots + \pi_K\sqrt{h} \times \text{VaR}_{1,\alpha}$$

Parameter estimate of the best model from PT. Astra International Tbk (ASII), PT. Telekomunikasi Indonesia Tbk (TLMK), and PT. Unilever Indonesia Tbk (UNVR) shown by Table 2.
In calculation VaR, we only need significant parameter estimate of $\pi$ and $\sigma$. Calculation VaR OF PT. Astra International Tbk (ASII), PT. Telekomunikasi Indonesia Tbk (TLMK), and PT. Unilever Indonesia Tbk (UNVR) shown by Table 3

Table 3 show that the 5% 1-day VaR of ASII is 3.56% of return value. With 100 million rupiah invested in the return the VaR is Rp. 100 million × 0.03561 = Rp. 3561000. In other words, we are 95% confident that we will lose no more than Rp. 3561000 from investing in this fund over the next day. The Interpretation Applicable to the VaR others.

Table 2. Estimate Parameter of Bayesian MLAR model

| Node | ASII MLAR(3;[3],[3,6], 0) | TLKM MLAR(3;[2,3],[3,4],[2,3,4]) | UNVR MLAR(3;2,[11],(2,[11])) |
|---|---|---|---|
| $\pi_1$ | 0.33320 | 0.33350 | 0.33450 |
| $\pi_2$ | 0.33280 | 0.33330 | 0.33280 |
| $\pi_3$ | 0.33400 | 0.33320 | 0.33280 |
| $\phi_{1,1}$ | | | -0.26520 |
| $\phi_{1,2}$ | | -0.09508 | -0.09611 |
| $\phi_{1,3}$ | -0.09142 | -0.03814 | |
| $\phi_{2,3}$ | -0.09649 | -0.08405 | |
| $\phi_{2,4}$ | | -0.07648 | |
| $\phi_{2,6}$ | -0.05720 | | |
| $\phi_{2,11}$ | | | 0.05798 |
| $\phi_{3,1}$ | | | -0.26480 |
| $\phi_{3,2}$ | | -0.01041 | -0.09634 |
| $\phi_{3,3}$ | | -0.08699 | |
| $\phi_{3,4}$ | | -0.08622 | |
| $\phi_{3,11}$ | | | 0.05125 |
| $\sigma_1$ | 0.02149 | 0.01786 | 0.02053 |
| $\sigma_2$ | 0.02151 | 0.01786 | 0.02032 |
| $\sigma_3$ | 0.02133 | 0.01788 | 0.02058 |

Table 3. VaR calculation

| Stock | Long Investment | 95% CI | 99% CI |
|---|---|---|---|
| ASII | 1-day | 0.03561 | 0.06924 |
| | 5-days | 0.07963 | 0.15482 |
| | 20-days | 0.15926 | 0.30964 |
| TLKM | 1-day | 0.02908 | 0.05769 |
| | 5-days | 0.06501 | 0.12900 |
| | 20-days | 0.13003 | 0.25800 |
| UNVR | 1-day | 0.03401 | 0.06612 |
| | 5-days | 0.07605 | 0.14786 |
| | 20-days | 0.15210 | 0.68878 |

## 5. Conclusion

In this paper show that analysis using MLAR approach was determined DIC value better than MNAR approach. So. estimation using MLAR more appropriate to calculate VaR. Furthermore. the higher the confidence interval is used. the VaR values generated will be greater and the longer the investment is done. then the risk will be received by investors are also getting bigger. In other words, the possibility of losses that come will be even greater.

## Reference


[1] Wong. C. S.. & Li. W. K. (2000). On a Mixture Autoregressive. *Journal of the Royal Statistical Society Series B*. 95-115.

[2] Le. N. D.. Martin. R. D.. & Raftery. A. E. (1996). Modelling Flat Stretches. Burst. and Outlier Time Series In Exponential Variable. *J. Am. Statist. Ass.*. 1504-1514.

[3] Nguyen. H. D.. & McLachlan. G. J. (2016). Laplace mixture of linear experts. *Computational Statistics and Data Analysis 93*. 177-191.

[4] Nguyen. H. D.. Geoffrey. M. J.. Ullmann. J. F.. & Janke. A. L. (2016). Laplace mixture autoregressive models. *Statistics and Probability Letters 110*. 18-24.

[5] Kotz. S.. Kozubowski. T. J.. & Podgorski. K. (2001). *The Laplace Distribution and Generalizations: A Revisit with Applications to Communications. Economics. Engineering. and Finance*. New York: Springer Science+Business Media. LLC.

[6] Dempster. A. P. (1974). The Direct Use of Likelihood for Significance Testing. *Proceedings of Conference on Foundational Questions in Statistical*. 335-352.

[7] Spiegelhalter. D. J.. Best. N. G.. Carlin. B. P.. & van der Linde. A. (2002). Bayesian Measures of Model Complexity and Fit" (with discussion). *Journal of the Royal Statistical Society. 64*. 583-639.

[8] Butler. C. (1999). *Mastering Value at Risk*. New York: Pretice Hall.

[9] Alexander. C. (2008). *Market Risk Analysis* (Vol. IV). West Sussex: John Wiley & Sons Ltd.

[10] Putri. U. M. (2016). *Analisis Risiko Investasi Saham Syariah Menggunakan Metode Value at Risk dengan Pendekatan Mixture Normal Autoregressive*. Surabaya: Institut Teknologi Sepuluh Nopember.

# Comparison of Grand Median and Cumulative Sum Control Charts on Shuttlecock Weight Variable in CV Marjoko Kompas dan Domas

**N Musdalifah[1], S S Handajani[2], E Zukhronah[3]**

[1] Department of Mathematics, Faculty of Mathematics and Natural Sciences. Sebelas Maret University.

[2,3] Department of Statistics, Faculty of Mathematics and Natural Sciences. Sebelas Maret University.

E-mail: nurulmdlfh@student.uns.ac.id, ssulistijowati@yahoo.com, etikzukhronah@yahoo.co.id

**Abstract.** Competition between the homoneous companies cause the company have to keep production quality. To cover this problem, the company controls the production with statistical quality control using control chart. Shewhart control chart is used to normal distributed data. The production data is often non-normal distribution and occured small process shift. Grand median control chart is a control chart for non-normal distributed data, while cumulative sum (cusum) control chart is a sensitive control chart to detect small process shift. The purpose of this research is to compare grand median and cusum control charts on shuttlecock weight variable in CV Marjoko Kompas dan Domas by generating data as the actual distribution. The generated data is used to simulate multiplier of standard deviation on grand median and cusum control charts. Simulation is done to get average run lenght (ARL) 370. Grand median control chart detects ten points that out of control, while cusum control chart detects a point out of control. It can be concluded that grand median control chart is better than cusum control chart.

***Keywords:*** *grand median control chart, cusum control chart, ARL.*

## 1. Introduction

Free markets make the trade transactions among the regions in the world grow easy and fast. It caused the homogeneous companies that offer their products on the same area on a competition. Quality assurance of the production is a way to compete with the homogeneous companies. Efforts used to keep and to improve the quality of the result product can be done with quality control process statistics (Ramadhani [6]).

A tool used in statistical quality control of the production process is the control chart. Control chart is used to determine the process of controlled or uncontrolled through the upper control limit (UCL) and the lower control limit (LCL). A control chart may indicate an out of control condition when one or more points fall beyond the control limits or when the plotted points exhibit some nonrandom pattern of behavior (Montgomery [5]).

Shewhart control chart is a control chart that used to the normal distribution. Production quality characteristics is often non-normal distributed. To overcome this problem, we can use

**Table 1.** Data structure of samples

| 1 | 2 | $\cdots$ | $n$ |
|---|---|---|---|
| $x_{1.1}$ | $x_{1.2}$ | $\cdots$ | $x_{1.n}$ |
| $x_{2.1}$ | $x_{2.2}$ | $\cdots$ | $x_{2.n}$ |
| $\vdots$ | $\vdots$ | $\ddots$ | $\vdots$ |
| $x_{m.1}$ | $x_{m.2}$ | $\cdots$ | $x_{m.n}$ |

grand median control chart. Acccording to (Altukife [1]), grand median control chart use the number of observation exceeding the grand median.

A production data often occur small process shift. Cusum control chart is used because it's sensitive in detecting small process shift. The comparison between grand median and cusum control charts is to get the sensitive control chart. We obtained it by detecting small process shift on non-normal distribution. This research use shuttlecock weight variable in CV Marjoko Kompas dan Domas.

## 2. Grand Median Control Chart

Grand median is a statistic used to measure the center of m sample each of size n observations (Altukife [1]). Let $x_{i1}, x_{i2}, x_{i3}, ..., x_{in}$ are a sample of independent observations of size $n$ taken at period-$i$, where $i = 1, 2, ...., m$. Data structure of samples can be seen in Table 1.

Let $e$ is the grand median calculated from the combined total of $N$ observations, $N$ is the $m$ samples each of size $n$. Assume that $O_{1i}$ is the number of observations that exceeds e in the $i^{th}$ sample. Hence, $O_{2i}$ is the number of observations less than or equal $e$ in the $i^{th}$ sample. The probability occurrence of $O_{1i}$ and $O_{2i}$ is 0.5. Therefore $O_{1i}$ for $i = 1, 2, ...., m$ is binomial(n,0.5). A mean and variance of random variable of binomial distribution are

$$\mu = \frac{n}{2} \quad (1)$$

$$\sigma^2 = \frac{n}{4}. \quad (2)$$

The lines for the grand median control chart are given as follows

$$LCL = \mu - D\sqrt{\sigma^2} \quad (3)$$

$$CL = \mu$$

$$UCL = \mu + D\sqrt{\sigma^2} \quad (4)$$

where $D$ is a multiplier of standard deviation and it is chosen to produce the desired type I error probability value (Altukife [1]). Equations (1) and (2) are substituted to equations (3) and (4) results control limit of grand median control chart

$$LCL = \frac{n}{2} - D\sqrt{\frac{n}{4}} \quad (5)$$

$$CL = \frac{n}{2}$$

$$UCL = \frac{n}{2} + D\sqrt{\frac{n}{4}}. \quad (6)$$

## 3. Cumulative Sum Control Chart

Cumulative sum (cusum) control chart collects all the information from the first sample until the last sample. Suppose that samples of sizes $n > 1$ are collected and $\bar{x}_i$ is the average of the $i^{th}$ sample, then $\mu_0$ is the target of the process mean. However, if the mean shifts upward to some value $\mu_1 > \mu_0$, then an upward or positive drift will be developed in the cumulative sum $C_i$. Conversely, if the mean shifts downward to some $\mu_1 < \mu_0$ then a downward or negative drift in $C_i$ will be developed. The size of the shift is measured in terms of the process mean and standard deviation given by

$$\delta = \frac{\mu_1 - \mu_0}{\sigma}.$$

Observation of cusum control chart can be formed using tabular cusum defined by Koshti [4] as

$$C_i{}^+ = max[0, \bar{x}_i - (\mu_0 + K) + C_{i-1}{}^+] \quad (7)$$

$$C_i{}^- = max[0, (\mu_0 - K) - \bar{x}_i + C_{i-1}{}^-] \quad (8)$$

where the initial value $C_i^+$ and $C_i^-$ is zero. Otherwise $K$ is reference value defined by

$$K = \frac{\delta}{2}\sigma = \frac{\mu_1 - \mu_0}{2}.$$

where $\delta$ is the size of the shift. Cusum control chart is a good control chart used to detect a small process shift. Usually the small process shift is approximately less than $1.5\sigma$ (Ferdinant [3]). Control limit of cusum control chart are defined as follows

$$LCL = -h\sigma \quad (9)$$

$$UCL = h\sigma. \quad (10)$$

## 4. Average Run Length

Average run lenght (ARL) is the average number of points that must be plotted before a point indicates an out of control condition. ARL is calculated from the average simulation run length. Run length was defined as the number of samples required to find samples out of the control limit. Calculating of the ARL is defined by

$$ARL = \frac{1}{Pr(rejected\ H_0|H_0\ is\ true)} = \frac{1}{\alpha}, \quad (11)$$

where $\alpha$ is probability of type I error or probability of production process is uncontrolled when the process is in control (Montgomery [5]).
Different technique that used to calculate the ARL is Siegmund approximation written as

$$ARL = \begin{cases} \frac{exp(-2\Delta b)+2\Delta b-1}{2\Delta^2}, & \text{for } \Delta \neq 0 \\ b^2, & \text{for } \Delta = 0 \end{cases} \quad (12)$$

with $\Delta = \delta - k$ for $C_i^+$ and $\Delta = -\delta - k$ for $C_i^-$,
$b = h + 1.166$,
$k = \frac{K}{\sigma}$.
To obtained ARL of the two sided cusum is written as

$$\frac{1}{ARL} = \frac{1}{ARL^+} + \frac{1}{ARL^-}.$$

Comparison between the two control chart can be done if the both of control chart have the same ARL value. ARL value is selected by setting alpha value. Simulation the multiplier of standard deviation ($D$ and $h$) is to get the expected of ARL. Simulation is done by generating data as the actual distribution data.

IOP Conf. Series: Journal of Physics: Conf. Series **855** (2017) 012027 doi:10.1088/1742-6596/855/1/012027

## 5. Data Distribution Test

One of data distribution test is Anderson-Darling test statistic. According to Fallo et al [2], the steps of the hypothesis test are as follows.

(i) Hypothesis
$H_0$ : Shuttlecock weight data followed certain distribution.
$H_1$ : Shuttlecock weight data unfollowed certain distribution.

(ii) Significance Level
$\alpha$ =0.05.

(iii) Critical area (CA)
CA = $\{A^*|A^* > A_\alpha\}$. $H_0$ is rejected if $A^* \in$ CA.

(iv) Test Statistics

$$AD = -N - \sum_{i=1}^{N} \frac{(2i-1)}{N}[lnF(Z_t + ln(1 - F(Z_{N+1-t}))],$$

where
$AD$ : test statistics Anderson Darling,
$N$ : number of observations,
$Z_t$ : standardized variable, and
$F(Z_t)$ : the probability of the standard normal cumulative distribution function in $Z_t$.

Test for normal distribution data using a modification of the test statistic AD is written as

$$A^* = AD(1 + \frac{0.75}{N} + \frac{0.25}{N^2}).$$

Distribution test statistic having the smallest value of AD indicated that the distribution is a representative distribution (Wibawati et al. [7]).

## 6. Research Method

This research used primary data of shuttlecock weight variable that produced by CV Marjoko Kompas dan Domas in district Sukoharjo. Data were taken 50 samples, where each sample size 20. Before applying the both of control charts, it is determined a multiplier value of standard deviation $D$ and $h$. Multiplier value is selected to result in expected ARL. The following steps are to determine the value of D .

(i) Generate data as the actual distribution data as much as 1000, then to calculate the grand median.
(ii) Set any limit value D and then to make the limit control chart.
(iii) Generate data as the actual data distribution with a sample size of 20, and then calculating the statistics $O_{1i}$.
(iv) Statistics $O_{1i}$ was plotted on limit control chart.
(v) If $O_{1i}$ was inside of the control chart then back to plot the statistics $O_{1i}$ on step 3 and if it was outside the control chart then defined as Run Length (RL).
(vi) Step 3 until step 5 was repeated 1.000 times.
(vii) Calculate ARL by sum the RL value and then it is divided in 1000.

The steps to determine the value of $h$ cusum control chart are defined as follows.

(i) Determine the mean process shift
(ii) Determine the initial value of h and calculating ARL of cusum control chart
(iii) Repeat steps 1 and 2 to get the expected ARL.

Next applying both of controls chart on observation data. The steps are

(i) Apply the grand median control chart
   (a) Determining the grand median then search for observations in each sample whose value exceeds the grand median
   (b) The number of observations which value exceeds the grand median in each sample was plotted on a control limits of grand median control chart.
   (c) Count the points are outside the control limit.
(ii) Apply the cusum control chart
   (a) Determine the mean process shift and the multiplier of standard deviation.
   (b) Calculate value of tabular cusum $C_i^+$ and $C_i^-$ as an observation point of cusum control chart.
   (c) Count the points are outside the control limit
(iii) Compare the both of control chart.

## 7. Result

Test of data distribution to determine the observations follow the certain distribution. Steps of hypothesis test using the test statistic AD was written as follows

(i) Hypothesis
$H_0$ : Shuttlecock weight data followed normal distribution.
$H_1$ : Shuttlecock weight data unfollowed normal distribution.
(ii) Significance level
$\alpha$ =0.05.
(iii) Critical area (CA)
CA $= \{A^*|A^* > A_\alpha = A_{0.05} = 0.752\}$. $H_0$ is rejected if $A^* \in$ CA.
(iv) Test statistics.

$$A^* = AD(1 + \frac{0.75}{N} + \frac{0.25}{N^2})$$

Table 2 showed $A^* = 1.068$.
(v) Conclusion
Because $A^*$ is in CA so $H_0$ is rejected, it can be stated that shuttlecock weight data followed non-normal distribution.

The results of the statistic test AD on some distributions are shown in Table 2. Distribution test having the smallest value AD was 3-parameter Weibull distribution. In order that, hypothesis test using $H_0$ 3-parameter Weibull distribution data got statistic test $A^* = 0.648 < A_{0.05} = 2.492$. Because $A \notin CA$, $H_0$ was not rejected which it stated data followed 3-parameter Weibull distribution

Data distribution test was used to generate data as the actual distribution. The generated data used to the simulation of multiplier standard deviation was to get ARL 370. The multiplier of standard deviation of $D$ value gave amount 2.71 by equations (11) and $h$ value amount 4.87 that calculated from equation (12). Observations of grand median control chart are defined by $O_1i$. Equation (5) and (6) gave LCL amount 3.9043 and UCL amount 16.0597.

IOP Conf. Series: Journal of Physics: Conf. Series **855** (2017) 012027 doi:10.1088/1742-6596/855/1/012027

**Table 2.** The result of statistics test on some distributions

| Distribusi | *AD* |
|---|---|
| Normal | 1.067 |
| Lognormal | 0.762 |
| Exponential | 442.195 |
| 2-p Exponential | 176.951 |
| Weibull | 17.243 |
| 3-p Weibull | 0.648 |
| Smallest Extreme Value | 18.927 |
| Largest Extreme Value | 5.705 |
| Gamma | 0.857 |
| Logistic | 1.384 |
| Loglogistic | 1.237 |

Observations of cusum control chart are defined from tabular cusum value using equation (7) and (8). Equation (9) and (10) gave LCL amount -0.2922 and UCL amount 0.2922 of cusum control chart. Grand median and cusum control charts can be seen in Figure 1 and Figure 2.

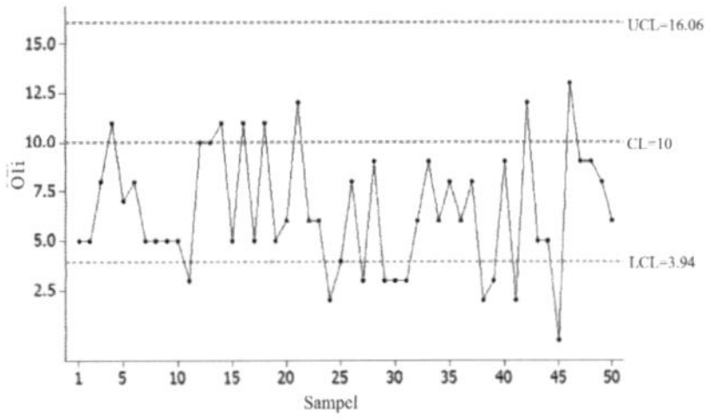


**Figure 1.** Grand median control chart

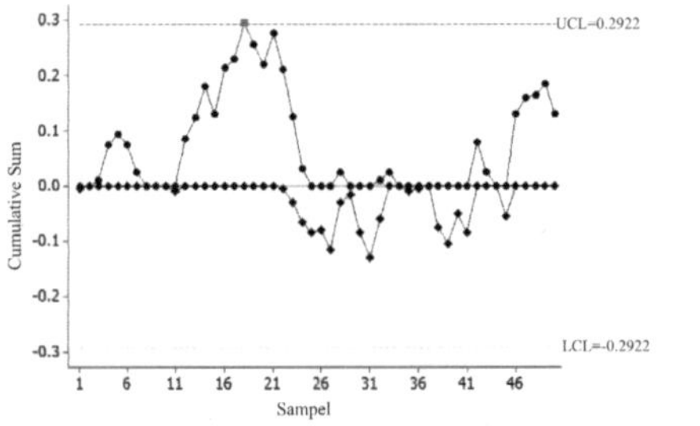


**Figure 2.** Cusum control chart

Figure 1 showed the control chart with LCL of grand median at 3.94 and UCL at 16.06 detect the 10 points that out of control. Figure 2 showed the cusum control chart with LCL at 0.2922 and UCL at 0.2922 detect a point that out of control. Observation points that out of control limit LCL and UCL on the grand median control chart were the sample 11, 24, 27, 29, 30, 31, 38, 39, 41, and 45, while on the cusum control chart was the sample to 18.

## 8. Conclusion

Simulation multiplier of standard deviation with the ARL 370 results $D$ value amount 2.71 and $h$ value amount 4.87. It's to get the LCL and UCL the both of control charts. From the number out of control points it can be concluded that the grand median was better than cusum control chart on weight shuttlecock data that 3-parameter Weibull distribution. It means that the company produces the shuttlecock that qualified the specification, so the consumers are satisfied. The satisfaction of consumers can cause the increasing number of sale production. So the income of the company will be increased.

## References


[1] Altukife, F.S., *A New Nonparametric Control Chart Based on Observations Exceeding The Grand Median*, Pakistan Journal of Statistic **19** (2003), no.3, 343-351.

[2] Fallo, J.O., Setiawan, and Susanto, B., *Perbandingan Uji Berdasarkan Anderson-Darling, Cramer, dan Liliefors*, Prosiding Seminar Nasional Mathematika, UNNES, Semarang, (2013), 152-158.

[3] Ferdinant, P.F., *Pengendalian Kualitas Proses dengan Peta Kendali CUSUM dan EWMA*, Prosiding Seminar Nasioal Industrial Services (SNIS) III, UNTIRTA, Banten, (2013), 147-152.

[4] Koshti, V.V., *Cumulative Sum Control Chart*, International Journal of Phisicsand Mathematical Siences **1(1)** (2011), 28-32.

[5] Montgomery, D.C., *Introduction to Statistical Quality Control*, $6^{nd}$ed., John Willey and Sons, Washington, 2009.

[6] Ramadhani, G.S., *Ananlisis Pengendalian Kualitas Menggunakan Diagram Kendali Demerit*, Skripsi, FMIPA Universitas Diponegoro, Semarang, 2014.

[7] Wibawati, Mashuri M and A.K. Gupta, *Diagram Kontrol Nonparametrik Mann Whitney untuk Target*, Skripsi, FMIPA Universitas Teknologi Surabaya, Surabaya, 2007.

IOP Conf. Series: Journal of Physics: Conf. Series **855** (2017) 012028 doi:10.1088/1742-6596/855/1/012028

# The $C$-integral of sequence-valued functions on the sequence space $\ell^\infty$

Muslich[1], Sutrima[1], Supriyadi Wibowo[1]
[1]Department Of Mathematics, Faculty of Mathematics and Natural Sciences, Sebelas Maret University, Surakarta, Indonesia
E-mail: muslich_mus@yahoo.com, zutrima@yahoo.co.id , supriyadi_w@yahoo.co.id

**Abstract.** In this paper the $C$-integral of sequence-valued functions, the new concept of integral theory, is constructed. The concept is based on the $M_\alpha$- integral. We report some of the fundamental properties of the $C$-integral including the uniqueness, linearity, the Cauchy criterion for integrability and the Henstock Lemma.

## 1. Introduction

It is well known that the Lebesgue integral is equivalent to the McShane integral (see, e.g., [2], [3], [5], [9], [10]). In 2010, based on the McShane integral, Park *et al.* [6] and Park *et al.* [7] defined the $M_\alpha$-integral, a Riemann-integral type, and constructed its properties. If a constant $\alpha = 1/\varepsilon$, then the $M_\alpha$-integral is called the $C$-integral, [1]. Some extensions of the $M_\alpha$-integral have been studied for real-functions, such as Park *et al.* [6] researched the integration by parts for the $M_\alpha$-integral and Park [4] introduced the $M_\alpha$-integral of Banach-Valued functions concept and obtained its properties and especially, Zhao D [13] has studied the relations among Henstock, McShane and $C$-integral. Zachriwan *et al.*[12] have investigated the Henstock integral of sequence-valued functions on the sequence space $\ell^p$, $1 \leq p < \infty$. As an extension of the $C$-integral, the authors are motivated by Zachriwan's paper for studying the $C$-integral for which the function has value in a sequence space $\ell^\infty$ and investigate some of their properties.

## 2. Basic Definitions

Before introducing the $C$-integral of sequence-valued functions concept, we start by some notations. Let $\delta$ be a positive function on the closed interval $[a,b]$ , i.e. $\delta: [a,b] \to \mathcal{R}^+$, and $\mathfrak{D} = \{([u_{i-1},u_i],\xi_i)\}_{i=1}^n$ is a finite collection of interval-point pairs, where $[u_{i-1},u_i], 1 \leq i \leq n$ are non-overlapping of subintervals of $[a,b]$. The following definitions can be found in [4], [7], [8], [11] and [ 13].

**Definition 2.1** *Given* $\mathfrak{D} = \{([u_{i-1},u_i],\xi_i)\}_{i=1}^n$. *We say that*
(i) $\mathfrak{D}$ *is a partial partition of* $[a,b]$ *if* $\bigcup_{i=1}^n [u_{i-1},u_i] \subset [a,b]$,
(ii) $\mathfrak{D}$ *is a partition of* $[a,b]$ *if* $\bigcup_{i=1}^n [u_{i-1},u_i] = [a,b]$,
(iii) $\mathfrak{D}$ *is a $\delta$-fine McShane partition of* $[a,b]$ *if* $[u_{i-1},u_i] \subset (\xi_i - \delta(\xi_i), \xi_i + \delta(\xi_i))$ *for all* $i = 1,2,3,\dots,n$,
(iv) $\mathfrak{D}$ *is a $\delta$-fine $M_\alpha$- partition of* $[a,b]$ *for a constant* $\alpha > 0$ *if it is a $\delta$-fine McShane partition of* $[a,b]$ *and satisfying the condition* $\sum_{i=1}^n dist(\xi_i,[u_{i-1},u_i]) < \alpha$, *where* $dist(\xi_i,[u_{i-1},u_i]) = inf\{|u-\xi_i| : u \in [u_{i-1},u_i]\}$,
(v) $\mathfrak{D}$ *is a $\delta$-fine $C$-partition of* $[a,b]$ *for a positive constant* $\alpha = \frac{1}{\varepsilon}$ *if it is a $\delta$-fine McShane partition of* $[a,b]$ *and satisfying the condition* $\sum_{i=1}^n dist(\xi_i,[u_{i-1},u_i]) < \frac{1}{\varepsilon}$.

If $\mathfrak{D} = \{([u_{i-1},u_i],\xi_i)\}_{i=1}^n$ is a $\delta$-fine partition of $[a,b]$ and $\delta(\xi) \leq \delta_1(\xi)$ then every $\delta$-fine partition $\mathfrak{D}$ of $[a,b]$ is a $\delta_1$-fine partition $\mathfrak{D}$ of $[a,b]$, and integral sum over $\mathfrak{D}$ for a function $\bar{f}: [a,b] \to \ell^\infty$ is defined as

$$S(\bar{f},\mathfrak{D}) = (\mathfrak{D})\sum_{i=1}^{n} \bar{f}(\xi_i)(u_i - u_{i-1})$$

where $\ell^\infty = \left\{\bar{u} = \{u_1, u_2, \ldots, u_j, \ldots\} : sup\{|u_j|\}_{j=1}^{\infty} < \infty\right\}$ and $\|\bar{u}\| = sup\{|u_j|\}_{j=1}^{\infty}$ for each $\bar{u} \in \ell^\infty$.

**Definition 2.2 (Zachriwan *et al.* [12])** *A function $\bar{f}: [a,b] \to \ell^\infty$, exists if and only if there exists a sequence of real valued functions $f_j: [a,b] \to \mathcal{R}, j = 1,2,3, \ldots$ such that $\bar{f}(u) = \{f_1(u), f_2(u), \ldots, f_j(u), \ldots\} = \{f_j(u)\}_{j=1}^{\infty} \in \ell^\infty \Leftrightarrow sup\{|f_j(u)|\}_{j=1}^{\infty} < \infty$ for all $\bar{f}(u) \in \ell^\infty$.*

**Definition 2.3** *Let $\bar{f}, \bar{g}: [a,b] \to \ell^\infty$, where $\bar{f}(u) = \{f_1(u), f_2(u), \ldots, f_j(u), \ldots\}$ and $\bar{g}(u) = \{g_1(u), g_2(u), \ldots, g_j(u), \ldots\}$, for all $u \in [a,b]$. We define*
*(i)* $\bar{f} = \bar{g}$ *if and only if* $f_j(u) = g_j(u)$, *for all* $j = 1,2,3, \ldots$ and $u \in [a,b]$,
*(ii)* $\bar{f} \leq \bar{g}$ *if and only if* $f_j(u) \leq g_j(u)$, *for all* $j = 1,2,3, \ldots$ and $u \in [a,b]$,
*(iii) the null function* $\bar{f} = \bar{\theta}$ *on* $[a,b]$ *if and only if* $f_j(u) = 0$ *on* $[a,b]$, *for all* $j = 1,2,3 \ldots$ and $u \in [a,b]$,
*(iv)* $\bar{f} = \bar{\theta}$ *almost everywhere on* $[a,b]$ *if and only if* $f_j(u) = 0$ *almost everywhere on* $[a,b]$, *for all* $j = 1,2,3, \ldots,$
*(v)* $\bar{f} \leq \bar{g}$ *almost everywhere on* $[a,b]$ *if and only if* $f_j \leq g_j$ *almost everywhere on* $[a,b]$, *for all* $j = 1,2,3, \ldots,$
*(vi)* $(\bar{f} + \bar{g})(u) = \{(f_1 + g_1)(u), (f_2 + g_2)(u), \ldots, (f_j + g_j)(u), \ldots\}$, *for all* $u \in [a,b]$,
*(vii)* $\propto \bar{f}(u) = \{\propto f_1(u), \propto f_2(u), \ldots, \propto f_j(u), \ldots\}$, *for all* $\propto \in \mathcal{R}, u \in [a,b]$.

From [1] and [8] we will define the $C$-integral of real function and of sequence-valued functions as follow. A function $f: [a,b] \to \mathcal{R}$ is $C$-integrable on $[a,b]$ if there exists a real number $L \in \mathcal{R}$ such that for each $\varepsilon > 0$ there exists a positive function $\delta(\xi): [a,b] \to \mathcal{R}^+$ such that $|S(f,\mathfrak{D}) - L| < \varepsilon$ for each $\delta$-fine $C$-partition $\mathfrak{D}$ of $[a,b]$.

**Definition 2.4** *A function $\bar{f}: [a,b] \to \ell^\infty$ is $C$-integrable on $[a,b]$, in short, $\bar{f} \in C([a,b], \ell^\infty)$, if there exists a sequence $\bar{L} = (L_1, L_2, \ldots, L_j, \ldots) \in \ell^\infty$ such that for each $\varepsilon > 0$ there exists a positive function $\delta(\xi): [a,b] \to \mathcal{R}^+$ such that*

$$\|S(\bar{f},\mathfrak{D}) - \bar{L}\| < \varepsilon$$

*for each $\delta$-fine $C$-partition $\mathfrak{D}$ of $[a,b]$.*

The sequence $\bar{L} \in \ell^\infty$ is called the $C$-integral of $\bar{f}$ on $[a,b]$ and it will be written as $\bar{L} = (C)\int_a^b \bar{f}$.

## 3. Main Results

Based on the above definitions, we will provide some properties of $C$-integral including the uniqueness, linearity, the Cauchy criterion for integrability, the Henstock Lemma and the modified Henstock Lemma.

**Theorem 3.1** *If $\bar{f} \in C([a,b], \ell^\infty)$, then the $C$-integral of $\bar{f}$ on $[a,b]$ is unique.*

*Proof.* Let $\bar{L}_1, \bar{L}_2$ be the $C$-integrals of $\bar{f}$ on $[a,b]$. Given any $\varepsilon > 0$. From the hypothesis, there exists a positive function $\delta_i(\xi): [a,b] \to \mathcal{R}^+, i = 1,2$ such that

$$\|S(\bar{f},\mathfrak{D}_i) - \bar{L}_i\| < \frac{\varepsilon}{2}$$

for each $\delta_i$-fine $C$-partition $\mathfrak{D}_i$ of $[a,b]$. Define a positive function $\delta(\xi): [a,b] \to \mathcal{R}^+$, where $\delta(\xi) = \min\{\delta_1(\xi), \delta_2(\xi)\}$. If $\mathfrak{D}$ is a $\delta$-fine $C$-partition of $[a,b]$, then we have

$$\|\bar{L}_1 - \bar{L}_2\| \leq \|S(\bar{f},\mathfrak{D}) - \bar{L}_1\| + \|S(\bar{f},\mathfrak{D}) - \bar{L}_2\| < \frac{\varepsilon}{2} + \frac{\varepsilon}{2} = \varepsilon.$$

Since $\varepsilon > 0$ is arbitrary this implies that $\bar{L}_1 = \bar{L}_2$. This result shows that $\bar{f}$ has at most one $C$-integral on $[a,b]$. Hence theorem is proved. □

Let $C([a,b],\ell^\infty)$ be the collection of all the $C$-integrable functions $\bar{f}$ on $[a,b]$. The following result states that the set $C([a,b],\ell^\infty)$ is a vector space over $\mathcal{R}$ .

**Theorem 3.2** *If $\bar{f},\bar{g} \in C([a,b],\ell^\infty)$, then $k\bar{f} + l\bar{g} \in C([a,b],\ell^\infty)$ for every $k,l \in \mathcal{R}$, and $(C)\int_a^b(k\bar{f}+l\bar{g}) = k(C)\int_a^b \bar{f} + l(C)\int_a^b \bar{g}$.*

*Proof.* Without lose of the generality, we assume $k$ and $l$ are nonzero. Given any $\varepsilon > 0$. Let $(C)\int_a^b \bar{f} = \bar{L}_1$ and $(C)\int_a^b \bar{g} = \bar{L}_2$, where $\bar{L}_1$, $\bar{L}_2 \in \ell^\infty$. There exists two positive functions $\delta_i: [a,b] \to \mathcal{R}^+$, $i = 1,2$ such that $\|S(\bar{f},\mathfrak{D}_1) - \bar{L}_1\| < \frac{\varepsilon}{2|k|}$ for each $\delta_1$-fine $C$-partition $\mathfrak{D}_1$ of $[a,b]$ and $\|S(\bar{g},\mathfrak{D}_2) - \bar{L}_2\| < \frac{\varepsilon}{2|l|}$ for each $\delta_2$-fine $C$-partition $\mathfrak{D}_2$ of $[a,b]$. Define a positive function $\delta(\xi): [a,b] \to \mathcal{R}^+$, where $\delta(\xi) = \min\{\delta_1(\xi),\delta_2(\xi)\}$, then for each $\delta$-fine $C$-partition $\mathfrak{D}$ of $[a,b]$, we have

$$\|S(k\bar{f}+l\bar{g}, P) - (k\bar{L}_1 + l\bar{L}_2)\| \le |k|\|S(\bar{f},\mathfrak{D}) - \bar{L}_1\| + |l|\|S(\bar{g},\mathfrak{D}) - \bar{L}_2\| < \varepsilon.$$

Definition 2.4 shows that $k\bar{f}+l\bar{g} \in C([a,b],\ell^\infty)$ and

$$(C)\int_a^b (k\bar{f}+l\bar{g}) = k\bar{L}_1 + l\bar{L}_2 = k(C)\int_a^b \bar{f} + l(C)\int_a^b \bar{g}.$$

Hence the theorem is proved. □

**Theorem 3.3 (Cauchy's Criterion Theorem)** *Let $\bar{f}: [a,b] \to \ell^\infty$. A function $\bar{f} \in C([a,b],\ell^\infty)$ if and only if for every $\varepsilon > 0$ there exsits a positive function $\delta(\xi): [a,b] \to \mathcal{R}^+$ such that*

$$\|S(\bar{f},\mathfrak{D}_1) - S(\bar{f},\mathfrak{D}_2)\| < \varepsilon$$

*for every two $\delta$-fine $C$-partitions $\mathfrak{D}_1$ and $\mathfrak{D}_2$ of $[a,b]$.*

*Proof.* ($\Rightarrow$) Let $\varepsilon > 0$ be given. Since $\bar{f} \in C([a,b],\ell^\infty)$, then there exists *a* positive function $\delta(\xi): [a,b] \to \mathcal{R}^+$ such that $\left\|S(\bar{f},\mathfrak{D}) - (C)\int_a^b \bar{f}\right\| < \frac{\varepsilon}{2}$ for every $\delta$-fine $C$-partition $\mathfrak{D}$ of $[a,b]$. Therefore, for two arbitrary $\delta$-fine $C$-partitions $\mathfrak{D}_1$ and $\mathfrak{D}_2$ of $[a,b]$, we have

$$\|S(\bar{f},\mathfrak{D}_1) - S(\bar{f},\mathfrak{D}_2)\| \le \left\|S(\bar{f},\mathfrak{D}_1) - (C)\int_a^b \bar{f}\right\| + \left\|S(\bar{f},\mathfrak{D}_2) - (C)\int_a^b \bar{f}\right\| < \frac{\varepsilon}{2} + \frac{\varepsilon}{2} = \varepsilon.$$

($\Leftarrow$) We choose $\varepsilon_r = 1/r$, $r \in \mathbb{N}$. There is a decreasing sequence $\{\varepsilon_r\}$ and a positive function $\delta_r^0(\xi): [a,b] \to \mathcal{R}^+$ such that for every $\delta_r^0$-fine $C$-partitions $\mathfrak{D}_r'$ and $\mathfrak{D}_r''$ of $[a,b]$ we have

$$\|S(\bar{f},\mathfrak{D}_r') - S(\bar{f},\mathfrak{D}_r'')\| < \frac{1}{r},$$

where $S(\bar{f},\mathfrak{D}_r') = \{(\mathfrak{D}_r')\sum_{i=1}^n f_j(\xi_i)(u_i - u_{i-1})\}_{j=1}^\infty$. Therefore, we have

$$\left\|\left\{(\mathfrak{D}_r')\sum_{i=1}^n f_j(\xi_i)(u_i - u_{i-1})\right\}_{j=1}^\infty - \left\{(\mathfrak{D}_r'')\sum_{i=1}^n f_j(\xi_i)(u_i - u_{i-1})\right\}_{j=1}^\infty\right\| < \frac{1}{r}.$$

This implies that

$\left|(\mathfrak{D}_r')\sum_{i=1}^n f_j(\xi_i)(u_i - u_{i-1}) - (\mathfrak{D}_r'')\sum_{i=1}^n f_j(\xi_i)(u_i - u_{i-1})\right| < \frac{1}{r}$, $j = 1,2,3,\ldots$. Now, we show that $f_j: [a,b] \to \mathcal{R}$ is $C$-integrable on $[a,b]$ for $j = 1,2,3,\ldots$. Given $\varepsilon_r = 1/r$, $r \in \mathbb{N}$, and so we have two consequences. First, there is a decreasing sequence $\{\delta_r(\xi)\}$ of positive functions on $[a,b]$, where $\delta_r(\xi) = \min\{\delta_1^0(\xi),\delta_2^0(\xi),\delta_3^0(\xi),\ldots,\delta_r^0(\xi)\}$. Second, there is a Cauchy sequence of real number $\{(\mathfrak{D}_r)\sum_{i=1}^n f_j(\xi_i)(u_i - u_{i-1})\}_{j=1}^\infty$, where $\mathfrak{D}_r$ is a $\delta_r$-fine $C$- partition of $[a,b]$. By the Cauchy sequence properties, we have $f_j$, $j = 1,2,3,\ldots$ is $C$-integrable on $[a,b]$. Therefore $\bar{f} = \{f_j\}_{j=1}^\infty \in C([a,b],\ell^\infty)$. This completes the proof of the Theorem. □ See Park *et al.* [7] Theorem 2.2 for details.

**Theorem 3.4** *Let $\bar{f}: [a,b] \to \ell^\infty$. If $\bar{f} \in C([a,b],\ell^\infty)$, then $\bar{f} \in C([c,d],\ell^\infty)$ for every $[c,d] \subseteq [a,b]$.*

*Proof.* Assume that $\bar{f} \in C([a,b],\ell^{\infty})$. Given $\varepsilon > 0$, there exists a positive function $\delta(\xi):[a,b] \to \mathcal{R}^+$ such that

$$\|S(\bar{f},\mathfrak{D}')-S(\bar{f},\mathfrak{D}'')\| < \varepsilon$$

for every $\delta$-fine $C$-partitions $\mathfrak{D}'$ and $\mathfrak{D}''$ of $[a,b]$. We define $\delta$-fine $C$-partitions $\mathfrak{D}_1$ and $\mathfrak{D}_2$ of $[c,d]$, $\mathfrak{D}_3$ of $[c,d] \cup [d,b]$. Let $\mathfrak{D}^1=\mathfrak{D}_1 \cup \mathfrak{D}_3$ and $\mathfrak{D}^2=\mathfrak{D}_2 \cup \mathfrak{D}_3$, then the unions of $\mathfrak{D}_1 \cup \mathfrak{D}_3$ and $\mathfrak{D}_2 \cup \mathfrak{D}_3$ form the $\delta$-fine $C$-partitions of $[a,b]$, where $c$ and $d$ are always a partition point of $\mathfrak{D}^i$, $i = 1,2$. So we have

$$\|S(\bar{f},\mathfrak{D}_1)-S(\bar{f},\mathfrak{D}_2)\| = \|S(\bar{f},\mathfrak{D}_1 \cup \mathfrak{D}_3)-S(\bar{f},\mathfrak{D}_2 \cup \mathfrak{D}_3)\|$$
$$= \|S(\bar{f},\mathfrak{D}^1)-S(\bar{f},\mathfrak{D}^2)\| < \varepsilon$$

By Theorem 3.3, then $\bar{f} \in C([c,d],\ell^{\infty})$. Hence the theorem is proved. □

**Theorem 3.5** *Let* $\bar{f}:[a,b] \to \ell^{\infty}$. *If* $\bar{f} \in C([a,c],\ell^{\infty})$ *and* $\bar{f} \in C([c,b],\ell^{\infty})$, *then* $\bar{f} \in C([a,b],\ell^{\infty})$ *and satisfying*

$$(C)\int_a^b \bar{f} = (C)\int_a^c \bar{f} + (C)\int_c^b \bar{f}.$$

*Proof.* Given any $\varepsilon > 0$. Let the sequence $\bar{L}_1$ and $\bar{L}_2$ be the $C$-integrals of $\bar{f}$ on $[a,c]$ and on $[c,b]$, respectively. From the hypothesis, there exists two positive functions $\delta_1(\xi):[a,c] \to \mathcal{R}^+$ and $\delta_2(\xi):[c,b] \to \mathcal{R}^+$ such that $\|S(\bar{f},\mathfrak{D}) - \bar{L}_1\| < \frac{\varepsilon}{2}$ for each $\delta_1$-fine $C$-partition $\mathfrak{D}$ of $[a,c]$ and $\|S(\bar{f},\mathfrak{D}) - \bar{L}_2\| < \frac{\varepsilon}{2}$ for each $\delta_2$-fine $C$-partition $\mathfrak{D}$ of $[c,b]$. We define a positive function $\delta(\xi) = \min\{\delta_1(\xi), c-\xi\}$ if $\xi \in [a,c)$, $\min\{\delta_1(\xi),\delta_2(\xi)\}$ if $\xi = c$ and $\min\{\delta_1(\xi),\delta_2(\xi)\}$ if $\xi \in (c,b]$. Therefore for each $\delta$-fine $C$-partition $\mathfrak{D}$ of $[a,b]$ we have

$$\|S(\bar{f},\mathfrak{D}) - (\bar{L}_1 + \bar{L}_2)\| \le \|S(\bar{f},\mathfrak{D}) - \bar{L}_1\| + \|S(\bar{f},\mathfrak{D}) - \bar{L}_2\| < \frac{\varepsilon}{2} + \frac{\varepsilon}{2} = \varepsilon.$$

Definition 2.4 shows that $\bar{f} \in C([a,b],\ell^{\infty})$ and

$$(C)\int_a^b \bar{f} = \bar{L}_1 + \bar{L}_2 = (C)\int_a^c \bar{f} + (C)\int_c^b \bar{g}.$$

Hence the theorem is proved. □

**Theorem 3.6 (Henstock Lemma 1)** *Let* $\bar{f} \in C([a,b],\ell^{\infty})$ *i.e. for every* $\varepsilon > 0$ *there exists a positive function* $\delta(\xi):[a,b] \to \mathcal{R}^+$ *such that for each* $\delta$*-fine* $C$*-partition* $\mathfrak{D} = \{([u_{i-1},u_i],\xi_i)\}_{i=1}^n$ *of* $[a,b]$ *satisfies*

$$\left\|S(\bar{f},\mathfrak{D})-(C)\int_a^b \bar{f}\right\| < \varepsilon.$$

*If* $\mathfrak{D}^* = \{([u_{i-1}^*,u_i^*],\xi_i^*)\}_{i=1}^k$ *is an arbitrary* $\delta$*-fine partial* $C$*-partition of* $[a,b]$, *then*

$$\left\|S(\bar{f},\mathfrak{D}^*)-\sum_{i=1}^{k}(C)\int_{u_{i-1}^*}^{u_i^*} \bar{f}\right\| < 2\varepsilon.$$

*Proof.* Let $\mathfrak{D}$ be a $\delta$-fine C -partition of $[a,b]$. Choose two arbitrary $\delta$-fine partial $C$-partitions of $\mathfrak{D}$, called $\mathfrak{D}^* = \{([u_{i-1}^*,u_i^*],\xi_i^*)\}_{i=1}^k$ and $\mathfrak{D}^{**} = \{([u_{i-1}^{**},u_i^{**}],\xi_i^{**})\}_{i=1}^l$ such that $cl([a,b]\backslash \bigcup_{i=1}^k[u_{i-1}^*,u_i^*]) = \bigcup_{j=1}^l[u_{j-1}^{**},u_j^{**}]$, where $k + l = n$. Since $\bar{f} \in C([a,b],\ell^{\infty})$ by Theorem 3.4, then $\bar{f} \in C([u_{j-1}^{**},u_j^{**}],\ell^{\infty})$.

Given $\varepsilon > 0$. There exists a positive function $\delta_j(\xi):[u_{j-1}^{**},u_j^{**}] \to \mathcal{R}^+$ such that for every $\delta_j$-fine $C$-partition $\mathfrak{D}_j$ *of* $[u_{j-1}^{**},u_j^{**}]$ we have $\left\|S(\bar{f},\mathfrak{D}_j)-(C)\int_{u_{j-1}^{**}}^{u_j^{**}} \bar{f}\right\| < \frac{\varepsilon}{l}$. We define $\mathfrak{D} = \mathfrak{D}^* \cup \mathfrak{D}^{**}$, then $\mathfrak{D}$ is a $\delta$-fine $C$-partition of $[a,b]$ and from Theorem 3.5, we have

$$\left\|S(\bar{f},\mathfrak{D})-(C)\int_a^b \bar{f}\right\| = \left\|S(\bar{f},\mathfrak{D}^*)+S(\bar{f},\mathfrak{D}^{**})-(C)\int_a^b \bar{f}\right\| < \varepsilon.$$

This implies that

$$\left\|S(\bar{f},\mathfrak{D}^*)-\sum_{i=1}^{k}(C)\int_{u_{i-1}^*}^{u_i^*}\bar{f})\right\| \leq \left\|S(\bar{f},\mathfrak{D})-(C)\int_a^b \bar{f}\right\|+\left\|S(\bar{f},\mathfrak{D}^{**})-\sum_{j=1}^{l}(C)\int_{u_{j-1}^{**}}^{u_j^{**}}\bar{f})\right\|$$
$$< \varepsilon + \sum_{j=1}^{l}\frac{\varepsilon}{l} = 2\varepsilon .$$

Hence the theorem is proved. □

**Theorem 3.7** *Let* $\bar{f}:[a,b]\to \ell^{\infty}$. *If* $\bar{f}=\bar{\theta}$ *almost everywhere on* $[a,b]$, *then* $\bar{f}\in C([a,b],\ell^{\infty})$ *and* $(C)\int_a^b \bar{f}=\bar{\theta}$.

*Proof.* By hipotesis $\bar{f}=\{f_1,f_2,f_3,\dots,f_j,\dots\}=\bar{\theta}$ almost everywhere on $[a,b]$, it implies that $f_j(u)=0$ almost everywhere on $[a,b]$, for all $j=1,2,3,\dots$We define $\mathcal{A}_j=\{u\in[a,b]: f_j(u)\neq 0\}$ and $\mathcal{A}_{jm}=\{u\in[a,b]: m-1\leq |f_j(u)|<m\}$, for every $m=1,2,3,\dots$, then $\mathcal{A}_j=\bigcup_{m=1}^{\infty}\mathcal{A}_{jm}$. Since $\mu(\mathcal{A}_j)=0$ then $\mu(\mathcal{A}_{jm})=0$, for all $m=1,2,3,\dots$.
Given any $\varepsilon>0$, then there exists an open sets $\mathcal{H}_m\subset[a,b]$ such that $\mathcal{H}_m\supset\mathcal{A}_{jm}$ and $\mu(\mathcal{H}_m)<\frac{\varepsilon}{m2^m}$.
Define a positive function $\delta(\xi):[a,b]\to\mathcal{R}^+$ such that $\delta(\xi)=1$, if $\xi\in[a,b]\setminus\mathcal{A}_j$, and $\delta(\xi)=dist(\xi,\mathcal{A}_{jm}^C)$ if $\xi\in\mathcal{A}_{jm}$, for all $j=1,2,3,\dots$.
If $\mathfrak{D}=\{(I_i,\xi_i)\}_{i=1}^{n}$, where $I_i=[u_{i-1},u_i]$ is a $\delta$-fine $C$-partition of $[a,b]$ we have

$$\|S(\bar{f},\mathfrak{D})\|=\|(\mathfrak{D})\sum_{i=1}^{n}\bar{f}(\xi_i)|I_i|\|=\sup\{|(\mathfrak{D})\sum_{i=1}^{n}f_j(\xi_i)|I_i||\}_{j=1}^{\infty},$$

Then there exists an index $j$ that is $j=j^*$ such that

$$\sup\{|(\mathfrak{D})\sum_{i=1}^{n}f_j(\xi_i)|I_i||\}_{j=1}^{\infty}=|(\mathfrak{D})\sum_{i=1}^{n}f_{j^*}(\xi_i)|I_i||,\text{ and we have}$$

$$\begin{aligned}\|S(\bar{f},\mathfrak{D})\|&=|(\mathfrak{D})\sum_{i=1}^{n}f_{j^*}(\xi_i)|I_i||\leq(\mathfrak{D})\sum_{i=1}^{n}|f_{j^*}(\xi_i)||I_i|\\&=(\mathfrak{D})\sum_{\xi\in\mathcal{A}_{j^*m}}|f_{j^*}(\xi_i)||I_i|+(\mathfrak{D})\sum_{\xi\notin\mathcal{A}_{j^*m}}|f_{j^*}(\xi_i)||I_i|\\&<\sum m.\mu(\mathcal{H}_m)+0<m\sum\frac{\varepsilon}{m2^m}=\varepsilon.\end{aligned}$$

This shows that $\bar{f}=\{f_j\}_{j=1}^{\infty}\in C([a,b],\ell^{\infty})$ and $(C)\int_a^b\bar{f}=\bar{\theta}$. □

**Corollary 3.8** *Let* $\bar{f},\bar{g}:[a,b]\to\ell^{\infty}$ *and* $\bar{f}\in C([a,b],\ell^{\infty})$. *If* $\bar{f}=\bar{g}$ *almost everywhere on* $[a,b]$, *then* $\bar{g}\in C([a,b],\ell^{\infty})$ *and* $(C)\int_a^b\bar{f}=(C)\int_a^b\bar{g}$.

*Proof.* Obviously the function $\bar{g}-\bar{f}=\bar{\theta}$ almost everywhere on $[a,b]$. By Theorem 3.7 $(C)\int_a^b(\bar{g}-\bar{f})=\bar{\theta}$. From Theorem 3.2, the function $\bar{g}=\bar{f}+(\bar{g}-\bar{f})$ is $C$-integrable on $[a,b]$ and satisfying $(C)\int_a^b\bar{g}=(C)\int_a^b\bar{f}+(C)\int_a^b(\bar{g}-\bar{f})=(C)\int_a^b\bar{f}$. □

**Theorem 3.9** *Let* $\bar{f}:[a,b]\to\ell^{\infty}$. *If* $\bar{f}\geq\bar{\theta}$ *almost everywhere on* $[a,b]$ *and* $\bar{f}\in C([a,b],\ell^{\infty})$, *then* $(C)\int_a^b\bar{f}\geq\bar{\theta}$.

*Proof.* Let $\bar{L}=(C)\int_a^b\bar{f}=\{(C)\int_a^b f_1,(C)\int_a^b f_2,\dots,(C)\int_a^b f_j,\dots\}$, where $\bar{f}=\{f_1,f_2,f_3,\dots,f_j,\dots\}\in C([a,b],\ell^{\infty})$. Given any $\varepsilon>0$, there exists a positive function $\delta(\xi):[a,b]\to\mathcal{R}^+$ such that for each $\delta$-fine $C$-partition $\mathfrak{D}=\{([u_{i-1},u_i],\xi_i)\}_{i=1}^{n}$ of $[a,b]$ satisfies

$$\|S(\bar{f},\mathfrak{D})-\bar{L}\|=\left\|(\mathfrak{D})\sum_{i=1}^{n}(\bar{f}(\xi_i)(u_i-u_{i-1})-\bar{L})\right\|.$$

$$=\sup\left\{\left|(\mathfrak{D})\sum_{i=1}^{n}f_j(\xi_i)(u_i-u_{i-1})-(C)\int_a^b f_j\right|\right\}_{j=1}^{\infty}<\varepsilon.$$

It implies that $\left|(\mathfrak{D})\sum_{i=1}^{n}f_j(\xi_i)(u_i-u_{i-1})-(C)\int_a^b f_j\right|<\varepsilon$ for all $j=1,2,3,\dots$ and

$$(C)\int_a^b f_j - \varepsilon < (\mathfrak{D})\sum_{i=1}^n f_j(\xi_i)(u_i - u_{i-1}) < (C)\int_a^b f_j + \varepsilon.$$

Since $f_j \geq 0$ for all $j = 1,2,3,\ldots$ and $(u_i - u_{i-1}) \geq 0$ we have

$$(C)\int_a^b f_j + \varepsilon > (\mathfrak{D})\sum_{i=1}^n f_j(\xi_i)(u_i - u_{i-1}) \geq 0.$$

This shows that $(C)\int_a^b f_j \geq 0$ for all $j = 1,2,3,\ldots$. Therefore $(C)\int_a^b \bar{f} \geq \bar{\theta}$. □

**Corollary 3.10** *Let* $\bar{f},\bar{g}:[a,b] \to \ell^\infty$ *and* $\bar{f},\bar{g} \in C([a,b],\ell^\infty)$. *If* $\bar{f} \geq \bar{g}$ *almost everywhere on* $[a,b]$, *then* $(C)\int_a^b \bar{f} \geq (C)\int_a^b \bar{g}$.

*Proof.* We construct the function $\bar{h} = \bar{f} - \bar{g}$. Then $\bar{h} \geq \bar{\theta}$ almost everywhere on $[a,b]$ and from Theorem 3.8, it is obvious that $(C)\int_a^b \bar{h} = (C)\int_a^b(\bar{f} - \bar{g}) = (C)\int_a^b \bar{f} - (C)\int_a^b \bar{g} \geq \bar{\theta}$. Consequently, $(C)\int_a^b \bar{f} \geq (C)\int_a^b \bar{g}$. □

Now we will define the primitive function of $C$-integrable function $\bar{f}$ on $[a,b]$ and discuss the following modified Henstock Lemma.

**Definition 3.11** *Let* $\bar{f} = \{f_1, f_2, \ldots, f_j, \ldots\} \in C([a,b],\ell^\infty)$. *The function* $\bar{F} = \{F_1, F_2, \ldots, F_j, \ldots\}: [a,$ $[a,b] \to \ell^\infty$ *such that*

$$\bar{F} = \int_a^x \bar{f}, \qquad a \leq x \leq b$$

*is called* $C$*-primitive function of* $\bar{f}$ *on* $[a,b]$.

**Theorem 3.12** The collection of all the $C$-primitive functions on $[a,b]$ is a vector space over $\mathcal{R}$.
*Proof.* The proof is omitted. □

From the Definition 3.11, if $\mathfrak{D} = \{([u_{i-1},u_i],\xi_i)\}_{i=1}^n$ we can write

$$(\mathfrak{D})\sum_{i=1}^n \bar{F}[u_{i-1},u_i] = (\mathfrak{D})\sum_{i=1}^n (\bar{F}(u_i) - \bar{F}(u_{i-1})) = \sum_{i=1}^n (C)\int_{u_{i-1}}^{u_i} \bar{f} = \bar{F}[a,b].$$

**Theorem 3.13** *Let* $\bar{f} \in C([a,b],\ell^\infty)$, *the* $\bar{F}$ *is a* $C$*-primitive function of* $\bar{f}$ *on* $[a,b]$ *if and only if for any* $\varepsilon > 0$ *there exists a positive function* $\delta(\xi): [a,b] \to \mathcal{R}^+$ *such that for each* $\delta$*-fine* $C$*-partition* $\mathfrak{D} = \{([u_{i-1},u_i],\xi_i)\}_{i=1}^n$ *of* $[a,b]$ *satisfies*

$$(\mathfrak{D})\sum_{i=1}^n \|\bar{f}(\xi_i)(u_i - u_{i-1}) - \bar{F}[u_{i-1},u_i]\| < \varepsilon.$$

*Proof.* (⇒) Given $\varepsilon > 0$. By the Henstock Lemma there exists a positive function $\delta(\xi): [a,b] \to \mathcal{R}^+$ such that for each $\delta$-fine $C$-partition $\mathfrak{D} = \{([u_{i-1},u_i],\xi_i)\}_{i=1}^n$ of $[a,b]$ satisfies

$$\left\|S(\bar{f},\mathfrak{D}) - (C)\int_a^b \bar{f}\right\| = \left\|(\mathfrak{D})\sum_{i=1}^n (\bar{f}(\xi_i)(u_i - u_{i-1}) - \bar{F}[u_{i-1},u_i])\right\| < \frac{\varepsilon}{4}.$$

Moreover, if $\mathfrak{D}^* = \{([u_{i-1}^*,u_i^*],\xi_i^*)\}_{i=1}^k$ is a $\delta$-fine partial $C$-partition of $[a,b]$ then it satisfies

$$\left\|S(\bar{f},\mathfrak{D}^*) - \sum_{i=1}^k (C)\int_{u_{i-1}^*}^{u_i^*} \bar{f}\right\| = \left\|(\mathfrak{D}^*)\sum_{i=1}^k (\bar{f}(\xi_i)(u_i - u_{i-1}) - \bar{F}[u_{i-1},u_i])\right\| < \frac{\varepsilon}{2}.$$

If $\mathfrak{D} = \{([u_{i-1},u_i],\xi_i)\}_{i=1}^n$ is a $\delta$-fine $C$-partition of $[a,b]$, then there exists an index $j$ that is $j = j^*$ such that $sup\{|f_j(\xi_i)(u_i - u_{i-1}) - F_j[u_{i-1},u_i]|\}_{j=1}^\infty = |f_{j^*}(\xi_i)(u_i - u_{i-1}) - F_{j^*}[u_{i-1},u_i]|$, then we have

$(\mathfrak{D})\sum_{i=1}^n \|\bar{f}(\xi_i)(u_i - u_{i-1}) - \bar{F}[u_{i-1},u_i]\|$

$$= (\mathfrak{D}) \sum_{i=1}^{n} \left\| \left\{ f_j(\xi_i)(u_i - u_{i-1}) - F_j[u_{i-1}, u_i] \right\}_{j=1}^{\infty} \right\|$$
$$= (P) \sum_{i=1}^{n} sup\left\{ \left| f_j(\xi_i)(u_i - u_{i-1}) - F_j[u_{i-1}, u_i] \right| \right\}_{j=1}^{\infty}$$
$$= (P) \sum_{i=1}^{n} \left| f_{j^*}(\xi_i)(u_i - u_{i-1}) - F_{j^*}[u_{i-1}, u_i] \right|,$$
$$= (P) \sum_{Nonneg} \left| (f_{j^*}(\xi_i)(u_i - u_{i-1}) - F_{j^*}[u_{i-1}, u_i]) \right| + (P) \sum_{Neg} \left| (f_{j^*}(\xi_i)(u_i - u_{i-1}) - F_{j^*}[u_{i-1}, u_i]) \right|,$$

where $\sum_{Nonneg}$ and $\sum_{Neg}$ are the partial Riemann sums of $(f_{j^*}(\xi_i)(u_i - u_{i-1}) - F_{j^*}[u_{i-1}, u_i] \geq 0$ and $(f_{j^*}(\xi_i)(u_i - u_{i-1}) - F_{j^*}[u_{i-1}, u_i] < 0$, respectively.
Furthermore,

$$(\mathfrak{D}) \sum_{Nonneg} sup\left\{ \left| f_j(\xi_i)(u_i - u_{i-1}) - F_j[u_{i-1}, u_i] \right| \right\}_{j=1}^{\infty} + (\mathfrak{D}) \sum_{Neg} sup\left\{ \left| f_j(\xi_i)(u_i - u_{i-1}) - F_j[u_{i-1}, u_i] \right| \right\}_{j=1}^{\infty}$$
$$= (\mathfrak{D}) \sum_{Nonneg} \left\| (\bar{f}(\xi_i)(u_i - u_{i-1}) - \bar{F}[u_{i-1}, u_i]) \right\| + (\mathfrak{D}) \sum_{Neg} \left\| (\bar{f}(\xi_i)(u_i - u_{i-1}) - \bar{F}[u_{i-1}, u_i]) \right\|.$$

By the Henstock Lemma, $(\mathfrak{D}) \sum_{Nonneg} \left\| (\bar{f}(\xi_i)(u_i - u_{i-1}) - \bar{F}[u_{i-1}, u_i]) \right\| < \frac{\varepsilon}{2}$ and $(\mathfrak{D}) \sum_{Neg} \left\| (\bar{f}(\xi_i)(u_i - u_{i-1}) - \bar{F}[u_{i-1}, u_i]) \right\| < \frac{\varepsilon}{2}$.
Therefore, $(\mathfrak{D}) \sum_{i=1}^{n} \left\| \bar{f}(\xi_i)(u_i - u_{i-1}) - \bar{F}[u_{i-1}, u_i] \right\| < \frac{\varepsilon}{2} + \frac{\varepsilon}{2} = \varepsilon$.
($\Leftarrow$) Conversely. This is a consequence of Theorem 3.3 and triangle inequality. This completes the proof of the theorem. □

By Definition 3.11 and Theorem 3.13 we will modify the Henstock Lemma in Theorem 3.6 to the following corollary.

**Corollary 3.14 (Henstock Lemma 2)** *Let $\bar{f} \in C([a,b], \ell^{\infty})$ with $\bar{F}$ is C-primitive function $\bar{f}$ on $[a,b]$ i.e. for every $\varepsilon > 0$ there exists a positive function $\delta: [a,b] \to \mathcal{R}^+$ such that for each $\delta$-fine C-partition $\mathfrak{D} = \{([u_{i-1}, u_i], \xi_i)\}_{i=1}^{n}$ of $[a,b]$ satisfies*

$$(\mathfrak{D}) \sum_{i=1}^{n} \left\| \bar{f}(\xi_i)(u_i - u_{i-1}) - \bar{F}[u_{i-1}, u_i] \right\| < \varepsilon.$$

*If $\mathfrak{D}^* = \{([u_{i-1}^*, u_i^*], \xi_i^*\}_{i=1}^{k}$ is an arbritrary $\delta$-fine partial C-partition of $[a,b]$, then*

$$(\mathfrak{D}^*) \sum_{i=1}^{k} \left\| \bar{f}(\xi_i)(u_i^* - u_{i-1}^*) - \bar{F}[u_{i-1}^*, u_i^*] \right\| < 2\varepsilon.$$

*Proof.* The result can be derived directly from the Definition 3.11 and Theorem 3.13. □

## 4. Conclusion

We have defined the $C$-integral of sequence-valued functions and discussed some of its properties. Based on the $C$-integral of sequence-valued functions concept, we are able to define the primitive function of $C$-integrable function for modifying the Henstock Lemma. Finally, there is a possibility to investigate the $ACG$ property of $C$-integral of sequence-valued functions.

## 5. Acknowledgement

The authors wish to thank Sebelas Maret University Indonesia for funding this research in the year of 2016.

## References

[1] Bongiorno, B. On the Minimal Solution of the Problem of Primitives. *Journal of Mathematical Analysis and Applications*. 2000; 251: 479-487.
[2] Gordon, R. A. *The Integrals of Lebesgue Denjoy Perron and Henstock*. Graduate Studies in Mathematics, The American Mathematical Society; 1994.
[3] Paluga, R. N. and Lonoy, S. R. Jr. On The McShane Integral in Topological Vector Spaces. *Chamchuri Journal of Mathematics*. 2010; 2: 45-56.
[4] Park, J. M., Kim, B. M., Kim, Y. K., and Lee, H. K. The-$M_\alpha$ integral of Banach-Valued Functions. *Journal of The Chungcheong Mathematical Society*. 2012; 25: 115-125.
[5] Park, J. M. and Lee D. H. The Denjoy Extension of The McShane Integral. *Bull. Korean Math Soc*. 1996; 33: 411-416.
[6] Park, J. M., Lee, D. H.,Yoon, J. H., and Lee, H.K. The Integration by Parts for the-$M_\alpha$ Integral. *Journal of The Chungcheong Mathematical Society*. 2010; 23: 861-870.
[7] Park, J. M., Ryu, H. W. And Lee, H. K. The -$M_\alpha$ integral. *Journal of The Chungcheong Mathematical Society*. 2010; 23: 99-108.
[8] Tuo, Y. L. 2006 Multipliers for Generalized Riemann Integrals in The Real Line. *Mathematica Bohemica*. 2006; 2: 161-166.
[9] Vyborny, R. Kurzweil's PU as The Lebesgue Integral. *Mathematica Bohemica*. 2006; 131: 11-14.
[10] Yoon, J. H., Eun, G. S. and Lee, Y. C. On McShane-Stieltjes Integral. *Kangweon-Kyungki Math Jour*. 1997; 5: 217-225.
[11] You, X., Cheng, J. and Zhao, D. On Strong $M_\alpha$- Integral of Banach Valued Functions. *Journal of The Chungcheon Mathematical Society*. 2013; 26: 259-268.
[12] Zachriwan, Soeparna, D. and Daru Unoningsih. *Integral Henstock Fungsi Bernilai di Dalam Ruang Barisan* $\ell^p$ $1 \leq p < \infty$. Prosiding Seminar Nasional Matematika dan Statistika VI ITS Surabaya. 2003; 231-240
[13] Zhao, D. and Ye, G. C-Integral and Denyoy-C Integral. *Commun Korean Math Soc*. 2007; 22: 27-39.

IOP Conf. Series: Journal of Physics: Conf. Series **855** (2017) 012029 doi:10.1088/1742-6596/855/1/012029

# Model reduction of unstable systems using balanced truncation method and its application to shallow water equations

**Kiki Mustaqim, Didik Khusnul Arif, Erna Apriliani and Dieky Adzkiya**

Department of Mathematics, Institut Teknologi Sepuluh Nopember, Surabaya 60111, Indonesia

E-mail: kiki.mustaqim14@mhs.matematika.its.ac.id, didik@matematika.its.ac.id, april@matematika.its.ac.id, dieky@matematika.its.ac.id

**Abstract.** In this paper, we discuss the reduction of unstable systems. Suppose we have a large-scale unstable system. The reduction of such systems are done in order to obtain a simpler model that has a similar behavior with the original model. In reality, a higher order model is not preferred because the computation takes more time. Thus we need a method to reduce the order of a model that is known as model reduction. Model reduction of a system is the method of approximation of a system with a lower order but the dynamic behavior is equal or almost equal to the original model. In this paper, we discuss model reduction of unstable systems. Finally this method is applied to shallow water equation that describes the flow of water in rivers: we obtain a procedure for model reduction of unstable systems and the reduced model of shallow water equations.

## 1. Introduction

Studying systems' behavior can be done through mathematical model. If we want to construct a model that is close to the real phenomenon, we need a large number of variables. As a consequence, the mathematical model has many state variables. Existing systems in the universe often has a big order. On the other hand, this affects the computational time to simulate the model, due to the huge number of variables in the system. Hence, we need a method to simplify the order of a system.

There are many methods in the literature for model order reduction, such as balanced truncation, modal analysis, krylov method, Hankel norm approximation, singular value decomposition [1, 2, 3]. The application of balanced truncation methods to heat conduction distribution has been discussed in [4, 5]. Those methods are suitable for stable, controllable and observable systems. However, in reality, there are many unstable systems. Thus we need to develop a model reduction method for unstable systems. In this paper, we discuss a model reduction procedure for unstable systems using balanced truncation method. In the model order reduction of an unstable system, first the system is decomposed into stable subsystem and unstable subsystems using decomposition algorithms. Then we reduce the stable subsystem by using balanced truncation method, i.e. constructing a balanced system, partition the systems and truncating the least controllable or least observable state. Finally, we add the unstable

IOP Conf. Series: Journal of Physics: Conf. Series **855** (2017) 012029 doi:10.1088/1742-6596/855/1/012029

system into the reduced stable subsystems. Then, we apply the model reduction of unstable systems to shallow water equations.

The paper is structured as follows. Section 2 discusses the shallow water equations that will be reduced later. Then Section 3 discretizes the original model into a discrete-time system. In Section 4, we discuss the model reduction of unstable systems and its application to shallow water equations.

## 2. Shallow Water Equations

In this section, we discuss the shallow water equation that describes the flow of water in rivers [6]:

$$\frac{\partial h}{\partial t} + D\frac{\partial v}{\partial x} = 0 \tag{1}$$

$$\frac{\partial v}{\partial t} + g\frac{\partial h}{\partial x} + C_f u = 0 \tag{2}$$

with initial and boundary conditions:

$$h(x,0) = 1, v(x,0) = 0, h(0,t) = \psi_b(t), v(L,t) = v_N(t) \tag{3}$$

where $h(x,t)$ is the water level above the reference plane at position $x$ and time $t$, $v(x,t)$ is the average current velocity at position $x$ and time $t$, $t$ is the time variable, $x$ is the position along the river, $D$ is the water depth, $g$ is the gravitational acceleration and $c$ is a friction constant.

## 3. Discretization

In this section, we discretize the Shallow Water Equation so that we can obtain a discrete-time system that is suitable for model reduction. We use implicit scheme called Preissman discretization [7] as follows:

$$\frac{\partial v}{\partial x} = \theta\left(\frac{v_{i+1}^{k+1} - v_i^{k+1}}{\Delta x}\right) + (1-\theta)\left(\frac{v_{i+1}^{k} - v_i^{k}}{\Delta x}\right) \tag{4}$$

$$\frac{\partial h}{\partial t} = \theta\left(\frac{h_{i+1}^{k+1} - h_i^{k+1}}{2\Delta t}\right) + (1-\theta)\left(\frac{h_{i+1}^{k} - h_i^{k}}{2\Delta t}\right) \tag{5}$$

then, from equations (1) and (2), we obtain

$$\frac{1}{2\Delta t}h_i^{k+1} - \frac{D\theta}{\Delta x}v_i^{k+1} + \frac{1}{2\Delta t}h_{i+1}^{k+1} + \frac{D\theta}{\Delta x}v_{i+1}^{k+1} = \frac{1}{2\Delta t}h_i^{k} + \frac{D(1-\theta)}{\Delta x}v_i^{k} + \frac{1}{2\Delta t}h_{i+1}^{k} - \frac{D(1-\theta)}{\Delta x}v_{i+1}^{k} \tag{6}$$

$$\begin{aligned}\frac{-g\theta}{\Delta x}h_i^{k+1} + \left(\frac{1}{2\Delta t} + \frac{C_f\theta}{2}\right)v_i^{k+1} + \frac{g\theta}{\Delta x}h_{i+1}^{k+1} + \left(\frac{1}{2\Delta t} + \frac{C_f\theta}{2}\right)v_{i+1}^{k+1} = \frac{g(1-\theta)}{\Delta x}h_i^{k} + \\ \left(\frac{1}{2\Delta t} - \frac{C_f(1-\theta)}{2}\right)v_i^{k} - \frac{g(1-\theta)}{\Delta x}h_{i+1}^{k} + \left(\frac{1}{2\Delta t} - \frac{C_f(1-\theta)}{2}\right)v_{i+1}^{k}\end{aligned} \tag{7}$$

Using the initial and boundary condition (3) and generate terms (6) and (7) for $i = 0$ to $N = 5$, then, the following values for the parameters are assumed

$$L = 30km, D = 10m, C_f = 0.0002, \Delta t = 10, \theta = 0.3, g = 9.8m/s^2$$

Thus, we can write the equations (6) and (7) in matrix notation as follows

$$A_1 x_{k+1} = A_2 x_k + B u_k \tag{8}$$

$$x_{k+1} = A_1^{-1} A_2 x_k + A_1^{-1} B_1 u_k \tag{9}$$

The above system is a discrete-time linear-time-invariant system

$$x_{k+1} = A x_k + B u_k$$

where $A = A_1^{-1} A_2$ *and* $B = A_1^{-1} B_1$. By using the parameters described above, those matrices are given by:

$$A = \begin{pmatrix} 0.8465 & 0 & 0 & 0 & 0 & 0 & 0 & 0 & 0 & 0 & 0 & 0 \\ 0.0311 & 0.9980 & -0.0653 & -0.0007 & 0.0653 & 0.0020 & -0.0654 & -0.0033 & 0.0655 & 0.0046 & -0.0327 & 0.9972 \\ 0 & 0 & 1.0003 & 0.0333 & -0.0013 & -0.0666 & 0.0026 & 0.0667 & -0.0039 & -0.0668 & 0.0023 & -0.0231 \\ 0 & 0 & 0.0327 & 0.9983 & -0.0653 & -0.0013 & 0.0654 & 0.0026 & -0.0654 & -0.0039 & 0.0327 & -0.9970 \\ 0 & 0 & 0 & 0 & 1.0003 & 0.0333 & -0.0013 & -0.0666 & 0.0026 & 0.0667 & -0.0016 & 0.0032 \\ 0 & 0 & 0 & 0 & 0.0327 & 0.9983 & -0.0653 & -0.0013 & 0.0654 & 0.0026 & -0.0327 & 0.9968 \\ 0 & 0 & 0 & 0 & 0 & 0 & 1.0003 & 0.0333 & -0.0013 & -0.0666 & 0.0010 & 0.0167 \\ 0 & 0 & 0 & 0 & 0 & 0 & 0.0327 & 0.9983 & -0.0653 & -0.0013 & 0.0327 & -0.9969 \\ 0 & 0 & 0 & 0 & 0 & 0 & 0 & 0 & 1.0003 & 0.0333 & -0.0003 & -0.0367 \\ 0 & 0 & 0 & 0 & 0 & 0 & 0 & 0 & 0.0327 & 0.9983 & -0.0327 & 0.9974 \\ 0 & 0 & 0 & 0 & 0 & 0 & 0 & 0 & 0 & 0 & 1.0000 & 0.0233 \\ 0 & 0 & 0 & 0 & 0 & 0 & 0 & 0 & 0 & 0 & 0 & 0 \end{pmatrix} \tag{10}$$

$$B = \begin{pmatrix} 0 \\ 0 \\ 0 \\ 0 \\ 0 \\ 0 \\ 0 \\ 0 \\ 0.0005 \\ 0.0500 \\ -0.0005 \\ 1.0000 \end{pmatrix}, x_{k+1} = \begin{pmatrix} u_0^{k+1} \\ h_0^{k+1} \\ u_1^{k+1} \\ h_1^{k+1} \\ u_2^{k+1} \\ h_2^{k+1} \\ u_3^{k+1} \\ h_3^{k+1} \\ u_4^{k+1} \\ h_4^{k+1} \\ u_5^{k+1} \\ h_5^{k+1} \end{pmatrix}, x_k = \begin{pmatrix} u_0^k \\ h_0^k \\ u_1^k \\ h_1^k \\ u_2^k \\ h_2^k \\ u_3^k \\ h_3^k \\ u_4^k \\ h_4^k \\ u_5^k \\ h_5^k \end{pmatrix}, u_k = \begin{pmatrix} 0 \\ 0 \\ 0 \\ 0 \\ 0 \\ 0 \\ 0 \\ 0 \\ 0 \\ 0 \\ 0 \\ 1 \end{pmatrix} \tag{11}$$

then, we construct a measurement equation at time $k$, as follows

$$y_k = C_k x_k \tag{12}$$

where

$$C = \begin{pmatrix} 0 & 1 & 0 & 0 & 0 & 0 & 0 & 0 & 0 & 0 & 0 & 0 \end{pmatrix} \quad D = 0 \tag{13}$$

Thus, we have system $G = (A, B, C, D)$ as the original system in the model order reduction. The properties of the preceding system are unstable, uncontrollable and unobservable. In the next section, we discuss a model reduction method for such unstable systems.

## 4. Model Reduction of Unstable Systems

In this section, we discuss model reduction of unstable systems and its applications to shallow water equations. The steps in model order reduction of unstable systems are as follows. First, we decompose the original system into stable and unstable subsystems (Section 4.1). Then, we reduce the stable subsystem using balanced truncation method (Section 4.2). Finally, we combine the reduced stable subsystem and the unstable subsystem (Section 4.3).

### *4.1. Unstable System Decomposition*

In this section, the unstable system $G$ is decomposed into stable subsystem and unstable subsystem using a decomposition algorithm [8, 9]. The decomposition algorithm consists of two steps:

**Step 1** Transform the system $G$ using a unitary matrix $U$ in block diagonal upper Schur form [10]. Thus, in the first step, the transformed system becomes:

$$G_t = \left(\begin{array}{c|c} U'AU & U'B \\ --- & --- \\ CU & D \end{array}\right) = \left(\begin{array}{cc|c} A_{t11} & A_{t12} & B_{t1} \\ 0 & A_{t22} & B_{t2} \\ --- & --- & --- \\ C_{t1} & C_{t2} & D \end{array}\right) \tag{14}$$

**Step 2** The transformed system $G_t$ in the first step, contains a coupling term $A_{t12}$. Using transformation $X_t = WX$, we obtain completely decoupled system $G_d$ as follows:

$$G_d = \left(\begin{array}{c|c} W^{-1}A_tW & W^{-1}B_t \\ --- & --- \\ C_tW & D \end{array}\right) = \left(\begin{array}{cc|c} A_{11} & 0 & B_1 \\ 0 & A_{22} & B_t \\ --- & --- & --- \\ C_1 & C_2 & D \end{array}\right) \tag{15}$$

The transformed model can be decomposed into stable and unstable subsystems as follows

$$\begin{aligned} G_d &= \left(\begin{array}{c|c} A_{11} & B_1 \\ --- & --- \\ C_1 & D \end{array}\right) + \left(\begin{array}{c|c} A_{22} & B_2 \\ --- & --- \\ C_2 & 0 \end{array}\right) \\ &= G_s(Stable\ subsystem) + G_u(Unstable\ subsystem) \end{aligned} \tag{16}$$

By using the parameters defined in the preceding section, the stable and unstable subsystems are given by

$$Gs = \left(\begin{array}{ccccccc|c} 0.9980 & -0.0000 & -0.0022 & 0.0014 & 0.0002 & -0.0009 & -0.0044 & -0.0041 \\ 0 & 0.8465 & 0 & 0 & 0 & 0 & 0 & 0 \\ 0 & 0 & 0.9663 & -0.0353 & -0.0172 & -0.0155 & -0.0071 & 0.0002 \\ 0 & 0 & 0 & 0.9663 & -0.0325 & -0.0294 & -0.0147 & -0.0004 \\ 0 & 0 & 0 & 0 & 0.9663 & -0.0593 & -0.0272 & 0.0008 \\ 0 & 0 & 0 & 0 & 0 & 0.9663 & -0.0330 & -0.0245 \\ 0 & 0 & 0 & 0 & 0 & 0 & 0 & 16.000 \\ ---- & ---- & ---- & ---- & ---- & ---- & ---- & ---- \\ -17.3644 & -0.0000 & -7.0040 & 2.4502 & -0.9750 & 1.4097 & 0.0156 & 0 \end{array}\right) \tag{17}$$

$$G_u = \left(\begin{array}{ccccc|c} 1.0323 & -0.0620 & 0.0321 & -0.0174 & 0.0009 & -0.2121 \\ 0 & 1.0323 & -0.0342 & 0.0185 & -0.0006 & 0.2363 \\ 0 & 0 & 1.0323 & -0.0368 & -0.0009 & -0.5164 \\ 0 & 0 & 0 & 1.0323 & 0.0014 & 0.9656 \\ 0 & 0 & 0 & 0 & 1.0000 & 0.0950 \\ ---- & ---- & ---- & ---- & ---- & ---- \\ -3.7625 & -3.2642 & -1.5663 & -0.8370 & -3.2687 & 0 \end{array}\right) \tag{18}$$

### *4.2. Reduction Stable Subsystem*

*4.2.1. Non-minimal system* If the system $G_s$ is stable and non-minimal, we can determine the minimal order of system $G_s$ by using "minreal.m" command of Matlab [11]. It eliminates uncontrollable or unobservable state in state-space models, The output has minimal order ($m$) and has the same response characteristics as the stable subsystem. The order of minimal system $G_s$ is $6^{th}$.

*4.2.2. Balanced Realized Model Reduction* The steps of reduction [10] are as follows:

**Step 1** Determining controllability and observability gramians. We solve the Lyapunov equations for discrete-time system, and obtain the controllability Gramian, $P$ and Observability Gramian, $Q$:

$$APA - P + BB = 0 \text{ and } A^TQA - Q + CC = 0$$

where in our case

$$P = \begin{pmatrix} 1.3940 & -0.2355 & 0.4021 & -0.8253 & 1.0620 & -0.0660 \\ -0.2355 & 0.2550 & -0.0109 & 0.0735 & -0.0881 & 0.0035 \\ 0.4021 & -0.0109 & 0.8939 & -0.0909 & 0.1600 & -0.0063 \\ -0.8253 & 0.0735 & -0.0909 & 3.7160 & -0.3654 & 0.0122 \\ 1.0620 & -0.0881 & 0.1600 & -0.3654 & 4.5936 & -0.3927 \\ -0.0660 & 0.0035 & -0.0063 & 0.0122 & -0.3927 & 256.0000 \end{pmatrix} \quad (19)$$

$$Q = 10^4 \begin{pmatrix} 7.5501 & -0.1140 & 0.2893 & -0.1134 & -0.2496 & -0.0255 \\ -0.1140 & 0.0818 & -0.0799 & 0.0310 & -0.0164 & 0.0008 \\ 0.2893 & -0.0799 & 0.1052 & -0.0514 & 0.0226 & -0.0016 \\ -0.1134 & 0.0310 & -0.0514 & 0.0345 & -0.0231 & 0.0009 \\ -0.2496 & -0.0164 & 0.0226 & -0.0231 & 0.0379 & 0.0003 \\ -0.0255 & 0.0008 & -0.0016 & 0.0009 & 0.0003 & 0.0001 \end{pmatrix} \quad (20)$$

**Step 2** Determination of hankel singular values (HSV). We obtain HSV of the system by finding the square root of the eigen values of product of $P$ and $Q$:

$$HSV = \sqrt{\lambda_i(PQ)} = \begin{pmatrix} 327.1102 & 45.1700 & 20.2872 & 9.9500 & 3.4579 & 0.8132 \end{pmatrix}$$

**Step 3** Performing singular value decomposition (SVD) of the gramians

$$P = U_p \Sigma_p V_p^T \text{ and } Q = U_q \Sigma_q V_q^T \quad (21)$$

**Step 4** Finding Truncation Matrices ($S_L$ and $S_R$). Let $V_R = U_p\sqrt{\Sigma_p}$, $V_L = U_q\sqrt{\Sigma_q}$ and $E = V_L^T V_R$, Perform singular value decomposition of E such that

$$E = U_E \Sigma_E V_E^T \quad (22)$$

Then, we get the truncation matrices are obtained as $S_L = V_L\ U_E\ V_E^{-1/2}$ In our problem, the matrices $S_L$ and $S_R$ are given by

$$S_L = \begin{pmatrix} 15.0779 & 4.9136 & -1.3724 & 0.7625 & -0.3531 & 0.0465 \\ -0.3677 & 2.3781 & -4.0138 & 4.0882 & -2.6846 & 0.3729 \\ 0.7671 & -3.5190 & 3.7912 & -0.7231 & -0.9586 & 0.1790 \\ -0.3449 & 2.5361 & -0.4330 & -1.0507 & -0.3756 & 0.1085 \\ -0.3961 & -2.5627 & -1.1332 & -0.7168 & -0.0245 & 0.0957 \\ -0.0529 & 0.0289 & 0.0272 & 0.0317 & 0.0422 & 0.0512 \end{pmatrix}$$

and $S_R = V_R\ U_E\ V_E^{-1/2}$

$$S_R = \begin{pmatrix} 0.0651 & 0.0013 & -0.0144 & -0.0087 & 0.0103 & 0.0709 \\ -0.0111 & -0.0022 & -0.0332 & 0.0861 & -0.1782 & 0.1007 \\ 0.0205 & -0.0407 & 0.1350 & -0.0406 & -0.2717 & 0.2210 \\ -0.0418 & 0.1506 & -0.0346 & -0.3925 & -0.3485 & 0.4200 \\ 0.0443 & -0.1830 & -0.2738 & -0.2600 & -0.0821 & 0.5227 \\ -0.0440 & 0.1801 & 0.3677 & 0.8396 & 3.1280 & 16.0818 \end{pmatrix}$$

**Step 5** Obtaining Balanced Realized Minimal Model. BT model of $G_s$ using the truncation matrices $S_L$ and $S_R$ is as follows,

$$G_d = \left(\begin{array}{c|c} A_{bal} & B_{bal} \\ --- & --- \\ C_{bal} & D_{bal} \end{array}\right) + \left(\begin{array}{c|c} S_L' A S_R & S_L' B \\ --- & --- \\ C S_R & D \end{array}\right) \tag{23}$$

In our case, the balanced realized minimal model is given by:

$$G_{bal} = \left(\begin{array}{cccccc|c} 0.9988 & -0.0016 & 0.0012 & -0.0015 & 0.0018 & -0.0022 & -0.8988 \\ 0.0016 & 0.9971 & 0.0095 & -0.0048 & 0.0082 & -0.0091 & 0.5083 \\ 0.0012 & -0.0095 & 0.9944 & 0.0234 & -0.0133 & 0.0190 & 0.4663 \\ 0.0015 & -0.0048 & -0.0234 & 0.9850 & 0.0538 & -0.0414 & 0.5221 \\ 0.0018 & -0.0082 & -0.0133 & -0.0538 & 0.9217 & 0.1819 & 0.6760 \\ 0.0022 & -0.0091 & -0.0190 & -0.0414 & -0.1819 & -0.0335 & 0.8174 \\ ---- & ---- & ---- & ---- & ---- & ---- & ---- \\ -0.8988 & -0.5083 & 0.4663 & -0.5221 & 0.6760 & -0.8174 & 0 \end{array}\right)$$

**Step 6** Obtaining $r^{th}$ order BT Model. We select the reduced order number $r = 1$ of the system $G_s$ on the basis of higher magnitudes of HSV. So, we get $1^{st}$ order BT model as

$$G_{sr-bt} = \left(\begin{array}{c|c} 0.9988 & -0.8988 \\ ----- & ----- \\ -0.8988 & 0 \end{array}\right) \tag{24}$$

*4.3. Reduced order model for original unstable system*

The balanced truncated reduced model of unstable system is obtained by adding the unstable subsystem into the reduced stable subsystem:

$$G_{r-bt} = G_{sr-bt} + G_u$$

In our case, it is given by

$$G_{r-bt} = \left(\begin{array}{cccccc|c} 0.9988 & 0 & 0 & 0 & 0 & 0 & -0.8988 \\ 0 & 1.0323 & -0.0620 & 0.0321 & -0.0174 & 0.0009 & -0.2121 \\ 0 & 0 & 1.0323 & -0.0342 & 0.0185 & -0.0006 & 0.2363 \\ 0 & 0 & 0 & 1.0323 & -0.0368 & -0.0009 & -0.5164 \\ 0 & 0 & 0 & 0 & 1.0323 & -0.0014 & 0.9656 \\ 0 & 0 & 0 & 0 & 0 & 1.0000 & 0.0950 \\ ---- & ---- & ---- & ---- & ---- & ---- & ---- \\ -0.8988 & 3.7625 & 3.2642 & -1.5663 & -0.8370 & -3.2687 & 0 \end{array}\right) \tag{25}$$

## 5. Conclusion

In this work, we have developed a model reduction for unstable systems. Even our procedure works for non-minimal systems. Furthermore we have successfully obtained a reduced model for shallow water equations. We can use the results in this paper to design a low-dimensional controller, observer and some other applications. In this approach, we have not analyzed the procedure formally. We are planning to work on such issue in the future.

**Acknowledgments**

This work has been supported by:

- Penelitian Doktor Baru number 31040/IT2.11/PN.08/2016 with the title of Konstruksi Algoritma Filter Kalman Tereduksi Pada Sistem Tidak Stabil Dan Aplikasinya Pada Aliran Air Sungai;
- Penelitian Unggulan Perguruan Tinggi year 2017 with the title of Estimasi Ketinggian dan Kecepatan Aliran Sungai dengan Filter Kalman Tereduksi Metode Residual.

## References

[1] Djukić S and Sarić A 2012 *Serbian journal of electrical engineering* **9** 131–169
[2] Sari Y, Arif D and Adzkiya D 2016 *International Conference on Mathematics: Pure, Applied and Computation*
[3] Kartika D, Arif D and Adzkiya D 2016 *International Conference on Mathematics: Pure, Applied and Computation*
[4] Arif D, Widodo, Apriliani E and Salmah 2011 *3rd International Conferences and Workshops on Basic and Applied Sciences* p M007
[5] Arif D, Widodo, Salmah and Apriliani E 2012 *Proceedings of the 8th International Conference on Mathematics, Statistics and Its Applications*
[6] Verlaan M 1998 *Efficient Kalman Filtering Algorithms for Hydrodynamic Models* Ph.D. thesis Delft University of Technology
[7] Arif D, Widodo, Salmah and Apriliani E 2014 *International Journal of Control and Automation* **7** 257–270
[8] Kumar D, Tiwari J and Nagar S 2011 *International Journal of Engineering Science and Technology* **3**
[9] Kumar D, Tiwari J and Nagar S 2011 *International Journal of Engineering Science and Technology* **3** 2825–2831
[10] Singh S, Nagar S and Pal J 2006 *IEEE International Conference on Industrial Technology* pp 1522–1527
[11] MATLAB and Robust Control Toolbox *Release 2010a* (Natick, Massachusetts: The MathWorks Inc.)

IOP Conf. Series: Journal of Physics: Conf. Series **855** (2017) 012030 doi:10.1088/1742-6596/855/1/012030

# Conjecturing and Generalization Process on The Structural Development

**Khomsatun Ni'mah[1], Purwanto[2], Edy Bambang Irawan[2], Erry Hidayanto[2]**
[1]Student of Doctoral Program of State University of Malang – Indonesia
[2]Lecturer of State University of Malang – Indonesia

e-mail: denayu.khomsah@gmail.com, purwanto.fmipa@um.ac.id, edy.bambang.fmipa@um.ac.id, erry.hidayanto.fmipa@um.ac.id

**ABSTRACT.** This study aims to describe how conjecturing process and generalization process of structural development to thirty children in middle school at grade 8 in solving problems of patterns. Processing of the data in this study uses qualitative data analysis techniques. The analyzed data is the data obtained through direct observation technique, documentation, and interviews. This study based on research studies Mulligan et al (2012) which resulted in a five - structural development stage, namely prestructural, emergent, partial, structural, and advance. From the analysis of the data in this study found there are two phenomena that is conjecturing and generalization process are related. During the conjecturing process, the childrens appropriately in making hypothesis of patterns problem through two phases, which are numerically and symbolically. Whereas during the generalization of process, the childrens able to related rule of pattern on conjecturing process to another context.

## 1. BACKGROUND

Patterning is critical to the abstraction of mathematical ideas and relationships, and the development of mathematical reasoning in young children (English, 2004; Mulligan, Prescott & Mitchelmore, 2004; Waters, 2004). The integration of patterning in early mathematics learning can promote the development of mathematical modelling, representation and abstraction of mathematical ideas It seems advantageous then, that initiatives in mathematics curricula and assessment in Australia and at international level are promoting the development of early mathematical patterning and reasoning (Clements, 2004; Doig, 2005).

In mathematics education there is a general consensus that patterning involves "observing, representing and investigating patterns and relationships in social and physical phenomena and between mathematical objects themselves" (Australian Education Council, 1991, p 4). Mathematical patterns encountered in school range from number sequences and spatial arrays to algebraic generalisations and geometrical theorems A pattern may be defined as a numerical or spatial regularity, and the relationship between the various components of a pattern constitute its structure Pattern and structure are thus at the heart of school mathematics Early algebraic thinking in the elementary school may involve the development of thinking skills where the letter symbol is assigned to describe patterns. Other activities, for example analysing relationships among quantities, noticing structure, studying change, generalising, problem solving, modelling, justifying, proving, and predicting, can be engaged without using the letter symbol (Keiran, 2004). The conceptual development of 'pattern' in a variety of situations engages these processes.

Pattern generalization is the activity to make a general rule pattern based on specific examples. Specific examples can be graphic, numeric, verbal and algebraic pattern (Janvier, 1987). In particular, patterns are seen by some researchers as path to transition to algebra because they are a fundamental step to build the generalization that is the essence of mathematics (Zazkis and Lijedahl, 2002). In generalizing patterns, it is not enough to declare a general rule and order patterns verbally but must also state the general rule of pattern with symbol.

International studies about the approach to algebra that involves the processes of generalization concern the study of: patterns, algebraically representable functional correspondences between pairs of variables, equations, structural aspects of arithmetic operations, simple numerical theorems (formulation of conjectures and their justification).

A conjecture is a logical statement, but whose truth has not been confirmed (Cañadas and Castro, 2005; Ontario Ministry of Education, 2005; Mason et al., 2010; Reid, 2002). The process of producing conjecture is called the conjecturing process. Conjecturing process is the mental activity used in building a conjecture based on one's knowledge (Sutarto et al., 2015). Mental activity in building conjecture is a process that occurs in the mind that can be seen through the behaviour of students in problem solving.

Associated with the conjecturing in the school mathematics is a process of building new knowledge for the students. This is in accordance with the statement by Lee and Sriraman (2010) that conjecturing in mathematical problem solving in learning process is to construct new knowledge for the students according to the existing knowledge that students already have.

**Indicators of Conjecturing Process**

The most common strategy used in organizing cases is registering or sorting the data; (3) Searching for and predicting patterns is one's activity when observing repetitive and regular situations, one naturally imagines that the pattern may apply to the next cases of unknown; (4) Formulating a conjecture is making a statement about all possible cases, based on empirical facts, but with an element of doubt or in other words conjecture is a statement that has not been validated; (5) Validating the conjecture is the activity performed to justify the conjecture generated based on specific cases but not in general; (6) Generalizing the conjecture is an activity on changing confidence related to the generated conjecture, that the conjecture is valid in general; (7) Justifying generalization is the activity performed to justify generalizations. Justifying the generalization involves giving reasons that explain conjecture with the intention of convincing others that the resulting conjecture is correct.

**Table 1. Stages of Conjecturing Process**

| Stages of Conjecturing Process | Descriptions |
|---|---|
| Observing the case | Initial activities against specific cases of problems:<br>• Observing and counting the number of green square, white square, green or white square<br>• Observing and counting the number of green rectangle |
| Organizing the case | Activities that involve the use of strategies that facilitate work in certain cases:<br>• Making a list or table to associate symbol 1x1 with the image number 1, symbol 2x2 with the image number 2, symbol 3x3 with the image number 3, and so on |
| Searching for and predicting the pattern | Activities of observing certain objects either organized or disorganized and thinking about the next object that is not yet known:<br>• Calculating the difference between a square to the first, second, third, fourth and fifth<br>• Calculating the difference between a white square and green square to the first, second, third, fourth and fifth |
| Formulating the conjecture | Making a statement about all possible cases, based on empirical facts but have not been validated:<br>• Declaring the n formula of green square as 4n+4, white square as $n^2$, and the formula for n green or white square as $(n+2)^2$ |

| | |
|---|---|
| Validating the conjecture | Activities carried out to establish the truth of the conjecture produced by certain new cases but not in general:<br>• To sketch the next object that represents the next pattern to establish the truth of the conjecture generated |
| Generalizing the conjecture | Changes related to the confidence of the generated conjecture, that the conjecture is valid in general:<br>• Believing the formula of n green square as 4n+4, white square as $n^2$, and the formula for n green or white square as $(n+2)^2$ |
| Justifying the generalization | Giving reason that explain the generalization is correct:<br>• Justifying generalization based on specific cases |

**Indicators of Generalization Process**

Successful strategies seemed to proceed through the following sequence of stages: a direct modelling stage, the stage of identification of a pattern, the stage of proof testing of the pattern, and the final stage for finding a rule for the general case (Dindyal, 2007).

The direct modelling stage involved the focus students actually using strategies such as drawing, counting, and writing down the first few cases systematically. For example, drawing five white arrangement of square tiles measuring nxn (n = 1, 2, 3, 4, 5) which is surrounded by tiles green square measuring 1x1 tiles green-colored tiles are connected directly with its sides and its vertices. Counting the number of white square for the n formula, the number of green square for the n formula, the number of white or green square for the n formula, and the number of green rectangle for the n formula.

The second stage (identification of the pattern) was the stage during which the students were actually able to identify some useful pattern. Which pattern one chooses depends on the particular aspect of the pattern that one wishes to observe (Phillips, 1993), and this depended considerably on the students' systematic counting, drawing, or writing/recording from the first stage. For examples, identified the pattern of number of white square as 1, 4, 9, 16, 25. The pattern of number of green square as 8, 12, 16, 20, 24. The pattern of number of white or green square as 9, 16, 25, 36, 49. For the first pattern, the number of white square was square number. For the second pattern, the number of green square was multiple of 4. For the third pattern, the number of green or white square was square number.

In the third stage (proof testing of the pattern), the successful focus students tested their conjectures about the patterns by using a particular case beyond the range for them to model directly. For examples, the student were asked about larger grids where counting the number of green rectangle on strategies were not very practical that these students looked for alternative strategies. Counting the number of green rectangle where 1x2, 1x3, 1x4, ... 1x(n+2) grid for every 1x1, 2x2, 3x3, 4x4, and 5x5 grid.

In the final stage (finding a rule for the general case), the students had to come up with a generalisation. Swafford and Langrall (2000) had claimed that the generalisation of a tiling problem situation might be presented symbolically. For the symbolic, the student wrote the number of green square as 4n+4 which similar to 8, 12, 16, 25, 36 strategy for the 1x1, 2x2, 3x3, 4x4, and 5x5 grid. For the second pattern, the student wrote the number of green rectangle (where 1x2, 1x3, 1x4, ..., 1x(n+2) grid) as $4(n+2)\left(\frac{n+1}{2}\right)$ which similar to 4+8, 4+8+12, 4+8+12+16, 4+8+12+16+20, 4+8+12+16+20+24 strategy for the 1x1, 2x2, 3x3, 4x4, and 5x5 grid.

## 2. METHODOLOGY

### Subject

Researchers asked 30 students at junior high school (SMPN 1 Pace) to complete tiling problem. After experiencing saturation data in the subject, seven students able to make conjecture simbolically and then the only one student able to make generalization simbolically.

### Instrument

There are two types of instruments used, main and auxiliary instruments. The main instrument is the researchers themselves who act as planners, data collectors, data analysts, interpreters, and reporters of research results. The auxiliary instrument used in this study is a Tiling Problem (TP) and interviews. The problem given aims to obtain a description of the process of conjecturing and generalization of students, while the interview used was unstructured interview. The TP is presented in Figure 1.

### Procedure

In the first stage, students completed TP. In the second stage, the researchers conducted work-based interviews to understand the process of conjecturing and generalization and then the researchers recorded them by using a handy cam.

### Data analysis

This study is a qualitative research with descriptive exploratory approach. At the data analysis stage, the activities conducted by researchers were (1) transcribing the data obtained from interviews, (2) data reduction, including explaining, choosing principal matters, focusing on important things, removing the unnecessary ones, and organizing raw data obtained from the field, (4) describing the conjecturing and generalization process in the solving of tiling problem, and (5) conclusion.

## 3. RESULTS AND DISCUSSION

Based on the results of the analysis of the TP answer sheets and interviews, we obtained data on the conjecturing and generalization process undertaken by students in solving tiling problem. Of the 10 students who did conjecturing process and produced a formula or symbolic general rule, 2 students did generalization process and produced a new combine formula which resulted from conjecturing process and other formula symbolically. The results of the process of conjecturing are presented in Table 1. After experiencing saturation in the process of data collection, there were 10 students that did the conjecturing and then the only 2 student that did the generalization. Of the 2 students, we chose one subject that did conjecturing and generalization process, that is the subject S1.

**Table 2. Prosentase of Ability to Make Conjecturing and Generization of Pattern**

| Unable to make conjecturing or generalization of pattern | Able to make | |
|---|---|---|
| | Conjecturing Pattern | Generalization Pattern |
| 18 | 10 | 2 |
| 60% | 33,3% | 6,7% |

### Conjecturing Process

IOP Conf. Series: Journal of Physics: Conf. Series **855** (2017) 012030 doi:10.1088/1742-6596/855/1/012030

At the stage of action, subject S1 realized that the images (first, second, third, fourth and fifth image) formed a pattern. S1 observed cases by observing and counting the number of square section with distinguishing green square and white square of the first, second, third, fourth and fifth image. Here are excerpts of the interview and the work of S1.

I : What did you think when reading this issue?
S1 : The pattern, off course.
I : What kind of patterns?
S1 : The pattern of addition, the pattern of multiples of 4, the pattern of exponential number

It is always 4, namely from to the first, second, third, fourth, and fifth and onwards (pointing to the results of the work). Based on the number of squares on the first, second, third, fourth and fifth images, subject S1 organized the case by writing a symbol indicating a similar pattern by circling the first, second, third, fourth and fifth images. This was confirmed by the transcript of the interview S1 and the work of S1 in Figure 1.

S1 : I see a pattern. many green tiles are always multiples of 4, while many white tiles are squared numbers

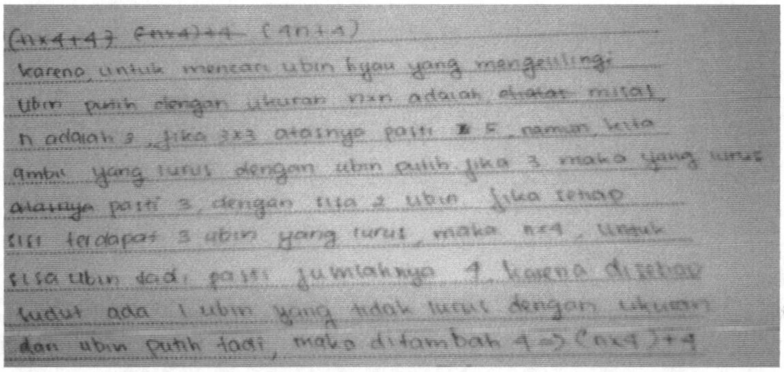
(4n+4)
karena untuk mencari ubin hijau yang mengelilingi
ubin putih dengan ukuran nxn adalah misal
n adalah 3, jika 3x3 atasnya pasti 5, namun kita
ambil yang lurus dengan ubin putih jika 3 maka yang lurus
atasnya pasti 3, dengan sisa 2 ubin jika setiap
sisi terdapat 3 ubin yang lurus, maka nx4, untuk
sisa ubin tadi pasti jumlahnya 4, karena disetiap
sudut ada 1 ubin yang tidak lurus dengan ukuran
dan ubin putih tadi, maka ditambah 4 => (nx4)+4

Figure 1. the work of S1 to seek green square tiles that surround the white tiles

Translate in English:

Because to find the green tiles that surround the white tiles with size nxn is suppose that n is 2, 3x3 if it would have been 5, but we take that straight to the white tile if 3 then straight definite 3, with the remaining 2 tiles. If each side there are three tiles straight then nx4. for the rest of the tiles definite number 4, because in every corner there is one tile that is not proportional to the size and white tiles just now, then increase 4(nx4)+4

At the process stage, S1 internalized the action by finding and predicting patterns. The activities were carried out by calculating number of green square of the first, second, third, fourth and fifth image, that was multiple of 4, as well as thinking about the fourth, fifth, sixth image and so on.

At the object stage, S1 encapsulated process to formulate a conjecture by connecting between the first, second, third, fourth and fifth image to the difference and the number of the green square of the image given. For example, for the first image, the student wrote eight for the first image, because there were four square which 1x1 grid, wrote 4 for the number of square which on its four corner points. For the second image, S1 wrote twelve, because there were four square which 1x2 grid, wrote 4 for the number of square which on its four corner points. The third image, S1 wrote sixteen, The fourth image, S1 wrote twenty and the fifth image, S1 wrote twenty. Furthermore, S1 formulated the general formula for *n* image, that was 4n+4) . This was confirmed by the results of the work of S1

(Figure 4) and interview transcript S1. S1 validated conjecture by calculating conformity of the formula with the number of green square of the sixth image. The formula obtained was correct as the number of green square at the fifth image plus 4 equal to 24. This is evidenced by the following interview transcript of S1 and the work of S1.

**S1** : So for the first image I created a formula; for example, the first image was multiplied by the addition, and I got 4, then right, up, left which parallel, down was fifth, and I added all, and the result was 4. Once multiplied, then added with four the green square which on its four corner points.

**S1** : I'm sure, I tried to count in the sixth image, the result is the same as for the fifth image, 24 plus 4 is 28.

At the scheme stage, S1 believed the general formula 4n+4 produced was correct, based on the results of the validation. By believing that formula, S1 was generalizing. Next on justify generalization stage, S1 pinpointed the specific examples as done in validating the conjecture in order to convince others that the conjecture was generated correctly. This can be shown from the following interview excerpts.

**I** : Ok. How did you explain to others that the resulting formula is true (pointing to the work of S1).

**S1** : I'm going to show you an example.

**I** : Examples like what?

**S1** : Suppose the first image is true, the second image is true, the third image is true, fourth image is true, and fifth image is true, then is true and so on.

**Generalization Process**

In Tiling Problems focus students had to identify a general pattern starting from few specific cases. It was expected that reasoning inductively from a few cases the focus students would be able to generate a general rule or formula. Successful strategies seemed to proceed through the following sequence of stages: a direct modelling stage, the stage of identification of a pattern, the stage of proof testing of the pattern, and the final stage for finding a rule for the general case.

The direct modelling stage involved the focus students actually using strategies such as drawing and counting. For example, in Tiling Problem most of the students drew a white squares in the 1×1 grid, 2x2 grid, 3x3 grid, 4x4 grid and 5x5 grid which surrounded and connected directly with green tiles and measuring 1x1 grid, and then counted the number of white squares in the 1×1 grid, 2x2 grid, 3x3 grid, 4x4 grid and 5x5 grid and again counted the number of green rectangle which measured 1x2, 1x3, 1x4, ... 1x(n+2) of white square in the 1x1 grid, 2x2 grid, 3x3 grid, 4x4 grid and 5x5 grid before identifying any pattern. For example, the number of green rectangle which measured 1x2 and 1x3 in the white square which 1x1 grid. The number of green rectangle which measured 1x2, 1x3 and 1x4 in the white square which 2x2 grid.

The second stage was the stage during which the students were actually able to identify some useful pattern. Which pattern one chooses depends on the particular aspect of the pattern that one wishes to observe (Phillips, 1993), and this depended considerably on the students' systematic drawing and counting from the first stage. For example, in Tiling Problem for the number of white square in 5×5 grid, some students identified emergent some pattern: (i) the pattern of the number white square as 1, 4, 9, 16, 25 and (ii) the pattern of the number green square as 8, 12, 16, 20, 24, (iii) the pattern of the number green rectangle as 4 + 8 + 12 + 16 + ... + 4(n+1). The generalisation was $n^2$ for the first pattern, multiple of 4 for the second pattern which was later simplified to $4n$ for $n$ = 2, 3, 4, 5,... etc. For the third pattern, can be modified with 4 (1 + 2 + 3 + ... + (n+1)), and then

IOP Conf. Series: Journal of Physics: Conf. Series **855** (2017) 012030 doi:10.1088/1742-6596/855/1/012030

$4\,(n+2)\left(\dfrac{n+1}{2}\right)$, thus a systematic way of counting the number of white squares, green square and green rectangle helped the students to generalise. The generalisation was fairly easy when there were sufficient examples to make the pattern quite evident. In this problem, the successful students were able to identify a connection between the number of white square and green square, connection between the number of green rectangle and green square. The systematic way of recording the number of green rectangle and green square in a table helped the successful students to identify a pattern in the results. In problems where this was not the case, the students had more difficulties in coming up with a useful pattern. For example, the problem of determining the number of rectangular tiles are colored green in the n pattern, the most students had to come up with a generalisation based on only one initial case. This proved to be hard for the students. Lee (1996) has pointed out that the problem for many students is not the inability to see a pattern but the inability to see an algebraically useful pattern.

In the third stage, the successful focus students tested their conjectures about the patterns by using a particular case beyond the range for them to model directly. For example, in the students were asked to find the number of green rectangle in the n pattern. Generally, the students who were able to attain this stage were able to get to the algebraic generalisation later. "Counting on" was a common strategy for some of the focus students to reach a solution for the pattern to - n, but this was not very helpful as an overall strategy. It was when these students were asked about larger grids such as 7×7 grid where counting on strategies were not very practical that these students looked for alternative strategies. So, they used their earlier patterns such as 4(1+2+3+ .. +(7+1)) = 4 (1+2+3+ .. +8), in this stage, the student identified there are four pairs which sum of the two number are the same numerically, there are 1+8 = 2+7 = 3+6 = 4+5. It is differences if larger grids such as 10×10 grid where counting on strategies were not very practical that these students looked for alternative strategies. So, they used their earlier patterns such as 4(1+2+3+ .. +(10+1)) = 4 (1+2+3+ .. +11), in this stage, the student identified there are five pairs numerically which sum of the two number are the same, there are 1+11 = 2+10 = 3+9 = 4+8 = 5+7 and then number 6 which reminder. They used the patterns they had identified to do so.

In the final stage, the students had to come up with a generalisation. Swafford and Langrall (2000) had claimed that the generalisation of a problem situation might be presented verbally or symbolically. In the problems that were used in this study, the focus students avoided a verbal generalisation and all of them tried to give symbolic generalisations. For the symbolic, this involved constructing an algebraic relation for the pattern they had noticed. Their success in the first three stages of the solution process helped them to come to the right conclusion. The students used the pattern that they had identified earlier to come up with the generalisation. For example, S1 from school SMPN 1 Pace, grades 7 wrote $(4n+4)+4(n+2)\left(\dfrac{n+1}{2}\right)$ which was similar to her 16+4(5)(2) strategy for n = 3.

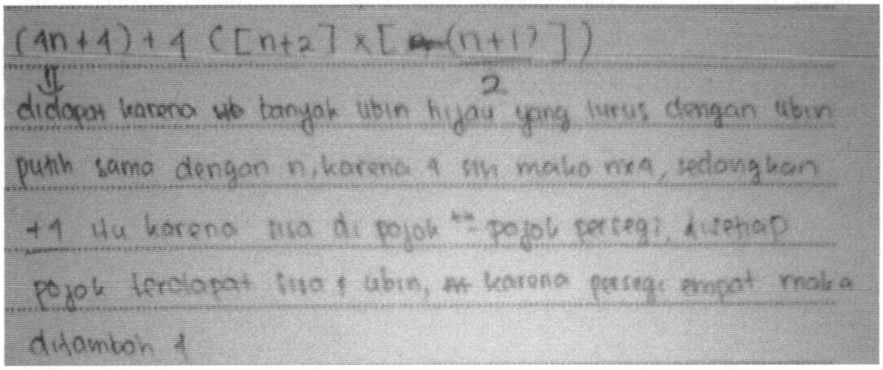

Figure 2. The work of S1 to seek green tiles are rectangular

Translate in English:

$$(4n+4)+4(n+2)\left(\frac{n+1}{2}\right)$$

$(4n+4)$, obtained because many straight green tiles with white tiles equal to n, because four sides then nx4, while + 4 it for the rest of the corners of the square at each corner are the remaining 4 tiles, as rectangular then plus 4.

Based on figure 3, S1 find a general rule of the number of green rectangle which 1x1 grid, (4n+4). 4n is obtained from the number of rectangular tiles measuring 1x2 that is located right on the top, bottom, right and left white square tiles as much as 4, while n represents the size of the white square tiles. while coupled with 4, was obtained from the number of green tiles are located at the ends. More detail, description for answer S1 in generalization problem in Figure 3.

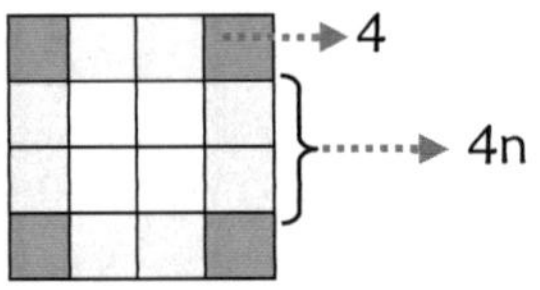


Figure 3. General rule of the number of green rectangle which 1x1 grid is (4n+4)

Some of the difficulties encountered by the students, such as producing variables on their own, and writing down the relations algebraically, hampered the students' progress. For instance, the students found it very difficult to come up with a symbolic generalisation for this problem. Generally, the students were able to fill up the table, but their search was for a linear symbolic relationship. Most of them were able to identify a recursive relationship in the table but only S1, from school SMPN 1 Pace, gave an explicit recursive formula. His formula was $(4n+4)+4(n+2)\left(\frac{n+1}{2}\right)$, where $(4n+4)$ stood for the number of green rectangle which 1x1 grid, and then $4(n+2)\left(\frac{n+1}{2}\right)$ stood for the number of green rectangle which 1x2, 1x3, 1x4,... 1x(n+2) grid. This was confirmed by the transcript of the work of S1 to determine how she get this general formula in Figure 4.

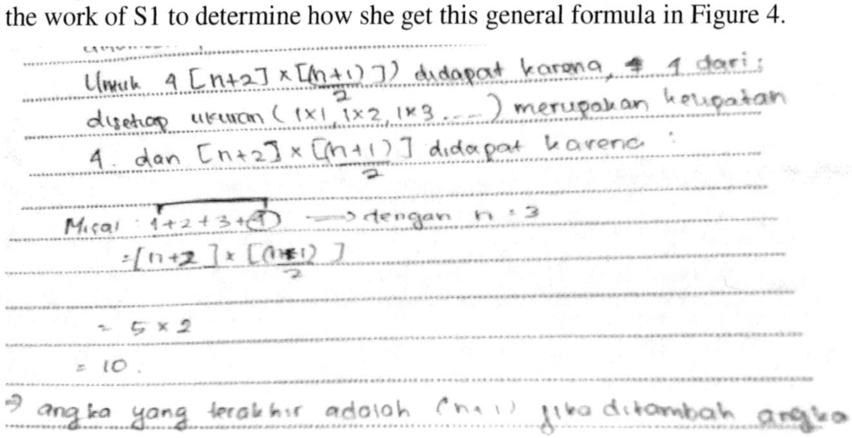

Figure 4. The work of S1 to seek green tiles are rectangular is $(n+2)\left(\frac{n+1}{2}\right)$

Translate in English:

For $4(n+2)\left(\frac{n+1}{2}\right)$ comes from because 4 from: every measured (1x1, 1x2, 1x3,...) which multiple by 4 and then $(n+2)\left(\frac{n+1}{2}\right)$ comes from because:

For example:

$= 1 + 2 + 3 + 4$ with $n = 3$

$= (n+2)\left(\frac{n+1}{2}\right)$

$= 5 \times 2$

$= 10$

The last number is (n+2) if addition with number

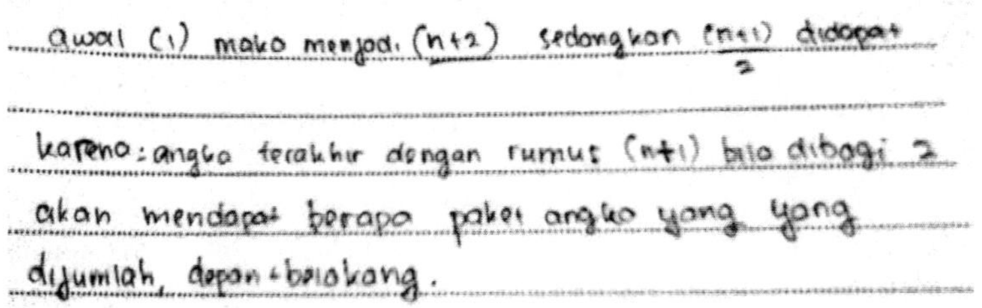


Figure 5. The work of S1 to seek green tiles are rectangular is $(n+2)\left(\frac{n+1}{2}\right)$

Translate in English:

The first (1) then (n+2). While $\left(\frac{n+1}{2}\right)$ comes from because the last number with formula (n+1) if divided y 2 will get the number which adding, the first number adding the last number.

Based on the Figure 4 and 5 above, S1 get formula $(n+2)\left(\frac{n+1}{2}\right)$, where (n+2) it is mean that how much the sum of pairs two number (n + 2 = 3 + 2 = 5), and then S1 get general formula $\left(\frac{n+1}{2}\right)$ comes from how many pairs number (the fisrt number adding to the last number, the second number adding to the third number) which result of adding the two pairs number is same. There are two pairs number $\left(\frac{3+1}{2}=2\right)$ added, 1+4 = 2+3 which the result is same.

However, she was able to give an explicit symbolic representation of the Tiling Problem in generalisation pattern. Some authors caution that, in their attempt to write symbolic representations, students often focus on inappropriate aspects of a number pattern – particularly the recursive relationship between successive terms in a sequence (MacGregor & Stacey, 1993; Orton & Orton, 1994). Thus, in this problem, it might be possible that the students' focus on the recursive relationship was responsible for their inability to produce an explicit generalisation. Even this problem was problematic for some of the students. The students had difficulties in coming up with the generalisation about the combine between number of green rectangle which 1x1 grid and number of green rectangle which 1x2, 1x3,..., 1x(n+2) grid because inability to determine the number of green rectangles. It seemed that a limited number of initial cases might not be enough for the students to find a pattern and hence a generalisation from the pattern, although Dreyfus (1991) had claimed that sometimes it is better to abstract from a single case.

## 4. CONCLUSIONS AND IMPLICATIONS

The conjecturing process in the solving of tiling problem is composed of conjecturing process based by contrast. The conjecturing process based on contras happens at the action stage, in which the subject builds a conjecture by observing and counting the number of square separately based on differentiating a green square and a white square; and at the process stage, object and scheme is done perfectly.

The identification of a useful pattern by the students was a significant factor in their successful symbolic generalisation, which seemed to proceed in four sequential stages. However, the students had difficulties with symbolic generalisations. The students generally avoided verbalising their generalisations. Students with a weaker background in algebra, more difficulties generalising compared to the other students. Even the students with a stronger background in algebra displayed some misconceptions in handling algebraic expressions. In this study, the tiling problems had some connections to geometry, which may have added to the students' difficulties. In future studies, a broader range of problems with similar generalising activities may provide a more complete image.

## REFERENCES

Australian Education Council (1991). *A national statement on mathematics for Australian schools* Melbourne: Curriculum Corporation

Cañadas MC, Castro E (2005). A proposal of categorisation for analysing inductive reasoning. In M. Bosch (Ed.), Proceedings of the CERME 4 International Conference. Sant Feliu de Guíxols, Spain. pp. 401-408

Clements, D H (2004). Major themes and recommendations In D H Clements & J Sarama (Eds ) *Engaging young children in mathematics: Standards for early childhood mathematics education* Mahwah, NJ: Lawrence Erlbaum Associates

Dindyal, J.(2007). High School Students' Use of Patterns and Generalisations. Proceedings of the 30th annual conference of the Mathematics Education Research Group of Australasia. Australia: MERGA.

Doig, B (2005). Developing formal mathematical assessment for 4-to 8-year-olds *Mathematics Education Research Journal, 16*, 100-119

Dreyfus, T. (1991). Advanced mathematical thinking processes. In D. Tall (Ed.), *Advanced mathematical thinking* (pp. 25-41). Boston: Kluwer Academic Publishers.

English, L. D. (2004). *Mathematical and Analogical Reasoning of Young Learning*. Mahwah, NJ: Lawrence Erlbaum.

Janvier C (1987). Translation processes in mathematics education. In C. Janvier (Ed.), Problems of representation in the teaching and learning of mathematics. Hillsdale, New Jersey:LEA. pp. 27-32.

Kieran, C (2004, July). The development of algebraic thinking and symbolization Paper presented to the PME Research Session, *10th International Congress on Mathematical Education,* Danish Technical University, Copenhagen

Lee, L. (1996). An initiation into algebraic culture through generalization activities. In N. Bednarz, C. Kieran, & L. Lee (Eds.), *Approaches to algebra: Perspectives for research and teaching* (pp. 87-106). Boston: Kluwer Academic Publishers.

Lee KH, Sriraman B (2010). Conjecturing via reconceived classical analogy. Educ. Stud. Math. 76(2):123-140.

MacGregor, M., & Stacey, K. (1993). Cognitive models underlying students' formulation of simple linear equations. *Journal for Research in Mathematics Education, 24*(3), 217-232.

Mason J, Burton L, Stacey K (2010). Thingking mathematically second edition, England: Pearson Education Limited.

Mulligan, J. T., Prescott, A., & Mitchelmore, M. C. (2004). Children's development of structure in early mathematics. In Høines, M.J & Fuglestad, A. (Eds.), *Proceedings of the 28th PME International Conference*, 3, 393-401).

Ontario Ministry of Education (2005). The Ontario Curriculum: Grades 9 and 10 mathematics. Toronto, ON: Queen's Printer for Ontario.

Orton, A., & Orton, J. (1994). Students' perception and use of pattern and generalization. In J. P. da Ponte & J. F. Matos (Eds.), *Proceedings of the 18th annual conference of the International Group for the Psychology of Mathematics Education* (Vol. 3, pp. 404-414). Lisbon, Portugal: PME.

Phillips, E (1993). Issues surrounding algebra. In C. Lacampagne, W. Blair, & J. Kaput (Eds.), *The algebra initiative colloquium*,(Vol. 2, pp. 67-79). Washington, DC: US Department of Education.

Reid D (2002). Conjectures and refutations in grade 5 matematics. J. Res. Math. Educ. 33(1):5-29.

Sutarto Toto N, Subanji S (2015). Indicator of conjecturing process in a problem solving of the pattern generalization. Proceding International Conference on Educational Research and Development (ICERD), UNESA Surabaya. pp. 32-45.

Swafford, J. O., & Langrall, C. W. (2000). Grade 6 students' preinstructional use of equations to describe and represent problem situations. *Journal for Research in Mathematics Education, 31*(1), 89-112.

Waters, J (2004). Mathematical patterning in early childhood settings In I Putt, R Faragher & M Mclean (Eds ) *Mathematics Education for the Third Millennium: Towards 2010* (Proceedings of the 27th Annual Conference of the Mathematics Education Research Group of Australasia, Townsville, Vol 2, pp 565-572) Sydney: MERGA

Zazkis R, Liljedahl P (2002). Generalization of patterns: the tension between algebraic thinking and algebraic notation. Educ. Stud. Math. 49:379-402.

IOP Conf. Series: Journal of Physics: Conf. Series **855** (2017) 012031 doi:10.1088/1742-6596/855/1/012031

# Software Piracy Detection Model Using Ant Colony Optimization Algorithm

**Nor Astiqah Omar[1], Zeti Zuryani Mohd Zakuan[2] and Rizauddin Saian[1]**

[1]Faculty of Computer and Mathematical Sciences, Universiti Teknologi MARA, Malaysia
[2]Faculty of Law, Universiti Teknologi MARA, Malaysia

E-mail: astiqah93lavenda@ymail.com, zeti@perlis.uitm.edu.my and rizauddin@perlis.uitm.edu.my

**Abstract.** Internet enables information to be accessible anytime and anywhere. This scenario creates an environment whereby information can be easily copied. Easy access to the internet is one of the factors which contribute towards piracy in Malaysia as well as the rest of the world. According to a survey conducted by Compliance Gap BSA Global Software Survey in 2013 on software piracy, found out that 43 percent of the software installed on PCs around the world was not properly licensed, the commercial value of the unlicensed installations worldwide was reported to be $62.7 billion. Piracy can happen anywhere including universities. Malaysia as well as other countries in the world is faced with issues of piracy committed by the students in universities. Piracy in universities concern about acts of stealing intellectual property. It can be in the form of software piracy, music piracy, movies piracy and piracy of intellectual materials such as books, articles and journals. This scenario affected the owner of intellectual property as their property is in jeopardy. This study has developed a classification model for detecting software piracy. The model was developed using a swarm intelligence algorithm called the Ant Colony Optimization algorithm. The data for training was collected by a study conducted in Universiti Teknologi MARA (Perlis). Experimental results show that the model detection accuracy rate is better as compared to J48 algorithm.

## 1. Introduction

University is a sacred place to nurture students so that they will become a better person and later can contribute to the society. However this is not the case nowadays. Malaysia as well as other countries in the world is faced with issues of piracy committed by the students in universities. Easy access to the internet is one of the factors which contributes towards piracy in Malaysia as well as the rest of the world. Piracy in universities has gained much attention due to recent high profile lawsuits faced by the universities [1]. The Software Alliance in 2013 has conducted Compliance Gap BSA Global Software Survey on software piracy. It was found that unlicensed software usage continued to be a major problem in 2013. According to the survey, 43 percent of the software installed on PCs around the world was not properly licensed and the commercial value of the unlicensed installations worldwide was $62.7 billion. Universities become concerned about the issue as they might face negative publicity and also they may face legal actions.

Piracy in the university can be in the form of software piracy, music piracy, movies piracy and piracy of intellectual materials such as books, articles and journals piracy. Students in university play a critical role in piracy due to the advances in peer-to-peer file sharing programs which have made it easy for students to illegally download and share copyrighted content via campus network [2]. According to [1], piracy on campus has gained abundant attention due to recent high profile lawsuits against developers of file sharing software. Universities become concerned as they can face negative publicity along with the risk of large fines. Software piracy has been literally defined as "the unauthorized copying or distribution of copyrighted software" [3]. However, software piracy has no definite definition under the law. According to [2], software piracy is a form of copyright infringement which can happen in higher institutions where peers usually share information with each other. Thus, the act of software piracy can be dealt with under the provision of copyright infringement. For example, in Malaysia copyright works is governed by the Copyright Act 1987.

The Copyright Act 1987 provides protection for copyrightable works in Malaysia. Section 7(1) Copyright Act 1987 provides for works which are eligible for copyright which includes literary works; musical works; artistic works; films; sound recording; and broadcasts. Literary works according to Section 3 of the Copyright Act 1987 includes work in the form of computer program. Thus, software as a form of computer program is a copyrightable work and is protected under the Malaysian copyright law. These works according to Section 7(2) "shall be protected irrespective of their quality and the purpose for which they were created". This provision shows that the law emphasized on the importance of protecting one's work. Copyright infringement is defined under Section 36 Copyright Act 1987, which provides that copyright is infringed by a person when he does something with the work without the consent of the owner of the work. If software is pirated, the pirate according to Section 36 has infringed the copyright of the software. For any act of copyright infringement, Section 37(1) of the Copyright Act 1987 provides that it "shall be actionable at the suit of the owner of the copyright", which means that the owner of the copyright has the right to sue the infringer. The offences and punishments relating to copyright infringement are provided under Section 41 Copyright Act 1987 which is illustrated in table 1.

This phenomenon requires attention from academics and researchers as the act of software piracy is detrimental to students. By having the law in place, students who commit software piracy might be facing legal consequences by the manufacturer seeking to thwart the theft of their commercial products. If the students are later found guilty, they might be dismissed from the university. Even though software piracy is common among students, it is actually a serious matter. The punishment imposed on the students might be able to act as a deterrence so that students will not do the same thing when they venture into the working environment. This study which involves predicting software piracy tendency among students, is important because after predicting their tendency, prevention can be done. Prevention is important for the students so that they are more sensitive to the likelihood of conviction and punishment which may act as a deterrent against software piracy.

Thus, this study has developed a classification model for detecting software piracy. The model was developed using an implementation of the Ant Colony Optimization (ACO) algorithm [4, 5] for data classification called the Ant-Miner [6, 7, 8].

## 2. Methodology

Three phases involved in developing the classification model as depicted by figure 1. The data pre-processing step will convert the data in an appropriate for Ant-Miner. In the second step, Ant-Miner trained the data to produce the classification model. The last step validates the classification model using predictive accuracy.

The first phase is the Data Pre-Procesing. The data was extracted from a survey conducted among the students at UiTM Perlis. Figure 2 shows an example of a question in the questionnaire. Each question was labelled by a question number in the most left column. The next column represents the question itself. Finally, the rest of the columns represents a number

**Table 1.** Offences and punishments relating to copyright infringement.

| Offences | Punishment |
|---|---|
| Sales, hire, distribution, possession, exhibition, imports of any infringing copies | • First offence: a fine not less than RM2,000 and not more than RM 20,000 for each infringing copy or imprisonment for a term not exceeding 5 years or both.<br>• Subsequent offence, a fine not less than RM4,000 and not more than RM 40,000 for each infringing copy or imprisonment for a term not exceeding 10 years or both. |
| Contrivance of infringing copies | • First offence, a fine not less than RM4,000 and not more than RM40,000 for each contrivance or imprisonment for a term not exceeding 10 years or both.<br>• Subsequent offence, a fine not less than RM8,000 and not more than RM80,000 for each contrivance or imprisonment for a term not exceeding 20 years or both. |
| Circumvention, removal, alteration, distribution, import of the electronic works without authority | • First offence, a fine not more than RM250,000 or imprisonment not exceeding 5 years or both.<br>• Subsequent offence, a fine not less than RM500,000 or imprisonment not exceeding 10 years or both. |

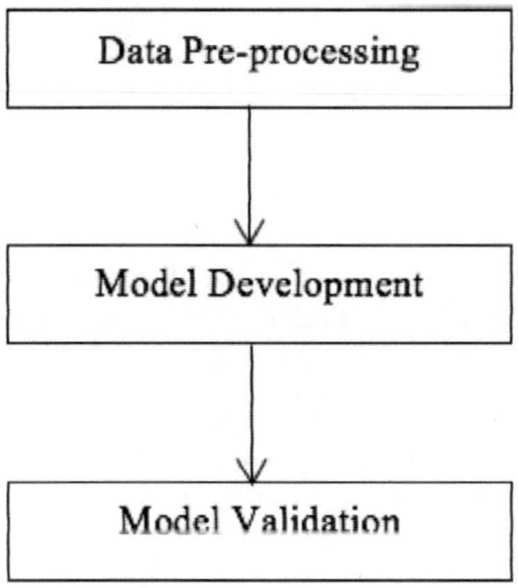


**Figure 1.** Research phases

of choices to be chosen by an individual which are strongly disagree (1), disagree (2), somewhat disagree (3), neither agree nor disagree (4), somewhat agree (5), agree (6) and strongly agree (7). The choices was divided into two parts. First, choices 1-4 reflects good attitude labeled as "Good" and second, choices 5-7 reflects bad attitude labeled as "Bad". Table 2 presents data

IOP Conf. Series: Journal of Physics: Conf. Series **855** (2017) 012031 doi:10.1088/1742-6596/855/1/012031

description of this study.

| | | | | | | | | |
|---|---|---|---|---|---|---|---|---|
| 2 | I have allowed people to copy digital contents I have purchased. | 1 | 2 | 3 | 4 | 5 | 6 | 7 |

**Figure 2.** Example of Question in the Questionnaire.

**Table 2.** Data Description.

| Attribute | Values |
|---|---|
| Gender | Male, Female |
| Attitude | Good, Bad |
| Income | High, Low |
| Ethics | Yes, No |
| Law | Yes, No |
| Perception | Will Piracy, Will Not Piracy |

The second phase is the model development. The classification model was developed using ACO. ACO is a metaheuristic algorithm inspired by the foraging behaviour of ant colonies introduced in the early 1990s [9, 5]. The purpose of ACO is to find the best solution using a set of artificial ants that communicates indirectly using an item called the pheromone. The amount of pheromone on one of the trails used by the majority ants will increase as time passed by and will decrease on the less used trails. As a consequent, since ants tend to follow the pheromones, all or most of the ants will finally converge to the best trail, with the high density of pheromones, which happened to be the shortest trail from the nest to the food source. However, there are a small number of ants still using the longer branch. This is the effect of "path exploration", where the density of the pheromone will not bias some of the ants. ACO has also been applied to various fields such as the travelling salesman problem, sequential ordering, flow shop scheduling and the graph colouring problems.

The first ACO implementation for data classification is called the Ant-Miner. Ant-Miner follows a divide-and-conquer approach to discover classification rules called the rule induction. Rule induction is a technique to extract set of rules from a data set [10, 11]. Each rule is in the form of IF <term1 AND term2 AND ... > THEN <class>. A rule contains two parts: the rule ancedent and the class. The rule antecedent is a combinations of one or many terms. Each term is a triple <attribute = value>, such as <Attitude = Good>. The rule consequent specifies the class predicted.

First, an ant starts with an empty rule. The ant will add one term at a time to its current partial rule. The rule addition will stop if adding any term to the rule would make the rule cover less than a pre-defined minimum number of cases, or all attributes have already been used by the ant for the rule antecendent.

Ant-Miner uses a heuristic measure as evaluation measure to fill in the antecedent part of the rule, by selecting the best term to be included into the partial rule. The heuristic measure (2) is the normalization of entropy measures between terms (1). The algorithm selects one best rule from a set of discovered rules, based on a quality measure using some fitness function.

$$H\left(W|A_i = V_{ij}\right) = -\sum_{w=1}^{k} P\left(w|A_i = V_{ij}\right) \cdot \log_2 P\left(w|A_i = V_{ij}\right) \tag{1}$$

where $W$ is the class attribute, $k$ is the number of classes and $P\left(w|A_i = V_{ij}\right)$ is the number of tuples for $A_i = V_{ij}$ with class $w$.

$$\eta_{ij} = \frac{\log_2 k - H\left(W|A_i = V_{ij}\right)}{\sum_{i=1}^{a} x_i \cdot \sum_{j=1}^{b_i} \left(\log_2 k - H\left(W|A_i = V_{ij}\right)\right)} \tag{2}$$

where $a$ is the total number of attributes, $x_i$ is set to one if atttribute $A_i$ was not yet used by the current ant or to zero otherwise, and $b_i$ is number of values for attribute $A_i$.

Ant-Miner has an exploration behaviour and exploitation behaviour. The exploration behaviour is contributed by a value called the pheromone level. Ant-miner uses a probability (3) that is proportional to the product of heuristic value and pheromone level for that term, to add terms to a rule. Dorigo in his book, called this transition rule, random proportional transition rule [4]. Detail algorithms are described in [7].

$$P_{ij} = \frac{\eta_{ij} \times \tau_{ij}}{\sum_{i=1}^{a} x_i \cdot \sum_{j=1}^{b_i} \left(\eta_{ij} \cdot \tau_{ij}\right)} \tag{3}$$

where $\tau_{ij}$ is the current amount of pheromone for term $A_i = V_{ij}$.

Table 3 shows an example of entropy, heuristic and probability calculation for each terms.

**Table 3.** An example of entropy, heuristic and probability calculation.

| Terms | Entropy | Heuristic | Probability |
|---|---|---|---|
| gender=male | 0.937 | 0.097 | 0.100 |
| gender=female | 0.943 | 0.087 | 0.101 |
| attitude=good | 0.936 | 0.099 | 0.100 |
| attitude=bad | 0.950 | 0.077 | 0.102 |
| income=high | 0.890 | 0.168 | 0.095 |
| income=low | 0.968 | 0.049 | 0.104 |
| ethics=yes | 0.931 | 0.106 | 0.100 |
| ethics=no | 0.946 | 0.082 | 0.101 |
| law=yes | 0.884 | 0.179 | 0.095 |
| law=no | 0.964 | 0.056 | 0.103 |

In this study, the number of ants was set to 100, the maximum uncovered cases = 10, rule for convergence = 10 and number of iterations = 100 are constant use but minimum case per rule will change. The second column in table 5 shows the result of accuracy for minimum case per rule from 6 to 10 with 10 folds cross validation. The percentage shows that the interval of accuracy in between 64.38% to 64.94% at every different minimum case per rule. Thus, these results show more transparent in a graph which is illustrated in figure 5.

The final phase figure 1 is the model validation. This study calculate the predictive accuracy of the classification model to know the dataset quality using $k$-fold cross validation with $k = 10$[12]. The accuracy was calculated using formula (4).

$$Accuracy = \left( \frac{predicted - actual}{actual} \right) \times 100\% \tag{4}$$

Finally, this study makes a comparison between Ant-Miner and J48 based on the percentage of accuracy. J48 is an implementation of C4.5 algorithm [13] in Weka software [14]. Classification model that has a higher percentage of accuracy is better.

## 3. Findings and Discussion

This study tested the effect of changing the value of min. cases per rule parameter. First, this study has a look at the simplicity of the discovered rules. The simplicity of discovered rules is measured by the number of discovered rules and the average number of terms in a rule.

Table 4 shows the number of rules and the number of terms for different values of min. cases per rule. The lower the number of rules and terms, the better the classification model is. The number of rules decreased as the value of min. cases per rule is decreased. The lowest number of rules is 5.7 when min. cases per rule is set to 10 as depicted by figure 3. The lowest number of rules is 6.4 when min. cases per rule is set to 7 as depicted by figure 4.

**Table 4.** Predictive accuracy, number of rules and number of terms for different values of min. cases per rule.

| Min. Cases per rule | Number of rules | Number of terms |
|---|---|---|
| 6 | 6 | 7.6 |
| 7 | 5.5 | 6.4 |
| 8 | 5.8 | 6.7 |
| 9 | 5.9 | 7 |
| 10 | 5.7 | 6.5 |

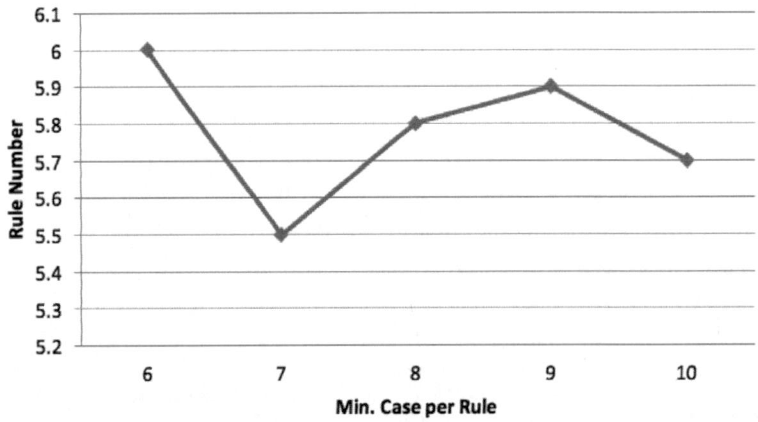


**Figure 3.** Number of rules for different values of min case per rule.

Table 5 shows the predictive accuracy. The higher the value of the accuracy, the better the classification model is. According to figure 3, the highest accuracy is 64.94% when the min. cases

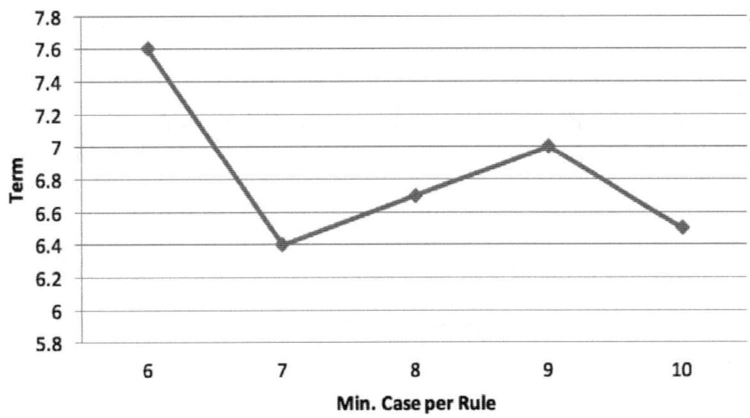


**Figure 4.** Number of terms for different values of min case per rule.

**Table 5.** Predictive accuracy for different values of min. cases per rule.

| Min. Cases per rule | Accuracy (%) |
|---|---|
| 6 | 64.38 |
| 7 | 64.88 |
| 8 | 64.94 |
| 9 | 64.90 |
| 10 | 64.70 |

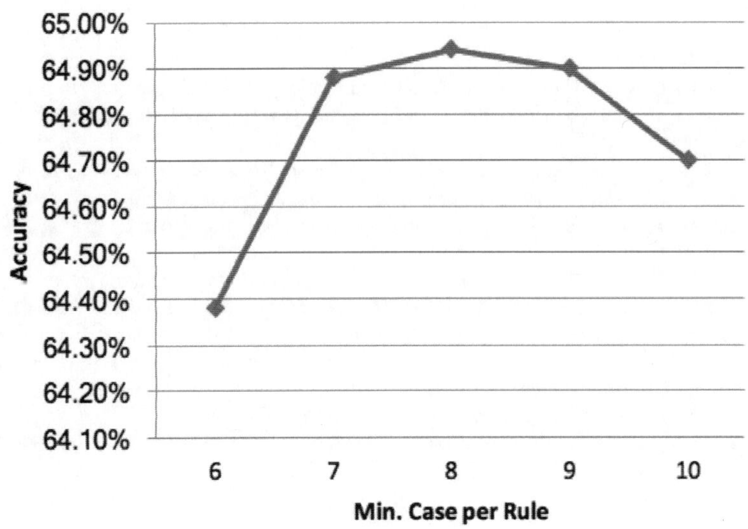


**Figure 5.** Accuracy for different values of min case per rule.

per rule is set to the value of 8. However, the accuracy starts to decrease when the parameter value was increased.

The results comparing the predictive accuracy of Ant-Miner and J48 are reported in table 6. As shown in table 6, the accuracy of classification model by J48 is 64.10% while the accuracy of classification model by Ant-Miner is 64.94%. Ant-Miner discovered rules with a better predictive accuracy than J48.

**Table 6.** Predictive accuracy (%) of Ant-Miner and J48.

| J48 | Ant-Miner |
|---|---|
| 64.10 | 64.94 |

## 4. Conclusion

The main objective of this study is to develop a classification model for detecting software piracy tendency using a variant of ACO algorithm called the Ant-Miner. It is found that the Ant-Miner produced a more accurate classification model for detecting software piracy tendency as compared to J48.

## Acknowledgments

The authors wish to thank Universiti Teknologi MARA for funding this study under DKCP (600-UiTMPs/PJIM&A/ST/DKCP (02/2012)).

## References

[1] Chiang E P and Assane D 2008 *The Journal of Socio-Economics* **37** 1371–1380
[2] Gerlach J H, Kuo F Y B and Lin C S 2009 *Journal of the American Society for Information Science and Technology* **60** 1687–1701
[3] BSA 2014 The compliance gap: Bsa global software survey. Tech. rep. BSA — The Software Alliance
[4] Dorigo M and Sttzle T 2004 *Ant colony optimization* (the MIT Press) ISBN 0-262-04219-3
[5] Dorigo M, Maniezzo V and Colorni A 1996 *Systems, Man, and Cybernetics, Part B: Cybernetics, IEEE Transactions on* **26** 29–41
[6] Parpinelli R S, Lopes H S and Freitas A A 2001 An ant colony based system for data mining: applications to medical data *Proceedings of the genetic and evolutionary computation conference (GECCO-2001)* pp 791–797
[7] Parpinelli R S, Lopes H S and Freitas A A 2002 *IEEE transactions on evolutionary computation* **6** 321–332
[8] Parpinelli R S, Lopes H S and Freitas A A 2002 *Data mining: A heuristic approach* 191–208
[9] Dorigo M, Maniezzo V and Colorni A 1991 Positive feedback as a search strategy Tech. Rep. 91-016 Dipartimento di Elettronica, Politecnico di Milano Milan, Italy
[10] Han J, Pei J and Kamber M 2011 *Data mining: concepts and techniques* (Elsevier)
[11] Tan P N, Steinbach M and Kumar V 2006 *Introduction to data mining* (Pearson Addison Wesley Boston)
[12] Witten I H, Frank E and J Pal C 2016 *Data Mining: Practical Machine Learning Tools and Techniques, Second Edition (Morgan Kaufmann Series in Data Management Systems)* (San Francisco, CA, USA: Morgan Kaufmann Publishers Inc.) ISBN 0128042915
[13] Quinlan J R 2014 *C4.5: Programs for machine learning* (Elsevier)
[14] Hall M, Frank E, Holmes G, Pfahringer B, Reutemann P and Witten I H 2009 *ACM SIGKDD explorations newsletter* **11** 10–18

IOP Conf. Series: Journal of Physics: Conf. Series **855** (2017) 012032 doi:10.1088/1742-6596/855/1/012032

# Error-correcting pairs for a public-key cryptosystem

**Ruud Pellikaan**

Dept. of Mathematics and Computing Science, Eindhoven University of Technology
P.O. Box 513, 5600 MB Eindhoven, The Netherlands

E-mail: g.r.pellikaan@tue.nls

**Irene Márquez-Corbella**

Dept. of Mathematics, Statistics and O. Research, University of La Laguna, Spain

E-mail: irene.marquez.corbella@ull.es

**Abstract.** Code-based Cryptography (CBC) is a powerful and promising alternative for quantum resistant cryptography. Indeed, together with lattice-based cryptography, multivariate cryptography and hash-based cryptography are the principal available techniques for post-quantum cryptography. CBC was first introduced by McEliece where he designed one of the most efficient Public-Key encryption schemes with exceptionally strong security guarantees and other desirable properties that still resist to attacks based on Quantum Fourier Transform and Amplitude Amplification.

The original proposal, which remains unbroken, was based on binary Goppa codes. Later, several families of codes have been proposed in order to reduce the key size. Some of these alternatives have already been broken.

One of the main requirements of a code-based cryptosystem is having high performance $t$-bounded decoding algorithms which is achieved in the case the code has a $t$-error-correcting pair (ECP). Indeed, those McEliece schemes that use GRS codes, BCH, Goppa and algebraic geometry codes are in fact using an error-correcting pair as a secret key. That is, the security of these Public-Key Cryptosystems is not only based on the inherent intractability of bounded distance decoding but also on the assumption that it is difficult to retrieve efficiently an error-correcting pair.

In this paper, the class of codes with a $t$-ECP is proposed for the McEliece cryptosystem. Moreover, we study the hardness of distinguishing arbitrary codes from those having a $t$-error correcting pair.

## 1. Introduction

In 1978 [17] McEliece presented the first PKC system based on the theory of error-correcting codes. In 1986 Niederreiter [19] presented a dual version of McEliece cryptosystem which is equivalent in terms of security. Their main advantages are its fast encryption and decryption schemes. It is an interesting candidate for post-quantum cryptography.

## 2. Code-based cryptography

A *linear code* $C$ is a subspace of $\mathbb{F}_q^n$. The *weight* of $\mathbf{x} \in \mathbb{F}_q^n$ is the number of nonzero entries of $\mathbf{x}$ and is denoted by $\mathrm{wt}(\mathbf{x})$. The *(Hamming) distance* between $\mathbf{x}$, $\mathbf{y} \in \mathbb{F}_q$ is the number of entries where $\mathbf{x}$ and $\mathbf{y}$ differ and is denoted by $d(\mathbf{x}, \mathbf{y})$. The *minimum distance* of $C$ is the minimal value of $d(\mathbf{x}, \mathbf{y})$ where $\mathbf{x},\ \mathbf{y} \in C$ and $\mathbf{x} \neq \mathbf{y}$. Since $C$ is linear it is equal to the minimum weight of $C$, that is the minimal value of $\mathrm{wt}(\mathbf{x})$ where $\mathbf{x} \in C$ and $\mathbf{x} \neq \mathbf{0}$.

The *parameters* of the code are denoted by $[n, k, d]$, where $n$ is its *length*, $k$ its *dimension* and $d$ its *minimum distance*. The *(information) rate* of $C$ is defined by $R = k/n$.

Let $C$ be an $\mathbb{F}_q$-linear code of length $n$ and dimension $k$. A *generator matrix* $G$ of $C$ is a $k \times n$ matrix with entries in $\mathbb{F}_q$ such that its rows are a basis of $C$. A *parity check matrix* $H$ is $(n-k) \times n$ matrix with entries in $\mathbb{F}_q$ such that $\mathbf{c}H^T = \mathbf{0}$ if and only if $\mathbf{c} \in C$.

The problem of *minimum distance decoding* has as input $(G, \mathbf{y})$, where $G$ is a generator matrix of a code $C$ over $\mathbb{F}_q$ of parameters $[n, k, d]$ and $\mathbf{y} \in \mathbb{F}_q^n$ is a received word. The output is a codeword $\mathbf{c} \in C$ of minimal distance to $\mathbf{y}$. One can phrase the problem equivalently in terms of a parity check matrix $H$ of the code. Then the input is $(H, \mathbf{s})$, where $\mathbf{s} \in \mathbb{F}_q^{n-k}$. The output is an $\mathbf{e} \in \mathbb{F}_q^n$ of minimal weight such that $\mathbf{e}H^T = \mathbf{s}$. The relation of the two versions is given by $\mathbf{s} = \mathbf{y}H^T$ the *syndrome* and $\mathbf{e} = \mathbf{y} - \mathbf{c}$ the error vector of the received word $\mathbf{y}$.

The security of code-based cryptosystems is based on the hardness of decoding up to half the minimum distance. The minimum distance decoding problem was shown by Berlekamp-McEliece-Van Tilborg [2] to be NP-hard. The status of the hardness of decoding up to half the minimum distance is an open problem. McEliece proposed to use binary Goppa codes for his PKC system.

In the McEliece PKC system a collection $\mathcal{K}$ of generator $k \times n$ matrices is chosen for which an efficient decoding algorithm is available that corrects all patterns of $t$ errors. The *encryption map*

$$E_G : \mathcal{P} \rightarrow \mathcal{C}$$

for a given key $G \in \mathcal{K}$ is defined by $E_G(\mathbf{m}, \mathbf{e}) = \mathbf{m}G + \mathbf{e}$. An *adversary* $A$ is a map from $\mathcal{C} \times \mathcal{K}$ to $\mathcal{P}$. This adversary is successful for $(x, G) \in \Omega$ if $A(E_G(x), G) = x$.

Let $\mathcal{C}$ be a class of codes such that every code $C$ in $\mathcal{C}$ has an efficient decoding algorithm correcting all patterns of $t$ errors. Let $G \in \mathbb{F}_q^{k\times n}$ be a generator matrix of $C$. In order to mask the origin of $G$, take a $k \times k$ invertible matrix $S$ over $\mathbb{F}_q$ and an $n \times n$ permutation or monomial matrix $\Pi$. Then for the McEliece PKC the matrices $G$, $S$ and $\Pi$ are kept secret while $G' = SG\Pi$ is made public. Furthermore the (trapdoor) one-way function of this cryptosystem is usually presented as follows:

$$\mathbf{x} = (\mathbf{m}, \mathbf{e}) \mapsto \mathbf{y} = \mathbf{m}G' + \mathbf{e},$$

where $\mathbf{m} \in \mathbb{F}_q^k$ is the plaintext and $\mathbf{e} \in \mathbb{F}_q^n$ is a random error vector with Hamming weight at most $t$.

## 3. Error-correcting pairs

From now on the dimension of a linear code $C$ will be denoted by $k(C)$ and its minimum distance by $d(C)$. Given two elements $\mathbf{a}$ and $\mathbf{b}$ in $\mathbb{F}_q^n$, the *star product* is defined by coordinatewise multiplication, that is $\mathbf{a} * \mathbf{b} = (a_1b_1, \ldots, a_nb_n)$ while the *standard inner multiplication* is defined by $\mathbf{a} \cdot \mathbf{b} = \sum_{i=1}^{n} a_ib_i$.

Let $A$, $B$ and $C$ be subspaces of $\mathbb{F}_q^n$. Then $A * B$ is the subspace generated by $\{\mathbf{a} * \mathbf{b} \mid \mathbf{a} \in A \text{ and } \mathbf{b} \in B\}$. And $C^\perp = \{\mathbf{x} | \mathbf{x} \cdot \mathbf{c} = 0 \text{ for all } \mathbf{c} \in C\}$ is the *dual* code of $C$. Furthermore $A \perp B$ means $\mathbf{a} \cdot \mathbf{b} = 0$ for all $\mathbf{a} \in A$ and $\mathbf{b} \in B$.

**Definition 3.1.** Let $C$ be a linear code in $\mathbb{F}_q^n$. The pair $(A, B)$ of linear codes over $\mathbb{F}_{q^m}$ of length $n$ is called a *t-error-correcting pair* (ECP) over $\mathbb{F}_{q^m}$ for $C$ if the following properties hold:

E.1 $(A * B) \perp C$,

E.2 $k(A) > t$,

E.3 $d(B^\perp) > t$,

E.4 $d(A) + d(C) > n$.

**Remark 3.2.** In the above definition $A$ and $B$ are $\mathbb{F}_{q^m}$-linear codes and $C$ is an $\mathbb{F}_q$-linear code. So by $k(A)$ the dimension of $A$ over $\mathbb{F}_{q^m}$ is meant. And $d(B^\perp)$, $d(A)$ and $d(C)$ mean the minimum distances of $B^\perp$ and $A$ over $\mathbb{F}_{q^m}$ and of $C$ over $\mathbb{F}_q$.

**Remark 3.3.** The notion of an error-correcting pair for a linear code was introduced in 1988 by Pellikaan [22] and independently by Kötter in [11, 12] in 1992. It is shown that a linear code in $\mathbb{F}_q^n$ with a $t$-error-correcting pair has a decoding algorithm which corrects up to $t$ errors with complexity $\mathcal{O}(n^3)$.

**Remark 3.4.** Note that if $(A, B)$ is a pair of codes that satisfies Conditions E.1, E.2, E.3 and the following two conditions:

*E.5* $d(A^\perp) > 1$, *that means A is a non-degenerate code,*

*E.6* $d(A) + 2t > n$,

then $d(C) \geq 2t + 1$ and $(A, B)$ is a $t$-ECP for $C$ by [23, Corollary 3.4].

In the following we consider eight collections of pairs.

**Example 3.5.** Let $\mathbf{a}$ be an $n$-tuple of mutually distinct elements of $\mathbb{F}_q$ and $\mathbf{b}$ be an $n$-tuple of nonzero elements of $\mathbb{F}_q$. Then the *generalized Reed-Solomon* code $\mathrm{GRS}_k(\mathbf{a}, \mathbf{b})$ is defined by

$$\mathrm{GRS}_k(\mathbf{a}, \mathbf{b}) = \{(f(a_1)b_1, \ldots, f(a_n)b_n) \mid f(X) \in \mathbb{F}_q[X] \text{ and } \deg(f(X)) < k\}.$$

If $k \leq n \leq q$, then $\mathrm{GRS}_k(\mathbf{a}, \mathbf{b})$ is an $[n, k, n-k+1]$ code. Furthermore the dual of a GRS code is again a GRS code, in particular $\mathrm{GRS}_k(\mathbf{a}, \mathbf{b})^\perp = \mathrm{GRS}_{n-k}(\mathbf{a}, \mathbf{b}^\perp)$ for some $\mathbf{b}^\perp$ that is explicitly known.

Let $A = \mathrm{GRS}_{t+1}(\mathbf{a}, \mathbf{u})$, $B = \mathrm{GRS}_t(\mathbf{a}, \mathbf{v})$ and $C = \mathrm{GRS}_{2t}(\mathbf{a}, \mathbf{u} * \mathbf{v})^\perp$. Then $(A, B)$ is a $t$-ECP for $C$. Conversely let $C = \mathrm{GRS}_k(\mathbf{a}, \mathbf{b})$, then $A = \mathrm{GRS}_{t+1}(\mathbf{a}, \mathbf{b}^\perp)$ and $B = \mathrm{GRS}_t(\mathbf{a}, \mathbf{1})$ is a $t$-ECP for $C$ where $t = \left\lfloor \frac{n-k}{2} \right\rfloor$.

So GRS codes are the prime examples of codes that have a $t$-error-correcting pair. GRS codes are not suited for a coded-based PKC by the attack of Sidelnikov-Shestakov [26].

**Example 3.6.** Let $C$ be a subcode of a code $D$ that has $(A, B)$ as a $t$-ECP. Then condition (E.1) holds for $(A, B)$ with respect to $D$. So $\mathbf{a} * \mathbf{b} \cdot \mathbf{d} = 0$ for all $\mathbf{d}$ in $D$. Hence $\mathbf{a} * \mathbf{b} \cdot \mathbf{c} = 0$ for all $\mathbf{c}$ in $C$, since $C \subseteq D$. Conditions (E.2), (E.3) and (E.4) hold. Therefore $(A, B)$ is also a $t$-ECP for $C$.

In particular, let $C$ be a subcode of the code $\mathrm{GRS}_{n-2t}(\mathbf{a}, \mathbf{b})$. This GRS code has a $t$-error-correcting pair by Example 3.5 which is also a $t$-ECP for $C$.

The class of subcodes of GRS codes was proposed by Berger and Loidreau in [1] for code-based PKC to resist precisely the Sidelnikov-Shestakov attack. But for certain parameter choices this proposal is also not secure as shown by Márquez et al. [14].

**Example 3.7.** The Goppa code $\Gamma(L, g(X))$ associated to a Goppa polynomial $g(X)$ of degree $r$ and an $n$-tuple $L$ of points in $\mathbb{F}_{q^m}$ can be viewed as an alternant code, that is a subfield subcode of a GRS code of codimension $r$. Therefore such a code has an $\lfloor r/2 \rfloor$-error-correcting pair. If the Goppa polynomial is square free of degree $r$ in an extension of $\mathbb{F}_2$, then the binary Goppa code has an $r$-ECP, since $\Gamma(L, g(X)) = \Gamma(L, g(X)^2)$.

Goppa codes were proposed by McEliece [17] for his PKC system. Sidelnikov-Shestakov made a claim [26] that their method for GRS codes could be extended to attack Goppa codes as well, but this had never been substantiated by a paper in the public domain. A binary Goppa code using elements in the extension $\mathbb{F}_{2^m}$ and with a square free Goppa polynomial of degree $t$ over $\mathbb{F}_{2^m}$ has parameters $[n, k, d]$ with $n \leq 2^m$, $k \geq n - mt$ and $d \geq 2t + 1$. For these codes efficient decoding algorithms are known that decode all patterns with $t$ errors.

A binary Goppa code with parameters $[1024, 524, 101]$ as proposed by McEliece is no longer secure with nowadays computing power due to recent improvements in the decoding algorithms.

**Example 3.8.** Algebraic geometry (AG) codes were introduced in 1977 by V.D. Goppa. Recall that GRS codes can be seen as the class of AG codes on the projective line, that is the algebraic curve of genus zero. We refer the interested reader to [24, 25].

Let $\mathcal{X}$ be an algebraic curve defined over $\mathbb{F}_q$ with genus $g$. By an algebraic curve we mean a curve that is absolutely irreducible, nonsingular and projective. Let $\mathcal{P}$ be an $n$-tuple of $\mathbb{F}_q$-rational points on $\mathcal{X}$ and let $E$ be a divisor of $\mathcal{X}$ with disjoint support from $\mathcal{P}$ of degree $e$. Then the algebraic geometry code $C_L(\mathcal{X}, \mathcal{P}, E)$ is the image of the Riemann-Roch space $L(E)$ of rational functions with prescribed behavior of zeros and poles at $E$ under the evaluation map $\mathrm{ev}_{\mathcal{P}}$. If $e < n$, then the dimension of the code $C_L(\mathcal{X}, \mathcal{P}, E)$ is at least $e + 1 - g$ and its minimum distance is at least $n - e$. If $e > 2g - 2$, then its dimension is $e + 1 - g$. The dual code $C_L(\mathcal{X}, \mathcal{P}, E)^{\perp}$ is again AG. If $e > 2g - 2$, then the dimension of the code $C_L(\mathcal{X}, \mathcal{P}, E)^{\perp}$ is at least $n - e - 1 + g$ and its minimum distance is at least $d^* = e - 2g + 2$. If $e < n$, then its dimension is $n - e - 1 + g$.

If $A = C_L(\mathcal{X}, \mathcal{P}, E)$ and $B = C_L(\mathcal{X}, \mathcal{P}, F)$, then $A * B \subseteq C_L(\mathcal{X}, \mathcal{P}, E + F)$. So there are abundant ways to construct error-correcting pairs of an AG code. An AG code on a curve of genus $g$ with designed minimum distance $d^*$ has a $t$-ECP over $\mathbb{F}_q$ with $t = \lfloor (d^* - 1 - g)/2 \rfloor$ by [21, Theorem 1] and [22, Theorem 3.3]. If $m$ is sufficiently large, then there exists a $t$-ECP over $\mathbb{F}_{q^m}$ with $t = \lfloor (d^* - 1)/2 \rfloor$ by [23, Proposition 4.2].

Algebraic geometry codes were proposed by Niederreiter [19] and Janwa-Moreno [10] for code-based PKC systems. This system was broken for low genus zero [26], one and two [9, 18]. For arbitrary genus it was shown by Márquez et al. [13, 15] that these codes are not secure for rates $R$ in the intervals $[\gamma, \frac{1}{2} - \gamma]$, $[\frac{1}{2} + \gamma, 1 - \gamma]$, $[\frac{1}{2} - \gamma, 1 - 3\gamma]$ and $[3\gamma, \frac{1}{2} + \gamma]$, where $R = k/n$ is the information rate and $\gamma = g/n$ the relative genus. Recently Couvreur et al. [6] showed that it is not necessary to retrieve the triple $(\mathcal{X}, \mathcal{P}, E)$ and the Riemann-Roch space $L(E)$ but that one can stay in the realm of $\mathbb{F}_q^n$ and its subspaces in order to find an error-correcting pair.

**Example 3.9.** Geometric Goppa codes are subfield subcodes of algebraic geometry codes generalizing the classical Goppa codes that are subfield subcodes of GRS codes. Geometric Goppa codes were proposed by Janwa-Moreno [10]. Couvreur et al. [5] showed that certain geometric Goppa codes are not secure for a PKC system.

**Example 3.10.** Let $(A, B)$ be a pair of codes with parameters $[n, t+1, n-t]$ and $[n, t, n-t+1]$, respectively, and $C = (A * B)^{\perp}$, then the minimum distance of $C$ is at least $2t + 1$ and $(A, B)$ is a $t$-error-correcting pair for $C$ by [23, Corollary 3.4]. The dimension of $A * B$ is at most $t(t+1)$. So the dimension of $C$ is at least $n - t(t+1)$. In Appendix A it is shown that this is almost always equal to $n - t(t+1)$ for random choices of $A$ and $B$.

If $q$ is considerably larger than $n$, then a random linear code is MDS with very high probability. So taking random codes $A$ and $B$ of length $n$ and dimensions $t + 1$ and $t$, respectively, this gives a very large class of codes for the McEliece PKC. However with large field the key size becomes larger and recall that the main obstacle for coded-based cryptosystems was the key size.

## 4. The ECP one-way function

Let $\mathcal{P}(n, t, q)$ be the collection of pairs $(A, B)$ such that there exist a positive integer $m$ and a pair $(A, B)$ of $\mathbb{F}_{q^m}$-linear codes of length $n$, that satisfy Conditions E.2, E.3, E.5 and E.6. Let $C$ be the $\mathbb{F}_q$-linear code of length $n$ that is the subfield subcode that has all elements of $A * B$ as parity checks. So

$$C = \mathbb{F}_q^n \cap (A * B)^{\perp}.$$

Then the minimum distance of $C$ is at least $2t + 1$ and $(A, B)$ is a $t$-ECP for $C$ as was noted in Remark 3.4. Let $\mathcal{F}(n, t, q)$ be the collection of $\mathbb{F}_q$-linear codes of length $n$ and minimum distance

$d \geq 2t+1$. Consider the following map

$$\begin{array}{rccc} \varphi_{(n,t,q)}: & \mathcal{P}(n,t,q) & \longrightarrow & \mathcal{F}(n,t,q) \\ & (A,B) & \longmapsto & C. \end{array}$$

The question is whether this map is a one-way function.

Let $U$ and $V$ be generator matrices of the codes $A$ and $B$, with rows denoted by $\mathbf{u}_i$ and $\mathbf{v}_i$, respectively. Let $U * V$ be the matrix with the rows $\mathbf{u}_i * \mathbf{v}_j$ ordered lexicographically. Let $(U*V)(q^l)$ be the matrix with entries the $q^l$-power of the entries of $U*V$. Let $W$ be the reduced row echelon form (with the zero rows deleted) of the matrix with rows all the rows of $(U*V)(q^l)$ for $l = 0, 1 \dots, m-1$. Then $W$ has entries in $\mathbb{F}_q$ and is a parity check matrix of $C$. In this way

$$(U,V) \longmapsto \; W$$

is an implementation of the map $\varphi_{(n,t,q)}$.

If the map $\varphi_{(n,t,q)}$ is indeed difficult to invert, then we will call it the *ECP one-way function* and the code $C$ with parity check matrix $W$ might be used as a public-key in a coding based PKC. Otherwise it would mean that the PKC based on codes that can be decoded by error-correcting pairs is not secure.

**Remark 4.1.** Note that $\mathbf{u}\Pi * \mathbf{v}\Pi = (\mathbf{u} * \mathbf{v})\Pi$ for every permutation or monomial matrix $\Pi$. Thus, if $(A,B)$ is a $t$-ECP for $C$, then $(A\Pi, B\Pi)$ is a $t$-ECP for $C\Pi$.

Furthermore, if $S$ and $T$ are invertible matrices of the correct sizes to be multiplied on the left of the matrices $U$ and $V$, respectively, then $U*V$ generates the same code as $(SU)*(TV)$ since $(SU)*\mathbf{v} = S(U*\mathbf{v})$ and $\mathbf{u}*(TV) = T(\mathbf{u}*V)$ for all vectors $\mathbf{u}$ and $\mathbf{v}$. Therefore the usual masking $SHP$ of a parity check matrix $H$ by means of an invertible matrix $S$ and a permutation matrix $P$ is already incorporated in the choice of the pair of generator matrices $(U,V)$.

## 5. Distinguishing a code with an ECP

Let $\mathcal{K}$ be a collection of generator matrices of codes that have a $t$-error-correcting pair and that is used for a coded-based PKC system. In this section we address assumption A.2 whether we can distinguish arbitrary codes from those coming from $\mathcal{K}$.

Let $C$ be a $k$ dimensional subspace of $\mathbb{F}_q^n$ with basis $\mathbf{g}_1, \dots, \mathbf{g}_k$ which represents the rows of the generator matrix $G \in \mathbb{F}_q^{k \times n}$. We denote by $S^2(C)$ the *second symmetric power* of $C$, or equivalently the *symmetrized tensor product* of $C$ with itself. If $\mathbf{x}_i = \mathbf{g}_i$, then $S^2(C)$ has basis $\{\mathbf{x}_i\mathbf{x}_j \mid 1 \leq i \leq j \leq n\}$ and dimension $\binom{k+1}{2}$. Furthermore we denote $C * C$ by $C^{(2)}$ the *square* of $C$, that is the linear subspace in $\mathbb{F}_q^n$ generated by $\{\mathbf{a} * \mathbf{b} | \mathbf{a}, \mathbf{b} \in C\}$. See [3, §4 Definition 6] and [4, 14]. Now, following the same scheme as in [13], we consider the linear map

$$\sigma: \; S^2(C) \longrightarrow C^{(2)},$$

where the element $\mathbf{x}_i\mathbf{x}_j$ is mapped to $\mathbf{g}_i * \mathbf{g}_j$. The kernel of this map will be denoted by $K^2(C)$. Then $K^2(C)$ is the solution space of the following set of equations:

$$\sum_{1 \leq i \leq i' \leq k} g_{ij} g_{i'j} \mathbf{X}_{ii'} = 0, \quad 1 \leq j \leq n.$$

There is no loss of generality in assuming $G$ to be systematic at the first $k$ position, making a suitable permutation of columns and applying Gaussian elimination, if necessary. Then $G = \begin{pmatrix} I_k & P \end{pmatrix}$ where $I_k$ is the $k \times k$ identity matrix and $P$ is an $k \times (n-k)$ matrix formed by the last $n-k$ columns of $G$. Now $H = \begin{pmatrix} P^T & -I_{n-k} \end{pmatrix}$ is a parity check matrix of $C$, or equivalently $H$ is a generator matrix of the $[n, n-k]$ code $D = C^\perp$.

In [8, III] and [20, Ch. 10] a system $\mathcal{L}_P$ associated to the matrix $P$ of $k$ linear equations involving the $\binom{n-k}{2}$ variables $Z_{jl}$, with $k+1 \leq j < l \leq n$, is defined as

$$\mathcal{L}_P = \left\{ \sum_{j=k+1}^{n-1} \sum_{j'>j}^{n} p_{ij} p_{ij'} \mathbf{Z}_{jj'} = 0 \mid 1 \leq i \leq k. \right\}$$

This system differs from the system of equations obtained for the kernel $K^2(C)$ in interchanging indices $i$ and $j$ and the strict inequality $j < j'$ in the summation, instead of $i \leq i'$. Denote the kernel of $\mathcal{L}_P$, that is the space of all solutions of $\mathcal{L}_P$, by $K(\mathcal{L}_P)$.

**Proposition 5.1.**

$$\dim K(\mathcal{L}_P) = \dim K^2(D)$$

*Proof.* Let $M$ be the $\binom{k+1}{2} \times n$ matrix with entries $(g_{ij} g_{i'j})_{\substack{1 \leq i \leq i' \leq k \\ 1 \leq j \leq n}}$. Then a basis of $K^2(C)$ can be read of directly as the kernel of $M$. Note also that the dimension of $C^{(2)}$ is equal to the rank of $M$. Furthermore, since $C^{(2)}$ is the image of the linear map $\sigma$, by the first isomorphism theorem we get:

$$\dim K^2(C) + \dim C^{(2)} = \dim S^2(C) = \binom{k+1}{2}.$$

Let $\mathbf{h}_i$ be the $i$-th row of the parity check matrix $H$, $\mathbf{e}_i$ be the $i$-th vector in the canonical basis of $\mathbb{F}_q^{n-k}$ and $\mathbf{q}_i$ be the $i$-th row of the matrix $P^T$. Then $q_{ij} = p_{j,i+k}$ and $\mathbf{h}_i = (\mathbf{q}_i | -\mathbf{e}_i)$. Therefore

$$\mathbf{h}_j * \mathbf{h}_{j'} = \begin{cases} \left( \mathbf{q}_j * \mathbf{q}_j \mid \mathbf{e}_j \right) & \text{if } j = j', \\ \left( \mathbf{q}_j * \mathbf{q}_{j'} \mid \mathbf{0} \right) & \text{if } j < j'. \end{cases}$$

Let $M_1$ be the $k \times \binom{n-k}{2}$ matrix with entries $(p_{ij} p_{ij'})_{\substack{1 \leq i \leq k \\ k < j < j' \leq n}}$, then

$$\dim K(\mathcal{L}_P) = \binom{n-k}{2} - \operatorname{rank}(M_1)$$

Now let $M_2$ be the $\binom{n-k+1}{2} \times n$ matrix with entries $(h_{ij} h_{i'j})_{\substack{1 \leq i \leq i' \leq n-k \\ 1 \leq j \leq n}}$. Then

$$\dim D^{(2)} = \operatorname{rank}(M_2) = n - k + \operatorname{rank}(M_1)$$

Therefore

$$\begin{aligned} \dim K(\mathcal{L}_P) &= \binom{n-k}{2} - \operatorname{rank}(M_1) \\ &= \binom{n-k}{2} + n - k - \dim D^{(2)} \\ &= \dim K^2(D) \end{aligned}$$

□

The dual statement of Proposition 5.1 gives: $\dim K(\mathcal{L}_{P^T}) = \dim K^2(C)$.
For every $[n, k]$ code $C$ over $\mathbb{F}_q$ the following inequality holds:

$$\dim C^{(2)} \leq \min\{n, \binom{k+1}{2}\}.$$

However if the entries of the matrix $P$ are taken independently and identically distributed, then the inequality holds with equality with high probability what is actually proved in the next proposition.

**Proposition 5.2.** *Let $C$ be an $[n,k]$ code with $n > \binom{k+1}{2}$ chosen at random. Then*

$$\Pr\left(\dim(C^{(2)}_{random}) = \binom{k+1}{2}\right) = o(1)$$

*Proof.* Let $C$ be a linear code with parameters $[n,k]$ over $\mathbb{F}_q$ with $n > \binom{k+1}{2}$.

We have seen in the proof of Proposition 5.1, with the role of $C$ and $D = C^\perp$ interchanged that the linear system $\mathcal{L}_{PT}$ associated with $C$ consists of $n-k$ linear equations and $\binom{k}{2}$ unknowns. In case $n-k > \binom{k}{2}$ or equivalently $n > \binom{k+1}{2}$ Faugère et al. [8] proved that the dimension of the solution space of $\mathcal{L}_{PT}$ is 0 with high probability. Therefore under the same hypothesis we have that the dimension of $C^{(2)}_{\text{random}}$ is $\binom{k+1}{2}$ with high probability. □

**Remark 5.1.** See [4, Theorem 2.3] for an improved version of Proposition 5.2.

**Example 5.2.** Let $C$ be a GRS code with parameters $[n,k]$, take for instance $C = \text{GRS}_k(\mathbf{a},\mathbf{b})$ where $\mathbf{a}$ is an $n$-tuple of mutually distinct elements of $\mathbb{F}_q$ and $\mathbf{b}$ is an $n$-tuple of nonzero elements of $\mathbb{F}_q$. Then $C^{(2)}$ is the code $\text{GRS}_{2k-1}(\mathbf{a},\mathbf{b}*\mathbf{b})$ if $2k-1 \leq n$ and $\mathbb{F}_q^n$ otherwise. Hence $\dim C^{(2)} = \min\{2k-1, n\}$. Therefore

$$\dim K^2(C) = \binom{k+1}{2} - (2k-1) = \binom{k-1}{2} \text{ if } 2k-1 \leq n.$$

**Example 5.3.** Let $C$ be a $k$-dimensional subcode of the code $\text{GRS}_l(\mathbf{a},\mathbf{b})$. Then $C^{(2)}$ is a subcode of the code $\text{GRS}_{2l-1}(\mathbf{a},\mathbf{b}*\mathbf{b})$, if $2l-1 \leq n$. Thus

$$\dim C^{(2)} \leq \min\{2l-1, n\}.$$

Moreover if $4l-3k-1 < q$ and $2l-1 \leq \binom{k+1}{2}$, then it was shown in [14] that $C^{(2)}$ is equal to $\text{GRS}_{2l-1}(\mathbf{a},\mathbf{b}*\mathbf{b})$ with high probability so, under this hypothesis,

$$\Pr\left(\dim C^{(2)} = 2l-1\right) = 1 - o(1).$$

The dual code $D = C^\perp$ contains the code $\text{GRS}_l(\mathbf{a},\mathbf{b})^\perp = \text{GRS}_{n-l}(\mathbf{a},\mathbf{b}^\perp)$. That is, $D^{(2)}$ contains the square of $\text{GRS}_{n-l}(\mathbf{a},\mathbf{b}^\perp)$ which is equal to $\text{GRS}_{2n-2l-1}(\mathbf{a},\mathbf{b}^\perp*\mathbf{b}^\perp)$ if $2n-2l-1 \leq n$, or equivalently if $n \leq 2l+1$. Recall that the star product of the rows of a generator matrix of any linear code gives a generating set for its square code, that is the square of any $[n,s]$ linear code is generated by $\binom{s+1}{2}$ elements. In particular $D^{(2)}$ is generated by $\binom{n-k+1}{2}$ elements but since $\text{GRS}_{2n-2l-1}(\mathbf{a},\mathbf{b}^\perp*\mathbf{b}^\perp) \subseteq D^{(2)}$ there are at least $\binom{n-l+1}{2} - (2n-2l+1)$ dependent elements of this generating set. Thus

$$\dim D^{(2)} \leq \binom{n-k+1}{2} - \binom{n-l+1}{2} + 2n - 2l - 1 = \binom{n-k+1}{2} - \binom{n-l-1}{2}.$$

**Example 5.4.** The problem of distinguishing Goppa, alternant and random codes from each other was studied by Faugère et al. in [8]. Their experimental results give rise to a conjecture on the dimension of $K(\mathcal{L}_P)$ for Goppa and alternant codes of high rate.

**Example 5.5.** Let $C = \mathcal{C}_L(\mathcal{X},\mathcal{P},E)$ where $\mathcal{X}$ is an algebraic curve over $\mathbb{F}_q$ of genus $g$, $\mathcal{P}$ is an $n$-tuple of mutually distinct $\mathbb{F}_q$-rational points of $\mathcal{X}$ and $E$ is a divisor of $\mathcal{X}$ with disjoint support from $\mathcal{P}$ of degree $e$. Then $C^{(2)} \subseteq \mathcal{C}_L(\mathcal{X},\mathcal{P},2E)$.

Assume moreover that $2g-2 < e < n/2$. Then $C$ has dimension $k = e+1-g$ and $\mathcal{C}_L(\mathcal{X},\mathcal{P},2E)$

has dimension $2e + 1 - g = k + e$. Hence $\dim C^{(2)} \leq k + e$.
Let $G$ be a generator matrix of an algebraic geometry code $C$. Take the columns of $G$ as homogeneous coordinates of points in $\mathbb{P}^{e-g}$, this gives a projective system $\mathcal{Q} = (Q_1, \dots, Q_n)$ of points in the projective space $\mathbb{P}^{e-g}(\mathbb{F}_q)$. Since $e > 2g$ there exists an embedding of the curve $\mathcal{X}$ in $\mathbb{P}^{e-g}$ of degree $e$

$$\begin{array}{cccc} \varphi_E: & \mathcal{X} & \longrightarrow & \mathbb{P}^{e-g} \\ & P & \longmapsto & \varphi_E(P) = (f_0(P), \dots, f_{e-g}(P)) \end{array}$$

where $\{f_0, \dots, f_{e-g}\}$ is a basis of $L(E)$ such that $\mathcal{Q} = \varphi_E(\mathcal{P})$ lies on the curve $\mathcal{Y} = \varphi_E(\mathcal{X})$. The space $I_2(\mathcal{Q})$ of quadratic polynomials that vanish on $\mathcal{Q}$ can be identified with $K^2(C)$. Furthermore if $2g + 2 \leq e < \frac{1}{2}n$, then $I_2(\mathcal{Y}) = I_2(Q)$ and $I(\mathcal{Y})$, the vanishing ideal of $\mathcal{Y}$, is generated by $I_2(Q)$. Now

$$\dim K^2(C) = \binom{k+1}{2} - \dim C^{(2)} \geq \binom{k}{2} - e.$$

Therefore $\mathcal{Y}$ is given as the intersection of at least $\binom{k}{2} - e$ quadrics in $\mathbb{P}^{e-g}$. For more details we refer the reader to [13, 15].

**Example 5.6.** Let $t(t+1) < n$. Let $(A, B)$ be a pair of random codes of dimension $t+1$ and $t$, respectively. Take $C = (A * B)^\perp$ as in Example 3.10. Let $D = C^\perp = A * B$. Then $D^{(2)} = A^{(2)} * B^{(2)}$. Hence

$$\dim D^{(2)} \leq \binom{t+2}{2}\binom{t+1}{2}$$

which is about half the expected value $\binom{t(t+1)}{2}$ in case $\binom{t(t+1)}{2} < n$ by Proposition 5.2, since $D$ has dimension $t(t+1)$ with high probability by Appendix A.

## 6. Acknowledgement
An earlier version of this work was presented for the first time by the second author at the *Code-Based Cryptography Workshop*, May 2012 at the Technical University of Denmark, Lyngby and posted at arXiv [16] and furthermore at the conferences Applications of Computer Algebra 2013 and 2014 at Malaga and Fordham, respectively, and finally at the 3rd IndoMS International Conference on Mathematics and Its Applications, Depok in 2015.

## References
[1] Berger, T., Loidreau, P.: How to mask the structure of codes for a cryptographic use. Designs, Codes and Cryptography 35, 63–79 (2005)
[2] Berlekamp, E., McEliece, R., van Tilborg, H.: On the inherent intractability of certain coding problems. IEEE Transactions on Information 24, 384–386 (1978)
[3] Cascudo, I., Chen, H., Cramer, R., Xing, X.: Asymptotically good ideal linear secret sharing with strong multiplication overy any fixed finite field. In: Halevi, S. (ed.) Advances in Cryptology - CRYPTO 2009, Lecture Notes in Computer Science. vol. 5677, pp. 466–486. Springer, Berlin (2009)
[4] Cascudo, I., Cramer, R., Mirandola, D., Zémor, G.: Squares of random linear codes. IEEE Trans. Inform. Theory 61(3), 1159–1173 (2015)
[5] Couvreur, A., Márquez-Corbella, I., Pellikaan, R.: Cryptanalysis of public-key cryptosystems that use subcodes of algebraic geometry codes. In: 4ICMCTA, Coding Theory and Application, CIM Series in Mathematical Sciences 3, pp. 133–140 (2014)
[6] Couvreur, A., Márquez-Corbella, I., Pellikaan, R.: A polynomial time attack against algebraic geometry code based public key cryptosystems. In: IEEE International Symposion on Information Theory ISIT 2014, p. 1446 (2014)
[7] Faugère, J., Gauthier-Umana, V., Otmani, A., Perret, L., Tillich, J.: A distinguisher for high rate McEliece cryptosystems. In: Proceedings IEEE Information Theory Workshop 2011. Paraty, Brazil (October 16-20, 2011)

[8] Faugère, J.C., Gauthier-Umaña, V., Otmani, A., Perret, L., Tillich, J.P.: A distinguisher for high-rate McEliece cryptosystems. IEEE Trans. Inform. Theory 59(10), 6830–6844 (2013)
[9] Faure, C., Minder, L.: Cryptanalysis of the McEliece cryptosystem over hyperelliptic codes. In: Proceedings 11th Int. Workshop on Algebraic and Combinatorial Coding Theory, ACCT 2008. pp. 99–107 (2008)
[10] Janwa, H., Moreno, O.: McEliece public crypto system using algebraic-geometric codes. Designs, Codes and Cryptography 8, 293–307 (1996)
[11] Kötter, R.: A unified description of an error locating procedure for linear codes. In: Proceedings of Algebraic and Combinatorial Coding Theory, pp. 113–117. Voneshta Voda (1992)
[12] Kötter, R.: On algebraic decoding of algebraic-geometric and cyclic codes. Ph.D. thesis, Linköping University of Technology, Linköping Studies in Science and Technology, Dissertation no. 419 (1996)
[13] Márquez-Corbella, I., Martínez-Moro, E., Pellikaan, R.: On the unique representation of very strong algebraic geometry codes. Designs, Codes and Cryptography pp. 215–230 (2012)
[14] Márquez-Corbella, I., Martínez-Moro, E., Pellikaan, R.: The non-gap sequence of a subcode of a generalized Reed-Solomon code. Designs, Codes and Cryptography 66(1-3) (2013)
[15] Márquez-Corbella, I., Martínez-Moro, E., Pellikaan, R., Ruano, D.: Computational aspects of retrieving a representation of an algebraic geometry code. J. Symbolic Computation 64, 67–87 (2014)
[16] Márquez-Corbella, I., Pellikaan, R.: Error-correcting pairs for a public-key cryptosystem. Preprint arXiv:1205.3647 (2012)
[17] McEliece, R.: A public-key cryptosystem based on algebraic coding theory. DSN Progress Report 42–44, 114–116 (1978)
[18] Minder, L.: Cryptography based on error correcting codes. Ph.D. thesis, 3846 EPFL (2007)
[19] Niederreiter, H.: Knapsack-type crypto systems and algebraic coding theory. Problems of Control and Information Theory 15(2), 159–166 (1986)
[20] Otmani, A.: Contribution to the cryptanalysis of code-based primitives. Université de Caen Basse Normandie, Caen (2011)
[21] Pellikaan, R.: On a decoding algorithm of codes on maximal curves. IEEE Trans. Inform. Theory 35, 1228–1232 (1989)
[22] Pellikaan, R.: On decoding by error location and dependent sets of error positions. Discrete Math. 106–107, 369–381 (1992)
[23] Pellikaan, R.: On the existence of error-correcting pairs. Statistical Planning and Inference 51, 229–242 (1996)
[24] Stichtenoth, H.: Algebraic function fields and codes. Springer, Berlin (1993)
[25] Tsfasman, M., Vlădut, S.: Algebraic-geometric codes. Kluwer Academic Publishers, Dordrecht (1991)
[26] V.M. Sidelnikov, V., Shestakov, S.: On the insecurity of cryptosystems based on generalized Reed-Solomon codes. Discrete Math. Appl. 2, 439–444 (1992)

## Appendix A. The dimension of $A * B$

Let $A$ and $B$ be two linear codes over $\mathbb{F}_q$ with parameters $[n, s]$ and $[n, t]$, generated by the set $\{\mathbf{a}_1, \ldots, \mathbf{a}_s\}$ and $\{\mathbf{b}_1, \ldots, \mathbf{b}_t\}$ of vectors in $\mathbb{F}_q^n$, respectively. Let $M$ be an $st \times n$ matrix over $\mathbb{F}_q$ whose rows consist on the vectors $\mathbf{a}_i * \mathbf{b}_j = (a_{i,1}b_{j,1}, \ldots, a_{i,n}b_{j,n}) \in \mathbb{F}_q^n$ with $i \in \{1, \ldots, s\}$ and $j \in \{1, \ldots, t\}$ ordered lexicographically. Then the rows of $M$ form a generating set of the code $A * B$.

Indeed $M$ is a block-matrix consisting of $s$ blocks $M_i = (\mathbf{a}_i * \mathbf{b}_j)_{1 \leq j \leq t}$ with $i \in \{1, \ldots, s\}$ of size $t \times n$. We define the support of a codeword $\mathbf{c} = (c_1, \ldots, c_n)$ by $\text{supp}(\mathbf{c}) = \{i \mid c_i \neq 0\}$. Note that if $i \notin \text{supp}(\mathbf{a}_j)$ then the $i$-th column of $M_j$ consists of zeros.

In the following lines, assuming that $st < n$ we will prove that $M$ has full rank with high probability. We proceed by a similar procedure as in [8, VI Theorem 2] where it is proved that the solution space of the linear system associated to an arbitrary random linear code is zero with high probability.

Let $E_1 = \text{supp}(\mathbf{a}_1)$. Suppose $|E_1| \geq t$. Let $F_1$ be a subset of $E_1$ with cardinality $t$. To simplify notation and without loss of generality, we can always assume that $F_1$ corresponds to the first $t$ elements in $\mathbf{a}_1$, by permuting the elements if necessary. Let $M^{(1)}$ be a square submatrix of $M_1$ formed by its first $t$ columns, i.e.

$$M^{(1)} = (a_{1,j}b_{i,j})_{\substack{j \in F_1 \\ 1 \leq i \leq t}} \in \mathbb{F}_q^{t \times t}$$

Now we define by induction $E_i := \text{supp}(\mathbf{a}_i) \setminus F_{i-1}$ and the subset $F_i$ as the first $t$ elements of the subset, assuming that $|E_i| \geq t$. The square matrix $M^{(i)} \in \mathbb{F}_q^{t\times t}$ is obtained from $M_i$ by taking the $F_i$-indexed columns, for $i \in \{1, \ldots, s\}$. Then clearly the following Lemma holds.

**Lemma Appendix A.1.** *If $|E_i| \geq t$ for all $i \in \{1, \ldots, s\}$ then*

$$\text{rank}(M) \geq \sum_{i=1}^{s} \text{rank}(M^{(i)}).$$

**Lemma Appendix A.2.** *If $|E_i| \geq t$ for all $i = 1, \ldots, s$ then*

$$\Pr\left(\sum_{i=1}^{s} d(M^{(i)}) \geq u\right) \leq K^s q^{\frac{-u^2}{s}}$$

*where $d(M^{(i)}) = t - \text{rank}(M^{(i)})$ for $i = 1, \ldots, s$ and $K$ is a constant depending only on $q$.*

*Proof.* See [8, Lemma 5]. □

**Lemma Appendix A.3.** *Let $u_i = n - (i-1)t$ with $i = \{1, \ldots, s\}$, then*

$$\Pr\left(|E_i| < t, \mid |E_1| \geq t, \ldots, |E_{i-1}| \geq t\right) \leq e^{-2\frac{\left(\frac{q-1}{q}u_i - t + 1\right)^2}{u_i}}$$

*Proof.* See [7, Lemma 6] and [8]. □

**Theorem Appendix A.4.** *Assume that $st < n$. Then for any function $w(x)$ tending to infinity as $x$ goes to infinity we have*

$$\Pr\left(D \geq w(t)\right) = o(1),$$

*where $D = st - rank(M)$.*

*Proof.* Note that if $|E_i| \geq t$ for $i \in \{1, \ldots, s\}$ then $D \leq \sum_{i=1}^{s} d(M^{(i)})$.

Let $S_1$ be the event $\sum_{i=1}^{s} d(M^{(i)}) \geq w(t)$ then using Lemma Appendix A.2 we have that $\Pr(S_1) = o(1)$. And let $S_2$ be the event of having at least one $E_i$ with $i \in \{1, \ldots, s\}$ such that $|E_i| < t$. Then the probability of the complement of event $S_2$ is given by

$$\Pr\left(\overline{S_2}\right) = \Pr\left(\bigcap_{i=1}^{s} |E_i| \geq t\right) = \prod_{i=1}^{s} \Pr\left(|E_1| \geq t, \ldots, |E_i| \geq t\right) = 1 - o(1)$$

by Lemma Appendix A.3. Then we deduce that the sought probability is

$$\Pr\left(D \geq w(t)\right) \leq \Pr\left(S_1 \cup S_2\right) \leq \Pr\left(S_1\right) + \Pr\left(S_2\right) = o(1).$$

□

# Self-exciting point process in modeling earthquake occurrences

**H. Pratiwi[1], I. Slamet[2], D. R. S. Saputro[3], Respatiwulan[4]**

[1,2,4]Statistics Study Program, Universitas Sebelas Maret, Jl. Ir. Sutami 36A, Surakarta 57126, Indonesia

[3]Mathematics Study Program, Universitas Sebelas Maret, Jl. Ir. Sutami 36A, Surakarta 57126, Indonesia

E-mail: hpratiwi@mipa.uns.ac.id, isnandarslamet@staff.uns.ac.id, dewiretnoss@staff.uns.ac.id, respatiwulan@staff.uns.ac.id

**Abstract.** In this paper, we present a procedure for modeling earthquake based on spatial-temporal point process. The magnitude distribution is expressed as truncated exponential and the event frequency is modeled with a spatial-temporal point process that is characterized uniquely by its associated conditional intensity process. The earthquakes can be regarded as point patterns that have a temporal clustering feature so we use self-exciting point process for modeling the conditional intensity function. The choice of main shocks is conducted via window algorithm by Gardner and Knopoff and the model can be fitted by maximum likelihood method for three random variables.

## 1. Introduction

A spatio-temporal point process is a random collection of points, where each point represents the time and location of an event. Examples of events include incidence of disease, sightings or births of a species, or the occurrences of fires, earthquakes, lightning strikes, tsunamis, or volcanic eruptions. Typically the spatial locations are recorded in three spatial coordinates, e.g. longitude, latitude, and height or depth, though sometimes only one or two spatial coordinates are available or of interest. One particularly useful approach to modeling spatio-temporal point processes is through the conditional intensity function which depend on the location and times of past events occurring in the time interval $(0, t)$ [9]. In this paper we analyze earthquake model in Java island using a spatial-temporal point process approach. We present self exciting point process where the earthquake magnitude as the mark.

Point process and its notation are defined in second section. Third section defines the truncated exponential distribution and fourth section defines the spatio-temporal point precess and the self-exciting point process. We consider the distributions of earthquake magnitudes, mainshocks, and aftershocks in estimating the intensity function. Finally we apply the model to the earthquake data in Java island in the last section.

## 2. Point process

Consider a sequence $(X_n)$ of random vectors in the state space $E$ and define, for $A \subset E$,

$$N(A) = \#\{i \geq 1 : X_i \in A\},$$

i.e., $N(A)$ counts the number of $X_i$'s falling into $A$. The state space $E$, where the points live, is a Borel subset of a finite-dimensional Euclidean space and $E$ is equipped with the $\sigma$-field $\mathcal{E} = \mathcal{B}(E)$ of the Borel sets generated by the open sets of $E$. It is convenient to write a point process using the Dirac measure $\varepsilon_x$ at $x \in E$:

$$\varepsilon_x(A) = I_A(x) = \begin{cases} 1 & \text{if } x \in A, \\ 0 & \text{if } x \notin A, \end{cases} \qquad A \in \mathcal{E}$$

For a given sequence $(x_i)_{i \geq 1}$ in $E$,

$$m(A) = \sum_{i=1}^{\infty} \varepsilon_{x_i}(A) = \#\{i \geq 1 : x_i \in A\}, \qquad A \in \mathcal{E}$$

define a counting measure on $\mathcal{E}$ which is called point measure if $m(K) < \infty$ for all compact sets $K \subset E$. This means that any compact set $K$ must not contain infinitely many points $x_i$.

Let $M_p(E)$ be the space of all point measures on $E$ equipped with the smallest $\sigma$-field $\mathcal{M}_p(E)$ which contains all sets of the form

$$\{m \in M_p(E) : m(A) \in B\}$$

for any $A \in \mathcal{E}$ and any Borel set $B \subset [0, \infty]$, i.e., it is the smallest $\sigma$-field making the maps $m \to m(A)$ measurable for all $A \in \mathcal{E}$ [6].

**Definition 1.** A point process $N$ with state space $E$ is a measurable map from the underlying outcome space $\Omega$ equipped with a $\sigma$-field $\mathcal{F}$ to $(M_p(E), \mathcal{M}_p(E))$.

In other words, a point process $N$ is random element or a random function which assumes point measures as values: for every $\omega \in \Omega$ the value $m(\cdot) = N(\cdot, \omega)$ is a point measure. In particular, $N(K) < \infty$ for compact sets $K \subset E$. The next result gives a justification of the fact that we may interpret a point process $N$ as a collection $(N(A))_{A \in \mathcal{E}}$ of the random variables $N(A)$ with values in $\{0, 1, \ldots, \infty\}$.

**Lemma 1.** *The mapping N from $(\Omega, \mathcal{F})$ to $(M_p(E), \mathcal{M}_p(E))$ is a point process on E if and only if, for every $A \in \mathcal{E}$, N(A) is a random variable with values in $\{0, 1, \ldots, \infty\}$ such that $N(A) < \infty$ for compact $A \subset E$.*

## 3. Truncated Exponential Distribution

The truncated exponential model is based on Gutenberg-Richter magnitude recurrence relation. The Gutenberg-Richter recurrence relation is expressed as:

$$log\ N(M) = a - bM \tag{1}$$

where *N(M)* is the mean number of earthquakes per unit volume and per unit time having magnitude greater than $M$ [4]. The $a$-value represents the number of events per unit time with magnitudes greater than zero, which describes the seismic activity rate in the area of interest. The $b$-value defines the ratio of small and large magnitude events and typically takes a value between 0.8-1.0 [3], [5]. Equation (1) describes the frequency of earthquakes as a power of

IOP Conf. Series: Journal of Physics: Conf. Series **855** (2017) 012033 doi:10.1088/1742-6596/855/1/012033

magnitude. Therefore, exponential probability distribution would describe the probability of occurrence rate of earthquake.

Although the standard Gutenberg-Richter recurrence relation can be applied to an infinite range of magnitudes, it is common to apply bounds at minimum and maximum magnitude values. This is because seismic sources are usually associated with a capacity to produce some maximum magnitude and for engineering purposes earthquakes of very small magnitudes that do not cause damage to structures are not of interest. If parameter of magnitude distribution is $\beta$ then the corresponding probability density function using these minimum and maximum values is expressed below in its bounded form [3]:

$$f(m) = \frac{\beta e^{-\beta m}}{e^{-\beta M_{min}} - e^{-\beta M_{max}}} \tag{2}$$

$M_{min}$ is the smallest magnitude considered in the model (here $M_{min} = 5$) and $M_{max}$ is the largest magnitude event which can be expected in the region. Different $M_{max}$ values may be assigned in different regions.

## 4. Self Exciting Point Process

A spatio-temporal point process $N$ is mathematically defined as a random measure on a region $S \subseteq \mathrm{R} \times \mathrm{R}^3$ of space and time, taking values in the non-negative integers $\mathrm{Z}^+$. In this framework the measure $N(A)$ represents the number of points falling in the subset $A$ of $S$. Any analytical spatiotemporal point process is characterized uniquely by its associated conditional intensity process $\lambda$[6]. $\lambda(t, x, y|H_t)$ may be thought of as the frequency with which events are expected to occur around a particular location $(t, x, y)$ in spacetime, conditional on the prior history, $H_t$, of the point process up to time $t$ [11].

A self exciting point processes can be used for modeling events that are clustered together in time and space. Consider events $\{(t_i, x_i, y_i, m_i) : i = 1 \ldots, N\}$, where $(x_i, y_i) \in D$ is location of $i$-th earthquake, $D \subset \mathrm{R}^2$, $t_i \in [0, T]$ is time of $i$-th earthquake, $m_i > m_0$ is magnitude of $i$-th earthquake, $m_0$ is cut off magnitude. The self-exciting point process captures the tendency for earthquakes to spawn aftershocks, clustered about their parents [10]. The conditional intensity function is

$$\lambda(t, x, y|H_t) = \mu(x, y) + \sum_{i:t_i<t} k(m_i)g(t - t_i)v(x - x_i, y - y_i|m_i) \tag{3}$$

where

(i) the function $\mu(x, y)$ represents the deterministic background rate and is assumed to take the semi-parametric form
$$\mu(x, y) = \mu u(x, y)$$
where $\mu$ is an unknown constant and $u(x, y)$ is an unknown function

(ii) mainshocks of magnitude $m$ produce a number of aftershock with distribution $k(m)$

(iii) aftershocks are spatially distributed about their parents according to the bivariate density function $v(\cdot)$

(iv) aftershocks are temporally distributed about their parents according to a nonnegative function $g(\cdot)$.

In application we make use of the specific function

$$k(m) = Ae^{\alpha(m-m_0)} \tag{4}$$

**Table 1.** Window Parameters

| $M$ | $\triangle R$ (km) | $\triangle T$ (days) |
|---|---|---|
| 2.5 | 19.5 | 6 |
| 3.0 | 22.5 | 11.5 |
| 3.5 | 26 | 22 |
| 4.0 | 30 | 42 |
| 4.5 | 35 | 83 |
| 5.0 | 40 | 155 |
| 5.5 | 47 | 290 |
| 6.0 | 54 | 510 |
| 6.5 | 61 | 790 |
| 7.0 | 70 | 915 |
| 7.5 | 81 | 960 |
| 8.0 | 94 | 985 |

for number of aftershocks. The temporal aftershocks follow Omori's Law: the frequency of aftershocks at time $t$ after the main shock decays hyperbolically [12]:

$$g(t) = (p-1)c^{p-1}(t+c)^{-p} \tag{5}$$

and the spatial distribution of aftershock is determined by the Gaussian density function [7]:

$$v(x,y|m) = \frac{1}{2\pi d e^{\alpha(m-m_0)}} exp\{-\frac{1}{2}\frac{x^2+y^2}{de^{\gamma(m-m_0)}}\} \tag{6}$$

For determining the intensity function (3) the declustering algorithm is applied. We focus on the window algorithms, where magnitude-dependent windows in space and time are used to identify earthquakes of the same cluster. Here we use the window method by Gardner and Knopoff to identify the mainshocks (Table 1) [1], [2]. Furthermore we should find the background intensity $u(x,y)$ first to estimate the conditional intensity function. An estimator of the background intensity can be estimated using thinning procedure and the model parameters $\theta = (\mu, A, c, \alpha, p, d, q, \gamma)$ are estimated simultaneously using an iterative approach [13], [14], [15]. The parameter vector $\theta$ is estimated by maximizing the log-likelihood function

$$l(\theta) = \sum_i \lambda(t_i, x_i, y_i|H_{t_i}) - \int \lambda(t,x,y|H_t)dxdydt.$$

## 5. Earthquake modeling

In this research we use the earthquake data in Java Island obtained from U.S. Geological Survey. The data was restricted to events with time period January 1973 - December 2010 from the rectangular area $5.46^o - 11^o$S and $104.98^o - 114.87^o$E. The data are shorted in ascending magnitude order and the cumulative number of events $N$, of magnitude $M$ or greater is found. A straight line fit to the plot of *log N* versus $M$ yields Gutenberg-Richter recurrence model with $R^2 = 99\%$ and $p$-value $= 0$. Figure 1 describes the frequency of earthquakes as a power of magnitude. Therefore, exponential probability distribution would describe the probability of occurrence rate of earthquakes. We take $M_{min} = 5$ and $M_{max} = 7.9$ and estimate $\beta$ in equation (2) using maximum likelihood method. Hence we have $\beta = 3.235$ and we obtain the magnitude distribution. Furthermore we apply Gardner-Knopoff window to identify mainshocks

IOP Conf. Series: Journal of Physics: Conf. Series **855** (2017) 012033 doi:10.1088/1742-6596/855/1/012033

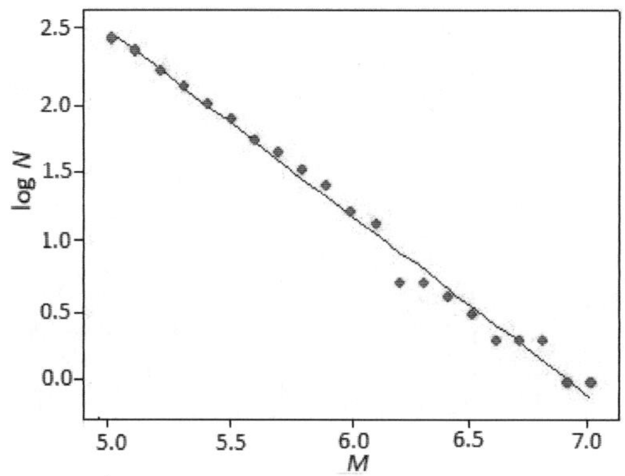


**Figure 1.** Plot of recurrence model, log $N = 8.890 - 1.285M$

and aftershocks. Here we consider shallow deep (depth $\leq$ 70 km)[8] with magnitude $\geq 5$ for mainshocks.

**Algorithm 1**

(i) Identify an event with magnitude $\geq 5$ and depth $\leq$ 70 km. Record this event by $i$ with magnitude $m_i$, location $(x_i, y_i)$ and occurrence time $t_i$, $i = 1, 2, \ldots$

(ii) For every event after $i$, say $ij$, $j = 1, 2, \ldots$ calculate

(a) $\triangle t_{ij} = |t_{ij}| - |t_i|$

(b) $\triangle x_{ij} = |x_{ij}| - |x_i| \times 111.3195$

(c) $\triangle y_{ij} = |y_{ij}| - |y_i| \times 111.3195$

(d) $\triangle r_{ij} = \sqrt{\triangle x_{ij}{}^2 + \triangle y_{ij}{}^2}$

Calculation is done for events $ij$ having magnitude $\leq m_i$ and satisfy window parameters (Table 1), i.e. $\triangle r_{ij} \leq \triangle R$ and $\triangle t_{ij} \leq \triangle T$.

(iii) Stop step (ii) if Gardner-Knopoff criteria is not satisfied, go to step (i).

**Algorithm 2**

(i) Input the length of the time intervals, number of events, starting points of time, the magnitude threshold, the starting point of time target interval, the number of points for the polygon region, number of grids for background.

(ii) Input an initial $u_0(x, y)$.

(iii) Input initial parameters.

(iv) Fit the conditional intensity function $\lambda(t, x, y|H_t)$.

(v) Calculate the background rate at the position of all events $\hat{\mu}(x, y)$.

(vi) Calculate probability of the $j$th earthquake is trigger by $i$th earthquake $\rho_{i,j}$.

(vii) Calculate probability of the $j$th event belongs to the background $\varphi_j$.

(viii) Steps (iv) - (vii) are repeated enough times such that the results converge.

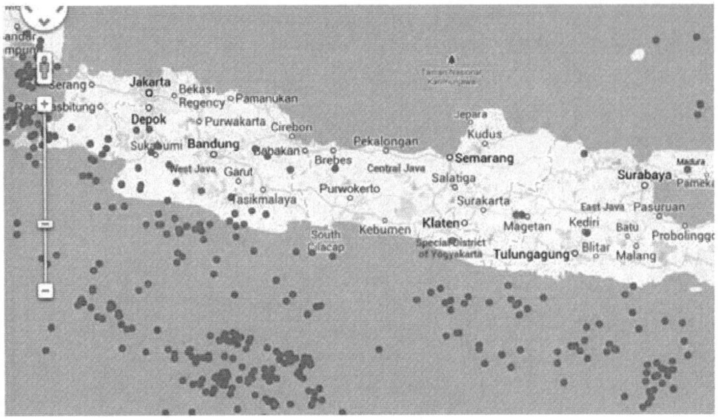


**Figure 2.** Epicentral locations of mainshocks

**Table 2.** Maximum Likelihood Estimates for the Java earthquake

| parameter | estimate |
|---|---|
| $\mu$ | 0.822674974 |
| $A$ | 0.325757138 |
| $c$ | 0.026135075 |
| $\alpha$ | 1.408691215 |
| $p$ | 1.325195590 |
| $d$ | 0.004627986 |
| $q$ | 1.501570185 |
| $\gamma$ | 1.020769545 |

By using Algorithm 1 we can identify 45 mainshocks from 491 events including three great earthquakes, the 2006 Yogyakarta earthquake, the 2006 Ciamis-Cilacap earthquake, and the 2009 Tasikmalaya earthquake. The epicentral locations of mainshocks is presented in Figure 2 while Table 2 shows the maximum likelihood estimates for parameters of the self-exciting point process obtained from Algorithm 2. This estimates have log likelihood = -1589.4533 and AIC = 3194.9. From the probability $\rho_j$ we can see that most of the events are either independent events or aftershocks. About 36.46% of the events have probability of being an aftershock. Figure 3 shows that high values of conditional intensity function are located in $112.019^o E, 6.969^o S$ (Java Sea, north of Tuban, East Java), $113.247^o E, 7.15^o S$ (Sampang, Madura Island), $114.521^o E, 7.27^o S$ (Madura Strait), $105.756^o E, 10.429^o S$ (Hindia Ocean, south of Banten Province).

Although we obtain maximum likelihood estimates for the parameters of the model, there remains some room for further improvement. In this model, information on shocks less than 5 in magnitude was not available. Modifications of (4) - (6) can be made to model variability in mean aftershock number and the spatio-temporal clustering of aftershocks. However we should be care to avoid over parametrization of model.

## 6. Conclusion

A probabilistic approach for earthquake data in this research is self-exciting point process. This model consist of three variables and eight parameters to be estimated simultaneously. Four high values of the conditional intensity function in Java Island are located in the north of Tuban regency, Sampang regency, Madura strait, and the south of Banten Province. From the

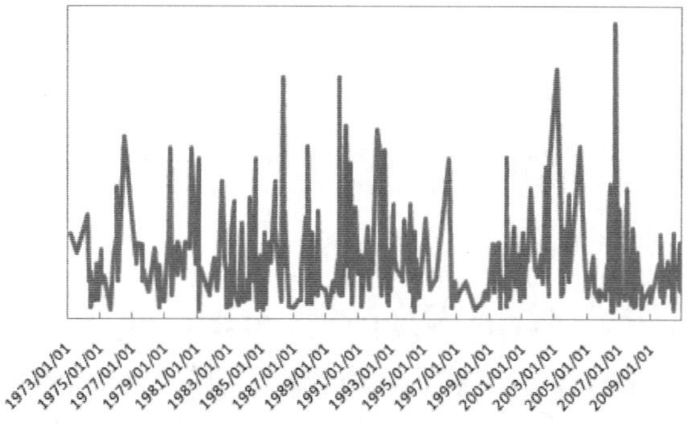


**Figure 3.** Plot of conditional intensity function

algorithm, we also obtain the probability of the event being an aftershock.

## 7. Acknowledgments

We are grateful to Universitas Sebelas Maret for financial support through Grant of Maintenance Research Group 2015. We also thank referees for their critical and helpful comments.

## 8. References

[1] Christophersen, A., M.C. Gerstenberger, D.A. Rhoades, and M.W. Stirling, Quantifying the Effect of Declustering on Probabilistic Seismic Hazard, *Proceedings of the Ninth Pacific Conference on Earthquake Engineering Building an Earthquake-Resilient Society*, Auckland, New Zealand, 2011.

[2] Gardner, J.K. and L. Knopoff, Is the Sequence of Earthquakes in Southern California, with Aftershocks Removed, Poissonian?, *Bulletin of the Seismological Society of America*, vol. **64**, No. 5, pp. 1363-1367, 1974.

[3] Godinho, J., Probabilistic Seismic Hazard Anaylisis - An Introduction to Teeoritical Basis and Applied Methodology, *Dissertation*, University of Patras, Greece, 2007.

[4] Knopoff, L., The Magnitude Distribution of Declustered Earthquakes in Southern California, *Proceedings of the National Academy of Sciences of the United States of America*, Vol. **97**, No. 22, pp. 11880-11884, 2003.

[5] Marzocchi, W. and L. Sandri, A Review and New Insights on the Estimation of the b-value and Its Uncertainty, *Annals of Geophysics*, Vol. **46**, No. 6, pp. 1271-1282, 2003.

[6] Mikosch, T., *Non-Life Insurance Mathematics*, 2$^{nd}$ ed., Berlin: Springer-Verlag, 2009.

[7] Musmeci, F. and D. Vere-Jones, A Space-Time Clustering Model for Historical Earthquakes, *Annals of the Institute of Statistical Mathematics*, Vol. **44**, No. 1, pp. 1-11, 1992.

[8] NEIC, *M7.5 Northern Peru Earthquake of 26 September 2005*. National Earthquake Information Center, U.S. Geological Survey, October 17, 2005.

[9] Pratiwi, H., Subanar, and Danardono, Analysis of Earthquake Data Based on Space-Time Epidemic Type Aftershocks Sequence Model, *International Journal of Applied Mathematics and Statistics*, vol. **52**, No. 5, 2014.

[10] Rathbun, S. L., Asymptotic Properties of the Maximum Likelihood Estimator for Spatio-Temporal Point Process, *Journal of Statistical Planning and Inference*, vol. **51**, pp. 55-74, 1996.

[11] Schoenberg, F. P., D. R. Brillinger, and P. Guttorp, Point Processes, Spatial-Temporal, *Encyclopedia of Environmetrics*, Vol. 3, pp. 1573-1577, 2002.

[12] Vere-Jones, D., Seismology - A Statistical Vignette, *Journal of the American Statistical Association*, Vol. **95**, No. 451, pp. 975-978, 2000.

[13] Zhuang, J., Second-order Residual Analysis of Spatio-temporal Point Processes and Applications in Model Evaluation, *Journal of the Royal Statistical Society: Series B (Statistical Methodology)*, Vol. **4**, No. 68, pp. 635-653, 2006.

[14] Zhuang, J., Y. Ogata, and D. Vere-Jones, Analyzing Earthquake Clustering Features by Using Stochastic Reconstruction, *Journal of Geophysical Research*, Vol. **109**, No. B5, B05301, 2004.

[15] Zhuang, J., Y. Ogata, and D. Vere-Jones, Stochastic Declustering of Space-Time Earthquake Occurrences, *Journal of the American Statistical Association*, Vol. **97**, No. 458, pp. 369-380, 2002.

# The Completion of Non-Steady-State Queue Model on The Queue System in Dr. Yap Eye Hospital Yogyakarta

**Arum Helmi Manggala Putri[1], Retno Subekti[2], and Nikenasih Binatari[3]**
Department of Mathematics Education, Faculty of Mathematics and Natural Science, Yogyakarta State University, Indonesia

E-mail: Arumhelmi.manggalaputri@gmail.com [1], retnosubekti@uny.ac.id [2], nikenasih@uny.ac.id [3]

**Abstract.** Dr Yap Eye Hospital Yogyakarta is one of the most popular reference eye hospitals in Yogyakarta. There are so many patients coming from other cities and many of them are BPJS (Badan Penyelenggara Jaminan Sosial, Social Security Administrative Bodies) patients. Therefore, it causes numerous BPJS patients were in long queue at counter C of the registration section so that it needs to be analysed using queue system. Queue system analysis aims to give queue model overview and determine its effectiveness measure. The data collecting technique used in this research are by interview and observation. After getting the arrival data and the service data of BPJS patients per 5 minutes, the next steps are investigating steady-state condition, examining the Poisson distribution, determining queue models, and counting the effectiveness measure. Based on the result of data observation on Tuesday, February 16th, 2016, it shows that the queue system at counter C has (M/M/1):(GD/∞/∞) queue model. The analysis result in counter C shows that the queue system is a non-steady-state condition. Three ways to cope a non-steady-state problem on queue system are proposed in this research such as bounding the capacity of queue system, adding the servers, and doing Monte Carlo simulation. The queue system in counter C will reach steady-state if the capacity of patients is not more than 52 BPJS patients or adding one more server. By using Monte Carlo simulation, it shows that the effectiveness measure of the average waiting time for BPJS patients in counter C is 36 minutes 65 seconds. In addition, the average queue length of BPJS patients is 11 patients.

*Keywords: queue system, (M/M/1):(GD/∞/∞), non-steady-state, Monte Carlo simulation*

## 1. Introduction

The queue is a waiting line that usually happens when the demand for service is over capacity. In daily life, the queue case can be seen and even felt directly by society especially the queue of public facilities. One of the examples is in Dr. Yap Eye Hospital which is the one and only eye hospital in Yogyakarta. Dr. Yap Eye Hospital has cooperated in medical services especially the eye medical with the government instance. The newest medical service from the government is *BPJS* (Badan Penyelenggara Jaminan Sosial, Social Security Administrative Bodies). BPJS is an Indonesian government-owned enterprise which is obliged to provide protection for socio-economic. That corporation will minimize the cost of medicine since it has been guaranteed by *BPJS* company. Thus there are many *BPJS* patients coming to Dr. Yap Eye Hospital. Since there are so many *BPJS* patients which are over capacity, it causes patients hoarding. The patients hoarding that mostly occurs is in the counter C of the registration section. This is the result of arrival rate that exceeds the service rate of *BPJS* patients. Therefore, it requires a perfect decision to increase the quality of the service for the patients especially the *BPJS* ones. Queue theory used to calculate systematically, so that it can solve the problems of queue system. This theory is also use to predict the important parameter such as the average waiting time and the average queue length of *BPJS* patients [1].

Several research that relevant to queue system were done by Ratna [2], William, Lawrence & John [3], Toshiba, Sanjay & Anil [4] and so on. Ratna [2] applied on Jamkesmas patients data in RSUD Dr. Karyadi Semarang. The analysis of non-steady-state queue system, or transient case, has been done by William, Lawrence & John [3] and the result was achieved by developing computation formula from both symbolic and numeric exact where result are checked against Monte Carlo simulation. Another simple queue model *(M/M/1/∞)* was implemented by Toshiba, Sanjay & Anil [4] in Bank service to improve the optimal service rate.

Complex queue case can be solved by using simulation. Simulation can be used to represent a reality that occurs in the queue system. Simulation can be defined as a process of designing a model of a real system and conducting experiments with this model for the purpose of understanding the behaviour of the system [5]. One of the applications of Monte Carlo queueing system can be found in the patient queue [1]. Monte Carlo simulation is a technique that converts uncertainties of input variables in the model into probability distributions [6]. In that simulation, it requires a random number generator to re-form the opportunity distribution [7]. Monte Carlo simulation is a very effective technique, widely accepted for true result. Other ways to solve the problems of complex queue system are by restricting the queue system capacity or adding the number of the servers. Thus, the queue system is able to reach stability or steady-state.

Steady-state condition is a stable state where the arrival rate is less than the service rate and it can be determined by searching utilization. Utilization refer to the proportion of time that a server (or system of servers) is busy handling customers [8]. Utilization should be strictly less than one for the system to function well (steady-state condition). Utilization is usually represented by the symbol $\rho$ [9]. If $\rho > 1$ then the arrival rate is faster than it can be served. This shows that the queue length is expected to increase unlimited so it is non-steady-states. Also, if $\rho = 1$ then the arrival rate is equal to the service rate so that the queue cannot occur. In other words, steady-state will not occur in the queue system.

## 2. Research Methodology

The research methodology used in this research was descriptive. The descriptive methodology aims to acquire information related to what recently happens and to see the relevancy among the available variables [10]. It has several steps: collecting data, analyzing data, and drawing a conclusion.

### *2.1 Research Location and Time*

This research was done in Dr. Yap Eye Hospital Yogyakarta located in Teuku Cik Ditiro 5 Yogyakarta. The data collection was taken in counter C which is in registration section. This research was taken on Tuesday, February 16th, 2016 at 06.00 am to 13.00 pm.

### *2.2 Data Collecting Technique*

The data used in this research was primary and secondary data. The primary data was gained by interview and observation. The interview resulted in queue system design for *BPJS* patients while the observation resulted in the arrival data and the service data of *BPJS* patients. Next, the secondary data was gained by medical record installation document. The document was in the form of the printout of the *BPJS* patients queue number in counter C.

*2.3 Data Analysis and Non-Steady-state Queue System Problem Solving*

*2. 3. 1 Data Analysis*
Data analysis was performed in counter C of the *BPJS* patients registration section. The steps to analyze the data were as follows:

a. Categorized the primary data on Tuesday in counter C consisting of the arrival and service data of *BPJS* patients every 5 minutes. The data grouping taken in this research was the data at 08.00 am to 11.00 am.

b. Sought the average arrival rate ($\lambda$) and the average service rate ($\mu$), then investigated the steady-state solution. The queue system will reach the steady-state when $\rho = \lambda/\mu < 1$. It means the average arrival rate was less than the average service rate or in other words the queue system reached the steady-state in which then was performed Monte Carlo simulation to seek the effectiveness measure.

c. Examined the Poisson distribution of arrival time data and service time data was gained by Kolmogorov-Smirnov test.

With the objective to improve the case BPJS queue system, the following assumptions are considered:

1) The arrival of patients is identically independent and follow the Poisson probability distribution with parameter $\lambda$ [11] and given by:
$$P_n(t) = \frac{e^{-\lambda t}(\lambda t)^n}{n!}; \qquad n = 1,2,3 \ldots t > 0$$
Where $P_n(t)$ indicates the probability of arriving $n$ patients in time interval $t$
2) The service time of patients is identically independent which meets negative distribution with parameter $\mu$ and is given by:
$$F(t) = 1 - e^{-\mu t}; \qquad t > 0$$

d. Determined the appropriate model using queue system in the counter C of registration section.

e. Counted the effectiveness measure of *BPJS* patients queue system including counting the average waiting time on the queue and the average queue length.

Queue system analysis can be depicted in the flow chart, Figure 1.

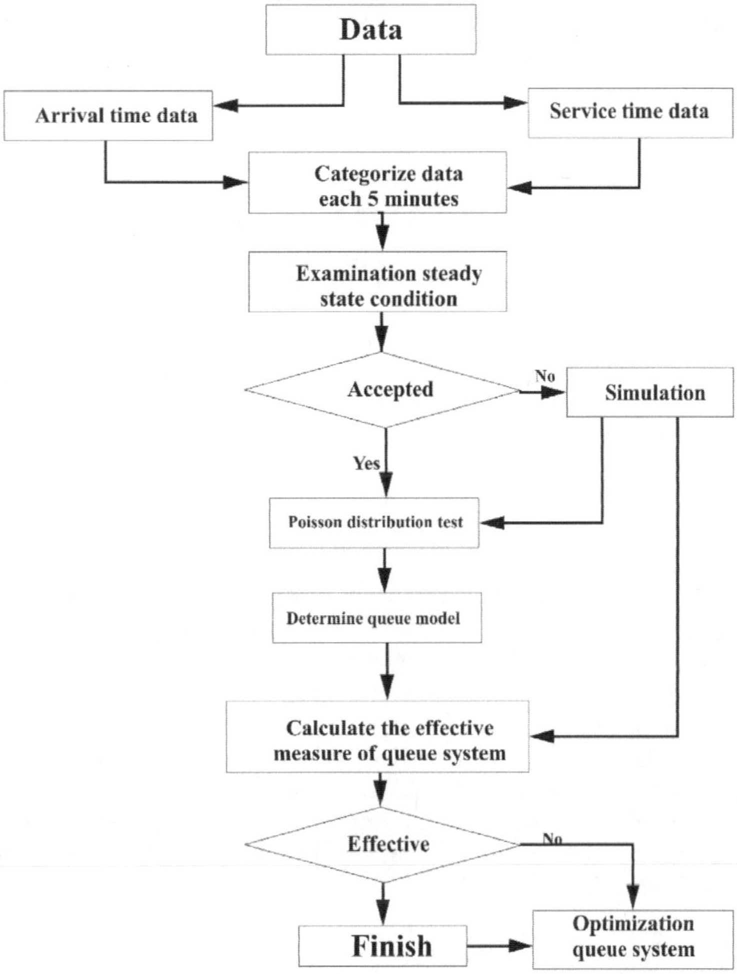


Figure 1. Analysis Data Flow

*2.3.2 Solution of non-steady-state queue system problems*

Characteristic and assumption of queue model in counter C can be summarized into Kendall notation from $(M/M/1)$: $(GD/\infty/\infty)$. In queue model $(M/M/1)$: $(GD/\infty/\infty)$, the inter-arrival times and service times are assumed that they are distributed Exponential [12]. Moreover, queue system in counter C only has one server and unlimited capacity of calling source. Figure 2 is the diagram which shows the model of $(M/M/1)$: $(GD/\infty/\infty)$.

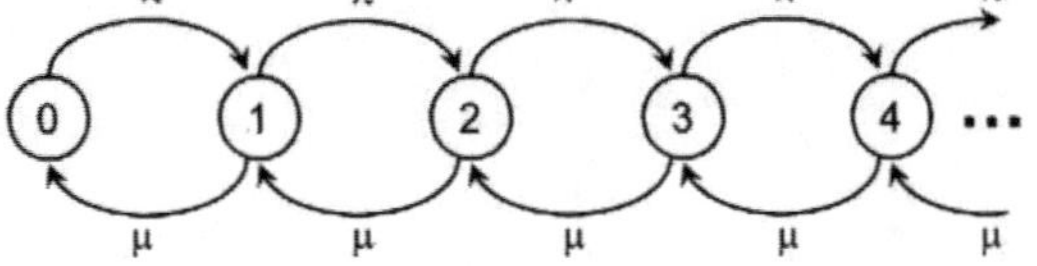


Figure 2. Rate transition diagram for the $(M/M/1)$ queue

Furthermore, for example, $n$ shows the number of customer. Arrival system is assumed as a birth and a departure/service is assumed as a death of the queue system. The average arrival rate ($\lambda$) and the average service rate ($\mu$) do not rely on the number of customer in queue system. Therefore, the queue model $(M/M/1):(GD/\infty/\infty)$ is the process of a birth and a death, where $\lambda_n = \lambda\ (n \geq 0)$ and $\mu_n = \mu\ (n \geq 1)$. Based on this process, we have :

$$P_n = \left(\frac{\lambda}{\mu}\right)^n P_0$$

Assumed that $\rho = {}^{\lambda}/_{\mu}$ , then

$$P_n = (\rho)^n P_0$$

On the queue model $(M/\ M/1):(GD/\infty/\infty)$, it is obtained the value of $P_0$ and $P_n$:

$$P_0 = 1 - \rho$$

$$P_n = (1-\rho)\rho^n \qquad n \geq 0$$

assumed if $\rho = {}^{\lambda}/_{\mu} < 1$. In some cases that are found, it shows that the average arrival rate ($\lambda$) is very large, thus $\rho > 1$. This is a non-steady-state case $(M/M/1):(GD/\infty/\infty)$. As a result, the analysis of queue system cannot be solved by analytic model. The solving problem of non-steady-state queue system can be run by some ways, we propose that three ways to cope the problem:

a. Monte Carlo simulation

One of the alternative solutions of the non-steady-state queue system is Monte Carlo simulation. It is caused by non-steady-state queue system that cannot be solved by using analytic model, so that Monte Carlo simulation is the best way. Monte Carlo simulation is an approach for reforming probability distribution based on a choice or a random number. Figure 3 depicts the steps to do Monte Carlo simulation in the flowchart:

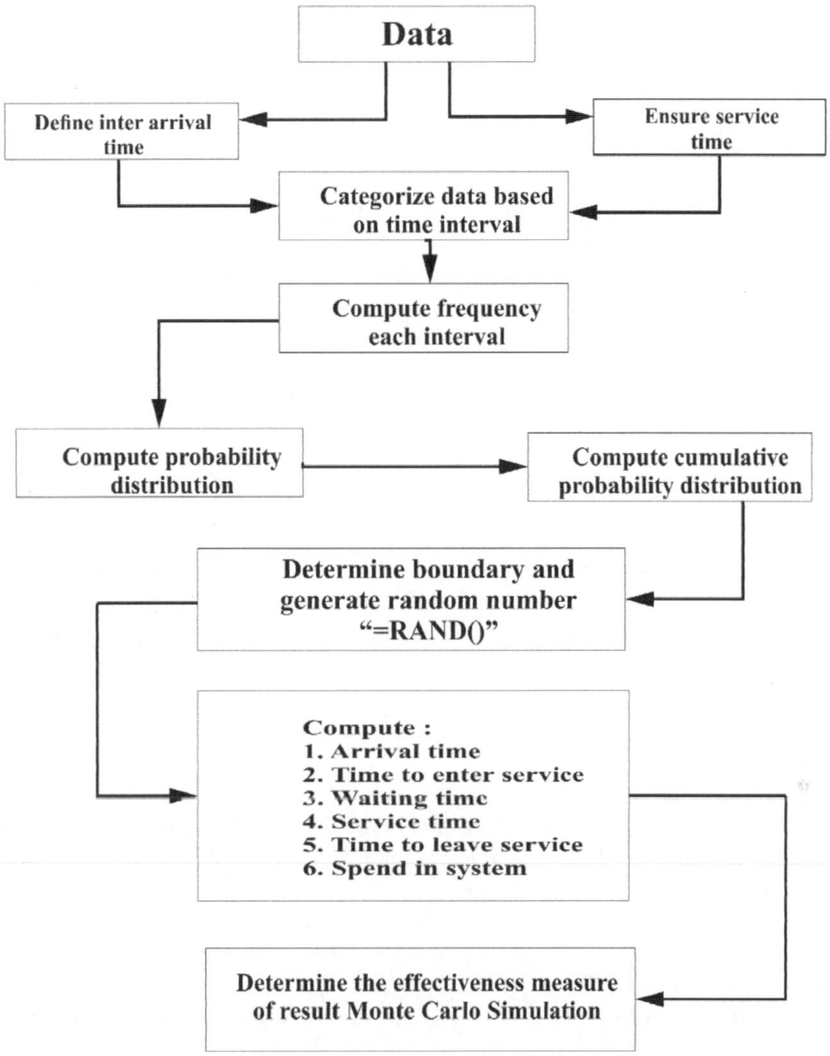


Figure 3. Monte Carlo simulation

b. Bounding the capacity of the queue system

Besides doing Monte Carlo simulation, another way that can be run is bounding the capacity of queue system. This restriction of the capacity of queue system was carried out by reducing the service data of *BPJS* patients one by one that has classified per 5 minutes. After it was reduced one by one, then the service data classified again per 5 minutes and counted again until the number of service rate ($\mu$) less than the arrival rate ($\lambda$). Therefore, the average service rate ($\mu$) will be smaller and queue system reach the steady-state condition.

c. Add the number of a server at queue system.

Besides the two ways described previously, the next way to solve the queue system that is non-steady-state is adding the number of servers. Adding the number of servers can be done until the queue system reached steady-state condition. Thus, after the queue system reach the steady-state condition then we can count the effectiveness by using an analytical model.

## 3. Result and Discussion

### *3.1 Result of research*

The results of research which done in counter C of registration section is the average arrival rate and the average service rate of *BPJS* patients. In addition, the Kolmogorov-Smirnov test using *SPSS* software can be known that the arrival data and service data of *BPJS* patients are Poisson distribution. The result of research the average arrival rate and the average service rate after data are grouped per 5 minutes can be shown in Table 1.

Table 1. Steady-state condition

| Name | $\lambda$ | $\mu$ | $\rho = \frac{\lambda}{\mu}$ | *Steady State* |
|---|---|---|---|---|
| Counter C | 0.4056 | 0.3388 | 1.1972 | Not yet |

Based on Table 1 the queue system at counter C has not reached steady-state conditions because of $\rho > 1$. This cause the effectiveness measure cannot be calculated using analytical models. Therefore, to solve the non-steady-state queue system by using Monte Carlo simulation, bounding of capacity queue system and added the servers.

### *3.2 Discussion*

#### *3. 2. 1 Model of queue system*

The discussion of the queue system analysis is obtained appropriate queue model at counter C of registration section that follows *(M/M/1):(GD/∞/∞)* model. The *(M/M/1):(GD/∞/∞)* queue model have one server that is a computer and have discipline service called General Discipline or First Come First Served. In addition, the capacity of queue system and calling sources of *BPJS* patients is unlimited.

#### *3. 2. 2 The effectiveness measure of non-steady-state queue system*

The problem of non-steady-state queue system to calculate the effectiveness measure can be solve in the following way:

a. Monte Carlo simulation

Procedure of Monte Carlo simulation using MS. Excel [13] as follows:

- Determine the inter-arrival time, the cumulative probability, and random number interval.

| Inter-Arrival Time (minutes), $x$ | Frequency $f(x)$ | $P(x) = \frac{f(x)}{\sum_x f(x)}$ | Cumulative Probability | Random Number Interval |
|---|---|---|---|---|
| 0 | 12 | 0.1644 | 0.0000 | 0.0000-0.1643 |
| 1 | 20 | 0.2740 | 0.1644 | 0.1644-0.4383 |
| 2 | 16 | 0.2192 | 0.4384 | 0.4384-0.6574 |
| 3 | 9 | 0.1233 | 0.6575 | 0.6575-0.7807 |
| 4 | 3 | 0.0411 | 0.7808 | 0.7808-0.8218 |

| **Inter-Arrival Time (minutes), $x$** | **Frequency $f(x)$** | $P(x) = \frac{f(x)}{\sum_x f(x)}$ | **Cumulative Probability** | **Random Number Interval** |
|---|---|---|---|---|
| 5 | 7 | 0.0959 | 0.8219 | 0.8219-0.9177 |
| 6 | 2 | 0.0274 | 0.9178 | 0.9178-0.9451 |
| 8 | 1 | 0.0137 | 0.9452 | 0.9452-0.9588 |
| 10 | 1 | 0.0137 | 0.9589 | 0.9589-0.9725 |
| 12 | 1 | 0.0137 | 0.9726 | 0.9726-0.9862 |
| 13 | 1 | 0.0137 | 0.9863 | 0.9863-0.9999 |
| | ∑=73 | ∑=1.0000 | ∑=1.0000 | |

- Determine the service time, cumulative probability and random number interval

| **Service Time (minutes),$y$** | **Frequency $f(y)$** | $P(y) = \frac{f(y)}{\sum_y f(y)}$ | **Cumulative Probability** | **Random Number Interval** |
|---|---|---|---|---|
| 0 | 3 | 0.0492 | 0.0000 | 0.0000-0.0491 |
| 1 | 16 | 0.2623 | 0.0492 | 0.0492-0.3114 |
| 2 | 14 | 0.2295 | 0.3115 | 0.3115-0.5409 |
| 3 | 9 | 0.1475 | 0.5410 | 0.5410-0.6884 |
| 4 | 11 | 0.1803 | 0.6885 | 0.6885-0.8688 |
| 5 | 2 | 0.0328 | 0.8689 | 0.8689-0.9015 |
| 6 | 1 | 0.0164 | 0.9016 | 0.9016-0.9179 |
| 8 | 2 | 0.0328 | 0.9180 | 0.9180-0.9507 |
| 14 | 2 | 0.0328 | 0.9508 | 0.9508-0.9835 |
| 23 | 1 | 0.0164 | 0.9836 | 0.9836-0.9999 |
| | ∑=61 | | ∑=1.0000 | |

- After we obtained random number interval for each cumulative probability of arrival and services data of *BPJS* patients, then we make 73 trial rows of arrival *BPJS* patients as shown in Table 2.

IOP Conf. Series: Journal of Physics: Conf. Series **855** (2017) 012034 doi:10.1088/1742-6596/855/1/012034

Table 2. Monte Carlo Simulation

| | Time) R1 | - (Minutes) | | | | | (Service Time) R2 | (Minutes) | Leave Service s |
|---|---|---|---|---|---|---|---|---|---|
| 1 | | | 0 | 0 | 0 | 0 | 0.1484 | 1 | 1 |
| 2 | 0.5819 | 2 | 2 | 2 | 0 | 0 | 0.5421 | 3 | 5 |
| 3 | 0.1413 | 0 | 2 | 5 | 3 | 1 | 0.8849 | 5 | 10 |
| 4 | 0.2517 | 1 | 3 | 10 | 7 | 2 | 0.3384 | 2 | 12 |
| 5 | 0.2166 | 1 | 4 | 12 | 8 | 3 | 0.9336 | 8 | 20 |
| 6 | 0.1505 | 0 | 4 | 20 | 16 | 4 | 0.9269 | 8 | 28 |
| 7 | 0.7623 | 3 | 7 | 28 | 21 | 4 | 0.7954 | 4 | 32 |

| | | | | | | | | | |
|---|---|---|---|---|---|---|---|---|---|
| . | . | . | . | . | . | . | . | . | . |
| . | . | . | . | . | . | . | . | . | . |
| . | . | . | . | . | . | . | . | . | . |
| 71 | 0.2827 | 1 | 184 | 230 | 46 | 10 | 0.8206 | 4 | 234 |
| 72 | 0.7865 | 4 | 188 | 234 | 46 | 10 | 0.1079 | 1 | 235 |
| 73 | 0.6873 | 3 | 191 | 235 | 44 | 11 | 0.7825 | 4 | 239 |

After doing the computation, the result can be used to calculate the average waiting time in queue ($W_q$) and length of queue ($L_q$). Here are the results of the calculation of average waiting times and length queues with repetition as much as 10 times:

Table 3. Repetition result of Monte Carlo Simulation

| **Repetition** | **The effectiveness measure** | |
|---|---|---|
| | **Average waiting time in queue ($W_q$)** | **Average queue length ($L_q$)** |
| 1 | 32.74 | 10.59 |
| 2 | 32.29 | 9.55 |
| 3 | 33.26 | 9.49 |
| 4 | 38.51 | 11.07 |
| 5 | 31.41 | 10.10 |
| 6 | 42.42 | 12.70 |
| 7 | 40.44 | 13.75 |
| 8 | 30.85 | 7.75 |
| 9 | 34.48 | 13.40 |
| 10 | 42.45 | 13.30 |

Furthermore, we make a graph to find the midpoint of both the average waiting time in queue and the average queue length. The values are currently being sought by using the lower and upper limits. In Table 3 it can be seen that the lower limit value W$q$ currently on the 8[th] repetition and the upper limit is at the 10[th] repetition. The lower limit value contained in the L$q$ replicates the 8[th] and the upper limit on the repetition 7[th]. Two graphs in Figure 4 and Figure 5 are the visualization of Table 3:

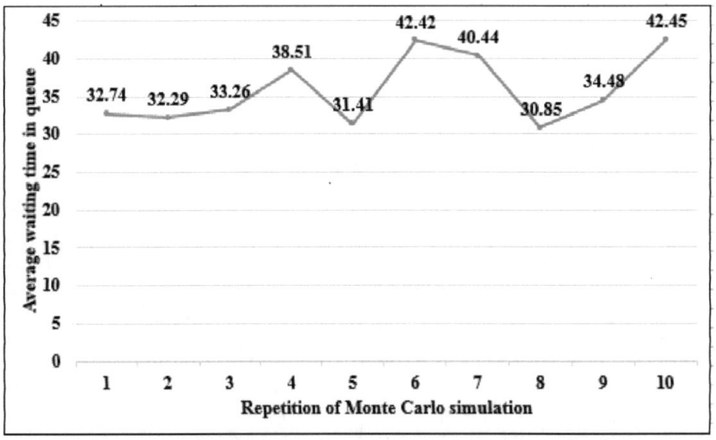


Figure 4. Repetition of average waiting time ($W_q$)

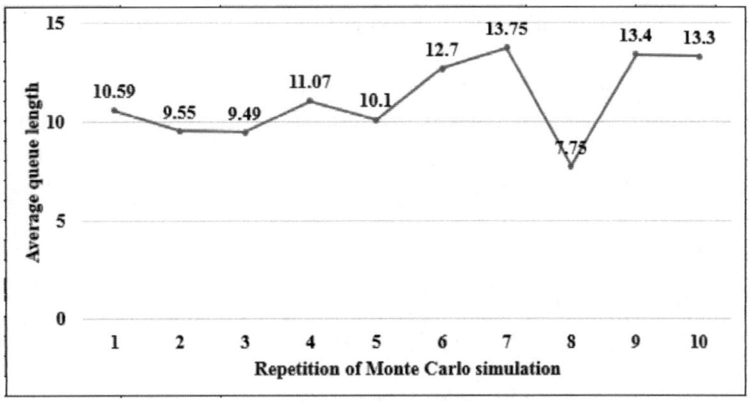


Figure 5. Repetition of average queue length ($L_q$)

Based on the previous chart, it can be seen that the average waiting time and length of queue are:

$$W_q = \frac{30.85 + 42.45}{2} = 36.65\ minutes$$

$$L_q = \frac{7.75 + 13.75}{2} = 10.75 \approx 11\ patients$$

b. Bounding capacity of the queue system

Bounding capacity of the queue system is the second way to solve a problem of non-steady-state queue system. The following service data of *BPJS* patients when it was reduced one by one and regrouped per 5 minutes:

Table 4. Service data of *BPJS* patients based on the interval per 5 minutes

| **Interval with *I* services** | **The number of *BPJS* patients services at intervals $I_i$ ($K_i$)** | **The frequency or the number of intervals $I_i$ ($f(I_i)$)** | **The number of *BPJS* patients who served during the period $I_i$ ($K_i \times f(I_i)$)** |
|---|---|---|---|
| $I_0$ | 0 | 9 | 0 |
| $I_1$ | 1 | 8 | 8 |
| $I_2$ | 2 | 13 | 26 |
| $I_3$ | 3 | 6 | 18 |
| $I_4$ | 4 | 0 | 0 |
| | | $\sum I = 36$ | $\sum N = 52$ |

Based on the service data of *BPJS* patients regrouping per 5 minutes, obtained value of the average service rate is less than previous one:

$$\mu = \frac{\sum N}{\sum I} = \frac{52}{36} = 1.4444\ \frac{patients}{5\ minutes} = 0.2889\ \frac{patients}{minutes}$$

So that utility level of queue system at counter C can be recalculate as follows:

Table 5. Steady-state condition after bounding capacity

| **Name** | $\lambda$ | $\mu$ | $\rho = \frac{\lambda}{\mu}$ | ***Steady State*** |
|---|---|---|---|---|
| Counter C | 0.2722 | 0.2889 | 0.9422 | Yes |

Based on Table 5, the queue system at counter C will reach steady-state condition when queue capacity is bounded for 52 *BPJS* patients. Because it bounded for 52 *BPJS* patients, so that queue model becomes $(M/M/1):(GD/52/\infty)$. Then, the effectiveness measure of queue system at counter C can be calculate using analytic formula $(M/M/1):(GD/N/\infty)$ as follow:

$$W_s = \frac{1}{\mu\left(1-\frac{\lambda}{\mu}\right)} \tag{1}$$

$$= \frac{1}{0.2889(1-\frac{0.2722}{0.2889})} = 59.8802\ minutes$$

$$L_s = \frac{\frac{\lambda}{\mu}}{\left(1-\frac{\lambda}{\mu}\right)} \tag{2}$$

$$= \frac{\frac{0.2722}{0.2889}}{(1-\frac{0.2722}{0.2889})} = 16.3010 = 16\ patients$$

So from capacity bounding of queue system generate the average waiting time is 59.8802 minutes and the average queue length is 16 patients.

c. Adding servers of queue system

The another way to solve non-steady-state queue system problem is adding servers. After adding one server, the queue model is $M/M/C$ and the utility level for each server is $r = \lambda/_{c\mu}$ [12]. The Table 6 bellow is steady-state condition of queue system at counter C:

Table 6. Steady-state condition after added server

| **Name** | $c$ | $\lambda$ | $\mu$ | $r = \frac{\lambda}{c\mu}$ | ***Steady State*** |
|---|---|---|---|---|---|
| Counter C | 2 | 0.4056 | 0.3388 | 0.5986 | Yes |

Based on the Table 6 the queue system at counter C will reach steady-state condition when adding one more server. Therefore, the model of queue system at counter C become $(M/M/2):(GD/\infty/\infty)$. So that, the effectiveness measure at counter C can be calculate using analytic formula $(M/M/c):(GD/\infty/\infty)$ as follows :

$$\lambda = 0.4056\ ;\ \mu = 0.3388\ ;\ c = 2$$

$$\rho = \frac{\lambda}{\mu} = \frac{0.4056}{0.3388} = 1.1972$$

$$P_0 = \left\{ \sum_{n=0}^{c-1} \frac{\rho^n}{n!} + \frac{\rho^c}{c!} \left( \frac{1}{1 - \frac{\rho}{c}} \right) \right\}^{-1} \tag{3}$$

$$= \left\{ \sum_{n=0}^{2-1} \frac{1.1972^n}{n!} + \frac{1.1972^2}{2!} \left( \frac{1}{1 - 1.1972/2} \right) \right\}^{-1} = 0.2511$$

$$L_s = \left[ \frac{\rho^{c+1}}{(c-1)!\,(c-\rho)^2} \right] P_0 + \rho \tag{4}$$

$$= \left[ \frac{1.1972^{2+1}}{(2-1)!\,(2-1.1972)^2} \right] 0.2511 + 1.1972 = 1.8657 \approx 2\, patients$$

$$W_s = \frac{\rho^c}{\mu(c-1)!\,(c-\rho)^2} P_0 + \frac{1}{\mu} \tag{5}$$

$$= \frac{1.1972^2}{0.3388(1)!\,(2-1.1972)^2} 0.2511 + \frac{1}{0.3388} = 4.5998\, minutes$$

So, the queue system resolution can be done by adding the number of servers results the average queue length is 2 patients and the average waiting time is 4.5998 minutes.

## 4. Conclusion

Based on the results of this research and discussion of the queue system at the counter C, *BPJS* patients Dr. Yap Eye Hospital Yogyakarta is following *(M/M/1):(GD/∞/∞)* model. We obtained that queue system at the counter C was non-steady-state. It means the average arrival rate is more than the average service rate. Therefore the effectiveness measure of the queue system in this research can be solved by three ways such as doing Monte Carlo simulation, bounding the capacity of queue system and adding the number of servers.

First, Monte Carlo simulation works with new trials based on random numbers generated by Ms. Excel and referred to the real data. The average waiting time and length of queue are calculated using the these new trials. By using Monte Carlo simulation, it shows that the average waiting time in queue for *BPJS* patients at counter C is 36 minutes 65 seconds and the average queue length is 11 patients. Second, bounding the capacity of the new system means we reduced the number of customers one by one until it reachs steady-state condition. By using this way, the arrival capacity of patients is less than 52 *BPJS* patients. Last, the third solution is by adding the number of server one by one until it reachs steady-state condition. Since the system will follow steady-state condition by using two server, it means that we only need to add one server.

### Acknowledgment

The authors would like to thank you to Dr. Yap Eye Hospital Yogyakarta which gave permission for this research.

### References

[1] Ishan P, V P Sakhare, M S Shelke, P B Sawaitul 2015 Reduction of Waiting Time by Using Simulation and Queuing Analysis *International Journal on Recent and Innovation Trends in Computing and Communication* **3** 55-59
[2] Ratna N, Rochmad & Kartono 2014 Analisis Proses Antrian Multiple Channel Single Phase di Loket Administrasi dan Rawat Jalan RSUP Dr. Kariadi Semarang *Unnes Journal of Mathematics* **3** 1-6
[3] William H K, Lawrence M L, John H D 2012 Transient Queueing Analysis *INFORMS Journal on Computing* **24** 10-28
[4] Toshiba Sheikh, Sanjay Kumar S, Anil Kumar K 2013 Application of Queuing Theory for the Improvement of Bank Service *International Journal of Advanced Computational Engineering and Networking* **1** 15-18
[5] Syed Shujauddin Sameer 2014 Simulation: Analysis of Single Server Queuing Model *International Journal on Information Theory* **3** 47-54
[6] H I Okagbue, S O Edeki, A A Opanuga 2014 A Monte Carlo Simulation Approach in Assessing Risk and Uncertainty Involved in Estimating the Expected Earnings of an Organization: A Case Study in Nigeria *American Journal of Computational and Applied Mathematics* **4** 161-166
[7] S Shanmugasundaram, S Punitha 2014 A Study on Multi Server Queuing Simulation *International Journal of Science and Research* **3** 1519-1521
[8] Ajay Kumar S, Rajiv Kumar, Girish Kumar S 2013 Queueing Theory Approach with Queueing Model: A Study *International Journal of Engineering Science Invention* **2** 01-11
[9] EC Emmanuel Chinwuko, Ezeliora Chukwuemeka D, Okoye Patrick U, Obiafudo Obiora J 2014 Analysis of a Queuing System in an Organization (a Case Study of First Bank PLC, Nigeria) *American Journal of Engineering Research (AJER)* **3** 63-72
[10] Mardalis 2003 *Metode Penelitian Suatu Pendekatan Proposal* Jakarta: Bumi Aksara
[11] Eshetie Berhan 2015 Bank Service Performance Improvements using Multi-Server Queue System *IOSR Journal of Business and Management* **17** 65-69
[12] Donald G, John F, James M, Carl M 2008 *Fundamental of Queueing Theory: Fourth Edition* New Jersey & Sons Inc: John Wiley.
[13] Bernard W 2008 *Introduction to Management Science: Eight Edition* Jakarta: Salemba Empat

# On (local) metric dimension of graphs with *m*-pendant points

**Rinurwati[1,2], Slamin[3] and H Suprajitno[1]**
[1]Mathematics Department, Universitas Airlangga, Indonesia
[2]Mathematics Department, Sepuluh Nopember Institute of Technology, Indonesia
[3]Information System of Study Program, Universitas Jember, Indonesia

E-mail: rinur@matematika.its.ac.id, rinurwati@gmail.com

**ABSTRACT.** An ordered set of vertices *S* is called as a (local) resolving set of a connected graph $G=(V_G,E_G)$ if for any two adjacent vertices $s \neq t \in V_G$ have distinct representation with respect to *S*, that is $r(s\mid S) \neq r(t\mid S)$. A representation of a vertex in *G* is a vector of distances to vertices in *S*. The minimum (local) resolving set for *G* is called as a (local) basis of *G*. A (local) metric dimension for *G* denoted by dim(*G*), is the cardinality of vertices in a *basis* for *G*, and its local variant by $\dim_l(G)$.

Given two graphs, *G* with vertices $s_1$, $s_2$, ..., $s_p$ and edges $e_1$, $e_2$, ..., $e_q$, and *H*. An edge-corona of *G* and *H*, $G\Diamond H$ is defined as a graph obtained by taking a copy of *G* and *q* copies of *H* and for each edge $e_j = s_is_h$ of *G* joining edges between the two end-vertices $s_i$, $s_h$ of $e_j$ and each vertex of *j*-copy of *H*.

In this paper, we determine and compare between the metric dimension of graphs with *m*-pendant points, $G\Diamond mK_1$, and its local variant for any connected graph *G*. We give an upper bound of the dimensions.

## 1. Introduction

Metric dimension is an attractive parameter in Graph Theory. This is shown by many applications and research results about it, in the last two decade recently. One of them is related to the structure of chemistry compound, such that there are some researchs by using notion in Chemistry such as: metric dimension of corona graph [5, 10] and its local variant [1, 6, 7, 8].

The development of the last result is local metric dimension of edge-corona graphs [9], that is for operation of graphs, *G* and *H*, with *H* of order greater than equal to two. It is important to motivate researcher in Graph Theory to produce the metric dimension of graphs.

Although edge-corona graph has appeared in [4] for case spectrum of graphs, at that time there is no result about metric dimension of this graph. Moreover, specially for the other variant.

In this paper, we go on our research about (local) metric dimension edge-corona graph *G* and *H*, with *H* of order one. We call this graph as graph with pendant points. We determine and compare between the metric dimension of graphs with *m*-pendant points and its local variant for any connected graph *G*. We give an upper bound of the dimensions.

## 2. Material and method

In this paper, we use a simple, finite and connected graph $G=(V_G,E_G)$, with $V_G$ is a set of vertices in *G* and $V_G$ is a set of edges in *G*. An ordered set of vertices $S=\{s_1,s_2,\ldots,s_k\}\subset V_G$ is called a resolving set of $G$ if for every two distinct vertices $u,v \in V_G$ hold $r(u|S) \neq r(v|S)$. The representation of a vertex $x \in V_G$ with respect to $S$, denoted by $r(x|S)$, is defined as $r(x|S) = (d(x,s_1), d(x,s_2), ..., d(x,s_k))$, where $d(x,s_i)$ is the distance between vertices $x, s_i \in V_G$. If $u,v \in V_G$ are adjacent then $S$ is called a local resolving set of $G$. A minimal (local) resolving set of $G$ is a (local) basis of $G$ and its cardinality, $\dim_l(G)$, called by a (local) metric dimension of $G$.

We use the following results that have been obtained from some researchs before.

**Theorem 2.1.** [3] *Let G be a connected graph of order* $n \geq 2$.
(*i*). $\dim(G) = 1$ *if only if* $G = P_n$.
(*ii*). $\dim(G) = n - 1$ *if only if* $G = K_n$.
(*iii*). *For* $n \geq 4$, $\dim(G) = n - 2$ *if only if* $G = K_{r,s}$ $(r, s \geq 1)$, $G = K_r + \overline{K_s}$ $(r \geq 1, s \geq 2)$ *or* $G = K_r + (K_1 \cup K_s)$, $(r,s \geq 1)$.
(*iv*). *For* $n \geq 3$, $\dim(C_n) = 2$.

**Theorem 2.2.**[6] *Let G be a connected graph of order* $n \geq 2$ *and diameter k, then*
(*i*). $\dim_l(G) = 1$ *if only if G is a bipartite graph..*
(*ii*). $\dim_l(G) = n - 1$ *if only if* $G = K_n$.

**Theorem 2.3.**[9] *For any connected-graph which is not an incomplete-bipartite graph G of order* $n_1 \geq 3$ *with* $|E(G)| = m$ *and any graph complement of complete graph* $H = \overline{K_{n_2}}$ *hold* $\dim_l(G \Diamond H) = \dim_l(G)$.

In this study, we determine and compare the metric dimension of graphs with pendant points and their local variant. We use a simple, finite and connected graph $G = (V_G, E_G)$, with $V_G$ is a set of vertices in $G$ and $V_G$ is a set of edges in $G$. We start with basic concepts and notations. Notation $u \sim v$, we mean that $u$ adjacent to $v$.

Given two graphs, $G$ with vertices $s_1, s_2, ..., s_p$ and edges $e_1, e_2, ..., e_q$, and $H$. An edge-corona $G \Diamond H$ graph of $G$ and $H$ graphs, is defined as a graph obtained by taking a copy of $G$ and $q$ copies of $H$ and for each edge $e_j = s_i s_h$ of $G$ joining edges between the two end-vertices $s_i$, $s_h$ of $e_j$ and each vertex of $j$-copy of $H$ [4]. If we $x$ take $H \cong mK_1, m \geq 1$, we called a $G \Diamond H$ graph as graph with pendant points. The set of vertices of $G$ will denoted by $\{x_1, x_2, ..., x_p\}$ and the vertex of a trivial graph $K_1$ (that is a pendant point) will be denoted by $y_j$ such that $y \sim x_i$ and $y \sim x_h$ too, with $e_j = x_i x_h$, $i \neq h \in \{1,2,...,p\}, j \in \{1,2,...,q\}$.

## 3. Main Results

In this part, we want to determine the metric dimension of $G \Diamond H$ for $H \cong K_1$. We determine the exact values of $\dim(G \Diamond K_1)$ for some simple connected graphs $G$. We obtain general bound of $\dim(G \Diamond mK_1)$.

### 3.1. The Value of $\dim(G \Diamond mK_1)$

Here, as $G$, we use $P_n, C_n, K_n, S_n$, and $K_{r,s}; r + s = n, r \geq 2, s \geq 1$.

#### 3.1.1. The exact value of $\dim(G \Diamond mK_1)$ when G is a path graph

Let $P_p$ is a path of order $p$ and measure $q = p$-1, and $K_1$ is a trivial graph. A $G$ graph, is an edge-corona graph of $P_p$ and $K_1$ denoted by $G \cong P_p \Diamond K_1$, has $q$ copies of a trivial graph $K_1$. For a simple way, a copy of a $j^{th}$ trivial graph $K_1$ denoted by $H_j = y_j$ for some $j \in \{1,2,...,q\}$. An edge-corona graph $G \cong P_p \lozenge K_1$ has a set of vertices $V_G = \{x_1, x_2, ..., x_p\} \bigcup \{y_1, y_2, ..., y_q\}$, and a set of edges $E_G = \{x_i x_{i+1} | 1 \le i \le p\text{-}1\} \bigcup \{x_i y_j, x_{i+1} y_j \,|1 \le i=j \le p\text{-}1\}$. We consider $\langle e_j \rangle \lozenge H_j \subset P_p \lozenge K_1$. The metric dimension of $P_p \lozenge K_1$, with $p \ge 3$ given by the following Theorem 3.1.1.

**Theorem 3.1.1.** *Given a path of order* $p = |V_{P_p}|$ *with measure* $q = |E_{P_p}|$, *and a trivial graph* $K_1$.*For* $p \ge 3$, *the metric dimension of edge-corona of* $P_p$ *and* $K_1$, $\dim(P_p \lozenge K_1)$, *is two.*

*Proof.* 1). According to Theorem 2.1., because $G \cong P_p \lozenge K_1$ is not a path then $\dim(G) \ge 2$.

2). We will show $\dim(G) \le 2$.

Choose a set $W=\{y_1, y_q\} \subset V_G$ and a vertex $v \in V_G$. The representation of all vertices in $V_G \backslash \{y_1, y_q\}$ are:

$r(x_1|W) = (1, p\text{-}1)$
$r(x_p|W) = (p\text{-}1, 1)$
$r(x_i|W) = (i\text{-}1, p\text{-}i);\ 2 \le i \le p\text{-}1$
$r(y_h|W) = (h, q\text{-}h\text{+}1);\ 2 \le h \le q\text{-}1$

We can show that all of the representation above are distinct. For example, $(s\text{-}1, p\text{-}s) = (t, q\text{-}t\text{+}1)$ for some $s$ and $t$ fixed. Then $s = t\text{+}1$ and $s = n\text{-}q\text{+}t\text{-}1$, a contradiction. So $W$ is a resolving set of $G$ with minimal cardinality. Hence $\dim(P_p \lozenge K_1) = 2$. □

### 3.1.2. The exact value of $\dim(G \lozenge mK_1)$ when $G$ is a cycle graph

Given a cycle $C_p$ of order $p$ and measure $q=p$, and a trivial graph $K_1$. A graph $G$, is an edge-corona graph of $C_p$ and $K_1$ denoted by $G \cong C_p \lozenge K_1$, has $q$ copies of a trivial graph $K_1$. For a simplification, a copy of a $j^{th}$ trivial graph denoted by $H_j = y_j$ for some $j \in \{1,2,...,q\}$.An Edge-corona graph $G \cong C_p \lozenge K_1$ has a set of vertices $V_G = \{x_1, x_2, ..., x_p\} \bigcup \{y_1, y_2, ..., y_p\}$and a set of edges $E_G = \{x_i x_{i+1} | 1 \le i \le p\text{-}1, x_p x_1\} \bigcup \{x_i y_h, x_{i+1} y_h \,|1 \le i \le n\text{-}1, 1 \le h \le n\}$. We consider $\langle e_j \rangle \lozenge H_j \subset C_p \lozenge K_1$. The metric dimension of $C_p \lozenge K_1$, with $p \ge 4$ given by the following Theorem 3.1.2.

**Teorema 3.1.2.** *Given a cycle* $C_p$ *of order p and measure q and a trivial graph* $K_1$. *For* $p \ge 4$, *the metric dimension of an edge-corona graph of* $C_p$ *and* $K_1$,

$$\dim(C_p \lozenge K_1) = \begin{cases} 2 & , \textit{for even } p \\ 3 & , \textit{for odd } p. \end{cases}$$

*Proof.* a). When $p = 2k$ for $k \in N$.

1). According to Theorem 2.1., because $G \cong C_p \lozenge K_1$ is not a path then $\dim(G) \ge 2$.

2). We will show $\dim(G) \le 2$.

Choose a set $W=\{y_1, y_k\} \subset V_G$ and a vertex $v \in V_G$. Representations of all vertices in $V_G \backslash \{y_1, y_q\}$ are:

$$r(x_i\,|\,W)=\begin{cases}(1,k) & ,i=1\\(i-1,k-i+1) & ,2\le i\le k\\(k,1) & ,i=k+1\\(p-i+2,i-k) & ,k+2\le i\le p.\end{cases}$$

All of the above representations are distinct. We can check for some fixed $s$ and $t$ as follows ($s$-1, $k$-$s$+1) = ($p$-$t$+2, $t$-$k$). Then $s$-1 =$p$-$t$+2 and $k$-$s$+1= $t$-$k$, a contradiction. Thus, $W$ is a resolving set for $G$ and obviously from minimal cardinality. Hence, $\dim(C_p\Diamond K_1) = 2$. □

Since $K_n\Diamond K_1 \cong K_{n+1}$, then $\dim(K_n\Diamond K_1)=(n+1)-1=n$ (Theorem 2.1.). Since $S_n\Diamond K_1 \cong f_n$ and $\dim(f_n)$ = $n$ then $\dim(S_n\Diamond K_1)=n.$ For $r=2, s=1$, $K_{2,1}\cong P_3$ so that $\dim(K_{2,1}\Diamond K_1)=2$ (Theorem 3.1.1.). For $r=2, s=2$, $K_{2,2}\cong C_4$ so that $\dim(K_{2,2}\Diamond K_1)=2$ (Theorema 3.1.2.). Moreover, we just need to show $\dim(K_{r,s}\Diamond K_1)$ for $r\ge 2, s\ge 3$, similar to the proof in the following theorem.

#### 3.1.3. The exact value of $\dim(G\Diamond mK_1)$ when $G$ is a bipartite graph

Given a bipartite graph $K_{r,s}$, $r\ge 2, s\ge 3$, of order $p$=$r$+$s$ and measure $q$, and a trivial graph $K_1$. The set of vertices of $K_{r,s}$ is $V_{K_{r,s}}=V_1\cup V_2=\{x_{1i}\,|\,i=1,2,...,r\}\cup\{x_{2h}\,|\,h=1,2,...,s\}$ and the set of edges is $E_{K_{r,s}}=\{x_{1i}x_{2h}\,|\,i=1,2,...,r;h=1,2,...,s\}$. A $G$ graph, is an edge-corona graph of $K_{r,s}$ and $K_1$ denoted by $G\cong K_{r,s}\Diamond K_1$, has $q$ copies of a trivial graph $K_1$. As above, a copy of a $j^{\text{th}}$ trivial graph denoted by $H_j = y_j = y_{ih}$, $i=1,2,...,r;h=1,2,...,s$. An Edge-corona graph $G\cong K_{r,s}\Diamond K_1$ has a set of vertices $V_G = V_{K_{r,s}}\cup\bigcup_{i=1}^{r}(\bigcup_{h=1}^{s}\{y_{ih}\})$ and a set of edges $E_G = E_{K_{r,s}}\bigcup\{x_iy_h,\ x_{i+1}y_h\ |1\le i\le p\text{-}1,\ 1\le h\le p\}$.

We consider $\langle e_j\rangle\Diamond H_j\subset K_{r,s}\Diamond K_1$. The metric dimension of $K_{r,s}\Diamond K_1$, with $p\ge 4$ given by the following Theorem 3.1.3.

**Theorem 3.1.3.** *Given a bipartite graph $K_{r,s}$ of order $p$=$r$+$s$ and measure $q$ and a trivial graph $K_1$.For any integer r, s with $r\ge 2$, $s\ge 3$, the metric dimension of an edge-corona graph of $K_{r,s}$ and $K_1$,* $\dim(K_{r,s}\Diamond K_1)=p-2.$

The following results is the metric dimension of graphs with $m$-pendant points.

**Theorem 3.1.4.** *Given a connected graph G of order $p\ge 4$ and measure $q\ge 2$. Then,*

$$\dim(G\Diamond mK_1)=\begin{cases}mq-1 & ,\mathit{for}\ G\cong S_n\\ q(m-1) & ,\mathit{otherwise}.\end{cases}$$

It is mentioned in [2] that if we add a point at a graph $G$ such that there are greater than one edge incident to this point then metric dimension of this graph maybe fixed, decrease or increase from metric dimension of $G$. From this statement and the results in Theorem 3.1.1. until Theorem 3.1.3., we give the following lemma.

**Lemma 3.1.5.** *If $G'$ is a graph obtained by adding a pendant point to every edge of a connected graph G of order $p\ge 2$, then* $\dim(G)\le\dim(G')\le\dim(G)+1.$

IOP Conf. Series: Journal of Physics: Conf. Series **855** (2017) 012035 doi:10.1088/1742-6596/855/1/012035

Lemma 3.1.5. useful to obtain the following theorem.

**Theorem 3.1.6.** *For every connected graph $G$ of order $p \geq 3$ with $|E(G)| = q \geq 2$,* $\dim(G\Diamond mK_1) \leq q(m-1)$.

The above results is an upper bound of $\dim(G\Diamond mK_1)$. Now, we determine the local metric dimension of $G\Diamond H$ for $H \cong mK_1$.

### 3.2. The Exact Value of $\dim_l(G\Diamond mK_1)$

As mentioned above, a graph $G$ that we use in this paper are $P_n$, $C_n$, $K_n$, $S_n$ and $K_{r,s}$, $r \geq 2, s \geq 1$. In this part, we determine the exact values of $\dim_l(G\Diamond mK_1)$ for some simple connected graphs $G$. By Theorem 2.3., we only show $\dim_l(G\Diamond mK_1)$ with $G$ is a bipartite graph that is $P_n$, $S_n$ and $K_{r,s}$, $r \geq 2, s \geq 1$.

#### 3.2.1. The exact value of $\dim_l(G\Diamond mK_1)$ when G is a path graph

Let $P_p$ be a path with vertex set $V_{P_p} = \{x_i \mid 1 \leq i \leq p, i \in N\}$ and edge set $E_{P_p} = \{x_i x_h \mid 1 \leq i \leq p-1, h = i+1; i \in N\}$ and $K_1$ is a trivial graph. A graph $G \cong P_p\Diamond mK_1$, has $mq$ copies of a trivial graph $K_1$. We denote by $y_{jk}$, the pendant point of edge $e_j = x_i x_h$, $1 \leq i = j \leq p-1; h = i+1$. Here, a copy of $j^{th}$-$mK_1$ denoted by $H_j = (mK_1)_j = \bigcup_{k=1}^{m}\{y_{jk}\}$ for some $j \in \{1,2,\ldots,q\}$. A graph $G \cong P_p\Diamond mK_1$ has a set of vertices $V_G = V_{P_p} \cup \bigcup_{k=1}^{m} y_{jk}$ and a set of edges $E_G = E_{P_p} \cup \{x_i y_{jk}, x_{i+1} y_{ik} \mid 1 \leq i = j \leq p-1, 1 \leq k \leq m\}$. We prove the following theorem.

**Theorem 3.2.1.** *Let $P_p$ be a path graph of order $p \geq 4$, and $K_1$ is a trivial graph. For all integer $m \geq 1$,* $\dim_l(P_p\Diamond mK_1) = 2$.

*Proof.* Let $G \cong P_p\Diamond mK_1$. Since $G$ is not a bipartite graph, then by Theorem 2.3., we obtain $\dim_l(G) \geq 2$. Moreover, we just need to show $\dim_l(G) \leq 2$. We will show that $W = \{x_1, x_m\}$ is a resolving set of $G$. The representation of all vertices in $V_G - W$ are

$$r(x_i \mid W) = (i-1, p-i), \quad 2 \leq i \leq p-2$$
$$r(y_{jk} \mid W) = (j, p-j), \quad 1 \leq j \leq p-1; \quad 1 \leq k \leq m.$$

All the above representations are distinct. Suppose that $(s-1, p-s) = (t, p-t)$ for some fixed $s$ and $t$. Then $s = t+1$ and $s = t$, a contradiction. Thus, representation of any two adjacent vertices are distinct, too. Therefore, $W$ is a local resolving set of $G$. Obviously cardinality of $W$ is minimal. So, $\dim_l(G) \leq 2$. □

#### 3.2.2. The exact value of $\dim_l(G\Diamond mK_1)$ when G is a star graph

Let $S_p$ be a star graph with vertex set $V_{S_p} = \{c, x_i \mid 1 \leq i \leq p, i \in N\}$ and edge set $E_{S_p} = \{cx_i \mid 1 \leq i \leq p, i \in N\}$ and $K_1$ is a trivial graph. A graph $G \cong S_p\Diamond pK_1$, has $mp$ copies of a trivial graph $K_1$. We denote by $y_{jk}$, the pendant point of edge $e_i = cx_i$, $1 \leq i \leq p$. We denote a copy of $i^{th}$-$mK_1$ by $H_i = (mK_1)_i = \bigcup_{k=1}^{m}\{y_{ik}\}$ for some $i \in \{1,2,\ldots,p\}$. A graph $G \cong S_p\Diamond mK_1$ has a set of vertices $V_G = V_{S_p} \cup$

$\bigcup_{k=1}^{m} y_{ik}$, $1 \le i \le p$ and a set of edges $E_G = E_{S_p} \cup \{cy_{ik}, x_i y_{ik} \mid 1 \le i \le p,\ 1 \le k \le m\}$. The following is Theorem 3.2.2. and its proof.

**Theorem 3.2.2.** *Let $S_p$ be a star graph of order $n \ge 4$, and $K_1$ is a trivial graph. For all integer $m \ge 1$, $\dim_l(S_p \Diamond mK_1) = p-1$.*

*Proof.* Let $G \cong S_p \Diamond mK_1$. By Theorem 2.2., since $G$ is not a complete graph, then $\dim_l(G) \ge p-1$. We will show that $W = \{x_i \mid 1 \le i \le p, p \in N\}$ is a local resolving set of $G$.The representation of all vertices in $V_G - W$ are

$$r(c \mid W) = (1,1,\ldots,1)$$
$$r(y_{ik} \mid W) = (2,2,\ldots,\underbrace{1}_{i-term},2,\ldots,2),\quad 1 \le i \le p;\quad 1 \le k \le m.$$

From direct observation, we can see that all the above representations are distinct. Such that for any two adjacent vertices in $G$ also have distinct representations. So, $W$ is a local resolving set of $G$ and $|W| \le p-1$.

Suppose that $B$ is a local basis of $G$ with $|B| < |W|$. Because $H$ is a null graph, if there is $z \in B \cap V_{H_i}$, for some $i$, then the pairs of vertices of $G$ which are distinguished by $z$ can be distinguised also by $x_i$. Let the set $B_1$ obtained from $B$ by replacing $x_i$ by $z \in B \cap V_{H_i}$, with $i \in \{1,2,\ldots,p\}$. So, $B_1$ is a local resolving set of $G$ and $|B_1| \le |B| < |W| = \dim_l(G)$, a contradiction. Thus, $W$ is a local basis of $G$. □

### 3.2.3. The exact value of $\dim_l(G \Diamond mK_1)$ when $G$ is a bipartite graph

Let $K_{r,s}, r \ge 2, s \ge 1$ be a bipartite graph of order $p = r + s$ and measure $q = r.s$ with vertex set $V_{K_{r,s}} = V_1 \cup V_2$, where $V_1 = \{x_i \mid 1 \le i \le r,\ r \in N\}$, $V_2 = \{y_h \mid 1 \le h \le s;\ s \in N\}$, and edge set $E_{K_{r,s}} = \{x_i y_h \mid 1 \le i \le r,\ 1 \le h \le s;\ i,h \in N\}$. $K_1$ is a trivial graph. A graph $G \cong K_{r,s} \Diamond mK_1$ has $mq$ copies of a trivial graph $K_1$. We denote by $z_{i(h(k))}$, $1 \le k \le m$; the pendant point $K_1$ of edge $e_{i(h)} = x_i x_h$, For some $1 \le h \le s$; $1 \le i \le r$. We denote a copy of $j^{th}$-$mK_1$ by $H_j = ((mK_1)_h)_i = \bigcup_{k=1}^{m} \{z_{i(h(k))} = z_{jk}\}$ for some $j \in \{1,2,\ldots,q\}$, with $j = \begin{cases} h & ,i = 1 \\ (i-1)r + h & ,i \ge 2. \end{cases}$ A graph $G \cong K_{r,s} \Diamond mK_1$ has a set of vertices $V_G = V_{K_{r,s}} \cup \bigcup_{k=1}^{m} z_{jk}$, $1 \le j \le q$ and a set of edges $E_G = E_{K_{r,s}} \cup \{x_{1i} z_{jk}, x_{2i} z_{jk} \mid 1 \le i \le r,\ 1 \le j \le q$ $1 \le k \le m\}$.The following is Theorem 3.2.3. and its proof.

**Theorem 3.2.3.** *Let $K_{r,s}$ with $r \ge 2$, $s \ge 1$, is a bipartite graph of order $n \ge 4$, and $K_1$ is a trivial graph. For any integer $m \ge 1$, $\dim_l(K_{r,s} \Diamond mK_1) = \max\{r,s\}$.*

*Proof.* Let $G \cong K_{r,s} \Diamond mK_1$. We suppose that $s \ge t$. We will prove that $W = \{z_{1(h(1))} \mid 1 \le h \le s\}$ is a local resolving set of $G$.The representation of all vertices in $V_G - W$ are

$$r(x_i \mid W) = \begin{cases} (1,1,\ldots,1) & ,i = 1 \\ (2,2,\ldots,2) & ,2 \le i \le r. \end{cases}$$

$$r(y_h \mid W) = (2,2,\ldots,\underbrace{1}_{h-term},2,\ldots,2),\quad 1 \le h \le s.$$

$$r(z_{1(h(k))}) \mid W) = (2,2,\ldots,2),\quad 1 \le h \le s;\ 2 \le k \le m.$$

$$r(z_{i(h(k))}) \mid W) = (3,3,\ldots,\underbrace{2}_{h-term},3,3,\ldots,3),\quad 2 \le i \le r;\ 1 \le h \le s;\ 1 \le k \le m.$$

Among pendant points $z_{i(h(k))}$ are not adjacent. For every $2 \le j \ne l \le r$, $x_j \nsim x_l$, and for every $1 \le j \ne l \le s$, $y_j \nsim y_l$. So, for every pairs of adjacent vertices in $G$ have distinct representations. Therefore, $W$ is a local resolving set of $G$.

Suppose that $A$ is a local basis of $G$ with $|A| < |W|$. Because $H$ is a null graph, if there is $v \in A \cap V_{H_j}$, for some $j$, then the pairs of vertices of $G$ which are distinguished by $v$ can be distinguised also by $x_j$ or $y_j$. Let the set $A_1$ obtained from $A$ by replacing $x_j$ or $y_j$ by $v \in A \cap V_{H_j}$, with $j \in \{1,2,\ldots,q\}$. So, $A_1$ is a local resolving set of $G$ and $|A_1| \le |A| < |W| = \dim_l(G)$, a contradiction. Thus, $W$ is a local basis of $G$. □

From the results that have disscused in part 3.1 and 3.2, we can conclude as follows.

**Theorem 3.2.4.** *Let G be a connected graph of order $p \ge 4$. For any integer $m \ge 1$,*

$$\dim_l(G \Diamond mK_1) \le \dim(G \Diamond mK_1).$$

## 4. Conclussions.

We have discussed the (local) metric dimension of edge-corona of some connected graphs $G$ and null graph $\overline{K_m}$ or $mK_1$ that is a graph with pendant points. As metric dimension, if we add a point at a graph $G$ such that there are greater than one edge incident to this point, then the local metric dimension of this graph can be fixed, decrease or increase from local metric dimension of $G$ and $\dim_l(G \Diamond mK_1) \le \dim(G \Diamond mK_1)$.

**Acknowledgement**

Authors would like to tank the anonymous referee for all of their comments which is very useful to improve this paper.

**References**

[1] Barrgan-Ramirez G A, Gomez C G and Rodriguez-Velazquez J A 2014 Close formula for the local metric dimension of corona product graphs *Electronic Notes in Discrete Mathematics*
[2] Buczkowski P T, Chartrand G, Poisson C and Zhang P 2003 On *k*-dimensional of graphs and their basis *Peoridica Mathematica Hungarica* **46** **1** pp 9-15
[3] Chartrand G, Eroh L, Johnson M A and Oellermann O R 2000 Resolvability in graphs and the metric dimension of a graph, *Discrete Applied Mathematics* **105** 99–113
[4] Hou Y and Shiu W C 2010 The spectrum of the edge corona of two graphs *Electronic Journal of Linear Algebra* **20** 586-59
[5] Iswadi H, Baskoro E T, and Simanjuntak R 2010 On the metric dimension of corona product of graphs *Far East J. Math. Sci.* (FJMS) **40** **1** 23-36.
[6] Okamoto F, Phinezy B and Zhang P 2010 The local metric dimension of a graph *Mathematica Bohemica* **135** **3** 239-255
[7] Rodriguez-Velazquez J A, Barragan-Ramirez G A, and Gomez C G 2013 On the local metric dimension of corona product graphs, *ArXiv.*1308.6689*v*1[math.CO].

[8] Rodriguez-Velazquez J A, Gomez C G and Barragan-Ramirez G A 2014 Computing the local metric dimension of a graph from the local metric dimension of primary subgraphs *International Journal of Computer Mathematics*
[9] Rinurwati, Slamin and Herry Suprajitno General Results of local metric dimensions of edge-corona of graphs 2016 *International Mathematical Forum* **11** **16** 739-799
[10] Yero I G, Kuziak D and Rodriguez-Velazquez J A 2011 On the metric dimension of corona product graphs *Computers and Mathematics with Applications* **61** **9** 2793-2798

# Survival Bayesian Estimation of Exponential-Gamma Under Linex Loss Function

**S. W Rizki, M. N Mara, and E Sulistianingsih**
Mathematics Department of Mathematics and Natural Sciences Faculty of Tanjungpura University. Indonesia

E-mail: setyo.wirarizki@math.untan.ac.id

**ABSTRACT.** This paper elaborates a research of the cancer patients after receiving a treatment in cencored data using Bayesian estimation under Linex Loss function for Survival Model which is assumed as an exponential distribution. By giving Gamma distribution as prior and likelihood function produces a gamma distribution as posterior distribution. The posterior distribution is used to find estimatior $\hat{\lambda}_{BL}$ by using Linex approximation. After getting $\hat{\lambda}_{BL}$, the estimators of hazard function $\hat{h}_{BL}$ and survival function $\hat{S}_{BL}$ can be found. Finally, we compare the result of Maximum Likelihood Estimation (MLE) and Linex approximation to find the best method for this observation by finding smaller MSE. The result shows that MSE of hazard and survival under MLE are 2.91728E-07 and 0.000309004 and by using Bayesian Linex worths 2.8727E-07 and 0.000304131, respectively. It concludes that the Bayesian Linex is better than MLE.

## 1. Introduction

The Exponential distribution is a known distribution which can be used in many studies like biology, economics and demography. It is a very popular one-parameter distribution which is often used in survival model study, see [1]. Let $T$ is defined to be the time of failure of the entity known to exist at time $t = 0$, and is therefore frequently called the failure time random variable. Now, if $T$ is the time to failure, then the probability of still functioning at time $t$ is the same as the probability that the failure is later (mathematically greater) than the value of $t$. The survival density function (SDF), probability density function (PDF) and hazard rate function (HRF) are defined as below,

$$S(t) = e^{-\lambda t} \quad . \quad t \geq 0. \ \lambda \geq 0 \tag{1}$$

$$f(t) = -\frac{d}{dt}S(t) = \lambda e^{-\lambda t} \tag{2}$$

$$\lambda(t) = \frac{f(t)}{S(t)} = \lambda \tag{3}$$

Besides, using Maximum Likelihood to estimate the parameters, Bayesian method is required in this research. It builds a posterior distribution by formulating a likelihood function of the exponential distribution and a prior. The Gamma distribution is an appropriate prior for it because if we switch the parameter $\alpha$ with 1, it will change to be an Exponential distribution [2]. This research will observe survival probability of cancer patients after receiving the treatment in a cencored data. The data is taken from R versi 3.3.0 which has an exponential distribution. Cencoring is a feature that recurrent in lifetime and reliability data analysis, it occurs when exact lifetimes or run-outs can only be collected for a portion of the inspection units [3]. The survival study can predict probability of survival and hazard rate by some methods. There are some estimation methods in statistics. Bayesian is one of them which use likelihood function and prior distribution to find posterior distribution. The data is an Exponential distributed which will be composed with a Gamma distribution as its prior to construct a posterior distribution. It gives the relative weights to each parameter value after analizing the data [4]. The Bayesian inference has some approximations such as Generalised Non-informative Prior. Linear Exponential Loss Function, Lindley Approximation, General Entropy Loss Function and Squared Error Loss Function.

## 2. Maximum Likelihood Estimation on Cencored Data

Cencoring is a way to handle an uncomplete data which is caused some events like death. loss or out from observation. According to [5], variables $T_1 \ldots . T_n$ represent $n$ individual lifetimes. A time $t_i$ is the lifetime or a cencoring time. The variable $\delta_i = 1$ if $T_i = t_i$ and 0 if $T_i > t_i$ is called the cencoring or status indicator for $t_1$. Value $t_1$ is obtained from $\min(T_i, C_i), i = 1,2,3,\ldots,n$ where $T_i$ is the duration of their remission measured from time of entry to study and $C_i$ is the time between their date of entry and the end of study. The likelihood function of cencored data for observation $(t_i, \delta_i)$ $i = 1,2,..,n$ can be calculated is defined as,

$$L(t_i;\lambda,\delta) = \prod_{i=1}^{n} [f(t_i;\lambda)]^{\delta_i}[S(t_i;\lambda)]^{1-\delta_i} \tag{4}$$

The likelihood function of exponential distribution for observation $(t_i, \delta_i)$ $i = 1,2,\ldots,n$ can be calculated by

$$L(t_i;\lambda,\delta) = \prod_{i=1}^{n} \left[\lambda e^{-\lambda t_i}\right]^{\delta_i}\left[e^{-\lambda t_i}\right]^{1-\delta_i}$$

$$= \left[\lambda^{\delta_1} \cdot \lambda^{\delta_2} \cdots \lambda^{\delta_n}\right]\left[e^{-\lambda t_1} \cdot e^{-\lambda t_2} \cdots e^{-\lambda t_n}\right]$$

$$= \lambda^{\sum_{i=1}^{n}\delta_i} e^{-\lambda(\sum_{i=1}^{n} t_i)} \tag{5}$$

then we find a natural logaritme of likelihood function above,

$$l = \ln L(t_i;\lambda,\delta) = \left(\sum_{i=1}^{n} \delta_i\right)\ln\lambda - \left(\sum_{i=1}^{n} t_i\right)\lambda$$

by deriving $l$ to parameter $\lambda$. we obtain,

$$\frac{dl}{d\lambda} = 0$$

$$\frac{d}{d\lambda}\left[\left(\sum_{i=1}^{n} \delta_i\right)\ln\lambda - \left(\sum_{i=1}^{n} t_i\right)\lambda\right] = 0$$

$$\frac{\sum_{i=1}^{n}\delta_i}{\lambda} - \sum_{i=1}^{n} t_i = 0$$

$$\hat{\lambda} = \frac{\sum_{i=1}^{n}\delta_i}{\sum_{i=1}^{n} t_i} \tag{6}$$

We get $\hat{\lambda}$ is a Maximum Likelihood Estimation of $\lambda$. Later, composing $\hat{\lambda}$ into both of survival model and hazard function are

$$\hat{S}_{ML}(t_i;\hat{\lambda}) = e^{-\hat{\lambda} t_i} = e^{-\left(\frac{\sum_{i=1}^{n}\delta_i}{\sum_{i=1}^{n} t_i}\right)t_i} \tag{7}$$

$$\hat{h}_{ML}(t_i;\hat{\lambda}) = \hat{\lambda} = \frac{\sum_{i=1}^{n}\delta_i}{\sum_{i=1}^{n} t_i} \tag{8}$$

$\hat{S}_{ML}(t_i;\hat{\lambda})$ and $\hat{h}_{ML}(t_i;\hat{\lambda})$ are Maximum Likelihood Estimation of survival model and hazard function.

## 3. Formulating Posterior Distribution

In this case, gamma distribution is assigned as a conjugate prior distribution to exponential distribution with parameter λ where $0 < \lambda < \infty$. The exponential distribution is a special form of gamma distribution. The gamma distribution is defined by

$$\Gamma(\alpha) = \int_0^{\infty} \lambda^{\alpha-1} e^{-\lambda}\, d\lambda \tag{9}$$

where $\alpha > 0$ and $0 < \lambda < \infty$

Selecting gamma distribution with parameter $(\alpha, \beta)$ as prior distribution for exponential distribution is based on closed form of gamma distribution. Let $T$ is a continous random varible of gamma distribution with parameter $\alpha$ and $\beta$, we can describe the probability density function as below.

$$f(t) = \frac{\beta^{\alpha}}{\Gamma(\alpha)} t^{\alpha-1} e^{-\beta t} \ ; t > 0, \alpha > 0, \beta > 0 \tag{10}$$

The posterior distribution is a formula of likelihood function and prior distribution. It is defined by

$$f(\lambda \mid t_i) = \frac{f(\lambda) f(t_i; \lambda)}{\int_0^{\infty} f(\lambda) f(t_i; \lambda)\, d\lambda} \tag{11}$$

The density function $f(\lambda \mid t_i)$, $f(\lambda)$ and $f(t_i; \lambda)$ show the posterior distribution, the prior distribution and likelihood function, respectively. Let $T \sim Exponential\,(\lambda)$ and prior density function $\lambda \sim Gamma(\alpha, \beta)$ then we can formulate the posterior distribution which can be stated as a conditional function of $\lambda$ and knowing $t$ as,

$$\begin{aligned}
f(\lambda \mid t_i) &= \frac{f(\lambda; t_i)}{f(t_i)} \\
&= \frac{f(\lambda) f(t_i; \lambda)}{\int_0^{\infty} f(\lambda) f(t_i; \lambda)\, d\lambda} \\
&= \frac{\left(\frac{\beta^{\alpha}}{\Gamma(\alpha)} \lambda^{u-1} e^{-\beta\lambda}\right)\left(\lambda^{\sum_{i=1}^{n} \delta_i} e^{-\lambda(\sum_{i=1}^{n} t_i)}\right)}{\int_0^{\infty} \left(\frac{\beta^{\alpha}}{\Gamma(\alpha)} \lambda^{\alpha-1} e^{-\beta\lambda}\right)\left(\lambda^{\sum_{i=1}^{n} \delta_i} e^{-\lambda(\sum_{i=1}^{n} t_i)}\right) d\lambda} \\
&= \frac{\frac{\beta^{\alpha}}{\Gamma(\alpha)} \lambda^{\sum_{i=1}^{n} \delta_i + \alpha - 1} e^{-\lambda(\sum_{i=1}^{n} t_i + \beta)}}{\int_0^{\infty} \frac{\beta^{\alpha}}{\Gamma(\alpha)} \lambda^{\sum_{i=1}^{n} \delta_i + \alpha - 1} e^{-\lambda(\sum_{i=1}^{n} t_i + \beta)}\, d\lambda} \\
&= \frac{\frac{\beta^{\alpha}}{\Gamma(\alpha)} \lambda^{\sum_{i=1}^{n} \delta_i + \alpha - 1} e^{-\lambda(\sum_{i=1}^{n} t_i + \beta)}}{\frac{\beta^{\alpha}}{\Gamma(\alpha)} \Gamma\left(\sum_{i=1}^{n} \delta_i + \alpha\right)\left(\frac{1}{\sum_{i=1}^{n} t_i + \beta}\right)^{\sum_{i=1}^{n} \delta_i + \alpha}} \\
&= \frac{\lambda^{\sum_{i=1}^{n} \delta_i + \alpha - 1} e^{-\lambda(\sum_{i=1}^{n} t_i + \beta)}}{\Gamma\left(\sum_{i=1}^{n} \delta_i + \alpha\right)\left(\frac{1}{\sum_{i=1}^{n} t_i + \beta}\right)^{\sum_{i=1}^{n} \delta_i + \alpha}} \\
&= \frac{\left(\sum_{i=1}^{n} t_i + \beta\right)^{\sum_{i=1}^{n} \delta_i + \alpha}}{\Gamma\left(\sum_{i=1}^{n} \delta_i + \alpha\right)} \lambda^{\sum_{i=1}^{n} \delta_i + \alpha - 1} e^{-\lambda(\sum_{i=1}^{n} t_i + \beta)}
\end{aligned} \tag{12}$$

From the result above we obtain the posterior distribution is a Gamma distribution or it can be expressed by $Gamma(\sum_{i=1}^{n} \delta_i + \alpha\,, \sum_{i=1}^{n} t_i + \beta)$ where $\lambda$ is a variable and $t_i$ is sample.

## 4. Bayesian Linex Loss Function

Bayesian Method is a well known estimation method in whole statistical methodology studies. There are three loss functions of the Bayes estimator. Linex is one of loss function in Bayesian estimation. The posterior expectation of the Linex loss function is according to Zellner in [3]. The Linex loss function is one of bayesian approaches. Using result of Zellner (1986) in [5], the parameter estimator of $\lambda$ which is denoted by $\hat{\lambda}_{lin}$ under Linex loss function is defined by,

$$\hat{\lambda}_{lin} = -\frac{1}{c}\ln\left[E\left(e^{-c\lambda}\right)\right] \tag{13}$$

We found the posterior formula in (12) that can be used to find a parameter estimator $\hat{\lambda}_{BL}$ under Bayes Linex loss function,

$$\begin{aligned}
E\left[e^{-c\lambda}\right] &= \int_0^{\infty} e^{-c\lambda} f(\lambda \mid t_i)\, d\lambda \\
&= \int_0^{\infty} e^{-c\lambda}\left[\frac{\left(\sum_{i=1}^n t_i+\beta\right)^{\sum_{i=1}^n \delta_i+\alpha}\lambda^{\sum_{i=1}^n \delta_i+\alpha-1}e^{-\lambda\left(\sum_{i=1}^n t_i+\beta\right)}}{\Gamma\left(\sum_{i=1}^n \delta_i+\alpha\right)}\right] d\lambda \\
&= \frac{\left(\sum_{i=1}^n t_i+\beta\right)^{\sum_{i=1}^n \delta_i+\alpha}}{\Gamma\left(\sum_{i=1}^n \delta_i+\alpha\right)}\int_0^{\infty} \lambda^{\sum_{i=1}^n \delta_i+\alpha-1}e^{-\lambda\left(\sum_{i=1}^n t_i+\beta+c\right)}\, d\lambda \\
&= \frac{\left(\sum_{i=1}^n t_i+\beta\right)^{\sum_{i=1}^n \delta_i+\alpha}\Gamma\left(\sum_{i=1}^n \delta_i+\alpha\right)}{\left(\sum_{i=1}^n t_i+\beta+c\right)^{\sum_{i=1}^n \delta_i+\alpha}\Gamma\left(\sum_{i=1}^n \delta_i+\alpha\right)} \\
&= \left(\frac{\sum_{i=1}^n t_i+\beta}{\sum_{i=1}^n t_i+\beta+c}\right)^{\sum_{i=1}^n \delta_i+\alpha}
\end{aligned} \tag{14}$$

from (14) we get the parameter estimation under Linex loss function as following

$$\begin{aligned}
\hat{\lambda}_{BL} &= -\frac{1}{c}\ln\left[E\left(e^{-c\lambda}\right)\right] \\
&= -\frac{1}{c}\ln\left[\left(\frac{\sum_{i=1}^n t_i+\beta}{\sum_{i=1}^n t_i+\beta+c}\right)^{\sum_{i=1}^n \delta_i+\alpha}\right]
\end{aligned} \tag{15}$$

Futhermore, the survival and hazard function under Linex loss function are expressed by equation (16) and (17),

$$\begin{aligned}
\hat{S}_{BL}(t_i;\hat{\lambda}) &= e^{-\hat{\lambda}t_i} \\
&= e^{-\left(-\frac{1}{c}\ln\left[\left(\frac{\sum_{i=1}^n t_i+\beta}{\sum_{i=1}^n t_i+\beta+c}\right)^{\sum_{i=1}^n \delta_i+\alpha}\right]\right)t_i}
\end{aligned} \tag{16}$$

$$\begin{aligned}
\hat{h}_{BL}(t_i;\hat{\lambda}) &= \hat{\lambda} \\
&= -\frac{1}{c}\ln\left[\frac{\left(\sum_{i=1}^n t_i+\beta\right)^{\sum_{i=1}^n \delta_i+\alpha}}{\left(\sum_{i=1}^n t_i+\beta+c\right)^{\sum_{i=1}^n \delta_i+\alpha}}\right]
\end{aligned} \tag{17}$$

IOP Conf. Series: Journal of Physics: Conf. Series **855** (2017) 012036 doi:10.1088/1742-6596/855/1/012036

## 5. Result

There are many diseases around our life. Some of them are dangerous and making a high death risk like diabetes mellitous, cancer, stroke and heart attack. Researchers are interested in study about cancer. In the last decades, cancer patients are growing bigger in the world. This disease has suffered people in all ages. Science makes some contributions to detect the risk which is caused by the diseases especially in Mathematics and Statistics. Survival study is derived from both mathematics and statistics. It relates to Actuarial Science which has developed in some countries. The data is taken from R versi 3.3.0. It describes about patients of lung cancer which has an exponential distribution. The result of survival value under MLE and Bayes Linex is presented as below,

**Table 1. Calculation of Survival Probability under MLE and Bayes Linex Loss Function.**

| The lenghth of Cencoring Time | Survival Probability | Estimation of Survival Probability under MLE | Estimation of Survival Probability under Bayesian Linex |
|---|---|---|---|
| 1 | 0.991811898 | 0.992347739 | 0.992343628 |
| 1 | 0.991811898 | 0.992347739 | 0.992343628 |
| 2 | 0.983690841 | 0.984754035 | 0.984745875 |
| 3 | 0.975636280 | 0.977218440 | 0.977206294 |
| 4 | 0.967647670 | 0969740509 | 0.969724439 |
| 7 | 0.944072173 | 0.947648307 | 0.947620826 |
| 8 | 0.936342013 | 0.940396655 | 0.940365488 |
| 10 | 0.921071062 | 0.926059399 | 0.926021035 |
| 11 | 0.913529238 | 0.918972951 | 0.918931073 |
| 12 | 0.906049167 | 0.911940730 | 0.911895395 |
| 13 | 0.898630344 | 0.904962322 | 0.904913584 |
| 15 | 0.883974439 | 0.891165298 | 0.891109920 |
| 16 | 0.876736366 | 0.884345868 | 0.884287250 |
| 18 | 0.862437533 | 0.870863161 | 0.870798222 |
| 19 | 0.855375806 | 0.864199089 | 0.864131067 |
| 20 | 0.848371902 | 0.857586012 | 0.857514958 |
| 21 | 0.841425346 | 0.851023540 | 0.850949504 |
| 22 | 0.834535670 | 0.844511285 | 0.844434318 |
| 24 | 0.820925094 | 0.831635896 | 0.831553211 |
| 25 | 0.814203275 | 0.825272001 | 0.825186530 |
| 27 | 0.800924304 | 0.812689932 | 0.812599032 |
| 29 | 0.787861902 | 0.800299690 | 0.800203545 |
| 30 | 0.781410809 | 0.794175588 | 0.794076889 |
| 31 | 0.775012537 | 0.788098348 | 0.787997140 |
| 33 | 0.762372734 | 0.776083028 | 0.775976934 |
| 35 | 0.749939076 | 0.764250893 | 0.764140085 |
| 36 | 0.743798498 | 0.758402646 | 0.758289544 |
| 42 | 0.707996613 | 0.724241065 | 0.724115058 |
| 44 | 0.696449783 | 0.713199311 | 0.713069317 |
| 45 | 0.690747181 | 0.707741724 | 0.707609793 |
| 48 | 0.673918010 | 0.691618263 | 0.691480743 |
| 51 | 0.657498860 | 0.675862120 | 0.675719335 |
| 52 | 0.652115192 | 0.670690246 | 0.670545776 |
| 53 | 0.646775606 | 0.665557949 | 0.665411828 |
| 54 | 0.641479742 | 0.660464926 | 0.660317187 |
| 56 | 0.631017746 | 0.650395501 | 0.650244626 |
| 59 | 0.615643806 | 0.635578476 | 0.635423141 |
| 61 | 0.605603173 | 0.625888469 | 0.625730318 |
| 63 | 0.595726294 | 0.616346195 | 0.616185350 |
| 72 | 0.553236206 | 0.575174574 | 0.575003032 |
| 73 | 0.548706251 | 0.570773187 | 0.570600595 |
| . | . | . | . |
| . | . | . | . |
| . | . | . | . |
| 999 | 0.000270947 | 0.000465751 | 0.00046575 |

The table 1 presents the survival probability of cancer patients after getting treatment under MLE and Bayesian Linex Loss function. Both of results give the value are greater than the real value but the estimation value under Bayesian Linex is closer to the survival value than the MLE's. The result shows that the all values are inversely proportional with the time. The survival probability of patients is getting smaller and converges to zero after less 3 years later. It means that the effect of treatment will be fading by the time. Decreasing point of survival probability is around 10% in about 10 days of different range. The degenerate rate is associated to hazard function of survival study. The hazard value estimation under both MLE and Bayesian Linex Loss function is given in table 2 as following,

**Table 2. The Hazard Values of MLE and Bayes Linex Loss Function Estimation.**

| Hazard | Estimation of Hazard Value under MLE | Estimation of Hazard Value under Bayesian Linex |
|---|---|---|
| 0.008221809 | 0.007681689 | 0.007685833 |

The hazard value serves a failure rate in reliability. From table 2, the failure rate of the real value, using MLE and Bayesian Linex are 0.8221809%, 0.7681689% and 0.7685833%, respectively. It means the Bayesian's is nearer to the real value than MLE's. Calculation of Mean Squared Error (MSE) is one way to find the best method by viewing the smallest MSE of both results. Those are shown by table 3 as below,

**TABLE 3. MSE Values of Survival and Hazard under MLE and Bayes Linex Loss Function.**

| MSE | Hazard | Survival |
|---|---|---|
| **MLE** | 2.91728E-07 | 0.000309004 |
| **Bayesian Linex Loss Function** | 2.8727E-07 | 0.000304131 |

The result shows that MSE value of Hazard and survival under Bayes Linex Loss Function is less than MSE value of Hazard and survival under MLE.

**Graph 1. The Comparison Plot of The Real Survival Probability and Using MLE.**

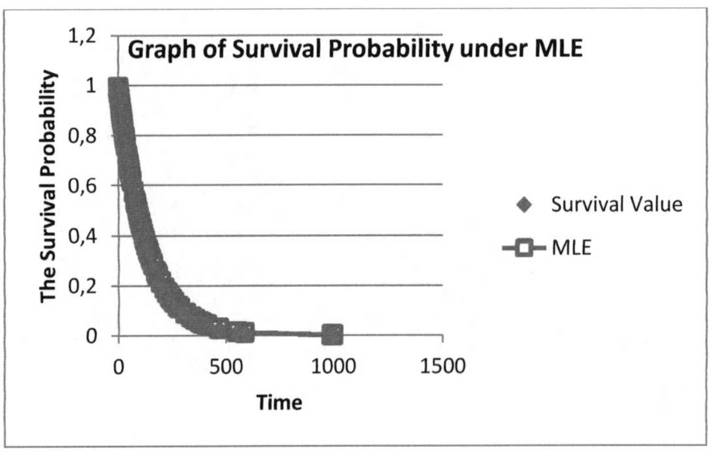

**Graph 2. The Comparison Plot of The Real Survival Probability and Using Bayesian Linex.**

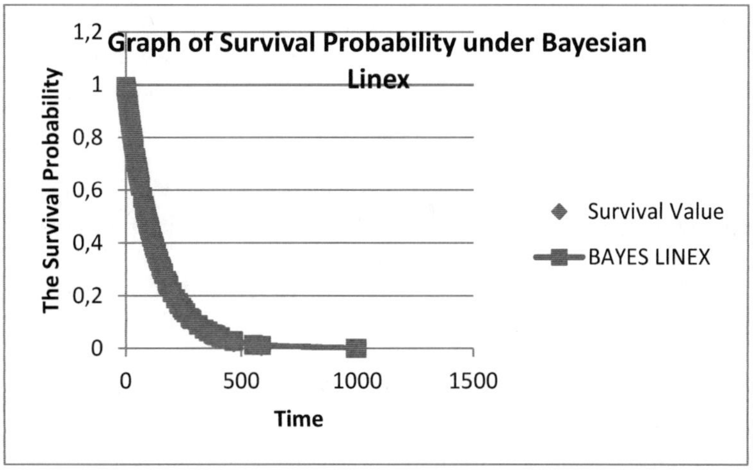


Although, both of graph 1 and graph 2 show that they have a slightly different value with the real value of survival probability, the line of graph 2 has a more rapid distant between the real's and Bayesian's.

## 6. Conclusion

To sum up, the research tells some points. First, the cancer is a high risk disease because the survival opportunity is decreasing rapidly by the time. It is not longer than 3 years the survival probability limits to zero. It means the cancer is almost impossible to be reaped. Second, the hazard rate and length of post treatment observation have a particular role to survival value. Those are inversely proportional with the survival probability. The last, the MSE shows that the Bayesian Linex Loss function is better than MLE.

## 7. Acknowledgements

We wish to thank the Chief of LPPKM and Dean of Mathematics and Natural Science Faculty Tanjungpura University who fund and support this research. By doing the research, we can improve the creativity and enrich knowledge in our field study.

## REFERENCES

[1] London. D,. 1988. *Survival Models and Their Estimation.* Third Edition, Connecticut: Actex Publications.

[2] Fitria. S,. Helmi & Rizki. S. W.. 2016. Survival Model Parameter Estimation Exponential Distribution of Data Censored by the Method of Maximum Likelihood and Bayesian SELF. *Bimaster,* 5(03), pp. 213-220.

[3] Guure. C. B. & Ibrahim. N. A,. 2012. Bayesian Analysis of the Survival Function and Failure Rate of Weibull Distribution with Censored Data. *Mathematical Problems in Engineering.*

[4] Bolstad. W. M,. 2007. *Introduction to Bayesian Statistics.* Second Edition, New Jersey: John Wiley & Sons, Inc.

[5] Metiri. F.. Zeghdoudi. H. & Remita. M. R,. 2016. On Bayes Estimatesof Lindley Disitribution under Linex Loss Function: Informative and Non Informative Priors. *Global Journal of Pure and Applied Mathematics,* 12(1), pp. 391-400.

# Super $(a, d)$-Cycle-Antimagic Total Labeling on Triangular Ladder Graph and Generalized Jahangir Graph

**Mania Roswitha, Anna Amanda, Titin Sri Martini and Bowo Winarno**

Combinatorial Research Group, Department of Mathematics, Sebelas Maret University, Surakarta, Indonesia

E-mail: mania_ros@yahoo.co.id, mynameisanna.27@gmail.com, titinsmartini@gmail.com and bowowinarno.bw@gmail.com

**Abstract.** Let $G(V(G), E(G))$ be a finite simple graph with $|V(G)| = \nu_G$ and $|E(G)| = e_G$. Let $H$ be a subgraph of $G$. The graph $G$ is said to be $(a, d)$-$H$-antimagic covering if every edge in $G$ belongs to at least one of the subgraphs $G$ isomorphic to $H$ and there is a bijective function $\xi : V \cup E \to \{1, 2, \ldots, \nu_G + e_G\}$ such that all subgraphs $H'$ isomorphic to $H$, the $H'$-weights

$$w(H') = \Sigma_{v \in V(H')}\xi(v) + \Sigma_{e \in E(H')}\xi(e)$$

constitutes an arithmetic progression $\{a, a + d, a + 2d, \ldots, a + (t - 1)d\}$, where $a$ and $d$ are positive integers and $t$ is the number of subgraphs $G$ isomorphic to $H$. Such a labeling is called *super* if the vertices contain the smallest possible labels. This research provides super $(a, d)$-$C_3$-antimagic total labelng on triangular ladder $TL_n$ for $n \geq 2$ and super $(a, d)$-$C_{s+2}$-antimagic total labeling on generalized Jahangir $J_{k,s}$ for $k \geq 2$ and $s \geq 2$.

## 1. Introduction

Wallis [13] defined a graph labeling as a mapping from the set of vertices, edges, or both vertices and edges to the positive or non-negative integers. The types of graph labeling which is still widely studied today are magic labelings and antimagic labelings. Magic labelings were introduced by Sedláček [12] in 1963. It was followed by Kotzig and Rosa [6] who developed magic labelings into an edge-magic total labeling. Furthermore, Gutiérrez and Lladó [2] generalized the concept of an edge-magic total labeling into an $H$-magic covering.

Let $G = (V, E)$ be a finite and simple graph. An edge-covering of $G$ is a family of different subgraphs $H_1, H_2, \ldots, H_k$ such that every edge in $G$ belongs to at least one of the subgraphs $H_i$ for $1 \leq i \leq k$. If each $H_i$ is isomorphic to $H$, then $G$ admits an $H$-covering. A graph $G$ is said to be an $H$-magic if it admits an $H$-covering and there is a bijective function $f : V \cup E \to \{1, 2, \ldots, \nu_G + e_G\}$ such that for any subgraph $H'(V', E')$ of $G$ isomorphic to $H$, $\Sigma_{v \in V(H')}\xi(v) + \Sigma_{e \in E(H')}\xi(e) = m(f)$, where $m(f)$ is a magic sum. We call an $H$-supermagic if its vertices contain the smallest possible labels.

There are a large number of research problems on $H$-magic total labeling that have been applied on some classes of graphs. Lladó and Moragas [7] showed that a wheel graph, a windmill

IOP Conf. Series: Journal of Physics: Conf. Series **855** (2017) 012037 doi:10.1088/1742-6596/855/1/012037

graph, and a book graph are cycle-magic. Ngurah et al. [10] studied cycle-supermagic covering on a triangular ladder graph, a book graph, and grids $P_m \times P_n$ for $m \geq 3$ and $n = 3, 4, 5$, then Roswitha et al. [11] proved that a generalized Jahangir graph $J_{k,s}$, a complete bipartite graph $K_{2,n}$ for $n \geq 2$, and a wheel graph $W_n$ for $n$ odd and $n \geq 4$, respectively, are $C_{s+2}$-supermagic, $C_4$-supermagic, and $C_3$-supermagic.

In 2009, Inayah et al. [3] introduced an $(a, d)$-$H$-antimagic covering. A graph $G$ is said to be an $(a, d)$-$H$-antimagic covering if there exists a bijective function $\xi : V \cup E \to \{1, 2, \ldots, v_G + e_G\}$ such that for every subgraph $H'$ isomorphic to $H$, $w(H') = \Sigma_{v\in V(H')}\xi(v) + \Sigma_{e\in E(H')}\xi(e)$ constitutes an arithmetic progression $\{a, a+d, a+2d, \ldots, a+(t-1)d\}$ where $a$ and $d$ are positive integers and $t$ is the number of subgraphs $G$ isomorphic to $H$. Inayah et al. [3] studied an $(a, d)$-$H$-antimagic covering on fan graph for some $d$, then Inayah et al. [5] proved that shackles of a connected graph $H$ is a super $(a, d)$-$H$-antimagic.

This research attempts to study super $(a, d)$-$C_3$-antimagic on triangular ladder graph $TL_n$ for $n \geq 2$ and super $(a, d)$-$C_{s+2}$-antimagic on generalized Jahangir graph $J_{k,s}$ for $k \geq 2$ and $s \geq 2$.

## 2. Main Results

### *2.1. k-balanced multisets*

Maryati [8] and Maryati et al. [9] defined a multiset as a set that allows the existence of the same elements in it. Let $k \in \mathbb{N}$ and $Y$ be a multiset that contains positive integers. Y is said to be *k-balanced* if there exists $k$-subsets of $Y$, namely $Y_1, Y_2, \ldots, Y_k$, such that for every $i \in [1, k]$, $|Y_i| = \frac{|Y|}{k}$, $\sum Y_i = \frac{\sum Y}{k} \in \mathbb{N}$, and $\biguplus_{i=1}^{k} Y_i = Y$. If those are the cases for every $i \in [1, k]$, then $Y_i$ is called a *balanced subset* of $Y$.

**Lemma 2.1** [Roswitha et al. [11]] *Let $x$ and $k$ be nonnegative integers. Let $X = [x + 1, x + k]$ with $|X| = k$ and $Y = [x + k + 1, x + 2k]$ where $|Y| = k$. Then, the multiset $K = X \uplus Y$ is $k$-balanced for $j \in [1, k]$.*

Next, we have the following lemma.

**Lemma 2.2** *Let $x$ be a nonnegative integer and $k \geq 2$ be an even integer. If $Y = [x + 1, x + \frac{k}{2}] \uplus [x + 2, x + 2k + 1] \uplus [x + \frac{3k}{2} + 2, x + 2k + 1]$, then $Y$ is $k$-balanced for $i \in [1, k]$.*

*Proof.* For every $i \in [1, k]$ we define a multiset $Y_i = \{a_i, b_i, c_i\}$, where

$$a_i = \begin{cases} x + \lceil \frac{i}{2} \rceil, & \text{for } i \text{ odd;} \\ x + \lceil \frac{i+1}{2} \rceil, & \text{for } i \text{ even.} \end{cases}$$

$$b_i = x + \frac{3k}{2} + 2 - i$$

$$c_i = \begin{cases} x + \frac{3k}{2} + 2 + \lfloor \frac{i}{2} \rfloor, & \text{for } i \text{ odd;} \\ x + \frac{3k}{2} + 1 + \frac{i}{2}, & \text{for } i \text{ even.} \end{cases}$$

Furthermore, we defined

$$A = \{a_i \mid 1 \leq i \leq k\} = [x + 1, x + \frac{k}{2}] \uplus [x + 2, x + \frac{k}{2} + 1];$$

$$B = \{b_i \mid 1 \leq i \leq k\} = [x + \frac{k}{2} + 2, x + \frac{3k}{2} + 1];$$

$$C = \{c_i \mid 1 \leq i \leq k\} = [x + \frac{3k}{2} + 2, x + 2k + 1] \uplus [x + \frac{3k}{2} + 2, x + 2k + 1].$$

If $A \uplus B \uplus C = Y$ and $\biguplus_{i=1}^{k} Y_i = Y$, then for every $i \in [1, k]$ we obtained $|Y_i| = 3$ and $\sum Y_i = 3x + 3k + 4$. Therefore, $Y$ is $k$-balanced.

*2.2. $(k, \delta)$-anti balanced multisets*

Inayah [4] defined $(k,\delta)$-anti balanced multiset as follows. Let $k, \delta \in \mathbb{N}$ and $X$ be a set containing the elements of positive integers. A multiset $X$ is said to be $(k,\delta)$-anti balanced if there exists $k$ subsets from $X$, i.e. $X_1, X_2, \ldots, X_k$ such that for every $i \in [1,k]$, $|X_i| = \frac{|X|}{k}$, $\biguplus_{i=1}^{k} X_i = X$, and for $i \in [1, k-1]$, $\sum X_{i+1} - \sum X_i = \delta$ is hold. Here, we give several lemmas on $(k,\delta)$-anti balanced multisets.

**Lemma 2.3** *Let $k, s \geq 2$ be integers. If $X = [k+2, 2k+1], Y = [2, ks+1]$, and $Z = [ks+2, 2ks+1]$, then the multiset $K = (Y - X) \uplus Z$ is $(k,1)$-anti balanced.*

*Proof.* Let $k, s \geq 2$ be integers. We define a multiset $K_i = \{a_j^i \mid 1 \leq j \leq s\} \uplus \{b_j^i \mid 1 \leq j \leq s-1\}$ for $i \in [1,k]$, where

$$a_j^i = \begin{cases} (s+j)k + 2 - i, & \text{for } j \text{ odd;} \\ (s+j-1)k + 1 + i, & \text{for } j \text{ even.} \end{cases}$$

$$b_j^i = \begin{cases} j + i, & \text{for } j = 1; \\ jk + 1 + i, & \text{for } j \text{ even;} \\ (j+1)k + 2 - i, & \text{for } j \text{ odd and } j \geq 3. \end{cases}$$

It is obvious that for every $i \in [1,k]$, $|K_i| = 2s-1$, $K_i \subset K$, and $\biguplus_{i=1}^{k} K_i = K$. If $s$ is odd, then the sum of elements in $K_i$ is

$$\begin{aligned} \sum K_i =& \sum_{z=1}^{\frac{s-1}{2}} ((s+2z-1)k + 2 - i) + \sum_{z=1}^{\frac{s-1}{2}} ((s+2z-1)k + 1 + i) + 2ks + 2 - i + 1 + \\ & i + \sum_{z=1}^{\frac{s-1}{2}-1} ((2z)k + 1 + i) + \sum_{z=1}^{\frac{s-1}{2}-1} (((2z+1)+1)k + 2 - i) + ks - k + 1 + i \\ =& 2k(s^2-1) + 3s - 2 + i. \end{aligned}$$

If $s$ is even, then the sum of elements in $K_i$ is

$$\begin{aligned} \sum K_i =& \sum_{z=1}^{\frac{s}{2}} ((s+2z-1)k + 2 - i) + \sum_{z=1}^{\frac{s}{2}} ((s+2z-1)k + 1 + i) + 1 + i + \sum_{z=1}^{\frac{s}{2}-1} ((2z)k + \\ & 1 + i) + \sum_{z=1}^{\frac{s}{2}-1} (((2z+1)+1)k + 2 - i) \\ =& 2k(s^2-1) + 3s - 2 + i. \end{aligned}$$

Since $\sum K_i = 2k(s^2-1) + 3s - 2 + i$ for every $i \in [1,k]$ and for all $s \geq 2$ and $\sum K_{i+1} - \sum K_i = 1$ for every $i \in [1, k-1]$, then $K$ is $(k,1)$-anti balanced.

**Lemma 2.4** *Let $k, s \geq 2$ be integers. If $X = [k+2, ks+1]$ and $Y = [ks+2, 2ks+1]$, then the multiset $Z = X \uplus Y$ is $(k,3)$-anti balanced.*

*Proof.* Suppose $k, s \geq 2$ be integers. We defined a multiset $Z_i = \{a_j^i \mid 1 \leq j \leq s\} \uplus \{b_j^i \mid 1 \leq j \leq s-1\}$ for $i \in [1,k]$, where

$$a_j^i = \begin{cases} ks + k(j-1) + i + 1, & \text{for } j \text{ odd;} \\ k(s+j-1) + i + 1, & \text{for } j \text{ even.} \end{cases}$$

$$b_j^i = \begin{cases} k + j + i, & \text{for } j = 1; \\ 2 + 3k + k(j-2) - i, & \text{for } j > 1. \end{cases}$$

Hence, $|Z_i| = 2s-1$, $Z_i \subset Z$, and $\biguplus_{i=1}^{k} Z_i = Z$ for every $i \in [1,k]$. Since $\sum Z_i = 2k(s^2-1) + 3(s-1) + 3i$ for every $i \in [1,k]$ and for all $s \geq 2$ and $\sum Z_{i+1} - \sum Z_i = 3$ for every $i \in [1,k-1]$, then $Z$ is $(k,3)$-anti balanced.

**Lemma 2.5** *Let $k, s \geq 2$ be integers. If $X = [1, k(s-1)]$ and $Y = [k(s+1)+2, k(2s+1)+1]$, then the multiset $Z = X \uplus Y$ is $(k,5)$-anti balanced.*

*Proof.* Let $k, s \geq 2$ be integers. We define the multiset $Z_i = \{a_j^i \mid 1 \leq j \leq s\} \uplus \{b_j^i \mid 1 \leq j \leq s-1\}$ for $i \in [1,k]$, where

$$a_j^i = \begin{cases} k(s+1)+j-1+2i, & \text{for } j = 1,2; \\ k(s+1)+k(j-1)+1+i, & \text{for } j \text{ odd and } j \geq 3; \\ k(s+1)+kj+2-i, & \text{for } j \text{ even and } j \geq 4. \end{cases}$$

$$b_j^i = \begin{cases} k(j-1)+i, & \text{for } j \text{ odd;} \\ kj+1-i, & \text{for } j \text{ even.} \end{cases}$$

It is easy to verify that $|Z_i| = 2s-1$, $Z_i \subset Z$, and $\biguplus_{i=1}^k Z_i = Z$ for every $i \in [1,k]$. Since $\sum Z_i = 2k(s^2-1)+2s-3+5i$ for every $i \in [1,k]$ and for all $s \geq 2$ and $\sum Z_{i+1} - \sum Z_i = 5$ for every $i \in [1,k-1]$, then $Z$ is $(k,5)$-anti balanced.

**Lemma 2.6** *Let $x$ be a nonnegative integer and $k \geq 2$ be an even integer. If $X = [x+1, x+k] \uplus [x+2, x+k+1] \uplus [x+3, x+k+2]$, then $X$ is $(k,3)$-anti balanced.*

*Proof.* Let $k \geq 2$ be an even integer. For every $i \in [1,k]$, we define the multiset $X_i = \{x+i, x+1+i, x+2+i\}$. It can be verified that $|X_i| = 3$, $X_i \subset X$, and $\biguplus_{i=1}^k X_i = X$ for every $i \in [1,k]$. Since $\sum X_i = 3x+3+3i$ for every $i \in [1,k]$ and $\sum X_{i+1} - \sum X_i = 3$ for every $i \in [1,k-1]$, $X$ is $(k,3)$-anti balanced.

**Lemma 2.7** *Let $x$ be a nonnegative integer and $k \geq 2$ be an even integer. If $Y = \{x+1+2j, j = 0,1,\ldots,k\} \uplus \{x+2+j, j = 0,1,2,\ldots,2(k-1)\}$, then $Y$ is $(k,2)$-anti balanced.*

*Proof.* Let $k \geq 2$ be an even integer. For every $r \in [1,k]$, we define the multiset $Y_r = \{a_r, b_r, c_r\}$, where $a_r = 2k+2+x-2r$, $b_r = x+1+2r$, $c_r = x-1+2r$. It can be verified that $|Y_r| = 3$, $Y_r \subset Y$, and $\biguplus_{r=1}^k Y_r = Y$ for every $r \in [1,k]$. Since $\sum Y_r = 3x+2k+2+2r$ for every $r \in [1,k]$ and $\sum Y_{r+1} - \sum Y_r = 2$ for every $r \in [1,k-1]$, $Y$ is $(k,2)$-anti balanced.

**Lemma 2.8** *Let $x$ be a nonnegative integer and $k \geq 2$ be an even integer. If $Y = [x+1, x+k] \uplus [x+2, x+2k+1]$, then $Y$ is $(k,1)$-anti balanced.*

*Proof.* Let $k \geq 2$ be an even integer. For every $r \in [1,k]$, we define the multiset $Y_r = \{a_r, b_r, c_r\}$, where $a_r = x+r$, $b_r = x+1+r$, $c_r = x+2k+2-r$. It can be verified that $|Y_r| = 3$, $Y_r \subset Y$, and $\biguplus_{r=1}^k Y_r = Y$ for every $r \in [1,k]$. Since $\sum Y_r = 3x+2k+3+r$ for every $r \in [1,k]$ and $\sum Y_{r+1} - \sum Y_r = 1$ for every $r \in [1,k-1]$, $Y$ is $(k,1)$-anti balanced.

**Lemma 2.9** *Let $x$ be a nonnegative integer and $k \geq 2$ be an even integer. If $Y = [x+1, x+2k] \uplus [x+k+2, x+2k+1]$, then $Y$ is $(k,1)$-anti balanced or $(k,3)$-anti balanced.*

*Proof.* Let $k \geq 2$ be an even integer. In this proof, we have two separated cases.
**Case 1.** For every $i \in [1,k]$, we define the multiset $Y_i = \{a_i, b_i, c_i\}$, where $a_i = x+k+1-i$, $b_i = x+k+i$, and $c_i = x+k+1+i$. It is easy to check that $|Y_i| = 3$, $Y_i \subset Y$, and $\biguplus_{i=1}^k Y_i = Y$ for every $i \in [1,k]$. Since $\sum Y_i = 3x+3k+2+i$ and $\sum Y_{i+1} - \sum Y_i = 1$ for every $i \in [1,k-1]$, $Y$ is $(k,1)$-anti balanced.
**Case 2.** For every $i \in [1,k]$, we define the multiset $Y_i = \{a_i, b_i, c_i\}$, where $a_i = x+i$, $b_i = x+k+i$, and $c_i = x+k+1+i$. It is easy to check that $|Y_i| = 3$, $Y_i \subset Y$, and $\biguplus_{i=1}^k Y_i = Y$ for every $i \in [1,k]$. Since $\sum Y_i = 3x+2k+1+3i$ and $\sum Y_{i+1} - \sum Y_i = 3$ for every $i \in [1,k-1]$, $Y$ is $(k,3)$-anti balanced.

**Lemma 2.10** *Let $x$ be a nonnegative integer and $k \geq 2$ be an even integer. If $Y = [x+1, x+2k+1] \uplus \{x+1+2j, j = 1, 2, \dots, k-1\}$, then $Y$ is $(k,2)$-anti balanced.*

*Proof.* Let $k \geq 2$ be an even integer. For every $i \in [1,k]$, we define the multiset $Y_i = \{a_i, b_i, c_i\}$, where

$$a_i = \begin{cases} x-1+2i, & i \in [1, \frac{k}{2}]; \\ x+2k+2-2i, & i \in [\frac{k}{2}+1, k]. \end{cases}$$

$$b_i = \begin{cases} x+1+2i, & i \in [1, \frac{k}{2}]; \\ x-1+2i, & i \in [\frac{k}{2}+1, k]. \end{cases}$$

$$c_i = \begin{cases} x+2k+2-2i, & i \in [1, \frac{k}{2}]; \\ x+1+2i, & i \in [\frac{k}{2}+1, k]. \end{cases}$$

Clearly, $|Y_i| = 3$, $Y_i \subset Y$, and $\biguplus_{i=1}^{k} Y_i = Y$ for every $i \in [1,k]$. Since $\sum Y_i = 3x+2k+2+2i$ for every $i \in [1,k]$ and $\sum Y_{i+1} - \sum Y_i = 2$ for every $i \in [1, k-1]$, $Y$ is $(k,2)$-anti balanced.

**Lemma 2.11** *Let $x$ be a nonnegative integer and $k \geq 2$ be an even integer. If $Y = [x+1, x+2k] \uplus \{x+1+2j, j = 1, 2, \dots, k\}$, then $Y$ is $(k,6)$-anti balanced.*

*Proof.* Let $k \geq 2$ be an even integers. For every $i \in [1,k]$, we define the multiset $Y_i = \{a_i, b_i, c_i\}$, where $a_i = x-1+2i$, $b_i = x+2i$, and $c_i = x+1+2i$. It is easy to check that $|Y_i| = 3$, $Y_i \subset Y$, and $\biguplus_{i=1}^{k} Y_i = Y$ for every $i \in [1,k]$. Since $\sum Y_i = 3x+6i$ for every $i \in [1,k]$ and $\sum Y_{i+1} - \sum Y_i = 6$ for every $i \in [1, k-1]$, $Y$ is $(k,6)$-anti balanced.

### *2.3. Triangular ladder graph $TL_n$*

Jeyanthi and Maheswari (Gallian [1]) defined a triangular ladder graph $TL_n$ as a graph obtained from the ladders $L_n = P_n \times P_2$ $(n \geq 2)$ with additional edges $u_i v_{i+1}$ for $1 \leq i \leq n-1$, where the consecutive vertices of two copies of $P_n$ are $u_1, u_2, \dots, u_n$ and $v_1, v_2, \dots, v_n$ and the edges are $u_i v_i$ for $1 \leq i \leq n$. A triangular ladder graph $TL_n$ has $|V(TL_n)| = 2n$ and $|E(TL_n)| = 4n-3$.

**Theorem 2.12** *For $n \geq 2$, a triangular ladder graph $TL_n$ admits a super $(14n, 1)$-$C_3$-antimagic labeling.*

*Proof.* Let $G$ be a triangular ladder graph, $G - TL_n$ and $V(G)$ and $E(G)$ be the sets of vertices and edges of of $G$, respectively. Here, we define a bijective function $\xi_1 : V(G) \cup E(G) \rightarrow \{1, 2, \dots, 6n-3\}$ and $\xi_1(V(G) = \{1, 2, \dots, 2n\}$. Let $W = [1, 6n-3]$ be the set of labels for all vertices and edges of $G$. Partition $W$ into two sets, i.e. $K = [1, 2n]$ and $L = [2n+1, 6n-3]$. Let $H_i$ be the arbitrary subgraphs $C_3$ of $G$ with $V(H_i) = \{v_1, u_1, v_2, v_1\}$, $\{u_1, v_2, u_2, u_1\}$, $\{v_2, u_2, v_3, v_2\}$, $\{u_2, v_3, u_3, u_2\}$, ..., $\{v_{n-1}, u_{n-1}, v_n, v_{n-1}\}$, $\{u_{n-1}, v_n, u_n, u_{n-1}\}$. The number of subgraphs $C_3$ of $G$ is $(2n-2)$.

Next, we use the elements of $K$ to label the entire vertices of $G$, namely $v_1, u_1, v_2, u_2, \dots, v_n, u_n$, respectively. Such labeling is applied based on Lemma 2.6 with $x = 0$ and $k = 2n-2$. The elements of $X_i$ are used to label all vertices in every subgraph $H_i$ of $G$, i.e. $V(H_i)$. Since $\sum X_i = 3+3i$ for every $i \in [1, 2n-2]$ and $\sum X_{i+1} - \sum X_i = 3$ for every $i \in [1, 2n-3]$, so $X$ is $((2n-2), 3)$-anti balanced.

Now, we label all edges of $TL_n$ using the elements of $L$. This labeling is applied according to Lemma 2.7 with $x = 2n$ and $k = 2n-2$. The elements of $Y_r$ are used to label all edges in every subgraph $H_i$ of $G$, i.e. $\{u_{n-1}u_n, u_{n-1}v_n, u_n v_n\}$, $\{v_{n-1}v_n, u_{n-1}v_{n-1}, u_{n-1}v_n\}$, ..., $\{u_1u_2, u_1v_2, u_2v_2\}$, $\{v_1v_2, u_1v_1, u_1v_2\}$, respectively. Since $\sum Y_r = 10n-2+2r$ for every $r \in [1, 2n-2]$ and $\sum Y_{r+1} - \sum Y_r = 2$ for every $r \in [1, 2n-3]$, so $Y$ is $((2n-2), 2)$-anti balanced.

After all the vertices and edges of $G$ are labeled, we obtain $w(H_i) = \sum X_i + \sum Y_r$ as the sum of labels from each subgraph $H_i$. If $r = 2n-1-i$, then $w(H_i) = \sum X_i + \sum Y_{2n-1-i} =$

$3+3i+14n-4-2i = 14n-1+i$ for all $i \in [1, 2n-2]$. Since $w(H_{i+1}) - w(H_i) = 1 = d$ and $w(H_1) = 14n = a$, we can deduce that triangular ladder graph $TL_n$ is a super $(14n, 1)$-$C_3$-antimagic for $n \geq 2$.

As a consequence of this result, a triangular ladder graph $TL_n$ is a super $(a, d)$-$C_3$-antimagic covering for $d = 2, 3, 4, 5, 6, 9$. It can be proved by using Lemma 2.8, Lemma 2.2, Lemma 2.9 ($(k, 1)$-anti balanced), Lemma 2.10, Lemma 2.9 ($(k, 3)$-anti balanced), and Lemma 2.11 to $L$, respectively.

**Corollary 2.13** *For $n \geq 2$, a triangular ladder graph $TL_n$ admits*

*(i) a super $(12n+3, 2)$-$C_3$-antimagic total labeling,*
*(ii) a super $(12n+4, 3)$-$C_3$-antimagic total labeling,*
*(iii) a super $(12n+3, 4)$-$C_3$-antimagic total labeling,*
*(iv) a super $(10n+6, 5)$-$C_3$-antimagic total labeling,*
*(v) a super $(10n+6, 6)$-$C_3$-antimagic total labeling,*
*(vi) a super $(6n+12, 9)$-$C_3$-antimagic total labeling.*

### *2.4. Generalized Jahangir graph $J_{k,s}$*

Gallian [1] defined a generalized Jahangir graph $J_{k,s}$ as a graph contains $ks+1$ vertices consisting of a cycle $C_{ks}$ and one additional vertex that is adjacent to $k$ vertices of $C_{ks}$ at distance $s$ to each other on $C_{ks}$. A generalized Jahangir graph $J_{k,s}$ has $|V(J_{k,s})| = ks+1$ and $|E(J_{k,s})| = ks+k$.

**Theorem 2.14** *For $k, s \geq 2$, a generalized Jahangir graph $J_{k,s}$ is a super $(2ks^2+4ks+2k+3s+6, 1)$-$C_{s+2}$-antimagic.*

*Proof.* Let $G$ be a generalized Jahangir graph. Let the set of vertices of $G$, $V(G) = \{c, v_!, v_2, ..., v_s, v_{s+1}, ..., v_{2s}, v_{2s+1}, ..., v_{ks}\}$ and the set of edges of $G$, $E(G) = \{cv_1, cv_{s+1}, ..., cv_{ks}\}$. We define a bijective function $\xi_3 : V(G) \cup E(G) \to \{1, 2, \ldots, 2ks+k+1\}$ and we set $\xi_3(V(G) = \{1, 2, \ldots, ks+1\}$. Let $W = [1, 2ks+k+1]$, the set of labels for all vertices and edges of $G$. Partition $W$ into five sets, i.e. $P = \{1\}$, $X = [k+2, 2k+1]$, $Y - X = [2, ks+1] \setminus X$, $Z = [ks+2, 2ks+1]$, and $L = [2ks+2, 2ks+k+1]$. Next, we define $H_i$ as any subgraph $C_{s+2}$ of $G$, where $V(H_i) = \{c, v_1, v_2, \ldots, v_{s+1}, c\}$, $\{c, v_{s+1}, v_{s+2}, \ldots, v_{2s+1}, c\}$, $\{c, v_{2s+1}, v_{2s+2}, \ldots, v_{3s+1}, c\}$, $\ldots$, $\{c, v_{(k-1)s+1}, v_{(k-1)s+2}, \ldots, v_1, c\}$. The number of subgraphs of $G$ is $kC_{s+2}$.

Next, we label all of the vertices and edges from every subgraph $H_i$ of $G$. First, we label the central vertex of $G$ by the element of $P$, so $\xi_3(c) = 1$. Furthermore, we put the elements of $X$ and $L$ as the labels of vertices and edges that adjacent and incident to c, respectively. Let $k \geq 2$ and $M = X \uplus L$ with $|X| = k$ and $|L| = k$. Define $M_i = \{\{a_i, b_i\} \mid 1 \leq i \leq k\}$, where $a_i = 2(k+1) - i$ and $b_i = 2ks+1+i$ for every $i \in [1, k]$. Furthermore, we define the sets

$$\begin{aligned} A = \{a_i \mid 1 \leq i \leq k\} &= [k+2, 2k+1]; \\ B = \{b_i \mid 1 \leq i \leq k\} &= [2ks+2, 2ks+k+1]. \end{aligned}$$

If $A \uplus B = M$ and $\biguplus_{i=1}^{k} M_i = M$, then by Lemma 2.1, for $1 \leq i \leq k$, $|M_i| = 2$ and $\sum M_i = 2ks+2k+3$. Therefore, $M$ is $k$-balanced.

Next, we put the elements of $Y - X$ and $Z$ as the labels of vertices not adjacent to $c$ and edges which not incident to $c$. If $(Y - X) \uplus Z = K$, then for $k \geq 2$ and $i \in [1, k]$, by Lemma 2.3, we set $K_i = \{a_j^i \mid 1 \leq j \leq s\} \uplus \{b_j^i \mid 1 \leq j \leq s-1\}$ . Use the elements $b_j^i$ of $K_i$ for every $i \in [1, k]$ and $j \in [1, s-1]$ to label every vertex not adjacent to $c$. Meanwhile, the elements $a_j^i$ of $K_i$ for every $i \in [1, k]$ and $j \in [1, s]$ are used to label the edge not incident to $c$. Since $K_i = k(2s^2-2) + 3s - 2 + i$ for every $i \in [1, k]$ and $K_{i+1} - K_i = 1$ for every $i \in [1, k-1]$, so $K$ is $(k, 1)$-anti balanced.

We obtain the sum of labels from each subgraph $H_i$, $w(H_i) = 1 + 2(\sum M_i) + \sum K_i = 2ks^2 + 4ks + 2k + 3s + 5 + i$ for every $i \in [1, k]$. Since $w(H_{i+1}) - w(H_i) = 1 = d$ and $w(H_1) = 2ks^2 + 4ks + 2k + 3s + 6 = a$, it can be concluded that a generalized Jahangir graph $J_{k,s}$ is a super $(2ks^2 + 4ks + 2k + 3s + 6, 1)$-$C_{s+2}$-antimagic for $k, s \geq 2$.

The following corollary can be proved by using Lemma 2.1 to $K \uplus L = [2, k+1] \uplus [2ks + 2, 2ks + k + 1]$ and Lemma 2.4 to $X \uplus Y = [k + 2, ks + 1] \uplus [ks + 2, 2ks + 1]$ for Corollary 2.15(1), then Lemma 2.1 to $K \uplus L = [k(s-1) + 1, ks] \uplus [ks + 2, k(s+1) + 1]$ and Lemma 2.5 to $X \uplus Y = [1, k(s-1)] \uplus [k(s+1) + 2, k(2s+1) + 1]$ for Corollary 2.15(2).

**Corollary 2.15** *For $k, s \geq 2$, a generalized Jahangir graph $J_{k,s}$ admits*

*(i) a super $(2ks^2 + 4ks + 3s + 7, 3)$-$C_{s+2}$-antimagic covering,*

*(ii) a super $(2ks^2 + 5ks - 2k + 2s + 7, 5)$-$C_{s+2}$-antimagic covering.*

### *2.5. Acknowledgments*
The authors would like to thank DIKTI Indonesia for the funding under PUPT 2016.

## References
[1] Galian, J. A., *A Dinamic Survey of Graph Labeling*, The Electronic Journal of Combinatorics, **17** (2014), 1-384.#DS6
[2] Gutiérrez, A and A. Lladó, *Magic Covering*, J.Combin. Math. Combin. Comput., **55** (2005), 43-46.
[3] Inayah, N., A. N. M., Salman, and R. Simanjuntak, *On $(a, d)$-$H$-Antimagic Coverings of Graphs*, J. Combin. Math. Combin. Comput. **71** (2009), 273-281.
[4] Inayah, N., *Pelabelan Selimut $(a, d)$-$H$-Anti Ajaib pada Beberapa Kelas Graf*, Disertasi, ITB Bandung (2013).
[5] Inayah, N., R. Simanjuntak, and A.N.M. Salman, *Super (a, d)-H-Antimagic Total Labelings for Shackles of a Connected Graph* , Australasian Journal of Combinatorics, **57** (2013),127-138.
[6] Kotzig, A. and A. Rosa, *Magic Valuations of Finite Graphs*, Canad. Math. Bull (1970), 451–461.
[7] Lladó, A. and J. Moragas, *Cycle-magic Graphs*, Discrete Mathematics **307** (2007), 2925–2933.
[8] Maryati, T. K., *Karakteristik Graf H-Ajaib dan Graf H-Ajaib Super, Disertasi*, Institut Teknologi Bandung, Bandung, (2011).
[9] Maryati, T. K., A. N. M. Salman, E. T. Baskoro, J. Ryan, and M. Miller, *On H Supermagic Labelings for Certain Shackles and Amalgamations of a Connected Graph*, Utilitas Mathematica **83** (2010), 333-342.
[10] Ngurah, A. A. G. and E. T. Baskoro, *On Magic and Antimagic Total labeling of Generalized Petersen Graph*, Utilitas Mathematica **63** (2003), 97–107.
[11] Roswitha, M., E. T. Baskoro, T. K. Maryati, N. A. Kurdhi, and I. Susanti, *Further Results on Cycle-Supermagic Labeling*, AKCE Int. J. Graphs Comb. **10** (2013), 211-220.
[12] Sedlacek, J., *Theory of Graphs and Its Applications*, House Czechoslovak Acad. Sci. Prague, Scotland, 1964.
[13] Wallis, W. D., *Magic Graph*, Birkhäuser, Basel, Berlin, 2001.

# The Development of e-tutorial on Implementation National Curriculum 2013 for Mathematics Teacher

**YenitaRoza[1), Gita Satria[2), Syarifah Nur Siregar[3)**

Universitas of Riau

[1)yenita.roza@lecturer.unri.ac.id, [2)gita.satria@lecturer.unri.ac.id,
[3) nurhafirays@yahoo.com

**Abstract.** Curriculum 2013 is the new national Curriculum in Indonesia that is targeted to be used in all Indonesian schools in 2019. At this time the teacher training continues but the number and locations of teachers very diffuse and time constraints to be an obstacle for the government to be able to conduct training for teachers. This research resulted in the e-tutorial which is designed for mathematics teachers in studying the process of Curriculum implementation. This product will assist the government in accelerating the preparation of teachers in implementation of Curriculum 2013. This e-tutorial contains the dynamics of Curriculum development, learning model, learning assessment, lesson plan, curriculum stages of implementation and government regulation that is relevant to the implementation of Curriculum 2013. The product development started with a needs analysis through discussions with mathematics teachers about their difficulties in the implementation of the Curriculum 2013. This e-tutorial was developed using *Application of Adobe Director 11*. This paper discusses the results of need analysis, process development and results of product revisions made based on input from teachers during the FGD. From the discussion, it can be concluded that this e-tutorial easily understood by teachers and help them to understand the implementation of Curriculum 2013

**Keywords: Curriculum 2013; e-tutorial; mathematics teacher**

## 1. Introduction

The Curriculum is a set of plans and arrangements regarding the objectives, content and learning materials as well as the tools used to guide the implementation of learning activities to achieve specific educational objectives (UU no 20 tahun 2003). The Curriculum is designed in accordance with the times in which the Curriculum is used. In line with the shifting paradigms of learning government of Indonesia through Department of Education announced the change of National Curriculum in 2013. This new curriculum were built based-on understanding of the competence to be achieved in the learning process and realizing the importance of being trained for high-level thinking skills. The government published the new Curriculum named Curriculum 2013 that officially used on July 2013. In the beginning, the Curriculum only implemented at schools designated by the government which began with the first semester at grade 1 and grade 4of elementary school, grade 7 junior of high school and grade 10 of senior high school.

One of changes that occurred in Curriculum 2013 were the demands to the learning process that should lead to creativity. This demand is facilitated by scientific approach containing step of observing, questioning, experimenting, reasoning and networking. This is a student center approach that required both student and teacher to be creative in learning environment. These changes requires shift of the mindset both teachers and students in implementing the learning process. Teacher need to facilitate student facing through those five steps. Another significant change to the Curriculum 2013 is the assessment of learning outcomes. In the previous Curriculum put more emphasis on knowledge assessment, but the Curriculum 2013 make teachers must assess the improvement of knowledge, skills and attitude, as each of these components require different evaluation techniques. Many teacher faced difficulties to do those types of assessment.

In 2014 the government issued a policy that Curriculum 2013 applied across the school for grades 2,3,4 and 5th of Elementary Schools, 7th and 8th grade of junior high and and grades 10th and 11th of senior high school. Curriculum 2013 is applied in 6,221 schools since the year 2013/2014. In the year 2014/2015, this Curriculum is implemented in schools at 295 districts and cities across Indonesia. Applying new curriculum at the same time for all level schools definitely a big work to do for all people involve in education.

In order to prepare the implementation of the Curriculum 2013, the government has been planning and carrying out massive training of the teachers who will implement this Curriculum. Training strategy used is a leveling training. There are three levels of training for National Trainer, National Instructror and Teachers as the end-user. In the initial phase the government prepare and conduct a short training (3 days) for the National Trainer. National Trainer recruited from lecturer of Educational faculty, senior trainer from government office of Education, and selected teachers. National Trainers responsible to train the National Instructors that of representatives designated teacher from each district. Selected candidates for National Instructors are trained for 72 hours in the provincel level and then given the responsibility to train teachers who will implement Curriculum 2013 directly. End-user teachers given exercises for 52 hours in their respective districts. By taking training teachers are assumed ready to implement the Curriculum 2013 with all the demands of change from the previous Curriculum.

From researcher direct observations as the National Trainer, training that is provided for teachers limited to introduction level. Limited activities at training for teachers and time constraints did not make them ready for implementation. The training conducted so far only reached the stage of knowing yet not preparing teachers to implement curriculum. From the facts of various observation, the main problem in implementation the Curriculum 2013 are the lack of teacher capability. Teachers need more drill and practice on designing and implementing lesson plan. Another problem is the delays on supplying teacher books and the books for students. For that reason, at the beginning of 2015, the Ministry of Education and Culture decided to temporarily suspend the use of the Curriculum 2013 at some schools.

Government Regulation *No 160*, published December 11th, 2014 stated that the units of primary and secondary education that have implemented the Curriculum 2013 for three semesters can continue to use the them, the school that implemented curriculum 2013 just for one semester should re-implement the Curriculum 2006 untill there is a provision of the Ministry to implement the Curriculum 2013 (Permendikbud 160, 2014). However, government also declared that if the school already implementing for three semesters, but was not ready they can request in writing for not using the Curriculum 2013. At this regulation also stated that schools could use the Curriculum 2006 up to 2019/2020. The Schools that have not implemented the Curriculum 2013 will be giving a training and assistance for school principals, teachers, an academic staff and educators supervisor.

Referring to the regulation above, the researchers saw the need to design a better and flexible way in preparing future teachers for implementation new curriculum for avoiding same failure. Preparation strategy for teachers should be able to accommodate the facts and circumstances that teachers' ability in Indonesia is varied as teachers location spread from big cities to rural areas, including disadvantaged and outer regions. The existence of a teacher location also seriously affected the ability of teachers to obtain information. The teacher who was in town is closer to the sources of information are typically more agile of information and more adaptive to change. The use of technology as training media is one of the good choice at the moment.

According to Kopp (2012) The e-tutor plays a major role in supporting virtual collaborative learning, as he/she supervises learners in collaboratively solving tasks, acquiring new skills, and applying new knowledge. As mention by Feng (2016) Students very much appreciated e-learning that focused on the existing curriculum, but extended it so that they could do things in virtual reality that safety or cost factors prevent them from doing in real life at that stage in their studies.

Training is the giving of information and knowledge usually through speech, the written word or other methods of demonstration. In essence, through training, we are not looking to reshape the behaviour of an individual rather to teach the employee or learner how things are done so that they can then carry out a process on their own. This allows instructors/trainers to provide their learners with the tools to tackle current issues, develop life-long skills, improve on their problem-solving skills and utilize resources to the best of their ability (Epignosis,2014).

According to researcher one of the alternatives that can be provided suitably in preparing teachers for implement the Curriculum 2013 is the e-tutorial material. As stated by Paay (2000) O*nline learning* is defined as learning that takes place partially or entirely over the Internet. It was found that no difference in effects between the treatment and the control, between online learning and face-to-face learning. This e-tutorial material is used based on technology, currently technologies have been available widespread in Indonesia, so it is not expected to be a drag, but it is obvious that the availability of electricity is essential to be able to use this materials that are electronic-tuturial. Hay (2008) found that teaching material presented trough e-learning get more appreciation from students.

Material e-tutorial is designed to be used user-friendly, it can be used as a supplement for teachers who have had a training or use independently for teachers who have not had the opportunity to attend the training. By studying the e-tutorial teachers are expected to understand the concepts of the Curriculum 2013, the changed from the previous Curriculum, preparing lesson plans and learning assessment. The examples video of classroom activities that are designed based on the Curriculum 2013 and presented directly by the teacher with a real classroom setting. By watching this video teacher is expected can design their own class and practice to implement them. In this study, in accordance with the limitations of researchers academic background, product was developed only for mathematic subjects of middle and high school.

Another advantage of this tutorial material is training control in the hands of teachers as the user. Different from regular training, with this material teachers can use it in accordance with the availability of their time and situation. Teachers can also repeat the lesson in accordance with their needs and speed in understanding the material. Availability of e-tutorial will also help save transportation costs required from remote areas for attending many training.

There are several possibilities to use this e-tutorial for implementation Curriculum 2013. This e-tutorial can be used as a starting material, all mathematics's teachers at junior high and senior high levels are given material to be studied. After using this material government can send instructors to the region to conduct short training and discuss the problems faced by teachers. This material can also be used by the trainer or instructor as a presentation to provide training to teachers and distributed it to the a teacher as guidance material when they face an obstacle in implementing Curriculum 2013.

## 2. Process of Development

The E-tutorial is developed through design research that done in several phases as follows:

**a. Define Phase**

At this stage of need analysis has been determined that the user of this program is mathematics teachers from junior and senior high schools who will implement the Curriculum 2013. Through Concept analysis has been selected the material to be presented, the order of presentation and presentation technique based on the information of the user characteristic. Task analysis is done to determine the forms of activity given that the user can achieve the desired objectives through e-tutorial. Define phase ended by determining the specific purpose of training that will be generated through the e-tutorial.

**b. Design Phase**

Design stage is the stage of designing the product, the design is made for the contents and the flow of the program. Design activities are divided into several sub-activities, namely the selection of media, selection of media format, and a preliminary draft. Media format that will be used is a multimedia tutorial that will use text, audio and video. In accordance with the format to be used then it is determined that the media will be used is the application Adobe Director 11 for creating applications, Adobe Photoshop CS4 for image editing, Adobe Audition 1.5 for processing and sound editing, Adobe Premiere 1.5 for video processing and program Adobe Flash Professional CS6 for supporting applications. At the initial design stage we arranged sequence and story-board material from the program. This is the initial design phase to begin development.

**c. Development Phase**

The next stage is the stage of development, where materials and preliminary design of a product made, at this stage all of the program are assemblied, all content that has been prepared in accordance with the program put together into the design of the story board. In the process carried out at least 2 times of focus group discussion (FGD) with prospective users for discussing parts of the product that has been made. The first FGD conducted for Curriculum Content and Mathematics subjects, second FGD done for Materials of Learning Design. At this stage program tray out also done to get input for improving the program.

**d. Dissemination Phase**

Stages of dissemination done after the product was revised based on the try-out results. Dissemination conducted on two different parts, namely the prospective user for introducing the product and to funders as a form of accountability. Dissemination done to the prospective user that facilitated by Department of Education Pekanbaru City as a research partner, to University or Riau and Department Higher Education as a funder..

## 3. Main Result

The development of research conducted to produce the user-friendly e-tutorial materials for Implementation of Curriculum 2013 for mathematics subjects. This product can be used by middle and high school mathematics' teacher who uses the Curriculum 2013. Products produced in the form of teacher training material contained IT-based Curriculum 2013, steps on analysis core and basic competency, steps to prepare a lesson plan and develop the evaluation instruments that includes the attitudes, knowledge and skills. This material will gradually be equipped with video links that provide examples of learning uses a scientific approach in the model of Problem Based Learning, Discovery learning and Project Based Learning as suggested in the implementation of the Curriculum 2013. Video will be taken in the actual classroom setting once teachers have been trained to use the device that made by the research team. In the e-tutorial is also provided all the relevant government regulation relate to the implementation of the Curriculum 2013, and a summary of each regulation to help user in understanding the regulation easier.

This product will simplify and accelerate the Ministry of Education and Culture program in preparing teachers, especially mathematics teachers to implement the Curriculum 2013. This product is targeted to be use in two ways as trainer presentation on teacher training or can be used as learning material for teachers independently and learn the relevant and quality examples. This effort will help government to achieve their target that all schools in the year 2019 is already using the Curriculum 2013. The products made have been validated and tested by the mathematics teacher for its practicalities.

The main result of this research is digital application in form of an e-tutorial for implementation, Curriculum 2013 on mathematics. Parts of the products discussed in this section. Some appearance on these products can be seen in the following figure:

Figure 1. Product Cover and Menu

The picture above show the main page of e-Tutorial that had two choices *Materi* and Video. The secon figure showed main menu that has 7 parts namely Foreword, Dinamics of curriculum changes, Learning Model, Learning Assesment, Lesson Plan, Implementation Curriculum 2013 and Relevan government regulation. This attractive main cover expect to gain user interest for using the program. The user can choose any one of the main menu to begin they learning, but recommended to start from the first choice.

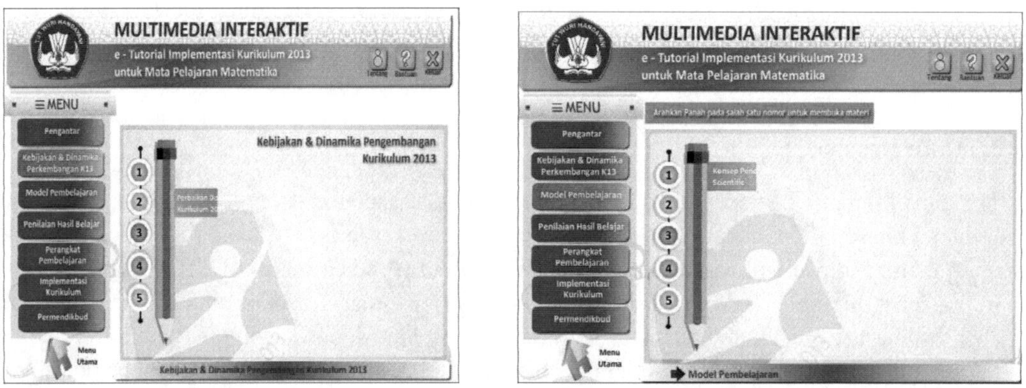


Figure 2. Product Sub Menu Dynamic Change of Curriculum

The Figure 2 give the sub- menu of Dinamics of Curriculum Change, it has five parts, they are the Background of changes, Mainframe of change, List of revision, and Implementation of change. This topics give the user the main idea or reason why they need the new curriculum. Understanding this background will help user to understand the change of curriculum.
The 3rd menu Learning model has 5 parts names Scientific approach, Discovery Model, Problem Based Model, Project Based Model and Inquiry Model. Those are model and approach that recommended to be used in implementing curriculum 2013. User will also see the video of those relevant model for their reference in developing their own program.

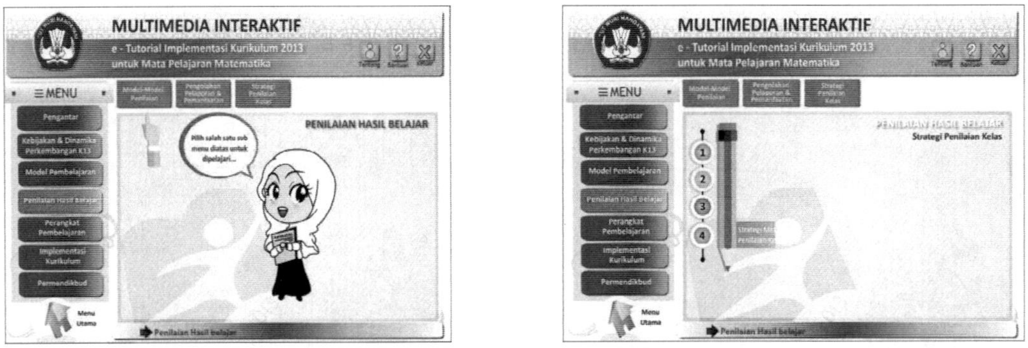


Figure 3. Menu Learning Assesment

The 4th Menu Learning Assesment shown in above figure. There are three sub menu in this part, they are Models of Assesment, Analisis and Usage of Assesment's data and Classroom Assesment Strategic. This Classroom Assesment Strategic presented in four parts as Assesment Guideline for Junior Highschool, Assesment guideline for Senior High School, Assesement Guideline for Teachers and Strategies in doing assessment.

Figure 4. Menu Lesson Instrument

At the Lesson Instrument Menu given 6 parts, they are Main Competency, Basic Competency, Students' activities sheets, Learning Media, Lesson Plan, Syllabus and Standard for Learning Process. Sub-Menu of students' activities sheets is divided into for 4 parts, Activities sheets based on Discovery model for grade 7 Junior High schools, Activities sheets based on Problem model for grade 7 Junior High schools. The other two are activities sheets for grade 10 of Senior High Schools student using Discovery and problem-based model.

Figure 5. Menu Syllabus and Standard of Process

There are five part at Syllabus menu, Main Syllabus for Senior High School, Specific Syllabus for Senior High School, Syllabus for Junior High Schools, Syllabus for year 7 of Junior High and year

10 of Senior High Schools. The Standard of Learning Proses giving a guideline for both learning process at Junior High School and Senior High Schools teachers.

Figure 6. Menu Learning Assessment

On the menu of Curriculum Implementation are given 3 submenus, textbooks, teaching material for curriculum implementation training as well as activity sheets for teacher in curriculum implementation training. On the sub-menu of textbook is provided the books for teacher and student at both level Junior and Senior High Schools. These books are standard textbook provided by government. The teacher and student can find supplement book at their own.

Figure 7. Menu Curriculum Implementation

Figure 7 show the menu for Curriculum Implementation, there are 6 different Activities sheet for teacher training. Those activities sheets will be used for drill and practice of Lesson Plan, Analysis of learning assessment, Analysis of Learning Outcome, Main Competency, Basic Competency, analysis of learning design, students and teacher textbooks. These activities are design by government for all training provided for teacher in preparing them to implement curriculum 2013.

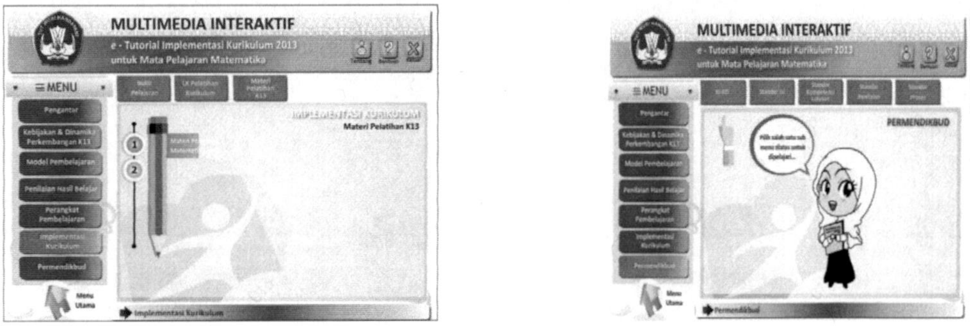


Figure 8. Menu Curriculum Implementation Process

On the above figure presented the menu on Curriculum Implementation process. The material is divided into two sub-menu, each for Junior and Senior High School. This content taken from the program teacher training that design by government in preparing teacher to use the Curriculum 2013.

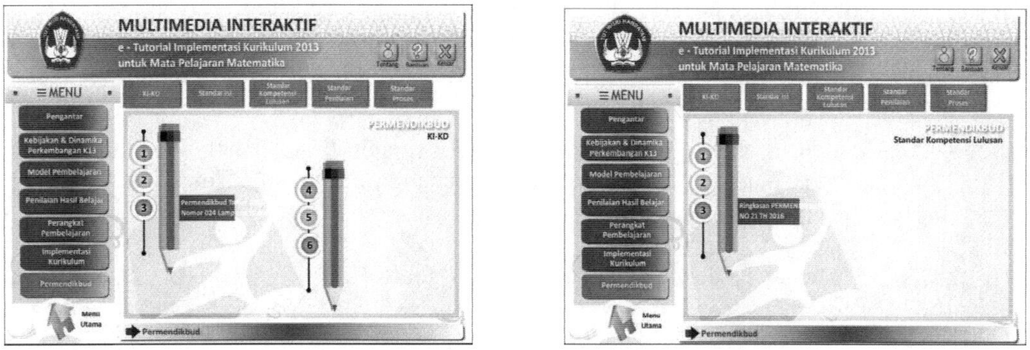


Figure 9. Menu Relevant Government Regulation

On this menu presented several government regulation that is relevant or use as basis on implementation of Curriculum 2013. Those regulation related to Learning Outcome, Standard of Learning Process, Standard for Learning Assessment. Each of them available for both Junior and Senior High School.

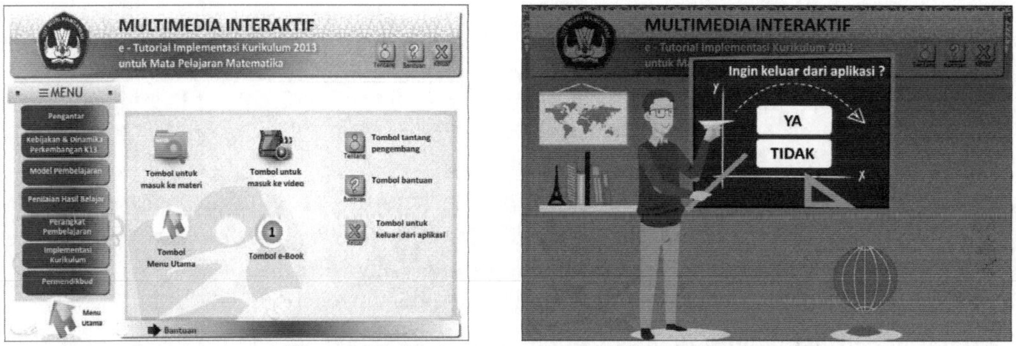


Figure 10. Menu program guideline and exit page

Figure 10 show the pages of program guideline and the exit page. The guideline page give the user the way to use the program as the whole. In this submenu been explained all the buttons being used. User get the reminder if they really want to close the program when the end of program reached.

## 4. Conclusion

From the development, the researcher tried to present the conclusions of the process and product research:

a. The product development process is determined by the availability of data and materials from the training that has been developed by the government. Therefore, change of the Curriculum of training materials at the last minute of research activity has affected the products development.

b. The e-tutorial product is an alternative and supplement material that will be used by teachers in preparing for the implementation of the Curriculum 2013, the materials used in this product at least equal to the material provided by the government training program.

c. In this product also added some material designed and developed by researchers as a support for teachers in understanding the various model of learning design.

**Acknowledgment**
This research was conducted by funding from the Higher Education Department through The Center of Research and Community Service University of Riau on competitive grants scheme. Researchers would like to acknowledge support of this financing. Researcher also addressed the appreciation to teachers, graduate students of Mathematics Education College of Education University of Riau who have shared their idea in the FGD to improve the e-tutorial contents.

## References

B.Koop, M.C. Matteucci, and C. Tomasetto. "e-Tutorial support for Collaborative online Learning: An explorative study on experienced and inexperience e-tutors". Computer. Edu, Vol 58. no.1. pp.12-20, 2012

D. B. Hay *et al.*, "Measuring the quality of e-learning," *Br. J. Educ. Technol.*, vol. 39, no. 6, pp. 1037–1056, 2008.

Depdiknas.2000.Panduan Pengembangan Multimedia Pembelajaran. Jakarta. Depdiknas

Depdiknas,2006. Kurikulum Tingkat Satuan Pendidikan (KTSP). Jakarta. Depdiknas

Epignosis LLC . 2014. *E-Learning, Concept, Trends and Application.* E-book. V1-1.2014.

Hofstetter, Fred T. 2001. Multimedia Literacy. 3rd Edition. McGraw-Hill. Int, Edition: New York

Kadir,A&Triwahyuni,T.C. 2014. Pengantar Teknologi Informasi Edisi Revisi. Andi Offset, Yogya.

L. Feng and J.-L. Ha, "Effects of Teachers' Information Literacy on Lifelong Learning and School Effectiveness," *Eurasia J. Math. Sci. Technol. Educ.*, vol. 12, no. 6, pp. 1653–1663, 2016.

Noesgaard, S,Signe. 2015. *The Effectivenes of E-Learning: An Explorative and Integrative Review of the Definitions, Methodologies and Factors that Promote E-Learning Effectiveness.* Electronic Journal of e-learning Voleume 13 Issued 4.

Paay, Jeni. *Learning by Interacting: Comparing the Effectivenes of an Interactive Tutorial with a Standard Electronic Book Interface.* School of Computing, Tasmania.

Perry,P.1994.*MultimediaDeveloper's Guide*.SamsPublishing.Indianapolis

Robin dan Linda.2001. *Perkembangan Multimedia danCDInteraktif.* [Online]. Tersedia:http://www.http://maroebeni.wordpress.com/2016/01/05/perkembangan-multimedia-dan-cd-interaktif/

Suyanto,M.2003. Multimedia Alat untuk Meningkatkan Keunggulan Bersaing. Andi Offset. Yogyakarta

Thiagarajan,S., Semmel,D.S&Semmel,M.I.1974.*Instructional Development for Teacher Training Of Expectional Children.* Minneapolis, Minnesota: Leadership Training Institute/ SpecialEducation, University of Minnesota.

Turban,dkk. 2002. Aplikasi Multimedia Interaktif, Paradigma: Yogyakarta

U.S. Department of Education. 2010. *Evalulation of Evidence-Based Practices in Online Learning: A Meta-Analysis and Review of Online Learning Studies.*

# Censored Hurdle Negative Binomial Regression (Case Study: Neonatorum Tetanus Case in Indonesia)

**[1]Riza Yuli Rusdiana, [2]Ismaini Zain, [3]Santi Wulan Purnami**

[1]Master's Program Student of Statistics Departement, Institut Teknologi Sepuluh Nopember, Surabaya, Indonesia. [2,3]Lecturer of Statistics Departement, Institut Teknologi Sepuluh Nopember, Surabaya, Indonesia.

[1]rizayuli.rusdiana @yahoo.co.id, [2] ismaini_z@statistika.its.ac.id,
[3]santi_wp@statistika.its.ac.id

**Abstract**. Hurdle negative binomial model regression is a method that can be used for discreate dependent variable, excess zero and under- and overdispersion. It uses two parts approach. The first part estimates zero elements from dependent variable is zero hurdle model and the second part estimates not zero elements (non-negative integer) from dependent variable is called truncated negative binomial models. The discrete dependent variable in such cases is censored for some values. The type of censor that will be studied in this research is right censored. This study aims to obtain the parameter estimator hurdle negative binomial regression for right censored dependent variable. In the assessment of parameter estimation methods used Maximum Likelihood Estimator (MLE). Hurdle negative binomial model regression for right censored dependent variable is applied on the number of neonatorum tetanus cases in Indonesia. The type data is count data which contains zero values in some observations and other variety value. This study also aims to obtain the parameter estimator and test statistic censored hurdle negative binomial model. Based on the regression results, the factors that influence neonatorum tetanus case in Indonesia is the percentage of baby health care coverage and neonatal visits.
**Keywords:** Hurdle negative binomial, neonatorum tetanus, right censored

## 1. Introduction

In the neonatal period, neonatorum tetanus is a major cause in the case of infant mortality. It is caused by Clostridium tetani which can lead to death. Stiff muscle pain is tetanus characteristic which is caused by neurotoxin and produced by Clostridium tetani in wound anaerobic. World Health Organization (WHO) in 1988 and UNICEF through the World Summit for Children in 1990 invited the world to eleminate neonatorum tetanus in 2000. However, this target was not achieved because it has not found an effective operational strategy. In 1999, developing countries were invited UNICEF, WHO and UNFPA to achieve the target of maternal and neonatal tetanus elemination in 2005, which later shifted to 2015. Elemination is achieved if the number of neonatorum tetanus case less than 1 case per 1000 live births. In 2008, WHO estimated 59000 death of newborns caused by neonatorum tetanus. There are 46 countries that have not been eleminated neonatorum tetanus, one of which is Indonesia[1].

The occurrence of neonatorum tetanus case is caused by non-medical factors and medical factors. Medical factors include standard prenatal care (lack of antenatal care to pregnant women, the lack of knowledge of pregnant women about the importance of immunization against tetanus toxoid), perinatal care (lack of availability of maternity facilities and medical personnel so many births do at home and use tools that are not sterile , including the handling of the umbilical cord) and neonatal care (neonatal born in a state of non-sterile, high prematurity, etc.), while for non-medical factors related to local customs [2].

Hurdle negative binomial model is two-component models with a truncated negative binomial component for positive counts and a hurdle component that models the zero counts. Among generalized linier models for handling underdispersion or overdispersion, the hurdle negative binomial regression model is one of the few that can accommodate both underdispersion and overdispersion [3]. In many applications, count data are often censored a specific point from above (right) or below (left) or a combination of them (interval). Right censored is those where observations are clustered at an upper threshold. The previous research [4] suggested the use of censored hurdle negative binomial model. They applied the hurdle negative binomial model to a data set on fish from UCLA Academic Technology Services website and compared it to the negative binomial regression models.

The number of neonatorum tetanus case, measured by the number of neonatorum tetanus case per province is an example of an event count. The basic model to establish is the Poisson regression model. However, this model rarely fits data because of overdispersion. Count data such as neonatorum tetanus case often show high incidence of zero counts. In such cases, hurdle negative binomial model may be more appropriate. Neonatorum tetanus case data contains zero value on most observations and some other value has non-negative integer value. This kind of data is called censored data. Type of censor that will be examined in this study is right censored. We argue that a two-part model is suited to describe the processes of (i) whether neonatorum tetanus case occurred; and (ii) the number of cases if neonatorum tetanus case is occured. This paper, therefore used hurdle negative binomial model with right censored to establish the determinants of neonatorum tetanus case among province in Indonesia. Our analysis also explored parameter estimation on hurdle negative binomial with dependent variable right censored using maximum likelihood estimation.

In this article, the main objective is to explain how we can use hurdle negative binomial model in right censored data. In section 2, the hurdle negative binomial regression model is defined and the likelihood function of hurdle negative binomial model in right censored data is formulated. In section 3, the parameter estimation is discussed using maximum likelihood method. In section 4, we applied hurdle negative binomial model in right censored data to an neonatorum tetanus data set.

## 2. Censored Hurdle Negative Binomial Model

In particular, a hurdle negative binomial model is mixed by a binary outcome of the count being below or above the hurdle, with a truncated negative binomial model for outcomes above the hurdle. That is why hurdle models sometimes are also called two-part models. The most important usage of a hurdle count data model is the hurdle at zero. The hurdle at zero formulation can account for excess zeros. It means that this model can be used in situations where there are many zeros at the dependent variable. In this case, the hurdle at zero defines a probability ($P(Y = 0)$) that is the first part of the two part models [4]. The truncated negative binomial model account for positive integer at the dependent variable. It is of the two part models defines a probability ($P(Y = y_i)$), $y_i = 1, 2, \ldots$

Suppose a count dependent variable is a random variable and affected by $p$ independent variables ($x_1, x_2, \ldots, x_p$). The hurdle negative binomial regression model derived by [3] is that the distribution of $Y_i$ , conditional on independent variables ($x_1, x_2, \ldots, x_p$), and it is defined by

$$P(Y_i = y_i) = \begin{cases} \dfrac{e^{x_i^T \boldsymbol{\delta}}}{1+e^{x_i^T \boldsymbol{\delta}}} & y_i = 0 \\ \dfrac{1}{1+e^{x_i^T \boldsymbol{\delta}}} \dfrac{g}{1-\left(1+e^{x_i^T \boldsymbol{\beta}}\kappa\right)^{-\kappa^{-1}}} & y_i > 0 \end{cases} \quad (1)$$

and

$$g = g(y_i;\kappa,\boldsymbol{\beta}) = \frac{\Gamma\left(y_i + \frac{1}{\kappa}\right)}{\Gamma(y_i+1)\Gamma\left(\frac{1}{\kappa}\right)} \left(1+e^{x_i^T\boldsymbol{\beta}}\kappa\right)^{-\kappa^{-1}-y_i} \left(e^{x_i^T\boldsymbol{\beta}}\kappa\right)^{y_i} \quad (2)$$

where $x_i = [1 \;\; x_{1i} \;\; x_{2i} \; \ldots \; x_{pi}]^T$ is $(p+1)\times 1$ dimensional vector, $\boldsymbol{\beta} = [\beta_0 \;\; \beta_1 \;\; \beta_2 \; \ldots \; \beta_p]^T$ and $\boldsymbol{\delta} = [\delta_0 \;\; \delta_1 \;\; \delta_2 \; \ldots \; \delta_p]^T$ is a $(p+1)\times 1$ dimensional vector of regression parameters. The parameter $\kappa$ is a measure of dispersion. When $\kappa = 0$, the HNB model reduces to the Poisson regression model. For $\kappa > 0$, the HNB model can used to fit overdispersed count data. When $\kappa < 0$, the HNB model can used to fit underdispersed count data.

Zero hurdle model with logit link function:

$$\text{logit}\,(\pi_i) = \hat{\delta}_0 + \sum_{j=1}^{p} x_{ij}\hat{\delta}_j \qquad i = 1,2,\ldots n \text{ and } j = 1,2,\ldots,p \quad (3)$$

Truncated negative binomial model with log link function:

$$\log(\mu_i) = \hat{\beta}_0 + \sum_{j=1}^{p} x_{ij}\hat{\beta}_j \qquad i = 1,2,\ldots n \text{ and } j = 1,2,\ldots,p \quad (4)$$

For some observations in data set, the value of $Y_i$ may be censored [4].Censored data is data that contains zero values in some observations whereas for most other observations has particular value that varies. No censoring occurs for the $i^{th}$ observation, $Y_i = y_i$ and censoring occurs for the $i^{th}$ observation, $Y_i$ is at least equal to $y_i$ ($Y_i \geq y_i$). When the data are censored, the distribution that applies to the sample data is derived by using binary variable $d_i$, this variable is defined as:

$$d_i = \begin{cases} 1 & \text{if } \; Y_i \geq y_i \; \text{(censored)} \\ 0 & \text{otherwise (uncensored)} \end{cases} \quad (5)$$

The probability function of observed counts $y$ for individual $i$ is equal to

$$\text{P}(Y = y_i, d_i, x_i) = P(Y_i = y_i)^{1-d_i} \, P(Y_i \geq y_i)^{d_i} \quad (6)$$

where

$$P(Y_i \geq y_i) = \sum_{\ell=y_i}^{\infty} \text{P}(Y_i = \ell) = 1 - \sum_{\ell=0}^{y_i-1} \text{P}(Y_i = \ell) \quad (7)$$

then HNB model in right censored data introduced by [4] is given by

$$\begin{aligned} \text{P}(Y = y_i, d_i, x_i) &= P(Y_i = y_i)^{1-d_i} \, P(Y_i \geq y_i)^{d_i} \\ &= \left[I_{y_i=0} f(y_i;\kappa,\boldsymbol{\delta},\boldsymbol{\beta}) I_{y_i>0} f(y_i;\kappa,\boldsymbol{\delta},\boldsymbol{\beta})\right]^{1-d_i} \left[\sum_{\ell=y_i}^{\infty} f(\ell;\kappa,\boldsymbol{\delta},\boldsymbol{\beta})\right]^{d_i} \end{aligned} \quad (8)$$

The first term on the right gives the likelihood contribution of the noncensored observations, while the second term on the right gives the contribution of the censored observations.

## 3. Parameter Estimation

CHNB can be estimated by using maximum likelihood method. The maximum likelihood method is used to estimate the parameter κ, **δ** and **β** . The likelihood function of CHNB model is given by

$$L(\kappa,\boldsymbol{\delta},\boldsymbol{\beta}\mid y_i)=\prod_{i=1}^{n}\left[I_{y_i=0}f\left(0;\kappa,\boldsymbol{\delta},\boldsymbol{\beta}\right)I_{y_i>0}f\left(y_i;\kappa,\boldsymbol{\delta},\boldsymbol{\beta}\right)\right]^{1-d_i}\left[\sum_{\ell=y_i}^{\infty}f(\ell;\kappa,\boldsymbol{\delta},\boldsymbol{\beta})\right]^{d_i} \tag{9}$$

The log-likelihood function of

$$l(\kappa,\boldsymbol{\delta},\boldsymbol{\beta}\mid y_i)=\sum_{i=1}^{n}\Big\{(1-d_i)\Big[I_{y_i=0}\ln(\pi_i)+I_{y_i>0}\Big(\ln(1-\pi_i)+\ln(g) -\ln\Big(1-\big(1+e^{x_i^T\boldsymbol{\beta}}\kappa\big)^{-\kappa^{-1}}\Big)\Big)\Big]+(d_i)\ln\Big[\sum_{\ell=y_i}^{\infty}f(\ell;\kappa,\boldsymbol{\delta},\boldsymbol{\beta})\Big]\Big\} \tag{10}$$

Estimation of κ, **δ** and **β** are obtained by taking the partial derivations of the log-likelihood function and setting them equal to zero. Thus, we obtain

$$\frac{\partial l(\kappa,\boldsymbol{\delta},\boldsymbol{\beta}\mid y_i)}{\partial\kappa}=\sum_{i=i}^{m}\left\{(1-d_i)I_{y_i>0}\left[\frac{g'_\kappa}{g}-\frac{\kappa^{-1}e^{x_i^T\boldsymbol{\beta}}\left(1+e^{x_i^T\boldsymbol{\beta}}\kappa\right)^{-1}-\kappa^{-2}\ln\left(1+e^{x_i^T\boldsymbol{\beta}}\kappa\right)}{1-\left(1+e^{x_i^T\boldsymbol{\beta}}\kappa\right)^{-\kappa^{-1}}}\left(1+e^{x_i^T\boldsymbol{\beta}}\kappa\right)^{-\kappa^{-1}}\right]+\frac{d_i}{F}F'_k\right\}=0 \tag{11}$$

$$\frac{\partial l(\kappa,\boldsymbol{\delta},\boldsymbol{\beta}\mid y_i)}{\partial\boldsymbol{\delta}^{\mathrm{T}}}=\sum_{i=i}^{n}\left\{(1-d_i)\left[I_{y_i=0}\frac{\pi_i'}{\pi_i}-I_{y_i>0}\frac{\pi_i'}{1-\pi_i}\right]+\frac{d_i}{F}F'_{\boldsymbol{\delta}^T}\right\}=0 \tag{12}$$

$$\frac{\partial l(\kappa,\boldsymbol{\delta},\boldsymbol{\beta}\mid y_i)}{\partial\boldsymbol{\beta}^{\mathrm{T}}}=\sum_{i=i}^{m}\left\{(1-d_i)\,I_{y_i>0}\left[\frac{g'_{\boldsymbol{\beta}^T}}{g}-\frac{x_i e^{x_i^T\boldsymbol{\beta}}\left(1+e^{x_i^T\boldsymbol{\beta}}\kappa\right)^{-\kappa^{-1}-1}}{1-\left(1+e^{x_i^T\boldsymbol{\beta}}\kappa\right)^{-\kappa^{-1}}}\right]+\frac{d_i}{F}F'_{\boldsymbol{\beta}^T}\right\}=0 \tag{13}$$

where

$$F=\sum_{\ell=y_i}^{\infty}f(\ell;\kappa,\boldsymbol{\delta},\boldsymbol{\beta})=\sum_{\ell=y_i}^{\infty}\frac{(1-\pi_i)g}{1-\left(1+e^{x_i^T\boldsymbol{\beta}}\kappa\right)^{-\kappa^{-1}}} \tag{14}$$

$g'_\kappa$ is first derivative $g$ against κ and $g'_{\boldsymbol{\beta}^T}$ is first derivative $g$ against $\boldsymbol{\beta}^{\mathrm{T}}$. $F'_k$, $F'_{\boldsymbol{\delta}^T}$ and $F'_{\boldsymbol{\beta}^T}$ are first derivative $F$ against parameters κ, $\boldsymbol{\delta}^{\mathrm{T}}$ and $\boldsymbol{\beta}^{\mathrm{T}}$. The likelihood equations from (12) until (14) are non-linear in parameters κ, **δ** and **β**. These equations are solved simultaneously by using an iterative algorithm. It can be used to carry out Newton-Raphson method for solving these equations. On taking the second partial derivatives, we obtained Hessian matrix which is a square matrix and given by

$$\boldsymbol{H}(\boldsymbol{\theta}^{(m)}) = \begin{bmatrix} \frac{\partial^2 l(\kappa,\boldsymbol{\delta},\boldsymbol{\beta})}{\partial \kappa^2} & \frac{\partial^2 l(\kappa,\boldsymbol{\delta},\boldsymbol{\beta})}{\partial \boldsymbol{\delta} \partial \kappa} & \frac{\partial^2 l(\kappa,\boldsymbol{\delta},\boldsymbol{\beta})}{\partial \boldsymbol{\beta} \partial \kappa} \\ & \frac{\partial^2 l(\kappa,\boldsymbol{\delta},\boldsymbol{\beta})}{\partial \boldsymbol{\delta} \partial \boldsymbol{\delta}^{\mathrm{T}}} & \frac{\partial^2 l(\kappa,\boldsymbol{\delta},\boldsymbol{\beta})}{\partial \boldsymbol{\beta} \partial \boldsymbol{\delta}^{\mathrm{T}}} \\ \textit{simetris} & & \frac{\partial^2 l(\kappa,\boldsymbol{\delta},\boldsymbol{\beta})}{\partial \boldsymbol{\beta} \partial \boldsymbol{\beta}^{\mathrm{T}}} \end{bmatrix}$$

second partial derivative against parameter κ, **δ** and **β** as follows:

$$\frac{\partial^2 l(\kappa,\boldsymbol{\delta},\boldsymbol{\beta} \mid y_i)}{\partial \kappa^2} = \sum_{i=i}^{m} \left\{ (1-d_i) I_{y_i>0} \left( \frac{g''_\kappa g - (g'_\kappa)^2}{g^2} - \left( \frac{(A'_\kappa - B'_\kappa)C - C'_\kappa (A-B)}{C^2} D \right) + D'_\kappa \left( \frac{A-B}{C} \right) \right) + d_i \left( -\frac{1}{F^2} (F'_\kappa)^2 + \frac{1}{F} F''_\kappa \right) \right\} \quad (15)$$

$$\frac{\partial^2 l(\kappa,\boldsymbol{\delta},\boldsymbol{\beta} \mid y_i)}{\partial \boldsymbol{\delta} \partial \boldsymbol{\delta}^{\mathrm{T}}} = \sum_{i=i}^{n} \left\{ (1-d_i) \left[ I_{y_i=0} \frac{\pi_i'' \pi_i - \pi_i' \pi_i'^{*}}{\pi_i^2} - I_{y_i>0} \frac{\pi_i'' (1-\pi_i) - \pi_i' \pi_i'^{*}}{(1-\pi_i)^2} \right] + d_i \left[ \frac{F''_{\boldsymbol{\delta}} F - F'_{\boldsymbol{\delta}^{\mathrm{T}}} F'_{\boldsymbol{\delta}}}{F^2} \right] \right\} \quad (16)$$

$$\frac{\partial^2 l(\kappa,\boldsymbol{\delta},\boldsymbol{\beta} \mid y_i)}{\partial \boldsymbol{\beta} \partial \boldsymbol{\beta}^{\mathrm{T}}} = \sum_{i=i}^{m} \left\{ (1-d_i) \left[ I_{y_i>0} \frac{g''_{\boldsymbol{\beta}} g - g'_{\boldsymbol{\beta}^T} g'_{\boldsymbol{\beta}}}{g^2} - \frac{E'_{\boldsymbol{\beta}} H + H'_{\boldsymbol{\beta}} E}{C} + \frac{C'_{\boldsymbol{\beta}} E H}{C^2} \right] + d_i \left[ \frac{F''_{\boldsymbol{\beta}} F - F'_{\boldsymbol{\beta}^T} F'_{\boldsymbol{\beta}}}{F^2} \right] \right\} \quad (17)$$

$$\frac{\partial^2 l(\kappa,\boldsymbol{\delta},\boldsymbol{\beta} \mid y_i)}{\partial \boldsymbol{\delta} \partial \kappa} = \sum_{i=1}^{m} d_i \left\{ \frac{F''_{\kappa\boldsymbol{\delta}} F - F'_{\boldsymbol{\delta}} F'_k}{F^2} \right\} \quad (18)$$

$$\frac{\partial^2 l(\kappa,\boldsymbol{\delta},\boldsymbol{\beta} \mid y_i)}{\partial \boldsymbol{\beta} \partial \kappa} = \sum_{i=i}^{m} \left\{ (1-d_i) I_{y_i>0} \left[ \frac{g''_{\kappa\boldsymbol{\beta}} g - g'_{\boldsymbol{\beta}} g'_\kappa}{g^2} - \left( \frac{(A'_{\boldsymbol{\beta}} - B'_{\boldsymbol{\beta}})C - C'_{\boldsymbol{\beta}} (A-B)}{C^2} \right) D + D'_{\boldsymbol{\beta}} \left( \frac{A-B}{C} \right) \right] + d_i \left[ \frac{F''_{\kappa\boldsymbol{\beta}} F - F'_{\boldsymbol{\beta}} F'_k}{F^?} \right] \right\} \quad (19)$$

$$\frac{\partial^2 l(\kappa,\boldsymbol{\delta},\boldsymbol{\beta} \mid y_i)}{\partial \boldsymbol{\beta} \partial \boldsymbol{\delta}^{\mathrm{T}}} = \sum_{i=1}^{n} d_i \left\{ \frac{F''_{\boldsymbol{\delta}^T \boldsymbol{\beta}} F - F'_{\boldsymbol{\delta}^T} F'_{\boldsymbol{\beta}}}{F^2} \right\} \quad (20)$$

The expressions for other function provided in the appendix.

## 4. Application

This study was based on Indonesian health profile 2015 [5]. Data and information on Indonesian health profile comes from technical units in the Ministry of Health and other institutions which related with health data such as BPS and BKKBN. Data included in this analysis comprised of 33 province in

Indonesia. The independent variables are baby health care coverage, TT immunization of pregnant women, neonatal visits and TT immunization of women of fertile age. Baby health care coverage are handled by trained health care personnel throughout the health care facility. Newborn services are care counseling include ASI and umbilical cord care. TT immunization of pregnant women is given to pregnant women with at least two doses of tetanus toxoid called TT2+. In addition to the baby health care coverage, indicators that describe the health service for neonatal period is neonatal visit which requires that every newborn visit at least three times as standard in region. TT immunization of women of fertile age is given to women of fertile age with at least one doses of tetanus toxoid.

The dependent variable, the number of neonatorum tetanus case, is non-negative integer ranging from zero to twenty one in the sample. The number of neonatorum tetanus case in Indonésia is 53 cases with case fatality as many as 27 people. The distribution of the number of neonatorum tetanus case is 0.4 percent or 13 province from 33 province in Indonesia which exposed neonatorum tetanus case.

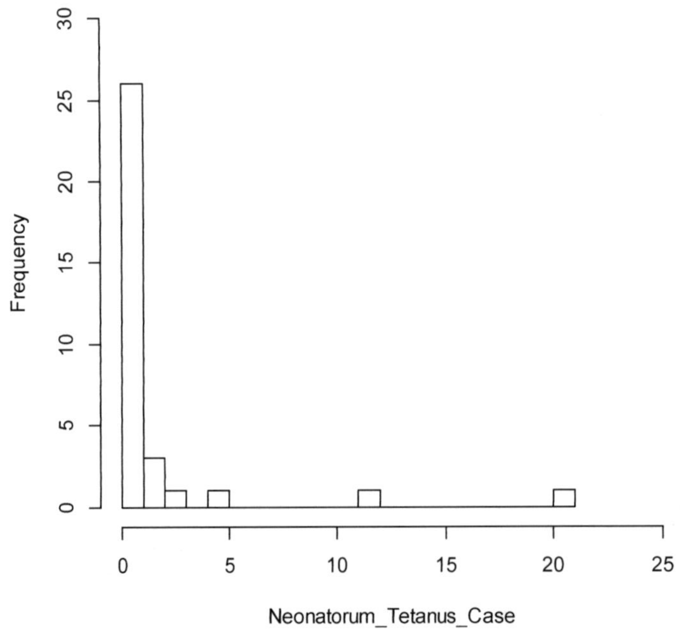


**Figure 1.** Histogram of the number of neonatorum tetanus case

Figure 1 shows frequency distribution of the number of neonatorum tetanus case. The graph indicates a high frequency of zeros and long right tail giving evidence of overdispersion in the data.

**Table 1.** Descriptives statistics of variables

| Variable | Mean | Variance |
|---|---|---|
| Number of neonatorum tetanus case | 1.61 | 17.25 |
| Baby health care coverage | 74.96 | 614.32 |
| TT immunization of pregnant woman | 57.57 | 481.04 |
| Neonatal visits | 68.95 | 612.69 |
| TT immunization of women of fertile age | 2.06 | 13.51 |

The sample variance of dependent variable is greater than the sample mean, therefore we expected that overdispersion occurs and dispersion parameter κ may be positive. In this study used right censored with censored point of 1, because we expected less than one case in each province or neonatorum tetanus case was eleminated in Indonesia. About 39.4% of the sample has dependent variable $y_i \geq 1$ , and then we consider as censored. The dispersion parameter κ is positive 3.145 and 0.188 which also indicate that modeling overdispersed data. The AIC values for hurdle negative binomial and censored hurdle negative binomial models are 109.5 and 50.7, respectively, which also indicate that using censored hurdle negative binomial model is more appropriate than hurdle negative binomial model.

**Table 2.** Parameter Estimation of Regression

| Independent Variables | Hurdle Negative Binomial | | | | | |
|---|---|---|---|---|---|---|
| | Censored | | | Not Censored | | |
| | Estimate | OR | P-value | Estimate | OR | P-value |
| $\delta_0$ | 0.108 | 1.114 | 0.941 | 0.009 | 1.009 | 0.994 |
| $\delta_1$ | 0.021 | 1.021 | 0.793 | 0.033 | 1.034 | 0.588 |
| $\delta_2$ | -0.015 | 0.985 | 0.564 | -0.013 | 0.987 | 0.517 |
| $\delta_3$ | 0.006 | 1.006 | 0.939 | -0.013 | 0.987 | 0.827 |
| $\delta_4$ | 0.266 | 1.305 | 0.624 | -0.176 | 0.839 | 0.374 |
| $\beta_0$ | -0.278 | 0.757 | 0.953 | -3.647 | 0.026 | 0.396 |
| $\beta_1$ | -0.569 | 0.566* | 0.044 | -0.033 | 0.968 | 0.844 |
| $\beta_2$ | -0.079 | 0.924 | 0.549 | 0.009 | 1.009 | 0.829 |
| $\beta_3$ | 0.432 | 1.540** | <0.0001 | 0.076 | 1.079 | 0.677 |
| $\beta_4$ | -0.159 | 0.853 | 0.834 | 0.074 | 1.077 | 0.514 |
| κ | 0.188 | 1.207 | 0.374 | 3.145 | 23.219 | 0.699 |
| -2LL | | 28.7 | | | 87.5 | |
| AIC | | 50.7 | | | 109.5 | |

*significant at 5% level, **significant at 1% level

Table 2 summarizes the results censored hurdle negative binomial regression. The number of neonatorum tetanus case was found to be associated with baby health care coverage ($X_1$) and neonatal visits ($X_3$). The number of neonatorum tetanus case decreased with increasing baby health care coverage ($X_1$) and neonatal visits ($X_3$). There was no significant difference in neonatorum tetanus case with TT immunization of pregnant woman ($X_2$) and TT immunization of women of fertile age ($X_4$).
Zero hurdle model:

$$\pi_i = \frac{\exp(0.108 + 0.021X_1 - 0.015X_2 + 0.006X_3 + 0.266X_4)}{1 + \exp(0.108 + 0.021X_1 - 0.015X_2 + 0.006X_3 + 0.266X_4)}$$

Truncated negative binomial model:

$$\mu_i = \exp(-0.278 - 0.5691X_1 - 0.079X_2 + 0.432X_3 + 0.159X_4)$$

The average of number of neonatorum tetanus cases expected will be decreased by 0.566 for a percent increase baby health care coverage and will be increased by 1.54 for a percent increase neonatal visit when other variables constant.

## 5. Conclusion

The result of parameter estimation by the Maximum Likelihood Estimation method is shaped so that the implicit Newton Raphson iteration process is required. Based on applied study CHNB regression is better than HNB regression. Results indicate that neonatorum tetanus cases in Indonesia is associated with baby health care coverage and neonatal visit.

## Appendix

From (11) until (20), the expression for other function given by

$$g'_{\kappa} = g\left[\frac{\Gamma'\left(y_i+\kappa^{-1}\right)}{\Gamma\left(y_i+\kappa^{-1}\right)} - \frac{\Gamma'\left(\kappa^{-1}\right)}{\Gamma\left(\kappa^{-1}\right)} + \kappa^{-2}\ln\left(1+e^{x_i^T\beta}\kappa\right) - \frac{\left(y_i+\kappa^{-1}\right)e^{x_i^T\beta}}{\left(1+e^{x_i^T\beta}\kappa\right)} + y_i\kappa^{-1}\right]$$

$$g'_{\beta^T} = g\left[\frac{\boldsymbol{x}_i\left(y_i - e^{x_i^T\beta}\right)}{\left(1+e^{x_i^T\beta}\kappa\right)}\right]$$

$$\pi_i' = \frac{x_i e^{x_i^T\delta}}{\left(1+e^{x_i^T\delta}\right)^2}$$

$$F'_{\kappa} = \sum_{\ell=y_i}^{\infty}(1-\pi_i)\left[\frac{g'_{\kappa}}{\left(1-\left(1+e^{x_i^T\beta}\kappa\right)^{-\kappa^{-1}}\right)} + g\left(1+e^{x_i^T\beta}\kappa\right)^{-\kappa^{-1}}\left(\frac{\ln\left(1+e^{x_i^T\beta}\kappa\right)^{\kappa^{-2}} - \kappa^{-1}e^{x_i^T\beta}\left(1+e^{x_i^T\beta}\kappa\right)^{-1}}{\left(1-\left(1+e^{x_i^T\beta}\kappa\right)^{-\kappa^{-1}}\right)^2}\right)\right]$$

$$F'_{\delta^T} = \frac{\partial F}{\partial \boldsymbol{\delta}^{\mathrm{T}}} = \sum_{\ell=y_i}^{\infty} g\frac{-\pi_i'}{\left(1-\left(1+e^{x_i^T\beta}\kappa\right)^{-\kappa^{-1}}\right)}$$

$$F'_{\beta^T} == \sum_{\ell=y_i}^{\infty}(1-\pi_i)\left[\frac{g'_{\beta^T}\left(1-\left(1+e^{x_i^T\beta}\kappa\right)^{-\kappa^{-1}}\right) - \left(x_i e^{x_i^T\beta}\left(1+e^{x_i^T\beta}\kappa\right)^{-\kappa^{-1}-1}\right)g}{\left(1-\left(1+e^{x_i^T\beta}\kappa\right)^{-\kappa^{-1}}\right)^2}\right]$$

$$A = \kappa^{-1}e^{x_i^T\beta}\left(1+e^{x_i^T\beta}\kappa\right)^{-1} \qquad B = \kappa^{-2}\ln\left(1+e^{x_i^T\beta}\kappa\right) \qquad C = 1-\left(1+e^{x_i^T\beta}\kappa\right)^{-\kappa^{-1}}$$

$$D = \left(1+e^{x_i^T\beta}\kappa\right)^{-\kappa^{-1}} \qquad E = \left(1+e^{x_i^T\beta}\kappa\right)^{-\kappa^{-1}-1} \qquad H = \boldsymbol{x}_i e^{x_i^T\beta}$$

$$P = \left(y_i+\kappa^{-1}\right)e^{x_i^T\beta} \qquad Q = 1+e^{x_i^T\beta}\kappa$$

## References

[1]. Kemenkes. (2012). *Buletin: Eleminasi Tetanus Maternal dan Neonatal.* [http://www.depkes.go.id/resources/download/pusdatin/buletin/buletin-mnte.pdf] diakses tanggal 11 Agustus 2016.
[2]. Handoko, M. (2011). Tetanus Neonatorum. Kepaniteraan Klinik Ilmu Penyakit Anak.[https://www.scribd.com/doc/51651896/Tetanus-neonatorum] diakses tanggal 10 Oktober 2016.
[3]. Pohlmeier, W. and Ulrich, V. (1995). *An Econometric Model of The Two-Part Decision-Making Process In The Demand For Health Care*. The Journal of Human Resources, 30, 339–361.

[4]. Saffari, S. E., Robiah, A. and Greene, W. (2012). *Hurdle Negative Binomial Regression Model with Right Cencored Count Data*. Malaysia: Journal of Statistics and Operations Research Transactions, Vol.36(2):181-194.

[5]. Kemenkes. (2016). *Profil Kesehatan Indonesia 2015*. Jakarta: Kemenkes RI.

IOP Conf. Series: Journal of Physics: Conf. Series **855** (2017) 012040 doi:10.1088/1742-6596/855/1/012040

# Predicting Flood in Perlis Using Ant Colony Optimization

**Syaidatul Nadia Sabri and Rizauddin Saian**

Faculty of Computer and Mathematical Sciences, Universiti Teknologi MARA, Malaysia

E-mail: nadyasabry@gmail.com and rizauddin@perlis.uitm.edu.my

**Abstract.** Flood forecasting is widely being studied in order to reduce the effect of flood such as loss of property, loss of life and contamination of water supply. Usually flood occurs due to continuous heavy rainfall. This study used a variant of Ant Colony Optimization (ACO) algorithm named the Ant-Miner to develop the classification prediction model to predict flood. However, since Ant-Miner only accept discrete data, while rainfall data is a time series data, a pre-processing steps is needed to discretize the rainfall data initially. This study used a technique called the Symbolic Aggregate Approximation (SAX) to convert the rainfall time series data into discrete data. As an addition, Simple K-Means algorithm was used to cluster the data produced by SAX. The findings show that the predictive accuracy of the classification prediction model is more than 80%.

## 1. Introduction

Flood has terrible impacts on people as it disrupts their daily activities [1]. Flood also gives impacts to the country. In 2010, Perlis had experienced the worst flood whereby the flood had effected 47,000 hectares of land in Perlis. It damaged property belonging 63,000 people and led to monetary loss of nearly RM200 million. The cause of the flood, among others,was due to the existing river was not able to cope with the overflow of water from Timah Tasoh Dam in Perlis.

Analysis of rainfall is an important task nowadays. Many researches [2, 3, 4, 5, 6] had been carried out to predict the occurrence of flood due to heavy rainfall. A good classification prediction model is needed to develop a good flood early warning system. This is to ensure that the government is ready to handle the situation and as a result be able to reduce the effect of flood disaster.

This study developed a classification prediction model using rainfall data from three rainfall gauging stations in Perlis from year 1993 to 2014. The gauging stations are Kaki Bukit, Tasoh dan Padang Besar. These three stations are located nearby the Timah Tasoh reservoir. Timah Tasoh reservoir is the largest multipurpose reservoir in Perlis which serves as a flood mitigation and water supply.

The research phases is divided into two main phases, data pre-processing and model development as depicted by figure 1.

## 2. Data pre-processing

Rainfall data is a time series data. The data pre-processing phase is a phase to prepare the data so that it can be used in the classification algorithm chosen for this study. The main output of

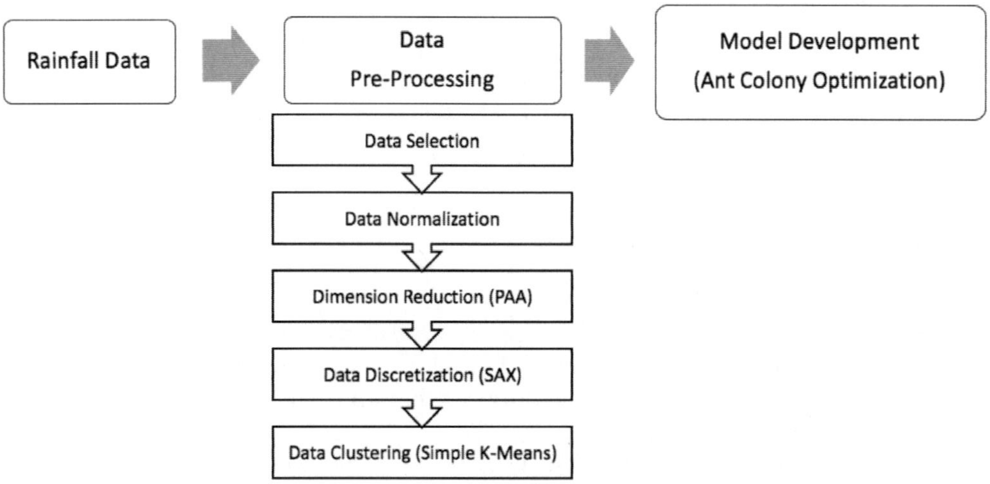


**Figure 1.** Research phases.

the pre-processing phase is a discrete rainfall data.

The data pre-processing phase consists of five steps: data selection, data normalization, dimension reduction, data discretization and data clustering.

The first step in the data pre-processing step is data selection. This step will select the rainfall data from three target months with the highest average amount of rainfall. Table 1 depicted the average rainfall distribution for each month from year 1993 until 2014. Three months that received high average amount of rainfall are September, October and November. Hence, this study selected rainfall data on September, October and November.

**Table 1.** Average rainfall distribution (Kaki Bukit, Tasoh and Padang Besar).

| Month | 01 | 02 | 03 | 04 | 05 | 06 | 07 | 08 | 09 | 10 | 11 | 12 |
|---|---|---|---|---|---|---|---|---|---|---|---|---|
| Average | 37.1 | 56.8 | 138.3 | 161.8 | 140.5 | 148.6 | 160.7 | 188.6 | 215.4 | 265.4 | 212.3 | 140.1 |

The second step is data normalization. This step prepares data input on a suitable format before the implementation of data representation. The normalization reduced the gap between the data. The rainfall data is normalized at range of 0 and 1.

The third step is dimension reduction. This step reduced the data dimensionality. This study used Piecewise Aggregate Approximation (PAA) [7] to convert the original rainfall data (day, rainfall value) to new segments (week, average rainfall). The data size of each of the segments is seven days. The average rainfall for seven days is the value for each of the segments. The rainfall data dimension is 91 days (September to November). Each segment size is equal to 7 days. Besides, each of the segments will represent weekly average rainfall. PAA approximates the rainfall data of length n into vector $\bar{C} = (\bar{c}_1, \ldots, \bar{c}_w)$ of any arbitrary length $w \leq n$ where each of $\bar{c}_i$ is calculated as in (1).

IOP Conf. Series: Journal of Physics: Conf. Series **855** (2017) 012040 doi:10.1088/1742-6596/855/1/012040

$$\overline{c}_i = \frac{w}{n} \sum_{j=\frac{n}{w}(i-1)+1}^{\frac{n}{w}i} c_j \tag{1}$$

where

- $\overline{c}_i$ is PAA data values
- $n$ is original data length (days)
- $w$ is the number of segments (weeks).

The forth step is data discretization. The discretization step converts PAA values into Symbolic Aggregate Approximation (SAX) symbol using Gaussian distribution table [8] depicted by table 2. This study divided the rainfall distribution into three classes ($\alpha = 3$): light, moderate and heavy) as shown in table 3. According to table 2, when $\alpha = 3$, $\beta_1 = -0.43$ and $\beta_2 = 0.43$. Table 4 displays the rainfall data in SAX alphabet symbol format. Finally, from the SAX alphabet symbol, the data is transformed into SAX integer symbol format as for 'a' to 1, 'b' to 2 and 'c' to 3 as shown in table 5.

**Table 2.** Gaussian distribution table [8].

| $\alpha$ | 3 | 4 | 5 | 6 | 7 | 8 | 9 | 10 |
|---|---|---|---|---|---|---|---|---|
| $\beta_1$ | −0.43 | −0.67 | −0.84 | −0.97 | −1.07 | −1.15 | −1.22 | 1.28 |
| $\beta_2$ | 0.43 | 0 | −0.25 | −0.43 | −0.57 | −0.67 | −0.76 | −0.84 |
| $\beta_3$ | | 0.67 | 0.25 | 0 | −0.18 | −0.32 | −0.43 | −0.52 |
| $\beta_4$ | | | 0.84 | 0.43 | 0.18 | 0 | −0.14 | −0.25 |
| $\beta_5$ | | | | 0.97 | 0.57 | 0.32 | 0.14 | 0 |
| $\beta_6$ | | | | | 1.07 | 0.67 | 0.43 | 0.25 |
| $\beta_7$ | | | | | | 1.15 | 0.76 | 0.52 |
| $\beta_8$ | | | | | | | 1.22 | 0.84 |
| $\beta_9$ | | | | | | | | 1.28 |

**Table 3.** Symbolic representation of rainfall rata for alphabet of size 3.

| Rainfall Distribution Class | Alphabet Symbol | Value |
|---|---|---|
| light | a | $\beta \leq -0.43$ |
| moderate | b | $-0.43 \leq \beta \leq 0.43$ |
| heavy | c | $\beta \geq 0.43$ |

The final step in the data pre-processing phase is data clustering. The discrete data is clustered into three classes: normal, alert and danger. This study used Simple K-Means with Euclidian distance [9] to perform the task using Weka software [10]. Table 6 depicted the rainfall data after the data clustering step.

**Table 4.** Rainfall distribution in SAX alphabet symbol.

| week1 | week2 | week3 | ... | week10 | week11 | week12 | week13 |
|---|---|---|---|---|---|---|---|
| b | b | b | ... | b | c | b | b |
| c | b | b | ... | a | c | a | b |
| c | b | c | ... | a | b | a | a |
| b | b | b | ... | b | a | b | b |
| b | b | a | ... | a | a | a | b |
| ... | ... | ... | ... | ... | ... | ... | ... |
| b | b | a | ... | b | b | c | a |
| b | b | b | ... | b | b | b | b |
| c | b | b | ... | b | b | b | b |
| c | b | b | ... | b | b | b | b |

**Table 5.** Rainfall distribution in SAX integer symbol.

| week1 | week2 | week3 | ... | week10 | week11 | week12 | week13 |
|---|---|---|---|---|---|---|---|
| 2 | 2 | 2 | ... | 2 | 3 | 2 | 2 |
| 3 | 2 | 2 | ... | 1 | 3 | 1 | 2 |
| 3 | 2 | 3 | ... | 1 | 2 | 1 | 1 |
| 2 | 2 | 2 | ... | 2 | 1 | 2 | 2 |
| 2 | 2 | 1 | ... | 1 | 1 | 1 | 2 |
| ... | ... | ... | ... | ... | ... | ... | ... |
| 2 | 2 | 1 | ... | 2 | 2 | 3 | 1 |
| 2 | 2 | 2 | ... | 2 | 2 | 2 | 2 |
| 3 | 2 | 2 | ... | 2 | 2 | 2 | 2 |
| 3 | 2 | 2 | ... | 2 | 2 | 2 | 2 |

**Table 6.** Rainfall data after data clustering step.

| week1 | week2 | week3 | ... | week10 | week11 | week12 | week13 | Class |
|---|---|---|---|---|---|---|---|---|
| 2 | 2 | 2 | ... | 2 | 3 | 2 | 2 | danger |
| 3 | 2 | 2 | ... | 1 | 3 | 1 | 2 | normal |
| 3 | 2 | 3 | ... | 1 | 2 | 1 | 1 | danger |
| 2 | 2 | 2 | ... | 2 | 1 | 2 | 2 | normal |
| 2 | 2 | 1 | ... | 1 | 1 | 1 | 2 | alert |
| ... | ... | ... | ... | ... | ... | ... | ... | ... |
| 2 | 2 | 1 | ... | 2 | 2 | 3 | 1 | danger |
| 2 | 2 | 2 | ... | 2 | 2 | 2 | 2 | danger |
| 3 | 2 | 2 | ... | 2 | 2 | 2 | 2 | danger |
| 3 | 2 | 2 | ... | 2 | 2 | 2 | 2 | danger |

## 3. Model development

The classification prediction model was developed by training the discrete rainfall data to construct classification rules using rule induction technique. Rule induction is a technique to extract set of rules from a data set [11, 12]. This study used a variant of Ant Colony Optimization (ACO) algorithm that was proposed by Parpinelli [13, 14, 15] named the Ant-Miner to perform the rule induction.

ACO Metaheuristic is a branch of an artificial intelligence technique called swarm intelligence, introduced in the early 1990s [16, 17]. It is a probabilistic technique for solving combinatorial optimizations problem, which was inspired by the behaviour of cooperating ants in finding path from its nest to the food source [18]. The purpose of ACO metaheuristic is to find the best solution using a set of artificial ants that communicates indirectly using an item called the pheromone. The amount of pheromone on one of the trails used by the majority ants will increase as time passed by and will decrease on the less used trails. As a consequent, since ants tend to follow the pheromones, all or most of the ants will finally converge to the best trail, with the high density of pheromones, which happened to be the shortest trail from the nest to the food source. However, there are a small number of ants still using the longer branch. This is the effect of "path exploration", where the density of the pheromone will not bias some of the ants.

Parpinelli proposed Ant-Miner algorithm in 2002 [13, 14, 15]. The Sequential Covering algorithm is the based for Ant-Miner algorithm, that extract rules directly from data. Sequential Covering algorithm [12] discovers rules in greedy fashion based on a certain evaluation measure. In other words, this algorithm select terms using some heuristic.

Ant-Miner uses a heuristic measure as evaluation measure to fill in the antecedent part of the rule, by selecting the best term to be included into the partial rule. The heuristic measure (3) is the normalization of entropy measures between terms (2). The algorithm selects one best rule from a set of discovered rules, based on a quality measure using some fitness function.

$$H\left(W|A_i = V_{ij}\right) = -\sum_{w=1}^{k} \left(P\left(w|A_i = V_{ij}\right) \cdot \log_2 P\left(w|A_i = V_{ij}\right)\right) \tag{2}$$

$$\eta_{ij} = \frac{log_2\left(k\right) - H\left(W|A_i = V_{ij}\right)}{\sum_{i}^{a} x_i \cdot \sum_{j}^{b_i} \log_2(k) - H\left(W|A_i = V_{ij}\right)} \tag{3}$$

where

- $W$ is the class attribute
- $a$ is the total number of attributes
- $x_i$ is set to one if the attribute $A_i$ was not yet used by the current ant, zero otherwise
- $k$ is the number of classes

Ant-Miner differs from other Sequential Covering algorithms implementation because this algorithm also depends on a value called the pheromone, which contributes to the behaviour of exploration of the algorithm. Hence, Ant-Miner uses a probability (4) that is proportional to the product of heuristic value and pheromone level for that term, to add terms to a rule. Dorigo in his book, called this transition rule, random proportional transition rule [18].

$$P_{ij}\left(t\right) = \frac{\tau_{ij} \cdot \eta_{ij}}{\sum_{i}^{a} \sum_{j}^{b_i} \tau_{ij} \cdot \eta_{ij}}, \forall i \in I \tag{4}$$

where

- $\tau_{ij}$ is the amount of pheromone at time $t$
- $I$ is the set of attributes that are not yet used by the ant.

Ant-Miner initialized equally the pheromone level, $\tau_{ij}$ using (5).

$$\tau_{ij}(t=0) = \frac{1}{\sum_{i=1}^{a} b_i} \tag{5}$$

where

- $a$ is the total number of attributes
- $b_i$ is the number of values in the domain of attribute $i$.

Ant-Miner updated the pheromone level after each ant colony has selected the best rule from a set of rules constructed by many ants in a colony using (6).

$$\tau_{ij}(t+1) = \tau_{ij}(t) + \tau_{ij}(t)Q, \forall i, j \tag{6}$$

where $Q$ is the quality of a rule.

For the next ant colony, terms in the rule antecedent that have been selected by the previous ant colony will have a higher level of pheromone and will probably are more favoured than other terms.

Ant-Miner selects the best rule after each ant colony has created a set of rules, based on rule's quality. This algorithm measures rule's quality using a fitness function that depends on the product of sensitivity and specificity, which were adapted from the field of information retrieval (IR). The quality of rules, $Q$ is calculated using (7).

$$Q = \frac{TP}{TP+FN} \times \frac{TN}{FP+TN} \tag{7}$$

where

- TP is the number of cases covered by the rule and having the same class as that being predicted by the rule
- FP is the number of cases covered by the rule and having a different class from that being predicted by the rule
- FN is the number of cases that are not covered by the rule while the class predicted by the rule
- TN is the number of cases that are not covered by the rule and having a different class predicted by the rule.

## 4. Findings and discussion

This study tested the rainfall data with different values of number of ants which are 15, 20, 25, 30, 40 and 45. In addition, the minimum cases per rule is fixed to 5, maximum uncovered cases is fixed to 10, rule for convergence is fixed to 10 and number of iterations is fixed to 10000.

Figure 2 depicted the predictive accuracy for different values of number of ants. When the number of ants is equal to 30, Ant-Miner produced a classification prediction model with the highest accuracy which is 84.29%. Meanwhile, the lowest accuracy reading is when the number of ants is equal to 25. The higher the accuracy, the better the classification prediction model is.

The classification prediction model produced by Ant-Miner has a higher predictive accuracy as compared to J48, an implementation of C4.5 algorithm [19] in Weka. The predictive accuracy for the classification prediction model produced by J48, using the same 22 years rainfall data, produced a much lower value which is 77.27% only.

Experiment was also done for the number of rules. The fewer the number of rules, the better the prediction model is. Figure 3 depicted the results from the experiments done to test the number of rules. It is found that when the number of ants is low, less than 35, the number of rules is small, but, increasing as the number of ants increased.

As for the number of terms, the results varies. However, the number of terms has a low value when the number of ants are 20 and 35 as depicted by figure 4.

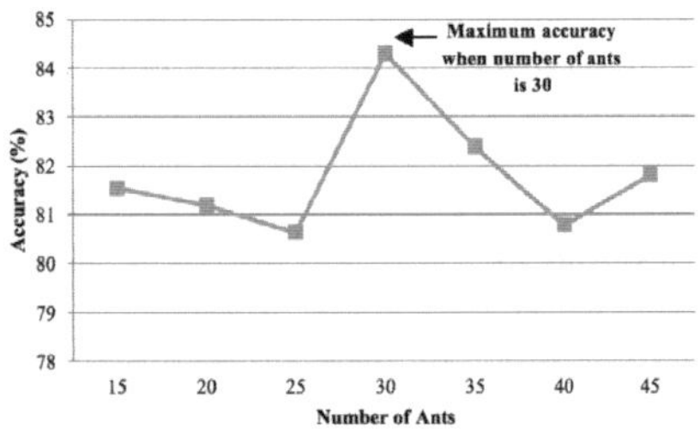


**Figure 2.** Predictive accuracy for various number of ants.

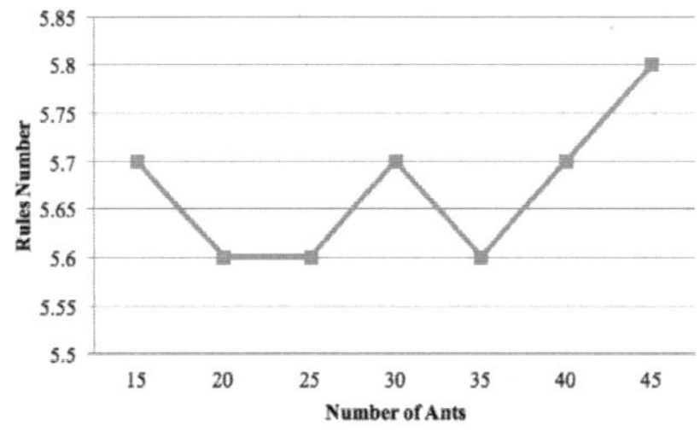


**Figure 3.** The Number of rules for various number of ants.

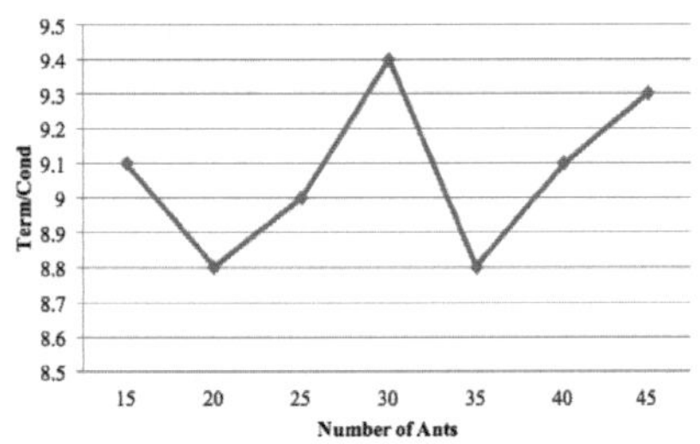


**Figure 4.** The number of terms for various number of ants.

## 5. Conclusion and future work

In this study the classification prediction model has been established for the rainfall recorded at three gauging stations in the north of Perlis. The findings show that predictive accuracy of the classification prediction model is promising at the value of more than 80%. As an addition, the predictive accuracy of the classification prediction model is better as compared to J48.

The classification prediction model established in this study can be used in predicting the occurrences of flood in Perlis. Hence, predicting can be used by the Perlis government to prepare the state as well as the people in order to reduce the effect of flood disaster.

One direction for future research is as follow. Since the class for rules constructed Ant-Miner will only be determined after the creation of each rule, the selected terms are not focused relevance. In other words, this algorithm will select a term that may be high in relevancy with the current set of previously selected terms, but not for the later assigned class. Therefore, like other ACO implementations for other fields, Ant-Miner might face the problem of stagnation, where optimized rule cannot be found, and thus make the program to run forever. So, it would be interesting to improve the terms selection strategy to reduce the stagnation problem.

## Acknowledgments

This study was funded by Long Term Research Grant Scheme (LRGS/b-u/2012/UUM/Teknologi Komunikasi dan Informasi) and DKCP (600-UiTMPs/PJIM&A/ST/DKCP (02/2012)). The authors would like to thank Department of Irrigation and Drainage Malaysia for supplying of required data.

## References

[1] Gasim M B, Toriman M E, Abdullahi M G *et al.* 2014 *International Journal of Interdisciplinary Research and Innovations* **2** 59–65

[2] Mokhtar S A, Ishak W H W and Norwawi N M 2014 Modelling of reservoir water release decision using neural network and temporal pattern of reservoir water level *2014 5th International Conference on Intelligent Systems, Modelling and Simulation* (IEEE) pp 127–130

[3] Ashaary N A, Wan-Ishak W and Ku-Mahamud K R 2015 Forecasting model for the change of reservoir water level stage based on temporal pattern of reservoir water level *Proceedings of the 5th International Conference on Computing & Informatics (ICOCI)* pp 692–697

[4] Othman M, Saian R, Nazri M N, Kamal N S A and Liyana N 2013 Fuzzy forecasting for water level of flood warning system in perlis *International Symposium on Mathematical Sciences and computing Research 2013 (iSMSC 2013)* pp 104–109

[5] Muhamad N S and Din A M 2015 Exponential smoothing techniques on time series river water level data *Proceedings of the 5th International Conference on Computing & Informatics (ICOCI)* pp 644–649

[6] Othman M, Azahari S N F and Massuut N A A 2016 Fuzzy spatial forecasting model of rainfall distribution for flood early warning *Regional Conference on Science, Technology and Social Sciences (RCSTSS 2014)* (Springer) pp 251–263

[7] Keogh E, Chakrabarti K, Pazzani M and Mehrotra S 2001 *ACM SIGMOD Record* **30** 151–162

[8] Lin J, Keogh E, Lonardi S and Chiu B 2003 A symbolic representation of time series, with implications for streaming algorithms *Proceedings of the 8th ACM SIGMOD Workshop on Research Issues in Data Mining and Knowledge Discovery* (ACM) pp 2–11

[9] Hall M, Witten I and Frank E 2011 *Data Mining: Practical Machine Learning Tools and Techniques* (Kaufmann, Burlington)

[10] Hall M, Frank E, Holmes G, Pfahringer B, Reutemann P and Witten I H 2009 *ACM SIGKDD Explorations Newsletter* **11** 10–18

[11] Han J, Pei J and Kamber M 2011 *Data mining: Concepts and Techniques* (Elsevier)

[12] Tan P N, Steinbach M and Kumar V 2006 *Introduction to Data Mining* (Pearson Addison Wesley Boston)

[13] Parpinelli R S, Lopes H S and Freitas A A 2002 An ant colony algorithm for classification rule discovery *Data Mining: A Heuristic Approach* pp 191–208

[14] Parpinelli R S, Lopes H S and Freitas A A 2002 *IEEE transactions on evolutionary computation* **6** 321–332

[15] Parpinelli R S, Lopes H S and Freitas A A 2001 An ant colony based system for data mining: Applications to medical data *Proceedings of the genetic and evolutionary computation conference (GECCO-2001)* pp 791–797

[16] Dorigo M, Maniezzo V and Colorni A 1991 Positive feedback as a search strategy Tech. Rep. 91-016 Dipartimento di Elettronica, Politecnico di Milano Milan, Italy

[17] Dorigo M, Maniezzo V and Colorni A 1996 *Systems, Man, and Cybernetics, Part B: Cybernetics, IEEE Transactions on* **26** 29–41

[18] Dorigo M and Sttzle T 2004 *Ant Colony Optimization* (the MIT Press) ISBN 0-262-04219-3

[19] Quinlan J R 2014 *C4.5: Programs for Machine Learning* (Elsevier)

# Conceptual or procedural mathematics for engineering students at University of Samudra

**Saiman** [1], **Puji Wahyuningsih** [2], **and Hamdani** [3]

[1] Departemen of mathematics education, University of Samudra, Aceh
[2] Departemen of chemistry, Faculty of Engineering, University of Samudra, Aceh
[3] Departemen of mechanical engineering , University of Samudra, Aceh

corresponding authors : saiman_mat@yahoo.com

**Abstract**. This study we investigate whether the emphasis in mathematics courses for engineering students would benefit from being more conceptually oriented than more procedurally oriented way of teaching. In this paper, we report in some detail from twenty-five engineering students comes from three departements ; mechanical engineering, civil engineering and industrial engineering. The aim was to explore different kinds of arguments regarding the role of mathematics in engineering courses, as well as some common across contexts. The result of interview showed that most of engineering students feel that conceptual mathematics is more important than procedural mathematics for their job the future.

**Keywords:** mathematics courses; engineering students; conceptual and procedural knowledge.

## Introduction

Several universities in Indonesian, especially faculty engineering in University of Samudra, the courses of mathematics is a subject of fundamental for engineering students in the future. Mathematics would became the basis for engineering students due to studied the others subjects in the curriculum. In the last two decades, several studies on mathematics at the workplace have been conducted. Most of these studies dealt with rather "lightweight" use of school mathematics. Just a few studies captured presumably heavy users of mathematics like engineers and scientists. Many engineering educators have proposed changes to the way that mathematics is taught to engineers: by having engineering professors teach mathematics rather than mathematics professors, by better integrating mathematics and engineering or by making large changes to the engineering curriculum. For example, uniting mathematics and engineering courses, or having mathematics and engineering educators collaborate in providing mathematics instruction to engineers. Her recommendations are based in a phenomenographic study of first year students. experiences learning mathematics (Cardella, 2008). Many of engineering students in the university are enrolled in the basic first-year courses mathematics, and to accommodate such large numbers, course enrollments are split into multiple large lectures and supplemented with smaller recitation sections.

According to Kent and Noss, mathematics 'is andwill remain crucial' for engineering education. It is not clear, however, how the mathematical education of engineering students should be organized. For example, the types of mathematical knowledge engineering students need, changes of practice and education suggested by the advance of computer technology, or how and when mathematics should be taught, are still discussed in the literature (Kent and Noss, 2003). This paper explores this issue by analysing interviews with twenty five of engineering students where the focus was on their views on the relevance of procedural and conceptual aspects of mathematics in the education. The aim of this study was to explore views and arguments concerning the role of conceptual versus procedural aspects of mathematics in engineering education.

**Blackground**

Correspondingly, the mathematical education of engineers has two major goals: It should enable students to understand, set up and use the mathematical concepts, models and procedures that are used in the application subjects like engineering mechanics, machine dynamics or control theory. In order to clarify this first goal, a lecturer has to investigate the use of mathematics in application textbooks or scripts of colleagues. Although this is also demanding and time-consuming for a non-engineer, the information is quite readily available. The second major goal of mathematics education is to provide students with a sound mathematical basis for their future professional life. For clarifying this rather nebulous goal, one needs information on the mathematical expertise or qualifications an engineer needs in his or her daily life (Alpers, 2010).

In his review of studies on the use of mathematics in engineering workplaces, Alpers identified several categories that 'capture the way mathematical thinking or activities occur during the work on practical problems' (Gainsburg, 2007). One category concerns the contextual embedding of mathematical models, concepts and procedures; another involves a recognition that the operation of technological tools still requires understanding of mathematical notations and graphics; while another category points at an appreciation that mathematics is important and necessary but not sufficient for engineering design, where also 'engineering judgement' and 'sceptical reverence' are critical.

Current experiences demonstrate that this belief is unfounded as engineering faculty members question the meaning of *mathematical competence*. Educators note, both through personal experience and in the research literature (Hiebert and Lefevre, 1986) that engineering students are not adapting the skills they learn in their math and physics classes to their engineering studies under traditional pedagogies. Often, the students just *don't see* how problem-solving strategies and skills learned in one setting apply to exercises and support thinking in another setting. Moreover, the difference of knowing about math, knowing about physics, and knowing about engineering concepts is both emphasized and encouraged by the curriculum in place in each of these disciplines. Indeed, transfer to new contexts of higher-level cognitive abilities from one domain to another is not automatic and may require special attention (Hiebert and Carpenter, 1992).

**Conceptual and procedural approaches in mathematics work**

Although conceptual and procedural knowledge cannot always be separated, it is useful to distinguish between the two types of knowledge to better understand knowledge development. After interviewing a number of mathematics education researchers, Baroody, *et al* (2007) and colleagues suggested that conceptual knowledge should be defined as 'knowledge about facts, generalizations, and principles', without requiring that the knowledge be richly connected (Rittle and Alibali, 1999). As with conceptual knowledge, the definition of procedural knowledge has sometimes included additional constraints. Within mathematics education, Star (2005) noted that sometimes: 'the term procedural knowledge indicates not only what is known (knowledge of procedures) but also one way that procedures (algorithms) can be known (e.g. superficially and without rich connections)' (p. 408).

At many universities, mathematics studies in engineering education begin with calculus and linear algebra courses, often taught at mathematics departments that treat their subject as valuable knowledge in its own right, rather than as a service subject for practical use outside mathematics. In this context, there is a strong tradition regarding the content of these courses as well as teaching approaches used. However, even within this 'traditional' approach to mathematics teaching, there is a difference between taking a mainly conceptual or procedural approach. Teaching with a focus on conceptual understanding may start by providing a contextual basis for the new mathematical knowledge requiring students to connect to their prior knowledge. Taking a more procedural approach, the lecturer may present definitions, notations and procedures without first providing meaningful contexts to concepts and methods involved (Hiebert and Lefevre, 1986).

While the distinction between these two aspects of doing, teaching and learning mathematics has prevailed, it has also been criticized. In mathematics education, this distinction between conceptual and procedural knowledge was much discussed in the 1980s, mainly within a psychological framework with reference to Piaget and mental schemes. Hiebert and Lefevre defined conceptual knowledge in mathematics as a connected network of knowledge in which 'the linking relationships are as prominent as the discrete pieces of information (Hiebert and Carpenter, 1992). As an example of conceptual knowledge, they note the construction of a relationship between the algorithm for multi-digit subtraction and knowledge of the positional place values of digits (Hiebert and Carpenter, 1992). Procedural knowledge was defined by two parts, constituted by step-by-step procedures for solving mathematical tasks, on one hand, and knowledge of the symbolic representations used in such procedures, on the other hand. To be competent in mathematics, then involves not only of knowledge of concepts and knowledge of procedures but also of relations between these two types of knowledge (Skemp, 1978).

Research suggests that there is a complex interplay between conceptual and procedural knowledge (Peled and Zaslavski, 2008; Rittle and Alibali, 1999); conceptual knowledge may influence or even become procedural with repeated exposure,while procedural knowledgemay support the development of conceptual knowledge (Artique, 2007; Baker and Czarnocha, 2002) Some authors see these two constructs as independent, while others have emphasized dynamic and evolutionary aspects between them. Gray and Tall (1994) coined the term 'procept' for an integrated knowledge of processes and concepts that is characteristic for successful learners of mathematics. Kieran (2013) claims, however, that the distinction is a false dichotomy, with reference to the nature of algebraic symbols. Also, in the context of more advanced mathematics, Wu argues from a number of examples that 'in mathematics, skills and understanding are completely intertwined (Engelbrecht, *et al.*, 2012).

In the context of mathematics in engineering education, the relation between conceptual and procedural approaches to solving mathematics tasks has been investigated within the larger project for the present study. That this relation is highly complex was illustrated by Engelbrecht, Bergsten and K agesten, who found that first-year engineering students tended to 'proceduralize' tasks designed to have a conceptual focus, sometimes requiring "quite sophisticated mathematical work". The study by Engelbrecht, Bergsten and K agesten on junior engineering students' achievement and views on conceptually versus procedurally focused mathematics tasks showed that the students considered procedural questions as more common in their mathematics curriculum, while conceptual (mathematical) questions were seen as more common in engineering subjects; they also expressed a higher confidence in their performance on the procedural tasks. However, both types of tasks were seen as relevant for their engineering studies. In (Engelbrecht, *et al.*, 2009), strong correlations between performance and confidence for the procedural items for junior and senior students were found, while for the conceptual items, these correlations were much stronger for the senior than for the junior students. This points to an increased familiarity with the conceptual aspects of mathematics through its use in applied subjects throughout the education.

**Methodology**

The sample of the research were twenty-five engineering students comes from three departements ; mechanical engineering, civil engineering and industrial engineering. We address various populations engineering students views on the role of mathematics in engineering education with a particular focus on conceptual/procedural approaches. We selected engineering students who had the highest and the lowest IPK of populations, using a combination of convenience and snowball sampling as sample of research (Geertz, 1973). In earlier parts of the project, we have problematized the distinction between conceptual and procedural approaches, and surveyed junior (second year) and senior (fourth year) engineering students from both departement.

In the overall project as well in this study, we use the following working definitions regarding the approach to the mathematical education of engineers: A *conceptual approach* includes translations between verbal, visual (graphical), numerical and formal/algebraic mathematical expressions (representations); linking relationships; and interpretations and applications of concepts (for example, by way of diagrams) to mathematical situations. A *procedural approach* includes (symbolic and numerical) calculations, employing (given) rules, algorithms, formulae and symbols.

The research question, *How do engineering students view the relevance and role of conceptual and procedural mathematical skills in the education ?*, was used as a guide to structure the interviews and the analysis of the interview protocols. In this paper, we report, as a case study, in some detail from three of these interviews for the full results from different engineering students, selected to represent three different 'poles' of engineering students, one being more on research and product development oriented and one more on site practice oriented (Bryman, 2004). By looking at these different cases, which may represent both, what Bryman calls a *unique case* and an *exemplifying case*, we hoped to be able to find different kinds of arguments regarding the role of mathematics in engineering work, as well as some that were common across contexts (Geertz, 1973).

IOP Conf. Series: Journal of Physics: Conf. Series **855** (2017) 012041 doi:10.1088/1742-6596/855/1/012041

The interviews were audio-recorded and transcribed. In the semi-structured interviews, (Geertz, 1973) the concepts of conceptual and procedural approaches were initially explained to the interviewees, using our working definitions and a few examples of items from the research instrument used with the students. To investigate the research question above, the interviews were organized by the following guiding questions:

1. What is your opinion about the subjects of mathematics in your departement?
2. Do the mathematics courses you could be relevant to the subject futher in your departement ? Why ?
3. If the subjects is relevant to your futher, which method is more useful for your project, conceptual or procedural mathematics ?

For the analysis of the two interview protocols, a matrix was set up where the answers from each interviewer were organized in columns sorted by questions. By identifying content categories of responses in a column across the two interviewees, similarities and differences between their views were possible to describe for each question. A picture of the overall view of each interviewee on the procedural/conceptual issue was obtained by a two-step coding procedure, with specific codes sorted within overarching themes.

**Interview 1 – The Civil Engineering Students**

Civil engineering students learn mathematics in the first semester, the second semester and the third semester. In studying subjects they need more advanced mathematics than the other sciences such as physics, chemistry, and biology. Mathematics plays a central part in the formation of civil engineering students, both in the form of an entry requirment to undergraduate courses, and as a core underpinning element of those course.

To the opening interview question, *What is your opinion about the subjects of mathematics in your departement*? Based on interviews, it can be concluded that mathematics is a subject which is very difficult to understanding. Mathematics as both a language and an art, emphasizing that he has not been much involved in computational activities. Firstly, it's a language to be learned, mathematical language, and then it it's also a little art, I mean you must have some, you have to have imagination to solve these problems sometimes, and that one can learn mechanically but some have the aptitude for this, and some may work hard on it and so.

While question, *Do the mathematics courses you could be relevant to the subject futher in your departement ? Why ?*. Many of engineering students answered that the mathematics courses could be relevant to the subjects futher in my department. In the practice of civil engineering, 'computational mathematics' is perceived as a tremendous opportunity, pushing forward the boundaries of civil engineering design. For example, in the structural design of buildings, there are high-profile, high-budget projects, pioneering the development of techniques that gradually become used across the whole range of buliding projects (Kent and Noss, 2003).

When they asked about the emphasis on procedural or conceptual aspects of mathematics for engineering education, they says the he believes a little on both parts should be studied. They want to learn mathematics procedurally as to compute than the concepts or theories. While they having some difficulties to explain why the procedural parts of his mathematics experience is needed, but they still have not realized the importance of the issue the sustainability of other subjects.

IOP Conf. Series: Journal of Physics: Conf. Series **855** (2017) 012041 doi:10.1088/1742-6596/855/1/012041

According to them, the courses of mathematics very difficult to learn because of less understanding of mathematics concepts and mathematical formulas that are difficult to remember. In addition, a simple counting techniques is also required by students in solving problems in mathematics. In education mathematics, engineering students are thus advised to encourage independence and confidence, and be able to collaborate with others on a challenging problem, do not sit with 'thumbin the answer key and looking'.

**Interview 2 – The mechanical engineering students**

Mechanical engineering students learn mathematics in the first semester and the second semester. To the opening interview question, *What is your opinion about the subjects of mathematics in your departement*? Based on interviews, it can be concluded that mathematics is a subject which is very difficult to understanding. Most of students come from vocational school, so many courses in they school not relevant to mathematics courses.

While question, *Do the mathematics courses you could be relevant to the subject futher in your departement ? Why ?*.Most of them answered that mathematics is relevant to the subject futher in our departement, such as technical mathematics, calculus, fluida mechanical, and etc. But, they have been difficulties to studied it. It happens because of the level of student knowledge that lower and mathematical formulas that are difficult to remember so the understanding of mathematical concepts and calculation in solving problems mathematical became less.

When they asked about the emphasis on the procedural aspects or conceptual mathematics for engineering education. They asked that conceptual mathematics is more important than procedural aspects for work futher. A *conceptual approach* includes translations between verbal, visual (graphical), numerical and formal/algebraic mathematical expressions (representations); linking relationships; and interpretations and applications of concepts (for example, by way of diagrams) to mathematical situations. A *procedural approach* includes (symbolic and numerical) calculations, employing (given) rules, algorithms, formulae and symbols. According to Baker (2002), in particular, engineering students must be able to express ideas that arise in engineering in the form of mathematical terms and then use their problem-sloving skills to understand the consequences. Based on my experineces in teaching ordinary differential equations to engineering students, students see their maathematical education as simply a vast collection of specific procedures. The question raised here is whether better coordination of the content in first-year math through conceptual and procedural approach could be improve student ability to use mathematics in subsequent engineering courses.

**Interview 3 – The Industrial Engineering Students**

To the opening interview question, *What is your opinion about the subjects of mathematics in your departement*? Based on interviews, it can be concluded that mathematics is a subject which is very difficult to understanding. Mathematics as both a language and an art, emphasizing that he has not been much involved in computational activities. Firstly, it's a language to be learned, mathematical language, and then it it's also a little art, I mean you must have some, you have to have imagination to solve these problems sometimes, and that one can learn mechanically but some have the aptitude for this, and some may work hard on it and so.

While question, *Do the mathematics courses you could be relevant to the subject futher in your departement ? Why ?*. Many of engineering students asked that the mathematic courses could be relevant to the subjects futher in my departement because some subjects require understanding of mathematical concepts and procedures, such as calculus, thermodinamic, industrial management, and the others.

They like to learn mathematics procedurally as to compute than the concepts or theories. While having some difficulties to pinpoint exactly why the procedural part of his mathematical experience is needed, They still hangs on to its importance even if he does not explicitly use it in his own work futher. Regarding the relevance of procedural versus conceptual mathematics for his present job, They continues to talk in rather general terms, acknowledging that he does not deal much with mathematics directly.

It's probably more about, creativity, and see the possibilities, the job they have today maybe a little more towards, instead of this calculating so maybe a little more to the formulation and problem-solving way, they would say. But perhaps it is far from that as well. Despite the uncertainty expressed here, he again emphasizes that 'one must be accurate and can't be sloppy' and that 'both are important', but that the key issue is problem solving, 'in order to formulate the problems and be able to see something and structure and sort out'. It is clear that a conceptual approach, expressed through seeing structure and order, is more relevant for his own work than managing the calculations.

## Discussion

The development of conceptual and procedural knowledge in mathematics was simultaneously. The conceptual knowledge is knowledge that learners have about something. The conceptual knowledge in mathematics is knowledge of concepts, facts, definitions, and formulas but the conceptual knowledge refers to knowledge about facts and concepts of mathematics of students or the factors that influence the thinking and interesting it learning (Anderson, 2006).

The conceptual knowledge is relatively static but a procedural knowledge is more dynamic. When the procedural knowledge was activated, the result is not as a retrieval of information, but a transformation of information. For example, the results work on the problems of 999/3 is 333. The input information (999/3) has been converted into an output (333) that is different from the input.

Another difference between conceptual knowledge and procedural knowledge is on the speed of activating. When procedural knowledge has been studied once, this knowledge will be working quickly and automatically. For example, a skilled reader review a passage (articles, etc.) quickly. Reviewers process is a procedural knowledge, because the reading material is converted into a meaning. Activation of conceptual knowledge is slower and is more conscious. Gagne (in Winkel, 2006), said that the difference between conceptual knowledge and procedural knowledge is follows. Someone had learned verbal information, if he can tell you about the information. The information is called as verbal, because we know it in the form of sentence. Someone had learned an intellectual skill, if he knows how to do something as opposed to know what that something was.

Mathematicals problems is a useful tool to study the conceptual and procedural knowledge, because problem solving in mathematics specifically requires the application of both types of knowledges. It was was due to problem solving requires an understanding of the problem situation to produce a settlement. Although it is generally argued that the problem solving

IOP Conf. Series: Journal of Physics: Conf. Series **855** (2017) 012041 doi:10.1088/1742-6596/855/1/012041

involves the understanding, and they often succes using the procedure without passing through the process of understanding.

For several years, researchers and educational practitioners were shocked by the systematic procedural errors committed by students. Observations on procedural errors that occur due to flaws in the conceptual knowledge. This shows that the procedural knowledge is based on the conceptual knowledge. A person who is given only the conceptual knowledge must construct the procedural knowledge. Two type of knowledges are often, and almost always that could not be separated each others (Shapiro, 2000).

The twenty five of engineering students interviewed for this study have very different department and are working in different areas and types of engineering. Already these examples thus highlight the potential complexity in designing what could be called 'engineering education', and in particular its mathematical component. The result of interview showed that most of engineering students feel that conceptual mathematics is more important than procedural mathematics for their job the future. But, if the students only understanding about the conceptual knowledge in mathematic education is not complete. It must construct the procedural knowledge. Based on the conceptual and procedural knowledge in mathematics education for engineering students especially student in university of Samudra is very important and could not be separated each others.

This message also echoes well the voices from the other twenty of five engineering students who were interviewed in the overall study described above (Bryman, 2002) Still, while the selection of the interviewees aimed to cover a wide spectrum of types of engineering work, the diversity and development of this professional field can never be 'covered' in any reasonable sense by one sample. By providing a 'thick description' of the arguments put forward in relaxed conversation settings by experienced engineers from different professional contexts, we hope to have pointed to some key concerns about the issue discussed relevant to their work. However, the intricacy of the relation between the procedural and conceptual aspects of mathematical work, as it has been described in these interviews, will need a continued empirical and epistemological attention in research to better understand the role of procedural knowledge in mathematics for engineering curriculum design.

There are complexities in the different uses of mathematics in engineering practice: not only the direct usefulness of mathematicals techniques and ideas to practice (for example, to perform a load case analysis of a certain kind of structure), but uses indirect usefulness of mathematics to practice. That is, it's formative role in development of an engineering students-the ways in which mathematics contributed to the development of technical expertise and judgement. For example, matrix alegbra role in developing an understanding of many engineering principles, but in the form of an explicit matrix may be of little practical 'use' for the working engineering. It is easy to see that the direct use of mathematics in practice has changed, but the changes indirectly uses are less apparent.

Futhermore, various kinds of mathematics may be relavant at different times in engineering careers. Take the case of calculus; it is important for formal technical engineering education to respect the principles of basic engineering principles, while in practice it may rarely have explicit for many engineering students but yet it still present, 'half remembered'as part of the appreciation of the analysis engineering education.

## Conclusion

The twenty five of engineering students have been interviewed for this study have very different departement and are working in different areas and types of engineering. The result of interview showed that most of engineering students feel that conceptual mathematics is more important than procedural mathematics for their job the future.

So, the math educators are very focused on improvements in teaching their discipline. Their purpose is to ensure that students understand the concepts relevant to the field. They also appreciate that most students enrolled in the first year in math classes will be continuing in engineering, and so acknowledge that the topics covered should be appropriate for engineering students. However, in math, the emphasis is on improving skills and simplifying conceptual development without explicit attention to its connections to engineering.

## References

[1] Alpers B 2010 Studies on the mathematical expertise of mechanical engineers *J. Math Model Appl.* 1(3):2–17

[2] Anderson O W & Krathwohl D R 2001 *A Taxonomy for Learning Teaching, and Assessing* (*A Revision of Blooms Taxonomy of Educational Objectives*). New York: Addision Wesley

[3] Artigue M 2007 Digital technologies: a window on theoretical issues in mathematics education. *Proceedings of CERME6*. Larnaca: University of Cyprus 68–82

[4] Baker W and Czarnocha B 2002 Written meta-cognition and procedural knowledge *Proceedings of the 2nd International Conference on the Teaching of Mathematics*.

[5] Baroody A J, Feil Y and Johnson A R 2007 An alternative reconceptualization of procedural and conceptual knowledge *J. Res Math Educ*. 38(2):115–131.

[6] Bryman A 2004 Social research methods Oxford University Press Oxford

[7] Cardella ME 2008 Which mathematics should we teach engineering students? An empirically grounded case for a broad notion ofmathematical thinking *Teach Math Appl*. 27(3):150– 159

[8] Engelbrecht J, Bergsten C and K°agesten O 2009 Undergraduate students' preference for procedural to conceptual solutions to mathematical problems *Int J.Math Educ Sci Technol.* 40(7):927– 940

[9] Engelbrecht J, Bergsten C and K°agesten O 2012 Conceptual and procedural approaches to mathematics in the engineering curriculum: student conceptions and performance *J. Eng. Educ*. 2012;101(1):138–162

[10] Gainsburg J 2007 The mathematical disposition of structural engineers *J. Res Math Educ*. 38:477–506

[11] Gray E and Tall D Duality 1994 ambiguity and flexibility: a proceptual view of simple arithmetic *J. Res Math Educ* 26(2):115–141

[12] Geertz C 1973 Thick description: toward an interpretive theory of culture New York Basic Books 3–30

[13] Hiebert J and Lefevre P 1986 Procedural and conceptual knowledge in mathematics: an introductory analysis Hillsdale (NJ) Erlbaum 1–27

[14] Hiebert J and Carpenter T 1992 Learning and teaching with understanding New York:Macmillan Publishing Company 65–97

[15] Kent P and Noss R 2003 Mathematics in the university education of engineers Available from: http://www.lkl.ac.uk/research/ REMIT/Kent-Noss-report-Engineering-Maths.pdf

[16] Kieran C 2013 The false dichotomy in mathematics education between conceptual understanding and procedural skills: an example from algebra Springer 153–171

[17] Peled I and Zaslavski O 2008 Beyond local conceptual connections: meta-knowledge about

procedures *Learn Math* 28(3):28–35
[18] Rittle J B and Alibali M W 1999 Conceptual and procedural knowledge of mathematics: does one lead to another? *J. Educ Psychol* 91(1):175–189.
[19] Shapiro Stewart 2000 Thinking about mathematics the philosophy of mathematics, New York. Oxford University Press
[20] Silver E 1986 Using conceptual and procedural knowledge: a focus on relationships. Hillsdale (NJ): Erlbaum. p. 181–198.
[21] Skemp R 1978 Relational and instrumental understanding *Arithmetic Teach* 26:9–15
[22] Star J R 2005 Reconceptualizing procedural knowledge *J. Res Math Educ* 36(5):404–411

IOP Conf. Series: Journal of Physics: Conf. Series **855** (2017) 012042 doi:10.1088/1742-6596/855/1/012042

# $H$-Supermagic Labeling on Coronation of Some Classes of Graphs with a Path

**H Sandariria[1], M Roswitha[2], T A Kusmayadi[3]**

Department of Mathematic, Faculty of Mathematics and Natural Sciences, Sebelas Maret University

E-mail: sandaririahardina@gmail.com, mania_ros@yahoo.co.id, tri.atmojo.kusmayadi@gmail.com

**Abstract.** A simple graph $G = (V, E)$ admits an $H$-covering if every edge in $E$ belongs to a subgraph on $G$ that isomorphic to $H$. A graph $G$ is $H$-magic if there is exists a total labeling $f : V(G) \cup E(G) \to \{1, 2, \ldots, |V|+|E|\}$, such that each subgraph $H' = (V'(H'), E'(H'))$ on $G$ satisfies $f(H') = \sum_{v \in V'} f(v) + \sum_{e \in E'} f(e) = m(f)$, where $m(f)$ is a constant magic sum. A graph $G$ is a $H$-supermagic labeling if $f(V) = \{1, \ldots, |V|\}$ and $s(f)$ is a constant supermagic sum. The research aims to study a $H$-supermagic labeling on corona product of a star graph with a path and a wheel graph with a path. We prove that the corona product of a star graph with a path $S_n \odot P_m$ is a $U_{m,2}$-supermagic for $m$ is odd or $m$, $n$ are even and $m \geq 3$ and the corona product of a wheel graph with a path $W_n \odot P_m$ is a $C_3 \odot P_m$-supermagic for $m \geq 3$.

## 1. Introduction

Graph labeling is an assignment of positive integers on vertices or edges, or both of them which fulfilled a certain condition (Gallian [2]). A labeling was first introduced by Sledl*áček* in 1963 (Gallian [2]). A magic labeling is a kind of a graph labeling that is the most studied. Suppose that a graph $G = (V(G), E(G))$ admits an $H$-covering. Consider a bijective function

$$f : V(G) \cup E(G) \to \{1, 2, \ldots, |V| + |E|\}.$$

Then $f$ is an $H$-magic labeling of $G$ if there is a positive integer $m(f)$, which we call them magic sum of $f$, such that for each subgraph $H' = (V(H'), E(H'))$ of $G$ isomorphic to $H$ we have,

$$f(H') = \sum_{v \in V(H')} f(v) + \sum_{e \in E(H')} f(e) = m(f).$$

In this case we say that the graph $G$ is $H$-magic labeling. When $f(V) = \{1, \ldots, |V|\}$ we say that graph $G$ is an $H$-supermagic labeling and $s(f)$ is a supermagic sum.

A magic covering was first introduced by Gutiérrez and Lladó [3], in which they found that a complete bipartite graph $K_{n,n}$ admitted a star magic covering $K_{1,n}$. Then Lladó and Moragas [5] proved in [3] the same graph containing a cycle. Later on Maryati et al. [6] obtained that some trees admit a path supermagic labeling and Ngurah et al. [8] worked on a path supermagic labeling on a cycle.

According to Lladó and Moragas [5], there are $C_3$-supermagic labelings on a wheel graph $W_n$ for $n \geq 5$ odd and $C_4$-supermagic labeling of a prism graph and a book graph. Roswitha and Baskoro [9] then proved that a cartepillar graph $S_{n_1,n_2,\ldots,n_k}$ is a $S_{n,n}$-supermagic for $n_1 = n_2 = \ldots = n_k = n$ and a *double star* $S_{n,n}$ has a $S_{n+1}$-supermagic labeling. A path $P_2$ and a cycle $C_n$ are a $P_h$-supermagic for $2 \leq h \leq n$.

In this paper, we study an $H$-supermagic labeling on the corona product of a star graph with a path and a wheel graph with a path. We show that the corona product of a star graph with a path $S_n \odot P_m$ is a $U_{m,2}$-supermagic for $m$ is odd or $m$, $n$ are even and $m \geq 3$ and the corona product of a wheel graph with a path $W_n \odot P_m$ is a $C_3 \odot P_m$-supermagic for $m \geq 3$.

## 2. $k$-Balanced Multiset

Maryati et al. [7] introduced $k$-balanced multiset as follows. Let $Y$ be a multiset of positive integers and $k \in N$. A multiset $Y$ is $k$-balanced if there are $k$ subsets from $Y$ where $Y_i = Y_1 = Y_2 = \ldots = Y_k$ then for each $i \in [1, k]$. We obtain $|Y_i| = |Y_k|$, $\sum Y_i = \frac{\sum Y}{k} \in N$ and $\biguplus_{i=1}^{k} Y_i = Y$.

**Lemma 2.1** [10] *Let $x$ and $y$ be nonegative integers. Let $X = [x+1, x+k]$ with $|X|$=$k$ and $Y$=$[x+k+1, x+2k]$ where $|Y|$=$k$. Then, the multiset $K$=$X \uplus Y$ is $k$-balanced for $j \in [1, k]$.*

**Lemma 2.2** [7] *Let $x$, $y$, and $z$ be positive integers. Then multiset $Y = [x+1, x+k] \uplus [y+1, y+k] \uplus [z+1, z+k]$ is $k$-balanced for $k \geq 3$ odd.*

**Lemma 2.3** [6] *Let $x$, $y$, and $k$ be integers, such that $1 \leq x \leq y$ and $k > 1$. If $X$=$[x, y]$ and $|X|$ is a multiple $2k$, then $X$ is $k$-balanced.*

## 3. A $U_{m,2}$-supermagic Labeling on A Star Graph Corona with Path $S_n \odot P_m$

Wallis [11] defined that a star $S_n$ is a graph which has $n+1$ vertices with one vertex has $n$ degrees and the others have 1 degree. A path $P_n$ is a graph constructed from a cycle $C_n$ with one edge removed. Kusmayadi [4] explained that an umbrella graph $U_{m,n}$ is a graph with a set of vertices $V(U_{m,n}) = \{x_1, x_2, \ldots, x_m, y_1, y_2, \ldots, y_n\}$ and a set of edges $E(U_{m,n}) = \{e_1, e_2, \ldots, e_{2m+n-2}\}$, with edges $e_i = x_i x_{i+1}$ for each $i = [1, m-1]$, edges $e_{i+m-1} = y_i y_{i+1}$ for each $i = [1, n-1]$, and edges $e_{i+m+n-2} = x_i y_1$ for each $i = [1, m]$. Frucht and Harary [1] defined that corona of graph $G$ with $H$, denoted by $G \odot H$ is a graph constructed from graph $G$ and $|V(G)|$ copies from $H$. i.e. there are $H_1, H_2, \ldots, H_{|V(G)|}$, then joined each vertex $v_i \in V(G)$ to all vertices from $V(H_i)$ for $i \in [1, k]$. A graph $S_n \odot P_m$ is a graph constructed from $S_n$ and $n$ copies $P_m$ called $P_m^i$ and joined every $v_i \in S_n$ to all vertices in $P_m^i$ for $i \in [1, n]$.

**Theorem 3.1** *Let $m$ and $n$ be positive integers with $m \geq 3$. For any $n$ and $m$ odd or $m$,$n$ even, a graph $S_n \odot P_m$ is $U_{m,2}$-supermagic.*

*Proof.* Let $G$ be a $S_n \odot P_m$ graph. Let $V(G) = \{a_j; 0 \leq j \leq (n+1)(m+1)\}$ and $E(G) = \{b_j; 0 \leq j \leq (n+1)(m+1)\} \uplus \{a_j^i; 0 \leq i \leq n+1, 1 \leq j \leq m-1\}$ where $|V(G)| = (n+1)(m+1)$ and $|E(G)| = 2nm + 2m + 1$. Let $f$ be a bijective function $f : V(G) \cup E(G) \to \{1, 2, \ldots, 3mn(n+1) + n\}$. Given a set of labels for all vertices and edges of $G$ denoted by $Z$ where $Z = [1, 3m(n+1) + n]$. Partition $Z$ into 2 sets: $A = [1, (n+1)(2m+1) + n]$ and $B = [(n+1)(2m+1) + n + 1, 3m(n+1) + n]$.

Then we define the total labeling $f$ on $G$ as follows. First, set $f(v_0) = 1$, the central vertex of $G$. Partition $A - \{1\}$ into 2 sets: $A_1 = [2, (n+1)(m+1)]$ and $A_2 = [(n+1)(m+1)+1, (n+1)(2m+1)+n]$. The vertices $a_j$ where $1 \leq j \leq (n+1)(m+1)$ are labeled by set $A_1$ and edges $b_j$ are labeled by set $A_2$. According to Lemma 2.1, if $x = 1$ and $k = m(n+1)+n$ we have $(m(n+1)+n)$-balanced. Let $A = A_1 \uplus A_2$, then $\uplus_{i=1}^{m(n+1)+n} = A$ and we have $\sum A_j = 2m(n+1) + 2n + 3$ for each $i \in [1, m(n+1)+n]$.

Next, we use $B$ to label edges $a_j^i$. $a_j^i$ defined a set of edges a path. The proof is divided into 2 cases.

**Case 1.** ***m* odd.** According to Lemma 2.3 if we set $x = (n+1)(2m+1) + n + 1$, $y = 3m(n+1) + n$, and $|B| = (n+1)(m-1)$, we obtain $(n+1)$-balanced where $\sum B_i = \frac{(m-1)}{2}(5m(n+1) + 3n + 2)$.

**Case 2.** ***m* even.** Partition $B$ into 2 sets: $B_1 = [(n+1)(2m+1) + n + 1, (2m+3)(n+1) + 2n + 1]$ and $B_2 = [(2m+3)(n+1) + 2n + 2, 3m(n+1) + n]$. Let $B = B_1$ for $m = 4$. According to Lemma 2.2 with $x = 2m(n+1) + 2n + 1$, $y = (2m+1)(n+1) + 2n + 1$, and $z = (2m+2)(n+1) + 2n + 1$, we have $B_1$ is $(n+1)$-balanced where $\sum B_{1i} = 6m(n+1) + 10n + \lceil\frac{n+1}{2}\rceil + 8$.

By Lemma 2.3, with $x = (2m+3)(n+1) + 2n + 2$, $y = 3m(n+1) + n$, and $|B_2| = (n+1)(m-4)$, we obtain $(n+1)$-balanced where $\sum B_{2i} = \frac{(m-4)}{2}(5m(n+1) + 6n + 5)$. Let $B = B_1 \uplus B_2$ for $m \geq 6$, then $\sum B_i = \frac{1}{2}(5m^2(n+1) - m(2n+3) - 4(n+1) + 2\lfloor\frac{n+1}{2}\rceil$.

Furthermore, the constant supermagic sum of a subgraph $U_{m,2}$:

$$f(U_{m,1}) = \begin{cases} \frac{1}{2}(9m^2(n+1) + m(6n+7) + n + 6), & \text{for } m \text{ odd;} \\ 2m^2(n+1) + m(10n+11) + 12(n+1) + \lceil\frac{n+1}{2}\rceil, & \text{for } m = 4\ ; \\ \frac{1}{2}(9m^2(n+1) + m(6n+7) + \lceil\frac{n+1}{2}\rceil + 4), & \text{for } m \text{ even, } m \geq 6.\square \end{cases}$$

Figure 1 illustrates an example of $U_{4,1}$-supermagic labeling on a $S_4 \odot P_4$ graph.

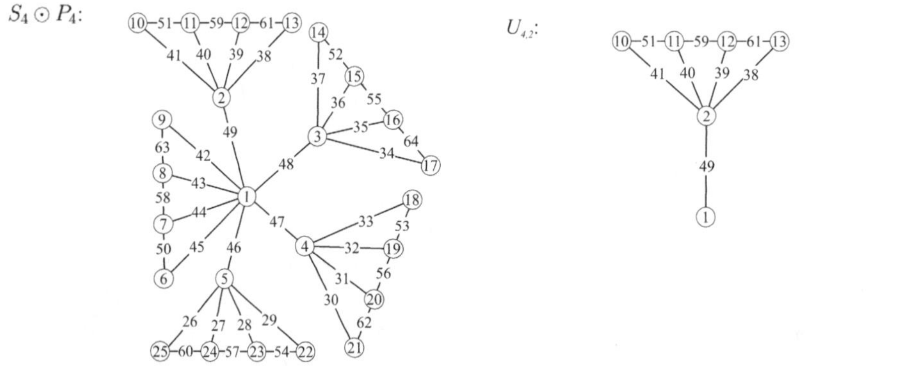


**Figure 1.** A $U_{4,2}$-supermagic labeling on a $S_4 \odot P_4$ graph

## 4. A $C_3 \odot P_m$-supermagic Labeling on a Wheel Graph Corona with Path $W_n \odot P_m$

Wallis et al. [11] defined a wheel $W_n$ is a graph obtained by joining $n$ vertices of $C_n$ with one central vertex. A graph $W_n \odot P_m$ constructed from $W_n$ and $n$ copies $P_m$ called $P_m^i$ and joined every $v_i \in W_n$ to all vertices $P_m^i$ for $i \in [1, n+1]$.

**Theorem 4.1** *A graph $W_n \odot P_m$ is $C_3 \odot P_m$-supermagic for $m \geq 3$.*

*Proof.* Let $G$ be a graph $W_n \odot P_m$. Let $V(G) = \{v_i; 0 \le i \le n\} \uplus \{a_j; 0 \le j \le (n+1)(m+1)\}$ and $E(G) = \{v_0v_1, v_0v_2, \dots, v_0v_n, v_1v_2, \dots, v_{n-1}v_n, v_nv_1\} \uplus \{b_j; 0 \le j \le m(n+1)\} \uplus \{a_j^i; 0 \le i \le n+1, 1 \le j \le m-1\}$ where $|V(G)| = (n+1)(m+1)$ and $|E(G)| = 2m(n+1)+n-1$. We have bijective function $f: V(G) \cup E(G) \to \{1, 2, \dots, 3m(n+1)+2n\}$. Given a set of label for all vertices and edges of $G$ denoted by $Z$ where $Z = [1, 3m(n+1)+2n]$. Partition $Z$ into 5 sets: $A = [1, n+1], B = [n+2, (n+1)(m+1)], C = [(n+1)(m+1)+1, (2m+1)(n+1)], D = [(2m+1)(n+1)+1, (2m+1)(n+1)+2n]$, and $E = [(2m+1)(n+1)+2n+1, 3m(n+1)+2n]$.

Set $A$ is used to label vertices $v_i$ where $0 \le i \le n$, $D$ is used to label edges $\{v_0v_1, v_0v_2, \dots, v_0v_n, v_1v_2, \dots, v_{n-1}v_n, v_nv_1\}$. The proof is divided into 2 cases, according to the values of $n$.

**Case 1. $n$ even.** (i) $\frac{n}{2}$ is even.

$$f(v_i) = \begin{cases} \lfloor \frac{i}{2} \rfloor + 1, & \text{for } i \text{ odd}; \quad i \in [1, n]; \\ \frac{1}{2}(i+n), & \text{for } i \text{ even}; \quad i \in [1, \frac{n}{2}]; \\ \frac{1}{2}(i+n)+1, & \text{for } i \text{ even}; \quad i \in [\frac{n}{2}+1, n]; \end{cases}$$

$$f(v_0v_i) = \begin{cases} 2(nm+n+m+1)-i, & \text{for } i \in [1, \frac{n}{2}]; \\ 2(nm+n+m)-i+1, & \text{for } i \in [\frac{n}{2}+1, n-1]; \\ 2(nm+m)+\frac{3}{2}n+1, & \text{for } i = n; \end{cases}$$

and

$$f(v_iv_{i+1}) = \begin{cases} 2(nm+n+m+1)+i, & \text{for } i \in [1, \frac{n}{2}-1]; \\ 2(nm+n+m)+i+3, & \text{for } i \in [\frac{n}{2}, n-2]; \\ 2(nm+m)+\frac{5}{2}n+2, & \text{for } i = n-1. \end{cases}$$

Next $f(v_0) = \frac{n}{4} + \lceil \frac{n-1}{2} \rceil + 1$ and $f(v_nv_1) = 2(nm+n+m)+2$. Let $f(C_3)$ be a constant sum in every subgraph $C_3$ of $G$. Hence, $f(C_3) = 6(nm+n+m+1) + \frac{3}{4}n + \lceil \frac{n-1}{2} \rceil + 1$.

(ii) $\frac{n}{2}$ is odd.

$$f(v_i) = \begin{cases} \lfloor \frac{i}{2} \rfloor + 1, & \text{for } i \text{ odd}; \quad i \in [1, n]; \\ \frac{1}{2}(i+n)+1, & \text{for } i \text{ even}; \quad i \in [1, \frac{n}{2}]; \\ \frac{1}{2}(i+n)+2, & \text{for } i \text{ even}; \quad i \in [\frac{n}{2}+1, n-1]; \\ \frac{i}{2}+1 & \text{for } i = n; \end{cases}$$

$$f(v_0v_i) = \begin{cases} 2(nm+n+m+1), & \text{for } i = 1; \\ 2(nm+n+m+1)-i, & \text{for } i \in [2, \frac{n}{2}]; \\ 2(nm+n+m)-i, & \text{for } i \text{ even}; \quad i \in [\frac{n}{2}+1, n-1]; \\ 2(nm+n+m+1)-i, & \text{for } i \text{ odd}; \quad i \in [\frac{n}{2}+1, n]; \\ 2(nm+m)+\frac{3}{2}n+1, & \text{for } i = n; \end{cases}$$

and

$$f(v_iv_{i+1}) = \begin{cases} 2(nm+n+m)+1, & \text{for } i = 1; \\ 2(nm+n+m)+i+1, & \text{for } i \in [2, \frac{n}{2}-1]; \\ 2(nm+n+m+1)+i, & \text{for } i \in [\frac{n}{2}, n-1]. \end{cases}$$

Then $f(v_0) = \frac{3}{2}(\frac{1}{2}n+1)$ and $f(v_nv_1) = 2(nm+m)+\frac{5}{2}n+1$. Let $f(C_3)$ be a constant sum in every subgraf $C_3$ of $G$. We have $f(C_3) = 6nm+\frac{29}{4}n+6m+\frac{15}{2}$.

**Case 2. $n$ odd.** First, set $f(v_0) = 1$. Next, label each vertex of $G$ other than central vertex by

$$f(v_i) = \begin{cases} \frac{n+i+2}{2}, & \text{for } i \text{ odd}; \\ \frac{i+2}{2}, & \text{for } i \text{ even}. \end{cases}$$

Then $f(v_0v_i) = 2(nm+n+m+1)-i$, $f(v_nv_1) = 2(nm+n+m)+2$, and

$f(v_iv_{i+1}) = 2(nm+n+m)+2+i$. Hence, the constant supermagic sum of $C_3$ is $f(C_3) = 6m(1+n) + \frac{(17+13n)}{2}$.

Furthermore, set $B$ is used to label $a_j$ where $1 \le j \le (n+1)(m+1)$ and $C$ is used to label $b_j$ where $1 \le j \le m(n+1)$. Let $M = B \uplus C$. According to Lemma 2.1 set $x = n+1$ and $k = m(n+1)$, then we have $m(n+1)$-balanced where $\sum M_i = 2(nm+n+m+1)+1$.

Next, we used $E$ to label edges $a_j^i$ where $1 \le i \le n+1$ and $1 \le j \le m-1$. The proof is divided into 2 cases, based on values of $m$.

**Case 1. *m* odd.** According to Lemma 2.3, if we put $x = 2m(n+1)+3n+2$, $y = 3m(n+1)+2n$, and $|E| = (n+1)(m-1)$, then we have $(n+1)$-balabced where $\sum E_i = \frac{1}{2}(m-1)(5m(n+1)+5n+2)$.

**Case 2. *m* even.** Partition $E$ into 2 sets: $E_1 = [(n+1)(2m+1)+2n+1, 2m(n+1)+6n+5]$ and $E_2 = [(2m(n+1)+6n+6, 3m(n+1)+2n]$. For $m = 4$ defined $E = E_1$. According to Lemma 2.2 with $x = (2m+1)(n+1)+2n$, $y = (2m+1)(n+1)+3n+1$, and $z = (2m+1)(n+1)+4n+2$, we obtain $(n+1)$-balanced where $\sum E_{1i} = 6m(n+1)+13n+\lceil\frac{n+1}{2}\rceil+8$.

Then, according to Lemma 2.3 for $x = 2m(n+1)+6n+6$, $y = 3m(n+1)+2n$ and $|E_2| = (n+1)(m-4)$, we have $(n+1)$-balanced where $\sum E_{2i} = \frac{1}{2}(m-4)(5m(n+1)+8n+6)$. Let $E = E_1 \uplus E_2$ for $m \ge 6$, then $\sum E_i = \frac{1}{2}(5m^2(n+1)-6n-2m-8)+\lceil\frac{n+1}{2}\rceil$.

Furthermore, the constant supermagic sum of a subgraph $C_3 \odot P_m$ for $n$ even and $\frac{n}{2}$ even are as follows.

$$f(C_3 \odot P_m) = \begin{cases} \frac{1}{4}(54m^2(n+1)+6m(8n+7)-3n+4\lceil\frac{n-1}{2}\rceil+16), & \text{for } m \text{ odd;} \\ (6m^2+30m)(n+1)+\frac{183n}{4}+3m+\lceil\frac{n-1}{2}\rceil+3\lceil\frac{n+1}{2}\rceil+31, & \text{for } m=4; \\ (\frac{27}{2}m^2+12m)(n+1)-\frac{9n}{4}+\lceil\frac{n-1}{2}\rceil+3\lceil\frac{n+1}{2}\rceil-5, & \text{for } m \text{ even, } m \ge 6. \end{cases}$$

Then, the constant supermagic sum of a subgraph $C_3 \odot P_m$ for $n$ even and $\frac{n}{2}$ odd are as follows.

$$f(C_3 \odot P_m) = \begin{cases} \frac{1}{4}(54m^2(n+1)+6m(8n+7)-n+18), & \text{for } m \text{ odd;} \\ 6m^2(n+1)+m(30n+33)+\frac{185n}{4}+3\lceil\frac{n+1}{2}\rceil+\frac{63}{2}, & \text{for } m=4\ ; \\ \frac{1}{4}(54m^2(n+1)+48m(n+1)-7n+12\lceil\frac{n+1}{2}\rceil-18, & \text{for } m \text{ even, } m \ge 6. \end{cases}$$

And the constant supermagic sum of a subgraph $C_3 \odot P_m$ for $n$ odd as follows.

$$f(C_3 \odot P_m) = \begin{cases} \frac{1}{2}(27m^2(n+1)+3m(8n+7)-2n+11), & \text{for } m \text{ odd;} \\ 6m^2(n+1)+m(30n+33)+\frac{13}{2}(7n+5)+3\lceil\frac{n+1}{2}\rceil, & \text{for } m=4\ ; \\ \frac{1}{2}(27m^2(n+1)+24m(n+1)-5n+6\lceil\frac{n+1}{2}\rceil-7, & \text{for } m \text{ even, } m \ge 6. \end{cases}$$ □

Figure 2 illustrates an example of $C_3 \odot P_3$-supermagic labeling on a $W_6 \odot P_3$ graph.

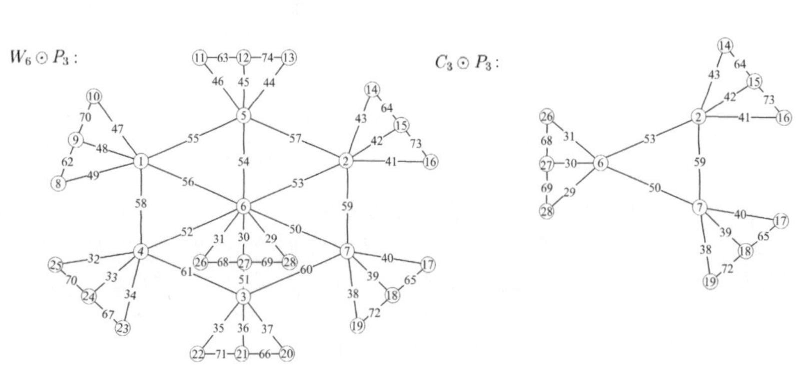


**Figure 2.** A $C_3 \odot P_3$-supermagic labeling on a $W_6 \odot P_3$ graph

## References

[1] Frucht, and F. Harary, *On The Corona of Two Graphs*, Aequationes Math **4** (1970), 322-325.

[2] Gallian, J. A., *A Dynamic Survey of Graph Labeling*, The Electronic Journal of Combinatorics **18**(2015).#DS6

[3] Gutiérrez, A. and A. Lladó, *Magic Coverings*, J. Combin. Math. Combin. Computing **55** (2005), 43-46.

[4] Kusmayadi, T. A., *Graf dan Digraf Eksentrik*, UNS Press, Surakarta, 2011.

[5] Lladó, A. and Moragas, *Cycle-magic Graph*, Discrete Mathematics **307** (2007), 2925-2933.

[6] Maryati, T. K., E. T. Baskoro, and A. N. M Salman, *$P_h$-supermagic Labelings of Some Trees*, J. Combin. Math. Combin. Computing **65** (2008), 182-189.

[7] Maryati, T. K., A. N. M. Salman, E. T. Baskoro, J. Ryan, and M. Miller, *On H-Supermagic Labelings for Certain Shackles and Amalgamations of a Connected Graph*, Utilitas Mathematica **83** (2010), 333-342.

[8] Ngurah A. A. G, A. N. M Salman, and L. Susilowati, *H-supermagic Labelings*, Discrete Mathematics (2010), 1293-1300.

[9] Roswitha, M. and E. T. Baskoro, *H-Magic Covering on Some Classes of Graphs*, The 5*th* International Conference on Research and Education in Mathematics (2012), 135-138.

[10] Roswitha, M., E. T. Baskoro, T. K. Maryati, N. A. Kurdhi, and I. Susanti, *Further Result on Cycle-Supermagic Labeling*, ACKE Int. J. Graphs Comb. **10** (2013), 211-220.

[11] Wallis, W. D., *Magic Graphs*, Birkhäuser, Boston, Basel, Berlin, 2001.

IOP Conf. Series: Journal of Physics: Conf. Series **855** (2017) 012043 doi:10.1088/1742-6596/855/1/012043

# Error Analysis Of Students Working About Word Problem Of Linear Program With NEA Procedure

**D.A. Santoso[1], Farid. A[2], and B. Ulum[3]**
Postgraduate, Semarang State University, Indonesia

E-mail: farid.akhmad01@gmail.com, denni_santoso@yahoo.com

**ABSTRACT.** Evaluation and assessment is an important part of learning. In evaluation process of learning, written test is still commonly used. However, the tests usually do not following-up by further evaluation. The process only up to grading stage not to evaluate the process and errors which done by students. Whereas if the student has a pattern error and process error, actions taken can be more focused on the fault and why is that happen. NEA procedure provides a way for educators to evaluate student progress more comprehensively. In this study, students' mistakes in working on some word problem about linear programming have been analyzed. As a result, mistakes are often made students exist in the modeling phase (transformation) and process skills (process skill) with the overall percentage distribution respectively 20% and 15%. According to the observations, these errors occur most commonly due to lack of precision of students in modeling and in hastiness calculation. Error analysis with students on this matter, it is expected educators can determine or use the right way to solve it in the next lesson.

## 1. Introduction

In designing mathematics learning, there are four important elements, ie who learned what was taught, how teaching, and how to evaluate it [1]. One important element in learning is evaluation. Without a thorough evaluation of the overall learning process will not be up in the process and outcome.

Evaluation has a very important role for learning in general and mathematics in particular. Kartono [4] stated that the measuring is an integral part of the learning process, students' performance-based measurements used to assess the student's ability to apply knowledge. In evaluating the necessary equipment / instrument specific for the purpose of evaluation and information to be extracted can be obtained accurately. Furthermore, Isman [1] states the evaluation and assessment tools play an important role because of that educators can find important information about the achievement of learning objectives and whether the student has successfully achieved or not.

In carrying out the evaluation, any curriculum, a written evaluation (paper and pencil test) is often used because it is relatively economical and relatively fast in the process of doubling and correction. However, the process often just end up in the stage of assessment and evaluation is not exhaustive. Whereas the matter of writing and analyze students' mistakes, we bids get the information that many of the students and the success of the learning process.

One does not go out of the assessment process is error analysis of the work done by the students. According Jabeen [2] the error analysis on student is to provide a deep understanding of the learning process. Errors students is an important tool to diagnose the difficulties experienced by students who require direct handling.

One of the analytical procedure that has been developed is NEA method. NEA analysis method (Newman Error Analysis) developed by Newman [7] and [8]. NEA procedure is designed as a simple diagnostic procedure. Newman [7] and [8], Prakitipong and Nakamura [9] states that when students try to solve the problems about the story written, so students must be able to reach certain levels. The levels are as follows: Level 1 Reading (Reading or Decoding), Level 2 Thorough understanding (Comprehension), Level 3 Modeling (Transformation), Level 4 Skills process (Process skills), and Level 5 Write down the answers (Encoding). When completing word problems there is the possibility of students to undertake processing errors and there were some who made a mistake in writing the final result.

Here are the details of the five stage of the procedure according to the NEA. Phase Reading (decoding), which is the stage where students read and understand the sentence in question. Errors that can occur because the students are not able to read and understand the terms in question, or not knowing what is known in terms of capital. This error characterized by the inability of students to write what they know. Stage of understanding (comprehension), is the next stage, ie when the student decides to understand the problem in question. This error characterized by the inability of students to write what was asked or errors in part on what is known. At this stage of modeling (transformation), the students demanded to be able to use methods, strategies, or the right formula to solve the problems. Error is observed when students are not able to use appropriate means and strategies or mistakes in using what is known into its strategy, it corresponds to a term introduced by Polya which device a plan for solving it. On the stage of process skill is the stage where students solve problems that have been modeled with the rules, procedures, or the appropriate algorithm although it has been able to write a mathematical model according to the questions asked.

The final stage of the procedure is to write down the answers NEA (encoding). This stage is the stage where students write the answers correctly and with the full unit or attribute is supposed to have the answers. Often at this stage of process skills, students have conducted a series of procedures and the exact algorithm, but are not able or less careful in rewriting what was requested or settlement in the matter of the story. Mistakes like these that will be observed in this study. If the last stage has been passed by the students, the students expressed absolutely perfect in solving the problem.

## 2. Problem on this research

Problem will be discussed on this article are as follows.

1) Based on the procedures NEA, what are common error made by the students and how they were distributed?
2) Why do these errors happen?
3) What is the theoretical solutions that can be done to solve it?

## 3. Research purposes

This study aims to provide information about various errors experienced by students in the work on the word problems task and common mistakes that is experienced by students. It is expected that educators or researchers can have this information and determine appropriate solutions to overcome the errors made by the students so that learning process becomes more effective and efficient. However, in this study deeper analysis on the cause or source of error are not included.

## 4. Method

In analyzing the problem in this research, descriptive analysis method is being used. It is used because we tries to convey is what is experienced by the subject of research in the form of behavior, perception, motivation, action, and others, in a holistic manner and by way of description in the form of words and language, in a specific context in which the natural and the using various scientific methods [6].

This study was conducted in SMK 1 Pekalongan in the academic year 2016/2017. The subjects were students in grade 11 as many as 32 students. Data is collected in the subject of mathematic when the topics is "linear programming". The data collection was done by tests, rubrics assessment, interview, and literature pertaining to this study. Rubrics assessment be a type of fault experienced by students and a description of the students' mistakes. Problem given test consists of three items a story that intends to explore the understanding of students. Problem 1, given a narrative and students are required to write / model of the narrative becomes a function constraints and objective function. On question 2, students are given a system of inequality and be required to determine the solution area. Problem 3, students are given a system of inequalities and the objective function in order to determine the optimum solution of system. It is assumed that question 1 is the easiest and 3 are the most difficult problems.

The results of the work students are then analyzed and classified by the NEA procedures. Data types of mistakes students then collected and tabulated to look overall data distribution. Of each type of error, then the error description is being reduced by process includes the selection process, simplification and abstraction in order to obtain a specific description of how poses the occurrence of errors of each type of error types. This reduction activities resulting classification is based on the type of error according to the procedure NEA.

## 5. Research Result

Here is ordered occurrence of mistakes made by the student for each of the items. It can be seen in question 1 that is assumed to be easiest, perfectly turned out students who answered only 19%, while the most common error is an error of type 3 that modeling (transformation). At about 2 and 3, students correctly answered respectively 59% and 41% when the matter is assumed to be more difficult than about 1. For the types of the most common mistakes made students is error type 4 is a skill process (process skill). In general the most common mistakes that students do is type 3 and 4, which experienced a 20% overall and 15% of the students.

Table 1. Distribution of the type of student errors

| **Student** | **question 1** | **question 2** | **question 3** | **Student** | **question 1** | **question 2** | **question 3** |
|---|---|---|---|---|---|---|---|
| 1 | 1 | 5 | 4 | 17 | 3 | 6 | 6 |
| 2 | 0 | 6 | 6 | 18 | 3 | 6 | 6 |
| 3 | 3 | 4 | 4 | 19 | 6 | 5 | 5 |
| 4 | 4 | 6 | 5 | 20 | 2 | 2 | 2 |
| 5 | 6 | 6 | 4 | 21 | 0 | 6 | 2 |
| 6 | 2 | 4 | 6 | 22 | 3 | 6 | 2 |
| 7 | 6 | 6 | 6 | 23 | 6 | 6 | 6 |
| 8 | 4 | 6 | 6 | 24 | 3 | 6 | 6 |
| 9 | 1 | 3 | 6 | 25 | 0 | 2 | 0 |
| 10 | 3 | 6 | 0 | 26 | 3 | 6 | 6 |
| 11 | 6 | 4 | 3 | 27 | 0 | 2 | 3 |
| 12 | 3 | 4 | 4 | 28 | 3 | 6 | 4 |
| 13 | 3 | 5 | 5 | 29 | 2 | 3 | 2 |
| 14 | 3 | 6 | 4 | 30 | 3 | 6 | 6 |
| 15 | 3 | 4 | 4 | 31 | 3 | 6 | 5 |
| 16 | 6 | 6 | 6 | 32 | 3 | 6 | 6 |

Information

0: Students do not work
1: Error phase 1 Reading (Reading or Decoding)
2: Error stage 2 Thorough understanding (Comprehension)
3: Error stage 3 Modeling (Transformation)
4: Error stage 4 Skills process (Process skills)
5: Error stage 5 Write down the answers (Encoding)
6: Done correctly

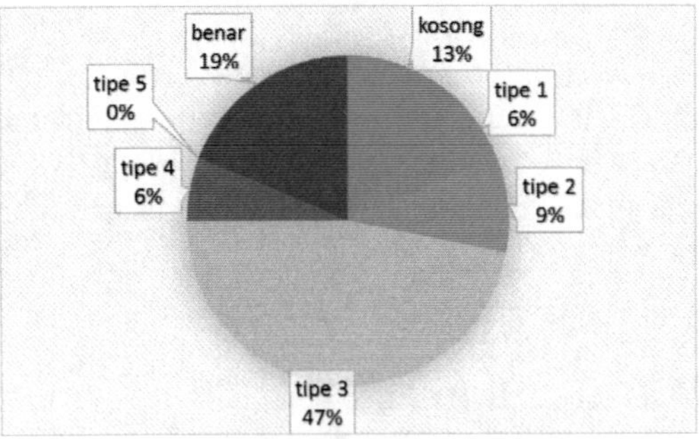


Figure 1

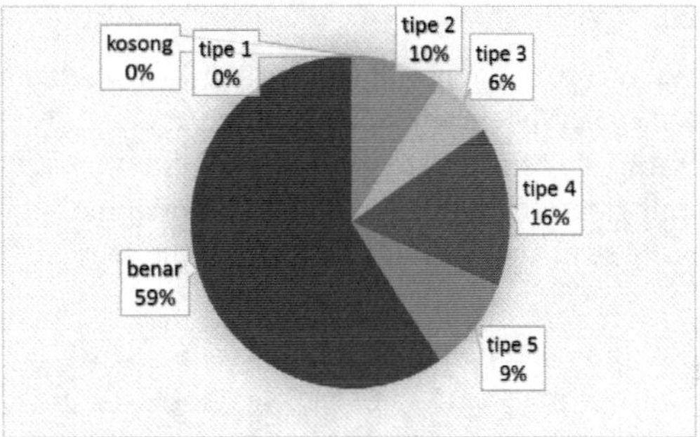


Figure 2

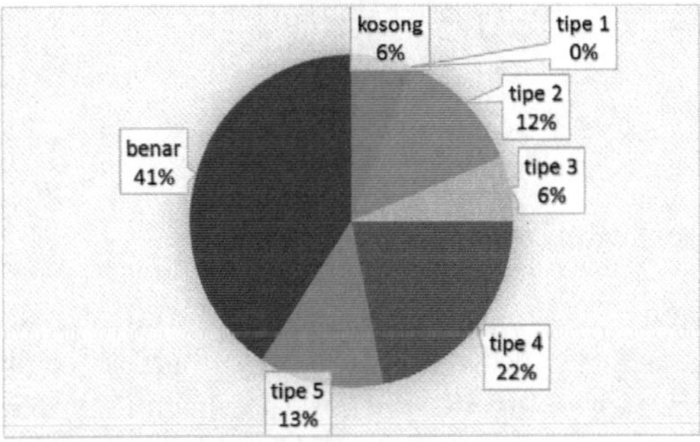


Figure 3

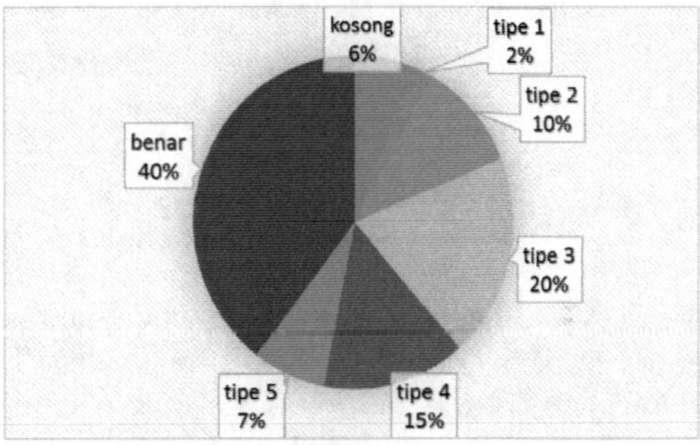


Figure 4

Information
Figure 1. Distribution of errors students to question 1
Figure 2. Distribution of errors students to question 2
Figure 3. Distribution of errors students to question 3
Figure 4. Distribution of errors students to question 4

Generally, error type 1 experienced only by 2% of the students. This error is only observed in question 1 error experienced by many as less conscientious students in reading about or even do not understand at all what is meant in the matter. Although a bit, it turns out, this type of error is experienced by students with higher grades. It shows the student hasty in reading the questions. Examples below are students who are mistaken in classifying information indicating students' mistakes in reading the questions correctly.

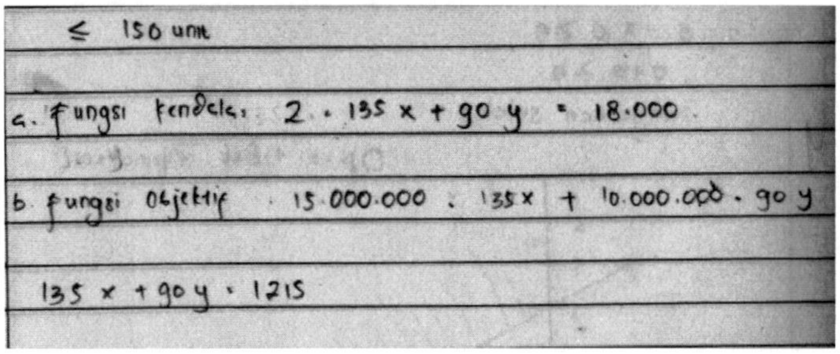


Figure 5. examples of student error type 1

Error type 2 experienced by 10% of students. Although the frequency of these errors is few, but it was observed in all items, which means an error occurs incidentally. In this type of error is generally the student fails to understand the problem as a whole and the students are less careful in sorting information. In the example below, the students reversed in letting variables known. This indicates an error in understanding about the students as a whole.

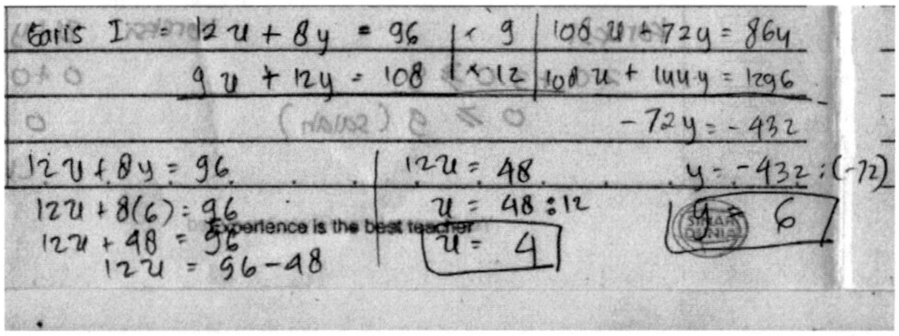


Figure 6. examples of student error type 2

Error type 3 happens most often when the research. This error occurs in a lot of students who have average grade but occur in all questions given. This means that this fault is a major barrier for the students. From the observation, this error is occurred because students are less conscientious, less careful, writing errors, and the rush of time. Specifically at students who have a high value, mistakes often occur because of a typing error. This is unfortunate because if these errors avoided, the work of these students is perfect.

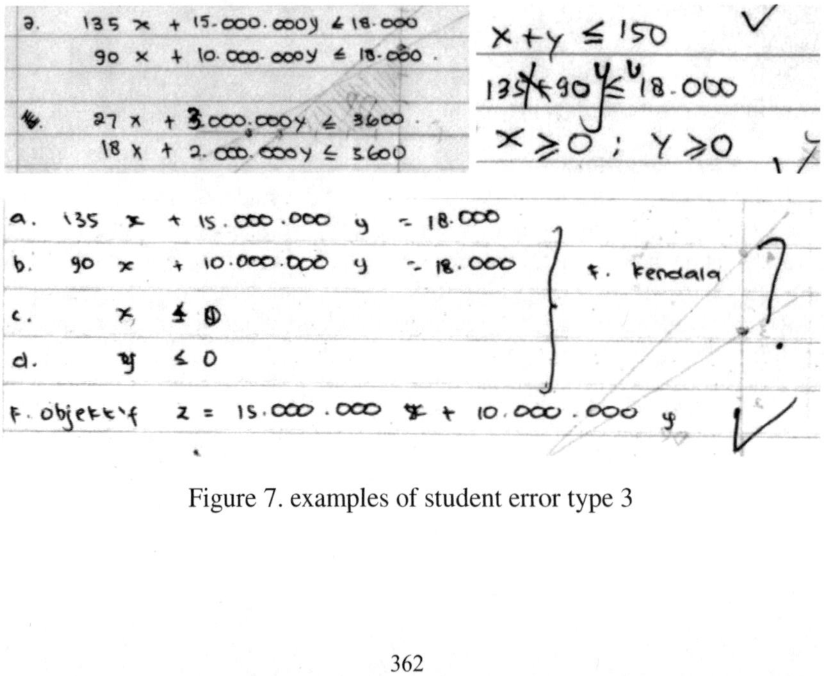


Figure 7. examples of student error type 3

Error type 4 is also very common. This error is observed on all question and performed by the students who have lower, average or higher grade. It is observed that the error occurred because of miscalculations, incorrect use of formula, or use of the inappropriate procedures. This means that the fault is not due to the incident, but it was an error processing of students.

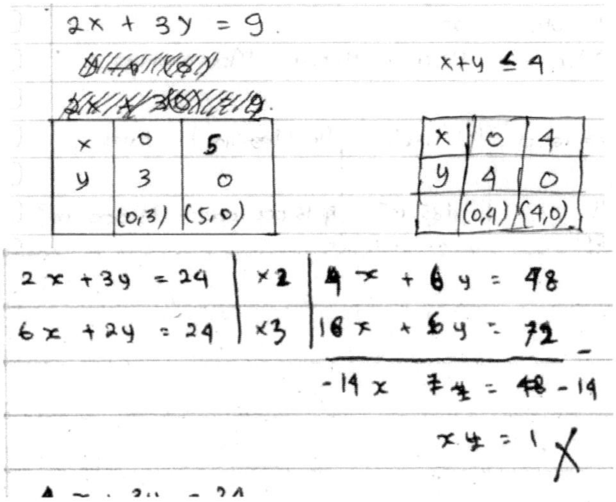

Figure 8. examples of student error type 4

Error type 5 occur quite infrequently with the frequency of occurrence of 7%. This error is very unfortunate because the student has successfully reached the stage of data processing but failed to write the final solution. These errors occur because of clerical errors and less understand students about issues in question.

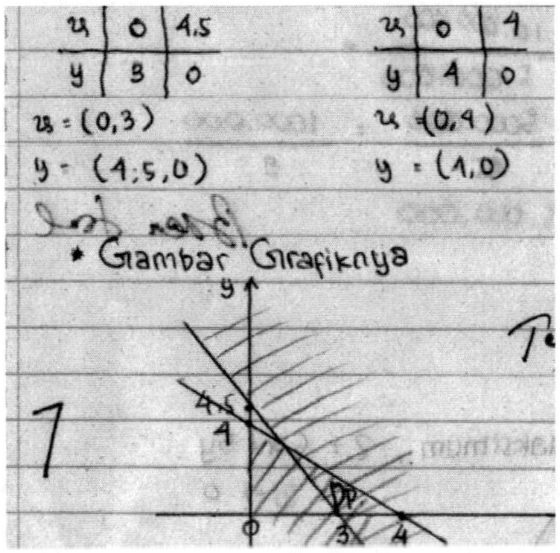


Figure 9. examples of student error type 5

One of the interested thing is the amount of the empty answer more than the type 1 error. The empty answers only contained in question 1, which is assumed about the story is the easiest for little calculation process. Although few, there are students who have a medium value but does not work on the matter. Based on the observations, students are lack of time due to an error in doing another matter that it took time to correct it.

## 6. Discussion

Sequentially, the most common mistakes made by students is 3rd stage that are modeling (transformation), process skills, thorough understanding (comprehension), 5th stage write down the answers (encoding), and phase 1st reading (reading or Decoding). This result is not too different from the research that has been done before in India (Kumar Jha) [5] and Malaysia (Singh) [10] . According to the findings Kumar Jha and Singh, mistake often made on the first three stages. They argued stage reading and comprehension as the most common mistakes. However, this is caused by the use of English as an language of math problems in their research, since in both countries the most widely used language is the local language. So, if the language factor ruled out errors were found to be similar to the findings in this study.

In another study conducted in Indonesia by Junaedi [3], the result is a little different. Junaedi find the most common errors is at the stage of encoding and comprehension. But it is natural to happen considering the studies were conducted at different levels of education. The occurrence of these errors are generally less rigorous because students will be writing, students will be less rigorous calculation, and the haste students in doing. These results are similar to results of research conducted [3].

These findings suggest still weak ability of students to understand and resolve the matter of the story. This shows students' literacy skills are lacking. As stated by Wardono [11] that mathematical literacy is the ability to formulate, implement, and interpret mathematics in a variety of situations including the ability to perform reasoning and use concepts, procedures, and facts to illustrate or explain a phenomenon. In his experiments, Wardono successfully tested the RME as a solution to improve this ability.

## 7. Conclusions and recommendations

Common error made by students consecutively are Transformation, Process skills, Comprehension, Encoding, and Reading (Decoding). Error of reading (decoding) mostly did by lower rank students. Error of comprehension and process skill were observed in entire group of students. Error of transformation and encoding mostly did by higher rank students.

Mistakes made by student are occurred due to hasty behavior, misclassify, misread, and misunderstand about task given. This suggest student are unable to understand the task and solving problem given. As stated above, to increase students understanding and reduce occurrence of error in student, Realistic Mathematics Educations (RME) may be the good solutions.

Deeper observations need to be done in order to know what treatment can reduce each of these errors. We recommend teacher to make an error analysis to understand what is the most error experienced by student and make accurate treatment on that problem.

## References

[1] Isman, Aytekin. 2011. Instructional Design In Education: New Model. *The Turkish Online Journal of Educational Technology*. Vol 10(1). PP: 136-142

[2] Jabeen, Ahmad. 2015. The Role of Error Analysis in Teaching and Learning of Second and Foreign Language. *Education and Linguistics Research*. Vol. 1,(2). PP: 52-61 Macro Institute. doi:10.5296/elr.v1i1.8189

[3] Junaedi, Iwan. 2015. Disclosure Causes of Students Error in Resolving Discrete Mathematics Problems Based on NEA as A Means of Enhancing Creativity. International Journal of Education. Macrothink Institute. Vol 7 (4). PP: 125 – 133

[4] Kartono. 2011. *Efektivitas Penilaian Diri Dan Teman Sejawat Untuk Penilaian Formatif Dan Sumatif Pada Pembelajaran Mata Kuliah Analisis Kompleks*. Prosiding Seminar Nasional Matematika Prodi Pendidikan Matematika, Universitas Muhammadiyah. Surakarta, 24 Juli 2011

[5] Kumar Jha, Shio. 2012. Mathematics Performance of Primary School Students in Assam (India): An Analysis Using Newman Procedure. *International Journal of Computer Applications in Engineering Sciences*. Vol 2(1). PP: 17-22

[6] Moleong, L.J. 2009. *Metodologi Penelitian Kualitatif.* Bandung: Remaja Rosdakarya

[7] Newman, M. A. 1977. An analysis of sixth-grade pupils' errors on written mathematical task. *Victorian Institute for Educational Research Bulletin*. Vol 39. PP: 31-43

[8] Newman, M. A. 1983. *Strategies for diagnosis and remediation*. Sydney: Harcourt, Brace

[9] Prakitipong, Natcha dan Nakamura, Satoshi. 2006. Analysis of Mathematics Performance of Grade Five Students in Thailand Using Newman Procedure. *Journal of International Cooperation in Education*. CICE Hiroshima University

[10] Singh, Parmjit *et al.* 2010. The Newman Procedure for Analyzing Primary Four Pupils Errors on Written Mathematical Tasks: A Malaysian Perspective. *Procedia Social and Behavioral Sciences*. PP: 264-271. doi:10.1016/j.sbspro.2010.12.036

[11] Wardono *et al.* 2015. The Realistic Scientific Humanist Learning With Character Education To Improve Mathematics Literacy Based On Pisa. *International Journal of Education and Research (IJERN)*. Vol 3 (1). PP: 349 – 362

IOP Conf. Series: Journal of Physics: Conf. Series **855** (2017) 012044 doi:10.1088/1742-6596/855/1/012044

# Rudder-roll stabilization using fgs-pid controller for sigma-e warship

**M Y Santoso**[1], **I Munadhif**[2], **A Wahidin**[3], **Ruddianto**[3], **Fathulloh**[3] **and R T Soelistijono**[3]

[1]Safety Engineering Study Program, Shipbuilding Institute of Polytechnic Surabaya, Indonesia

[2]Automation Engineering Study Program, Shipbuilding Institute of Polytechnic Surabaya, Indonesia

[3]Shipbuilding Engineering Study Program, Shipbuilding Institute of Polytechnic Surabaya, Indonesia

E-mail: yusuf.santoso@ppns.ac.id, iimunadhif.its@gmail.com, aangwahidin@gmail.com, ruddianto@yahoo.com, pakfathulloh@gmail.com, rachmad.tri@gmail.com

**ABSTRACT.** The aim of rudder-roll stabilization (RSS) is controlling ship heading and reducing roll motion simultaneously using one actuator, rudder. In this paper, RSS using FGS-PID for SIGMA-e warship was performed, both in normal and disturbed sea conditions. The fuzzy system for determining PID controller are constructed from SIGMA-e linear mathematical model. The wave disturbances are generated based on the WMO. The results show that FGS-PID has superior performance compared to conventional PID controller in heading control and roll damping. It means that the proposed control method can encounter the environmental changes.

## 1. Introduction

*Ship Integrated Geometric Modularity Approach* (SIGMA)-extended, or called SIGMA-e, is a warship designed by Indonesian Research and Technology Consortium in 2012. Some developments of the ship from the original SIGMA class warship are dimension, war reliability, and maneuverability [1]. One of the components of the development is ship automatic steering, or called autopilot.

The aim of rudder-roll stabilization (RSS) are controlling ship heading and reducing roll motion simultaneously using one actuator, i.e. rudder. It is in accordance with the autopilot objective: controlling heading without increasing roll motion. Excessive roll motion is a serious issue for a warship, especially in weapon operation, launching and recovery system, aircraft landing, and sonar operation [2]. Some researches were analyzed this technic [3]-[9]. As the result, roll reduction can be achieved to 50-70%, depends on ship dynamics [10].

Conventional RSS commonly works based on PID or LQR controller with constant parameters [11]. Those controllers give well damping performance for particular condition. However, in bad weather conditions or when the ship speed is changed, the autopilot system with fix parameters does not work well [3].

Recently, the tendency of combining some methods by exploiting individual advantages for achieving better controller is increasing (12-14). Fuzzy Gain Scheduling (FGS-PID) [15] is a nonlinear controller that utilize fuzzy rules to determine PID controller parameters. Fuzzy system will tune the parameters automatically, based on error signal and its derivative. Some studies, [8], [16] and [17], show the utilization of fuzzy system for control in container ship, oil tanker, and multirole naval vessel, respectively.

In this paper, FGS-PID application for RSS in a SIGMA-e warship was performed. First, mathematical model of the ship was derived from common ship mathematical model, regarding to SIGMA-e specification. Then, FGS-PID for the ship was designed. Finally, we discus about the performance of the proposed controller in several weather conditions, compared with conventional PID controller.

## 2. RRS control design

### *2.1 SIGMA-e modeling*

Ship has six degrees of freedom (DOF) in motion, consists of translation motion (position) in three directions: surge, sway, and heave; and rotation motion (orientation) in three axes: roll, pitch, and yaw – see Figure 1.

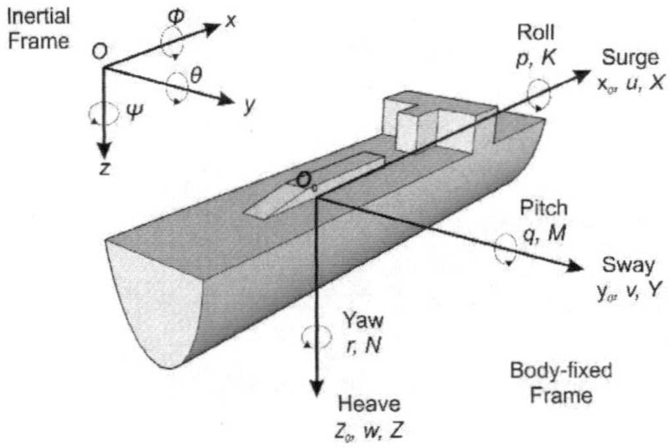


**Figure 1.** Standard notation and sign conventions for ship motion description based on SNAME standard [18].

Nonlinear marine vessel has the general state space form [19]:

$$\dot{x} = H^{-1} f(x,\delta) \tag{1}$$

where $\delta$ is the rudder angle,

$$x = \begin{bmatrix} u & v & p & r & \phi & \psi \end{bmatrix}^T \tag{2}$$

$$H = \begin{bmatrix} (m - X_{\dot{u}}) & 0 & 0 & 0 & 0 & 0 \\ 0 & (m - Y_{\dot{v}}) & -(mz_G + Y_{\dot{p}}) & (mx_G - Y_{\dot{r}}) & 0 & 0 \\ 0 & -(mz_G + K_{\dot{v}}) & (I_{xx} - K_{\dot{p}}) & -K_{\dot{r}} & 0 & 0 \\ 0 & (mx_G - N_{\dot{v}}) & -N_{\dot{p}} & (I_{zz} - N_{\dot{r}}) & 0 & 0 \\ 0 & 0 & 0 & 0 & 1 & 0 \\ 0 & 0 & 0 & 0 & 0 & 1 \end{bmatrix} \tag{3}$$

$$f(x,\delta) = \begin{bmatrix} X^*_{hyd}(x) + X_\delta + m\left(vr + x_G r^2 - z_G pr\right) \\ Y^*_{hyd}(x) + Y_\delta - mur \\ K^*_{hyd}(x) + K_\delta + mz_G ur \\ N^*_{hyd}(x) + N_\delta - mx_G ur \\ p \\ r \end{bmatrix} \tag{4}$$

For analyzing the linearized model, it is a common practice to decouple the surge equation from the others. Thus, the reduced state vector is $x = \begin{bmatrix} v & p & r & \phi & \psi \end{bmatrix}^T$. The linearized models are obtained straightforward from (1) as:

$$\dot{x} = H^{-1}\left[\frac{\partial f(z,u,\delta)}{\partial z}\bigg|_{\bar{z},\bar{u},\bar{\delta}} x + \frac{\partial f(z,u,\delta)}{\partial z}\bigg|_{\bar{z},\bar{u},\bar{\delta}} \delta\right] = H^{-1}[Ax + B\delta] \tag{5}$$

where the $A$ and $B$ are given in (6) and (7)

$$A = \begin{bmatrix} Y_{|u|v}|u| & Y_{|u|p}|u| & -mu + Y_{ur}u & Y_{uu\phi}u^2 & 0 \\ K_{|u|v}|u| & K_{up}u + K_p & K_{ur}u & W\overline{GM}_T + K_{uu\phi}u^2 & 0 \\ N_{|u|v}|u| & 0 & N_{|u|r}|u| - mx_G u & N_{|u|u\phi}|u|u & 0 \\ 0 & 1 & 0 & 0 & 0 \\ 0 & 0 & 1 & 0 & 0 \end{bmatrix} \tag{6}$$

$$\boldsymbol{B} = \begin{bmatrix} \boldsymbol{Y}_{ext} \\ \boldsymbol{K}_{ext} \\ \boldsymbol{N}_{ext} \\ 0 \\ 0 \end{bmatrix} \tag{7}$$

where:

$u^2$ and $|u|u$ : the square term of the surge speed which represent the forces and moments associated with the roll motion,

$Y_{ext}$, $K_{ext}$, $N_{ext}$ : possible contribution from external disturbance, rudders, propellers, bow thrusters and other devices.

Based on ship specification given in Table 1, it can be derived the mathematical model for SIGMA-e warship as:

$$\dot{x} = \begin{bmatrix} -0.066422 & 0.014261 & 0.666875 & 9.631995\times10^{-3} & 0 \\ 0.023389 & -0.139478 & -0.261273 & -0.168492 & 0 \\ 2.00943\times10^{-4} & 2.779796\times10^{-6} & 0.01108 & 2.889237\times10^{-4} & 0 \\ 0 & 1 & 0 & 0 & 0 \\ 0 & 0 & 1 & 1 & 0 \end{bmatrix} \tag{8}$$

**Table 1.** SIGMA-e specification

| No. | Parameters | Value |
|---|---|---|
| 1 | *Overall Length of the Ship* ($L_{OA}$) | 106 m |
| 2 | *Length of the Waterline* ($L_{WL}$) | 101,07 m |
| 3 | *Breadth* (B) | 14 m |
| 4 | *Mass of Ship* (m) | 2423 Ton |
| 5 | *Velocity* (U) | 15,43 m/s |
| 6 | *Draught* (D) | 3,7 m |
| 7 | *Depth* | 8,75 m |
| 8 | *Sea Water Density* ($\rho$) | 1014 Kg/m$^3$ |
| 9 | *Coefficient Block* (CB) | 0,441 |

*2.2 Rudder model*

As an actuator, the common widely used rudder model can be shown in [20]. Figure 2 shows the block diagram of rudder model, represent:

- Magnitude saturation: the move of rudder is limited within specific maximum angles, *i.e.* $-\delta_{\max} \le \delta \le \delta_{\max}$ $(rad)$,
- Slew rate saturation: the rudder rate motion is limited by a maximum value $\dot{\delta}_{\max}(rad/\sec)$
- Time delay: the main servo generates most of delay between the rudder command $\delta_c$ and the actual rudder angle $\delta$.

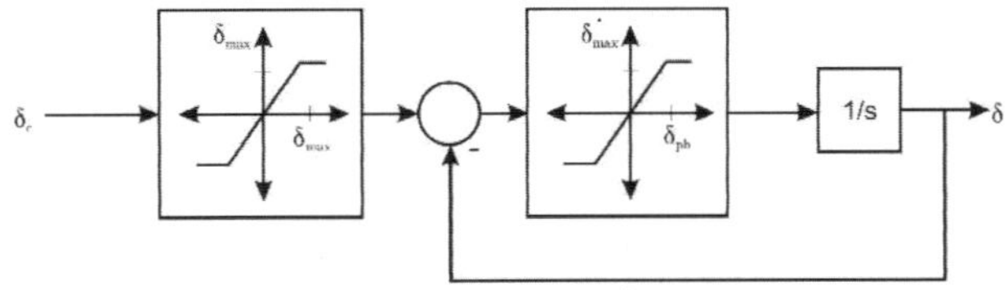


**Figure 2.** Simplified block diagram of the steering machine model [20].

*2.3 Wave model*

Sea waves are the main cause of roll motion. Waves can be described using frequency spectrum, e.g., Pierson-Moskowitz (PM) spectrum [21]. Linear approximation is commonly used by ship control engineer, because its simplicity and applicability. The approximation from PM spectrum can be determined by:

$$y(s) = h(s)\omega(s) \tag{9}$$

where $\omega(s)$ is a zero-mean Gaussian white noise with power spectrum and $h(s)$ is a transfer function to be determined.

A second order transfer function with damping for PM spectrum waves model was introduced in [22], and is written as:

$$h(s) = \frac{K_{\omega}s}{s^2 + 2\zeta\omega_0 s + \omega_0^2} \tag{10}$$

Where $K_{\omega} = 2\zeta\omega_0\sigma_{\omega}$ is gain constant, $\sigma_{\omega}$ is a constant that represent the wave intensity, $\zeta$ is damping coefficient, and $\omega_0$ is dominant wave frequency.

For PM spectrum, the dominant frequency spectrum can be defined as:

$$\omega_0 = 0.88\frac{g}{v} = 0.4\sqrt{\frac{g}{H_3}} \tag{11}$$

where $H_3 = 0.21V^2/g$ is the significant wave height in wind speed *V*.

*2.4 RSS with FGS-PID*

RSS utilizes the rudder for roll stabilization, which is controlling the ship heading and reducing its roll motion simultaneously. Figure 3 depicts a decoupled control system design can be used to obtain roll damping. FGS-PID is used for the yaw controller and roll damper.

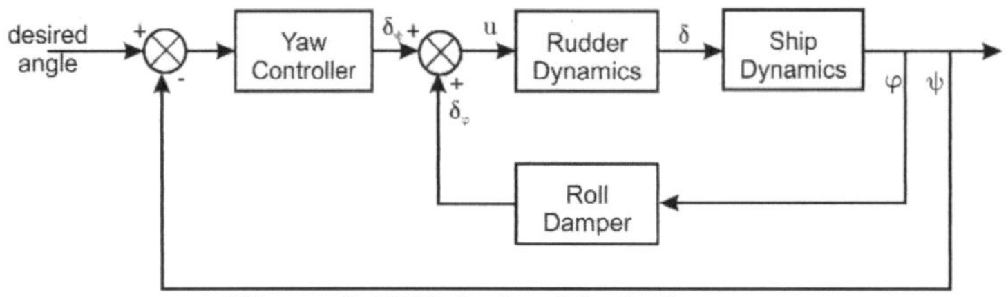


**Figure 3.** RSS design block diagram.

One of the methods which applied for adjusting gain values automatically based on environment conditions is FGS-PID. This method merging the advantages of fuzzy system in terms of adaptation and PID controller in terms of simplicity in control. The fuzzy system utilizes rule bases for tuning the three PID parameters. There are three constructed fuzzy system, i.e., fuzzy *Kp'* to get *Kp* value, fuzzy *Kd'* to get *Kd* value, and fuzzy alpha ($\alpha$) to get *Ki* value. Figure 4 shows input membership functions for fuzzy system, while Figure 5 depicts the outputs.

$$\mu_{small}(x) = -\frac{1}{4}\ln x \tag{12a}$$

$$\mu_{big}(x) = -\frac{1}{4}\ln(1-x) \tag{12b}$$

where $\mu$ is value of membership function, $0 \le \mu \le 1$.

In fuzzification process, the membership function for inputs is triangular membership function which consists of seven membership functions: N represents *negative*, P for *positive*, ZO for *approximately zero*, S for *small*, M for *medium*, and B for *big*. Then, NM represents *negative medium*, PB for *positive big*, and so on.

For the output, *fuzzy* $K_p$*'* and *fuzzy* $K_d$*'* have two membership function: *Big* and *Small*, defined in (12). Figure 2.9 shows the membership functions for *fuzzy* $K_p$*'* and *fuzzy* $K_d$*'*. While for *fuzzy* $\alpha$, the membership functions are singleton membership functions consists of *Small* (M), *Medium Small* (MS), *Medium* (M) and *Big* (B).

The value of $K_p'$, $K_d'$, and $\alpha$ are determined by a set of fuzzy rules. For the *i*th rule, it is:

if $e(t)$ is $A_i$ and $\Delta e(t)$ is $B_i$, then $K_p'$ is $C_i$, $K_d'$ is $D_i$, and $\alpha = \alpha_i$ (13)

where $i = 1,2,\ldots m$.

$e(t)$ is error signal, i.e. difference between the set-point and the measured value

$\Delta e(t)$ is error rate.

The rule base may be resulted from a desired process time response. For instance, if a big control signal must be generated to achieve a fast rise time, the PID controller should have a large proportional gain, a large integral gain, and a small derivative gain for generating a big control signal. As a result, $K_p'$ is set to Big, $K_d'$ is set to Small, and $\alpha$ is set to 2 or S. Another example, to avoid a large overshoot, a small control signal has to be produced. Then, $K_p'$ is set to Small, $K_d'$ is set to Big, and $\alpha$ is set to B. This rule represents a small proportional gain, a large derivative gain, and a small integral gain. Thus, the tuning rules for $K_p'$, $K_d'$, and $\alpha$ are given in Table 2.

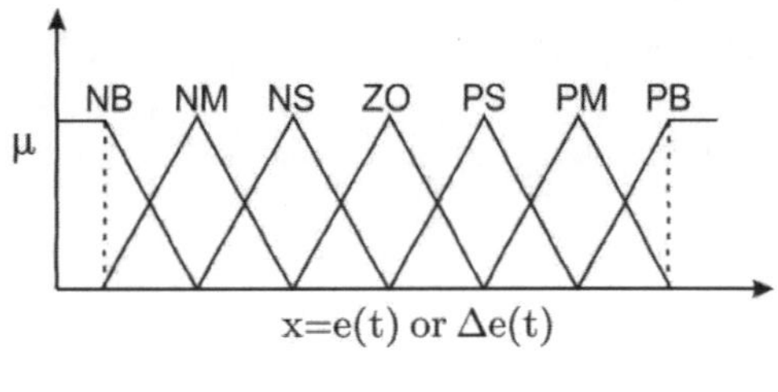


**Figure 4.** Fuzzy input membership function [15].

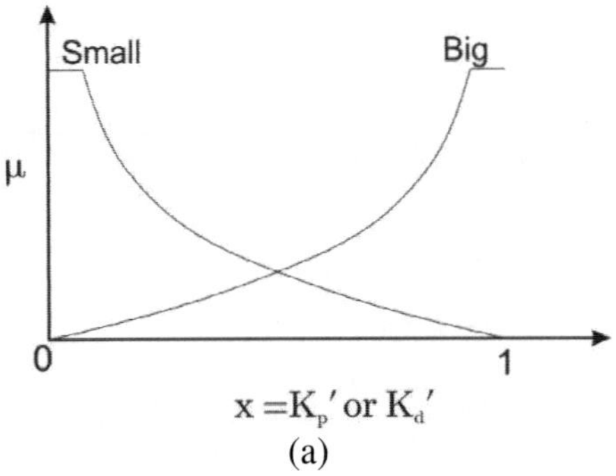


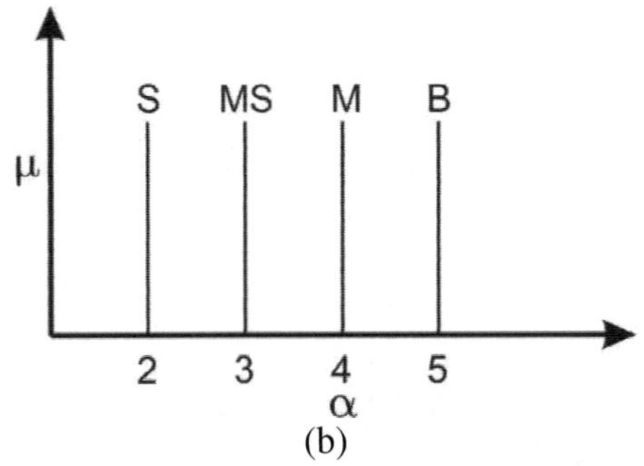


**Figure 5.** Fuzzy output membership function: (a) for *Kp'* and *Kd'*, (b) for *α* [15]

**Table 2.** Rule base for *Kp'*, *Kd'* and *α* [15]

| ***K$_p$', K$_d$', α*** | | ***error rate, Δe(t)*** | | | | | | |
|---|---|---|---|---|---|---|---|---|
| | | **NB** | **NM** | **NS** | **ZO** | **PS** | **PM** | **PB** |
| | **NB** | B,S,2 | B,S,2 | B,S,2 | B,S,2 | B,S,2 | B,S,2 | B,S,2 |
| | **NM** | S,B,3 | B,B,3 | B,S,2 | B,S,2 | B,S,2 | B,B,3 | S,B,3 |
| | **NS** | S,B,4 | S,B,3 | B,B,3 | B,S,2 | B,B3 | S,B,3 | S,B,4 |
| ***Error, e(t)*** | **ZO** | S,B,5 | S,B,4 | S,B,3 | B,B,3 | S,B,3 | S,B,4 | S,B,5 |
| | **PS** | S,B,4 | S,B,3 | B,B,3 | B,S,2 | B,B3 | S,B,3 | S,B,4 |
| | **PM** | S,B,3 | B,B,3 | B,S,2 | B,S,2 | B,S,2 | B,B,3 | S,B,3 |
| | **PB** | B,S,2 | B,S,2 | B,S,2 | B,S,2 | B,S,2 | B,S,2 | B,S,2 |

When the *Kp'*, *Kd'* and *α* are obtained, the PID parameters for FGS-PID can be determined using:

$$K_p = \left(K_{p.\max} - K_{p.\min}\right)K_p' + K_{p.\min} \tag{14a}$$

$$K_d = \left(K_{d.\max} - K_{d.\min}\right)K_d' + K_{d.\min} \tag{14b}$$

$$K_i = K_p^2 / \left(\alpha K_d\right) \tag{14c}$$

The ranges of $K_p$ and $K_d$ in (14) are determined using the rule:

$$K_{p.\min} = 0.32K_u\,,\quad K_{p.\max} = 0.6K_u$$

$$K_{d.\min} = 0.08K_uT_u\,,\quad K_{d.\max} = 0.15K_uT_u \tag{15}$$

where $K_u$ and $T_u$ are critical gain and period from oscillation when the P controller is applied to the system.

## 3. Main Results

In the following analysis, we consider the maneuverability of the ship and the implementation of FGS-PID to SIGMA-e warship in normal sea and disturbed sea conditions. There are three different disturbed sea conditions based on World Meteorological Organization (WMO), i.e., calm (rippled), moderate and rough sea. The results are simulated using block diagram code in Matlab Simulink®.

In order to find out the ship maneuverability, a turning circle test based on International Maritime Organization (IMO) Resolution MSC 137 (76) [23] was conducted. The test is applied to either port side and starboard side of the ship. Based on figure 6, it can be said that the designed ship complies with the standard. The tactical diameter of the turning circle is less than IMO criteria.

IOP Conf. Series: Journal of Physics: Conf. Series **855** (2017) 012044 doi:10.1088/1742-6596/855/1/012044

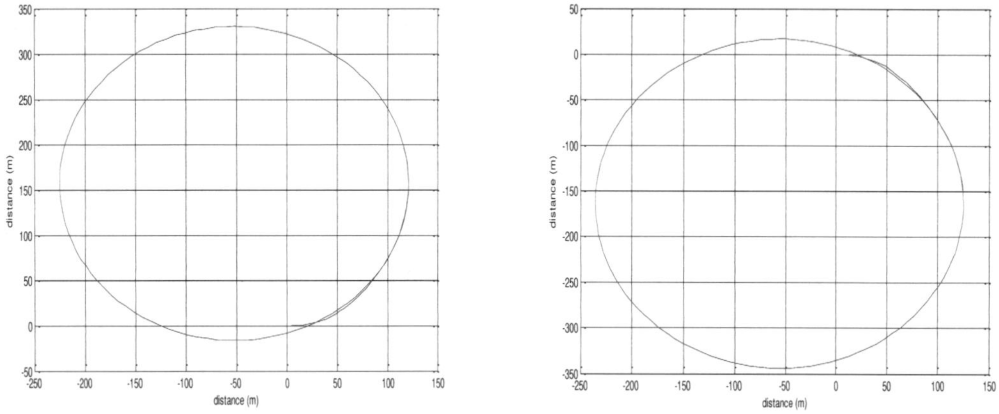


(a) (b)

**Figure 6.** Turning circle test result: (a) port side, (b) starboard side

Figure 7 and 8 depicts the simulation results in normal sea condition of yaw and roll motion, respectively. It can be shown that FGS-PID has better performance than conventional PID controller, in terms of overshoot and settling time. It corresponds to the data that given in Table 3.

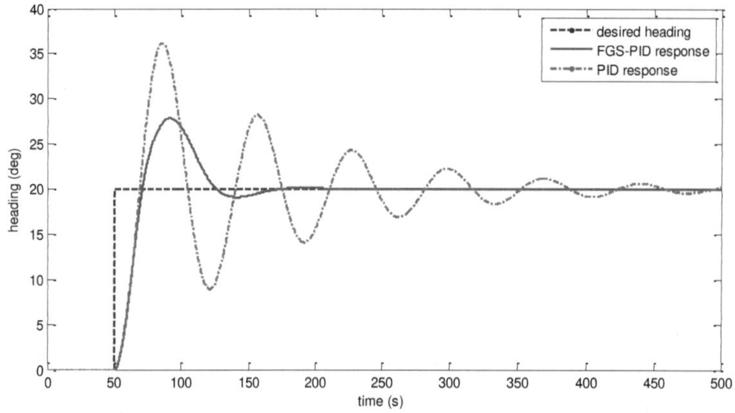


**Figure 7.** Yaw response for normal sea condition

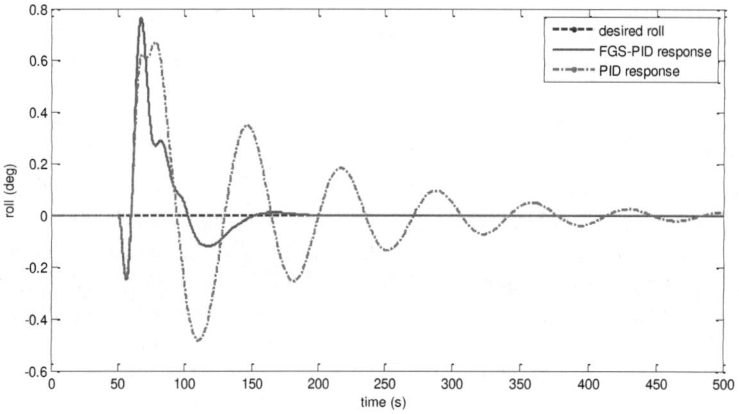


**Figure 8.** Roll response for normal sea condition

The simulation results in moderate sea condition for yaw and roll motion are depicted in Figure 9 and Figure 10, respectively. For calm (rippled) and rough sea condition, they have similar profile result with moderate condition. Calm (rippled) sea condition result has less ripples in the profile, while the rough sea condition result has more. Table 3 gives the detailed performances of controllers for all environment conditions.

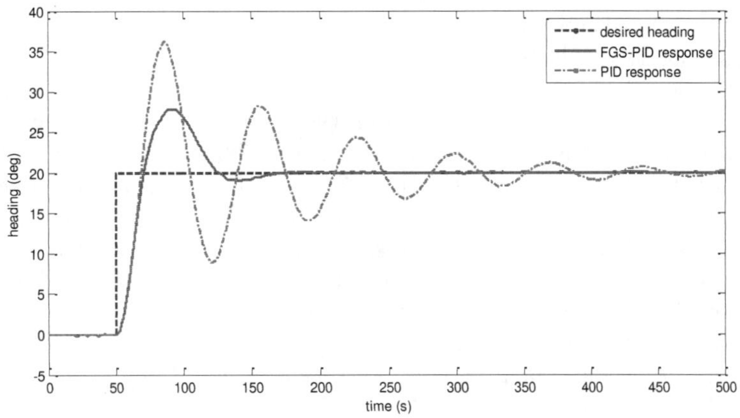


**Figure 9.** Yaw response for moderate sea condition

**Table 3.** Detailed simulation results

| Sea condition | Controller | Yaw maximum overshoot (deg) | Yaw settling time (s) | Roll RMSE |
|---|---|---|---|---|
| Normal | FGS-PID | 27.8555 | 120.5000 | 0.1129 |
| | PID | 36.1191 | 375.6000 | 0.1890 |
| Calm (rippled) | FGS-PID | 27.8559 | 120.5000 | 0.1130 |
| | PID | 36.1186 | 375.5000 | 0.1891 |
| Moderate | FGS-PID | 27.8299 | 120.6000 | 0.1139 |
| | PID | 36.1730 | 376.4000 | 0.1899 |
| Rough | FGS-PID | 27.8901 | 144.0000 | 0.1146 |
| | PID | 36.0530 | 371.1000 | 0.1906 |

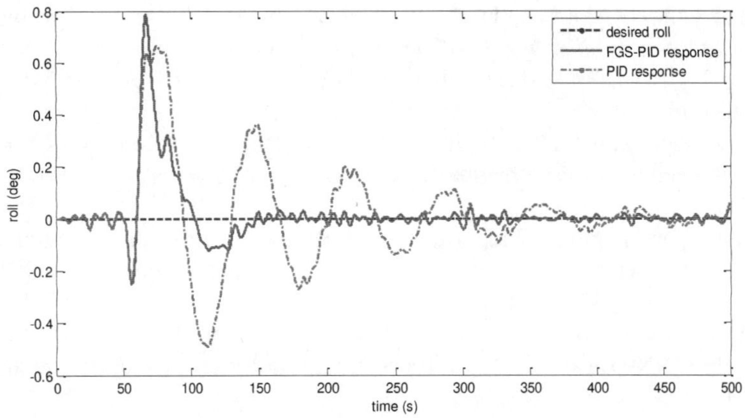


**Figure 10.** Roll response for moderate sea condition

## 4. Conclusions

In this paper simulation for applying FGS-PID to RSS of SIGMA-e warship is conducted. FGS-PID has superior performance in heading control and roll damping than conventional PID controller, either in normal or disturbed sea condition. It is shown that FGS-PID has more ability to adapt to the environmental changes compared to the conventional PID controller.

**Acknowledgment**

The authors would like to thank Shipbuilding Institute of Polytechnic Surabaya for funding to this research in the academic year of 2016.

**References**

[1] Akbar R, Aisjah A S 2014 Realtime Least Square Estimator Design for Sigma Class Extended 3 Meter Scale Model *National Maritime Seminar IX Hang Tuah University (Surabaya, Indonesia, 2014)*
[2] Perez T and Blanke M 2012 Ship roll damping control *Annual Reviews in Control* **36-1** 129-147
[3] Amerongen J and Lemke V N 1978 Optimum steering of ships with an adaptive autopilot *Fifth Ship Control Systems Symposium (Annapolis, USA)*
[4] Lauvdal T and Fossen T I 1998 Rudder roll stabilization of ships subject to input rate saturation using a gain scheduled control law *Conference on Control Applications in Marine Systems (CAMS'98) (Fukuoka, Japan)* 121-127
[5] Sgobbo J N and Parsons M 1999 Rudder/fin roll stabilization of the USCG WMEC 901 class vessel *Marine Technology and SNAME News* **36-3** 157-170
[6] Goodwin G, Perez T, Seron M, and Tzeng C Y 2000 On the fundamental limitations for rudder roll stabilization of ships *39th IEEE Conference on Decision and Control (Sydney, Australia)* 4705-4710
[7] Chun H H, Chun S H, and Kim S Y 2001 Roll damping characteristics of a small fishing vessel with a central wing *Ocean Engineering* **28-12** 1601–1619
[8] Santoso M Y, Su S F, Aisjah A S 2013 Nonlinear Rudder Roll Stabilization using Fuzzy Gain Scheduling – PID Controller for Naval Vessel *Proceedings of 2013 International Conference on Fuzzy Theory and Its Application (Taipei, Taiwan)* 94-99
[9] Munadhif I, Aisjah A S, and Masroeri A A 2015 Perancangan Sistem Kendali Kestabilan *Rolling* Kapal Perang Kelas SIGMA Saat Bermanuver Menggunakan *Fuzzy Gain Scheduling* - PID *Seminar Nasional Teknologi Informasi dan Multimedia 2015 (Yogyakarta, Indonesia)* 4.3-15 – 4.3-18
[10] Alarcin F and Gulez K 2007 Rudder roll stabilization for fishing vessel using neural network approach *Ocean Engineering* **34-12** 1811–1817
[11] van der Klugt P G M 1987 Rudder roll stabilization *Delft: Delft University of Techonlogy PhD Thesis Report*
[12] Syai'in M and Soeprijanto A 2010 Neural network optimal power flow (NN-OPF) based on IPSO with developed load cluster method *World Academy of Science, Engineering and Technology* **72** 48-53
[13] Syai'in M and Soeprijanto A 2012 Regular paper combination of generator capability curve constraint and statistic-fuzzy load clustering algorithm to improve NN-OPF performance *J. Electrical Systems* **8-2** 198-208
[14] Syai'in M, Soeprijanto A, and Yuniarno, E M 2011 New Algorithm for Neural Network Optimal Power Flow(NN-OPF) including Generator Capability Curve Constraint and Statistic-Fuzzy Load Clustering *International Journal of Computer Applications* **36-7**
[15] Zhao Z Y, Tomizuka M, and Isaka S 1993 Fuzzy gain scheduling of PID controllers *IEEE Transactions on System, Man, and Cyberetics* **23-5** 1392-1398
[16] Nejim S 2000 Rudder roll damping system for ships using fuzzy logic control *OCEANS 2000 MTS/IEEE Conference and Exhibition (Providence, RI)*1137-1143

[17] Aisjah A S, Masroeri A A, Efendi M A, Djatmiko E B, Ariyawan W D, and Iskandariyanto F A 2012 Fuzzy Control for Optimizing Ship Tracking in Karang Jamuang – Tanjung Perak *The Journal for Technology and Science* **23-4** 118-125

[18] SNAME 1950 Nomenclature for treating the motion of a submerged body through a fluid *(New York, USA)*

[19] Perez T and Blanke M 1988 Mathematical Ship Modeling for Control Application *Technical University of Denmark, Technical Report*

[20] van Amerongen J 1982 Adaptive steering of ships - A model reference approach to improved maneouvering and economic course keeping *Delft University of Techonlogy PhD Thesis Report*

[21] Pierson W J and Moskowitz L 1963 A proposed spectral form for fully developed wind seas based on the similarity theory of S. A. Kitaigorodsku *Technical Report U.S. Naval Oceanograhic Office Contract 62306-1042*

[22] Goodrich G 1969 Development and design of passive roll stabiliser *Transactions of The Royal Institution of Naval Architects* **111** 81–88

[23] International Maritime Organization (IMO) 2002 *Resolution: Standard for Ship Maneuverability* **MSC.137-76**

IOP Conf. Series: Journal of Physics: Conf. Series **855** (2017) 012045 doi:10.1088/1742-6596/855/1/012045

# The Preventive Control of a Dengue Disease Using Pontryagin Minimum Principal

**Eminugroho Ratna Sari, Nur Insani, Dwi Lestari**

Mathematics Education Department, Mathematics and Natural Sciences Faculty, Yogyakarta State University, Indonesia

eminugrohosari@gmail.com, nurinsani.utomo@gmail.com, dwilestari.math@gmail.com

**Abstract.** Behaviour analysis for host-vector model without control of dengue disease is based on the value of basic reproduction number obtained using next generation matrices. Furthermore, the model is further developed involving a preventive control to minimize the contact between host and vector. The purpose is to obtain an optimal preventive strategy with minimal cost. The Pontryagin Minimum Principal is used to find the optimal control analytically. The derived optimality model is then solved numerically to investigate control effort to reduce infected class.

## 1. Introduction

In the host-vector epidemic model, the population is divided into the host, in this case humans, and the vector, the Aedes aegypti mosquito as the carrier of the dengue virus. Research on the dynamic behaviour of the spread of dengue fever using a host-vector model by assuming that there is immunity after recovery has been performed by Esteva [1]. But in reality, dengue fever is caused by the dengue virus that has four serotypes, and a person can be infected by dengue fever again with a different serotype. The mathematical model that addresses this has been done by Soewono and Apriatna [2]. Their model has assumed of becoming susceptible again after recovering from a dengue fever, but has not considered the possibility of infected hosts that cannot spread the dengue fever.

Therefore, this paper discusses the extension of a dengue fever model where after recovery a host can be susceptible again. In addition, it is also assumed that there are some infected hosts but cannot spread the dengue fever. The proposed model (i.e. the first proposed model) and then will be analysed on its behaviour based on the value of the basic reproduction number ( [3], [4], [5]).

Furthermore, the first model will be expanded by adding a control variable. Research on the optimal control using control insecticide has been carried out by Rodrigues [6]. The results show that the control must be done repeatedly for the best result. But this leads to the increasing of the cost control. Dumont's research result [7] shows that using a combination treatment with chemicals, larvae control and adult mosquitoes can decrease the epidemic. Of course it is not easy to be done as it requires more data.

On the other hand, WHO [8] brings out some strategies to resolve the epidemic by prevention and control. Therefore, this paper also discusses the behaviour of our first proposed model if we give a preventive control variable to the model (i.e. second proposed model). This second model then will be analysed analytically by Pontryagin Minimum Principle. The behaviour of the first and second model (before and after being given controls) will be done numerically. At the end, we obtain the most effective control recommendations to minimize the dengue epidemic.

## 2. Mathematical Model of Dengue Disease without Control

The host population is divided into susceptible class, infectious class and recoverable (immune) class. Vector population are divided into susceptible class and infectious class. There is no recoverable (immune) class in the vector population as mosquitoes will always carry a dengue virus in their body for all their life. The spread of the dengue fever starts from the infected host is bitten by the vulnerable mosquito, resulting in the transmission of dengue virus to the mosquito, so the mosquito becomes infected. If the infected mosquito then bites the vulnerable host, then there will be a transmission which result on the host becomes infected. And so on.

Assume $S_H(t)$ represents the host population on the vulnerable class at $t$, $I_H(t)$ represents the host population for the infectious class at $t$, $R_H(t)$ represents the host population for the recoverable class at $t$, $S_V(t)$ denotes the vector population for the vulnerable class at $t$, and $I_V(t)$ denotes the vector population for infectious class at $t$. The population total for the host and the vector is denoted by $N_H$ and $N_V$, respectively, and assumed they have a constant value.

The birth and death rate in the host population are assumed to be equal, denoted by $\mu_H$. If $b$ denotes the number of mosquito bites per day, then for $N_V$ mosquitoes there will be will $bN_V$ mosquito bites per day. This means that the host will receive as much as $\frac{bN_V}{N_H}\frac{I_V}{N_V}$ infected mosquito bite per day. If $\beta_H$ denotes the transmission rate from the mosquito to the host, then the infection rate of the susceptible hosts is defined as $\beta_H\frac{bN_V}{N_H}\frac{I_V}{N_V}$ or $\beta_H\frac{bI_V}{N_H}$. We assumed that the recoverable host can be susceptible again to the dengue fever, and $\alpha$ denotes this state. The recovery rate is denoted by $\gamma_H$.

The growth rate for the population vector is denoted by $A$ and the mortality rate for the vector is denoted by $\mu_V$. As one mosquito will bite $\frac{b}{N_H}$ per day per person, then one mosquito will get the virus by $\frac{b}{N_H}I_H$ per day. If $\beta_V$ denotes the transmission rate from the host to the mosquito, the infection rate per susceptible mosquito is $\beta_V\frac{b}{N_H}I_H$. Furthermore, in this first proposed model, we also calculate the percentage of the infected hosts that cannot spread the dengue fever ($\varepsilon$). It means there will be $(1-\varepsilon)$ infected hosts that can spread the dengue fever. Thus, we obtain the infection rate per susceptible mosquito as $\beta_V\frac{b}{N_H}(1-\varepsilon)I_H$.

The following Fig.1 is the transmission diagram for the spread of the dengue fever by taking all the above assumptions.

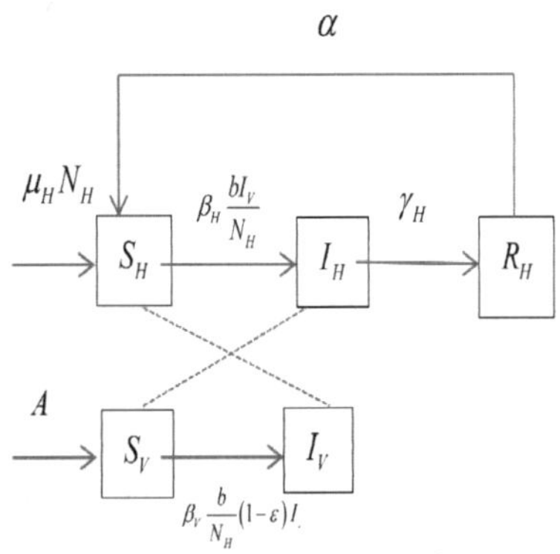


Figure 1. Dengue Fever Transmission Diagram

The mathematics model for the dengue fever transmission is given as follow.

$$\frac{dS_H(t)}{dt} = \mu_H N_H - \frac{\beta_H b S_H I_V}{N_H} - \mu_H S_H + \alpha R_H \quad (1a)$$

$$\frac{dI_H(t)}{dt} = \frac{\beta_H b S_H I_V}{N_H} - \mu_H I_H - \gamma_H I_H \quad (1b)$$

$$\frac{dR_H(t)}{dt} = \gamma_H I_H - \mu_H R_H - \alpha R_H \quad (1c)$$

$$\frac{dS_V(t)}{dt} = A - \frac{\beta_V b S_V (1-\varepsilon) I_H}{N_H} - \mu_V S_V \quad (1d)$$

$$\frac{dI_V(t)}{dt} = \frac{\beta_V b S_V (1-\varepsilon) I_H}{N_H} - \mu_V I_V \quad (1e)$$

where $S_H + I_H + R_H = N_H$ and $S_V + I_V = N_V$. As $\mu_V$ denotes the mortality transmission for the vector, thus the total of mortality for vector population is denotes by $\mu_V N_V$. In the other hand, the rate of change of the vector population is denoted by $\frac{dN_V}{dt}$, which is the recruitment rate subtracted by the total of mortality, means that $\frac{dN_V}{dt} = A - \mu_V N_V$. Hence, for $t \to \infty$, the solution of $N_V$ towards $\frac{A}{\mu_V}$

We obtain $S_V + I_V = \frac{A}{\mu_V}$.

Let the proportion for each class is denoted by

$$x_1 = \frac{S_H}{N_H}, x_2 = \frac{I_H}{N_H}, x_3 = \frac{R_H}{N_H}, \quad x_4 = \frac{S_V}{A/\mu_V}, \text{and } x_5 = \frac{I_V}{A/\mu_V}$$

and given that $x_3 = 1-(x_1 + x_2), x_4 = 1 - x_5$. Equation (1a) can be written as

IOP Conf. Series: Journal of Physics: Conf. Series **855** (2017) 012045 doi:10.1088/1742-6596/855/1/012045

$$\frac{1}{N_H}\left(\frac{dS_H(t)}{dt}\right)=\frac{1}{N_H}\left(\mu_H N_H-\frac{\beta_H b S_H I_V}{N_H}-\mu_H S_H+\alpha R_H\right)$$

$$\Leftrightarrow \frac{dx_1}{dt}=\mu_H-\frac{\beta_H b \, {}^{A}\!/_{\mu_V}}{N_H}x_1x_5-\mu_H x_1+\alpha\left(1-\left(x_1+x_2\right)\right) \tag{2a}$$

while for (1b), we obtain

$$\frac{1}{N_H}\left(\frac{dI_H(t)}{dt}\right)=\frac{1}{N_H}\left(\frac{\beta_H b S_H I_V}{N_H}-\mu_H I_H-\gamma_H I_H\right)$$

$$\Leftrightarrow \frac{dx_2}{dt}=\frac{\beta_H b \, {}^{A}\!/_{\mu_V}}{N_H}x_1x_5-\mu_H x_2-\gamma_H x_2 \tag{2b}$$

Using the same way, we derive (1e) as follow.

$$\frac{1}{{}^{A}\!/_{\mu_V}}\left(\frac{dI_V(t)}{dt}\right)=\frac{1}{{}^{A}\!/_{\mu_V}}\left(\frac{\beta_V b S_V(1-\varepsilon)I_H}{N_H}-\mu_V I_V\right)$$

$$\Leftrightarrow \frac{dx_5}{dt}=\beta_V b(1-\varepsilon)(1-x_5)x_2-\mu_V x_5.. \tag{2c}$$

The equilibrium point of System (2) is achieved when $\frac{dx_1}{dt}=0,\frac{dx_2}{dt}=0,\frac{dx_5}{dt}=0$. Using this condition, we have two equilibrium points that are $(1,0,0)$, which is called as disease free, and $\left(x_1^*,x_2^*,x_5^*\right)$, with

$$x_1^*=\frac{N_H(\mu_H+\gamma_H)\mu_V\left(\left((\mu_H+\gamma_H)+\alpha\right)\mu_V-(\mu_H+\alpha)(\varepsilon\beta_V b-\beta_V b)\right)}{(\beta_V b-\varepsilon\beta_V b)\left(\beta_H bA(\mu_H+\gamma_H)+\alpha\beta_H bA\right)-(\varepsilon\beta_V b-\beta_V b)(\mu_H+\alpha)\left(N_H(\mu_H+\gamma_H)\mu_V\right)}$$

$$x_2^*=\frac{(\mu_H+\alpha)\beta_H bA(\beta_V b-\varepsilon\beta_V b)-(\mu_H+\alpha)\left(N_H(\mu_H+\gamma_H)\mu_V^2\right)}{(\beta_V b-\varepsilon\beta_V b)\left(\left(\beta_H bA(\mu_H+\gamma_H)+\alpha\beta_H bA\right)\right)-(\varepsilon\beta_V b-\beta_V b)(\mu_H+\alpha)\left(N_H(\mu_H+\gamma_H)\mu_V\right)}$$

$$x_5^*=\frac{(\mu_H+\alpha)\beta_H bA(\beta_V b-\varepsilon\beta_V b)-(\mu_H+\alpha)\left(N_H(\mu_H+\gamma_H)\mu_V^2\right)}{\left(\left(\beta_H bA(\mu_H+\gamma_H)+\alpha\beta_H bA\right)\mu_V-(\mu_H+\alpha)\beta_H bA(\varepsilon\beta_V b-\beta_V b)\right)},$$

which is called as the endemic equilibrium point.

Then, we will discuss about the value of secondary infections as a result of the first infection. This value is called the basic reproduction number. To determine this value, we use the next generation matrix method [9].

Defined matrix $F$ as a matrix whose entries are the first derivative of all terms that "enters" the infectious class at $(1,0,0)$.

$$F\Big|_{(1,0,0)}=\begin{bmatrix}\dfrac{\partial\left(\dfrac{\beta_H bA}{N_H\mu_V}\right)x_1x_5}{\partial x_2} & \dfrac{\partial\left(\beta_V b(1-\varepsilon)\right)x_2}{\partial x_2}\\ \dfrac{\partial\left(\dfrac{\beta_H bA}{N_H\mu_V}\right)x_1x_5}{\partial x_5} & \dfrac{\partial\left(\beta_V b(1-\varepsilon)\right)x_2}{\partial x_5}\end{bmatrix}_{(1,0,0)}=\begin{bmatrix}0 & \beta_V b(1-\varepsilon)\\ \dfrac{\beta_H bA}{N_H\mu_V} & 0\end{bmatrix}$$

We also define matrix $V$ whose entries are the first derivatives from the terms that are "out" from the infectious class at $(1,0,0)$.

$$V\big|_{(1,0,0)} = \begin{bmatrix} \dfrac{\partial(\mu_H+\gamma_H)x_2}{\partial x_2} & \dfrac{\partial(\beta_V b x_2 x_5+\mu_V x_5)}{\partial x_2} \\ \dfrac{\partial(\mu_H+\gamma_H)x_2}{\partial x_5} & \dfrac{\partial(\beta_V b x_2 x_5+\mu_V x_5)}{\partial x_5} \end{bmatrix}_{(1,0,0)} = \begin{bmatrix} \mu_H+\gamma_H & \beta_V b x_5 \\ 0 & \beta_V b x_2+\mu_V \end{bmatrix}_{(1,0,0)} = \begin{bmatrix} \mu_H+\gamma_H & 0 \\ 0 & \mu_V \end{bmatrix}$$

Let $G$ is a product of matrix $F$ and $V^{-1}$, then

$$G = \begin{bmatrix} 0 & \dfrac{\beta_V b(1-\varepsilon)}{\mu_V} \\ \dfrac{\beta_H bA}{N_H \mu_V(\mu_H+\gamma_H)} & 0 \end{bmatrix}$$

The value of the basic reproduction number is obtained from the biggest eigen value from matrix G, which is $\overline{R}_0 = \sqrt{\dfrac{\beta_V b(1-\varepsilon)\beta_H bA}{N_H \mu_V^2(\mu_H+\gamma_H)}}$. If

$$R_0 = \frac{\beta_V b(1-\varepsilon)\beta_H bA}{N_H \mu_V^2(\mu_H+\gamma_H)}, \tag{3}$$

then $\overline{R}_0 = \sqrt{R_0}$.

Based on the value of the basic reproduction number (3), it is easy to show if $R_0 < 1$, then equilibrium point $(1,0,0)$ is asymptotically stable, and if $R_0 > 1$, the equilibrium point $(1,0,0)$ is not stable. Furthermore, the endemic equilibrium point $(x_1^*, x_2^*, x_5^*)$ is stable if $R_0 > 1$.

## 3. Optimal Control

This section discusses about the extension of the transmission model of dengue fever (1) controlled by the prevention control variable so that the contact between the host and the mosquito can be reduced, and denoted by $u_1$, where $0 \le u_1 \le 1$. We obtain the following model

$$\frac{dS_H(t)}{dt} = \mu_H N_H - (1-u_1)\frac{\beta_H b S_H I_V}{N_H} - \mu_H S_H + \alpha R_H \tag{4a}$$

$$\frac{dI_H(t)}{dt} = (1-u_1)\frac{\beta_H b S_H I_V}{N_H} - \mu_H I_H - \gamma_H I_H \tag{4b}$$

$$\frac{dR_H(t)}{dt} = \gamma_H I_H - \mu_H R_H - \alpha R_H \tag{4c}$$

$$\frac{dS_V(t)}{dt} = A - (1-u_1)\frac{\beta_V b S_V(1-\varepsilon)I_H}{N_H} - \mu_V S_V \tag{4d}$$

$$\frac{dI_V(t)}{dt} = (1-u_1)\frac{\beta_V b S_V(1-\varepsilon)I_H}{N_H} - \mu_V I_V \tag{4e}$$

It can be seen that the infection from the susceptible host occurs because there is a contact with the infected mosquito, thus the infection rate is $\beta_H \frac{bS_H I_V}{N_H}$. In the other hand, assume that the prevention control in order to reduce the contact between the host and the mosquito is denoted by $u_1$, thus $(1-u_1)$ is the possibility of control failure. Hence, the infection rate will become $(1-u_1)\frac{\beta_H bS_H I_V}{N_H}$ if the prevention control is given. Meanwhile, the infection of the susceptible mosquito happens because there is a contact with the host, thus the infection rate per susceptible mosquito is $\beta_V \frac{b}{N_H} S_V (1-\varepsilon) I_H$. If the prevention control is given, then the infection rate will become $(1-u_1)\beta_V \frac{b}{N_H} S_V (1-\varepsilon) I_H$. The prevention efforts in this case could be a counselling about the dengue fever, the provision of the abate powder or the fogging readiness if necessary.

To investigate the optimality of those efforts, given the objective function J, namely to minimize the infected host class and the cost to implement the $u_1$ control,

$$J(u_1) = \int_0^T \left( BI_H(t) + Cu_1^2(t) \right) dt \tag{5}$$

where $B$ is the weights for the infected hosts class, while C is the weights for prevention efforts. It means that we will find the solution from (5), which minimizes the number of the infected hosts with a minimal $u_1(t)$ control cost. In other words, will be sought so an $u_1^*$ optimal control such that

$$J(u_1^*) = \min \{ J(u_1) | u_1 \in \Gamma \} \tag{6}$$

with system (4) as a constraint and $\Gamma = \{ u_1 | u_1 \text{ measurable, with } 0 \le u_1 \le 1 \}$ is a control set.

Further, we define Hamiltonian as follow.

$$\begin{aligned} H &= BI_H(t) + Cu_1^2(t) \\ &+ \delta_1 \left( \mu_H N_H - (1-u_1)\frac{\beta_H bS_H I_V}{N_H} - \mu_H S_H + \alpha R_H \right) + \delta_2 \left( (1-u_1)\frac{\beta_H bS_H I_V}{N_H} - \mu_H I_H - \gamma_H I_H \right) \\ &+ \delta_3 (\gamma_H I_H - \mu_H R_H - \alpha R_H) + \delta_4 \left( A - (1-u_1)\frac{\beta_V bS_V (1-\varepsilon) I_H}{N_H} - \mu_V S_V \right) \\ &+ \delta_5 \left( (1-u_1)\frac{\beta_V bS_V (1-\varepsilon) I_H}{N_H} - \mu_V I_V \right) \end{aligned}$$

where $\delta_1, \delta_2, ..., \delta_5$ are costate variable. Furthermore, we use Minimum Pontryagin Principal to give an idea about the existence of optimal control through the following Lemma.

**Lemma 1**

Given $u_1^*$ as the optimal control that minimizes $J(u_1)$ for $\Gamma$ with System (4) as the constraint then there are costate variables $\delta_1, \delta_2, ..., \delta_5$ such that

IOP Conf. Series: Journal of Physics: Conf. Series **855** (2017) 012045 doi:10.1088/1742-6596/855/1/012045

$$\frac{d\delta_1}{dt} = \delta_1\left((1-u_1)\frac{\beta_H bI_V}{N_H} + \mu_H\right) - \delta_2(1-u_1)\frac{\beta_H bI_V}{N_H}$$
$$\frac{d\delta_2}{dt} = -B + \delta_2(\mu_H + \gamma_H) - \delta_3\gamma_H + \delta_4(1-u_1)\frac{\beta_V bS_V(1-\varepsilon)}{N_H} - \delta_5(1-u_1)\frac{\beta_V bS_V(1-\varepsilon)}{N_H}$$
$$\frac{d\delta_3}{dt} = -\alpha\delta_1 + \delta_3(\mu_H + \alpha)$$
$$\frac{d\delta_4}{dt} = \delta_4(1-u_1)\frac{\beta_V bI_H(1-\varepsilon)}{N_H} - \delta_5(1-u_1)\frac{\beta_V bI_H(1-\varepsilon)}{N_H}$$
$$\frac{d\delta_5}{dt} = \delta_1(1-u_1)\frac{\beta_H bS_H}{N_H} - \delta_2(1-u_1)\frac{\beta_H bS_H}{N_H} + \delta_5\mu_V$$

with transversality conditions $\delta_i(T)=0, i=1,...,5$, and the $u_1^*$ satisfies the optimality condition

$$u_1^* = \min\left\{\max\left(0, \frac{(\delta_2-\delta_1)\frac{\beta_H bS_H^* I_V^*}{N_H} + (\delta_5-\delta_4)\frac{\beta_V bS_V^*(1-\varepsilon)I_H^*}{N_H}}{2C}\right), 1\right\}.$$

**Proof.** The differential equations developing the costate variables are obtained by differentiation of the Hamiltonian function that evaluated at the optimal control. Then the costate system can be written as follow

$$\frac{d\delta_1}{dt} = -\frac{\partial H}{\partial S_H} = \delta_1\left((1-u_1)\frac{\beta_H bI_V}{N_H} + \mu_H\right) - \delta_2(1-u_1)\frac{\beta_H bI_V}{N_H}$$
$$\frac{d\delta_2}{dt} = -\frac{\partial H}{\partial I_H} = -B + \delta_2(\mu_H + \gamma_H) - \delta_3\gamma_H + \delta_4(1-u_1)\frac{\beta_V bS_V(1-\varepsilon)}{N_H} - \delta_5(1-u_1)\frac{\beta_V bS_V(1-\varepsilon)}{N_H}$$
$$\frac{d\delta_3}{dt} = -\frac{\partial H}{\partial R_H} = -\alpha\delta_1 + \delta_3(\mu_H + \alpha)$$
$$\frac{d\delta_4}{dt} = -\frac{\partial H}{\partial S_V} = \delta_4(1-u_1)\frac{\beta_V bI_H(1-\varepsilon)}{N_H} - \delta_5(1-u_1)\frac{\beta_V bI_H(1-\varepsilon)}{N_H}$$
$$\frac{d\delta_5}{dt} = -\frac{\partial H}{\partial I_V} = \delta_1(1-u_1)\frac{\beta_H bS_H}{N_H} - \delta_2(1-u_1)\frac{\beta_H bS_H}{N_H} + \delta_5\mu_V$$

with transversality conditions $\delta_i(T)=0, i=1,...,5$. On the interior of the control set, $0 \le u_1 \le 1$, the optimal control at the solutions $S_H^*, I_H^*, R_H^*, S_V^*, I_V^*$ of the corresponding state system (4) is derived by

$$\frac{\partial H}{\partial u_1} = 0$$

$$\Leftrightarrow 2Cu_1^* + \delta_1\frac{\beta_H bS_H^* I_V^*}{N_H} - \delta_2\frac{\beta_H bS_H^* I_V^*}{N_H} + \delta_4\frac{\beta_V bS_V^*(1-\varepsilon)I_H^*}{N_H} - \delta_5\frac{\beta_V bS_V^*(1-\varepsilon)I_H^*}{N_H} = 0$$

$$\Leftrightarrow u_1^* = \frac{(\delta_2-\delta_1)\frac{\beta_H bS_H^* I_V^*}{N_H} + (\delta_5-\delta_4)\frac{\beta_V bS_V^*(1-\varepsilon)I_H^*}{N_H}}{2C}$$

Hence, we obtain

$$u_1^* = \min\left\{\max\left(0, \frac{(\delta_2-\delta_1)\frac{\beta_H b S_H^* I_V^*}{N_H} + (\delta_5-\delta_4)\frac{\beta_V b S_V^* (1-\varepsilon) I_H^*}{N_H}}{2C}\right), 1\right\}.$$

## 4. Numerical Simulation

In this section, we present numerical simulation to describe the optimal solution of the host-vector model. The parameter values are given as below.

**Table 1.** Values of Parameters used in Host-Vector Model

| Parameter | Value | Reference |
|---|---|---|
| $\alpha$ | 0,9619 | [10] |
| $\mu_V$ | 0,071428 | [11] |
| $b$ | 0,5 | [1] |
| $\beta_V$ | 1 | [1] |
| $\beta_H$ | 0,75 | [1] |
| $\mu_H$ | 0,9619 | [12] |
| $\gamma_H$ | 0,1428 | [1] |

A susceptible host who is bitten by infected mosquito cannot transmit the virus directly, he/she needs 3-14 days in this condition [13], so the value of $\varepsilon$ is between $1/3$ and $1/14$. Using $A = 400$, $N_H = 10000$ and parameter in Table 1, we have $R_0 = 7 > 1$. It means without control, any solution will tend to the endemic equilibrium point. The initial values used were $S_H(0) = 0,6; I_H(0) = 0,3; I_V(0) = 0,1$. As shown in Figure 2, for susceptible hosts' class, the population rose initially, but dropped later on. This is because the population will go to the infectious class. While with the presence of the control, the population of the susceptible class increased over time.

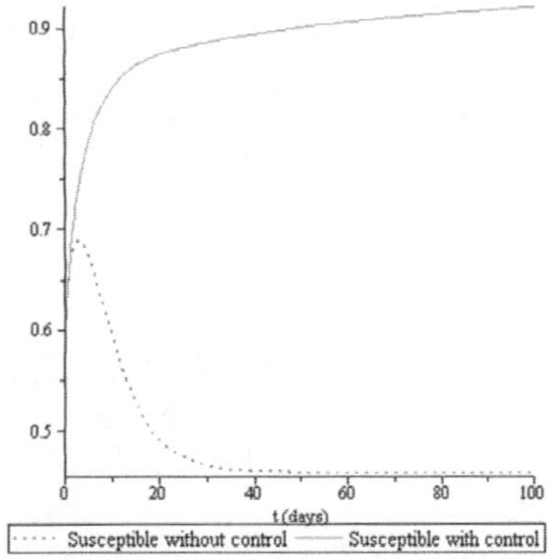


Figure 2. Susceptible human with and without control

Figure 3 explains that without control, the population of the infected hosts' increases. Conversely when the control is given, the population decreases sharply. In Figure 4, it can be seen that the vector population increases, both with and without controls. But, the population with control rose faster and higher than without control.

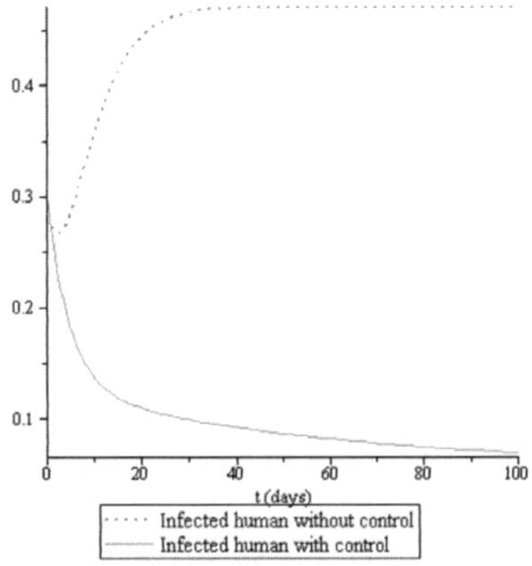


Figure 3. Infected human with and without control

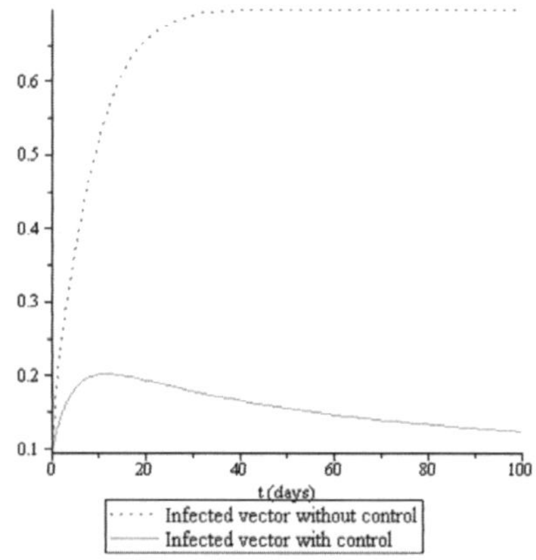


Figure 4. Infected vector with and without control

Figure 5 shows the optimal control function $u_1$ when the value in (5) are chosen $B = 1$ and $C = 50$. It can be seen that with preventive control can reduce the number of infectious host significantly.

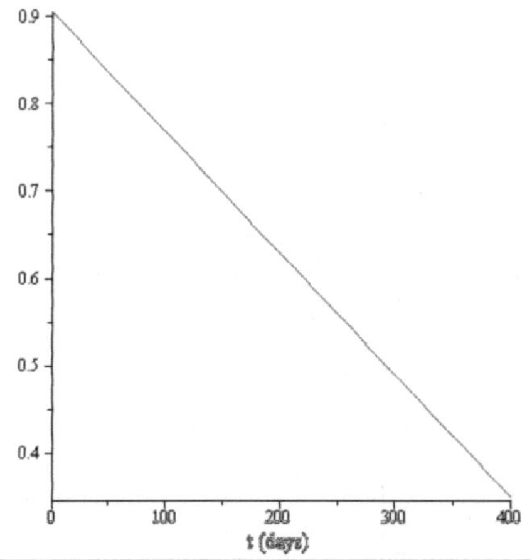


Figure 5. Optimal control function

## 5. Conclusion

This paper has discussed the development of a mathematical model for the dengue fever spread in the form of host-vector. The model then is extended by giving a prevention control for the spread. We described analytically that using Pontryagin Minimum Principle there is an optimal control. We also proved numerically that with a control strategy, we can decrease the population of infectious class both for the host and the vector.

IOP Conf. Series: Journal of Physics: Conf. Series **855** (2017) 012045 doi:10.1088/1742-6596/855/1/012045

**Acknowledgments**
The authors would like to thank to DIPA Faculty of Mathematics and Natural Sciences, Yogyakarta State University for the support and the research grant for this research.

**References**

[1] Lourdes Esteva and Cristobal Vargas, "Analysis of a Dengue Disease Transmission Model," *Mathematical Biosciences*, vol. 150, pp. 131-151, 1998.

[2] Edy Soewono and Asep K Supriatna, "A Two-dimensional Model for the Transmission of Dengue Fever Disease," *Bulletin of the Malaysian Mathematical Sciences Society*, vol. 24, pp. 49-57, 2001.

[3] R. M. Anderson and R. M. May, *Dynamics of Infectious Diseases*. Oxford UK: Oxford University Press, 1991.

[4] J. M. Hafferman, R. J. Smith, and L. M. Wahl, "Perspectives on the Basic Reproduction Number," *Journal of the Royal Society Interface*, vol. 2, pp. 281-293, 2005.

[5] Herbert W. Hethcote, "The Mathematics of Infectious Diseases," *SIAM review*, vol. 42, pp. 599-653, 2000.

[6] Helena Sofia Rodrigues, M. Teresa T Monteiro, and Delfim F.M. Torres, "Insecticide control in a Dengue epidemics model," in *Numerical Analysis and Applied Mathematics*, vol. 1281, 2010, pp. 979–982.

[7] Yves Dumont and Frederic Chiroleu, "Vector Control for the Chikungunya Disease," *Mathematical Biosciences and Engineering*, vol. 7, pp. 313–345, April 2010.

[8] WHO, *Global strategy for dengue prevention and control 2012-2020.*: WHO Press, 2012.

[9] P. van den Driessche and James Watmough, "Reproduction Numbers and Sub-Threshold Endemic Equilibria for Compartmental Models of Disease Transmission," *Mathematical Biosciences*, vol. 180, pp. 29-48, 2002.

[10] D. E. Mahmudah, A. Suryanto, and Trisilowati, "Optimal Control of a Vector-Host Epidemic Model with Direct Transmission," *Applied Mathematical Sciences*, vol. 7, pp. 4919-4927, 2013.

[11] Zhilan Feng and Jorge X. Velasco Hernandez, "Competitive Exclusion in a Vector-Host Model for the Dengue Fever," *Journal of mathematical Biology*, vol. 35, pp. 523-544, 1997.

[12] Badan Pusat Statistik. (2015) Angka Harapan Hidup Penduduk Beberapa Negara (tahun), 1995-2015. [Online]. http://www.bps.go.id/linkTabelStatis/view/id/1517

[13] Dengue Clinical Case Management (DCCM). (2014, March) [Online]. https://www.cdc.gov/dengue/training/cme/ccm/page45915.html

IOP Conf. Series: Journal of Physics: Conf. Series **855** (2017) 012046 doi:10.1088/1742-6596/855/1/012046

# Inflation data clustering of some cities in Indonesia

**Adi Setiawan, Bambang Susanto, and Tundjung Mahatma**
Department of Mathematics, Faculty of Science and Mathematics
Universitas Kristen Satya Wacana, Indonesia.

E-mail: adi.setiawan@staff.uksw.edu

**ABSTRACT.** In this paper, it is presented how to cluster inflation data of cities in Indonesia by using *k*-means cluster method and fuzzy *c*-means method. The data that are used is limited to the monthly inflation data from 15 cities across Indonesia which have highest weight of donations and is supplemented with 5 cities used in the calculation of inflation in Indonesia. When they are applied into two clusters with $k= 2$ for *k*-means cluster method and $c= 2$, $w = 1.25$ for fuzzy *c*-means cluster method, Ambon, Manado and Jayapura tend to become one cluster (high inflation) meanwhile other cities tend to become members of other cluster (low inflation). However, if they are applied into two clusters with $c=2$, $w=1.5$, Surabaya, Medan, Makasar, Samarinda, Makasar, Manado, Ambon dan Jayapura tend to become one cluster (high inflation) meanwhile other cities tend to become members of other cluster (low inflation). Furthermore, when we use two clusters with $k=3$ for *k*-means cluster method and $c=3$, $w= 1.25$ for fuzzy c-means cluster method, Ambon tends to become member of first cluster (high inflation), Manado and Jayapura tend to become member of second cluster (moderate inflation), other cities tend to become members of third cluster (low inflation). If it is applied $c=3$, $w = 1.5$, Ambon, Manado and Jayapura tend to become member of first cluster (high inflation), Surabaya, Bandung, Medan, Makasar, Banyuwangi, Denpasar, Samarinda dan Mataram tend to become members of second cluster (moderate inflation), meanwhile other cities tend to become members of third cluster (low inflation). Similarly, interpretation can be made to the results of applying 5 clusters.

## 1. Introduction

Monthly Consumer Price Index (CPI) of commodities in 82 cities across Indonesia is calculated and used to determine the monthly inflation of the cities and also nation-wide by the Central Statistics Agency (BPS). Each city has its own weighting based on the interests of the city. Jakarta, Surabaya, Bandung, Bekasi and Tangerang have the largest weight contribution i.e. 19.23%, 5.65%, 4.87%, 4.78% and 3.75%, respectively. In July 2016 monthly inflation rate is 0.69 per cent, while the inflation rate of the calendar year (January to July 2016) is amounted to 1.76 percent and the inflation rate year-on-year (July 2016 to July 2015) is 3.21 percent [1].

This paper presents the clustering of the inflation rates occurred in cities in Indonesia by using *k*-means cluster method and fuzzy *c*-means method. The data used is limited to the monthly inflation data from 15 cities across Indonesia which have the highest weight of contributions (61.96 % contribution to inflation in Indonesia) and is supplemented with 5 cities used in the calculation of inflation in Indonesia (3.97%). The data covers a period of November 2014 to July 2016.

## 2. Literature Review

The methods employed in this research –the *k*-means and fuzzy *c*-means clustering methods– are presented as follows.

The *k*-means algorithm is composed of the following steps:

1. Place *k* points into the space represented by the objects that are being clustered. These points represent initial group centers.
2. Assign each object to the group that has the closest centers.
3. When all objects have been assigned, recalculate the positions of the *k* centers.
4. Repeat steps 2 and 3 until the centers no longer move. This procedure a separation of the objects into groups from which the metric to be minimized can be calculated [2].

This algorithm aims at minimizing an objective function, in this case a squared error function i.e.

$$J=\sum_{j=1}^{k}\sum_{i=1}^{n}\left\|x_i^{(j)}-c_j\right\|^2$$

where $\left\|x_i^{(j)}-c_j\right\|^2$ is a chosen distance measure between a data point $x_i^{(j)}$ and the cluster center $c_j$, is an indicator of the distance of the *n* data points their respective cluster center.

The implementation of the *k*-means method using a small sample data is illustrated as follows. Suppose we have the monthly inflation data of Jakarta, Surabaya, Bandung and Bekasi as presented in Table 1.

Table 1. Table of monthly inflation of cities in Indonesia : Jakarta, Surabaya, Bandung and Bekasi.

| Month | Jakarta | Surabaya | Bandung | Bekasi |
|---|---|---|---|---|
| November 2014 | 1.43 | 1.27 | 1.27 | 1.99 |
| Desember 2014 | 2.74 | 2.23 | 2.34 | 1.99 |
| Januari 2014 | -0.41 | 0.41 | -0.05 | -0.17 |

The data may be expressed in matrix $Z$

$$Z=\begin{pmatrix}1.43 & 1.27 & 1.27 & 1.99\\ 2.74 & 2.23 & 2.34 & 1.99\\ -0.41 & 0.41 & -0.05 & -0.17\end{pmatrix}$$

and based on the *k*-means algorithm we obtain the center matrix, which consists of the first two points i.e.

$$V=\begin{pmatrix}1.43 & 1.27\\ 2.74 & 2.23\\ -0.41 & 0.41\end{pmatrix},$$

having the distance matrix

$$D=\begin{pmatrix}0 & 0.9788 & 0.5614 & 0.9663\\ 0.9788 & 0 & 0.4730 & 0.9552\end{pmatrix}$$

so that an objective function of 1.1361 is obtained. Furthermore, a clustering matrix can be expressed as

$$U=\begin{pmatrix}1 & 0 & 0 & 0\\ 0 & 1 & 1 & 1\end{pmatrix}.$$

This means that the monthly inflation data of Jakarta is in a separate cluster, while the monthly inflation data of Surabaya, Bandung and Bekasi in another cluster. By using the clustering matrix $U$ we obtain the center matrix:

$$V=\begin{pmatrix} 1.430 & 1.510 \\ 2.74 & 2.187 \\ -0.41 & 0.063 \end{pmatrix}.$$

In this case, the first column is the first point and the second column is the average of the second, third and fourth points. The distance matrix of each point to the corresponding centre can expressed by:

$$D=\begin{pmatrix} 0 & 0.9788 & 0.5614 & 0.9663 \\ 0.7325 & 0.4239 & 0.3065 & 0.5688 \end{pmatrix}$$

and an objective function of 0.5971. Other clustering matrices are then selected until the objective function value is not changed. The final result of this algorithm is the clustering matrix:

$$U=\begin{pmatrix} 0 & 0 & 0 & 1 \\ 1 & 1 & 1 & 0 \end{pmatrix},$$

the center matrix:

$$V=\begin{pmatrix} 1.3233 & 1.9900 \\ 2.4367 & 1.9900 \\ -0.0167 & -0.1700 \end{pmatrix},$$

a distance matrix:

$$D=\begin{pmatrix} 0.966 & 0.9552 & 0.8095 & 0 \\ 0.5080 & 0.4771 & 0.1153 & 0.8170 \end{pmatrix},$$

and an objective function of 0.4990.

Fuzzy *c*-means (FCM) is an algorithm to cluster data, in which the membership of each data point in the cluster is determined by its probability value. In *k*-means clustering method, in contrast, each point may only be a member of a cluster and not of any other cluster. Fuzzy *c*-means clustering is used to find the degree of membership in a cluster if the number of clusters $c$ and the weight $w$ are given, as in the examples found in [2], [3] and [4].

The algorithm of FCM may be described as follows [5] :

1. Suppose we have the data to be clustered in matrix $Z$ with the ordo $K \times N$, where $K$ = the number of variables/attributes/criteria, and $N$ = the number of sample points that will be clustered. Matrix $Z$ can be presented as follows

$$Z=\begin{pmatrix} z_{11} & z_{12} & \cdots & z_{1N} \\ z_{21} & z_{22} & \cdots & z_{2N} \\ \cdots & \cdots & \cdots & \cdots \\ z_{K1} & z_{K2} & \cdots & z_{KN} \end{pmatrix}.$$

2. The partition matrix or decision matrix $U$ depends on :
   a. The number of clusters $c$, where $N > c \geq 2$.
   b. The weight parameter $w$ that is used in clustering (as a rule, $w > 1$).
   c. Stopping criteria i.e. the threshold $\varepsilon$, a positive small number. Eg. $\varepsilon = 10^{-1}$; $10^{-3}$; $10^{-6}$, and so on. A maximum iteration may also be used, eg. an iteration of 100, 1000, or 10.000.
   d. Initial objective value ($P_0$).
3. An initial partition matrix $U_0$ is chosen, i.e. the degree of membership of clusters with the ordo of $c \times N$ such that the matrix $U_0$ can be presented as:

IOP Conf. Series: Journal of Physics: Conf. Series **855** (2017) 012046 doi:10.1088/1742-6596/855/1/012046

$$U = \begin{pmatrix} u_{11} & u_{12} & \ldots\ldots & u_{1N} \\ u_{21} & u_{22} & \ldots\ldots & u_{2N} \\ \ldots\ldots & \ldots\ldots & \ldots\ldots & \ldots\ldots \\ u_{c1} & u_{c2} & \ldots\ldots & u_{cN} \end{pmatrix}$$

having $\sum_{i=1}^{c} u_{ij} = 1$ for $j = 1, 2, \ldots., N$.

4. The cluster centre matrix $V$ is one with the ordo of $K \times c$, applying the formula:

$$v_i = \frac{\sum_{j=1}^{N} (u_{ij})^w z_j}{\sum_{j=1}^{N} (u_{ij})^w}$$

where $v_i$ is the $i$-th center of cluster for $i = 1, 2, \ldots., c$.

5. The objective value is obtained by using

$$P_n = \sum_{j=1}^{N} \sum_{i=1}^{c} (u_{ij})^w (d_{ij})^2$$

where $d_{ij}$= the function of Eucledian distance of the $i$-th cluster center and the $j$-th data, which is

$$d_{ij} = d(z_j, v_i) = \left[ \sum_{k=1}^{K} (z_{kj} - v_{ki})^2 \right]^{1/2}.$$

6. The partition matrix $U$ is modified so that each membership degree of the entry becomes $U_n$ which follows

$$u_{ij} = \left[ \sum_{h=1}^{c} \left( \frac{d_{ij}}{d_{hj}} \right)^{2/(w-1)} \right]^{-1}.$$

7. The iteration is stopped when the partition matrix does not change, i.e.

$$\| U_n - U_{n-1} \| < \varepsilon$$

or the objective function does not change, i.e. $| P_n - P_{n-1} | < \varepsilon$ or when the iteration limit is reached.

8. The final results are the partition matrix $U^*$ (each entry of which represents the degree of membership), and the cluster center matrix $V^*$.

The above algorithm may be described by employing a small data sample. Suppose we want to do clustering of data by using $c = 2$ cluster. Based on Table 1, the data can be expressed in matrix **Z**. With the number of cluster $c$ =2, the weight parameter $w = 2$, and epsilon $\varepsilon = 10^{-3}$, then the initial partition matrix $U$ is as follows:

$$U = \begin{pmatrix} 0.8 & 0.4 & 0.7 & 0.5 \\ 0.2 & 0.6 & 0.3 & 0.5 \end{pmatrix}.$$

This is to say that an initial matrix $U$ is used as a partition matrix, meaning that the monthly inflation data in Jakarta has a probability of 0.8 to become a member of the first cluster, and a probability of 0.2 to become a member of the second cluster. Meanwhile the monthly inflation data in Surabaya has a probability of 0.4 to become a member of the first cluster, and a probability of 0.6 to become the second cluster. In other words, the monthly inflation data in Jakarta tends to become a member of the first cluster, while the monthly inflation data in Surabaya tends to become a member of the second cluster. In this case –following the principle of fuzzy membership that uses the degree of membership– such that no data point is definite to become a member of the first cluster nor definite to become a member of the second cluster. The cluster centers based on the matrix $Z$ and partition matrix $U$ above is the matrix $V$ with ordo $3 \times 2$ as follows

$$V = \begin{pmatrix} 1.4534 & 1.5219 \\ 2.4380 & 2.1898 \\ -0.1713 & 0.1138 \end{pmatrix}.$$

The related objective function obtained is 0.4856.
The next step is fixing $U$ to become a new matrix $U_1$:

$$U_1 = \begin{pmatrix} 0.7974 & 0.2692 & 0.6607 & 0.4100 \\ 0.2026 & 0.7308 & 0.3393 & 0.5900 \end{pmatrix}.$$

The interpretation of the new partition matrix $U_1$ is analogous to the initial matrix $U$ as given above. By using the 8 steps to gain the final objective function of 0.4387, we get the final partition matrix $U^*$ as follows:

$$U^* = \begin{pmatrix} 0.9458 & 0.1089 & 0.2875 & 0.4338 \\ 0.0542 & 0.8911 & 0.7125 & 0.5662 \end{pmatrix}.$$

That means that for $c = 2$, $w = 2$ and $\varepsilon = 10^{-3}$, the monthly inflation data of Jakarta tends to become a member of the first cluster with a probability of 0.9458. The monthly inflation data of Surabaya, Bandung and Bekasi tends to become members of the second cluster with the probability of 0.8911, 0.7125 and 0.5662, respectively. The cluster center $V^*$ related to the final partition matrix $U^*$ may be expressed as follows:

$$V^* = \begin{pmatrix} 1.5067 & 1.4123 \\ 2.5869 & 2.2179 \\ -0.3381 & 0.1504 \end{pmatrix}.$$

By using the cluster center $V^*$, the Euclidean matrix $D^*$ is:

$$D^* = \begin{pmatrix} 0.1857 & 0.8620 & 0.4472 & 0.7862 \\ 0.7661 & 0.2963 & 0.2744 & 0.6988 \end{pmatrix}.$$

where $D_{ij}$ presents the distance between the $i$-th cluster center and the $j$-th data point. It appears that the greater the Euclidean distance, the less likely it is to become a member of the cluster.

Similarly, using $c = 2$, $\varepsilon = 10^{-3}$, with $w = 3$ and $w = 4$, the partition matrices will be respectively as follows:

$$U_2^* = \begin{pmatrix} 0.7633 & 0.2742 & 0.4264 & 0.4766 \\ 0.2367 & 0.7258 & 0.5736 & 0.5234 \end{pmatrix},$$

and

$$U_3^* = \begin{pmatrix} 0.6745 & 0.3553 & 0.4665 & 0.4810 \\ 0.3255 & 0.6447 & 0.5335 & 0.5190 \end{pmatrix}.$$

They required 5, 3 and 2 steps, and the objective functions are 0.2368, 0.1225 and 0.0623, respectively. Using $c = 3$, $\varepsilon = 10^{-3}$ and $w = 2$, the partition matrix is as follows:

$$U^* = \begin{pmatrix} 0.9992 & 0.0364 & 0.1934 & 0.0001 \\ 0.0003 & 0.0374 & 0.0887 & 0.9998 \\ 0.0005 & 0.9261 & 0.7179 & 0.0001 \end{pmatrix}.$$

It needs 10 steps, and the objective function is 0.0910. This means that when using three clusters, the monthly inflation data of Jakarta tends to become a member of the first cluster, while the monthly inflation data in Surabaya and Bandung tend to become members of the second cluster, and the data of Bekasi's tends to be in the third cluster.

## 3. Methods

The data used is the monthly inflation data of 15 cities across Indonesia. Jakarta, Surabaya, Bandung, Bekasi, Tangerang, Medan, Depok, Palembang, Makassar, Batam, Jember, Bogor, Banyuwangi, Denpasar, Pekanbaru give the highest contribution to the whole inflation in Indonesia. Five cities represent the regions in Indonesia, namely Samarinda, Mataram, Manado, Ambon and Jayapura. The data are taken at the beginning of Joko Widodo Administration, from November 2014 to July 2016. Clustering is performed by using the $k$-means and fuzzy $c$-means clustering methods, where $k = c = 2$, 3 and 5.

## 4. Results and Discussion

During the period of November 2014 to July 2016, the lowest average monthly inflation is achieved by Palembang, and the highest average is Medan. The lowest median is achieved by Bekasi and the highest median is Medan. The lowest standard deviation is achieved by Denpasar, while the highest is Jayapura. The lowest range is achieved by Denpasar, while the highest range is Jayapura. Fig. 1 presents the line diagram of the monthly inflation data for Jakarta, Medan, Denpasar, Ambon and Jayapura. Denpasar appears to have stable commodity prices as shown by the relatively small standard deviation compared to other cities. Contrarily, commodity prices in Jayapura are likely to fluctuate as shown by a relatively large standard deviation compared to other cities, and also supported by a relatively larger range.

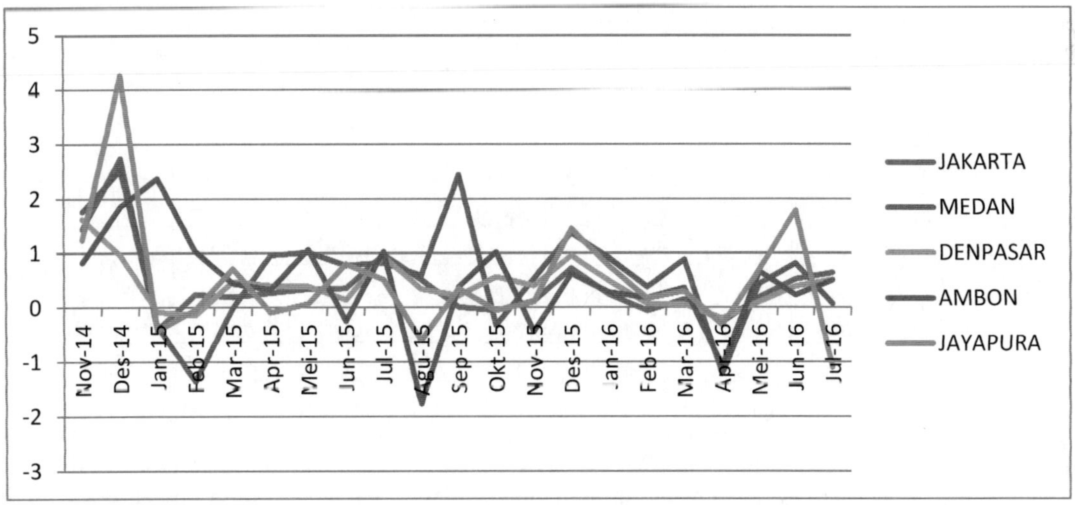


Fig. 1. Line diagram of monthly inflation for the period of November 2014 to July 2016 in Jakarta, Medan, Denpasar, Ambon and Jayapura.

Employing the $k$-means clustering method with $k = 2$ clusters, Manado, Ambon and Jayapura appear to become one cluster of high inflation rates, and other cities are in the other cluster of lower inflation rate. When using $k = 3$ clusters, Ambon is clustered by herself as having a high inflation rate,

IOP Conf. Series: Journal of Physics: Conf. Series **855** (2017) 012046 doi:10.1088/1742-6596/855/1/012046

Manado and Jayapura are in one cluster of a moderate inflationrate, and other cities are in another cluster of lower inflation rate. With $k = 5$ clusters, Ambon is in the very high inflation cluster, Manado and Jayapura in the high inflation cluster, Medan in the medium inflation cluster. Jakarta, Surabaya, Bandung, Bekasi, Makasar, Jember, Banyuwangi, Denpasar, Samarinda and Mataram are in the lower inflation cluster, while other cities become one cluster of very low inflation rates.

Table 2. The result of clustering for $c$=2 and $w$ = 1.25.

| | **$w$ = 1.25** | | | **$w$ = 1.25** | |
|---|---|---|---|---|---|
| **CITY** | **Cluster 1** | **Cluster 2** | **CITY** | **Cluster 1** | **Cluster 2** |
| **JAKARTA** | 0.0013 | 0.9987 | **JEMBER** | 0.0015 | 0.9985 |
| **SURABAYA** | 0.0007 | 0.9993 | **BOGOR** | 0.0027 | 0.9973 |
| **BANDUNG** | 0 | 1 | **BANYUWANGI** | 0.0008 | 0.9992 |
| **BEKASI** | 0.0019 | 0.9981 | **DENPASAR** | 0.0095 | 0.9905 |
| **TANGERANG** | 0.0016 | 0.9981 | **PEKANBARU** | 0.0117 | 0.9883 |
| **MEDAN** | 0.1728 | 0.8272 | **SAMARINDA** | 0.0604 | 0.9396 |
| **DEPOK** | 0.0007 | 0.9993 | **MATARAM** | 0.0485 | 0.9515 |
| **PALEMBANG** | 0.0109 | 0.9891 | **MANADO** | 0.9635 | 0.0365 |
| **MAKASAR** | 0.0044 | 0.9956 | **AMBON** | 0.9183 | 0.0817 |
| **BATAM** | 0.0108 | 0.9892 | **JAYAPURA** | 0.9336 | 0.0664 |

Table 3. The result of clustering for $c$=2 and $w$ = 1.50.

| | **$w$ = 1.50** | | | **$w$ = 1.50** | |
|---|---|---|---|---|---|
| **CITY** | **Cluster 1** | **Cluster 2** | **CITY** | **Cluster 1** | **Cluster 2** |
| **JAKARTA** | 0.3886 | 0.6114 | **JEMBER** | 0.3896 | 0.6104 |
| **SURABAYA** | 0.5013 | 0.4987 | **BOGOR** | 0.3222 | 0.6778 |
| **BANDUNG** | 0.2910 | 0.7090 | **BANYUWANGI** | 0.4514 | 0.5486 |
| **BEKASI** | 0.3587 | 0.6413 | **DENPASAR** | 0.4579 | 0.5421 |
| **TANGERANG** | 0.3449 | 0.6551 | **PEKANBARU** | 0.4098 | 0.5902 |
| **MEDAN** | 0.5111 | 0.4889 | **SAMARINDA** | 0.6639 | 0.3361 |
| **DEPOK** | 0.2519 | 0.7481 | **MATARAM** | 0.6206 | 0.3794 |
| **PALEMBANG** | 0.3418 | 0.6582 | **MANADO** | 0.5694 | 0.4306 |
| **MAKASAR** | 0.5060 | 0.4940 | **AMBON** | 0.5917 | 0.4083 |
| **BATAM** | 0.4009 | 0.5991 | **JAYAPURA** | 0.5589 | 0.4411 |

Employing the fuzzy $c$-means method with the parameter $w$ = 1.25 to create $c$=2 clusters, Manado, Ambon and Jayapura tend to become one cluster of high inflation rates, and other cities in the other cluster of low inflation rates. The clustering only yields cities which tend to form one cluster with the probabilities between 0 and 1. With the parameter $w$ = 1.50, and $c$ = 2, Manado, Ambon, Jayapura, Samarinda, Mataram, Surabaya, Medan and Makasar tend to be in one cluster, and the other cities tend in the other cluster.

With the parameter $w$ = 1.25 and $c$ = 3, Ambon tends to form a single cluster of high inflation, Manado and Jayapura in a medium inflation, and other cities in the lower inflation cluster. With the parameter $w = 1.5$ and $c = 3$, Manado, Ambon and Jayapura tend to become one cluster of high inflation. Surabaya, Bandung, Medan, Makasar, Banyuwangi, Denpasar, Samarinda and Mataram tend to form a medium inflation cluster, and other cities in a lower inflation cluster.

With the parameter $w = 1.25$ and $c = 5$, Ambon tends to form a very high inflation cluster. Jayapura and Manado in a high inflation cluster, Medan in a medium inflation cluster, and Jakarta, Surabaya, Bandung, Bekasi, Makasar, Jember, Banyuwangi, Denpasar, Samarinda and Mataram in a low inflation cluster, and the rest are in a very low inflation cluster. With the parameter $w = 1.5$ and $c = 5$, Ambon tends to form a very high inflation cluster. Manado and Jayapura tend to be in a high cluster, Surabaya, Bandung, Makasar, Samarinda and Mataram in a cluster. Tangerang, Medan, Batam, Bogor and Samarinda tend to be in a low inflation cluster, and the rest of the cities tend to be in a very low inflation cluster.

The fuzzy $c$-means method in this research is applied using the Eucledian distance, while the Mahalanobis distance can also be used as in paper [5]. There is another paper about fuzzy $c$-means clustering is [7], while ours also compares the results to those obtained by the $k$-means clustering method. It also cover wider geographical area of Indonesia. Fuzzy $c$-means method is also used in the paper [4], meanwhile $k$-means clustering method is also used in [8]. The comparison of the two methods may be found in the paper [2].

## 5. Conclusion

This paper has discussed the clustering of inflation data by using the $k$-means method and the fuzzy $c$-means method. This research may be extended by applying other clustering methods and longer periods of data.

## References

[1] BPS, 2016, *Berita Resmi Statistik No. 67/08 Tahun XIX, 01 Agustus 2016*, Badan Pusat Statistik, Jakarta.

[2] Sheshayee, A. & P. Sharmila, 2014, Comparative Study of Fuzzy C-Means and K-Means Algorithm for requirements Clustering, *Indian Journal of Science and Technology*, Vol 7 (6) page 853-857.

[3] Sanmorino, A., 2012, Clustering Batik Images using Fuzzy C-means Algorithm Based on Log Average Luminance, *Journal of Computer Engineering and Applications* Vol. 1, No. 1.

[4] Rustempasic, I. & M. Can, 2013, Diagnosis of Parkinson's Disease using Fuzzy C-Means Clustering and Pattern Recognition, *Southeast Europe Journal of Soft Computing*.

[5] Liu, Hsiang-Chuan, Bai-Cheng Jeng, Jeng-Ming Yih, Yen-Kuei Yu, 2009, Fuzzy $c$-means Algorthm Based on Standard Mahalanobis Distances, *Proceedings of the 2009 International Symposium on Information Processing*.

[6] Suganya, R., R. Shanthi, 2012, Fuzzy C-Means Algorithm A Review, *International Journal of Scientific and Research Publications* Vol. 2 Issue 11.

[7] Setiawan, A. & B. Susanto, 2016, Inflation data clustering of some cities in eastern Indonesia, *it was presented in Asian Mathematical Conference*, Denpasar, Bali 25-29 July 2016.

[8] Salman, Raied, V. Kecman, Q. Li, R. Strack, E. Test, Fast K-means Algorithm Clustering, *International Journal of Computer Networks & Communications* (IJCNC) Vol. 3 No. 4 page 17-31.

# Modelling tourists arrival using time varying parameter

[1] **Suciptawati P,** [2] **Sukarsa K G and** [3] **Eka N Kencana**

Department of Mathematics, Udayana University, Kampus Bukit Jimbaran–Badung 80361, ID

E-mail: [1] putusuciptawati@yahoo.co.id, [2] sukarsakomang@yahoo.com, [3] i.putu.enk@unud.ac.id

**Abstract.** The importance of tourism and its related sectors to support economic development and poverty reduction in many countries increase researchers' attentions to study and model tourists' arrival. This work is aimed to demonstrate time varying parameter (TVP) technique to model the arrival of Korean's tourists to Bali. The number of Korean tourists whom visiting Bali for period January 2010 to December 2015 were used to model the number of Korean's tourists to Bali (KOR) as dependent variable. The predictors are the exchange rate of Won to IDR (WON), the inflation rate in Korea (INFKR), and the inflation rate in Indonesia (INFID). Observing tourists visit to Bali tend to fluctuate by their nationality, then the model was built by applying TVP and its parameters were approximated using Kalman Filter algorithm. The results showed all of predictor variables (WON, INFKR, INFID) significantly affect KOR. For in-sample and out-of-sample forecast with ARIMA's forecasted values for the predictors, TVP model gave mean absolute percentage error (MAPE) as much as 11.24 percent and 12.86 percent, respectively.

## 1. Introduction

Tourism has been started attracts scholars and researchers to study this discipline and its related sectors in 1990s. At the beginning phase, tourism model regarding the visitors' demand utilised multiple regression techniques (MRTs) to model and forecast the growth of visitors to a destination. Qiu and Zhang [1] used multiple regression analysis to model tourist arrivals and their expenditures in Canada. However, this technique is criticised noting tourist arrivals had non-stationary properties that is not suitable for MRTs. One consequence of applying MRTs for non-stationaries data is spurious regression function relates the dependent and its predictors. Since this works, alternative methods such as co-integration (CI) technique, error correction model (ECM), vector autoregressive (VAR) has been tried and tested to model tourist arrivals. These methods utilise ordinary least squares (OLS) to estimate the parameter(s) by assuming its are constant over time. This assumption is not completely fulfilled noting the tourist arrivals tend to fluctuate because of internal and/or external destination factors.

The insufficiency of constant parameters assumption for tourism demand modelling is demonstrated by Witt and Song [2] who proved price elasticity of tourism demand decreased when cross-price elasticity increased over time. To overcome non-constant behaviour of tourism demand, both authors introduced time varying parameter (TVP) to model number of visitors as targeted variables and several macroeconomic indicators as predictor variables. In addition, Guo [5] stated by applying Kalman Filter algorithm thorough its capability to find the estimators adaptively, particularly by introducing an excitation condition, then the assumption regarding the stationarity or independence of predictors are not needed.

This work is aimed to model Korean tourists to Bali. According to Bali Provincial Statistics Office [3], the number of foreign tourists visiting Bali in year 2015 reached 4 million visitors, 6.65 percent growth compared to previous year. Classify according nationality, Australian tourists position in the first place while Korean visitors are in the fifth rank. To model Korean arrivals to Bali, TVP with Kalman Filter algorithm to 'purify' the state-spaces (time of visiting) is built. Refers to Welch and Bishop [4], a new state-space $x_{i+1}$ is a linear combination of the previous state $x_i$ and *process noise* $u_i$.

## 2. Research Method

Secondary data from Bali Tourism Office were used to model Korean arrivals to Bali at time $t$, and variables in the model are Monthly arrivals data (KOR$_t$), the exchange rate of WON to IDR (EXCHANGE$_t$), and inflation rate at Korea (INFKR$_t$) and Indonesia (INFID$_t$) for period January 2010–December 2015. Modelling steps are as follows:

(i) Defining state-space model that relates all of the variables under studied. Under normality assumption of errors in the model, refers to [6], TVP model in state-space form can be expressed by

$$KOR_t = x_t\alpha_t + u_t;\ u_t \sim NID(0, H_t). \tag{1}$$

In equation (1) – is also known as measurement equation – $x_t$ refers to a matrix of predictor variables, $\alpha_t$ is a vector denotes an unobserved parameters or state vector, and $u_t$ is a vector of temporary disturbances with mean 0 and variance matrix $H_t$.

(ii) Defining state or transition equation that represents the transition from state $t$ to state $t+1$. This transition can be expressed by

$$\alpha_{t+1} = \Phi_t\alpha_t + R_te_t;\ e_t \sim NID(0, Q_t). \tag{2}$$

In equation (2) $e_t$ represents permanent disturbances matrix with mean 0 and variance $Q_t$ while $\Phi_t$ and $R_t$ are transition matrices initially assumed to be known. Regarding the measurement and transition equations, two assumptions have to be made, namely (a) the initial vector $\alpha_0$ has mean 0 and variance $P_0$; and (b) $u_t$ and $e_t$ are assumed serially independent. If the $\Phi_t$ is an identity matrix, then $\alpha_{jt}$ will follows random walk process $\alpha_{jt} = \alpha_{jt-1} + e_t$, $j = 1, \cdots,$ *number of predictors.*

(iii) Estimating the parameters. To overcome the limitation of OLR in estimating non-constant parameters, it is common Kalman Filter (KF) algorithm is applied. According to [2], this algorithm uses iteration procedure to find optimal solution from all information available at state-space at the given time $t$. In brief, KF algorithm consists of three states:

(a) Initial state: at this state the initial values for $a_0$ and $P_0$ are determined. Whenever no information is available to init both values, then [2] suggested to use *diffuse priors* technique. This technique set $P_0$ as $kI$ where $k$ is an arbitrary large value range from $10^2$ to $10^7$ and I is an identity matrix. The initial values for $a_0$ is assign randomly;

(b) Prediction state: at this state the predicted values for $y_{t+1}$, $P_{t+1}$, and $a_{t+1}$ are counted. The initial values of $a_0$ and $P_0$ as the values for state $t$ is used to count

$$a_{t+1|t} = \Phi_t a_t. \tag{3}$$

$$P_{t+1|t} = \Phi_t P_t \Phi_t' + R_t Q_t R_t'. \tag{4}$$

Applying equation (3) and (4), we predict the value of $\hat{y}_{t+1}$ by equation (5)

$$\hat{y}_{t+1|t} = x_t a_{t+1|t}. \tag{5}$$

The prediction error and mean squared error (MSE) for $y_{t+1}$ can be examined by using the equations $r_{t+1} = y_{t+1} - \hat{y}_{t+1|t}$ and $F_{t+1} = x_{t+1}P_{t+1|t}x'_{t+1} + H_{t+1}$, respectively;

(c) Update state: this state is used to refine the $a_t$'s until optimal vector is reached. The update state utilises Kalman *gain* $K_t$ that can be expressed as $K_{t+1} = P_{t+1|t}x'_{t+1}F^{-1}_{t+1}$ and $a_{t+1}$ as well as $P_{t+1}$ are updated using these equations:

$$a_{t+1} = a_{t+1|t} + P_{t+1|t}x'_{t+1}F^{-1}_{t+1}(y_{t+1} - \Phi_{t+1}a_{t+1|t}). \tag{6}$$

and

$$P_{t+1} = P_{t+1|t} - P_{t+1|t}x'_{t+1}F^{-1}_{t+1}x_{t+1}P_{t+1|t}. \tag{7}$$

(iv) Conducting goodness-of-fit test. To validate the model is sufficient, two tests are conducted. Kolmogorov-Smirnov's test is used to check whether the measurement residuals follow normal distribution, and their correlation with residuals from transition matrix do not significant. The uncorrelated properties between measurement and transition residuals were checked by observing the correlation matrix.

## 3. Results and Discussion

### 3.1. Exploring the Data

Noting that TVP is suitable for time varying data – data that vary over time – it is necessary to explore before modelling is conducted. The plot for number of Korean tourists arrived at Bali is depicted in fig. 1. It is clear from fig. 1, the number of Korean tourists arrived at Bali vary

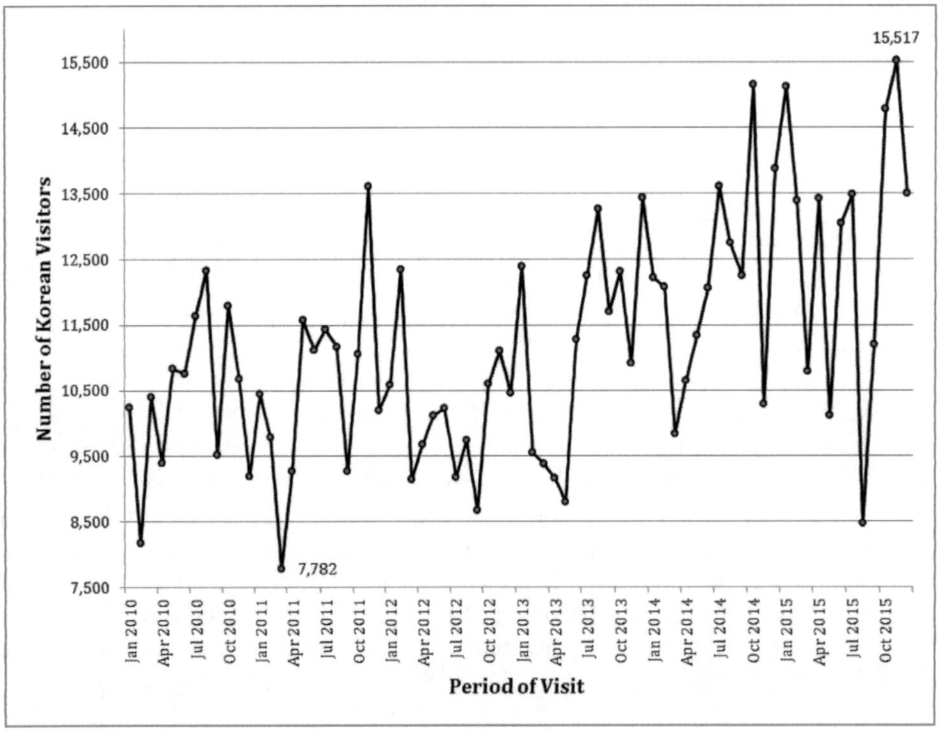


**Figure 1.** Number of Korean Arrivals at Bali for Jan. 2010–Dec. 2015

over time. In addition, it looks like the arrivals do not follow a specific pattern. Based on this figure, we conclude that Korean arrivals has time-varying properties.

IOP Conf. Series: Journal of Physics: Conf. Series **855** (2017) 012047 doi:10.1088/1742-6596/855/1/012047

For the predictor variables, plot of data show similar result. The exchange rate of WON to IDR (EXCHANGE) and the inflation rate at both countries (INFKR and INFID) demonstrate time-varying properties as show in fig. 2 and 3.

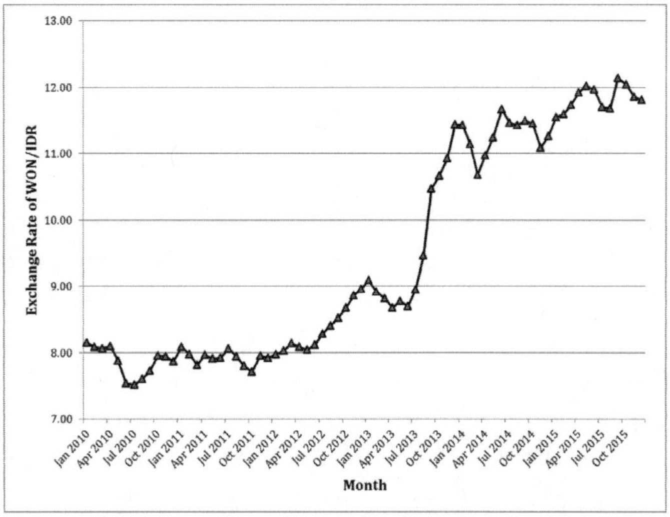


**Figure 2.** The EXCHANGE Rate of WON to IDR

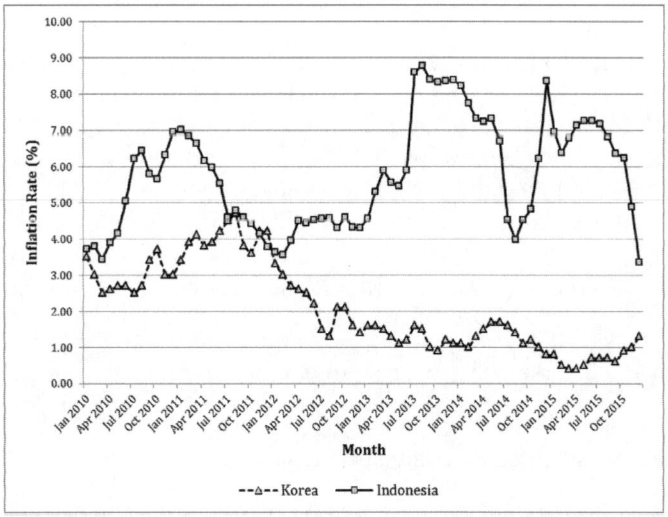


**Figure 3.** The Inflation Rates in Korea and Indonesia

By concluding the data matrix is appropriate to model using TVP, the state-space model were defined as follows:

$$\mathrm{KOR}_t = \alpha_{1t}\mathrm{EXCHANGE}_t + \alpha_{2t}\mathrm{INFKR}_t + \alpha_{3t}\mathrm{INFID}_t + u_t. \tag{8}$$

and

$$\alpha_{it+1} = \alpha_{it} + e_{it+1};\ i = 1, \cdots, 3. \tag{9}$$

### *3.2. Estimating the Parameters*

To estimate the model parameters, we applied maximum likelihood estimation (MLE) to know the covariance of measurement and transition equations $H_t$ and $Q_t$. By using EViews software, the value for $H_t$ and $Q_t$ are:

$$H_t = 2.05 \cdot 10^{-36}, \text{ and}$$

$$Q_t = \begin{pmatrix} 930.99 & 0 & 0 \\ 0 & 917.68 & 0 \\ 0 & 0 & 2.93 \cdot 10^{-22} \end{pmatrix}$$

The next step is to apply those values in estimating the variance of unobserved parameters $\alpha_t$ and its covariance state vector. KF algorithm is used with the initial value for $a_i$ and Kalman *gain* $k$ is set to 0 and $10^6$, respectively. The final state is listed on Table 1.

**Table 1.** Estimated the final state by using KF algorithm

| | Log(Variance) | Std. Error | z-Statistic | Prob. |
|---|---|---|---|---|
| Constant | 1.582 | 30.7 $10^{-5}$ | 5148.88 | 0.0000 |
| EXCHANGE | -0.046 | 8.6 $10^{-5}$ | -535.44 | 0.0000 |
| INFKR | 0.028 | 23.0 $10^{-5}$ | 121.25 | 0.0000 |
| INFID | 1.233 | 6.4 $10^{-5}$ | 19151.66 | 0.0000 |

Refers to Table 1, as noted by Bossche [7], the output of EViews for each of predictors shows the logarithmic of its variance. Taking the exponential of these values, we found the inflation rate of Indonesia (INFID) has the biggest variance as much as $exp(1.233) = 3.433$, and are followed by the inflation rate of Korea (INFKR) and the exchange rate between WON and IDR (EXCHANGE) as much as 1.028 and 0.955, respectively.

Our finding showed number of Korean tourists visiting Indonesia is more fluctuates on inflation of Indonesia and is less on the exchange rate between WON and IDR. Refers to Table 1, the measurement and state equation for Korean arrivals can be expressed as:

$$KOR_t = (EXCHANGE_t \quad INFKR_t \quad INFID_t) \begin{pmatrix} 0.955 \\ 1.028 \\ 3.434 \end{pmatrix} \tag{10}$$

and final covariance matrix for $\alpha_t$ is given by equation 12.

$$P_t = \begin{pmatrix} 1177.56 & -2911.28 & -8199.77 \\ -2911.28 & 1138650.00 & 30946.04 \\ -8199.77 & 30946.04 & 276626.30 \end{pmatrix} \tag{11}$$

### *3.3. Validating the Model*

To validate the resulted model, the residuals from measurement equation (10) is checked by using Kolmogorov-Smirnov normality test. The test showed p-value for test statistic is 0.056, slightly greater than type-I error ($\alpha$) 5 percent. We concluded, the residuals follow normal distribution. In addition, we also checked the correlation between the residuals and the predictors. Except for correlation between residuals and INFID as much as 0.266 (p-value = 0.024), the others were not significant with p-value as much as 0.864 and 0.862 for correlation between residuals–EXCHANGE and residuals–INFKR, respectively. Observing the residuals follow normal distribution and two of correlation values were not significant, than the resulted model as expressed in equation (10) is qualified to represent the relationship between Korean arrivals and its predictors.

The other technique to validate our model is to count the Mean Absolute Percentage Error (MAPE) value. MAPE can be understood as how good the model represents the original data, and its value can be counted using this formula:

$$\text{MAPE} = \frac{\sum_{t=1}^{m}\left(\frac{|e_t|}{y_t}\right)}{m} \tag{12}$$

For in-sample forecasting, we got MAPE's value is 11.24 percent. Although there is no consensus about the upper limit of MAPE to claim the model is good enough, we believe our model is sufficient to predict the number of Korean tourists come to Bali with the accuracy level as much as 88.76 percent. To validate, we make an out-of-sample forecast for six consecutive months in year 2016. In predicting Korean arrivals for January–June 2016, we forecast it on 2 different ways, i.e. (a) previously, we forecast all of the explanatory variables by using ARIMA models. Applying these forecasted values, then we make out-of-sample forecast of Korean arrivals for that period (Forecast$^1$); and (b) by utilising the real value for EXCHANGE, INFKOR, and INFID for that period (Forecast$^2$). The forecast's result is listed on Table 2. The MAPE's value for out-of-sample forecast using Forecast$^1$ and Forecast$^2$ as much as 12.86 percent and 12.28 percent, respectively.

**Table 2.** Out-of-Sample Forecast for Korean Arrivals

| Month | | EXCHANGE Rate | | KOR Inflation | | IND Inflation | | Korean Arrivals | | |
|---|---|---|---|---|---|---|---|---|---|---|
| | | Forecast | Real | Forecast | Real | Forecast | Real | Forecast$^1$ | Forecast$^2$ | Real |
| Jan. | 2016 | 11.80 | 11.43 | 1.42 | 6.8 | 2.96 | 4.13 | 13623 | 12827 | 14257 |
| Feb. | 2016 | 11.80 | 11.43 | 1.66 | 1.3 | 2.96 | 4.42 | 13775 | 13164 | 13247 |
| Mar. | 2016 | 11.80 | 11.49 | 1.84 | 1.0 | 3.05 | 4.45 | 13906 | 12975 | 8327 |
| Apr. | 2016 | 11.80 | 11.13 | 1.80 | 1.0 | 3.19 | 3.60 | 14880 | 12630 | 11900 |
| May | 2016 | 11.80 | 11.10 | 1.61 | 0.8 | 3.20 | 3.33 | 13749 | 12461 | 14498 |
| June | 2016 | 11.80 | 11.54 | 1.42 | 0.8 | 3.09 | 3.45 | 13625 | 12934 | 12419 |

Note:
Forecast$^1$: The forecasted values for explanatory variables were used
Forecast$^2$: The real values for explanatory variables were used

## 4. Conclusion

By demonstrating the use of time-varying parameter model to predict number of Korean tourists visit Bali, we found the inflation rate of Indonesia affects the Korean arrivals more fluctuate than the effects of inflation rate of Korea and the exchange rate between two countries. Furthermore, we found TVP model is qualified enough to model Korean demand to visit Bali with the in-sample MAPE's is 11.24 percent. By using the ARIMA's forecasted results for predictors' values, we got the out-of-sample forecast as much as 12.86 percent, slightly greater when the real values for predictor were used to with MAPE's as much as 12.28 percent.

## Acknowledged
We would like to thank the Udayana University for funding this research through Grant No. 486.64/UN14.2/PNL.01.03.00/2016.

## References
[1] Qiu H and Shang J 1995 *Journal of Travel Research* **34** 43–49
[2] Witt S F and Song H 2000 *Advances in Tourism Research: Tourism Demand Modelling and Forecasting. Modern Econometric Approaches* (New York: Routledge)
[3] Badan Pusat Statistik Bali 2015 *Statistik Wisatawan Mancanegara ke Bali* Denpasar: BPS Provinsi Bali
[4] Welch G and Bishop G 2001 *An Introduction to the Kalman Filter* (University of North Carolina at Chapel Hill)
[5] Lei GUO 1990 *IEEE Transaction on Automatic Control* **35 (2)** 141–147
[6] Gang Li, H Song and S F Witt 2005 *Journal of Travel Research* **44** 82–89
[7] Filip A. M. Van den Bossche 2011 *Journal of Statistical Software* **41(8)** 1–16

IOP Conf. Series: Journal of Physics: Conf. Series **855** (2017) 012048 doi:10.1088/1742-6596/855/1/012048

# Markov Switching Autoregressive Conditional Heteroscedasticity (SWARCH) Model to Detect Financial Crisis in Indonesia Based on Import and Export Indicators

**Sugiyanto, Etik Zukhronah, Yuliana Susanti and Sisca Rahma Dwi**

Study Program of Statistics, Faculty of Mathematics and Natural Sciences
Universitas Sebelas Maret

Sugiyanto61@staff.uns.ac.id

**Abstract**. A country is said to be a crisis when the financial system is experiencing a disruption that affects systems that can not function efficiently. The performance efficiency of macroeconomic indicators especially in imports and exports can be used to detect the financial crisis in Indonesia. Based on the import and export indicators from 1987 to 2015, the movement of these indicators can be modelled using SWARCH three states. The results showed that SWARCH (3,1) model was able to detect the crisis that occurred in Indonesia in 1997 and 2008. Using this model, it can be concluded that Indonesia is prone to financial crisis in 2016.

## 1. Introduction

Each country has different natural potentials that not available in other countries. This condition caused a country will need products that are not available in the country but available in other countries. Therefore, the country is trading with other countries, so that there are import and export activities. High imports and low exports may indicate a crisis in the country. Indonesia experienced the worst crisis that occurred in the middle of July 1997. So that it needs to hold a system of financial crisis detection. According to Kaminsky et al. [5], there are 15 indicators that can be used to detect crises such as import and export.

Some volatility models that ever introduced by experts are autoregressive conditional heteroscedasticity (ARCH), generalized autoregressive conditional heteroscedasticity (GARCH), and exponential generalized autoregressive conditional heteroscedasticity (EGARCH). In the volatility model the condition changes that occur in the data are ignored, but in the Markov model switching changes the condition is considered as an unobserved variable called state. Later Hamilton and Susmel [3] introduced the Markov switching ARCH (SWARCH) model. The model is able to explain the state changes and illustrate the data volatility well. The SWARCH model has been applied by some researchers, e.g. Chang et al. [1] uses SWARCH model assuming three states. The results show that SWARCH (3,2) model can identify crisis on stock market and Korean Won exchange rate crisis per US Dollar.

This study determines the appropriate model for the movement of import and export indicators. The model is used to detect early financial crisis in 2016. Furthermore, it is determined the relation of conditions of import and export indicators in detecting financial crisis in Indonesia.

## 2. ARMA Model

The ARMA model consists of two components, the autoregressive (AR) model with the p order and the moving average (MA) model with the q order (Tsay [7]). The ARMA model can be written as

$$r_t = \sum_{i=1}^{p} \emptyset_i r_{t-i} + a_t - \sum_{i=1}^{q} \theta_i a_{t-i},$$

where $r_t$ is the value of log return at time t, $\emptyset$ is parameter of AR model, θ is parameter of MA model, and $a_t$ is the residue of ARMA model at time t.

## 3. Volatility Model

Some models of volatility that ever introduced by experts are ARCH, GARCH and EGARCH.

### *3.1. ARCH Model*

The ARCH(m) model can be written as follows

$$\sigma_t^2 = \alpha_0 + \sum_{i=1}^{m} \alpha_i a_{t-i}^2,$$

where $\alpha_0 > 0, \alpha_i \geq 0,$, m is the order of the ARCH model and $\sigma_t^2$ is the residual variance of the ARMA model.

### *3.2. GARCH Model*

The GARCH(p, q) model can be written as follows

$$\sigma_t^2 = \alpha_0 + \sum_{i=1}^{p} \alpha_i a_{t-i}^2 + \sum_{j=1}^{q} \beta_j \sigma_{t-j}^2,$$

where $\beta_j \geq 0$ and $(p, q)$ is the order of the GARCH model.

### *3.3. EGARCH Model*

The EGARCH(r, s) model can be written as follows

$$\log \sigma_t^2 = \alpha_0 + \sum_{i=1}^{r} \alpha_i \left( \left| \frac{a_{t-i}}{\sqrt{\sigma_{t-i}^2}} \right| - \sqrt{\frac{2}{\pi}} \right) + \sum_{j=1}^{s} \beta_j \log \sigma_{t-j}^2 + \sum_{i=1}^{s} \gamma_i \frac{a_{t-i}}{\sqrt{\sigma_{t-i}^2}},$$

where $\gamma_j \geq 0$ and $(r, s)$ is the order of the EGARCH model.

## 4. SWARCH Model

According to Hamilton and Susmel [3], the SWARCH model can be written as

$$r_t = \mu_{s_t} + a_t, a_t = \sigma_t \varepsilon_t,$$

$$\sigma_{t,s_t}^2 = \alpha_{0,s_t} + \sum_{i=1}^{m} \alpha_{i,s_t} a_{t-i}^2,$$

where $\varepsilon_t \sim N(0,1)$, and $s_t = \{1,2, \dots, k\}$. The equation is said to be a SWARCH process with k state and m order and can be denoted as $a_t \sim SWARCH(k, m)$.

## 5. Filtered Probability

Filtered probability is the probability of a state in the t-period based on observational data until the t-period (Sopipan et al. [6]). According to Hermosillo and Hesse [4], the value of filtered probability of more than 0.6 is in state 3 with high volatility or can be said to occur the crisis in that data period. The value of filtered probability between 0.4 and 0.6 is in state 2 with medium volatility or can be interpreted as prone to crisis. While the filtered probability value less than 0.4 is in state 1 with low volatility or stable condition. According to Hamilton [2], filtered probability in state 3 can be written as

$$P_r[s_t=3|\psi_t] = 1 - P_r[s_t=1|\psi_t] - P_r[s_t=2|\psi_t],$$

where $\psi_t$ is an $a_t$ set until time t, $P_r[s_t=1|\psi_t]$ is filtered probability at state 1 and $P_r[s_t=2|\psi_t]$ is filtered probability at state 2.

## 6. Detection of Crisis

To detect the crisis can be seen from the value of filtered probability at state 3 in the period of the data. According to Sopipan et al. [6], filtered probability forecasting can be written as

$$\Pr(s_t = j\,|\psi_t) = \sum_{i=1}^{3} p_{ij} \Pr(s_{t-1} = i\,|\psi_{t-1}),$$

where $p_{ij}$ is the transition probability from state i to state j and $\Pr(s_{t-1} = i\,|\psi_{t-1})$ is filtered probability at state i and time t-1.

## 7. Research Methods

The data used in this research are monthly data of import and export from 1987 to 2015 that obtained from International Financial Statistics (IFS) CD-ROM issued by IMF and Bank Indonesia (BI). The steps of data analysis are

(1) Creating a data plot and then testing the stationary of data using a Dickey Fuller (ADF) augmented test.
(2) If the data are not stationary then it is transformed using log return.
(3) Create an Autocorrelation Function (ACF) and Partial Autocorrelation Function (PACF) plot of log return data to determine the appropriate ARMA model.
(4) Testing the effect of heteroscedasticity by using Lagrange multiplier test.
(5) Establishing an appropriate volatility model and combining volatility model with Markov Switching assuming three states.
(6) Calculating the value of filtered probability at state 3 to detect crisis.
(7) Forecasting the crisis conditions for one year ahead and determining the relation of conditions of import and export indicators.

## 8. Result and Discussion

Plot of import and export data can be seen in Figure 1 and Figure 2.

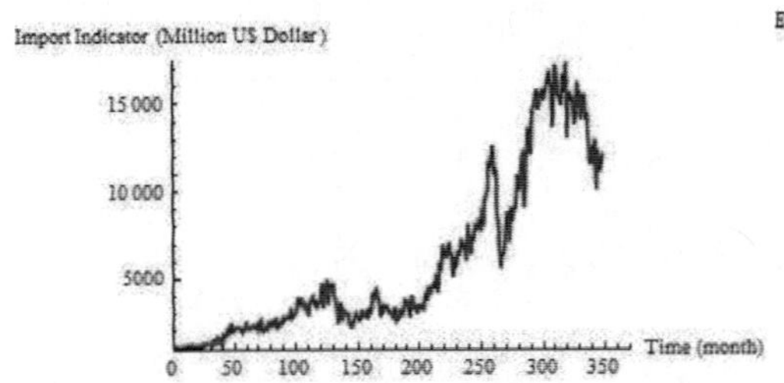


Figure 1. Plot of Import Data in 1987-2015

Figure 2. Plot of Export Data in 1987-2015

Figure 1 and Figure 2 show that the data fluctuates from time to time. This indicates that the data is not stationary, so that required data transformation using log return. Plot of log return of import and export can be seen in Figure 3 and Figure 4.

Figure 3.Plot of Log Return of Import

Figure 4. Plot of Log Return of Export

Figure 3 and Figure 4 show that the data are stationary and evidenced by an ADF test value of 0.001 less than $\alpha$ = 0.05. Since the log return of import and export data are stationary, it can be modeled using the ARMA model.

### *8.1. Establishment of ARMA Model*

The result of ARMA model parameter estimation based on plot of ACF and PACF got the appropriate model for import indicator that is ARMA (2,0) which can be written as

$$r_t = -0{,}552200\, r_{t-1} - 0{,}287786\, r_{t-2} + a_t\,.$$

For the export indicator, it is obtained ARMA(1,0) model which can be written as

$$r_t = -0{,}328037\, r_{t-1} + a_t$$

Based on Lagrange multiplier test, it is obtained probability value for imports of 0.0388 and exports of 0.0007, with both probabilities smaller than 0.05. This shows that the residue of ARMA model contains heteroscedasticity effect, so the volatility model is used.

### *8.2. Formation of Volatilty Model*

Volatility models include ARCH, GARCH, and EGARCH. The resuts of ARCH model parameter estimation for the import data are presented in Table 1 and for export data are presented in Table 2.

Table 1. The result of ARCH Model Parameter Estimation for Import Data

| Parameter | *ARCH*(1) | Prob. | *ARCH*(2) | Prob. |
|---|---|---|---|---|
| $\alpha_0$ | 0,010293 | 0.0000 | 0,010024 | 0.0000 |
| $\alpha_1$ | 0,289039 | 0.0042 | 0,283806 | 0.0057 |
| $\alpha_2$ | - | - | 0,023753 | 0.7427 |
| *AIC* | -1,442103 | - | -1,437041 | - |
| *SC* | -1,397541 | - | -1,381337 | - |

The corresponding model to the import data is ARCH (1) which can be written as

$$\sigma_t^2 = 0{,}010293 + 0{,}289039\, a_{t-1}^2$$

Table 2. The result of ARCH Model Parameter Estimation for Export Data

| Parameter | *ARCH*(1) | Prob. | *ARCH*(2) | Prob. | *ARCH*(3) | Prob. |
|---|---|---|---|---|---|---|
| $\alpha_0$ | 0,005780 | 0.0000 | 0,005108 | 0.0000 | 0,005098 | 0.0000 |
| $\alpha_1$ | 0,196269 | 0.0129 | 0,178101 | 0.0297 | 0,178505 | 0.0295 |
| $\alpha_2$ | - | - | 0,113774 | 0.0371 | 0,113503 | 0.0373 |
| $\alpha_3$ | - | - | - | - | 0,001399 | 0.9771 |
| *AIC* | -2,116821 | - | -2,122922 | - | -2,117146 | - |
| *SC* | -2,083471 | - | -2,078455 | - | -2,061561 | - |

The corresponding model to the export data is ARCH (1) which can be written as

$$\sigma_t^2 = 0{,}010293 + 0{,}289039\, a_{t-1}^2 \ ,$$

where $a_{t-1}^2$ is the quadratic residue at time t-1. Furthermore, the GARCH model is estimated and obtained that GARCH model has insignificant parameter. So the GARCH model is unusable, and the EGARCH model also can not be used. Therefore the volatility model used is ARCH (1) model.

A diagnostic test was then performed on the residual ARCH(1) model. Based on the Ljung-Box test, it is obtained the probability value more than 0.05. This indicates that the residual ARCH(1) model does not contain autocorrelation. From the result of Lagrange multiplier and Kolmogorov-Smirnov test, it is obtained probability value for imports of 0.8734 and 0.101 and probability for export of 0.4512 and 0.558, the probability value is greater than 0.05. Thus it can be concluded that the residual ARCH(1) model does not contain heteroscedasticity and normal distributed.

### *8.3. Formation of Combined Models of Volatility and Markov Switching*

The volatility model that can be used is ARCH, so the SWARCH model is formed. The SWARCH (3,1) estimation model for import data can be written as

$$r_t = \begin{cases} 0{,}000022 & \text{for } state\ 1 \\ 0{,}000026 & \text{for } state\ 2 \\ 0{,}000019 & \text{for } state\ 3 \end{cases} .$$

The value shows the average log return data on state 1 of 0.0000022, state 2 of 0.000026, and state 3 of 0.000019. The heteroscedasticity model of SWARCH (3,1) is

$$\sigma_t^2 = \begin{cases} 0{,}009127 + 0{,}009127\, \sigma_{t-1}^2 \text{for } state\ 1 \\ 0{,}009127 + 0{,}009127\, \sigma_{t-1}^2 \text{for } state\ 2 \\ 0{,}009127 + 0{,}009127\, \sigma_{t-1}^2 \text{for } state\ 3 \end{cases} .$$

While the probability of change from one state to another is explained by a transition probability matrix written as

$$P = \begin{pmatrix} 0{,}000000 & 0{,}499999 & 0{,}099881 \\ 0{,}000000 & 0{,}499999 & 0{,}099881 \\ 0{,}999999 & 0{,}000000 & 0{,}800238 \end{pmatrix}.$$

The value indicates the probability of changing state from low to high of 0.999999, the probability of survival in the medium state is equal to the change of state from medium to low by 0.499999, and the probability to survive in a high state of 0.800238.

Furthermore SWARCH(3,1) model for export data can be written as

$$r_t = \begin{cases} 0{,}000025 & \text{for } state\ 1 \\ 0{,}000018 & \text{for } state\ 2 \\ 0{,}000027 & \text{for } state\ 3 \end{cases} .$$

The heteroscedasticity model of SWARCH (3.1) for export data is

$$\sigma_t^2 = \begin{cases} 0{,}237830 + 0{,}226953\, \sigma_{t-1}^2 \text{for } state\ 1 \\ 0{,}229286 + 0{,}226953\, \sigma_{t-1}^2 \text{for } state\ 2 \\ 0{,}231370 + 0{,}226953\, \sigma_{t-1}^2 \text{for } state\ 3 \end{cases} .$$

The probability of change from one state to another can be explained by the transition probability matrix written as

$$P = \begin{pmatrix} 0{,}000000 & 0{,}000000 & 0{,}453617 \\ 0{,}000000 & 0{,}000000 & 0{,}453617 \\ 0{,}999999 & 0{,}999999 & 0{,}092766 \end{pmatrix}.$$

### *8.4. Filtered Probability*

Signal of financial crisis can be seen from filtered probability value at state 3. Plot of filtered probability of import data on period of January 1987 until December 2015 is presented in Figure 5. While the plot of filtered probability of export data on period of January 1987 to December 2015 is presented in Figure 6. In Figure 5 there are 44 months and in Figure 6 there are 28 months that have filtered probability value more than 0.6 so it indicates that indicator is in high volatility condition.

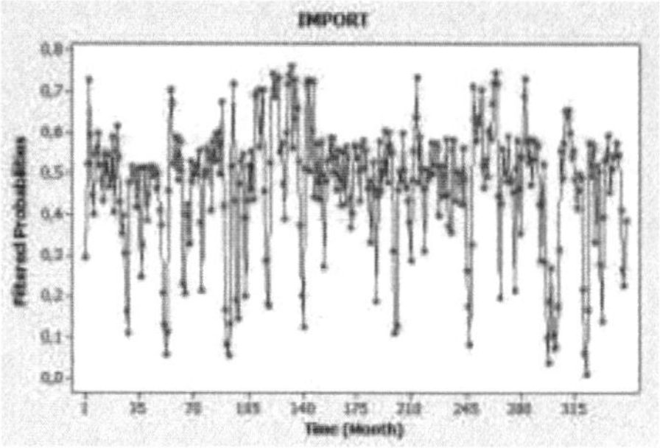


Figure 5. Filtered Probability of Import

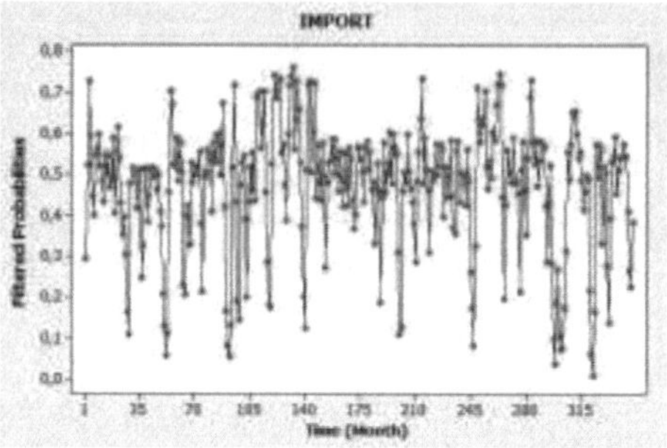


Figure 6. Filtered Probability of Export

### *8.5. Forecasting*

Forecasting of filtered probability based on import and export indicator is presented in Table 3.

Table 3. Forecasting of Filtered Probability of Import and Export Indicators in 2016

| Month | Import | Exsport |
|---|---|---|
| January | 0,581679 | 0,573237 |
| February | 0,474497 | 0,479944 |
| March | 0,542282 | 0,564582 |
| April | 0,562576 | 0,487791 |
| May | 0,568615 | 0,557458 |
| June | 0,570374 | 0,494254 |
| Juli | 0,570848 | 0,551595 |
| August | 0,570937 | 0,499573 |
| September | 0,570911 | 0,546769 |
| October | 0,570849 | 0,503951 |
| November | 0,570776 | 0,542797 |
| December | 0,570701 | 0,507555 |

Table 3 shows the value of filtered probability between 0.4 and 0.6. Therefore it can be concluded that Indonesia is prone to financial crisis in 2016 based on import and export indicators.

To determine the correlation of indicator condition used chi square test for independence. Since the value of $T = 22{,}70742 > \chi^2_{(0{,}95;4)} = 9{,}488$ and p-value = 0,000 less than $\alpha = 0.05$, it can be concluded that there is a condition relationship between import and export indicators in detecting financial crisis in Indonesia.

## 9. Conclusion

The appropriate model for the movement of import and export indicators from 1987 to 2015 is SWARCH (3,1). The SWARCH (3,1) model is able to provide information that 2016 is prone to financial crisis in Indonesia. There is a relationship of import and export indicators in detecting financial crisis in Indonesia.

IOP Conf. Series: Journal of Physics: Conf. Series **855** (2017) 012048 doi:10.1088/1742-6596/855/1/012048

**References**

[1] Chang K, Cho K Y and Hong M 2010 *J. of Economic Research* **15** 249-272.
[2] Hamilton J D 1989 *Econometrica* **57** 357-384.
[3] Hamilton J D and Susmel R 1994 *J. of Econometrics* **64** 307-333.
[4] Hermosillo B G and Hesse H 2009 *Global Market Condition and Systematic risk: IMF* IMF Working Paper.
[5] Kaminsky G, Lizondo S and Reinhart C 1998 *Leading Indicators of Currency Crises* IMF Staff Papers **45**.
[6] Sopipan N, Sattayatham P and Premanode B 2012 *J. of Mathematical Finance* **2** 121-131.
[7] Tsay R S 2005 *Analysis of Financial Time Series* (Canada: John Wiley and Sons).

# Application of Semiparametric Spline Regression Model in Analyzing Factors that Influence Population Density in Central Java

**Y D Sumantari [1], I Slamet [2], and Sugiyanto [2]**

[1] Department of Mathematics, Faculty of Mathematics and Natural Sciences, Universitas Sebelas Maret, Indonesia
[2] Department of Statistics, Faculty of Mathematics and Natural Sciences, Universitas Sebelas Maret, Indonesia

E-mail: isnandarslamet@staff.uns.ac.id

**Abstract.** Semiparametric regression is a statistical analysis method that consists of parametric and nonparametric regression. There are various approach techniques in nonparametric regression. One of the approach techniques is spline. Central Java is one of the most densely populated province in Indonesia. Population density in this province can be modeled by semiparametric regression because it consists of parametric and nonparametric component. Therefore, the purpose of this paper is to determine the factors that influence population density in Central Java using the semiparametric spline regression model. The result shows that the factors which influence population density in Central Java is Family Planning (FP) active participants and district minimum wage.

## 1. Introduction

Indonesia is an archipelago state which is the fourth most populated country in the world. The number of population is 255 million people which separate in about 2.342 islands (BPS, 2015). Due to the number of both population and island, each island in Indonesia has different population density. Java is an island which has the most densely populated in Indonesia. Central Java is the third most densely populated province in both Java and Indonesia after West Java and East Java. Population density is influenced by many factors. According to Sugiharyanto [8], the factors that influence population density are birth, climate and strategic places, economic, and social.

Semiparametric regression is a statistical analysis method that consists of parametric and nonparametric regression (Ruppert [7]). Parametric regression is used when the curve can be represented in terms of the parametric model (Hardle [4]). By contrast, nonparametric regression refers to flexible functional form of the regression curve (Eubank [2]). There are various approach techniques in nonparametric regression. One of the approach techniques is spline. Spline is work quite well in terms of giving a visually pleasing fit and has a definite computational advantage over the other methods (Eubank [2]).

The semiparametric spline regression has been applied to many cases. We may refer to Engle [1], who used semiparametric regression to estimates the relation between weather and electricity sales. Gimenez [3] applied semiparametric regression in capture-recapture modeling. Pambudi [6] also applied semiparametric spline regression to modeling Dengue Fever sickness in Madiun. Similarly, Marina [5] also applied semiparametric spline regression to determine the factors that influence criminality percentage in East Java. All of that papers provide a good result. Therefore, in this paper we use semiparametric spline regression to determine the factors that influence population density in Central Java.

## 2. Semiparametric Spline Regression

Semiparametric spline regression is a semiparametric regression that use spline approach. Assumed that a pairwise data set $(x_i, y_i, t_i)$ is following semiparametric regression, then the semiparametric regression equation can be modeled by

$$y_i = x'\beta_i + f(t_i) + \varepsilon_i, i = 1, 2, \ldots, n$$

where $y_i$ is the $i-$th response variable, $\beta_i$ is parameter of parametric function, $x'$ is predictor, $f(t_i)$ is nonparametric function, and $\varepsilon_i$ is residual which has $N(0, \sigma^2)$ distribution. Nonparametric function in this model can be formed by spline approach. Spline function with $p$ order and $k$ knot can be written as

$$f(t) = \Sigma_{j=0}^{p}\beta_i t_i^j + \Sigma_{k=1}^{K}\beta_{p+k}(t - K_k)_+^p$$

where $p$ is the order or degree of spline, and $(t - K_k)_+^p$ is piecewise of polynom with $(t - K_k)_+^p = \begin{cases} (t - K_k)^p, & t \geq K_k; \\ 0, & t < K_k. \end{cases}$

## 3. Research of Methodology

Data used in this research is secondary data of population density in Central Java. The factors are density population per $km^2$ $(y)$ as response variable, and the average age of first marriage $(x_1)$, the number of population $(t_1)$, the number of family planning (FP) active participants $(t_2)$, distance to province capital $(t_3)$, and district minimum wage $(t_4)$ as predictor variables. The steps to solve the problem of this paper are

(1) Draw the scatter plot between response variable and each predictor variables to determine the patterns in data.
(2) Determine the variable belongs to parametric or nonparametric component.
(3) Make the model of population density in Central Java using 1 knot, 2 knots, 3 knots of spline, and its combination.
(4) Choose the optimal knots by the minimum value of Generalized Cross Validation (GCV).
(5) Determine the best model by the optimal knots.
(6) Verify the significance of semiparametric spline regression parameter by $F$-test and $t$-test.
(7) Verify the residual assumption in the spline model.
(8) Interpret the model and make the conclusion.

## 4. Results

### *4.1. Scatter Plot and Knot*

First step is to apply the semiparametric spline regression in estimating the relationships among variables. It can be shown by plotting the data.

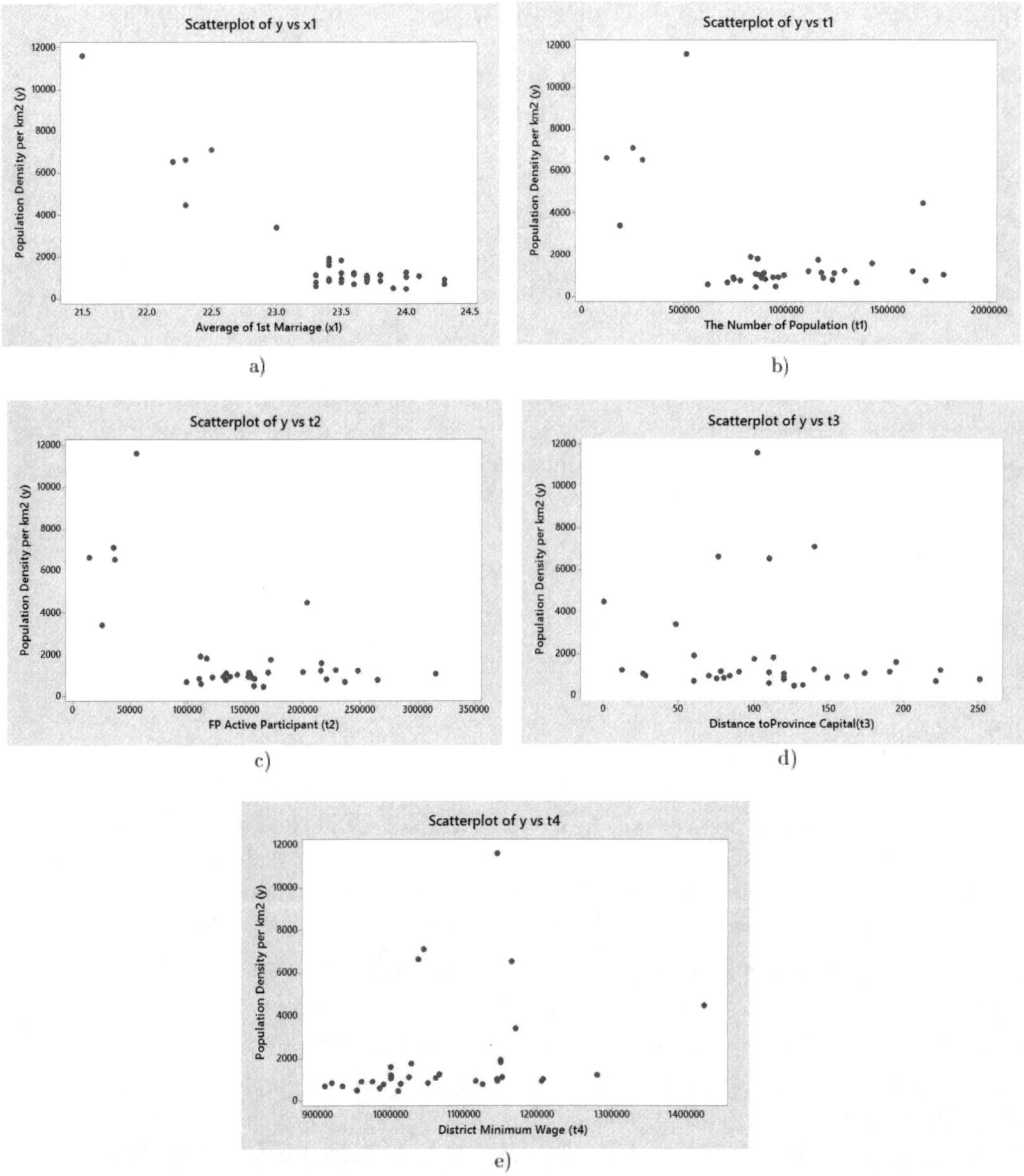


**Figure 1.** a) Scatter plot between $y$ vs $x_1$, b) Scatter plot between $y$ vs $t_1$, c) Scatter plot between $y$ vs $t_2$, d) Scatter plot between $y$ vs $t_3$, e) Scatter plot between $y$ vs $t_4$

From the Figure 1, it can be seen that scatter plot between density population per $km^2$ $(y)$ and the average of first marriage $(x_1)$ represents linear curve and scatter plot between density population per $km^2$ $(y)$ and the number of population $(t_1)$, density population per $km^2$ $(y)$ and the number of family planning (FP) active participants

($t_2$), density population per $km^2$ ($y$) and distance to province capital ($t_3$), and also density population per $km^2$ ($y$) and district minimum wage ($t_4$) do not represents any formed curve. For the variables that not represent any formed curve, it can be used spline method to approach the nonparametric function. Knot selection is important because it'll influence the result of model. The best knot is a knot which has minimum value of GCV. The knots and its GCV is shown in Table 1.

**Table 1.** Knots and its GCV

| No | Titik Knot | $t_1$ | $t_2$ | $t_3$ | $t_4$ | GCV |
|---|---|---|---|---|---|---|
| 1 | 1 | 576375 | 97622 | 62 | 1051655 | 728305 |
| 2 | 2 | 576375<br>975376 | 97622<br>170371 | 62<br>116 | 1051655<br>1175603 | 559577 |
| 3 | 3 | 234373<br>348374<br>633375 | 24872<br>76836<br>98999 | 39<br>87<br>90 | 991655<br>1101362<br>1161250 | 213616 |
| 4 | 1,2,3,3 | 576375 | 97622<br>170371 | 39<br>87<br>90 | 991655<br>1101362<br>1161250 | 664548 |
| 5 | 1,3,2,3 | 576375 | 24872<br>76836<br>98999 | 62<br>116 | 991655<br>1101362<br>1161250 | 198029 |
| 6 | 2,2,3,3 | 576375<br>975376 | 97622<br>170371 | 39<br>87<br>90 | 991655<br>1101362<br>1161250 | 729774 |
| 7 | 2,3,2,2 | 576375<br>975376 | 24872<br>76836<br>98999 | 62<br>116 | 1051655<br>1175603 | 193795 |
| 8 | 2,3,2,3 | 576375<br>975376 | 24872<br>76836<br>98999 | 62<br>116 | 991655<br>1101362<br>1161250 | 222076 |
| 9 | 3,3,2,2 | 234373<br>348374<br>633375 | 24872<br>76836<br>98999 | 62<br>116 | 1051655<br>1175603 | 203049 |
| 10 | 3,2,3,2 | 234373<br>348374<br>633375 | 97622<br>170371 | 39<br>87<br>90 | 1051655<br>1175603 | 648211 |

From Table 1, it can be shown that the minimum value of GCV is 193795. It means that the best knots is the combination of 2,3,2,2 knots.

*4.2. Semiparametric Spline Regression Model*

After we achieved the best knots, the next step is to determine the semiparametric spline regression model. Using *Minitab*, the result can be shown in Table 2.

**Table 2.** Result of Regression

| Variable | Parameter | Coefficient | $p - value$ |
|---|---|---|---|
| | $\beta_0$ | 9410 | 0.221 |
| $x_1$ | $\beta_1$ | -88 | 0.762 |
| $t_1$ | $\beta_2$ | -0.0163 | 0.131 |
| | $\beta_3$ | 0.0184 | 0.089 |
| | $\beta_4$ | -0.00023 | 0.861 |
| $t_2$ | $\beta_5$ | -0.2123 | 0.001 |
| | $\beta_6$ | 0.648 | 0.000 |
| | $\beta_7$ | -1.328 | 0.000 |
| | $\beta_8$ | 0.8825 | 0.000 |
| $t_3$ | $\beta_9$ | 16.7 | 0.243 |
| | $\beta_{10}$ | -16.9 | 0.315 |
| | $\beta_{11}$ | -1.17 | 0.852 |
| $t_4$ | $\beta_{12}$ | 0.00309 | 0.165 |
| | $\beta_{13}$ | -.0.00386 | 0.339 |
| | $\beta_{14}$ | 0.01432 | 0.006 |

From Table 2, it can be formed the semiparametric spline regression model. The model is $\widehat{y} = 9410 - 88x_1 - 0.0163t_1 + 0.0184(t_1 - 576375)_+ - 0.00023(t_1 - 975376)_+ - 0.2123t_2 + 0.648(t_2 - 24872)_+ - 1.328(t_2 - 76836)_+ + 0.8825(t_2 - 98999)_+ + 16.7t_3 - 16.9(t_3 - 62)_+ - 1.17(t_3 - 116)_+ + 0.00309t_4 - 0.00386(t_4 - 1051655)_+ + 0.01432(t_4 - 1175603)_+$, where the value of both $R^2$ dan MSE are 98.78% and 121250, respectively.

### *4.3. F-test and t-test*

(1) $F$-test
$F$-test is used to verify the significance of models with null hypothesis is no significant regression model. Using $\alpha = 0.05$, the critical region is to reject the null hypothesis if $p - value < \alpha$. The result of $p - value$ is 0.000. Therefore, the null hypothesis is rejected. It means that regression model is significant.

(2) $t$-test
$t$-test is used to verify the significance of parameters with null hypothesis is no significant parameter. Using $\alpha = 0.05$, the critical region is to reject the null hypothesis if $p - value < \alpha$. From Table 2, it can be shown that the significant parameter are $\beta_5$, $\beta_6$, $\beta_7$, $\beta_8$, and $\beta_{14}$.

### *4.4. Residual Assumption*

(1) Identically Distributed
A data is identical if its residual has homoscedasticity assumption. Homoscedasticity can be detected using Glejser test, with null hypothesis is residual has homoscedasticity assumption. Using $\alpha = 0.05$, the critical region is reject the null hypothesis if $p - value < \alpha$. The result of $p - value$ is 0.443. Therefore, the null hypothesis is not rejected. It means that residual has homoscedasticity assumption.

(2) Independent Distributed
Independent test is a test that expected the residuals have no correlation. Independent test can be check using autocorrelation function (ACF) plot. If

in ACF plot there is no lag which out of the margin level, then there is no autocorrelation in residuals. The result can be shown below.

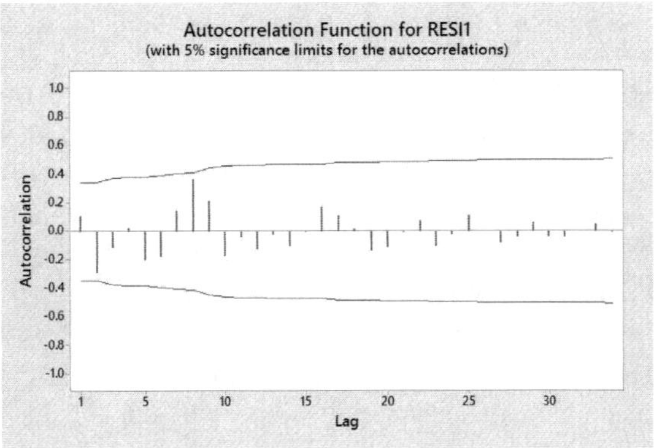


**Figure 2.** ACF Plot

From Figure 2, can be shown that there is no lag which out from confidence interval. Therefore, the residuals have no autocorrelation.

(3) Normality Test
Normality test is used to determine the normality of residual. This test is using Anderson-Darling test with null hypothesis is residual has normal distribution. For $\alpha = 0.05$ and $p - value$ is 0.623, residual has normal distribution.

*4.5. Interpretation of Model*
From the analysis above, it necessary to interpret the model. The coefficient of determination for this model is 98.78%. It means that 98.78% of population density can be explained by predictor variables. The high coefficient of determination shows that the regression line fits to data. The interpretation of semiparametric spline regression model that consists of factors that influence population density in Central Java shown below.

(1) Relationship between FP active participant ($t_2$) and population density per $km^2$ ($y$) with the assumption that other variables are constant is

$$\widehat{y} = -0.2123t_2 + 0.648(t_2 - 24872)_+ - 1.328(t_2 - 76836)_+ + 0.8825(t_2 - 98999)_+$$

where

$$\widehat{y} = \begin{cases} -0.2123t_2 & ; t_2 < 24872 \\ 0.4357t_2 - 16117 & ; 24872 \leq t_2 < 76836 \\ 85921 - 0.8923t_2 & ; 76836 \leq t_2 < 98999 \\ -1446 - 0.0098t_2 & ; t_2 \geq 98999 \end{cases}$$

When the FP active participants less than 24872, if it increases by 1 people then the model predicts that the population density per $km^2$ decreases by 0.2123 people. Area in this segment is Magelang City. When the FP active participants between 24872 and 76836, if it increases by 1 people then the model predicts that the population density per $km^2$ increases by 0.4357 people. Areas in this segment are Surakarta City, Salatiga City, Pekalongan City, and Tegal City. When the FP active participants between 76836 and 9899, if it increases by 1 people then the model predicts that the population density per $km^2$ decreases by 0.8923 people. Area in this segment is Purworejo Regency. When the FP active participants greater than 98999, if it increases by 1 people then the model predicts that the population density per $km^2$ decreases by 0.0098 people. Areas in this segment are Cilacap Regency, Banyumas Regency, Purbalingga Regency, Banjarnegara Regency, Kebumen Regency, Wonosobo Regency, Magelang Regency, Boyolali Regency, Klaten Regency, Sukoharjo Regency, Wonogiri Regency, Karanganyar Regency, Sragen Regency, Grobogan Regency, Blora Regency, Rembang Regency, Pati Regency, Kudus Regency, Jepara Regency, Demak Regency, Semarang Regency, Temanggung Regency, Kendal Regency, Batang Regency, Pekalongan Regency, Pemalang Regency, Tegal Regency, Brebes Regency, and Semarang City.

(2) Relationship between the district minimum wage ($t_4$) and population density per $km^2$ ($y$) with the assumption that other variables are constant is

$$\widehat{y} = 0.00309t_4 - 0.00386(t_4 - 1051655)_+ + 0.01432(t_4 - 1175603)_+$$

where

$$\widehat{y} = 0.01355t_4 - 12775; t_4 \geq 1175603$$

When the district minimum wage greater than 1175603, if the district minimum wage increases by Rp 1, then the model predicts that the population density per $km^2$ increases by 0.01355 peoples. Areas in this segment are Demak Regency, Semarang Regency, Kendal Regency, and Semarang City.

## 5. Conclusion

From the discussion, it can be concluded that

(1) the semiparametric spline regression model is given by

$$\begin{aligned}\widehat{y} = & \ 9410 - 88x_1 - 0.0163t_1 + 0.0184(t_1 - 576375)_+ - 0.00023(t_1 - 975376)_+ \\ & -0.2123t_2 + 0.648(t_2 - 24872)_+ - 1.328(t_2 - 76836)_+ + 0.8825(t_2 - \\ & 98999)_+ + 16.7t_3 - 16.9(t_3 - 62)_+ - 1.17(t_3 - 116)_+ + 0.00309t_4 \\ & -0.00386(t_4 - 1051655)_+ + 0.01432(t_4 - 1175603)_+\end{aligned}$$

where $y$ is population density per $km^2$, $x_1$ is the average age of first marriage, $t_1$ is the number of population, $t_2$ is family planning (FP) active participants, $t_3$ is distance to province capital, and $t_4$ is district minimum wage.

(2) The factors that influence population density in Central Java is FP active participants and district minimum wage.

## References

[1] Engle, R. F., C. W. J. Granger, John Rice, and Andrew Weiss, *Semiparametric Estimates of the Relation Between Weather and Electricity Sales*, Journal of the American Statistical Association, Vol. 81, No. 394 (Jun., 1986), pp. 310-320.

[2] Eubank, R. L., *Spline Smoothing and Nonparametric Regression*, New York: Marcel Dekker.Inc, 1988.

[3] Gimenez, O., C. Crainiceanu, C. Barbraud, S. Jenouvrier, and B. J. T. Morgan, *Semiparametric Regression in CaptureRecapture Modeling*, The International Biometric Society, No. 69, (2006) 691-698.

[4] Hardle, W., *Applied Nonparametric Regression*, New York, Cambridge University Press, 1990.

[5] Marina, S. M. T. dan I Nyoman Budiantara, *Pemodelan Faktor-Faktor yang Mempengaruhi Persentase Kriminalitas di Jawa Timur dengan Pendekatan Regresi Semiparametrik Spline*, Jurnal Sains dan Seni POMITS Vol. 2, No. 2, (2013) 2337-3520.
[6] Pambudi, F. A., *Pemodelan Jumlah Kasus Demam Berdarah Dengue (DBD) di Madiun Dengan Pendekatan Regresi Semiparametrik Spline di Kabupaten/Kota Madiun*, Thesis of Statistics Departement of FMIPA ITS, Surabaya, 2010.
[7] Ruppert, M. and R. Carroll, *Semiparametric Regression*, Cambridge University Press, New York, 2003.
[8] Sugiharyanto, *Geografi dan Sosiologi 2*, Yogyakarta: Yudhistira, 2007.
[9] http://www.jateng.bps.go.id

IOP Conf. Series: Journal of Physics: Conf. Series **855** (2017) 012050 doi:10.1088/1742-6596/855/1/012050

# Self-Intersection Local Times of Generalized Mixed Fractional Brownian Motion as White Noise Distributions

**Herry P. Suryawan, Boby Gunarso**

Department of Mathematics, Sanata Dharma University, Yogyakarta, Indonesia

E-mail: herrypribs@usd.ac.id, bobygunarso@hotmail.com

**Abstract.** The generalized mixed fractional Brownian motion is defined by taking linear combinations of a finite number of independent fractional Brownian motions with different Hurst parameters. It is a Gaussian process with stationary increments, posseses self-similarity property, and, in general, is neither a Markov process nor a martingale. In this paper we study the generalized mixed fractional Brownian motion within white noise analysis framework. As a main result, we prove that for any spatial dimension and for arbitrary Hurst parameter the self-intersection local times of the generalized mixed fractional Brownian motions, after a suitable renormalization, are well-defined as Hida white noise distributions. The chaos expansions of the self-intersection local times in the terms of Wick powers of white noises are also presented.

## 1. Introduction

The notion of mixed fractional Brownian motion was introduced by Cheridito in [2]. Let $a$ and $b$ be two real numbers such that $(a,b) \neq (0,0)$. A *mixed fractional Brownian motion* (MFBM) $M^H := (M_t^H)_{t\geq 0} := (M_t^{H,a,b})_{t\geq 0}$ of parameter $H$, $a$, and $b$ is a stochastic process defined on some probability space by $M_t^H = M_t^{H,a,b} := aB_t + bB_t^H$, where $(B_t)_{t\geq 0}$ is a Brownian motion and $(B_t^H)_{t\geq 0}$ is an independent fractional Brownian motion of Hurst parameter $H \in (0,1)$. This process was introduced to present a stochastic model of the discounted stock price in some arbitrage-free and complete financial markets. Some stochastic and analytic properties of MFBM was investigated by Zili in [14].

MFBM has been further generalized by Thäle in [13] to an arbitrary finite linear combinations of independent fractional Brownian motions. In this paper we shall name this process generalized mixed fractional Brownian motion. Let $\alpha_1, \ldots, \alpha_n$, $n \in \mathbb{N}$ be real numbers and not all $\alpha_k$ equals zero. A *generalized mixed fractional Brownian motion* (GMFBM) of parameter $H = (H_1, \ldots, H_n)$ and $\alpha = (\alpha_1, \ldots, \alpha_n)$ is a stochastic process $Z^H := (Z_t^H)_{t\geq 0} := (Z_t^{H,\alpha})_{t\geq 0}$ defined on some probability space by

$$Z_t^H = Z_t^{H,\alpha} := \sum_{k=1}^{n} \alpha_k B_t^{H_k},$$

where $(B_t^{H_k})_{t\geq 0}$, $k = 1, \ldots, n$ are independent fractional Brownian motions of Hurst parameter $H_k \in (0,1)$. Research in the internet traffic modeling using self-similar process shows the need of study of GMFBM, see e.g. [4]. Below we collect some basic properties of the GMFBM. For proofs and further comments on the importance of this process see [13] and references therein.

**Proposition 1.1.** *The GMFBM* $Z^H = (Z_t^{H,\alpha})_{t\geq 0}$ *is a centered Gaussian process with variance* $\sum_{k=1}^n \alpha_k^2 t^{2H_k}$ *and covariance function*

$$\mathbb{E}(Z_t^{H,\alpha} Z_s^{H,\alpha}) = \frac{1}{2}\sum_{k=1}^n \alpha_k^2 \left(t^{2H_k} + s^{2H_k} - |t-s|^{2H_k}\right).$$

*$Z^H$ has stationary increments and they are uncorrelated if and only if $H_k = \frac{1}{2}$ for all $k$. $Z^H$ is also $(c_1,\ldots,c_n; H_1,\ldots,H_n)$-self-similar in the sense that $\sum_{k=1}^n \alpha_k c_k^{-H_k} B_{c_k t}^{H_k} = \sum_{k=1}^n \alpha_k B_t^{H_k}$ in law. $Z^H$ is neither a Markov process nor a semimartingale, unless $H_k = \frac{1}{2}$ for all $k$. $Z^H$ exhibits a long-range dependence if and only if there exists $k$ with $H_k > \frac{1}{2}$. For all $T > 0$, with probability one $Z^H$ has a version, the sample path of which are Hölder continuous of order $\gamma < \min_{1\leq k\leq n} H_k$ on the interval $[0,T]$. Every sample path of $Z^H$ is almost surely nowhere differentiable.*

In this paper we focus on analyzing self-intersection local time of GMFBM using white noise theory. More precisely, we will study GMFBM on a specific probability space namely white noise space. The *self-intersection local time* of GMFBM $Z^H$ is formally defined by the Tanaka formula

$$L(T) := \int_0^T \int_0^T \delta(Z_t^H - Z_s^H)\, ds\, dt,$$

where $\delta$ is the Dirac delta function. Intuitively it measures the amount of time that the process spends intersecting itself on the time interval $[0,T]$. A rigorous definition of this object may be obtained by approximating the Dirac delta function by the heat kernel

$$p_\varepsilon(x) := \frac{1}{(2\pi\varepsilon)^{1/2}} e^{-\frac{1}{2\varepsilon}|x|^2},$$

as $\varepsilon > 0$ tends to zero. The first investigation related to this topic within white noise analysis framework was done by Faria et al [3] for the case $\alpha_1 = 1, \alpha_2 = \cdots = \alpha_n = 0$, and $H = \frac{1}{2}$ (Brownian motion). Drumond et al in [5] have extended this approach for the case $\alpha_1 = 1, \alpha_2 = \cdots = \alpha_n = 0$, and $H \in (0,1)$ (fractional Brownian motion). In this present work we are able to give meaning to $L^{(N)}$, i.e. a renormalized counterpart of $L(T)$, as a well-defined object in some white noise distribution space. Moreover, we find the chaos expansion of $L^{(N)}$ in terms of Wick powers of white noise. This allows us to define the kernel for $L^{(N)}$.

## 2. White Noise Analysis of GMFBM

We begin with some basic concepts of white noise theory used throughout this paper. For a more comprehensive explanation including various applications of white noise analysis, see [7, 8, 9]. In particular, for a survey on Gaussian white noise analysis and its application to Feynman's approach to quantum mechanics we refer to [12].

In the first part of this section we will recall a way to represent GMFBM as a generalized random variable on the white noise space. Firstly, we summarize the construction of a fractional Brownian motion in the white noise space. This approach was first introduced by Bender in [1]. Since GMFBM is a linear combination of independent fractional Brownian motions, its realization in the white noise space can be easily derived. For this purpose we follow closely the approach developed by Drumond et al in [5]. For a more detail on white noise approach of GMFBM see [10].

Let $(\mathcal{S}'_d(\mathbb{R}), \mathcal{C}, \mu)$ be the vector-valued white noise space, i.e., $\mathcal{S}'_d(\mathbb{R})$ is the space of vector-valued tempered distribution, $\mathcal{C}$ is the Borel $\sigma$-algebra generated by cylinder sets in $\mathcal{S}'_d(\mathbb{R})$, and the probability measure $\mu$ is uniquely determined by the Bochner-Minlos theorem such that its characteristic function given by

$$\int_{\mathcal{S}'_d(\mathbb{R})} e^{i\langle\vec{\omega},\vec{f}\rangle}\, d\mu(\vec{\omega}) = e^{-\frac{1}{2}|\vec{f}|_0^2}$$

for all vector-valued Schwartz test function $\vec{f} \in \mathcal{S}_d(\mathbb{R})$. Here $|\cdot|_0$ denotes the usual norm in the real Hilbert space of all vector-valued Lebesgue square-integrable functions $L^2_d(\mathbb{R})$. The dual pairing $\langle\cdot,\cdot\rangle$ on $\mathcal{S}'_d(\mathbb{R}) \times \mathcal{S}_d(\mathbb{R})$ can be considered as the bilinear extension of the inner product on $L^2_d(\mathbb{R})$, i.e.

$$\left\langle \vec{g}, \vec{f} \right\rangle = \sum_{j=1}^{d} \int_{\mathbb{R}} g_j(x) f_j(x)\, dx,$$

for all $\vec{g} = (g_1, \ldots, g_d) \in L^2_d(\mathbb{R})$ and $\vec{f} = (f_1, \ldots, f_d) \in \mathcal{S}_d(\mathbb{R})$. We should remark that we have the following Gel'fand triple

$$\mathcal{S}_d(\mathbb{R}) \subset L^2_d(\mathbb{R}) \subset \mathcal{S}'_d(\mathbb{R}).$$

It is well known that in the white noise space a version of the $d$-dimensional Brownian motion is given by the stochastic process $(B_t)_{t\geq 0}$ with

$$B_t := \left(\left\langle \cdot, \mathbf{1}_{[0,t]}\right\rangle, \cdots, \left\langle \cdot, \mathbf{1}_{[0,t]}\right\rangle\right),$$

such that

$$B_t(\vec{\omega}) := \left(\left\langle \omega_1, \mathbf{1}_{[0,t]}\right\rangle, \cdots, \left\langle \omega_d, \mathbf{1}_{[0,t]}\right\rangle\right), \quad \vec{\omega} = (\omega_1, \ldots, \omega_d) \in \mathcal{S}'_d(\mathbb{R})$$

where $\mathbf{1}_A$ denotes the indicator function on a set $A$.

A representation of fractional Brownian motion in term of indicator function is constructed using the concept of Weyl's fractional integral and Marchaud's fractional derivative. Precisely speaking, for an arbitrary Hurst parameter $H \in (0,1)$, a version of a $d$-dimensional fractional Brownian motion is given by

$$B_t^H(\vec{\omega}) := \left(\left\langle \omega_1, N_-^H \mathbf{1}_{[0,t]}\right\rangle, \cdots, \left\langle \omega_d, N_-^H \mathbf{1}_{[0,t]}\right\rangle\right), \quad \vec{\omega} = (\omega_1, \ldots, \omega_d) \in \mathcal{S}'_d(\mathbb{R})$$

where for a real-valued function $f$ the operator $N_-^H$ is defined by

$$N_-^H f := \begin{cases} \frac{(\frac{1}{2}-H)K_H}{\Gamma(H+\frac{1}{2})} \lim_{\varepsilon \searrow 0} \int_\varepsilon^\infty \frac{f(x)-f(x+y)}{y^{\frac{3}{2}-H}}\, dy & , \text{ if } H \in (0, \frac{1}{2}) \\ f & , \text{ if } H = \frac{1}{2} \\ \frac{K_H}{\Gamma(H-\frac{1}{2})} \int_x^\infty f(y)(y-x)^{H-\frac{3}{2}}\, dy & , \text{ if } H \in (\frac{1}{2}, 1) \end{cases}$$

provided the integrals exist for almost all $x \in \mathbb{R}$. Here $\Gamma$ denotes Gamma function and $K_H$ is the normalizing constant. We also consider the operator $N_+^H$ defined by

$$N_+^H f := \begin{cases} \frac{(\frac{1}{2}-H)K_H}{\Gamma(H+\frac{1}{2})} \lim_{\varepsilon \searrow 0} \int_\varepsilon^\infty \frac{f(x)-f(x-y)}{y^{\frac{3}{2}-H}}\, dy & , \text{ if } H \in (0, \frac{1}{2}) \\ f & , \text{ if } H = \frac{1}{2} \\ \frac{K_H}{\Gamma(H-\frac{1}{2})} \int_{-\infty}^x f(y)(x-y)^{H-\frac{3}{2}}\, dy & , \text{ if } H \in (\frac{1}{2}, 1). \end{cases}$$

For $H = (H_1, \ldots, H_n)$, and $\alpha = (\alpha_1, \ldots, \alpha_n)$, $H_k \in (0,1)$, $\alpha_k \in \mathbb{R}$, $n \in \mathbb{N}$ a version of a $d$-dimensional GMFBM of parameter $H$ and $\alpha$ in the white noise analysis setting is given by

$$Z_t^{H,\alpha}(\vec{\omega}) := \left( \left\langle \omega_1, \sum_{k=1}^{n} \alpha_k N_-^{H_k} \mathbf{1}_{[0,t]} \right\rangle, \cdots, \left\langle \omega_d, \sum_{k=1}^{n} \alpha_k N_-^{H_k} \mathbf{1}_{[0,t]} \right\rangle \right),$$

for $\vec{\omega} = (\omega_1, \ldots, \omega_d) \in \mathcal{S}'_d(\mathbb{R})$. Within this formalism we can choose a continuous version of the process according to Kolmogorov-Centsov theorem. At this point we should also emphasize that the white noise analysis approach in defining GMFBM has the advantage that the underlying probability space does not depend on the Hurst parameter under consideration.

It is known that for some type of functions, for example $f = \mathbf{1}_{[0,t]}, t > 0$ or $f \in \mathcal{S}_1(\mathbb{R})$, $N_-^H$ and $N_+^H$ are dual operators in the sense that the following integration by parts formula holds

$$\int_{\mathbb{R}} f(x) N_-^H g(x)\, dx = \int_{\mathbb{R}} (N_+^H f)(x) g(x)\, dx.$$

Moreover, the following estimation was proved in [5] and will be very useful for our purpose.

**Lemma 2.1.** *If $H \in (0,1)$ and $f \in \mathcal{S}_1(\mathbb{R})$, then there exists a nonnegative constant $C_H$, independent of $f$, such that*

$$\left| \int_{\mathbb{R}} f(x) N_-^H \mathbf{1}_{[s,t]}(x)\, dx \right| \leq C_H (t-s) \left( \sup_{x\in\mathbb{R}} |f(x)| + \sup_{x\in\mathbb{R}} |f'(x)| + |f|_0 \right)$$

*for all $s < t$.*

In the second part of this section we will review the concept of Hida distribution. Consider $L^2(\mu) := L^2(\mathcal{S}'_d(\mathbb{R}), \mathcal{C}, \mu)$ as a complex Hilbert space. This space is unitary isomorphic to the Fock space of symmetric square-integrable function, i.e.

$$L^2(\mu) \cong \left( \bigoplus_{k=0}^{\infty} \mathrm{sym} L^2(\mathbb{R}^k, k! d^k x) \right)^{\otimes d}.$$

This will lead to the chaos expansion of an element $F \in L^2(\mu)$, i.e.

$$F(\omega_1, \ldots, \omega_d) = \sum_{(m_1,\ldots,m_d)\in\mathbb{N}_0^d} \left\langle :\omega_1^{\otimes m_1}: \otimes \ldots \otimes :\omega_d^{\otimes m_d}:, \vec{f}_{(m_1,\ldots,m_d)} \right\rangle, \tag{1}$$

with kernel functions $\vec{f}_{(m_1,\ldots,m_d)}$ of the $m$-th chaos are in the Fock space. Here $:\omega^{\otimes n}:$ denotes the $m$-th Wick power of $\omega \in \mathcal{S}'_1(\mathbb{R})$, see e.g. [8] for details. For simplicity we introduce the following notations

$$\mathbf{m} = (m_1, \ldots, m_d) \in \mathbb{N}_0^d, \;\; m = \sum_{j=1}^{d} m_j, \;\; \mathbf{m}! = \prod_{j=1}^{d} m_j!,$$

which reduce (1) to

$$F(\vec{\omega}) = \sum_{\mathbf{m}\in\mathbb{N}_0^d} \left\langle :\vec{\omega}^{\otimes \mathbf{m}}:, \vec{f}_{\mathbf{m}} \right\rangle, \;\; \vec{\omega} \in \mathcal{S}'_d(\mathbb{R}).$$

Using, for example, the Wiener-Ito chaos expansion theorem and the second quantization operator of the Hamiltonian of harmonic oscillator, we can construct a Gel'fand triple

$$(\mathcal{S}) \subset L^2(\mu) \subset (\mathcal{S})^*$$

where $(\mathcal{S})$ is the space of white noise test functions such that its topological dual space is the space $(\mathcal{S})^*$, the space of generalized white noise functionals. Elements of $(\mathcal{S})$ and $(\mathcal{S})^*$ are also known as Hida test functions and Hida distributions, respectively. The reader not familiar with this subject may consult [7] or [8]. As we know with probability one sample paths of a GMFBM $Z^H$ are nowhere differentiable. However, it is possible to show that $Z^H$ is differentiable as a mapping from $\mathbb{R}$ into $(\mathcal{S})^*$, and the distributional derivative of $Z^H = Z_t^{H,\alpha}$ in $(\mathcal{S})^*$ is given by

$$W_t^{H,\alpha}(\vec{\omega}) := \left(\left\langle \omega_1, \sum_{k=1}^{n} \alpha_k \delta_t \circ N_+^{H_k} \right\rangle, \cdots, \left\langle \omega_d, \sum_{k=1}^{n} \alpha_k \delta_t \circ N_+^{H_k} \right\rangle\right),$$

where $\vec{\omega} = (\omega_1, \ldots, \omega_d) \in \mathcal{S}_d'(\mathbb{R})$ and $\delta_t$ denotes the Dirac delta function at $t$. The Hida distribution $W_t^{H,\alpha}$ is called *generalized mixed fractional white noise*. See [10] for details.

In the rest of this section we state an integrability criterion of a Hida distribution via the so-called S-transform. For a given $\vec{f} \in \mathcal{S}_d(\mathbb{R})$ and the corresponding Wick exponential

$$: e^{\langle \vec{\omega}, \vec{f} \rangle} ::= \sum_{\mathbf{n} \in \mathbb{N}_0^d} \left\langle : \vec{\omega}^{\otimes \mathbf{n}} :, \vec{f}^{\otimes \mathbf{n}} \right\rangle = e^{\langle \vec{\omega}, \vec{f} \rangle - \frac{1}{2}|\vec{f}|_0^2},$$

we define the *S-transform* of an element $\Phi \in (\mathcal{S})^*$ by

$$(S\Phi)(\vec{f}) := \left\langle\!\left\langle \Phi, : e^{\langle \cdot, \vec{f} \rangle} : \right\rangle\!\right\rangle, \quad \text{for all } \vec{f} \in \mathcal{S}_d(\mathbb{R}). \tag{2}$$

Here $\langle\langle \cdot, \cdot \rangle\rangle$ denotes the dual pairing between $(\mathcal{S})$ and $(\mathcal{S})^*$ which is defined as the bilinear extension of the sesquilinear inner product on $L^2(\mu)$. We observe that the multilinear expansion of (2) $S\Phi(\vec{f}) = \sum_{\mathbf{m} \in \mathbb{N}_0^d} \left\langle F_{\mathbf{m}}, \vec{f}^{\otimes \mathbf{m}} \right\rangle$, extends the chaos expansion to $\Phi \in (\mathcal{S})^*$ with distribution-valued kernels $F_{\mathbf{m}}$ such that $\langle\langle \Phi, \varphi \rangle\rangle = \sum_{\mathbf{m} \in \mathbb{N}_0^d} \mathbf{m}! \langle F_{\mathbf{m}}, \varphi_{\mathbf{m}} \rangle$, for every Hida test function $\varphi \in (\mathcal{S})$ with kernel functions $\varphi_{\mathbf{m}}$. The S-transform provides a quite useful way to study the Bochner integrability of a family of Hida distributions which depends on an additional parameter. For details and proofs see [6].

**Theorem 2.2.** *Let $(\Omega, \mathcal{F}, \nu)$ be a measure space and $\lambda \mapsto \Phi_\lambda$ be a mapping from $\Omega$ to $(\mathcal{S})^*$. If the S-transform of $\Phi_\lambda$ fulfils the following two conditions:*

*(1) the mapping $\lambda \mapsto (S\Phi_\lambda)(\vec{f})$ is measurable for all $\vec{f} \in \mathcal{S}_d(\mathbb{R})$, and*

*(2) there exist $C_1(\lambda) \in L^1(\Omega, \nu)$, $C_2(\lambda) \in L^\infty(\Omega, \nu)$ and a continuous norm $\|\cdot\|$ on $\mathcal{S}_d(\mathbb{R})$ such that $\left|(S\Phi_\lambda)(z\vec{f})\right| \le C_1(\lambda) e^{C_2(\lambda)|z|^2 \|\vec{f}\|^2}$, for all $z \in \mathbb{C}$, $\vec{f} \in \mathcal{S}_d(\mathbb{R})$,*

*then $\Phi_\lambda$ is Bochner integrable with respect to some Hilbertian norm which topologizing $(\mathcal{S})^*$. Hence $\int_\Omega \Phi_\lambda \, d\nu(\lambda) \in (\mathcal{S})^*$, and furthermore*

$$S\left(\int_\Omega \Phi_\lambda \, d\nu(\lambda)\right)(\vec{f}) = \int_\Omega (S\Phi_\lambda)(\vec{f}) \, d\nu(\lambda).$$

## 3. Self-Intersection Local Times of GMFBM

In several applications, for example in the context of polymer physics, Feynman's path integral, and financial mathematics, we need to pin a stochastic process at a particular spatial coordinate. Motivated by this fact we study the so-called Donskers delta function of the generalized mixed fractional Brownian motion. Therefore, in order to pin GMFBM at a point $c \in \mathbb{R}^d$ we consider the Donsker's delta function of GMFBM which is defined as the formal composition of the Dirac delta distribution $\delta_d \in \mathcal{S}'(\mathbb{R}^d)$ with a $d$-dimensional GMFBM $\left(Z_t^H\right)_{t \ge 0}$, i.e., $\delta_d\left(Z_t^H - c\right)$. We can give a precise meaning to the Donskers's delta function as a Hida distribution. First, we recall the Fourier-transform representation of Dirac delta function which is given by

$$\delta(x) = \frac{1}{2\pi}\int_{\mathbb{R}} e^{i\lambda x}\, d\lambda.$$

The following result has been proven in [11] and here we present a sketch of the proof for the sake of completeness.

**Proposition 3.1.** *The Bochner integral*

$$\delta_d\left(Z_t^H - Z_s^H\right) := \left(\frac{1}{2\pi}\right)^d \int_{\mathbb{R}^d} e^{i\lambda\left(Z_t^H - Z_s^H\right)}\, d\lambda,\ \ t \neq s$$

*is a Hida distribution with S-transform given by*

$$\begin{aligned}
&\left(S\delta_d\left(Z_t^H - Z_s^H\right)\right)(\vec{f}) \\
&= \frac{1}{\left(2\pi\sum_{k=1}^n \alpha_k^2|t-s|^{2H_k}\right)^{d/2}} \\
&\quad \times \exp\left(-\frac{1}{2\sum_{k=1}^n \alpha_k^2|t-s|^{2H_k}}\sum_{j=1}^d\left(\int_{\mathbb{R}}\sum_{k=1}^n \alpha_k f_j(x) N_-^{H_k}\mathbf{1}_{[s\wedge t, s\vee t]}(x)\, dx\right)^2\right),
\end{aligned}$$

*for all* $\vec{f} = (f_1, \ldots, f_d) \in \mathcal{S}_d(\mathbb{R})$*. In particular, its generalized expectation is given by*

$$\mathbb{E}_\mu\left(\delta_d\left(Z_t^H - Z_s^H\right)\right) = \frac{1}{\left(2\pi\sum_{k=1}^n \alpha_k^2 t^{2H_k}\right)^{d/2}}.$$

*Proof.* Without loss of generality we may assume $t > s$. For any $\vec{f} \in \mathcal{S}_d(\mathbb{R})$ we have

$$\begin{aligned}
F_\lambda(\vec{f}) &= Se^{i\lambda\left(Z_t^H - Z_s^H\right)}(\vec{f}) \\
&= \left\langle\left\langle \exp\left(i\lambda\left(\left\langle \cdot, \sum_{k=1}^n \alpha_k N_-^{H_k}\mathbf{1}_{[0,t]}\right\rangle - \left\langle \cdot, \sum_{k=1}^n \alpha_k N_-^{H_k}\mathbf{1}_{[0,s]}\right\rangle\right)\right), : e^{\langle\cdot,\vec{f}\rangle} : \right\rangle\right\rangle \\
&= \exp\left(-\frac{1}{2}\left|\vec{f}\right|_0^2\right)\int_{\mathcal{S}_d'(\mathbb{R})} \exp\left(\left\langle \vec{\omega}, i\lambda\sum_{k=1}^n \alpha_k N_-^{H_k}\mathbf{1}_{[s,t]} + \vec{f}\right\rangle\right) d\mu(\vec{\omega}) \\
&= \exp\left(-\frac{1}{2}|\lambda|^2\sum_{k=1}^n \alpha_k^2 (t-s)^{2H_k}\right)\exp\left(i\lambda\int_{\mathbb{R}} \vec{f}(x)\sum_{k-1}^n \alpha_k N_-^{H_k}\mathbf{1}_{[s,t]}(x)\, dx\right).
\end{aligned}$$

The mapping $\lambda \mapsto F_\lambda(\vec{f})$ is measurable for all $\vec{f} \in \mathcal{S}_d(\mathbb{R})$ and $\lambda \in \mathbb{R}^d$. To apply Theorem 2.2. it remains to show the boundedness. Let $z \in \mathbb{C}$ and by an application of Lemma 2.1 we obtain

$$\begin{aligned}
\left|F_\lambda(z\vec{f})\right| &\leq \exp\left(-\frac{1}{2}|\lambda|^2\sum_{k=1}^n \alpha_k^2(t-s)^{2H_k}\right) \\
&\quad \times \exp\left(|z||\lambda|\sum_{j=1}^d\sum_{k=1}^n |\alpha_k| C_{H_k}(t-s)\left(\sup_{x\in\mathbb{R}}|f_j(x)| + \sup_{x\in\mathbb{R}}|f_j'(x)| + |f_j|_0\right)\right).
\end{aligned}$$

IOP Conf. Series: Journal of Physics: Conf. Series **855** (2017) 012050 doi:10.1088/1742-6596/855/1/012050

Now define a continuous norm on $\mathcal{S}_d(\mathbb{R})$ as follow

$$\left\|\vec{f}\right\|_* := \sum_{j=1}^{d}\left(\sup_{x\in\mathbb{R}}|f_j(x)| + \sup_{x\in\mathbb{R}}|f_j'(x)| + |f_j|_0\right). \tag{3}$$

Thus,

$$\left|F_\lambda(z\vec{f})\right| \le \exp\left(-\frac{1}{4}|\lambda|^2\sum_{k=1}^{n}\alpha_k^2(t-s)^{2H_k}\right)\exp\left(\frac{\left(\sum_{k=1}^{n}K_{H_k}\right)^2}{\sum_{k=1}^{n}\alpha_k^2(t-s)^{2H_k}}|z|^2\left\|\vec{f}\right\|_*^2\right),$$

where $K_{H_k} := |\alpha_k|C_{H_k}(t-s)$. The first factor is an integrable function of $\lambda$, and the second factor is constant with respect to $\lambda$. Therefore, according to the Theorem 2.2, $\delta_d\left(Z_t^H - Z_s^H\right) \in (\mathcal{S})^*$. To obtain an explicit expression for the S-transform of $\delta_d\left(Z_t^H - Z_s^H\right)$ we calculate as follow.

$$\begin{aligned}
&S\left(\delta_d\left(Z_t^H - Z_s^H\right)\right)(\vec{f})\\
&= \left(\frac{1}{2\pi}\right)^d\int_{\mathbb{R}^d} Se^{i\lambda(Z_t^H - Z_s^H)}(\vec{f})\,d\lambda\\
&= \left(\frac{1}{2\pi}\right)^d\left(\frac{\pi}{\frac{1}{2}\sum_{k=1}^{n}\alpha_k^2(t-s)^{2H_k}}\right)^{d/2}\exp\left(\frac{\left|i\int_{\mathbb{R}}\vec{f}(x)\sum_{k=1}^{n}\alpha_k N_-^{H_k}\mathbf{1}_{[s,t]}(x)\,dx\right|_0^2}{2\sum_{k=1}^{n}\alpha_k^2(t-s)^{2H_k}}\right)\\
&= \frac{1}{\left(2\pi\sum_{k=1}^{n}\alpha_k^2(t-s)^{2H_k}\right)^{d/2}}\\
&\quad\times\exp\left(-\frac{1}{2\sum_{k=1}^{n}\alpha_k^2(t-s)^{2H_k}}\sum_{j=1}^{d}\left(\int_{\mathbb{R}}\sum_{k=1}^{n}\alpha_k f_j(x)N_-^{H_k}\mathbf{1}_{[s,t]}(x)\,dx\right)^2\right).
\end{aligned}$$

The generalized expectation is easily obtained by choosing $\vec{f} = 0$. □

Now we are ready to prove our main result on self-intersection local times $L(T)$ as well as their subtracted counterparts $L^{(N)}(T)$. For simplicity we assume $T = 1$ and write $L(T)$ and $L^{(N)}(T)$ simply by $L$ and $L^{(N)}$, respectively. In the sequel we fix the following notations $H_* := \max_{1\le k\le n} H_k$ , $\Delta := \{(s,t)\in\mathbb{R}^2 : 0 < s < t < 1\}$, $d^2(s,t)$ is the Lebesgue measure on $\Delta$, $\|\cdot\|_*$ is the norm on $\mathcal{S}_d(\mathbb{R})$ defined by Equation (3), and $\exp^{(N)}(x) := \sum_{m=N}^{\infty}\frac{x^m}{m!}$. We also define the truncated Donsker's delta function $\delta_d^{(N)}\left(Z_t^H - Z_s^H\right)$ via its $S$-transform, i.e.

$$\begin{aligned}
&S\left(\delta_d^{(N)}\left(Z_t^H - Z_s^H\right)\right)(\vec{f}) \qquad (4)\\
&= \frac{1}{\left(2\pi\sum_{k=1}^{n}\alpha_k^2(t-s)^{2H_k}\right)^{d/2}}\\
&\quad\times\exp^{(N)}\left(-\frac{1}{2\sum_{k=1}^{n}\alpha_k^2(t-s)^{2H_k}}\sum_{j=1}^{d}\left(\int_{\mathbb{R}}\sum_{k=1}^{n}\alpha_k f_j(x)N_-^{H_k}\mathbf{1}_{[s,t]}(x)\,dx\right)^2\right),
\end{aligned}$$

This is a unique Hida distribution whose existence is given via the characterization theorem of white noise distributions, see e.g. [6].

**Theorem 3.2.** *Let $\left(Z_t^{H,\alpha}\right)_{t\geq 0}$ be a GMFBM with parameter $H = (H_1,\ldots,H_n)$ and $\alpha = (\alpha_1,\ldots,\alpha_n)$, $H_k \in (0,1)$, $\alpha_k \in \mathbb{R}$, $n \in \mathbb{N}$. For any pair of integer $d \geq 1$ and $N \geq 0$ such that $2N(H_* - 1) + dH_* < 1$, the Bochner integral*

$$L^{(N)} := \int_\Delta \delta_d^{(N)}\left(Z_t^H - Z_s^H\right) d^2(s,t)$$

*is a Hida distribution.*

*Proof.* Note that the S-transform of $\delta_d^{(N)}\left(Z_t^H - Z_s^H\right)$ is a measurable function. Furthermore, for every $z \in \mathbb{C}$

$$\begin{aligned}
&\left|S\left(\delta_d^{(N)}\left(Z_t^H - Z_s^H\right)\right)(z\vec{f})\right| \\
&\leq \frac{1}{\left(2\pi\sum_{k=1}^n \alpha_k^2(t-s)^{2H_k}\right)^{d/2}} \exp^{(N)}\left(\frac{(t-s)^2\left(\sum_{k=1}^n|\alpha_k|C_{H_k}\right)^2}{2\sum_{k=1}^n\alpha_k^2(t-s)^{2H_k}}|z|^2\left\|\vec{f}\right\|_*^2\right) \\
&\leq \frac{1}{\left(2\pi\sum_{k=1}^n\alpha_k^2\right)^{d/2}}\cdot\frac{1}{(t-s)^{dH_*}}\exp^{(N)}\left(\frac{(t-s)^{2-2H_*}\left(\sum_{k=1}^n|\alpha_k|C_{H_k}\right)^2}{2\sum_{k=1}^n\alpha_k^2}|z|^2\left\|\vec{f}\right\|_*^2\right) \\
&\leq \frac{1}{\left(2\pi\sum_{k=1}^n\alpha_k^2\right)^{d/2}}.(t-s)^{2N(1-H_*)-dH_*}\exp\left(\frac{\left(\sum_{k=1}^n|\alpha_k|C_{H_k}\right)^2}{2\sum_{k=1}^n\alpha_k^2}|z|^2\left\|\vec{f}\right\|_*^2\right),
\end{aligned}$$

where $(t-s)^{2N(1-H_*)-dH_*}$ is integrable with respect to the Lebesgue measure on $\Delta$ if and only if $2N(1-H_*) - dH_* > -1$. Therefore we can conclude using Theorem 2.2 that $L^{(N)} \in (\mathcal{S})^*$. □

We are also able to derive the chaos expansion for the (truncated) self-intersection local times $L^{(N)}$.

**Theorem 3.3.** *Let $\left(Z_t^{H,\alpha}\right)_{t\geq 0}$ be a GMFBM with parameter $H = (H_1,\ldots,H_n)$ and $\alpha = (\alpha_1,\ldots,\alpha_n)$, $H_k \in (0,1)$, $\alpha_k \in \mathbb{R}$, $n \in \mathbb{N}$. For any pair of integer $d \geq 1$ and $N \geq 0$ such that $2N(H_* - 1) + dH_* < 1$, the kernel functions $F_{2\mathbf{m}}$ of $L^{(N)}$ are given by*

$$F_{2\mathbf{m}}(u_1,\ldots,u_{2m}) = \frac{\left(-\frac{1}{2}\right)^m}{\mathbf{m}!}\left(\frac{1}{2\pi}\right)^{d/2}\int_\Delta \frac{\prod_{l=1}^{2m}\left(\sum_{k=1}^n \alpha_k N_-^{H_k}\mathbf{1}_{[s,t]}\right)(u_l)}{\left(\sum_{k=1}^n\alpha_k^2(t-s)^{2H_k}\right)^{m+d/2}} d^2(s,t)$$

*for each $\mathbf{m} \in \mathbb{N}_0^d$ such that $m \geq N$. All other kernel functions $F_{\mathbf{m}}$ are identically equal to zero.*

*Proof.* By Theorem 2.2, the S-transform of self-intersection local times $L^{(N)}$ is obtained by integrating (4) over $\Delta$. Hence, given an element $\vec{f} = (f_1,\ldots,f_d) \in \mathcal{S}_d(\mathbb{R})$,

$$\begin{aligned}
SL^{(N)}(\vec{f}) &= S\left(\int_\Delta \delta_d^{(N)}\left(Z_t^H - Z_s^H\right)d^2(s,t)\right)(\vec{f}) \\
&= \int_\Delta S\left(\delta_d^{(N)}\left(Z_t^H - Z_s^H\right)\right)(\vec{f})\,d^2(s,t) \\
&= \int_\Delta \frac{1}{\left(2\pi\sum_{k=1}^n\alpha_k^2(t-s)^{2H_k}\right)^{d/2}} \\
&\quad\times\exp^{(N)}\left(-\frac{1}{2\sum_{k=1}^n\alpha_k^2(t-s)^{2H_k}}\sum_{j=1}^d\left(\int_\mathbb{R}\sum_{k=1}^n\alpha_k f_j(x)N_-^{H_k}\mathbf{1}_{[s,t]}(x)\,dx\right)^2\right)d^2(s,t)
\end{aligned}$$

$$= \left(\frac{1}{2\pi}\right)^{d/2} \int_\Delta \sum_{m=N}^{\infty} \left(-\frac{1}{2}\right)^m \frac{1}{\left(\sum_{k=1}^n \alpha_k^2 (t-s)^{2H_k}\right)^{m+d/2}}$$
$$\times \sum_{\substack{m_1,\dots,m_d \\ m_1+\dots+m_d=m}} \frac{1}{\mathbf{m}!} \prod_{j=1}^{d} \left( \int_{\mathbb{R}} f_j(x) \sum_{k=1}^{n} \alpha_k N_-^{H_k} \mathbf{1}_{[s,t]}(x)\, dx \right)^{2m_j} d^2(s,t).$$

Remember the general form of the chaos expansion

$$L^{(N)} = \sum_{\mathbf{m}\in\mathbb{N}_0^d} \langle :\vec{\omega}^{\otimes \mathbf{m}}:, F_{\mathbf{m}} \rangle \quad \text{and} \quad SL^{(N)}(\vec{f}) = \sum_{\mathbf{m}\in\mathbb{N}_0^d} \left\langle F_{\mathbf{m}}, \vec{f}^{\otimes \mathbf{m}} \right\rangle.$$

Thus, we conclude that

$$F_{\mathbf{m}} = \frac{\left(-\frac{1}{2}\right)^m}{\mathbf{m}!} \left(\frac{1}{2\pi}\right)^{d/2} \int_\Delta \frac{\left(\sum_{k=1}^n \alpha_k N_-^{H_k}\right)^{\otimes 2m}}{\left(\sum_{k=1}^n \alpha_k^2 (t-s)^{2H_k}\right)^{m+d/2}} d^2(s,t).$$

More precisely, for every $\mathbf{m} \in \mathbb{N}_0^d$ such that $m \geq N$ and $u_1, \dots, u_{2m} \in \mathbb{R}$ we have

$$F_{2\mathbf{m}}(u_1,\dots,u_{2m}) = \frac{\left(-\frac{1}{2}\right)^m}{\mathbf{m}!} \left(\frac{1}{2\pi}\right)^{d/2} \int_\Delta \frac{\prod_{l=1}^{2m} \left(\sum_{k=1}^n \alpha_k N_-^{H_k} \mathbf{1}_{[s,t]}\right)(u_l)}{\left(\sum_{k=1}^n \alpha_k^2 (t-s)^{2H_k}\right)^{m+d/2}} d^2(s,t),$$

while all other odd kernels $F_{\mathbf{m}}$ vanish. In particular, we obtain the generalized expectation of (truncated) self-intersection local times $L^{(N)}$, that is

$$\mathbb{E}_\mu(L^{(N)}) = F_{\mathbf{0}} = \left(\frac{1}{2\pi}\right)^{d/2} \int_\Delta \frac{1}{\left(\sum_{k=1}^n \alpha_k^2 (t-s)^{2H_k}\right)^{d/2}} d^2(s,t). \qquad \square$$

Theorem 3.2 shows that for one-dimensional GMFBM all self-intersection local times are well-defined as Hida distributions for all possible Hurst parameter $H_k \in (0,1)$, $k = 1\dots,n$. For $d \geq 2$, self-intersection local times are well-defined only for $\max_{1\leq k\leq n} H_k \leq \frac{1}{d}$. Informally speaking, if $H_k \geq \frac{1}{d}$ for some $k$ with $d \geq 2$, the local times only become well-defined once the divergent terms are removed.

## 4. Concluding Remarks

We proved under some conditions on the number of truncation of divergent terms, Hurst parameters, and the spatial dimension, that self-intersection local times of GMFBM are Hida distribution. Explicit expressions for the chaos expansions of the self-intersection local times are also presented. All of these results are obtained in the white noise analysis framework.

## References

[1] Bender, C, "An Ito formula for generalized functionals of a fractional Brownian motion with arbitrary Hurst parameter", *Stoc. Proc. Appl.* **104** (2003), 81-106.
[2] Cheridito, P, "Mixed fractional Brownian motion", *Bernoulli* **7(6)** (2001), 913-934.
[3] da Faria, M, Hida, T, Streit, L and Watanabe, H, "Intersection local times as generalized white noise functionals", *Acta Appl. Math.* **46** (1997), 351-362.
[4] Drakakis, K and Radulovic, D, "A discretized version of the self-similar model for internet traffic", *Appl. Math. Sci.* **2** (2008), 2743-2756.
[5] Drumond, C, Oliveira, M J and da Silva, J L, "Intersection local times of fractional Brownian motions with $H \in (0,1)$ as generalized white noise functionals", *AIP Conference Proceedings* **1021** (2008), 34-45.

[6] Kondratiev, Y, Leukert, P, Potthoff, J, Streit, L and Westerkamp, W, "Generalized functionals in Gaussian spaces: The characterization theorem revisited", *J. Funct. Anal.* **141** (1996), 301-318.
[7] Kuo, H-H, *White Noise Distribution Theory*, CRC Press, 1996.
[8] Hida, T, Kuo, H-H, Potthoff, J and Streit, L, *White Noise. An Infinite Dimensional Calculus*, Kluwer Academic Publishers, 1993.
[9] Obata, N, *White Noise Calculus and Fock Space*, Springer Verlag, 1994.
[10] Suryawan, H P, "Generalized mixed fractional Brownian motion as a generalized white noise functional", *MatStat* **11(1)** (2011), 10-19.
[11] Suryawan, H P, "Donsker's delta function of the generalized mixed fractional Brownian motion", *Proc. of the 1st Int. Sem. on Innovation in Mathematics and Mathematics Education* (2014).
[12] Suryawan, H P, "Gaussian white noise analysis and its application to Feynman path integral", AIP Conference Proceedings **1701** (2016), 030001 . http://dx.doi.org/10.1063/1.4940829
[13] Thäle, C, "Further remarks on mixed fractional Brownian motion", *Appl. Math. Sci.* **3(28)** (2009), 1885-1901.
[14] Zili, M, "On the mixed fractional Brownian motion", *J. Appl. Math. Stoch. Anal.* **2006** (2006), 1-9.

# Study on the factors affecting the quality of public bus transportation service in Bali Province using factor analysis

**M Susilawati**[1] **and D P E Nilakusmawati**[1]

[1]Mathematics Department, Faculty of Mathematics and Natural Sciences, Udayana University, Indonesia

E-mail: mdsusilawati@unud.ac.id

**Abstract**. The volume of mobility flows are increasing day by day and the condition of the number of people using private transport modes contribute to traffic congestion. With the limited capacity of the road, one of the alternatives solution to reduce congestion is to optimize the use of public transport. The purposes of this study are to determine the factors that influence user's satisfaction on the quality of public bus transportation service and determine variables that became identifier on the dominant factor affecting user's satisfaction. The study was conducted for the public bus transportation between districts in the province of Bali, which is among the eight regencies and one municipality, using a questionnaire as a data collection instrument. Service variables determinant of user's satisfaction in this study, described in 25 questions, which were analyzed using factor analysis. The results showed there were six factors that explain the satisfaction of users of public transport in Bali, with a total diversity of data that can be parsed by 61.436%. These factors are: Safety and comfort, Responsiveness, Capacity, Tangible, Safety, Reliability. The dominant factor affecting public transport user satisfaction is the safety and comfort, with the most influential variable is feeling concerned about the personal safety of users when on the bus.

## 1. Introduction

Transportation is a very important requirement for the community. One of the problems that often occur in all areas, especially urban areas is the need for transportation increases, caused by the increasing number and population growth. In addition, the Province of Bali as a favorite tourist destination and the rapid development of tourism industry contributed to the increase in the volume of mobility flows between districts. Volume of mobility flows vehicle or person will be increased each day, which became one of the causes of traffic congestion today, and potentially cause a bigger problem in the future if it does not handled properly.

With the limited capacity of the road, one of the alternatives solution to reduce congestion is to optimize the use of public transport. Real conditions in the field, interests and culture of the people using public transportation is still very low. Until now, only segments of society middle and lower level who use public transport services. According to Munawar [1], public transport is still less appealing because there are still shortcomings, particularly in terms of comfort, safety, speed, accuracy, ease, frequency and departures, and the facilities at the terminal and the bus stop.

In response to these circumstances, it is necessary to be handled more on the provision of public transport services. The study on the characteristics of users and the quality of public transport services is necessary to do. Beside that, Le-Klähn et al.[2] explained customer surveys are especially important nowadays considering the depth and amount of information they can provide so public transport providers can understand which service aspects play a more critical role in passenger satisfaction.

Traffic conditions in Bali, especially in urban areas showed conditions worsened caused by most people, especially the upper middle class people who use private transport in everyday activities. On the other hand, the condition of the low interest of the public to switch to using public transportation and low culture in the use of public transportation also contributed to the above-mentioned conditions.

The tendency of people traveling by private transport in urban areas will continue to increase if the condition of the transport system no more fundamentally improved. Tamin [3] states that the increased likelihood of a trip by private transport caused by: (1) Increased economic activity underserved by adequate public transportation; (2) The increasing purchasing power and the level of privacy that can not be served by public transport; (3) The increase in the price of land so that the addition of the highway less; (4) The opening of the new road further stimulate the use of private transport because usually in the new road network there has been no public transport services at the time; (5) Non-availability of transport feeder that bridge to travel to the main lines of public transport services; and (6) Less ensuring the security needs, the need for timely, the need for a long journey experienced in public transport services. According Ditjen.Hubdat (Ferdiansyah [4]), public transport problems commonly faced by urban areas in Indonesia can be classified into three levels of low accessibility, low level of service, and cost.

The purposes of this study are to determine the factors that influence user's satisfaction on the quality of public bus transportation service and determine variables that became identifier on the dominant factor affecting user's satisfaction on the public bus transportation service.

## 2. Research Methods

This research was conducted for the public bus transportation between districts in the province of Bali, which is among the eight regencies and one municipality. The data in this study is obtained from primary sources, which are taken directly by the researchers using a questionnaire. Method of measuring user satisfaction of the public bus transportation modes in this study using customer satisfaction surveys with interviews to public bus users using a questionnaire. The data collected is the passenger's perception of public bus service quality.

Sampling in this study conducted by purposive sampling is a sampling technique where the sample selection is done with certain subjective judgment based on multiple traits/characteristics of the sample, which is seen as closely linked to traits/ characteristics of the population that has been previously known. Considerations that are used in the selection of respondents, respondents who currently use/ride and/or currently use the public bus service as a means of mobility between districts in the province of Bali. Service variables determinant of user satisfaction in this study, described in 25 questions.

Data analysis techniques in this study, follow these steps:

(1) Conducting multivariate analysis techniques using Factor Analysis. Factor analysis steps refers to Hair et al.[5], as follows:

a. Test variables have been determined, using the method of Bartlett Test of Sphericity, KMO and measure of sampling adequacy (MSA).

b. Perform core processes in the factor analysis, a reduction of one or more factors of the variables that have pass the previous variable test. Conducting process rotation factor to the factors that have been formed. The purpose rotation to clarify the variables that enter into a particular factor. There are two different methods of rotation are orthogonal and oblique rotation. Known maintained orthogonal rotation if the axis perpendicular to one another (angle 90 degrees). Orthogonal rotation produce factors that are not correlated with each other. Conversely said oblique rotation if the rotation axis is not maintained to be perpendicular to each other and uncorrelated factors. Oblique rotation should be used if the factors in the population strongly correlate.

c. Interpretation of the factors that have been formed, by giving the name on the form factor, which is considered to represent variables members of these factors.

(2) Processing of the data in this study using SPSS version 19.0 For Window.

## 3. Results and Discussion

The data were obtained by questionnaires, it is necessary to test the instruments to ensure the questionnaire used is valid and reliable. The research result is said to be valid if there are similarities between the data collected with the actual data in the object studied. The method is often used to provide an assessment of the validity of the questionnaire is product-moment correlation, with the following formula:

$$r_{xy} = \frac{N\sum XY - (\sum X)(\sum Y)}{\sqrt{N\sum X^2 - (\sum X)^2}\sqrt{N\sum Y^2 - (\sum Y)^2}}. \quad (1)$$

Remarks:
X = Score test items
Y = Score total items
N = Number of respondents
Furthermore, t-test was calculated by using the formula (2).

$$t = \frac{r\sqrt{n-2}}{\sqrt{1-r^2}} \quad (2)$$

Wherein, $r$ is the correlation coefficient results $r_{xy}$, and $n$ is the number of respondents. Value $t$ obtained will then be compared with the value $t_{table}$ , if $t > t_{table}$ variables considered valid. The test results showed that the validity of 25 item questions to reveal more about the factors that affect the user satisfaction of public transport are all qualified validity so that it can be used for subsequent measurement.

The next test is to test reliability, the test is to indicate the extent to which an instrument is reliable or unreliable. If an instrument is used twice to measure the same phenomenon and the measurement results obtained by the same, then the instrument is said to have a high level of reliability.

**Table 1.** Value of reliability test

| Cronbach's Alpha | N of Items |
|---|---|
| .856 | 25 |

Sources: Primary data (2016)

Reliability testing table above were obtained Cronbach Alpha value of 0.856 which is greater than the prescribed criteria of 0.5, meaning all variables showed strong reliability. This means the entire test instrument meets the requirements for use in decision making further research. The analysis conducted in this study is factor analysis, this analysis is done to see the most dominant factor in determining the satisfaction of users of public transport in Bali. The steps are as follows:

### *3.1. Eligibility test data*

The first step is performed to see whether the data obtained deserve to be processed by factor analysis, is to look at the value of KMO.

**Table 2**. Value of KMO and Bartlett's

| | | |
|---|---|---|
| Kaiser-Meyer-Olkin Measure of Sampling Adequacy. | | .879 |
| Bartlett's Test of Sphericity | Approx. Chi-Square | 3346.591 |
| | df | 300 |
| | Sig. | .000 |

Sources: Primary data (2016)

KMO is an index to compare the magnitude of the correlation coefficient observations with a partial coefficient, which means that a large overall correlation coefficient of variables in the correlation matrix to be significant. KMO test value ranges between 0 and 1 which questioned the feasibility of factor analysis. Score KMO required to be more than 0.5, if the value of the index ranged from 0.5 to 1.0, a factor analysis feasible. Conversely, if the value of KMO below 0.5, a factor analysis was not feasible.

Based on calculations by the software SPSS 19.0 for windows obtained KMO value of 0.879 so worthy of factor analysis done by category is very good.

### *3.2. Barlett test*

Barlett Test is a statistical test to test whether the variables involved correlated. The null hypothesis ($H_0$) is no correlation between variables, whereas the alternative hypothesis (Ha) is a correlation between variables. Barlett Test value approximated by a chi-square value. Based on Table 2 obtained chi-square value is 3346.591 with significant 0,000, so that $H_0$ rejected and Ha accepted, meaning it is believed that there is a correlation between variables.

### *3.3. Factor extraction process*

In determining the number of the desired factor as a result of the extract in this study using eigenvalues which aims to determine the most dominant factor in determining the satisfaction of users of public transport in Bali, meaning that only those factors with eigenvalues> 1 were considered significant. From the results of data analysis using Principal Component Analysis, obtained six major components that have more than one eigenvalue indicating the number of factors that determine the satisfaction of users of public transport in Bali.

**Table 3**. Total variances explained major components

| Component | Initial Eigenvalues | | | Extraction Sums of Squared Loadings | | | Rotation Sums of Squared Loadings | | |
|---|---|---|---|---|---|---|---|---|---|
| | Total | % of Varian | Cumulativ % | Total | % of Varian | Cumulativ % | Total | % of Varian | Cumulativ % |
| **1** | 8.412 | 33.648 | 33.648 | 8.412 | 33.648 | 33.648 | 5.289 | 21.155 | 21.155 |
| **2** | 1.764 | 7.057 | 40.705 | 1.764 | 7.057 | 40.705 | 3.185 | 12.741 | 33.896 |
| **3** | 1.623 | 6.493 | 47.198 | 1.623 | 6.493 | 47.198 | 2.608 | 10.433 | 44.329 |
| **4** | 1.316 | 5.265 | 52.463 | 1.316 | 5.265 | 52.463 | 1.710 | 6.838 | 51.167 |
| **5** | 1.177 | 4.706 | 57.169 | 1.177 | 4.706 | 57.169 | 1.336 | 5.346 | 56.513 |
| **6** | 1.067 | 4.267 | 61.436 | 1.067 | 4.267 | 61.436 | 1.231 | 4.923 | 61.436 |

Sources: Primary data (2016)

Based on Table 3, from 25 variables proposed, formed six factors. Factor 1 has a total of eigenvalues of 8.412 or 33.648%, meaning that one factor able to explain 33.648% of the total of the factors that will affect the satisfaction of users of public transport in Bali. Factor 2 has a total of eigenvalues of 1.764 or 7.057%, which means Factor 2 is able to explain the 7.057% of the total factors that affect user satisfaction of public transportation in Bali. Factor 3 has a total value of eigenvalues of 1.623 or 6.493%, meaning that Factor 3 capable of explaining 6.493% of the total factors will affect the satisfaction of users of public transport in Bali. Factor 4, 5 and factor 6 with the same explanation can be seen in Table 3. The major cumulative contribution of six factors on consumer behavior amounted to 61.436%.

### *3.4. Factor rotation*

Results obtained by using the method of eigenvalues showed 6 factors formed in relation to the satisfaction of users of public transport in Bali. However, the results of the mapping using matrix components obvious that all variables are clustered in a factor of 1, 2, 3, and 4, which can be seen in Table 4.

**Table 4**. Component matrix

| | Component | | | | | |
|---|---|---|---|---|---|---|
| | 1 | 2 | 3 | 4 | 5 | 6 |
| $X_{25}$ | **.788** | .015 | .080 | .084 | .124 | -.105 |
| $X_{24}$ | **.754** | .136 | .029 | .148 | .126 | -.072 |
| $X_{21}$ | **.742** | .085 | .291 | .053 | .017 | -.069 |
| $X_{22}$ | **.733** | .352 | .149 | .056 | .055 | .007 |
| $X_{12}$ | **.656** | .221 | .234 | .063 | -.205 | .155 |
| $X_{20}$ | **.643** | .302 | .160 | .035 | .057 | .118 |
| $X_{11}$ | **.637** | .114 | .328 | .172 | .048 | .246 |
| $X_{19}$ | **.622** | .378 | .198 | .037 | -.075 | .007 |
| $X_{13}$ | **.599** | .305 | .246 | .025 | -.164 | .206 |
| $X_{14}$ | **.474** | .365 | .232 | -.148 | .200 | -.180 |
| $X_{17}$ | .241 | **.844** | .125 | .090 | .052 | .019 |
| $X^{18}$ | .220 | **.843** | .137 | .112 | .073 | .063 |
| $X_{16}$ | .305 | **.798** | .067 | .002 | .017 | .066 |
| $X_{15}$ | .101 | **.360** | .305 | -.075 | -.196 | .175 |
| $X_6$ | .410 | -.008 | **.697** | .073 | .022 | .088 |
| $X_7$ | .188 | .321 | **.636** | .039 | -.097 | .103 |
| $X_8$ | .330 | .120 | **.615** | .208 | .186 | .165 |
| $X_5$ | .261 | .279 | **.539** | .107 | .066 | -.270 |
| $X_4$ | .088 | .042 | **.445** | .156 | .397 | -.314 |
| $X_3$ | .142 | .084 | .264 | **.745** | -.118 | .002 |
| $X_2$ | .094 | .068 | -.194 | **.686** | .228 | -.064 |
| $X_1$ | .045 | -.030 | .258 | **.654** | -.159 | .219 |
| $X_{23}$ | .151 | .091 | .102 | -.044 | **.779** | .026 |
| $X_{10}$ | .182 | .277 | .173 | .197 | .019 | **.725** |
| $X_9$ | .327 | .166 | .222 | .099 | -.479 | **-.449** |

Extraction Method: Principal Component Analysis.
Rotation Method: Varimax with Kaiser Normalization.
a. Rotation converged in 10 iterations

The basis for deciding whether a variable is included in the factor 1, factor 2, or any other factor is to look at the value of loading factor. Prior grouped loading factor must meet the criteria of practical significance and statistical significance. Criteria practical significance is the loading factor is greater than 0:35, as in this study using 297 samples. Membership of each factor is as follows: Item questions $X_{25}$, $X_{24}$, $X_{21}$, $X_{22}$, $X_{12}$, $X_{20}$, $X_{11}$, $X_{19}$, $X_{13}$ and $X_{14}$ are clustered in factor 1, indicating that the items to these questions is a component of Factor 1. The next factor is named factor 1 safety and comfort.

Variables $X_{17}$, $X_{18}$, $X_{16}$ and $X_{15}$ clustered in factor 2 and is a component of the factor 2, factor 2 is named Factor Responsiveness. The components of Factor 3 consists of variables $X_6$, $X_7$, $X_8$, $X_5$, and $X_4$, and named Factor Capacity. Item questions $X_3$, $X_2$, and $X_1$ are clustered in factor 4, show items such questions are components of the Factor 4, Factor 4 is named Tangible. While the question items $X_{23}$ that are in the component Factor 5, this factor was named Safety, and the last is the item question $X_{10}$ and $X_9$ clustered on the components Factor 6, further Factor 6 is named by a Factor Reliability. The grouping of variables into a component of a factor can be seen in Table 5.

**Table 5.** Grouping variable become a component factor

| Factor | Number of question | Pernyataan |
|---|---|---|
| Factor 1 Safety and Comfort | $X_{25}$, $X_{24}$, $X_{21}$, $X_{22}$, $X_{12}$, $X_{20}$, $X_{11}$, $X_{19}$, $X_{13}$, $X_{14}$ | $X_{25}$ = feelings of self-users are concerned about security when in the bus<br>$X_{24}$ = feelings are concerned about security when the user luggage in the bus<br>$X_{21}$ = comfort in the bus during the trip<br>$X_{22}$= bus attendant quite responsive in helping users<br>$X_{12}$= bus attendant able to be patient and courteous in serving users<br>$X_{11}$ = Hygiene of bus<br>$X_{19}$ = bus attendant can be kind to each user<br>$X_{13}$ = bus attendant is able to communicate well with users<br>$X_{14}$ = treatments bus attendant in the serving regardless user's social status |
| Factor 2 *Responsiveness* | $X_{17}$, $X_{18}$, $X_{16}$, $X_{15}$ | $X_{17}$ = bus attendant quickly respond to complaints<br>$X_{18}$ = bus attendant willing to follow up on complaints from users<br>$X_{16}$ =bus attendant willing to listen to complaints from users<br>$X_{15}$ = user can submit a complaint easily |
| Factor 3 Capacity | $X_6$, $X_7$, $X_8$, $X_5$, dan $X_4$ | $X_6$ = Capacity of bus<br>$X_7$ = Adequacy of seating and standing<br>$X_8$ = complete facilities in bus<br>$X_5$= suitability travel expenses with service |
| Factor 4 *Tangibles* | $X_3$, $X_2$, dan $X_1$ | $X_3$ = Position stop, strategic and easy to reach, access speed<br>$X_2$ = stop position is easy to find<br>$X_1$ = Region coverage |
| Factor 5 Safety | $X_{23}$ | $X_{23}$ = misgivings users to the speed of the bus driver to drive the bus |
| Factor 6 Reliability | $X_{10}$ dan $X_9$ | $X_{10}$ = Timeliness arrived<br>$X_9$ = Completeness of facilities at bus stop |

Sources: Primary data (2016)

Table 5 shows the variables that are components in the sixth form factor based eigenvalues greater than one. In Factor 1, the safety and comfort factor having the greatest eigen value of 8.412 or can explain the variance data at 33.648%, with the most dominant variable is feeling concerned about the personal safety of users when on the bus. In addition to safety, comfort is also a variable that exists in this first factor. This indicates that people are very concerned about security and comfort in using public transport, especially buses.

The second factor is a factor of public transport user satisfaction is Responsiveness. The speed and sensitivity of officers in the bus respond to customer complaints is the main point in this second factor. The amount of variance can be explained by the second factor is 7.057%. Other factors that also affect user satisfaction in using public transport is the capacity, tangibles, safety and reliability.

The results of this study are in line with Redman et al.[6] presented a comprehensive review on service quality in public transportation and determined that reliability, frequency, price, speed, access, comfort, and convenience were the factors that attract car users to use public transportation. In a study by de Oña et al.[7], to determine the most important service quality factors explained that from a general perspective, comfort, personnel, information, and service were determined as the most important factors. Besides, Hassan et al.[8] explained that the most desirable service quality factors of public transportation services were reliability, frequency, capacity, price, cleanliness, comfort, security, staff, information, and ticketing system, with loading/ridership, travel time, travel distance, and service duration indicated as "efficiency" indicators

## 4. Conclusions

Based on the discussion in the previous chapter, the conclusions that can be drawn is: Results of factor analysis showed there are six factors that explain the satisfaction of users of public transport in Bali, with a total diversity of data that can be explained by 61.436%. These factors are: Safety and comfort; Responsiveness; Capacity; Tangible; Safety; Reliability. The dominant factor influencing customer satisfaction of public transport users is Safety and comfort, with the most influential variable is feeling concerned about the personal safety of customers while on the bus.

Based on the findings of this study suggested to the government to pay attention to the safety factor and passenger comfort when using public transport. The variables on these factors is the feeling of self-users are concerned about security when on the bus; feeling worried about the security of customer's belongings when in the bus; Comfort in the bus during the trip; bus attendant quite responsive in helping users; operator/supervisor of bus able to be patient and courteous in serving users; Cleanliness bus; bus attendant can be kind to each user; operator/supervisor able to communicate well with users; operator/supervisor treatment in the serving regardless of the user's social status.

## References

[1] Munawar A 2004 *Manajemen Lalu Lintas Perkotaan* (Yogyakarta: Beta Offset)

[2] Le-Klähn D L, Hall C M and Gerike R 2014 Analysis of visitor satisfaction with public transport in munich *J. Public Transportation* **17**(3)

[3] Tamin O Z 2000 *Perencanaan dan Pemodelan Transportasi* (Bandung: Penerbit ITB)

[4] Ferdiansyah R 2009 Kemungkinan peralihan penggunaan moda angkutan pribadi ke moda angkutan umum perjalanan Depok-Jakarta. *J. Perencanaan Wilayah dan Kota*, **20** 183-98

[5] Hair J F, Rolph E A, Ronald L T and William C B 1995 *Multivariate Data Analysis with Reading* (New Jersey: Prentice Hall International Editions)

[6] Redman L, Friman M, Gärling T and Hartig T 2013 Quality attributes of public transport that attract car users: a research review *J. Transport Policy* **25** 119-27.

[7] de Oña R, Eboli L and Mazzulla G 2014 Key factors affecting rail service quality in Northern Italy: a decision tree approach *Transport* **29**(1) 75-83.

[8] Hassan M N, Hawas Y E and Ahmed K 2013 A multi-dimensional framework for evaluating transit service performance *Transportation Research Part A*: *Policy and Practice* **50** 47-61

# The fundamental properties of quasi-semigroups

**Sutrima[1,2], Ch. Rini Indrati[1], Lina Aryati[1], and Mardiyana[3]**

[1)] Mathematics Department of Universitas Gadjah Mada
[2)] Mathematics Department of Universitas Sebelas Maret
[3)] Department of Mathematics Education of Universitas Sebelas Maret

E-mail: sutrima@mipa.uns.ac.id, rinii@ugm.ac.id, lina@ugm.ac.id, mardiyana@staff.uns.ac.id

**ABSTRACT.** In this paper we focus on the abstract Cauchy problems of the time-dependent evolution equation $\dot{x} = A(t)x(t)$. If the operator $A(t) = A$ is time-independent, we can use the $C_0$-semigroups theory to solve the abstract Cauchy problem. In this case $A$ is a infinitesimal generator of the $C_0$-semigroups. However, if $A(t)$ is time-dependent, we can not apply directly the $C_0$-semigroups theory to solve the problem. In this situation we can use the quasi semigroups theory as development of the two parameters semigroups. This semigroups is induced by bounded evolution operators $U(t,s)$ that satisfy some assumptions. In this paper we determine the fundamental properties of the quasi semigroups included its generator related to the time-dependent evolution equation.

## 1. Introduction

Let consider the time-independent abstract Cauchy problems

$$\dot{x}(t) = Ax(t), \quad t \geq 0 \tag{1}$$

on a Banach space $X$ and $A$ is a densely defined operator on $\mathcal{D}(A) \subseteq X$. Fattorini [6] gave necessary and sufficient conditions in order to (1) be well-posed in $t \geq 0$. Especially, if $A$ is the generator of $C_0$-semigroup of bounded linear operators on $X$, then the semigroups theory is a powerful tool for solving (1), (see [3], [5], [15], and [17]). We can get the comprehensively properties of $A$ and its application therein.

Next, we consider the time-dependent abstract Cauchy problems [7]

$$\dot{x}(t) = A(t)x(t) + f(t), \quad 0 \leq t \leq T \tag{2}$$

and the associated homogeneous equation

$$\dot{x}(t) = A(t)x(t). \tag{3}$$

Here $x$ is an unknown function from the real interval $[0,T]$ into a Banach space $X$, $f$ is a given function from $[0,T]$ into $X$, and $A(t)$ is a given, closed, linear operator in $X$ with domain $\mathcal{D}(A(t)) = \mathcal{D}$, independent of $t$ and dense in $X$. The higher order of parabolic type of (2) also was investigated by Obrecht [14].

The solution of (2) formally given by

$$x(t) = U(t,0)x_0 + \int_0^t U(t,s)f(s)ds, \quad x_0 = x(0) \tag{4}$$

where $U(t,s)$ is a linear operator on $X$ depending on $s$ and $t$, with $s \leq t$, [9]. The main problem is to find some sufficient conditions for the existence of $U(t,s)$.

If $A(t) = A$ is independent of $t$, then $U(t,s)$ is given formally by $U(t,s) = e^{(t-s)A}$. In this case $A$ is an infinitesimal generator of $C_0$-semigroup of bounded linear operators on $X$, and equation (2) and (3) admit a unique solution respectively. These suggest to generalize the results as $A(t)$ depend on $t$. It is a natural to take over some assumption on the infinitesimal generator $A$ for the $A(t)$ for each $t$. However, Ladas and Lashmikantham [10], Masuda [12] and Fattorini [6] gave the assumptions for $A(t)$ in order to each equation (2) and (3) admits a unique solution.

The family of operators $U(t,s)$ is also called evolution operator, propagator, solution operator or Green's function associated to (2) or (3). Let $B(X)$ be the set of all bounded linear operators from $X$ into $X$. Then each operator $U(t,s)$ belong to $B(X)$ for all $s,t \geq 0$, and satisfies the following conditions:

(a) $U(r,r) = I$, identity of $B(X)$,
(b) $U(t,r)U(r,s) = U(t,s)$, $0 \leq s \leq r \leq t$,
(c) $U(\cdot\,,\,\cdot)$ is strongly continuous,

(d) The operator $\frac{\partial U(t,r)}{\partial t}x$ exists, and continuous for each $x \in X$.

For detail properties of the evolution operator see [2], [3], [5], [6], [11] and [15].

The appearance of this the family $U(t,s)$ that urges the new semigroups theory of two parameters, is called quasi-semigroups. Leiva and Barcenas [11] have introduced the quasi-semigroups $R(t,s),\ s,t \geq 0$. They prove several properties of $R(t,s)$ and its generator. They also give the certain conditions in order to (2) has a unique solution. Furthermore, they can construct the necessary and sufficient conditions for exact and approximate controllability for the non-autonomous control systems. The sequel works about dual quasi-semigroups and controllability also was done by Barcenas *et.al* [1].

Megan and Cuc [13] study some stability concepts for linear systems the evolution which can be described by $C_0$-quasi-semigroup. The results obtained may be regarded as generalizations of well known results of Datko, Pazy, Littman and Neerven about exponential stability of $C_0$-semigroups. Moreover, the $C$-quasi-semigroups as generalization of $C$-semigroups also can be constructed [8]. Therein, some examples are given and the properties of $C$-quasi-semigroups are verified. Also some applications of $C$-quasi-semigroups in abstract evolution equations is considered.

In this paper we shall investigate the properties of $C_0$-quasi-semigroups which have not yet discussed in Leiva and Bercenas [11] inline to the properties of $C$-semigroups. Especially, we consider the Hille-Yosida theorem version in $C_0$-quasi-semigroups. The finally, we shall consider some applications of $C_0$-quasi-semigroups for solving the evolution equations that are described by PDE's.

## 2. Quasi-semigroups

The theory of quasi-semigroups of bounded linear operators, as a generalization of semigroups of operators, was introduced by Leiva and Barcenas [11].

**Definition 2.1**. Let $X$ be a Banach space. A two-parameter commutative family $\{R(t,s)\}_{s,t\geq 0} \subseteq B(X)$ is called a strongly continuous quasi-semigroup (or $C_0$-quasi-semigroup) of operators if for every $r,s,t \geq 0$ and $x \in X$:

($q1$) $R(t,0) = I$, the identity operator on $X$,

($q2$) $R(t,s+r) = R(t+r,s)R(t,r)$,

($q3$) $\lim_{s\to 0}\|R(t,s)x - x\| = 0$,

($q4$) there exists a continuous increasing mapping $M : [0,\infty) \to [0,\infty)$ such that

$$\|R(t,s)\| \leq M(t+s).$$

For a $C_0$-quasi-semigroup $\{R(t,s)\}_{s,t\geq 0}$ on a Banach space $X$, let $\mathcal{D}$ be the set of all $x \in X$ for which the following limits exist

$$\lim_{s\to 0^+}\frac{R(0,s)x - x}{s} \text{ and } \lim_{s\to 0^+}\frac{R(t,s)x - x}{s} = \lim_{s\to 0^+}\frac{R(t-s,s)x - x}{s},\ t > 0.$$

For $t \geq 0$ we define an operator $A(t)$ on $\mathcal{D}$ as

$$A(t)x = \lim_{s\to 0^+}\frac{R(t,s)x - x}{s}.$$

The family $\{A(t)\}_{t\geq 0}$ is called infinitesimal generator of the $C_0$-quasi-semigroups $\{R(t,s)\}_{s,t\geq 0}$.

Next, we say generator in short instead of infinitesimal generator. Throughout this paper we denote $T(t)$ and $R(t,s)$ as semigroups $\{T(t)\}_{t\geq 0}$ and quasi-semigroups $\{R(t,s)\}_{s,t\geq 0}$, respectively. We also denote $\mathcal{D}$ as domain of $A(t),\ t \geq 0$. We give some useful examples of $C_0$-quasi-semigroups that are summarized from [8] and [11].

**Example 2.2** *If $T(t)$ is a $C_0$-semigroup on a Banach space $X$ with its generator $A$, then $R(t,s)$, with*

$$R(t,s) = T(s),\quad s,t \geq 0,$$

*defines a $C_0$-quasi-semigroup on $X$ with its generator $A(t) = A,\ t \geq 0$, and $\mathcal{D} = \mathcal{D}(A)$.*

**Example 2.3** *Let $X$ be a Banach space of all bounded uniformly continuous functions on $[0,\infty)$ with the supremum-norm. The family of operators $R(t,s)$ in $B(X)$ defined by*

$$(R(t,s)x)(\xi) = x(s^2 + 2ts + \xi),\quad s,t \geq 0$$

*is a $C_0$-quasi-semigroup on $X$ with generator $A(t) = 2tx'$, $t \geq 0$, where $\mathcal{D} = \{x \in X : x' \in X\}$.*

We can verify easily the conditions $(q1) - (q4)$ of the Definition 2.1. For $t \geq 0$ fixed, and $x \in \mathcal{D}$ we have

$$\left(\frac{R(t,s)x - x}{s}\right)(\xi) = \frac{x(s^2 + 2ts + \xi) - x(\xi)}{s} = \frac{f(s) - f(0)}{s}, s > 0,$$

where $f(s) = x(s^2 + 2ts + \xi)$, for some $\xi > 0$. Consequently,

$$(A(t)x)(\xi) = \lim_{s \to 0^+} \left(\frac{R(t,s)x - x}{s}\right)(\xi) = f'(0) = 2tx'(\xi).$$

So, $R(t,s)$ is a $C_0$-quasi-semigroup on $X$ with its generator $A(t)x = 2tx'$ and $\mathcal{D} = \{x \in X : x' \in X\}$. Here $x'$ denotes the derivative of $x$ respect to $\xi$.

**Example 2.4** *If $T(t)$ be a $C_0$-semigroup on a Banach space $X$ with its generator $A$, then $R(t,s)$ with*

$$R(t,s) = e^{T(t+s) - T(t)}, \quad s, t \geq 0,$$

*is a $C_0$-quasi-semigroup on $X$ with its generator $A(t) = AT(t)$, $t \geq 0$, and $\mathcal{D} = \mathcal{D}(A)$.*

In fact, for $x \in \mathcal{D}$ and $T(t)$ is a $C_0$-semigroup we have

$$\begin{aligned} A(t)x = \lim_{s \to 0^+} \frac{R(t,s)x - x}{s} &= \lim_{s \to 0^+} \left(\frac{e^{T(t+s) - T(t)} - I}{s}\right)x \\ &= \frac{d}{ds}[T(t)]T(s)e^{T(s+t) - T(s)}\Big|_{s=0} x \\ &= AT(t)T(s)|_{s=0}x = AT(t)x. \end{aligned}$$

**Example 2.5** *Let $T(t)$ be a $C_0$-semigroup on a Banach space $X$ with its generator $A$. For $s, t \geq 0$,*

$$R(t,s) = T(g(t+s) - g(t)),$$

*where $g(t) = \int_0^t a(u)\, du$ and $a \in C[0, \infty)$ with $a(t) > 0$. The $R(t,s)$ is a $C_0$-quasi- semigroup on $X$ with its generator $A(t) = a(t)A$.*

For each $r, s, t \geq 0$ and $x \in X$ we have

$(q1)$ $R(t,0) = T(g(t) - g(t)) = I$,

$(q2)$
$$\begin{aligned} R(t, s + r) &= T\big(g(t + s + r) - g(t)\big) \\ &= T\big(g(t + s + r) - g(t + r) + g(t + r) - g(t)\big) \\ &= T\big(g(t + s + r) - g(t + r)\big)T\big(g(t + r) - g(t)\big) \\ &= R(t + r, s)R(t, r), \end{aligned}$$

$(q3)$ $\lim_{s \to 0^+} \|R(t,s)x - x\| = \lim_{s \to 0^+} \big\|T\big(g(t + s) - g(t)\big)x - x\big\| = 0$, since $g$ is continuous.

$(q4)$ Since $T$ is strongly continuous on $X$, there exists $\omega, M_\omega > 0$ such that $\|T(s)\| \leq M_\omega e^{\omega s}$. Therefore,

$\|R(t,s)\| = \|T(g(t + s) - g(t))\| \leq M(t + s)$, where $M(t + s) = M_\omega e^{\omega(g(t+s) - g(t))}$.

Moreover,

$$\begin{aligned} A(t)x = \lim_{s \to 0^+} \frac{R(t,s)x - x}{s} &= \lim_{s \to 0^+} \left(\frac{T(g(t + s) - g(t)) - I}{s}\right)x \\ &= g'(t + s)\frac{d}{ds}\big[T\big(g(t + s) - g(t)\big)x\big]\Big|_{s=0} = a(t)Ax. \end{aligned}$$

Thus, $R(t,s)$ is a $C_0$ quasi semigroup on $X$ with generator $A(t) = a(t)A$ and $\mathcal{D} = \mathcal{D}(A)$.

## 3. Main Results

In the following results we use in outline of results of [3], and some results because of reformulation from [1], [8], and [11].

**Theorem 3.1** *If $R(t,s)$ is a $C_0$-quasi-semigroup on a Banach space $X$, then*

(a) *For each $t \geq 0$, $R(t,.)$ is strongly continuous on $[0, \infty)$;*

(b) *For each $t \geq 0$ and $x \in X$ hold*

$$\lim_{s \to 0^+} \frac{1}{s} \int_0^s R(t,u)\, x\, du = x.$$

*Proof.* (a) For $s,t \geq 0$ fixed, by commutativeness and boundedness of $R(t,s)$, then for $r \geq 0$ and $x \in X$ we have

$$\begin{aligned}\|R(t,s+r)x - R(t,s)x\| &= \|R(t+s,r)R(t,s)x - R(t,s)x\| \\ &\leq \|R(t,s)\|\|R(t+s,r)x - x\|.\end{aligned}$$

By $(q3)$ we conclude that

$$\lim_{r\to 0^+} \|R(t,s+r)x - R(t,s)x\| = 0,$$

i.e $R(t,.)$ is strongly continuous on $[0,\infty)$.

(b) Let $x \in X$, $t \geq 0$, and $\varepsilon > 0$ be given. By strongly continuity of $R(t,.)$ there exists $\tau > 0$ such that for $s \in [0,\tau]$ we have

$$\|R(t,s)x - x\| < \varepsilon.$$

Therefore, for $s \in (0,\tau]$

$$\begin{aligned}\left\|\frac{1}{s}\int_0^s R(s,u)\,x\,du - x\right\| &= \left\|\frac{1}{s}\int_0^s [R(t,u)\,x - x]\,du\right\| \\ &\leq \frac{1}{s}\int_0^s \|R(t,u)x - x\|\,du < \frac{1}{s}\int_0^s \varepsilon\,du = \varepsilon.\end{aligned}$$

□

**Theorem 3.2** *If $R(t,s)$ is a $C_0$-quasi-semigroup on a Banach spce $X$ with its generator $A(t)$, then*

(a) *If $x \in \mathcal{D}$, $t \geq 0$, and $s_0, t_0 \geq 0$, then $R(t_0,s_0)x \in \mathcal{D}$ and*

$$R(t_0,s_0)A(t)x = A(t)R(t_0,s_0)x;$$

(b) *For each $s > 0$,*

$$\frac{\partial}{\partial s}(R(t,s)x) = A(t+s)R(t,s)x = R(t,s)A(t+s)x, \qquad x \in \mathcal{D};$$

(c) *If $A(\cdot)$ is locally integrable, then for every $x \in \mathcal{D}$ and $s \geq 0$*

$$R(t,s)x = x + \int_0^s A(t+u)R(t,u)\,x\,du;$$

(d) *If $f : [0,\infty) \to X$ is a continuous, then for every $t \in [0,\infty)$*

$$\lim_{h\to 0}\frac{1}{h}\int_s^{s+h} R(t,u)f(u)du = R(t,s)f(s).$$

*Proof.* (a) For $x \in \mathcal{D}$ states that

$$\lim_{s\to 0^+}\frac{R(t,s)x - x}{s} = A(t)x.$$

By strongly continuity of $R(t_0,s_0)$,

$$\begin{aligned}A(t)R(t_0,s_0)x &= \lim_{s\to 0^+}\frac{R(t,s)R(t_0,s_0)x - R(t_0,s_0)x}{s} \\ &= R(t_0,s_0)\lim_{s\to 0^+}\frac{R(t,s)x - x}{s} = R(t_0,s_0)\,A(t)x.\end{aligned}$$

So, $R(t_0,s_0)x \in \mathcal{D}$, and $R(t_0,s_0)A(t)x = A(t)R(t_0,s_0)x$.

(b) For $r > 0$, the commutativeness of $R(t,s)$ gives

$$\frac{R(t,s+r)x - R(t,s)x}{r} = R(t,s)\frac{(R(t+s,r) - I)x}{r}.$$

If $x \in \mathcal{D}$, then for $r \to 0^+$ the limit in the right hand exists. This implies the existence of limit in the left hand, i.e

$$\lim_{r\to 0^+}\frac{R(t,s+r)x - R(t,s)x}{r} = R(t,s)A(t+s)x. \tag{5}$$

For $r < 0$ such that $s + r \geq 0$,

$$\begin{aligned}\frac{R(t,s+r)x - R(t,s)x}{r} &= \frac{R(t,s)x - R(t,s+r)x}{-r} \\ &= \frac{R(t+s+r,-r)R(t,s+r)x - R(t,s+r)x}{-r} \\ &= R(t,s+r)\frac{R(t+s+r,-r)x - x}{-r} \\ &= R(t,s+r)\frac{R(t+s+r,-r)x - R(t+s,-r)x + R(t+s,-r)x - x}{-r}.\end{aligned}$$

The strongly continuity of $R(t,s)$ implies
$$\lim_{r\to 0^-} R(t+s+r,-r)x - R(t+s,-r)x = 0.$$
Therefore,
$$\lim_{r\to 0^-} \frac{R(t+s+r,-r)x - x}{-r} = \lim_{r\to 0^-} \frac{R(t+s,-r)x - x}{-r} = A(t+s)x.$$
Again, the strongly continuity of $R(t,s)$ implies
$$\lim_{r\to 0^-} \frac{R(t,s+r)x - R(t,s)x}{r} = R(t,s)A(t+s)x. \quad (6)$$
From (5) and (6) we have
$$\frac{\partial}{\partial s}(R(t,s)x) = A(t+s)R(t,s)x = R(t,s)A(t+s)x.$$
(c) By integrating the last equation from 0 to $s$, we have
$$R(t,s)x = x + \int_0^s R(t,u)A(t+u)xdu.$$
(d) For $t > 0$, by continuity of $f$ and the strongly continuity of $R(t,.)$ on $[0,\infty)$, then $y(.) = R(t,.)f(.)$ is continuous in $s$. If we define
$$F(\xi) = \int_s^{s+\xi} y(u)\ du,$$
then
$$F'(0) = \lim_{h\to 0} \frac{1}{h}\int_s^{s+h} y(u)du = \lim_{h\to 0} \frac{1}{h}\int_s^{s+h} R(t,u)f(u)du. \quad (7)$$
On the other hand, since
$$F'(\xi) = y(s+\xi)$$
then
$$F'(0) = y(s) = R(t,s)f(s). \quad (8)$$
From (7) and (8) we have
$$\lim_{h\to 0} \frac{1}{h}\int_s^{s+h} R(t,u)f(u)du = R(t,s)f(s).$$
□

In the semigroups theory, if $A$ is an infinitesimal generator of $C_0$-semigroup with domain $\mathcal{D}$, then $A$ is a closed operator and $\mathcal{D}$ is dense in $X$. These are is not always true for any $C_0$-quasi-semigroups.

We return to the $C_0$-quasi-semigroup $R(t,s)$ in Example 2.3, with the generator $A(t)x = 2tx'$ and its domain $\mathcal{D} = \{x \in X : x' \in X\}$. We can show that for every $t > 0$, $A(t)$ is not closed. In fact, we can choose a sequence $(x_n)$ in $\mathcal{D}$ with $x_n(\xi) = e^{-n\xi}/n$, for $\xi \geq 0$ and $n \in \mathbb{N}$. Obvious that
$$x_n \to x \quad \text{where} \quad x(\xi) = 0 \ \text{ for all } \xi \geq 0.$$
However,
$$A(t)x_n \to y$$
where $(A(t)x_n)(\xi) = -2te^{-n\xi}, \xi \geq 0$ and $y(\xi) = \begin{cases} -2t, & \xi = 0 \\ 0, & \xi > 0 \end{cases}$. We have $A(t)x \neq y$. Therefore, $A(t)$ is not closed.

**Example 3.3** *Let X be as in Example 2.3. For* $s > 0$ *given an initial value problem*
$$\frac{\partial x}{\partial t}(t,s) = 2(t+s)\frac{\partial x}{\partial s}(t,s)$$
$$x(0,s) = 0.$$
This initial value problem can be rewritten as
$$\dot{x}(t) = A(s+t)x(t)$$
$$x(0) = 0,$$
where $A(t)x = 2tx'$. By Theorem 3.2 (c), for each $x_0 \in \mathcal{D}$ the initial value problem has a solution $x(t) = R(s,t)x_0$ where $R(t,s)$ is a $C_0$-quasi-semigroup with generator $A(t)$ that satisfies $A(t)x(\xi) = 2tx'(\xi)$ on its domain
$$\mathcal{D} = \{x \in X : x' \in X, x(0) = 0\}.$$

The set $\mathcal{D}$ is not dense in $X$, since $\overline{\mathcal{D}} = \{x \in X : x(0) = 0\} \neq X$, [16]. This result also concludes that for $t \geq 0$ the operator $A(t)$ is not a generator for any $C_0$-semigroup.

□

The two examples previously state that the operator $A(t)$, $t \geq 0$, can be a generator for some $C_0$-quasi-semigroup but it must not satisfy the the necessary condition of the Hille-Yosida theorem [3]. Next, we define the resolvent operator of $A(t)$ similar to one in operator theory. For each $t \geq 0$ we define the resolvent operator of $A(t)$ as

$$\mathcal{R}_\lambda = \mathcal{R}(\lambda, A(t)) = (\lambda I - A(t))^{-1},$$

with its resolvent set $\rho(A(t))$.

Throughout this paper, we put $S$ as a set of all complex numbers $\lambda$ such that $-\theta \leq \arg\lambda \leq \theta$ with $\pi/2 < \theta < \pi$, see [10].

**Theorem 3.4** *Let $A(t)$ be a closed and densely defined generator of a $C_0$-quasi-semigroup $R(t,s)$ on a Banach space $X$ and resolvent $\mathcal{R}(\lambda, A(t))$ exists in $S$. If $\lambda \in \rho(A(t))$, then $\mathcal{R}(\lambda, A(t))R(t,s) = R(t,s)\mathcal{R}(\lambda, A(t))$ for all $s \geq 0$.*

*Proof.* For $\lambda \in \rho(A(t))$ and $x \in X$, we set $y = \mathcal{R}(\lambda, A(t))x$. Since $A(t)^{-1}$ exists, then we can write $y$ as

$$y = \mathcal{R}(\lambda, A(t))x = A(t)^{-1}(\lambda A(t)^{-1} - I)^{-1}x \quad \text{or} \quad A(t)y = (\lambda A(t)^{-1} - I)^{-1}x,$$

i.e $y \in \mathcal{D}$. From Theorem 3.2 (a) for $s \geq 0$ we have

$$R(t,s)(\lambda - A(t))y = (\lambda I - A(t))R(t,s)y$$

By applying $\mathcal{R}(\lambda, A(t))$ to the both sides, we conclude that $\mathcal{R}(\lambda, A(t))R(t,s)x = R(t,s)\mathcal{R}(\lambda, A(t))x$ as asserted.

□

**Theorem 3.5** *If $A(t)$ is a generator of a $C_0$-quasi-semigroup $R(t,s)$ on a Banach space $X$, then for $\lambda \in \mathbb{C}$, $A(t) - \lambda I$ is a generator of $C_0$-quasi-semigroup*

$$K(t,s) = e^{-\lambda s}R(t,s) \text{ for all } s,t \geq 0.$$

*Proof.* It is easy to show that $K(t,s)$ is a $C_0$-quasi-semigroup. Furthermore, for $t \geq 0$ and $x \in \mathcal{D}$,

$$\begin{aligned}\lim_{s\to 0^+} \frac{K(t,s) - I}{s}x &= \lim_{s\to 0^+} \frac{e^{-\lambda s}R(t,s) - I}{s}x \\ &= e^{-\lambda s}\left[-\lambda R(t,s) + \frac{\partial}{\partial s}R(t,s)\right]\Big|_{s=0} \\ &= e^{-\lambda s}[-\lambda R(t,s) + A(t+s)R(t,s)]\big|_{s=0} \\ &= A(t) - \lambda I.\end{aligned}$$

So, $A(t) - \lambda I$ is the generator of a $C_0$-quasi-semigroup $K(t,s)$.

□

The following we shall analyze the necessary and sufficient conditions of the Hille-Yosida theorem in order to an operator generates a quasi-semigroup. To achieve this aim we introduce the so called Yosida approximation, (see [3], [5], [17], and [18]). For $\lambda \in \rho(A)$ we set

$$A_\lambda = \lambda A\mathcal{R}(\lambda, A) = \lambda^2\mathcal{R}(\lambda, A) - \lambda I \in \mathcal{B}(X). \tag{9}$$

We note that

$$A\mathcal{R}(\lambda, A) = \lambda\mathcal{R}(\lambda, A) - I \quad \text{and} \quad A\mathcal{R}(\lambda, A)x = \mathcal{R}(\lambda, A)Ax \tag{10}$$

hold for all $x \in \mathcal{D}(A)$.

**Lemma 3.6** *Let $A(t)$ be a closed, densely defined and $M, \omega \geq 0$ such that $[\omega, \infty) \subseteq \rho(A(t))$ and*

$$\|\mathcal{R}(\lambda, A(t))\| \leq \frac{M}{\lambda}$$

*for all $\lambda \geq \omega$. Then $\lambda\mathcal{R}(\lambda, A(t))x \to x$ as $\lambda \to \infty$ for all $x \in X$ and $\lambda\mathcal{R}(\lambda, A(t))y \to A(t)y$ as $\lambda \to \infty$ for all $y \in \mathcal{D}$.*

*Proof.* Let $x \in \mathcal{D}$. Equation (10) and the assumptions yield that

$$\|\lambda\mathcal{R}(\lambda,A(t))x-x\|=\|\mathcal{R}(\lambda,A(t))A(t)x\|\leq\frac{M}{\lambda}\|A(t)x\|\rightarrow 0$$

as $\lambda\rightarrow\infty$. Because $\lambda\mathcal{R}(\lambda,A(t))$ is uniformly bounded, by $M$, the first assertion follows. The second one is a consequence of the first assertion, taking $x=A(t)y$ and using (10).

□

**Theorem 3.7** *If $A(t)$ is a closed, densely defined on $\mathcal{D}$ and $M,\omega\geq 0$ such that $[\omega,\infty)\subseteq\rho(A(t))$ and*

$$\left\|\mathcal{R}(\lambda,A(t))^r\right\|\leq\frac{M}{(\lambda-\omega)^r},\qquad r\geq 1,$$

*for all $\lambda>\omega$, then $A(t)$ generates a $C_0$-quasi-semigroup $R(t,s)$. Moreover, $\|R(t,s)\|\leq Me^{\omega s}$ for all $s,t\geq 0$.*

*Proof.* By using Yosida approximation we set

$$A_n(t):=nA(t)\mathcal{R}(n,A(t))=n^2\mathcal{R}(n,A(t))-nI$$

for all $n>\omega$, $n\in\mathbb{N}$. Lemma 3.6 and assumptions imply that

$$\lim_{n\rightarrow\infty}A_n(t)y=A(t)y \text{ for all } x\in\mathcal{D}.$$

Let $s\geq 0$. We define

$$R_n(s):=e^{sA_n(t)}=e^{-sn}\sum_{k=0}^{\infty}\frac{(n^2s)^k}{k!}\mathcal{R}(n,A(t))^k.$$

Hence

$$\|R_n(s)\|\leq e^{-sn}\sum_{k=0}^{\infty}\frac{(n^2s)^k}{k!}\frac{M}{(n-\omega)^k}$$
$$=Me^{-ns}e^{\left(\frac{n^2}{n-\omega}\right)s}=Me^{\left(\frac{n\omega}{n-\omega}\right)s}. \tag{11}$$

Next, take $m,n\in\mathbb{N}$, $s_0>0$, $y\in\mathcal{D}$ and $s\in[0,s_0]$. Obvious, it holds that $A_m(t)A_n(t)=A_n(t)A_m(t)$ and

$$A_n(t)e^{sA_m(t)}=A_n(t)\sum_{k=0}^{\infty}\frac{s^k}{k!}A_m^k(t)=\sum_{k=0}^{\infty}\frac{s^k}{k!}A_m^k(t)A_n(t)=e^{sA_m(t)}A_n(t).$$

Consequently, we can compute

$$e^{sA_n(t)}y-e^{sA_m(t)}y=\int_0^s\frac{d}{ds}e^{(s-u)A_m(t)}\,e^{uA_n(t)}y\,du=\int_0^s e^{(s-u)A_m(t)}e^{uA_n(t)}\,(A_n(t)-A_m(t))y\,du.$$

Equation (7) implies that

$$\|R_n(s)y-R_m(s)y\|=\|e^{sA_n(t)}y-e^{sA_m(t)}y\|\leq s_0\|A_n(t)y-A_m(t)y\|\rightarrow 0\quad\text{as } m,n\rightarrow\infty. \tag{12}$$

Therefore, $(R_n(s)y)$ is a Cauchy sequence, and so it converges in $\mathcal{D}$. Furthermore, we have that $(R_n(s))$ is equicontinuous on $X$. Since $\mathcal{D}$ is dense in $X$, then we can extend this convergence to every $x\in X$, (see Lemma 3.8 of [16]). Hence, we may define

$$R(t,s)x:=\lim_{n\rightarrow\infty}R_n(s)x$$

for all $s,t\geq 0$ and $x\in X$. In facts, $R(t,0)=I$ and

$$R(t,s+r)x=\lim_{n\rightarrow\infty}e^{(s+r)A_n}=\lim_{n\rightarrow\infty}e^{sA_n}.\lim_{n\rightarrow\infty}e^{rA_n}=R(t+r,s)R(t,r)x$$

for all $r,s,t\geq 0$. By letting $m\rightarrow\infty$ in (12), we deduce that

$$\|R_n(s)y-R(t,s)y\|\leq s_0\|A_n(t)y-A(t)y\|$$

for all $s\in[0,s_0]$. This states that $(R_n(s)y)$ uniformly converges to $R(t,s)y$ on $[0,s_0]$. Therefore, $R(t,\cdot)y$ is continuous for all $y\in\mathcal{D}$. Finally, by (11) there exists continuous increasing function $M_1$ on $[0,\infty)$ such that

$$\|R(t,s)\|\leq M_1(t+s)$$

where $M_1(t+s)=Me^{\omega s}$. Thus $R(t,s)$ is a $C_0$-quasi-semigroup.

Let $B(t)$ be the generator of $R(t,s)$. For $s>0$ and $y\in\mathcal{D}$, we have

$$\frac{1}{s}(R(t,s)y-y)=\lim_{n\rightarrow\infty}\frac{1}{s}\left(e^{sA_n(t)}y-y\right)=\lim_{n\rightarrow\infty}\frac{1}{s}\int_0^s e^{uA_n(t)}A_ny\,du=\frac{1}{s}\int_0^s e^{uA(t)}A(t)y\,du.$$

By letting $s\rightarrow 0^+$, it follows that $y\in\mathcal{D}(B(t))$ and $B(t)y=A(t)y$.

Now, if $\lambda>\omega$, then by proof of Theorem 3.4 we have

$$(\lambda I-A(t))\mathcal{D}=X.$$

Since we assume that $B(t)$ is a generator, then Theorem 3.2 (a) gives

$$(\lambda I-B(t))\mathcal{D}(B(t))=X.$$

However, $A(t)\mathcal{D}=B(t)\mathcal{D}$, and hence

$$(\lambda I-B(t))\mathcal{D}=(\lambda I-B(t))\mathcal{D}(B(t)).$$

Thus, $\mathcal{D} = \mathcal{D}(B(t))$, and this proves that $B(t) = A(t)$.

□

We call Theorem 3.7 as Hille-Yosida theorem in quasi-semigroups version. In fact, this is just the sufficient condition of Hille-Yosida theorem. The necessary condition of the theorem is not valid for any quasi-semigroups. Example 3.3 and comments thereon explain this condition. As consequence, Theorem 3.7 gives sufficient condition that guarantees the existence and uniqueness solution of the time-independent abstract Cauchy problems (2) and (3), see [8] and [11].

**Corollary 3.8** [8, 11] *If $A(t)$ is the generator of $C_0$-quasi-semigroup $R(t,s)$ on a Banach space $X$, then for each $x_0 \in \mathcal{D}$ and $r \geq 0$ the initial value problem*

$$\dot{x}(t) = A(r+t)x(t), \ x(0) = x_0, \tag{13}$$

*admits a unique solution.*

## 4. Applications

In this section we give some examples of application of $C_0$-quasi-semigroups to solve the partial differential equations completed by initial value or boundary conditions.

**Example 4.1** *For $r \geq 0$ consider the boundary condition problem*

$$\begin{aligned} &\frac{\partial x}{\partial t}(t,\xi) = a(r+t)\frac{\partial^2 x}{\partial \xi^2}(t,\xi), \qquad 0<\xi<1, \qquad t \geq 0 \\ &x(t,0) = x(t,1) = 0. \end{aligned} \tag{14}$$

*where $a$ is a continuous function with $a(t) > 0$ for $t \geq 0$.*

Let $X$ be a Hilbert space of $L_2[0,1]$ and the operator $A : \mathcal{D}(A) \to X$ given by

$$Ax = \frac{d^2x}{d\xi^2}$$

and

$$\mathcal{D}(A) = \{x \in X : x,\ x' \text{ absolutely continuous and } x'' \in X \text{ with } x(0) = x(1) = 0\}.$$

The boundary condition problem (14) can be written as

$$\dot{x}(t) = a(r+t)Ax(t), \ t \geq 0 \tag{15}$$

where $A$ is a generator of $C_0$-semigroup $T(t)$ that given

$$T(t)x = \sum_{n=1}^{\infty} e^{-\lambda_n t}\langle \phi_n, x\rangle x$$

where $\lambda_n = n^2\pi^2$ and $\phi_n(\xi) = \sin(n\pi\xi)$. From Example 2.5, the family of bounded operators

$$R(t,s) = T\big(g(t+s) - g(t)\big), \qquad t,s \geq 0$$

is a $C_0$-quasi-semigroup with the generator $A(t) = a(t)A$ and $\mathcal{D} = \mathcal{D}(A)$ which is dense in $X$. According to Theorem 3.2 (c) and Corollary 3.8, then for each $x_0 \in \mathcal{D}$ the problem (15) admits a unique solution

$$x(t) = R(r,t)x_0, \qquad x(0) = x_0.$$

Therefore, the initial value problem (14) has a unique solution

$$x(t,\xi) = T\big(g(t+r) - g(r)\big)\, x_0(\xi)$$

where $g(t) = \int_0^t a(s)\, ds$.

**Example 4.2** *Given a boundary condition problem*

$$\begin{aligned} &\frac{\partial x}{\partial t}(t,\xi) = \frac{1}{w(\xi)}\left(\frac{\partial}{\partial \xi}\left(p(\xi)\frac{\partial x}{\partial \xi}(t,\xi)\right) - q(\xi)x(t,\xi)\right), \qquad a<\xi<b, t \geq 0, \\ &\frac{\partial x}{\partial \xi}(a,t) = \frac{\partial x}{\partial \xi}(b,t) = 0. \end{aligned} \tag{16}$$

*where $w,\ p, q, x' = \frac{\partial x}{\partial \xi}$ are continuous functions on the interval $[a,b]$ with $p(\xi) > 0$ and $w(\xi) > 0$.*

Let $X$ be a Hilbert space of $L_2[a,b]$ and $A_{SL}$ is the Sturm-Liouville operator that defined as

$$A_{SL}x = \frac{1}{w}\left(-\frac{d}{d\xi}\left(p\frac{dx}{d\xi}\right) + qx\right)$$

on a domain

$$\mathcal{D}(A_{SL}) = \{x \in X: x, x' \text{ are absolutely continuous}, x'' \in X, x'(a) = 0, x'(b) = 0\}.$$

The boundary condition problem (16) can be written as

$$\dot{x}(t) = Ax(t), \ \ t \geq 0 \tag{17}$$

where $A = -A_{SL}$ is the generator of a $C_0$-semigroup $T(t)$ with $\mathcal{D}(A) = \mathcal{D}(A_{SL})$, see [3] and [4].

Example 2.2 states that the family of bounded operators

$$R(t,s) = T(s), \qquad t,s \geq 0$$

is a $C_0$-quasi-semigroup with its generator $A(t) = A$ and $\mathcal{D} = \mathcal{D}(A)$ which is dense in $X$. Furthermore, for each $x_0 \in \mathcal{D}$ (17) admits a unique solution

$$x(t) = R(s,t)x_0, \qquad x(0) = x_0.$$

Thus, the boundary condition problem (16) has a unique solution

$$x(t,\xi) = T(t)x_0(\xi).$$

## 5. Conclusions

In this paper we have discussed the fundamental properties of quasi-semigroups. This investigation has be done based on the existing properties of semigroups. The results show that some properties inline to the properties of semigroups, although the others do not. Especially, the Hille-Yosida theorem does not hold in $C_0$-quasi-semigroup except for the sufficient condition. On the final part, we have showed the powerfulness of $C_0$-quasi-semigroup to solve the problems related to PDE's. The next works, we shall investigate the contraction and Riesz-spectral in quasi-semigroups version.

## References

[1] Barcenas D, Leiva H and Moya A T 2006 The Dual Quasi-Semigroup and Controllability of Evolution Equations *J. Math. Anal. Appl.* **320** 691 – 702

[2] Curtain R F and Pritchard A J 1978 *Infinite-Dimensional Linear Systems Theory* (New York: Springer-Verlag)

[3] Curtain R F and Zwart H J 1995 *Introduction to Infinite-Dimensional Linear Systems Theory* (New York: Springer)

[4] Delattre C, Dochain D and Wilkin J 2003 Sturm-Liouville systems are Riesz-spectral systems *Int. J. Appl. Math. Comput. Sci.* **13** 481 – 484

[5] Engel K.J and Nagel R 2000 *One-Parameter Semigroups for Linear Evolution Equations* (New York: Springer-Verlag)

[6] Fattorini, H O 1983 *The Cauchy Problems* (Cambridge: Cambridge University Press)

[7] Hackman M 1968 The Abstract Time-Dependent Cauchy Problem *Amer. Math. Soc.* **133(1)** 1 – 50

[8] Janfada M 2010 On Regularized Quasi-Semigroups and Evolution Equations *Abstract and Applied Analysis. Hindawi Publishing Corporation.* Article ID 785428, 13 pages, doi:10.1155/785248

[9] Kato T 1953 Integration of the Evolution Equation in a Banach Space *J. Math. Soc. Japan.* **5(2)** 208 – 234

[10] Ladas G E and Lakshmikantham V 1972 *Differential equations in Abstract Spaces* (New York: Academic Press)

[11] Leiva H and Barcenas D 1991 *Quasi-semigroups, Evolution Equation and Controllability* (Merida: Facultad De Ciencias Departemento De Matematica)

[12] Masuda K 1972 On the Holomorphic Evolution Operators *J. Math. Anal. Appl.* **39** 706 – 711

[13] Megan M and Cuc V 1999 On Exponential Stability of $C_0$-Quasi-Semigroups in Banach Spaces. *Le Matematiche* **Liv (1999)** 229 – 241

[14] Obrecht E 1987 The Cauchy Problem for Time-Dependent Abstract Parabolic Equation of Higher Order *J. Math. Anal. Appl.* **125** 508 – 530

[15] Pazy A 1983 *Semigroups of Linear Operators and Applications to Partial Differential Equations* (New York: Springer)

[16] Royden H L 1989 *Real Analysis* (New York: Macmillan Publishing Company)

[17] Schnaubelt R 2012 *Lecturer Notes Evolution Equations* (Karlsruhe: -)

[18] Tucsnak M and Weiss G 2009 *Observation and Control for Operator Semigroups* (Berlin: Birkhauser)

IOP Conf. Series: Journal of Physics: Conf. Series **855** (2017) 012053 doi:10.1088/1742-6596/855/1/012053

# Classification of Status of the Region on Java Island using C4.5, CHAID, and CART Methods

**R A Syaraswati[1], I Slamet[2], and B Winarno[1]**

[1]Mathematics Department, Faculty of Mathematics and Natural Sciences, Universitas Sebelas Maret, Indonesia

[2]Statistics Department, Faculty of Mathematics and Natural Sciences, Universitas Sebelas Maret, Indonesia

E-mail: isnandarslamet@staff.uns.ac.id

**Abstract.** The indicator of region economic success can be measured by economic growth, presented by value of Gross Regional Domestic Product (GRDP). Java island has the biggest GDP contribution toward the Indonesian government, but not all of the region gives equality contribution. The C4.5, CHAID, and CART methods can be used for classifying the status of the region with nonparametric approach. The C4.5 and CHAID methods are non-binary decision tree, meanwhile the CART methods is binary decision tree. The purposes of this paper are to know how the classification and to determine the factors that influence on classification of the region. The dependent variable is status of the region which is divided into four categories based on Klassen typology. The result shows factors that have the biggest contribution on classification of status of the region on Java island based on C4.5 method are economic growth rate, electricity, gas, and water sector, and area. The factors that have the biggest contribution based on CHAID method are growth rate, manufacturing sector, and electricity, gas, and water sector, while based on CART method are growth rate, manufacturing sector, and electricity, gas, and water sector.

## 1. Introduction

Classification problems are very common in business, economic, health, and education. In economic, indicator of region economic success can be measured by economic growth, presented by value of Gross Regional Domestic Product (GRDP). Java island has the biggest GDP contribution toward the Indonesian government, but not all of the region gives equality contribution. To overview the growth pattern we can used Klassen typology. This technique divide the region into four quadrants, i.e quadrant I, II, III, and IV. There are many factors which's influence the division of the region, so that region is shifted into quadrant I, II, III, or IV. Therefore, overview is needed to solve classification problem. Classification method can be used to solve it.

There are two classification methods, parametric and nonparametric approach. Parametric approach has some assumptions that should be completed. Nonparametric approach does not need any assumption but it has high accuracy and easy to interpreted. One of classification method with nonparametric approach is classification tree. This classification is divided into two groups i.e binary and non-binary decision tree. C4.5 and CHAID (Chi-squared Automatic Interaction Detection) are non-binary decision tree, while CART (Classification and Regression Tree) is binary decision tree.

In this paper we use C4.5, CHAID, and CART methods. The C4.5 method uses entropy concept and can be used for discrete and continuous data. CHAID is an iterative technique that examines the predictors (e.g. classification variables) individually and utilizes them in the order dictated by their statistical significance (Gallagher *et al.* [5]). CART analysis is a powerful technique with significant potential (Lewis and Roger [10]).

The C4.5, CHAID, and CART methods have similar characteristic. Therefore, in this paper we applied classification of the status of the region on Java Island using C4.5, CHAID, and CART methods.

## 2. Decision Tree

### 2.1. C4.5 Method

Quinlan [12] made improvements to ID3 and extended to C4.5 in 1993. The C4.5 method can handle both of continuous and discrete attribute.

(1) Determine the root node. Root node is chosen by the biggest gain ratio. Let $S$ denote number of cases, $n$ is number of classes, and $p_i$ is the proportion of $S_i$ instances $S$. Therefore, the entropy of attribute $S$ is calculate as

$$Entropy(S) = \Sigma_{i=1}^{n} p_i \times log_2 p_i.$$

The information by attribute $A$ is defined as

$$Gain(S, A) = Entropy(S) - \Sigma_{i=1}^{n} \frac{|S_i|}{|S|} \times Entropy(S_i)$$

where $A$ is attribute, $|S_i|$ is number of cases in $i$ partition, and $|S|$ number of cases of $S$. Therefore, the information gain ratio of attribute $S$ is defined as

$$GainRatio(S, A) = \frac{(GainS, A)}{SplitInfo(S, A)}$$

where Split Info (S,A) is defined as

$$SplitInfo(S, A) = \Sigma_{j=1}^{n} p_j \times log_2 p_j.$$

with $p_j$ is the proportion of $S_j$ instances $S$.

(2) Determine the branch node.

(3) Iterate until all of the cases have a same class. The splitting of classification tree is stopped when all of the records have a same class, there is no more splitting attribute on record, and there is no empty record.

### 2.2. CHAID Method

CHAID is concerned with predicting a single variable, known as the dependent variable, based on a number of other variables, referred to as predictor variables (Gallagher [5]). The main part of CHAID method as follows.

(1) Chi-squared test (contingency table). Contingency table is data analyzing to defined relationship between several variables. Chi-squared test is used as follows

$$\chi^2 = \Sigma_{i=1}^{r} \Sigma_{j=1}^{c} \frac{(O_{ij} - e_{ij})^2}{e_{ij}}$$

where $O_{ij}$ is row $i$ and column $j$ observation, $e_{ij}$ is row $i$ and column $j$ expected frequency, $r$ is number of row, and $c$ is number of column.

(2) Bonferroni correction. Bonferroni correction is a multiple comparison correction when several dependent or independent statistical tests are being performed simultaneously.

### *2.3. CART Method*
CART can handle numerical data that are highly skewed or multi-modal, as well as categorical predictors with either ordinal or non-ordinal structure (Lewis and Roger[10]). The CART method consists of four basic step as follows.

(1) Tree constructing. Tree is constructed using recursive splitting of node. Let $k$, $k = 1, 2, ..., m$ is class where $m$ is number of classes. Gini splitting index is as follows.
$$i(t) = 1 - \Sigma_{i=1}^{m}[P(k|t)]^2,$$
where $[P(k|t)]$ is relative frequency of class $j$ in node $t$. Splitter is chosen by the biggest goodness of split ($\phi(s,t)$), as follows
$$\phi(s,t) = i(t) - p_L it_L - p_R it_R$$
where $p_L$ is left node proportion, $it_L$ is left node heterogenous function, $p_R$ is right node proportion, and $it_R$ is right node heterogenous function.

(2) Stopping the tree constructing process. At this step, maximal tree has been produced. The process is stopped when there is only one observation in each of the child nodes, all observations within each child node have the identical distribution of predictor variables, making splitting impossible, or an external limit on the number of levels in the maximal tree has been set by the user (depth option) (Lewis and Roger[10]).

(3) Tree pruning. In order to generate a sequence of simpler and simpler trees, each of which is a candidate for the appropriately-fit final tree. For tree pruning, the method of cost-complexity pruning is used (Lewis and Roger[10]).

(4) Optimal tree selection.

## 3. Research Methodology
This paper used C4.5, CHAID, and CART methods to classify the status of the region on Java island. This paper used 2014 BPS data. The dependent variable is status of the region which's divided into four categories (i.e K I, K II, K III, and K IV) based on Klassen typology. Agriculture sector ($X_1$), mining and quarrying sector ($X_2$), manufacturing sector ($X_3$), electricity, gas, and water sector ($X_4$), construction sector ($X_5$), commerce, hotel, and restaurant ($X_6$), transportation and communication ($X_7$), finance sector ($X_8$), services ($X_9$), economic growth rate ($X_{10}$), original local government revenue ($X_{11}$), the literacy rate ($X_{12}$), percentage of poor people ($X_{13}$), and area ($X_{14}$) are independent variables.

The first step is to divide the 113 region on Java island into four quadrant based on Klassen Typology. Next, construct a classification tree with C4.5. Starting with counting *Entropy* value, *Gain*, and then *GainRatio* to find a best splitting attribute. The higher attribute of gain ratio is a root node. Then, we make branches for every attribute until it on the same class. To construct a classification tree with CHAID method, we begin with merging the independent variables with *chi-square* test, splitting, and stopping when no more significant independent variable. To construct CART classification tree, we begin with split the node, stop splitting, prune, and choose the optimal classification tree. After that, the classification tree can be interpreted.

## 4. Result
### *4.1. Klassen Typology*
Klassen typology divided the status of the region into four quadrant. Quadrant I shows an advanced and fast growing region, quadrant II shows an advanced but pressured region, quadrant III shows a rapidly growing region, and quadrant IV shows a relatively underdeveloped region. There are 21 regions in quadrant I, 9 region in quadrant II, 42 region in quadrant III, and 41 regions in quadrant IV.

IOP Conf. Series: Journal of Physics: Conf. Series **855** (2017) 012053 doi:10.1088/1742-6596/855/1/012053

*4.2. C4.5 Method*

There are four steps to construct decision tree with C4.5 method.

(1) Determine the root node. To determine the root node, we begin with calculating *Entropy*, *Gain*, and then *Gain Ratio*. The result of calculating *Gain Ratio* is presented in Table 1.

**Table 1.** Determine root node

| Attribute | *Gain Ratio* | Attribute | *Gain Ratio* | Attribute | *Gain Ratio* |
|---|---|---|---|---|---|
| $X_1$ | 0.0876 | $X_6$ | 0.0781 | $X_{11}$ | 0.0870 |
| $X_2$ | 0.1075 | $X_7$ | 0.0787 | $X_{12}$ | 0.0391 |
| $X_3$ | 0.1182 | $X_8$ | 0.0626 | $X_{13}$ | 0.0851 |
| $X_4$ | 0.3190 | $X_9$ | 0.0315 | $X_{14}$ | 0.1628 |
| $X_5$ | 0.0728 | $x_{10}$ | 1.0000 | | |

Table 1 shows that $X_{10}$ has the biggest *Gain Ratio*, therefore $X_{10}$ is the root node.

(2) Determine the branch node. The branch node is the attribute that has the biggest *Gain Ratio* after the root node was deleted.

(a) First iteration. On $X_{10}$ first category, the biggest *Gain Ratio* is $X_4$ (0,1045). There are four categories in $X_4$, the fourth and fifth category classify the region into quadrant I. While first, second, and third category aren't.

(b) Second iteration. On $X_4$ first category, the biggest *Gain Ratio* is $X_{12}$ (0,3212). There are two categories in $X_{12}$, the first category classify the region into quadrant I. While second category classify the region into quadrant III.

(c) Third iteration. On $X_4$ second category, the biggest *Gain Ratio* is $X_2$ (0,3506). There are four categories in $X_2$, the first and second category classify the region into quadrant I. While third and fourth category classify the region into quadrant III.

(d) Fourth iteration. On $X_4$ third category, the biggest *Gain Ratio* is $X_{14}$ (0,3021). There are three categories in $X_{14}$, the first and second category classify the region into quadrant I. While third category classify the region into quadrant III.

(e) Fifth iteration. On $X_{10}$ second category, the biggest *Gain Ratio* is $X_{14}$ (0,1579). There are three categories in $X_{14}$, the second and third category classify the region into quadrant IV. While first category isn't.

(f) Sixth iteration. On $X_{14}$ first category, the biggest *Gain Ratio* is $X_5$ (0,3622). There are four categories in $X_{14}$, the first category classify the region into quadrant IV. While second, third, and fourth category classify the region into quadrant II.

All of branches node have classified the region, therefore process is stopped. The classification tree has 87,6% classification accuracy. The classification tree is presented in Figure 1.

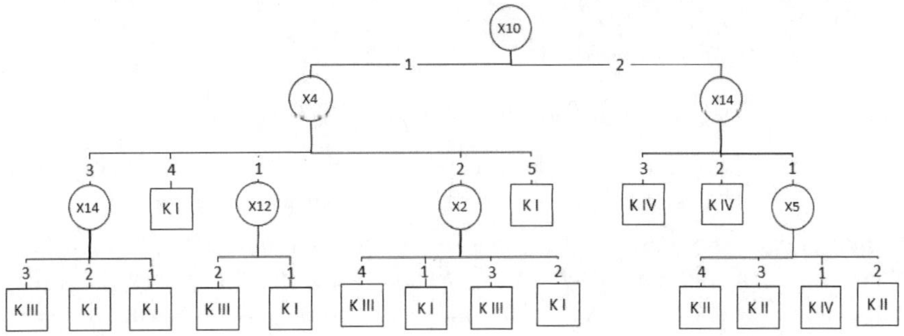


**Figure 1.** C4.5 Classification Tree

### *4.3. CHAID Method*

To construct decision tree with CHAID, we have to merge the independent variables to find the split values first. We apply chi-square test to merge the independent variables. After that, we determine which of the independent variables is the most significant in distinguishing among the dependent variable. The significant variables is given in Table 2.

**Table 2.** Primary Independent Variable

| Parent Node | Independent Variable | $\chi^2$ | Split Values |
|---|---|---|---|
| 1 | $X_{10}$ | 113 | 1 and 2 |
| 2 | $X_4$ | 14,032 | 2,3,4,5 and 1 |
| 3 | $X_3$ | 20,694 | 4 and 1,2,3 |
| 4 | $X_{14}$ | 9,808 | 3 and 1,2 |
| 5 | $X_{13}$ | 15,435 | 2,3and 1 |
| 6 | $X_3$ | 14,983 | 4 and 1,2,3 |

Variable $X_{10}$, $X_4$, $X_3$, $X_{14}$,and $X_{13}$ are parent node in tree classification with CHAID. Classification tree is shown in Figure 2.

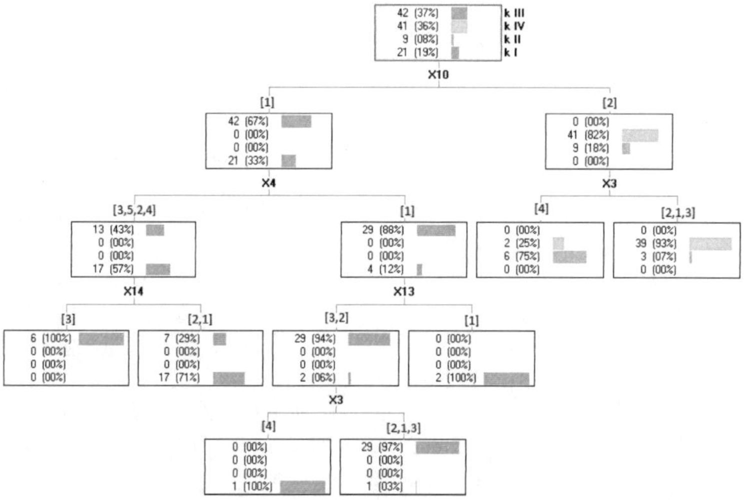


**Figure 2.** CHAID Classification Tree

### *4.4. CART Method*

There are four steps to construct decision tree with CART Method.

(1) Comparing classification accuracy of data combination. The best data combination is 88:12 with 85,86% : 85,71%. Therefore, this combination data is used.

(2) Tree constructing. Tree constructing begins at the root node, which includes all variable in the learning data. To construct a tree we have to split the node with splitting criteria, index gini. The best splitter is $X_{10}$ with improvement of 0.2500. Variable $X_{10}$ end up being the root node. It split into two terminal node, terminal node 1 formed by $X_{10}$ first category and terminal node 2 formed by $X_{10}$ second category. Splitting process stops on terminal node 6 and 7. After that we get the maximum tree classification which has 10 nonterminal node and 9 terminal node.

(3) Tree Pruning. In order to generate a sequence of optimum tree, the method of cost-complexity pruning is used.

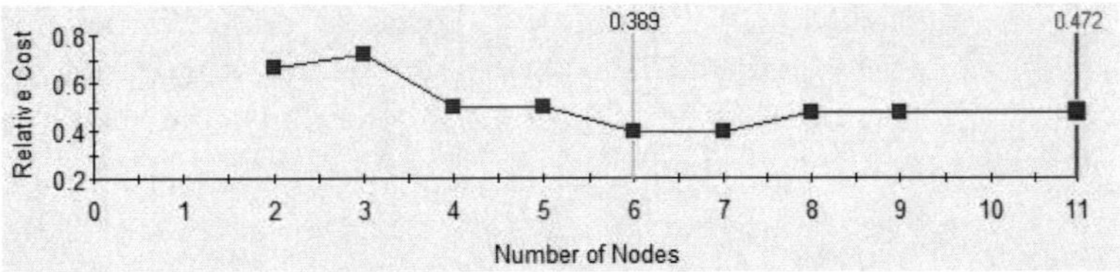


**Figure 3.** Pruning Maximal Tree

Optimal classification tree has resubtitution relative cost of 0,16683, complexity parameter of 0,031307, and Relative cost of 0,38889 +/- 0,05072.

(4) Optimal Tree Selection. We got optimal classification tree after prune the maximal tree classification. Optimal classification tree show by Figure 4.

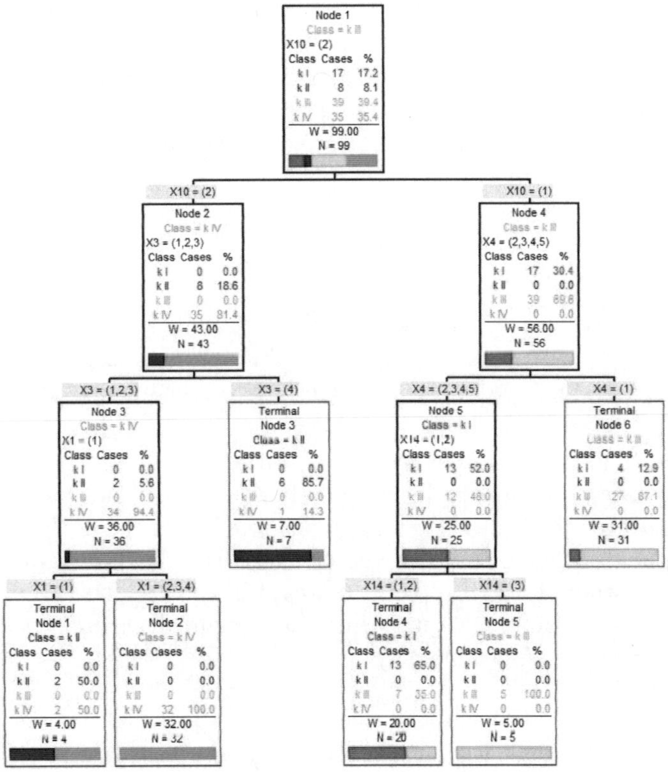


**Figure 4.** CART Classification Tree

The variables that include on optimal classification tree are agriculture sector, mining and quarrying sector, manufacturing sector, economic growth rate, and original local government revenue.

## 5. Interpretation

### 5.1. *C4.5 Algorithm*

Classification tree with C4.5 method has 7 nonterminal node and 17 terminal node. Classification tree can predict accurately of 87,6%. There are 7 terminal node predicted the region into quadrant I, 3 terminal node into quadrant II, 4 terminal node into quadrant III, and 3 terminal node into quadrant IV. The factors that have the biggest contribution are economic growth rate, electricity, gas, and water sector, and area.

### 5.2. *CHAID Method*

Classification tree with CHAID method has 6 nonterminal node and 7 terminal node. Classification tree can predict accurately of 88,47%. There are 1 region in quadrant I that misclassified into quadrant III, 3 regions in quadrant II that misclassified into quadrant IV, 7 regions in quadrant III that misclassified into quadrant I, and 2 regions in quadrant IV that misclassified into quadrant II. The factors that have the biggest contribution are economic growth rate, manufacturing sector, and electricity, gas, and water sector.

### 5.3. *CART Method*

Classification tree with CHAID method has 5 nonterminal node and 6 terminal node. Classification tree with testing data can predict accurately of 85,71%. There are no misclassification in quadrant I, 1 region in quadrant II that misclassified into quadrant IV, no misclassification in quadrant III, and 1 region in quadrant IV that misclassified into quadrant II. The factors that have the biggest contribution are economic growth rate, manufacturing sector, and electricity, gas, and water sector.

## 6. Conclusion

Based on the analysis, we can conclude.

(1) C4.5 classification tree can predict accurately of 87,6%, CHAID classification tree can predict accurately of 88,47%, and CART classification tree can predict accurately of 85,71%. Therefore, the CHAID method is the best method for classification of status of the region on Java island.

(2) The factors that influence the classification of status of the region on Java island based on C4.5 method are mining and quarrying sector, electricity, gas, and water sector, construction sector, economic growth rate, the literacy rate, and area. The factors influence based on CHAID method are manufacturing sector, electricity, gas, and water sector, economic growth rate, percentage of poor people, and area, while based on CART method the factors are agriculture sector, mining and quarrying sector, manufacturing sector, economic growth rate, and original local government revenue.

## References

[1] Agresti, A., *An Introduction to Categorical Data Analysis second edition*, New Jersey, Jon Wiley and Sons, Inc, 2007.

[2] Bagozzi, R.P., *Advanced Methods of Marketing Research.* Blackwell Publishers Ltd, Oxford, 1994.

[3] Badan Pusat Statistik, *Pertumbuhan Ekonomi Indonesia Triwulan III/2014 mencapai 2,96 Persen*, 2014. http://bps.go.id/Brs/view/id/225 (diakses tanggal 17 April 2016)

[4] Breiman, L., Friedman, J.H, Olshen, R.A, and Stone, C.J., *Classification and Regression Tree*, Capman and Hall, New York, 1984.

[5] Gallagher, C.A., Monroe, H.M., and Fish, J.L., *An Iterative Approach to Classification Analysis*, 2000. http://www.casact.org/pubs/dpp/dpp90/90dpp237.pdf (diakses tanggal 24 November 2015)

[6] Kasmir, *Analisis Laporan Keuangan*, Jakarta, Rajawali Pers, 2008.

[7] Kass, G.V., *An exploratory technique for investigating large quantities of categorical data.* Appl. Statist. 29, (1980), no.2, 119-127.

[8] Kusrini dan Taufiq, Emha. *Algoritma Data mining*, Yogyakarta, Andi Offsee, 2009.
[9] Lakshmi, B.N, Indumathi, T.S, and Ravi, N., *An Empherical Study on Decision Tree Classification Algorithms*, International Journal of Science, Engineering and Technology Research (IJSETR), (2015), vol. 4, issue 11.
[10] Lewis, M.D dan Roger, J., *An Introduction to Classification and Regression Tree (CART) Analysis*, Department of Emergency Medicine Harbor-UCLA Medical Center Torrance, California, 2000.
[11] Myers, J.H., S*egmentation and Positioning for Strategic Marketing Decisions*, Chicago, American Marketing Association, 1996.
[12] Quinlan, J. R.*Induction of Decision Trees*, Machine Learning, (1986), vol.1, 81-106.
[13] Ramani, R.G and Shanthi, S., *Classifier Prediction Evaluation in Modeling Road Traffic Accident Data.* Computational Inteligence & Computing Research (ICCIC), (2012), 1-4.
[14] Sadono, Sukirno, *Ekonomi Pembangunan Proses masalah dan Dasar Kebijakan*, Jakarta, Kencana, 2006.

# Online Statistical Modeling (Regression Analysis) for Independent Responses

I Made Tirta[1], Dian Anggraeni[2], Martinus Pandutama[3]

[1,2] Jurusan Matematika FMIPA Universitas Jember, Indonesia
[3] Fakultas Pertanian Universitas Jember, Indonesia
itirta.fmipa@unej.ac.id

**Abstract**. Regression analysis (statistical analmodelling) are among statistical methods which are frequently needed in analyzing quantitative data, especially to model relationship between response and explanatory variables. Nowadays, statistical models have been developed into various directions to model various type and complex relationship of data. Rich varieties of advanced and recent statistical modelling are mostly available on open source software (one of them is R). However, these advanced statistical modelling, are not very friendly to novice R users, since they are based on programming script or command line interface. Our research aims to developed web interface (based on R and shiny), so that most recent and advanced statistical modelling are readily available, accessible and applicable on web. We have previously made interface in the form of e-tutorial for several modern and advanced statistical modelling on R especially for independent responses (including linear models/LM, generalized linier models/GLM, generalized additive model/GAM and generalized additive model for location scale and shape/GAMLSS). In this research we unified them in the form of data analysis, including model using Computer Intensive Statistics (Bootstrap and Markov Chain Monte Carlo/ MCMC). All are readily accessible on our online Virtual Statistics Laboratory. The web (interface) make the statistical modeling becomes easier to apply and easier to compare them in order to find the most appropriate model for the data.

***Keywords:*** *additive models, linear models, MCMC Regression, online regression analysis, statistical models, web-interface*

## 1 Introduction

Regression analyss (statistical models) are among statistical methods which are frequently employed in analyzing quantitative data, especially to model dependences between response and several explanatory variables. Nowadays, statistical models have been developed into various directions to handle various type and complex relationship of data. Rich variety of advanced and recent statistical modelings are mostly available on open source software (one of them is R). However, these advanced statistical models, are mostly based on programming script or command line interface, which mean, that they are not easily accessed by applied or practical researchers. The gaps between developed and accessible statistical methods worried statisticians [1] that "practitioners continue to use inappropriate or suboptimal methods due to their being restricted to what is made available via GUIs".

Therefore it is essential to build interface to make advanced and most recent statistical methods, especially statistical model on R, becoming more user friendly and easier to access and to use. Several GUIs have been developed for various purposes. Explicet, is a GUI designed for management, analysis and visualization of microbiome data [2] and it is claimed has made the analysis of complex microbiome datasets becoming "much more accessible to the growing number of investigators". Microarray Я US, has been developed based on bioconductor R packages, mainly for researchers with no or little knowledge of R, to have a more reliable and accurate microarray data analysis [3]. Interactive web for statistics learning have also been developed. RwikiStat was developed by combining MediaWiki and Rweb [4] and combining theory with laboratory practice using Rweb, however user still need to have R scripting capabilities. Other types of statistics tutorial with combination of statistics theory and data analysis have been developed using R with shiny packages for spesific topic [5], [6]. This type of data analysis are accompanied with summary of theory and step by step choice analysis with example of interpretation to ensure users are doing analysis data with understanding but no need to master or understand R scripting.

Statistical models with general form $y_i = \mu_i + \varepsilon_i$ for $i$ = 1, 2,3, ..., $n$ have been extended into various directions. For model with independent errors, the model starti from (i) simple linear models (LM) having independent Gaussian errors, i.e., $\varepsilon_i \sim iid\ N(0, \sigma^2)$ and $\mu_i = \sum x_{ij}\beta_j$, for $j$ = 0,1,2,..., $p$ (with $p$ number of regressor/ predictors) [7], (ii) when outliers exist, there are several methods available using robust linear models approaches (RLM) [8][9] (iii) Generalized linear model (GLM) extends LM to accommodate independent errors with wider class of distributions known as the exponential family distributions (i.e, having continuous, count, or binary responses) and possibly nonlinear relationship between response means and the linear predictors, i.e., continuous and differentiable link function $g$ (such as log, logit, inverse/ reciprocal), such that $g(\mu_i) = \sum \beta_j\, x_{ij}$. [10][11]. Later, (iv) statistical model were again generalized to accommodate additive predictors (GAM) such that, $g(\mu_i) = f(x_i)$, for smooth function $f$ (parametric or nonparametric). One of the most frequently applied nonparametric smooth functions are the family of spline smoothers [12][13][14], and (v) perhaps the most recent development of statistical model with independent errors are extension of GAM into GAMLSS [15][16]. GAMLSS accommodates wider type of distribution (with 1, 2, 3, up to 4 parameters, such as the mean, variance, skewness and kurtosis). In addition to modeling the mean, with wider type of distributions, GAMLSS, can also model all other parameters of distributions, each may have its own link function. Recently GAMLSS is extended with variables selection capabilities [17]. In addition to those main statistical models, for small sample, the model are also extended to employ Computer Intensive Statistics (CIS) techniques, such as Bootstrap regression [7] and Markov Chained Monte Carlo (MCMC) regression [18].

All the statistical models mentioned above are already implemented in various packages on R. However, for novice R users, they are not easy to apply since they are all based on command line interface (script). Moreover, in addition to those packages, users may need to upload and call other functions from other R packages for drawing graph or calculating goodness of fit. In this paper we report the development Web-based-GUI interface that unifies most statistical models for independent responses using R and enriched by various options for data exploration, graphical visualisation and goodness of fit measures utilizing several selected R packages.

## 2 Methods

We develop an interface for unified online (web-based) statistical models for independent responses, which covers LM, RLM, GLM, GAM, GAMLSS, Bootsrap and MCMC regression based on various previously mentioned R- packages. We mainly utilize shiny toolkits [18] to build the interface. There are severeal main steps to follow in building the interface: (i) selecting main and related packages, including the primary functions for the models and secondary functions for graphical visualization, such as scatter plot and correlation plot matrices [20,21], and scatter plot with various smoother [22], and other regression visualization [23]; (ii) identifying the input parameters of the functions, (iii) defining input functions and their options in ui.r file and output functions server.r file; (iv) checking the compatibility of loaded packages and related functions; (v) uploading the files to the web (shiny server) so that they are readily accessible by users.

## 3 Results and discussion

### *3.1 General Features of the Web GUI interface*

At this stage, we have developed online statistical model fitting for independent responses, covering several models described previously with general features as follows (see Figure 1).

1) **Data Input**: internal database (for practical purposes), or import users' own data with csv or text format (for real data analysis). Users can load all chosen data, or only load small number of the

data (may be needed for illustration or practice with small size of data, such as CIS or Robust Linear Model)

2) **Data exploration** graphical representation on relationship among variables (correlation diagram and scatterplot diagram), graphical exploration of scatter plot with specific model (for examples distributions and link function for LM, RLM, GLM, and smoother for GAM, GAMLSS, see Figure 2, and Figure 3 as samples).
3) **Input Options** for statistics model, including choosing response, predictors (on mean for all models, and for shape and scale for GAMLSS), family and link function (for GLM, GAM and GAMLSS), smooth variable (for GAM and GAMLSS), number of simulations for CIS.
4) **Output Options**, including parameter estimates with their p-values, GOF (AIC, BIC, Adj-Rsquared), diagnostics and other visualization graphics and some selected detailed output (see Fig 4).

The web can be accessed at http://statslab-rshiny.fmipa.unej.ac.id/RProg/MSI/. The summary of features for each model fitting is given in Table 1 and the appearance of the web can be seen in Figure 1.

**Table 1.** Summary of features of the model fitting

| No | Parts | Input Option | Output Option |
|---|---|---|---|
| 1 | Input Data | • Internal database<br>• Import data (.csv, .txt)<br>• All or randomly select only small number of available data | • Summary of data<br>• List of data |
| 2 | Exploratory Data | • General exploration<br>• Smoother exploration (LM, RLM , GLM, GAM) | • Correlation matrix<br>• Correlation Diagram<br>• Scatter plot matrix<br>• Scatter plot with various smoother |
| | **Models Fitting** | | |
| 3 | LM | • Xs and Y<br>• Factor (dummy) | • Estimates (with p-val),<br>• Anova<br>• GOF (AIC, BIC, R-sqr, Adj-RSq)<br>• Diagnostik Graphic<br>• Stepwise regression (variable selection) |
| 4 | RLM | • Xs and Y<br>• Method M, MM, LTS<br>• Factor (Dummy) | • Estimates,<br>• Bonferroni test for outlier,<br>• MSE (mean square error)<br>• Graphic |
| 5 | GLM | • Xs and Y<br>• Family (link) exponential family including Negative Binomial (log)<br>• Factor (Dummy)<br>• NS or BS smoother | • Estimates (with p-val),<br>• Deviance Analysis<br>• GOF (AIC, BIC)<br>• Scatter plot<br>• Diagnostic Graphic<br>• Stepwise regression (variable selection) |
| 6 | GAM (based on mgcv package) | • Xs and Y<br>• Family (link): exponential family including Negative Binomial (log)<br>• Factor (Dummy)<br>• Spline smoother (Cubic Splines, Penalized Spline, Thin | • Estimates (with p-val),<br>• Deviance Analysis<br>• GOF (GCV, AIC, BIC)<br>• Scatter plot<br>• Diagnostics Graphic |

| No | Parts | Input Option | Output Option |
|---|---|---|---|
| | | Plate Spline) | |
| 7 | GAMLSS | • Xs and Y<br>• Family (link) including Nornmal Family, Zero Inflated Poisson (log)<br>• Spline smoothers (Cubic Splines, Penalized Spline, Thin Plate Spline) and limited local regression (loess)<br>• Linear and single Predictor for LSS<br>• Choice of Algorithms | • Responses distributions fit graphics<br>• Estimates (with p-val),<br>• Deviance Analysis<br>• GOF (AIC, BIC)<br>• Scatter plot<br>• Diagnostic Graphic |
| 8 | CIS | • Bootstrap regression<br>• MCMCRegress (for Gaussian responses)<br>• MCMCpoisson (for Poisson/ count responses)<br>• MCMClogit (for Binomial, especially binary responses) | Graphics of estimates and confidence interval based on 95% percentiles |

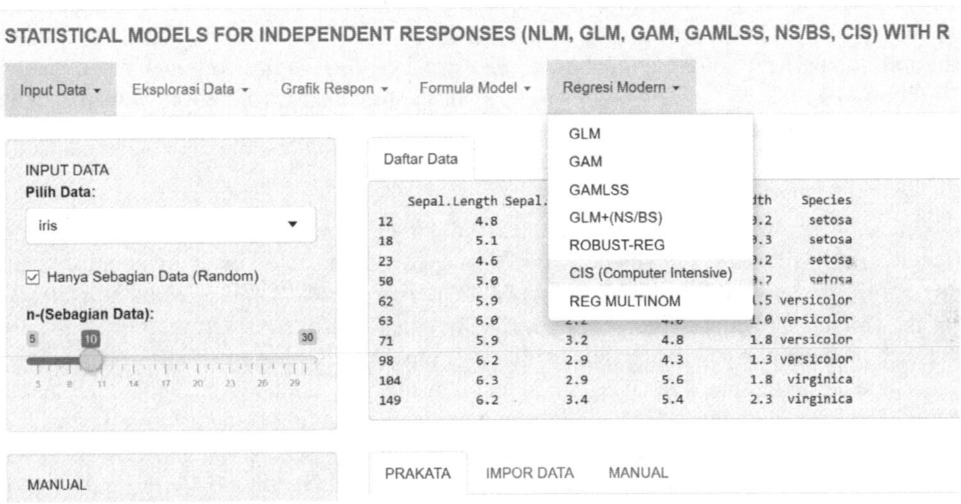


**Figure 1**. Web appearrance and the main menu (Navbar menu and sidebar menu)

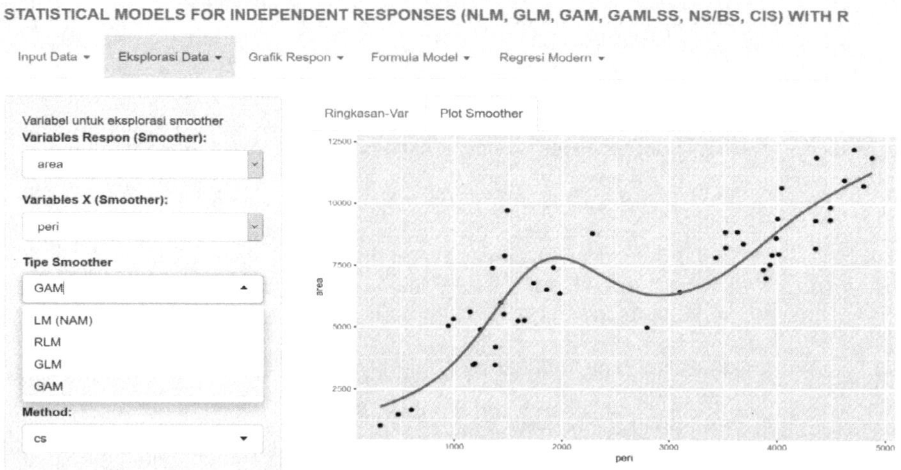


**Figure 2**. Data and model exploration using various type of smoothers (LM, RLM, GLM, GAM)

### 3.2 *Numerical Illustrations*

The following are numerical illustrations using iris data available on datasets package. The data were first published in 1935 [24]. The main purposes of the illustration are not to show the accuracy of the computation (results), since the results are the same if they are done via script, but to show that results (estimate and comparison among available model) can be done completely and more easily, using "point and click" on the web (see Figure 1). The fitting start from exploration to testing hypothesis about parameters of various alternatif models

#### 3.2.1 *Data Exploration*

From summary of data we see that the data consist of 1 factor and 4 variables, means that Species as factor may worth considering in the model. Graphical exploration can be made by creating scatter plot, with various type of smoother (available on menu). The first graphics explorations utilize scatter plot matrix of the variables, to check whether factor (i.e., Species) worth considering in the model. For the seek of clarity we only focus on Sepal.Length and Sepal.Width (Figure 3). The plot show that inclusion of Species in the model changes the regression lines directions (regression coefficients or the lines' slope) significantly from negative and may near zero (Figure 3a), to positive for each Species (Figure 3b). The next exploration using various smoother (with ggplot2 package), give us idea that the data may be better fitted using more advanced regression (such as GLM or GAM). Figure 4 shows that applying other continuous distributions (Gamma families) with nonidentity link seem improve the fitness of model. These graphics appearance suggests that Species should be included in the model and more advance model (such as GLM, GAM, GAMLSS, should be considered). Graphics explorations are very beneficial for giving rough idea. However, for more accurate results, user should check and compare goodness of fit measures such as AIC or BIC which are calculated and informed for every choice of model. For illustration or practice with Robust and CIS type regressions, users may randomly load only small amout of the data, however in this paper the illustration for Robust and CIS are excluded. The following are the summary output of all the iris data.

```
 Sepal.Length    Sepal.Width     Petal.Length    Petal.Width          Species
Min.   :4.30   Min.   :2.00   Min.   :1.00   Min.   :0.1   setosa    :50
1st Qu.:5.10   1st Qu.:2.80   1st Qu.:1.60   1st Qu.:0.3   versicolor:50
Median :5.80   Median :3.00   Median :4.35   Median :1.3   virginica :50
Mean   :5.84   Mean   :3.06   Mean   :3.76   Mean   :1.2
3rd Qu.:6.40   3rd Qu.:3.30   3rd Qu.:5.10   3rd Qu.:1.8
Max.   :7.90   Max.   :4.40   Max.   :6.90   Max.   :2.5
```

IOP Conf. Series: Journal of Physics: Conf. Series **855** (2017) 012054 doi:10.1088/1742-6596/855/1/012054

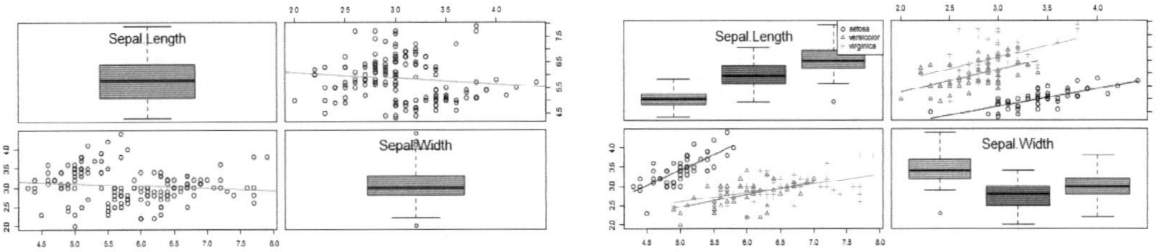


(a) (b)

**Figure 3**. Scatter plot for Sepal.Length vs Sepal.Width. (a). Without and (b). with factor/group separations)

### *3.2.2 Alternatives of model fittings*

Using our online data analysis, users can easily employ various types of modelings and various combinations of model parameters. For illustration, we set Sepal.Length as response and other variables or factor as explanatory variabels. We fit several models (i) Gaussian distribution (LM) with and without factor, (ii) Gamma with Log link (GLM) and (iii) GAM (by giving smoother on some variables, (iv) GAMLSS (by modeling the scale parameter), and (v) GLM with Natural or B-Splines. All the models are easily set in our interface and the GOF are informed for each model. We describe some of the fitting and summarise the results of all fittings.

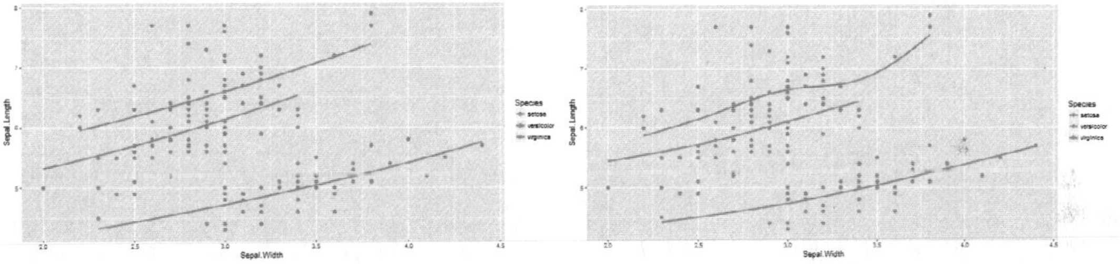

(a) (b)

**Figure 4**. Scatter plot for Sepal.Length vs Sepal.Width with Species as factor . (a) GLM with Gamma distribution and log-link; (b) GAM with Gamma distribution and log-link and additional cubic spline smoother

*(i) Fiiting Linear Model without Species*

```
Sepal.Length~Sepal.Width+Petal.length+Petal.Width

Coefficients:
             Estimate Std. Error t value Pr(>|t|)
(Intercept)    1.8560     0.2508    7.40  9.9e-12 ***
Sepal.Width    0.6508     0.0666    9.77  < 2e-16 ***
Petal.Length   0.7091     0.0567   12.50  < 2e-16 ***
Petal.Width   -0.5565     0.1275   -4.36  2.4e-05 ***
---
Signif. codes:  0 '***' 0.001 '**' 0.01 '*' 0.05 '.' 0.1 ' ' 1

Residual standard error: 0.315 on 146 degrees of freedom
Multiple R-squared:  0.859,      Adjusted R-squared:  0.856
F-statistic:  296 on 3 and 146 DF,  p-value: <2e-16
```

```
AIC  BIC   RSq    AdjRsq
84.6 99.7  0.859  0.856
```

*(ii) Fitting pararel Linear Model with Species as dummy*

```
Sepal.Length~Sepal.Width+Petal.length+Petal.Width+Species-1

Coefficients:
                  Estimate Std. Error t value Pr(>|t|)
Speciessetosa       2.1713     0.2798    7.76  1.4e-12 ***
Speciesversicolor   1.4477     0.2815    5.14  8.7e-07 ***
Speciesvirginica    1.1478     0.3536    3.25   0.0015 **
Sepal.Width         0.4959     0.0861    5.76  4.9e-08 ***
Petal.Length        0.8292     0.0685   12.10  < 2e-16 ***
Petal.Width        -0.3152     0.1512   -2.08   0.0389 *
---
Signif. codes:  0 '***' 0.001 '**' 0.01 '*' 0.05 '.' 0.1 ' ' 1

Residual standard error: 0.307 on 144 degrees of freedom
Multiple R-squared:  0.997,    Adjusted R-squared:  0.997
F-statistic: 9.22e+03 on 6 and 144 DF,  p-value: <2e-16

AIC  BIC  RSq    AdjRsq
79.1 100  0.997  0.997
```

*(iii) Fitting GLM with Gamma(log)*

```
Sepal.Length~Sepal.Width+Petal.length+Petal.Width+Species,
family=Gamma(link=log)

Coefficients:
                  Estimate Std. Error t value Pr(>|t|)
(Intercept)         1.1122     0.0476   23.35   <2e-16 ***
Speciesversicolor  -0.0746     0.0409   -1.83    0.070 .
Speciesvirginica   -0.1220     0.0568   -2.15    0.033 *
Sepal.Width         0.0939     0.0147    6.41    2e-09 ***
Petal.Length        0.1283     0.0117   11.00   <2e-16 ***
Petal.Width        -0.0491     0.0257   -1.91    0.058 .
---
Signif. codes:  0 '***' 0.001 '**' 0.01 '*' 0.05 '.' 0.1 ' ' 1

(Dispersion parameter for Gamma family taken to be 0.00273)

    Null deviance: 2.97256  on 149  degrees of freedom
Residual deviance: 0.39372  on 144  degrees of freedom
AIC: 74.95
                link  AIC BIC RSq   AdjRsq
family "Gamma" "log"  75  96  0.996 0.995
```

*(iv). Fitting GAM with Cubic Splines*

```
Family: Gamma
Link function: log

Formula:
Sepal.Length ~ s(Sepal.Width, bs = "cs", k = 5) + Petal.Length +
    Species
```

```
Parametric coefficients:
                   Estimate Std. Error t value Pr(>|t|)
(Intercept)          1.3973     0.0166   84.14  < 2e-16 ***
Petal.Length         0.1243     0.0109   11.45  < 2e-16 ***
Speciesversicolor   -0.1257     0.0361   -3.48  0.00067 ***
Speciesvirginica    -0.1970     0.0480   -4.11  6.7e-05 ***
---
Signif. codes:  0 '***' 0.001 '**' 0.01 '*' 0.05 '.' 0.1 ' ' 1

Approximate significance of smooth terms:
                 edf Ref.df    F p-value
s(Sepal.Width) 1.87      4 9.92 7.9e-10 ***
---
Signif. codes:  0 '***' 0.001 '**' 0.01 '*' 0.05 '.' 0.1 ' ' 1

R-sq.(adj) =  0.862   Deviance explained = 86.8%
GCV = 0.0028247  Scale est. = 0.0027034  n = 150
```

*(v) Fitting GAMLSS with Two Parameters Gamma and log link*

We have variety of choices of parameters for GAMLSS (sunch as type of distributions; formula for mean, sigma, nu and Tau, and, type of smoothers). We only choose Gamma with two parameter (mu and sigma), so we only have choices to model mu ($\mu$) and sigma ($\sigma$) (neither tau and nor nu). Apparently (for some rough choices) sigma does not dignificantly dependend upen some predictor. Therefore we only report model with constant sigma. Which mean interm of parameter model our GAMLSS does not differ significantly from GAM.

```
Family:  c("GA", "Gamma")
Fitting method: RS()

-------------------------------------------------------------------
Mu link function:  log
Mu Coefficients:
                    Estimate Std. Error t value Pr(>|t|)
(Intercept)           1.1147     0.0460   24.26  < 2e-16 ***
cs(Sepal.Width, 3)    0.0912     0.0141    6.47  1.5e-09 ***
Petal.Length          0.1305     0.0112   11.65  < 2e-16 ***
Petal.Width          -0.0400     0.0248   -1.61   0.1093
Speciesversicolor    -0.0924     0.0393   -2.35   0.0200 *
Speciesvirginica     -0.1454     0.0546   -2.66   0.0087 **
---
Signif. codes:  0 '***' 0.001 '**' 0.01 '*' 0.05 '.' 0.1 ' ' 1

-------------------------------------------------------------------
Sigma link function:  log
Sigma Coefficients:
            Estimate Std. Error t value Pr(>|t|)
(Intercept)  -2.9907     0.0577   -51.8   <2e-16 ***
---
Signif. codes:  0 '***' 0.001 '**' 0.01 '*' 0.05 '.' 0.1 ' ' 1

-------------------------------------------------------------------
Global Deviance:     55.2
            AIC:     75.2
            SBC:     105
```

### 3.2.3 *Comparing the models*

The estimate of each model and its GOF are summarized and compared in Table 2. We consider the best model (in term of number of parameters and value of likelihood) being the model with the smallest AIC or the biggest BIC.

**Tabel 2**. Comparison of models

| No | Model | Parameters | Estimates | GOF |
|---|---|---|---|---|
| 1 | LM without Factor | Intercept | 1.8560 (***) | AIC=84.6 |
| | | Sepal.Width | 0.6508 (***) | BIC=99.7 |
| | | Petal.Length | 0.7091 (***) | AdjRsq=0.856 |
| | | Petal.Width | -0.05565 (***) | |
| | LM with Factor (paralel model) | InterceptSetosa | 2.1713 (***) | AIC=79.1 |
| | | InterceptVersicolor | 1.4477 (***) | BIC=100 |
| | | InterceptVirginica | 1.1478 (**) | AdjRsq=0.997 |
| | | Sepal.Width | 0.4959 (***) | |
| | | Petal.Length | 0.8292 (***) | |
| | | Petal.Width | -0.3152 (*) | |
| 2 | GLM (with Gamma, log-link) | Intercept | 1.1122 (***) | AIC=74.95 |
| | | Versicolor | -0.0746 (NS) | BIC=96 |
| | | Virginica | -0.1220 (*) | RSq=0.996 |
| | | Sepal.Width | 0.0939 (***) | AdjRSq=0.995 |
| | | Petal.Length | 0.1283 (***) | |
| | | Petal.Width | -0.0491 (NS) | |
| 3 | GAM with cubic spline smoother on Sepal.Width | Intercept | 1.3950 (***) | AIC=73.1 AdjRsq=0.864 |
| | | SpeciesVersicolor | -0.0964 (*) | Deviance exlpained-= 87% |
| | | SpeciesVirginica | -0.1501 (**) | (When Petal.Width excluded, |
| | | Petal.Length | 0.1308 (***) | AIC=73.7) |
| | | Petal.Width | -0.0396 (NS) | |
| | | s(Sepal.Width) | Edf=1.77 (***) | |
| 4 | GAMLSS with cubic spline and constant sigma | (Intercept) | 1.1147 (***) | AIC=75.2 |
| | | Speciesversicolor | -0.0924 (*) | |
| | | Speciesvirginica | -0.1454 (**) | |
| | | cs(Sepal.Width, 3) | 0.0912 (***) | |
| | | Petal.Length | 0.1305 (***) | |
| | | Petal.Width | -0.0400(NS) | |
| | | Log(sigma) | -2.9907 (***) | |
| 5 | GLM (with Gamma, log-link) and natural spline on Sepal.Width | Intercept | 1.3273 (***) | AIC=74.9 |
| | | SpeciesVersicolor | -0.0939 (*) | BIC=102 |
| | | SpeciesVirginica | -0.1477 (*) | (When Petal.Width excluded |
| | | Petal.Length | 0.1308 (***) | AIC=109) |
| | | Petal.Width | -0.0393 (NS) | |
| | | ns(Sepal.Width, df = 3)1 | 0.0845 (**) | |
| | | ns(Sepal.Width, df = 3)2 | 0.1948 (**) | |
| | | ns(Sepal.Width, df = 3)3 | 0.2153 (***) | |

Notes:
(***) : p-val ≤ 0.1%
(**) : 0.1 < p-val ≤ 1%
(*) : 1% < p-val ≤ 5%
(NS) : p-val >5%

There are some remarks can be drawn from the results.

(i) It is worth to consider including factors (grups) in the model, when data do not heve observed group, users can perform cluster analysis and take the clusters as group (buliding cluster using Kmeans is also available in our online analysis)

(ii) The significance of individual parameter depends upon the combination of other parameters in the model. The parameters of some variabels may not be significant, but removing them from model can worsen the model (increase the AIC). Therefore, the parameters or variabels may be retained in the model.

(iii) To include spline smoothers in the model (with explonential family distributions), users can choose GAM or GLM+Natural or B-Spline, where the later are easier to interprete in term of using the model for prediction. (In this illustration GLM with natural spline has the smallest AIC value).

### *3.3 Advantages and disadvantages using online data analysis*

The method are placed in the web as part of Virtual Statistics Laboratory. The method can be accessed at http://statslab-rshiny.fmipa.unej.ac.id/RProg/MSI/. There are some adventages for users in using this online model fitting including (i) no need to install R, (ii) no need to master R scripting, (iii) users are easier to surf from one model to another, checking the graphical appearance and the GOF of the model (iv) user can access (do data analysis) using various type of gadgets (hp, tablet notebook, etc) and do simple to advanced statistical modeling with R. The main discomfort in using online data analysis is related to the speed of the available internet network and the number of users accessing the web at the same time. At this stage, web performances (the speed on various gadgets and various web browsers) have not been critically examined. However for local lectures or laboratory practices, students experience no noticable disruptions.

### *3.4 Future developments*

Some features have not been currently implemented namely (i) loess smoother for GAM (since they are conflicted with MGCV), (ii) nonlinear and multiple predictors model for the scale, shape and tau parameters in GAMLSS (iii) testing multicolinearity and models alternatives when it occurs in the predictors. These features, in near future, will be gradually included and tested.

## 4 Conclusion

Our online statistical model for independent responses, for LM, RLM, GLM, GAM , GAM LSS and CIS, has covered all main features (options) generally done using CLI (script programming), although for CIS types they are not illustrated. It enables users easier to do and compare various types of statistical modellings and choose the most appropriate model. In addition, user is also able to do various data explorations (scatter plot matrix, correlation plot matrix, and other visualization grafik). For GAM and GAMLSS more features are still to be added, and possibly extend the models to include multicollinearity.

**Acknowledgement**

This research is partly supported by PUPT Grant 2016, from DGHE Ministry of Research Technology and Higher Education Republic of Indonesia

**References**

[1]. Wallace B C, Dahabreh I J, Trikalinos TA, Lau J, Trow P, Schmid C H. Closing the Gap between Methodologists and End-Users: R as a Computational Back-End *Journal of Statistics Softwares.* Vol. 49, Issue 5, Jun 2012

[2]. Charles E. Robertson1, J. Kirk Harris, Brandie D. Wagner, David Granger, Kathy Browne, Beth Tatem 5, Leah M. Feazel 6, Kristin Park 1, Norman R. Pace 1 and Daniel N. Frank. Explicet: graphical user interface software for metadata-driven management, analysis and visualization of microbiome data. *Bioinformatics.* Vol. 29 no. 23 2013, pages 3100–3101

[3]. Dai, Y, Ling Guo, Meng Li and Yi-Bu Chen. Microarray Я US: a user-friendly graphical interface to Bioconductor tools that enables accurate microarray data analysis and expedites comprehensive functional analysis of microarray results. *BMC Research Notes* 2012, 5:282.

[4]. Subianto, M and H Sofyan. 2010. Interactive Statistics Learning with RwikiStat. *International Conference on Networking and Information Technology*

[5]. Tirta, IM. and D. Anggraini. 2015. Clustering: Analysis and Validation Using R-shiny web based interface. *ICOLIB: International Conference On Life Secience and Biotechnology*. 28-29 September 2015 URL: http://statslab-rshiny.fmipa.unej.ac.id/JORS/Cluster/
[6]. Tirta, IM. and D. Anggraini, L.C.Octaviani. 2016. Online and Interactive Web For Fitting GEE With Natural Splines For Longitudinal Data. *IBSC:International Basic Scince Conference*. FMIPA Universitas Jember. 26-27 September 2016. URL: http://statslab-rshiny.fmipa.unej.ac.id/JORS/GEE/
[7]. Fox, J. and S Weisberg. 2011. *An R Companion to Applied Regression,* Second Edition. Thousand Oaks CA: Sage. URL: http://socserv.socsci.mcmaster.ca/jfox/Books/Companion
[8]. Venables, W. N. & Ripley, B. D. 2002. *Modern Applied Statistics with S*. Fourth Edition. Springer, New York. ISBN 0-387-95457-0
[9]. Rousseeuw P., C. Croux, V. Todorov, A. Ruckstuhl, M. Salibian-Barrera, T. Verbeke, M. Koller, M. Maechler. 2016. *robustbase: Basic Robust Statistics*. R package version 0.92-6. URL http://CRAN.R-project.org/package=robustbase
[10]. McCullagh & Nelder. 1989. *Generalized Linear Models*. Chapman & Hall
[11]. Marschner,I. 2014. *glm2: Fitting Generalized Linear Models*. R package version 1.1.2. https://CRAN.R-project.org/package=glm2
[12]. Trevor Hastie 2016. gam: Generalized Additive Models. R package version 1.14. https://CRAN.R-project.org/package=gam
[13]. Wood, S.N. 2006. Generalized Additive Models: An Introduction with R. Chapman and Hall/CRC.
[14]. Wood, S.N. 2011. Fast stable restricted maximum likelihood and marginal likelihood estimation of semiparametric generalized linear models. *Journal of the Royal Statistical Society (B)* 73(1):3-36
[15]. Rigby R.A. and Stasinopoulos D.M. 2005. Generalized additive models for location, scale and shape,(with discussion), *Appl. Statist*., 54, part 3, pp 507-554.
[16]. Stasinopoulos D.M and Rigby R.A. 2008. Generalized Additive Models for Location Scale and Shape (GAMLSS) in R. *Journal of Statistitics Software*. Vol **23** (no 7). 1-46
[17]. Benjamin, H., Andreas Mayr, Matthias Schmid. 2016. gamboostLSS: An R Package for Model Building and Variable Selection in the GAMLSS Framework. *Journal of Statistitics Software*. Vol **74** (no 1). 1-31
[18]. Andrew D. Martin, Kevin M. Quinn, Jong Hee Park. 2011. MCMCpack: Markov Chain Monte Carlo in R. *Journal of Statistical Software*. **42(9)**: 1-21. URLhttp://www.jstatsoft.org/v42/i09/.
[19]. Chang, W., J. Cheng, JJ Allaire, Y. Xie and J McPherson. 2015. *shiny: Web Application Framework for R*. R package version 0.11.1. http://CRAN.R-project.org/package=shiny
[20]. Fox,J. 2016. *RcmdrMisc: R Commander Miscellaneous Functions*. R package version 1.0-5. https://CRAN.R-project.org/package=RcmdrMisc
[21]. Revelle, W. 2016. *psych: Procedures for Personality and Psychological Research*, Northwestern University, Evanston, Illinois, USA, https://CRAN.R-project.org/package=psych Version = 1.6.9.
[22]. Wickham. H. 2009. *ggplot2: Elegant Graphics for Data Analysis*. Springer-Verlag New York.
[23]. Patrick Breheny and Woodrow Burchett (2016). *visreg: Visualization of Regression Models. R package version 2.3-0*. https://CRAN.R-project.org/package=visreg.
[24]. Anderson, E. 1935. The irises of the Gaspe Peninsula, *Bulletin of the American Iris Society*, **59**, 2–5.

Appendix:

Table 3. Menu Structure of Online Statistical Model for Independent responses

| No | NavBar | Sub Menu | SideBar | Output |
|---|---|---|---|---|
| 1 | Input Data | - | Data Selection (internal & Import) | List of Data |
| 2 | Exploration | General | • Variable selection<br>• Type of Diagonal plot (histogram, boxplot, qqplot, density)<br>• Check and set for dummy | • Summary statistics<br>• Correlation matrix<br>• Correlation diagram |

| No | NavBar | Sub Menu | SideBar | Output |
|---|---|---|---|---|
| | | | • Form Kmean clustering | • Scatterplot matrix |
| | | Smoother | • response and predictor selection<br>• type of smoother and parameters (LM, RLM, GLM, GAM, Natural Spline) | • scatter plot with smoother |
| 3 | Response variable | • Respon Variable<br>• Histogram<br>• QQplot<br>• Box-Plot | • Respon Variable Slection<br>• For histogram<br>✓ Check for density estimate<br>✓ Check for density estimate<br>✓ Check for mean and median | Qqplot<br><br><br>BoxPlot<br>Histogram (with density estimate, mean and median) |
| 4 | Model Formula | Respon-predictors (for LM, GLM, GAM) | • Setting Response<br>• Setting Predictor<br>• Checking & Setting Dummy<br>• Setting Predictor for scatter plot | OLS output<br>• Summary<br>• GOF<br>• Anova<br>• Scatter Plot<br>• Visual Plot (from visreg packages)<br>• Stepwise output<br>• X matrix |
| | | Family and link (For GLM, GLM+NS and GAM) | • Family selection<br>• Link selection | • Scatter plot with smoother<br>• GOF of GLM |
| | | Formula for GAMLSS | • Response selection<br>• Distributionsfamily<br>• Predictor Selection<br>• Predictor for Sigma<br>• Predictor for Nu<br>• Predictor for Tau | • Histogram of data and fitness of theoretical density |
| 5 | Detail of Modern Reression(output and further selection) | GLM | - | • Summary of fit<br>• Deviance analysis<br>• Fiited Plot (2D)<br>• Stepwise<br>• Diagnostic plot |
| | | GAM | • Variable for nonparametric<br>• Type of smoother<br>• Df<br>• Spesific object of GAM | • Summary<br>• Fitted plot (2D)<br>• GOF<br>• Detail output of GAM |
| | | GLM+NS | • Response<br>• Predictor<br>• Nonparametric<br>• Df for NS | • Smoother plot (2D)<br>• Summary of fit<br>• Deviance analysis<br>• GOF<br>• Diagnosticplot<br>• X matrix |
| | | GAMLSS | (extension from Formula for GAMLSS)<br>• Type of Smoother | • Summary<br>• Plot |

| No | NavBar | Sub Menu | SideBar | Output |
|---|---|---|---|---|
| | | | • Df<br>• Estimation method (RS, CG, Mixed)<br>• Type of Plot (diagnostic, term, worm) | • GOF |
| | | Robust-Reg | Method (M,MM,MF, LTS) | • Summary ofOLS<br>• Summary of Robust<br>• Bonferroni Test<br>• GOF<br>• Plot (OLS and Robust) |
| | | CIS | • Response for CIS<br>• Predictor for CIS<br>• Type of CIS (Bootstrap, MCMC)<br>• Type of MCMC (Gaussian, Poisson, Logit-Binomial)<br>• Number of bootstraps<br>• number of burned in (MCMC) | • Estimate<br>• Plot of estimate<br>• Bootstrap CI<br>• Bootstrap Jakknife |

IOP Conf. Series: Journal of Physics: Conf. Series **855** (2017) 012055 doi:10.1088/1742-6596/855/1/012055

# $H$-Supermagic Labeling on Shrubs Graph and $L_m \odot P_n$

**Risala Ulfatimah[1], Mania Roswitha[2], Tri Atmojo Kusmayadi[3]**

Department of Mathematics, Faculty of Mathematics and Natural Sciences, Sebelas Maret University, Surakarta, Indonesia 57126

E-mail: rulfatimah@gmail.com[1], mania_ros@yahoo.co.id[2], tri.atmojo.kusmayadi@gmail.com[3]

**Abstract.** A finite simple graph $G$ admits an $H$-covering if every edge of $E(G)$ belongs to a subgraph of $G$ isomorphic to $H$. We said the graph $G = (V, E)$ that admits $H$-covering to be $H$-magic if there exists a bijection function $f : V(G) \cup E(G) \to \{1, 2, \ldots, |V(G)| + |E(G)|\}$ such that for each subgraph $H'$ of $G$ isomorphic to $H, f(H') = \sum_{v \in V'} f(v) + \sum_{e \in E'} f(e) = m(f)$ is constant.
Furthermore, if $f(V) = 1, 2, \ldots, |V(G)|$ then $G$ is called $H$-supermagic. In this research we defined $S_{2,2}$-supermagic labeling on shrub graph $\check{S}(m_1, m_2, \ldots, m_n)$ and fish-supermagic labeling on $L_m \odot P_n$ for $m, n \geq 2$.

## 1. Introduction

Gallian [2] defined a graph labeling as an assignment of integers to the vertices or edges, or both, subject to certain condition. Magic labeling is a type of graph labeling that the most often to be studied. In [2], magic labelings were first introduced in 1963 by Sedláček.

We consider finite and simple graphs. The vertex and edge sets of a graph $G$ are denoted by $V(G)$ and $E(G)$, respectively. An edge-covering of $G$ is a family of subgraphs $H_1, \ldots, H_k$ such that each edge of $E(G)$ belongs to at least one of the subgraphs $H_i, 1 \leq i \leq k$. If every $H_i$ is isomorphic to a given graph $H$, then $G$ admits and $H-$covering. An edge total labeling on $G$ is a bijective function $f : V(G) \cup E(G) \to 1, 2, \ldots, |V(G)| + |E(G)|$, with the property that, given any edge $xy$, we have $f(x) + f(xy) + f(y) = k$.

In 2005, Gutiérrez and Lladó [4] generalized the concept of an edge-magic total labeling into an $H$-magic covering as follows. Let $G = (V, E)$ be a finite simple graph that admits $H$-covering. A bijection function $f : V(G) \cup E(G) \to \{1, 2, \ldots, |V(G)| + |E(G)|\}$ is called $H$-magic labeling of $G$ if for every subgraph $H' = (V', E')$ of $G$ isomorphic to $H, f(H') = \sum_{v \in V'} f(v) + \sum_{e \in E'} f(e) = m(f)$ is constant. The graph $G$ is called $H$-supermagic if $f(V(G)) = \{1, 2, \ldots, |V(G)|\}$. In [4], it is proved that a complete bipartite graph $K_{n,n}$ is $K_{1,n}$-magic for $n \geq 1$.

Selvagopal and Jeyanthi in Gallian [2] proved that for a positive integer $n$, the $k$-polygonal snake of length $n$ is $C_4$-supermagic; for $m \geq 2, n = 3$, or $n > 4$, $C_n \times P_m$ is $C_4$-supermagic; $P_2 \times P_n$ and $P_3 \times P_n$ are $C_4$-supermagic for all $n \geq 2$. Roswitha et al. [8] proved $H$-magic covering on some classes of graphs. In this research we investigate that a shrub graph $\check{S}(m_1, m_2, \ldots, m_n)$ admits a double star $S_{2,2}$-supermagic labeling and a $L_m \odot P_n$ admits a fish-supermagic labeling for $m, n \geq 2$.

## 2. Main Results

### 2.1. *k-balanced multiset*

Maryati et al. [6] defined a multiset as a set that allows the same elements in it. A multiset $\{a_1^{k_1}, a_2^{k_2}, a_3^{k_3}, ..., a_n^{k_n}\}$ is defined as a set of $a_i$ as many as $k_i$ for any integers $n$ and $k_i, i \in [1, n]$. Let $V = \{a_1, a_2, a_3, a_4\}$ and $W = \{a_2, a_5, a_6\}$ be the multiset, then we have $V \uplus W = \{a_1, a_2, a_2, a_3, a_4, a_5, a_6\}$. In [6] was introduced a technique of partitioning a multiset, called $k-$balanced multiset, as follows. Let $k \in N$ and $Y$ be a multiset that contains positive integers. $Y$ is said to be $k-$balanced if there exists $k$ subsets of $Y$, say $Y_1, Y_2, \ldots, Y_k$, such that for every $i \in [i, k], |Y_i| = \frac{|Y|}{k}, \Sigma Y_i = \frac{\Sigma Y}{k} \in N$, and $\biguplus_{i=1}^{k} Y_i = Y$, then $Y_i$ is called a balanced subset of $Y$.

**Lemma 2.1** (Roswitha and Baskoro [7]) *Let $x$ and $y$ be non-negative integers. Let $X = [x+1, x(y+1)]$ with $|X| = xy$ and $Y = [x(y+2), 2x(y+1)-1]$ where $|Y| = xy$. Then, the multiset $K = X \uplus Y$ is $xy$-balanced with all its subsets are 2-sets.*

### 2.2. *$(k, \delta)$-anti balanced multiset*

Inayah [5] defined $(k, \delta)$-anti balanced multiset as follows. Let $k, \delta \in \mathbb{N}$ and $X$ be a set containing the elements of positive integers. A multiset $X$ is said to be $(k, \delta)$-anti balanced if there exists $k$ subsets from $X$, say $X_1, X_2, \ldots, X_k$ such that for every $i \in [1, k]$, $|X_i| = \frac{|X|}{k}$, $\biguplus_{i=1}^{k} X_i = X$, and for $i \in [1, k-1]$, $\sum X_{i+1} - \sum X_i = \delta$.

Here, we give several lemmas on $(k, \delta)$-anti balanced multiset.

**Lemma 2.2** Let $x, y, z$ and $k$ be non-negative integers, $k \geq 2$. Let $R = [x, x + \lfloor \frac{k}{2} \rfloor] \uplus [x+1, x + \lfloor \frac{k}{2} \rfloor] \uplus [y, y + \lfloor \frac{k-1}{2} \rfloor] \uplus [y, y + \lfloor \frac{k-1}{2} \rfloor - 1] \uplus [z, z+k-1]$, then $R$ is $(k, 2)$-anti balanced.

*Proof.*
For every $j \in [1, k]$ we define the multisets $R_j = \{x + \lfloor \frac{j}{2} \rfloor, y + \lfloor \frac{j-1}{2} \rfloor, z + (j-1)\}$. It is obvious that for each $j \in [1, k], |R_j| = 3, R_j \subset R$, and $\biguplus_{j=1}^{k} R_j = R$. Since $\sum R_j = x + y + z + \lfloor \frac{2j-1}{2} \rfloor + j - 1$ for every $j \in [1, k]$, then $\sum R_{j+1} - \sum R_j = 2$, $R$ is $(k, 2)$-anti balanced.

□

**Lemma 2.3** Let $x$ and $k$ be non-negative integers, $k \geq 2$. Let $Y = [x, x+k] \uplus [x+1, x+k-1]$, then $Y$ is $(k, 2)$-anti balanced.

*Proof.*
For every $j \in [1, k]$ we define the mutisets $Y_j = \{x+j-1, x+j\}$. Now we have for every $j \in [1, k], |Y_j| = 2, Y_j \subset Y$ and $\biguplus_{j=1}^{k} Y_j = Y$. Since $\sum Y_j = 2(x+j) - 1$ for every $j \in [1, k]$, then $\sum Y_{j+1} - \sum Y_j = 2$, $Y$ is $(k, 2)$-anti balanced.

□

*2.3. $S_{2,2}$-supermagic Labeling on Shrubs Graphs.*
Maryati [6] defined a shrub graph $\check{S}(m_1, m_2, \ldots, m_n)$ as a graph that is obtained from a star graph $K_{1,n}$ for $n \geq 2$ with central vertex $c$, and adding some vertices and edges so that for every $i \in [1, n], v_i$ is related with the new $m_i \geq 1$ vertices. According to Grossman [3], a double star $S(n, m)$ is a graph consisting of the union of two stars $K_{1,n}$ and $K_{1,m}$ together with a line joining their centers.

In Figure 1 and Figure 2 we show examples of a shrub graph $\check{S}(m_1, m_2, \ldots, m_n)$ and a double star $S_{2,2}$.

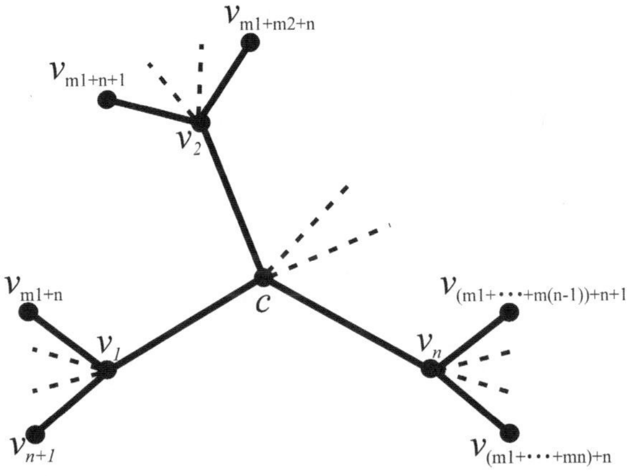


**Figure 1.** A shrub graph $\check{S}(m_1, m_2, \ldots, m_n)$.

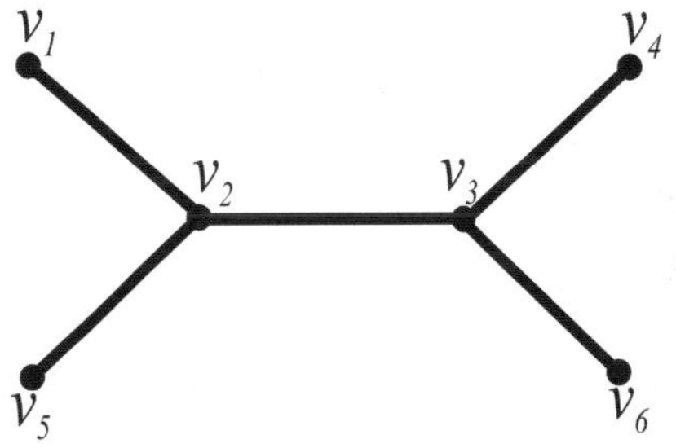


**Figure 2.** A double star $S_{2,2}$.

**Theorem 2.4** *Any shrub graph $\check{S}(m_1, m_2, \ldots, m_n)$ is $S_{2,2}$-supermagic for any integer $n$ and $m_i \geq 2, i \in [1, n]$.*

*Proof.*
Let $G$ be a shrub graph $\check{S}(m_1, m_2, \ldots, m_n)$ for any integer $n$ and $m_1, m_2, \ldots, m_n \geq 2$. Then $|V(G)| = m_1 + m_2 + \ldots + m_n + n + 1$ and $|E(G)| = m_1 + m_2 + \ldots + m_n + n$. Let $A = [1, 2(n + m_1 + m_2 + \ldots + m_n) + 1]$. Partition $A$ into 3 sets, $A = K \cup X$, with $X = L \uplus M$ where $K = \{1\}, L = [2, n + 1 + m_1 + \ldots + m_n], M = [n + 2 + m_1 + \ldots + m_n, 2n + 1 + 2(m_1 + \ldots + m_n)]$. Let $f$ be a total labeling of $G$ and $s(f)$ be the supermagic sum in every subgraph $H$ of $G$ that is isomorphic to double star $S_{2,2}$.

Now we define a total labeling $f$ on $G$ as follows. Label the center vertex with 1. Apply Lemma 2.1 to partition $X$ into $\{X_i, x \leq i \leq y\}$ with $x = 1$ and $y = n+m_1+m_2+\ldots+m_n$. We obtain that $X$ is $n+m_1+m_2+\ldots+m_n$-balanced with all its subsets are 2-sets and $\sum X_i = 2(n + m_1 + m_2 + \ldots + m_n) + 3$. Use the smaller labels in every $X_i$ for the vertices. Thus, the sum of labels of each double star $S_{2,2}$ is $s(f) = 10(n + m_1 + m_2 + \ldots + m_n) + 16$. Hence, the shrubs graph is $S_{2,2}$-supermagic. □

The example of $S_{2,2}$-supermagic labeling on $\check{S}(3, 2, 4, 2)$ is shown in Figure 3.

IOP Conf. Series: Journal of Physics: Conf. Series **855** (2017) 012055 doi:10.1088/1742-6596/855/1/012055

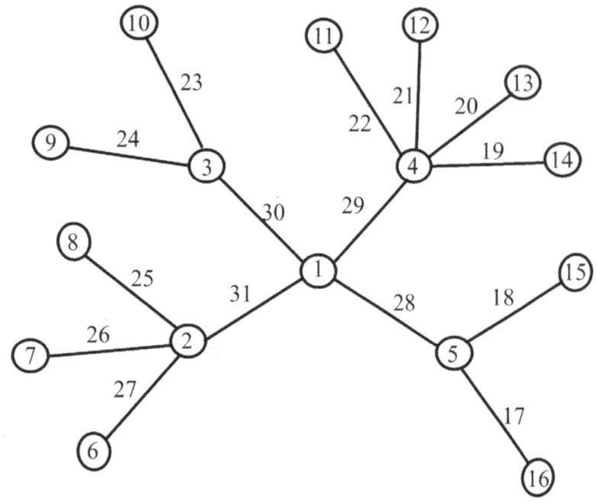


**Figure 3.** $S_{2,2}$-supermagic labeling on $\check{S}(3, 2, 4, 2)$.

*2.4.* ***A Fish-Supermagic Labeling on $L_m \odot P_n$.***

The corona $G_1 \odot G_2$ of two graphs $G_1$ and $G_2$ (where $G_i$ has $p_i$ points and $q_i$ lines) is defined as the graph $G$ obtained by taking one copy of $G_1$ and $p_i$ copies of $G_2$, and then joining by a line $i$'th point of $G_1$ to every point in the $i$'th copy $G_2$.

$L_m \odot P_n$ graph is obtained by taking one copy of $L_m$ and $2m$ copies of $P_n$ and then joining by a line the $i$'th vertex of $L_m$ to every vertex in the $i$'th copy of $P_n$.

Figure 4 is an example of $L_m \odot P_n$ graph.

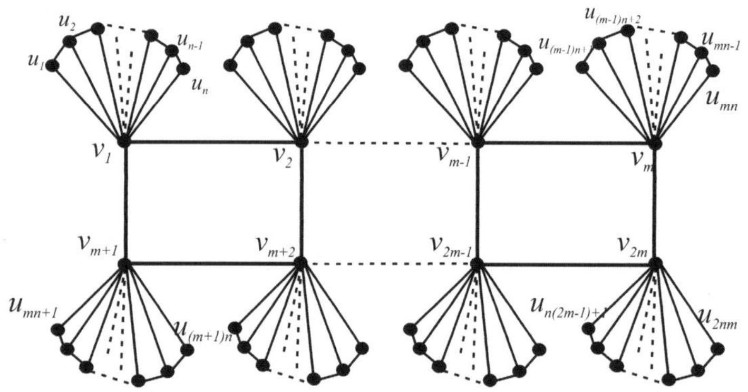


**Figure 4.** A $L_m \odot P_n$ graph.

Brandstadt [1] defined a fish graph as a graph on 6 vertices illustrated below.

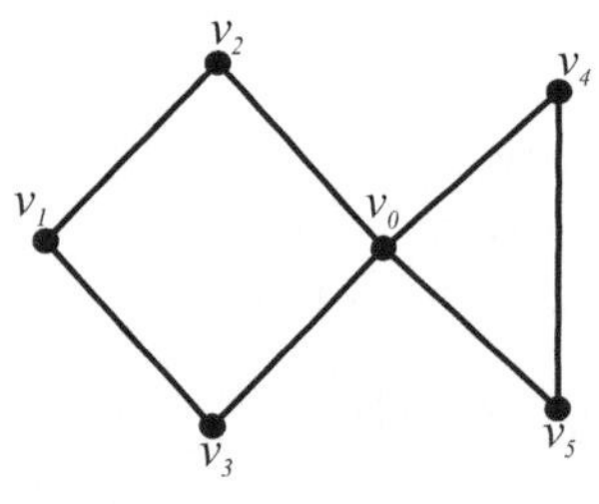


**Figure 5.** A fish graph.

**Theorem 2.5** *Any $L_m \odot P_n$ graph for $m, n \geq 2$ is a fish-supermagic.*

*Proof.*

Let $G$ be a $L_m \odot P_n$ graph for any integer $m, n \geq 2$. Then $|V(G)| = 2m(n + 1)$ and

$|E(G)| = m(4n+1) - 2$. Let $A = [1, 2m(3n-1) + 5(m-1) + 3]$. We define a bijective function $f : V(G) \cup E(G) \rightarrow \{1, 2, \ldots, 2m(3n-1) + 5m - 2\}$. Partition $A$ into 4 sets, $A = V \uplus W \uplus X \uplus Y$, where $V = [1, 2mn], W = [m(2n+3), m(4n+3) - 1], X = [m(4n+3), m(4n+3)+n-2] \biguplus [m(4n+3)+n, m(4n+3)+(2n-2)] \biguplus \ldots \biguplus [m(4n+3)+(2m-1)n, m(4n+3)+2mn-2]$, and $Y = [2mn+1, 2m(n+1)+(m-1)] \biguplus \{m(4n+3)+(n-1), m(4n+3)+(n-1)+n, m(4n+3)+(n-1)+2n, \ldots, m(4n+3)+(n-1)+2n(m-1)\}$.

Let $f$ be a total labeling of $G$ and $s(f)$ be the supermagic sum in every subgraph $H$ of $G$ that is isomorphic to fish graph. It is easier if a fish graph is divided into 2 subgraphs, $F_2 - \{v_0\}$ (a fan graph $F_2$ without a central vertex) and $C_4$.

There are two steps to label the $L_m \odot P_n$ graph. The first step is to prove that $2m(F_n - \{v_0\})$ is a $(F_2 - \{v_0\})$-supermagic using the elements of multisets $V \uplus W \uplus X$, and the second one is to prove that $L_m$ is $C_4$-supermagic using the element of set $Y$.

**Step 1.** The first step is to prove that $2m(F_n - \{v_0\})$ is $(F_2 - \{v_0\})$-supermagic using the elements of multisets $V \uplus W \uplus X$. We consider two cases on this proof.

**Case 1.** $n = 2$. For $n = 2$, the number of $(F_2 - \{v_0\})$ subgraphs on graph $G$ is $2m$ as shown in Figure 6.

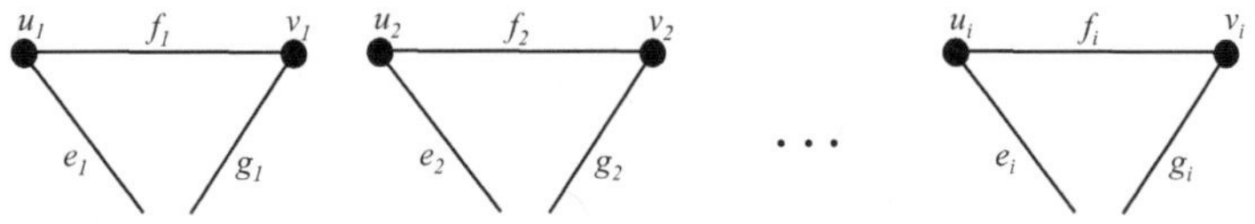


**Figure 6.** $F_2 - \{v_0\}$ subgraph on $L_m \odot P_2$.

Now, we prove that $2m(F_n - \{v_0\})$ on $L_m \odot P_2$ is $(F_2 - \{v_0\})$-supermagic using the element of multisets $V \uplus W \uplus X$. Let $H$ be $2m(F_n - \{v_0\})$. We have $V(H) = \{u_i, v_i : 1 \leq i \leq 2m\}$ and $E(H) = \{e_i \cup f_i \cup g_i : 1 \leq i \leq 2m\}$. Let $f$ be the total labeling of $H$ and $s(f)$ be the supermagic sum in every $F_2 - \{v_0\}$ subgraph on $2m(F_n - \{v_0\})$. Then for every $1 \leq i \leq 2m$ we define

$$f(x) = \begin{cases} i, & \text{if } x = v_i, \\ 2m+i, & \text{if } x = u_i, \\ m(4n+3)+1-2i, & \text{if } x = e_i, \\ m(4n+3)+2(i-1), & \text{if } x = f_i, \\ m(4n+3)-2i, & \text{if } x = g_i. \end{cases}$$

For every $1 \leq i \leq 2m$, let $(F_2 - \{v_0\})^i$ be the subgraph of $L_m \odot P_2$ with $V((F_2 - \{v_0\})^i) = \{u_i, v_i\}$ and $E((F_2 - \{v_0\})^i) = \{e_i, f_i, g_i\}$. It can be checked that for every $1 \leq i \leq 2m$, we have $\sum f((F_2 - \{v_0\})^i) = 3m(4n+3) + 2m - 1$. Hence $2m(F_2 - \{v_0\})$ is $(F_2 - \{v_0\})$-supermagic.

**Case 2.** $n > 2$. Now we prove that $2m(F_n - \{v_0\})$ is $(F_2 - \{v_0\})$-supermagic using the elements of multisets $V \uplus W \uplus X$. Let $f$ be the total labeling of $2m(F_n - \{v_0\})$ and $s(f)$ be the supermagic sum in every $F_2 - \{v_0\}$ of $2m(F_n - \{v_0\})$. The total labeling of $(F_n - \{v_0\})^i$ for every $1 \leq i \leq 2m$ will be executed within two steps. Firstly, we label the subgraph of $(F_n - \{v_0\})^i$ called $P_n^i$, then after that we label the edges other than $P_n^i$ that are incident with the vertices of $P_n^i$.

Lemma 2.2 is applied for every $i$ of $P_n^i$ where $1 \leq i \leq 2m$ with $x_i = 1 + (i-1)\lceil \frac{n}{2} \rceil, y_i = 2m\lceil \frac{n}{2} \rceil + (i-1)\lfloor \frac{n}{2} \rfloor + 1, z = m(4n+3) + n(i-1)$, and $k = n-1$ using the element of $V \uplus X$. We have that $V \uplus X$ in every $P_i$ is $(k, 2)$-anti balanced.

Next, by applying Lemma 2.3 we label the edges other than $P_n^i$ that are incident with the vertices of $P_n^i$ using the elements of $W$ with $x = m(4n+3) - ni$ and $k = n$, starting from the edge that incident with a vertex $v_n$ until the edge that incident with a vertex $v_1$ in every $P_n^i$. Now we have that $W$ for every $i \in [1, 2m]$ is $(k, 2)$-anti balanced. This implies that multiset $V \uplus W \uplus X$ is $2km$-balanced. Hence $2m(F_n - \{v_0\})$ of $L_m \odot P_n$ is $(F_2 - \{v_0\})$-supermagic with $\sum f((F_2 - \{v_0\})^i) = m(12n + 2\lceil \frac{n}{2} \rceil + 2) + 7(m-1) + 6$. We conclude that for $n \geq 2, 2m(F_n - \{v_0\})$ is $(F_2 - \{v_0\})$-supermagic.

**Step 2.** The next step is to prove that $L_m$ is $C_4$-supermagic using the element of set $Y$. Let $L_m \cong P_m \times P_2$ be a graph with $V(L_m) = \{u_i, v_i : 1 \leq i \leq m\}$ and $E(L_m) = \{u_iv_i : 1 \leq i \leq m\} \cup \{u_iu_{i+1}, v_iv_{i+1} : 1 \leq i \leq m-1\}$. Let $f$ be a total labeling of $V(L_m) \cup E(L_m)$ using the element of set $Y$. The total labeling of $L_m$ is defined as follows.

(i) Label each vertex of $L_m$ as follows.

$$f(u_i) = m(2n+1) - (i-1), \quad \text{for } i \in [1, m].$$

$$f(v_i) = \begin{cases} m(2n+1) + \frac{i+1}{2}, & \text{for } i \text{ odd}, i \in [1, m], \\ m(2n+1) + \lceil \frac{m}{2} \rceil + \frac{i}{2}, & \text{for } i \text{ even}, i \in [1, m]. \end{cases}$$

(ii) Next, label each edge of $L_m$ for every $i \in [1, m-1]$ as follows.

$$f(u_iu_{i+1}) = m(4n+3) + ni - 1$$
$$f(v_iv_{i+1}) = m(2n+2) + i$$

(iii) For every $i \in [1, m]$, label each edge of $L_m$ as follows.
**For $m$ odd.**

$$f(u_iv_i) = \begin{cases} m(4n+3) + n(\frac{m+1}{2}) + n(m-1) - n(\frac{i-1}{2}) - 1, & \text{for } i \text{ odd}, \\ m(4n+3) + mn + n(m-1) - n(\frac{i-2}{2}) - 1, & \text{for } i \text{ even}. \end{cases}$$

**For $m$ even.**

$$f(u_iv_i) = \begin{cases} m(4n+3) + mn + n(m-1) - n(\frac{i-1}{2}) - 1, & \text{for } i \text{ odd}, \\ m(4n+3) + n(\frac{m}{2}) + n(m-1) - n(\frac{i-2}{2}) - 1, & \text{for } i \text{ even}. \end{cases}$$

For every $1 \leq i \leq m-1$, let $C_4^i$ be as subgraph of $L_m \odot P_n$ with $V(C_4^i) = \{u_i, u_{i+1}, v_i, v_{i+1}\}$ and $E(C_4^i) = \{u_iu_{i+1}, v_iv_{i+1}, u_iv_i, u_{i+1}v_{i+1}\}$. It can be checked that for every $1 \leq i \leq m-1, \sum f(C_4^i) = m(22n+14) + n(2 + \lceil \frac{m+1}{2} \rceil) + \lceil \frac{m+1}{2} \rceil + (m-1)(3n+1)$. Hence $L_m$ is $C_4$-supermagic.

It has been proved that $2m(F_n - \{v_0\})$ is $(F_2 - \{v_0\})$-supermagic and $L_m$ is $C_4$-supermagic. Hence, $L_m \odot P_n$ is a fish-supermagic with its supermagic sum is $s(f) = m[(12n + 2 + 2\lceil \frac{n}{2} \rceil) + (22n + 14)] + n(2 + \lceil \frac{m+1}{2} \rceil) + \lceil \frac{m+1}{2} \rceil + (m-1)(3n+8) + 6$.

□

Figure 7 illustrates a fish-supermagic labeling on $L_3 \odot P_4$.

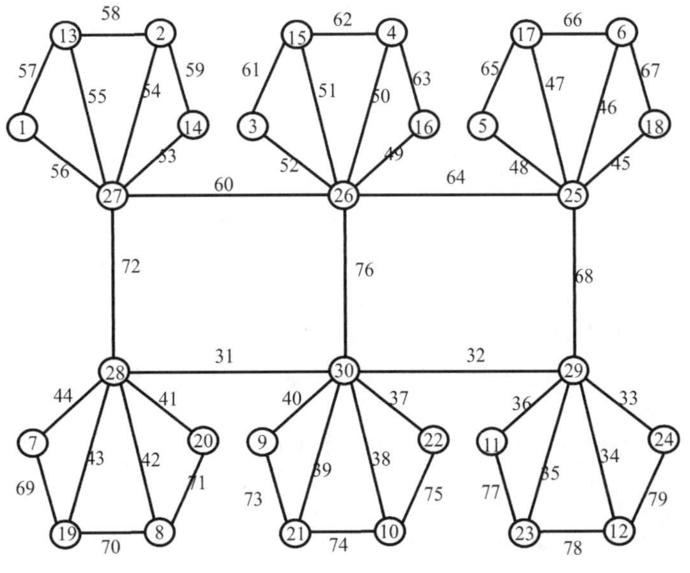

**Figure 7.** A fish-supermagic labeling on $L_3 \odot P_4$.

## References

[1] Brandstadt, A.,V.B.Le, and J.P. Spinrad, *Graph Classes: A Survey*, Philadelphia, SIAM Monographs on Disrete Mathematics and Applications, Society for Industrial Mathematics, 1987.

[2] Gallian, J. A., *A Dynamic Survey of Graph Labeling*, The Electronic Journal of Combinatorics. **18**(2015), 1-389 #DS6.

[3] Grossman, J.W., F. Harary, and M.Klawe, *Generalized Ramsey Theory for Graphs, x: Double Stars*, Discrete Mathematics **28** (1979), 247-254.

[4] Gutiérrez, A. and A. Lladó , *Magic Coverings*, J. Combin. Math. Combin. Comput **55** (2005), 43-56.

[5] Inayah, N., *Pelabelan $(a,d)-H-$Anti Ajaib pada Beberapa Kelas Graf, Disertasi, Institut Teknologi Bandung, Bandung, 2013.*

[6] Maryati, T.K.,*Karakteristik Graf H-Ajaib dan Graf H-Ajaib Super, Disertasi, Institut Teknologi Bandung, Bandung, 2011.* Cochin University of Science and Technology, Cochin, 2009.

[7] Roswitha, M. and E. T. Baskoro, *H-Magic Covering on Some Classes of Graphs*, American Institute of Physics Conference Proceedings on ICREM5 , **1450** (2012), 135-138.

[8] Roswitha, M., E.T.Baskoro, T.K.Maryati, N.A.Kurdhi, and I.Susanti, *Further Results on Cycle-Supermagic Labeling*, AKCE Int, J.Graphs Comb **10** (2013), 211-220.

IOP Conf. Series: Journal of Physics: Conf. Series **855** (2017) 012056 doi:10.1088/1742-6596/855/1/012056

# Internet messenger based smart virtual class learning using ubiquitous computing

**K Umam[1,2], S N S Mardi[1], M Hariadi[1]**

[1] Department of Electrical Engineering, Institut Teknologi Sepuluh Nopember (ITS) Surabaya, 60111, Indonesia.
[2] Walisongo State Islamic University, 50185, Semarang, Indonesia.

Khothibul@gmail.com, smardi88@gmail.com, mochar@ee.its.ac.id

**Abstract**. Internet messenger (IM) has become an important educational technology component in college education, IM makes it possible for students to engage in learning and collaborating at smart virtual class learning (SVCL) using ubiquitous computing. However, the model of IM-based smart virtual class learning using ubiquitous computing and empirical evidence that would favor a broad application to improve engagement and behavior are still limited. In addition, the expectation that IM based SVCL using ubiquitous computing could improve engagement and behavior on smart class cannot be confirmed because the majority of the reviewed studies followed instructions paradigms. This article aims to present the model of IM-based SVCL using ubiquitous computing and showing learners' experiences in improved engagement and behavior for learner–learner and learner–lecturer interactions. The method applied in this paper includes design process and quantitative analysis techniques, with the purpose of identifying scenarios of ubiquitous computing and realize the impressions of learners and lecturers about engagement and behavior aspect and its contribution to learning

## 1. Introduction

The idea of a smart campus today is incredibly powerful. Smart campus has been widely accepted and used in strategy of education reformation in several countries in the world for improving the quality of the campus learning environment[1]. With the adopted of technological development of smartphone technology, smarter campuses are built to benefit the lecturer and learner, manage the available resources and enhance user experience with proactive services anytime and anywhere[2].

Within the context of smart campus, technology of smartphone devices can be used to support teaching and learning in ubiquitous environment through Interactive internet messenger (IM). Some examples of IM in educational activities are content generating, interacting, collaborating and sharing [3][4], Content generating occurs when affective interaction and developing Question and Answer sessions. This application is designed to provide a platform for learners to share their knowledge, broadcast news, and asks for help [3][5][6]. Member are able to share or publish their work and ideas by message for others to view and download. For example, multimedia and documents (like pdf, ms. Office product) files can be shared with message[3][7]. IM can be used by learners to collaboratively learn how to solve problems with members of a group, or to organize collaborative learning and study groups[3][4]. By collaborating, learner extends their interactions by establishing active communication with their peers, with the aim of working towards particular outcomes.

Many researchers have discussed and focused on the infrastructure required for constructing smart campuses and applications that related to learning at ubiquitous environment, in [8] have proposed a framework specification for u-learning in a smart campus model that integrates real world learning resources with a social network of campus. This model aims to embed of learners within a smart campus environment that provides context based personalized learning and interaction or feedback, in [5] has designed and implemented a prototype system based OnCampus scheme. The model developed in three functional modules, namely Group, Transaction, and Forum modules to provide services related to learning. The model OnCampus can significantly contribute in the following: through the creation of social environment based on interests mining, the equipping of educational guidance based on emotion analysis, as an information sharing platform, and by developing a secondary platform that aimed at the optimal allocation of campus resources, in [9] has proposed an adaptive contents delivery model for context aware ubiquitous learning according to three level service models, which create the adaptive contents for learners to get a all access in learning according to learners' interest and contexts. The evaluation of a Ubiquitous learning system shows that the learners may not only study in media on mobile device at anytime and at any place and also get a better learning experience, e.g. Learners may get instant help from other participates with ubiquitous device and access the rich contents.

The majority of mobile and u-learning model studies showed positive effects. However, the IM based SVCL using ubiquitous computing and empirical proof that would favor a broad application of mobile technology devices and ubiquitous learning in smart virtual class settings to improve engagement and behavior is still limited. In addition, the expectation that mobile learning could improve engagement and behavior on smart virtual class cannot be confirmed because the majority of the reviewed of studies followed instructions paradigms.

This paper aims to design and implement and showing empirical evidence about IM based SVCL using ubiquitous computing to improve engagement and behavior for learner–learner and learner–lecturer interactions on smart campus. The method applied in this paper includes model design process and quantitative analysis techniques, with the purpose of identifying scenario of the project and realize the impressions of learners and teachers about the engagement and behavioral aspects, and its contribution to learning.Literatur review

## 2. Literatur review

### *2.1. Ubiquitous Computing on Smart Vitual Class Learning (SVCL)*

Education in a smart virtual class supported by mobile and smart technologies, making use of smart tools and mobile devices, can be considered smart education with ubiquitous environment. In this respect, we observe that novel technologies have been widely adopted in campus for provides connectivity between learners and their surrounding environments. For learners, learning goals is inherently identified to trigger didactic models which guide their instruction around real world data, based on unique learning contexts and delivered with the right information, at the right time, at the right place, at the right way, to the right person[10]. This is a new enhancement for academic from pedagogical processes through which learning is diffused just in one time like a process of production, when learner are ready to achieve a targeted level of instruction.

The smart campus environment provides such settings of ubiquitous learning in which virtual and electronic resources are made available through smart and mobile devices that help learners navigate the real and the surrounding physical world[5] learning happens all around the learners without any conscious effort on the part of the learner. The learning process as ubiquitous learning doesn't restrict to be in the classroom or formal learning environments. On the contrary, the learning involves situating learners in both the real and the virtual world to extend learners' learning experiences[11]. Therefore, ubiquitous learning on smart campus must provide learners with opportunities to learn in their own environment, with another term, selecting and using the proper equipment for their work and with their living experiences.

The Development concept of SVCL and ubiquitous computing in educational scenarios is very broad and new challenge for research. This approach can also be defined as: processes supported by the use of smart and mobile device technology to integrate learners with their learning context and its surrounding environment, so that it can support an ongoing, contextualized and useful learning to the learner[12]. As for systems designed to support learning should have the following characteristics[13]:

- Permanency: The content of learning information remains except the learners purposely remove it.
- Accessibility: The content of learning information has been always available whenever the learners need to use it.
- Immediacy: The content of learning information can be retrieved with immediately by the learners.
- Interactivity: The learners can communicate with peers, lecturer, and experts with efficiently and effectively.
- Context-awareness: The learning environment can adapt to the learners like a real situation for providing adequate information for the learners.

Finally, the challenges of a ubiquitous computing system go beyond providing the information in any time or place, as they also aim to provide the right information at the right time and the right way. Considering these aspects, the system must be able to survive and play a key role.

*2.2. IM based smart virtual class learning using ubiquitous computing*

The Enhancement of smart devices based on mobile group communication technologies becomes one of the important factors that brings into new various services on campus. Internet Messenger (IM) can be used an alternative approach to smart campus based learning environment, IM has a series of attractive features, including multiform messaging (e.g., Document, image, audio, and video), instant access to messages (e.g., Timely transmission, recordation, storage media and retrieval of messages over the Internet), move quickly cross device sharing (e.g., PC, tablet, and smart phone)[6].

Interaction in IM provides a mobile group communication service among members of the campus community and divides the campus into small regions, if a group member send a message to a group, the member sends the message to the IM server and then the server forwards it towards the each of a group members. This interaction makes the IM as an essential component of the learning process. In the u-learning environment, interaction among participants is a crucial concern to lecturer and learner[14]. As a complex and multifaceted component, interaction provides a diverse range of functions in u-learning, and is vital for effectively improving learning[14]. Some researches indicated a positive effect of interaction on learners' academic achievement and performance as well as their satisfaction[15].

*2.3. IM to improve engagement and behavior*

The engagement concept of learner and lecturer is an important concern in in the learning process at the campus. Some literature used term of engagement to indicate readiness for devotes considerable energy to studying, desire and compulsion to actively participate in everyday campus activities involving things like attending to classes, participates actively in leaner organizations, adherence to lecturer's directives in the class like submit a report of progress or assignment[16][17]. Learner engagement includes both the academic and non academic activities and when get more engaged in campus experience, it tends to generate high quality learning[18].

The rapid development of IM like whatsapp, Catfiz, facebook messenger has sparked the creative incorporation of media into current pedagogical application and processes, this technology are rapidly moving beyond their original purpose and significantly increased, even impact on leaner's collaboration and engagement[19], At the lecture course, the lecturers devide instruction between a direct presentation and via IM, which is physically closer, provides a more individualized way of displaying information for learners, thus contributing to the improvement of learning efficiency and learner engagement when the specific instruction is aimed to encourage learner-centered activities[20].

This literature review examines the link between interactive internet messenger group to be used as a medium for ubiquitous learning in order to improve learner engagement and behavior on a smart campus. The concepts of cognitivism and constructivism were introduced as the foundation paradigms are important to this study. Secondary theories embodied the issues of learner motivation; learning styles; the special needs of learner on campus; and interactive learning in the campus setting. In research has shown collaboration, active participation and learner-centered and most importantly for this study is the model of ubiquitous learning using IM stimulates learner interest and attention leading to increased motivation and engagement during lessons. The college learner engagement, as evidenced by behavior during lectures, is an essential component in learning. Incorporation of technology into campus instruction not only kindles attentiveness learner, satisfies the accommodation of learner needs, and use instructional strategies that are consistent with the application of innovative technologies available today, but also complies country mandates about use of technology in education

## 3. Approach and methodology

In this section will be discussed about methodical approaches to modeling of IM based SVCL using ubiquitous computing to improve engagement and behavior for smart campus. In describing the methods used to conduct this research, the model design of IM based SVCL using ubiquitous computing, participants, data collection and analysis will be discussed and follows the guidelines of quantitative research technique[21]. In general, the main purpose of this research was designed and implement and showing empirical evidence about IM based SVCL using ubiquitous computing to improve engagement and behavior for learner–learner and learner–lecturer interactions on smart campus. To further enhance the credibility of the research, standardized observations of classes were conducted, Data collected from the source using a questionnaire that includes a semi-structured questions and express an opinion, and threats to validity were managed as closely as possible.

### *3.1. IM based SVCL using ubiquitous computing Model*

This modeling technique adopted to purpose development of scenarios and conceptual design of the u-learning environment as illustrated in fig. 1, the scenario shows that a learner can learn in the ubiquitous learning environment, Notifications and adaptive learning support messages are sent to learners by IM on the group, like illustrated in fig. 2. The learner can use the smartphone or mobile devices to perform learning tasks quickly. Moreover, the model of learning system can be recommended by lecturers to learner to consult on their learning problems by IM.

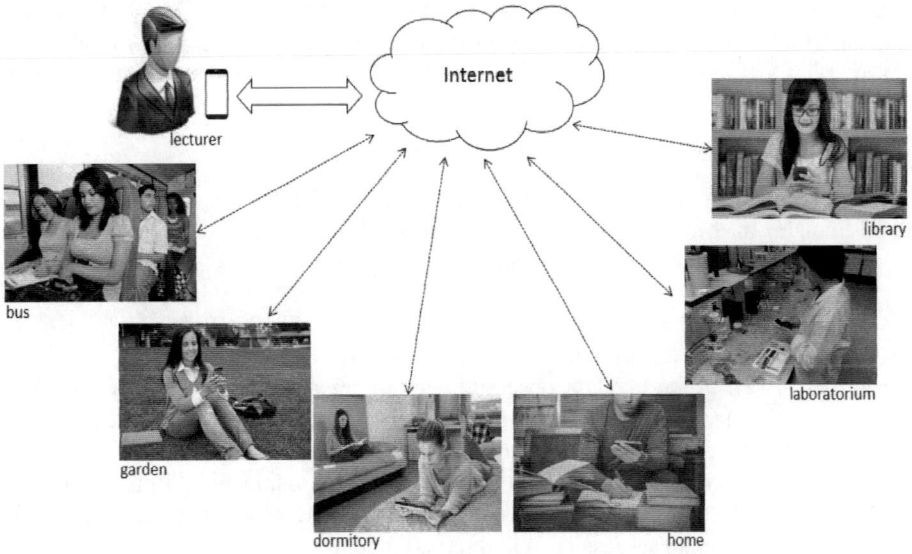


**Figure 1.** Scenario and conceptual design of the u-computing environment

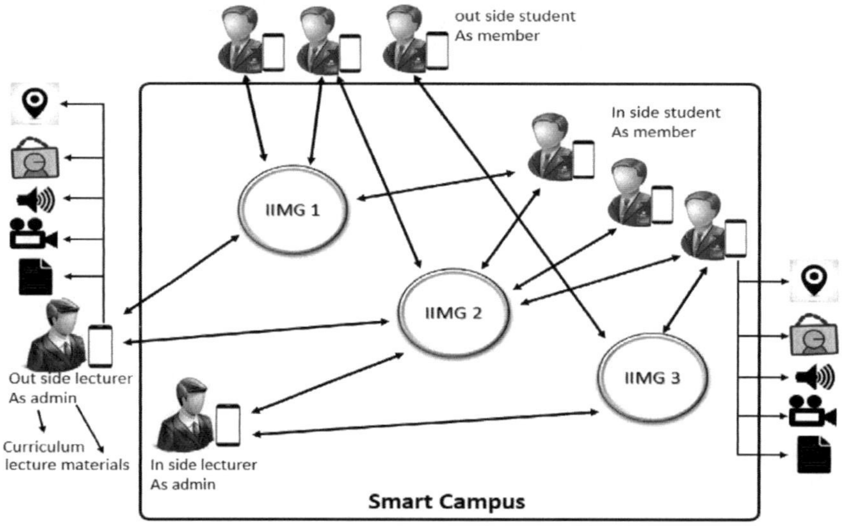


**Figure 2.** IM based SVCL using ubiquitous computing model

The platform based on IM provides learners and lecturer with an interactive learning environment using a combination of multimedia communications (e.g., Document, instant text, image, audio and video messages). It allows them to access, exchange and share their ideas so as to provide sufficient opportunities of interacting anywhere and anytime. In this model we designed a four-mode interactive mechanism for the platform. The first mode is to change a class for certain subjects into groups, lecturer as admin can assemble, organize and coordinate those groups by themselves, and the learner can join to group after registration and approved by the lecturer. The second mode is the scenario to post and circulate information using this platform, for example: lecturer transmit and distribute information like curriculum, syllabus, college contract and learning material. During online course, the learner can control what is distributed by the lecturer, the discussion pace and posting and demonstrate their own understanding or the result of the test or homework via text or visual media. The third mode is to track and record. When interaction exists between lecturer and learner, The IM platform automatically recorded all of feedback on every work from all learners and lecturer. The final mode is to conclude with brief remarks about the learners' knowledge acquirement and achievement via self-report based on the feedback.

### *3.2. Participant*

In 2015/2016, participants in this study were two lecturers of UIN Walisongo Semarang and 147 learners (80 Male, 67 Female) from five classes, they were enrolled form dakwah and communication faculty. The ages of learners ranged from 20 to 23 years and the class sizes ranged from 25 to 35 learners.

### *3.3. Data collection and analysis*

The investigation of this research using a quantitative same-subjects approach for data collection through systematic direct group or class observations combined with a post study survey to assist in addressing the results. Because this was a same group class, subjects were exposed to both conditions in the research, instruction without model use (Control class), followed by model use during lesson instruction (experiment class), and use of the IM based SVCL using ubiquitous computing was the independent variable in this investigation, the dependent variable, student engagement and behaviour evidenced by their behavior during lessons and interaction.

After the socialization of learning model for students in the mid lecture on December 2015, next, an initial questionnaire was given to the whole class of 147 learners before the start of the experiment to find out if they owned mobile devices and whatsapp. Those who did not have smart phones were encouraged to to buy or merge with a nearby friend while those without whatsapp applications were encouraged to download application and install them. While initially some students were slow in opening installs whatsapp application when they learnt from colleagues how they were easily accessed and communication with the group or their peer friend, they managed to create one for themselves. During the pilot program, daily activities of the group were facilitated by the lecturer that as a group administrator, moderator and driver to set the discussion or communication within the group, and researchers join in the group to monitor and determine the extent of use of the IM based SVCL using ubiquitous computing for engagement and collaboration between learner's communities, learner and their lecturer. Just before the end of the semester, the evaluation questionnaire was used to collect data from the students to measure the impact of model on their engagement and behavior. Each participant was provided a questionnaire consisting of 22 items was designed. The survey contained questions regarding learner self-perception of their engagement during model use. The results of attitude survey were converted into a numeric system allowing calculation of mean scores for each question and for each learner. There were five-point Likertscale items were used that ranged from 1 (strongly disagree) to 5 (strongly agree). Data aggregated into tables, from this information, an analysis was conducted to determine whether use of the model significantly affected learner engagement behaviors during classroom instruction, and whether the observed behavioral scores differed from the learners' perceptions of the influence of IM based SVCL using ubiquitous computing use on their behavior.

Measurement of learner attitudes towards IM based SVCL using ubiquitous computing was obtained using an existing attitude survey modified by the researcher. There were a number of questionnaires about learner attitudes toward technology. There is the Computer Attitude Questionnaire (CAQ) a 65-item Likert-scale type instrument for measuring of learners' attitudes was chosen by the researcher. Same item for measuring the importance and enjoyment of computer use were modified into an attitude survey to measure learner perceptions of the impact of the instructional technology being investigated, that is IM, and their perceptions about their enjoyment and its importance in groups instruction. This is a free instrument available online at https://iittl.unt.edu/content/computer-attitude-questionnaire-caq developed by The Technology Applications Center for Educator Development. There are 7 subscales measuring various learner attitude components, but only part 1 was used and it was modified by changing the word "computer" to " IM based SVCL using ubiquitous computing.

This instrument has been tested and used extensively by researchers. The reliability and validity data of CAQ are provided as part of the survey packages conducted in 1995 by Knezek and Christensen with a population of five hundred eightylearner and eight seventh and eighth grades in Texas to determine stability of measurement for the instrument. This attitude survey was modified by the researcher for comparative analysis of actual actual learner behavior and their own perception of their attitudes towards the model.

## 4. Result

This section presents the results obtained from the model defined and implemented in order to develop u-learning model and analyze its impact in real environment, the results of the research based on the data collected and presented. Results are analyzed and presented in various sections below;

- Do you have a smart phone?

This question required for detection of participants to indicate if they had a cell phone and if so which type (for example, smart phone or normal phone). As shown in Table 1, 100% of the participants had a mobile phone and 85% of those devices were smart phone with the capabilities of accessing the internet, sending instant or multimedia messages, and other applications.

**Table 1.** Mobile device ownership.

| Mobile device ownership | |
|---|---|
| 100% of the Students Own a Phone | |
| Smart phone 85% | Normal phone 15% |

- Do you have a whatsapp application?

With our research, it was important to find out the average number of participants that had whatsapp applications on their mobile phone. As represented in Table 2, the data show that 95% of the participants who have a smartphone, they also have a whatsapp application and 5% did not have and did not use whatsapp for personal reasons.

**Table 2.** Whatsapp application ownership.

| Mobile device ownership | |
|---|---|
| Yes | No |
| 87% | 13% |

- How often do you use your smart phone to chat with whatsapp application?

This question was directed at finding out how often students used their mobiles to access whatsapp for class-related matters like discussion with peer friends or consult with their lecturer before and after learning models included. It was important to find out the contribution of the learner while accessing whatsapp to ascertain the number of the learner who were active participation, or sometimes or those who were not participating at all. The results are presented in Table 3.

**Table 3.** Chatting with peer learner or lecture

| | | Regularly | Sometimes | Never did |
|---|---|---|---|---|
| Before | With lecturer | 5% | 11% | 84% |
| | With Peer learner | 66% | 23% | 11% |
| After | With lecturer | 71% | 21% | 8% |
| | With Peer learner | 76% | 17% | 7% |

- Learns perceptions about benefit using whasapp

The learner was asked to specify their level of agreements or disagreements on a scale to measure of subject attitude. Learner self-perception of personal attitudes of IM based SVCL using ubiquitous computing was collected using the modified CAQ. The results are presented in Table 4.

On the information accessed on the whtasapp as shown in table 4. 74% either agreed or strongly agreed that they are enjoy classroom instruction in the model of IM based SVCL using ubiquitous computing, while 11% neutral and 15% either disagreed or strongly disagreed, Similarly, 79% of participants agreed or strongly agreed if IM based SVCL using ubiquitous computing can make concentrate better on the lesson and 67% agreed or strongly agreed if they are know how to interaction in model they will be able to get a good job. It is inversely with they are who tired in the IM based SVCL using ubiquitous computing as shown in table 4, 14% either agreed or strongly agreed, while 8% neutral and 78% either disagreed or strongly disagreed.

On the issue of usefulness of IM based SVCL using ubiquitous computing for engagement (Table 4 Item 1, 3 to 11, 13 to 17, 19, 21 and 22) a large number of 78% of the participants believed that the IM based SVCL using ubiquitous computing helped them to engage and be informed about the module in and outside the classroom. 11% neutral position and 11% disagreed.

The results of this observation and investigation showed that the IM based SVCL using ubiquitous computing use has a positive effect on the behavior of all learners, thus, on their engagement with the ubiquitous learning environment. Overall, learners were aware of the positive impact of IM based SVCL using ubiquitous computing use on their engagement in ubiquitous classroom instruction. They regard positively, and this was evidenced by their task behavioral improvement. The investigation data showed general improvement in student behavior which translates into improved student engagement

**Table 4.** IM based SVCL using ubiquitous computing Attitude Survey Results

| No | QUIZ | Strongly agree | Agree | Normal | Disagree | Strongly disagree |
|---|---|---|---|---|---|---|
| 1 | I enjoy classroom instruction in the IM based SVCL model using ubiquitous computing | 32% | 42% | 11% | 7% | 8% |
| 2 | I am tired in the the IM based SVCL model using ubiquitous computing | 7% | 7% | 8% | 46% | 32% |
| 3 | I will be able to get a good job if I know how to interaction in the IM based SVCL model using ubiquitous computing | 33% | 34% | 21% | 9% | 3% |
| 4 | I can concentrate better on the lesson in the IM based SVCL model using ubiquitous computing | 38% | 41% | 12% | 7% | 2% |
| 5 | I would work harder if in the IM based SVCL model using ubiquitous computing | 25% | 49% | 13% | 7% | 6% |
| 6 | I know that the IM based SVCL model using ubiquitous computing gives me more opportunities to learn new things | 42% | 37% | 12% | 7% | 2% |
| 7 | I understand the lesson better when the the IM based SVCL model using ubiquitous computing is used | 36% | 39% | 12% | 9% | 4% |
| 8 | I believe that if more teachers used the the IM based SVCL model using ubiquitous computing, I would enjoy school more | 38% | 38% | 11% | 7% | 6% |
| 9 | I believe that it is important for me to be able to use technologies such as the the IM based SVCL model using ubiquitous computing | 47% | 35% | 9% | 6% | 3% |
| 10 | I can learn new things when the the IM based SVCL model using ubiquitous computing is used | 41% | 36% | 9% | 9% | 5% |
| 11 | I feel comfortable in the IM based SVCL model using ubiquitous computing | 53% | 32% | 5% | 8% | 2% |
| 12 | I think lessons take longer in the IM based SVCL model using ubiquitous computing | 3% | 8% | 3% | 36% | 50% |
| 13 | In the IM based SVCL model using ubiquitous computing does not scare me | 43% | 35% | 10% | 8% | 4% |

| | | | | | | |
|---|---|---|---|---|---|---|
| 14 | In the IM based SVCL model using ubiquitous computing does not make me nervous | 51% | 31% | 9% | 6% | 3% |
| 15 | In the IM based SVCL model using ubiquitous computing not difficult | 38% | 46% | 8% | 5% | 3% |
| 16 | In the IM based SVCL model using ubiquitous computing, I want to work and learn whenever I can | 53% | 31% | 5% | 8% | 3% |
| 17 | I work very hard on my schoolwork | 33% | 34% | 21% | 8% | 4% |
| 18 | I do not try hard in school | 5% | 9% | 14% | 46% | 26% |
| 19 | I pay attention in class | 27% | 47% | 13% | 8% | 5% |
| 20 | When I am in class, I just act as if I'm working | 4% | 20% | 46% | 21% | 9% |
| 21 | It is important to do my best in campus | 48% | 36% | 5% | 8% | 3% |
| 22 | I always try to complete my assignments | 47% | 35% | 8% | 7% | 3% |

*The scoring scale for questions 2, 12, 18, and 20 was reversed

## 5. Conclusion and Recommendations

The result of this research shows that the IM based SVCL using ubiquitous computing emphatically positive and can improve significantly on learner's engagement and behavior. Our analysis indicates that learners and lecturer, highly engaged in the learning process in ways transcended traditional classroom activities. The experimental evidence shows that IM based SVCL using ubiquitous computing can be used as an educational tool in smart campus to help learners to engage and collaborate. Therefore, we conclude that, IM based SVCL using ubiquitous computing could be effective for student to engage as well as succeed in their campus activities.

For future work, the opportunities exist for further studies in this area. Corporation between u-learning model with mobile learning or e-learning it is possible to be done to produce models more smart in order to create a smart campus dream, so the learning activities between students and professors increasingly interwoven dynamic and output is expected to have high competitiveness.

## References

[1] I Tsai and C Yeh 2015 Using a Novel Model to Measure Smart Campus Service Quality : An Empirical Case *IJEETDM* vol. 1 no. 4 pp. 175–181
[2] G D Chen C K Chang and C Y Wang 2008 Ubiquitous learning website: Scaffold learners by mobile devices with information-aware techniques *Comput. Educ*, vol. 50 no. 1 pp. 77–90
[3] I. Smit 2015 WhatsApp with learning preferences? *Proc. - Front. Educ. Conf. FIE*, vol. 2015 Decem
[4] Y T Chuang 2015 SSCLS: A smartphone-supported collaborative learning system *Telemat. Informatics* vol. 32 no. 3, pp. 463–474
[5] X Dong X Kong F Zhang Z Chen and J Kang 2016 OnCampus: a mobile platform towards a smart campus *Springerplus* vol. 5 no. 1 p. 974
[6] R G R Gu S Z S Zhang and J H J He 2009 Instant Messaging Tool Based Online Question-and-Answer System: An Effort for a New Group Learning Environment *Int. Conf. Comput. Intell. Softw. Eng.* pp. 1–4
[7] Y Atif and S Mathew and A Lakas 2015 Building a smart campus to support ubiquitous learning *J. Ambient Intell. Humaniz. Comput.* vol. 6 no. 2 pp. 223–238
[8] Y Atif and S Mathew 2013 A Social Web of Things Approach to a Smart Campus Model *IEEE Int. Conf. Green Comput. Commun. IEEE Internet Things IEEE Cyber Phys. Soc. Comput*

[9] Z Xinyou and O Toshio 2011 Adaptive multimedia content delivery for context aware u-learning *J.Mobile Learn. Organ.* vol. 5 no. 1 pp. 46–63
[10] G Fischer 2012 Context-Aware Systems : The ‘ Right ’ Information , at the ‘ Right ’ Time , in the ‘ Right ’ Place , in the ‘ Right ’ Way , to the ‘ Right ’ Person, *ACM*
[11] J L Shih H C Chu G J Hwang and Kinshuk 2011 An investigation of attitudes of students and teachers about participating in a context-aware ubiquitous learning activity *Br. J. Educ. Technol.* vol. 42 no. 3 pp. 373–394
[12] L H Wong M Milrad and M Specht 2015 Seamless learning in the age of mobile connectivity *Seamless Learn. Age Mob. Connect.* pp. 1–500
[13] J Yahya Ahmad 2010 The definition and characteristics of ubiquiteous learning *Int. J. Educ. Dev. using Inf. Commun. Technol.* vol. 6 no. 1 pp. 117–127
[14] D S Bental E Jennings E Papadopoulou N K Taylor M H Williams F R Blackmun I S Ibrahim M Y Lim I Mimtsoudis and S W Whyte 2015 Smartening Up the Student Learning Experience with Ubiquitous Media *ACM Trans. Multimed. Comput. Commun. Appl.* vol. 12 no. 1s pp. 1–23
[15] S Molomer and S Trausan-Matu 2014 Analysis of interactions in a collaborative learning session of instant messaging *Proc. - RoEduNet IEEE Int. Conf.* pp. 5–8
[16] H Coates 2006 student Engagement in Campus-Based and Online Education *Student Engagem. Campus-Based Online Educ. Univ. Connect.* vol. 9780203969 pp. 1–212
[17] E Ivala D Gachago J Condy and A Chigona 2013 Enhancing student engagement with their studies: A digital storytelling approach *Creat. Educ.* vol. 4 no. 10 pp. 82–89
[18] K Nelson J Clarke I Stoodley and T Creagh 2014 *Establishing a framework for transforming student engagement, success and retention in higher education institutions.* http://studentengagementmaturitymodel.net/
[19] J L Badge N F W Saunders and A J Cann 2012 Beyond marks: New tools to visualise student engagement via social networks *Res. Learn. Technol.* vol. 20 no. 1 pp. 1–14
[20] Y Kim S Jeong Y Ji S Lee K H Kwon and J W Jeon 2015 Smartphone response system using twitter to enable effective interaction and improve engagement in large classrooms *IEEE Trans. Educ.*, vol. 58, no. 2, pp. 98–103
[21] J W Creswell 2014 *Research Design: Qualitative, Quantitative, and Mixed Methods Approaches*, 4 Edition. SAGE Publications Inc. United State of America

IOP Conf. Series: Journal of Physics: Conf. Series **855** (2017) 012057 doi:10.1088/1742-6596/855/1/012057

# Satisfiability modulo theory and binary puzzle

**Putranto Utomo**

Department of Mathematics and Computer Science, Eindhoven University of Technology; and Faculty of Mathematics and Natural Science, Sebelas Maret University

E-mail: p.h.utomo@tue.nl

**Abstract.** The binary puzzle is a sudoku-like puzzle with values in each cell taken from the set $\{0, 1\}$. We look at the mathematical theory behind it. A solved binary puzzle is an $n \times n$ binary array where $n$ is even that satisfies the following conditions:

(1) No three consecutive ones and no three consecutive zeros in each row and each column,
(2) Every row and column is balanced, that is the number of ones and zeros must be equal in each row and in each column,
(3) Every two rows and every two columns must be distinct.

The binary puzzle had been proven to be an NP-complete problem [5].

Research concerning the satisfiability of formulas with respect to some background theory is called satisfiability modulo theory (SMT). An SMT solver is an extension of a satisfiability (SAT) solver. The notion of SMT can be used for solving various problem in mathematics and industries such as formula verification and operation research [1, 7].

In this paper we apply SMT to solve binary puzzles. In addition, we do an experiment in solving different sizes and different number of blanks. We also made comparison with two other approaches, namely by a SAT solver and exhaustive search.

## 1. Introduction

Recall the definition of binary puzzle in the abstract. Figure 1 is an example of initial setting of a binary puzzle. The unique solution satisfying all three conditions is given in Figure 2. Binary puzzles can be seen as constrained arrays [13]. One can also see this array from an erasure correcting point of view [12]. This paper focuses on solving binary puzzles.

| | | | | | | | |
|---|---|---|---|---|---|---|---|
| | 0 | | | | | | |
| | | | 1 | | 1 | | 0 |
| | | 0 | | | | | |
| | 1 | | | | | | |
| | | | | 1 | | | |
| | 0 | | | | 1 | | |
| | 0 | | | 0 | | | |
| | | | | 0 | | 0 | |

Figure 1: Unsolved Puzzle

| | | | | | | | |
|---|---|---|---|---|---|---|---|
| 1 | 0 | 1 | 0 | 1 | 0 | 1 | 0 |
| 0 | 1 | 0 | 1 | 0 | 1 | 1 | 0 |
| 1 | 0 | 0 | 1 | 1 | 0 | 0 | 1 |
| 0 | 1 | 1 | 0 | 0 | 1 | 0 | 0 |
| 0 | 1 | 0 | 1 | 1 | 0 | 1 | 1 |
| 1 | 0 | 1 | 0 | 1 | 1 | 0 | 0 |
| 1 | 0 | 0 | 1 | 0 | 0 | 1 | 1 |
| 0 | 1 | 1 | 0 | 0 | 1 | 0 | 1 |

Figure 2: Solved Puzzle

We have solved the binary puzzle by three methods: exhaustive search, system of polynomials, and satisfiability (SAT) theory [14]. In this paper, we consider a new approach by means of satisfiability modulo theory (SMT). Using this approach, we improve a lot in terms of the size of the puzzle.

## 2. Satisfiability modulo theory (SMT)

One of the most fundamental problems in mathematics and computer science is the satisfiability problem, which deals with checking and finding a solution of a formula with some constraints [6]. Some of these problems can be expressed as a propositional satisfiability (SAT) problem which can be solved efficiently by a modern SAT solver. But, some other problems are too complex to be expressed in propositional SAT problem and more simple and easily expressed in first or higher order logic, which can include functions, and non-Boolean variables. The more complex a problem, the more difficult to be solved.

One way to deal with this problem is usually by focusing on fragments of first order logic that are restricted semantically, by constraining the interpretation of certain symbols to some logical background theory, such as the theory of equality, integer number, real number, bitvector, and so on. The satisfiability problem with these restrictions is called satisfiability modulo theory (SMT) [3].

### *2.1. SMT solver*

Unlike SAT solvers, which are more mature and mostly have the same input and the same underlying theory, an SMT solver has more variance in terms of background theory behind the screen, depending on the problem it intends to solve. One of the reasons to make a restriction is to boost the performance of the solver. Nevertheless, there is an international initiative on SMT called SMT-LIB [2]. Two of its main purposes are to create standard descriptions of theories in SMT and to develop common input and output languages for SMT solvers.

In general, there are two approaches in designing an SMT solver, the lazy and the eager approach. The lazy approach is characterized by a combination of one or more theory solvers with a conflict-driven clause learning (CDCL) SAT engine, meanwhile the eager approach uses any technique so that the SMT problem is reduced to a propositional satisfiability (SAT) problem [3].

### *2.2. Related theories*

We consider several specific theories that are often used in SMT.

The theory of equality with uninterpreted function (EUF), also known as the empty theory is the most general case of theory because there are no restrictions on the model. This problem can be solved in polynomial time using the congruence closure algorithm [1].

The next theory we would like to introduce is the arithmetic theory. The arithmetic theory is often divided into three categories, integer arithmetic, real arithmetic, and difference logic. The symbols used in the real and integer arithmetic are in $\{+, -, *, \geq\}$. The linear integer arithmetic problem, also known as Presburger arithmetic, can be solved using the Presburger algorithm. For real arithmetic, The simplex based algorithm is often used in practice for solving the satisfiability problem [1]. As stated in [3], difference logic expressions are in the form $x - y \bowtie c$, where $\bowtie$ is an element of $\{=, \leq, \geq\}$, $c$ is a constant, and $x$ and $y$ are variables.

Another theory we would like to mention is the theory of bitvector. The function and predicate symbol in this theory usually include extraction, concatenation, bitwise logical operators, and arithmetic operations [1].

For one who is interested in other theories such as arrays, strings, finite sets, we refer to [3] or any other survey on satisfiability modulo theory.

### *2.3. Combining theory*

We often deal with a problem involving many theories at once. In solving this kind of problem, we need to combine the theory solver for the theories $\mathcal{T}_1, \ldots, \mathcal{T}_n$ into one theory solver that solves the satisfiability modulo combination theory $\mathcal{T}_1 \oplus \cdots \oplus \mathcal{T}_n$.

The idea for combination methods can be traced back to Nelson and Oppen. We refer to [3] for its procedure. One of the requirements using this kind of combining methods is that $\mathcal{T}_1, \ldots, \mathcal{T}_n$ should be disjoint in the sense that all theories do not share function symbols.

One example of a framework for combining theories is the delayed theory combination (DTC). It is based on the conflict-driven clause learning (CDCL) SAT engine, which is also known as DPLL($T$)[1] algorithm, where $T$ is a theory solver being used. The idea of DTC is to make an independent theory solver for each theory and extend the interface of the SAT engine so that it can interact with each theory solver using the Nelson-Oppen style [3].

Another way to combine one or more theories is by using Ackermann's expansion. In this method, every function in the input formula is replaced by a fresh variable and their congruence constraints. The new formula will then be equisatisfiable and contain no uninterpreted function symbols [1].

We choose Yices 2 as the solver for the binary puzzle, since it focuses on satisfiability and model generator and that is exactly our problem we want to solve. The comparison and performance of Yices can be seen in [10, 9, 8, 4].

The design of the SMT solver in Yices 2 follows the DTC model. One can find the details in [9].

## 3. Solving a binary puzzle with SMT

Since the bitvector representation provides a more compact representation and often allows a more efficient solution [1], we use the bitvector representation for the problem of the binary puzzle. Suppose we want to solve an $n \times n$ binary puzzle. First, we define $2n$ bitvector variables for the columns and the rows, where each bitvector consists of $n$ bits.

For the $1^{st}$ constraint, it is more natural to express every three consecutive cells as a number in their binary representation, for example, 011 is equal to 3 and 101 is equal to 5. Then we express the first constraint using bitvector arithmetic operator `less than` and `greater than` as explained in [9] as follows. Let $\mathbf{x} = x_1x_2x_3$ be any three consecutive cells, then $\mathbf{x}$ must be less than $111 = 7$ and $\mathbf{x}$ must be greater than $000 = 0$.

For the $2^{nd}$, balancedness constraint, we use integer arithmetic, since we need to count the number of ones in each row and each column to check whether the number of ones is equal to the number of zeros.

[1] DPLL is an abbreviation of Davis, Putnam, Logemann, and Loveland

For the $3^{rd}$ constraint, we use the operator `distinct` defined in [9] for all the columns and for all the rows so that there is no identical pair of columns and also no identical pair of rows.

Finally, we also need to impose the additional constraint to the row and column variables, so that they are consistent in the sense that the $i^{th}$ bit on column $j$ should be equal to the $j^{th}$ bit on row $i$.

## 4. Experiment

In this section, we will briefly discuss the simulation of solving the binary puzzle. Simulation has been done under 64 bit Debian 8, with hardware specification: Intel(R) Core(TM) i7-4700MQ CPU @ 2.40GHz, 10GB RAM. The software we used for the experiment are Sagemath 7.2 as the wrapper and front-end programming language; and Yices 2.5 as the SMT solver.

We prepared a database of solved binary puzzle with size ranging from $4 \times 4$ until $100 \times 100$. For each size, we have 6 distinct solved puzzles. In this research, we also interested whether there is a differences in the computation time for different puzzles. In our experiment, we hardly see any difference in the computation time, as we can see in Figure 3. In this figure, different colors represent different puzzles.

The procedure for acquiring a partially solved binary puzzle is as follow. Suppose we want an $n \times n$ puzzle with $k$ blanks, where $0 < k \leq n^2$ and $n = 2m$. First, we pick a particular puzzle from the database and then choose $k$ random cells from the $n \times n$ arrays to be erased.

We compare the execution time with the two other approaches for solving the binary puzzle, namely exhaustive search and propositional SAT solver. One can refer to [14] for the details of this approach.

In this research, we also did an experiment regarding the distribution of uniqueness. For each number of blanks, we calculate the ratio of uniqueness, that is the number of experiment having a unique solution divided by the total experiment. For $40 \times 40$, we refer to Figure 6. If the ratio is above 0.5, then the puzzle is more likely to have a unique solution. Table 1 shows when the ratio hits 0.5 for different sizes.

## 5. Conclusion

The result from this experiment is quite surprising, since it can solve binary puzzles beyond our expectation.

We have several remarks:

(i) The limit of the SAT solver with the current hardware specification is $14 \times 14$ since the precomputation blows up exponentially.

(ii) The computation time using the SMT solver is an exponential function $y = e^{-3.5597+0.0028x}$ in terms of the number of blanks. See Figure 5.

(iii) The comparison of different solvers and different sizes with 70% of blanks is given in Table 2. In this case, exhaustive search will fail to find a solution for $16 \times 16$ puzzle in a reasonable time compared to SMT solver

(iv) The exhaustive search methods can catch up with the SMT solver for very small portion of blanks, but as the blanks get more, the exhaustive search cannot find a solution in a reasonable time compared to the SMT solver. For the case of 10% of blanks, the limit for exhaustive search is $76 \times 76$, and comparison can be seen in Figure 4.

(v) Since the number of blanks does not significantly affect the size of the solver input, it is obvious that the precomputation time is not affected by the increase of the blanks, as we can see in Figure 7.

| Size | Number of blanks |
|---|---|
| $4 \times 4$ | 9 |
| $6 \times 6$ | 19 |
| $8 \times 8$ | 30 |
| $10 \times 10$ | 44 |
| $12 \times 12$ | 53 |
| $14 \times 14$ | 67 |
| $16 \times 16$ | 80 |
| $18 \times 18$ | 92 |
| $20 \times 20$ | 110 |
| $22 \times 22$ | 126 |
| $24 \times 24$ | 133 |
| $26 \times 26$ | 150 |
| $28 \times 28$ | 164 |
| $30 \times 30$ | 188 |
| $32 \times 32$ | 209 |
| $34 \times 34$ | 238 |
| $36 \times 36$ | 238 |
| $38 \times 38$ | 248 |
| $40 \times 40$ | 279 |

Table 1: Number of blanks where the ratio of uniqueness is equal to 0.5.

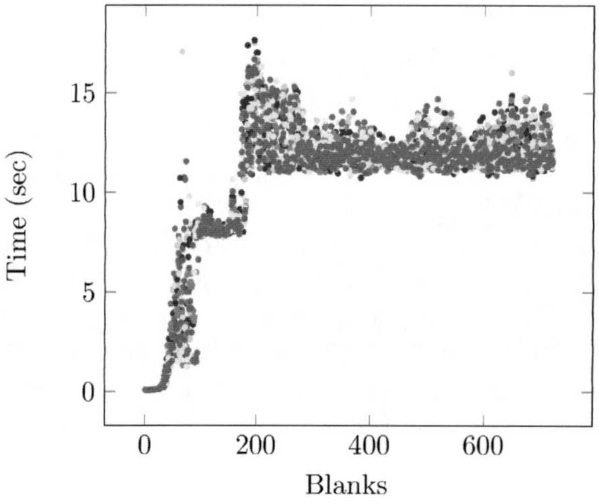


Figure 3: Comparisson time between different puzzle for $40 \times 40$ puzzle.

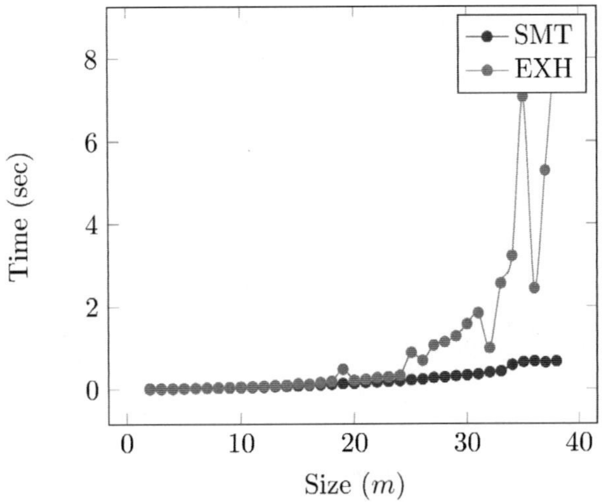


Figure 4: Comparison time between SMT solver and exhaustive search.

| Size | SAT (sec) | EXH (sec) | SMT (sec) |
|---|---|---|---|
| $4 \times 4$ | 0.007 | 0.022 | 0.015 |
| $6 \times 6$ | 0.075 | 0.179 | 0.017 |
| $8 \times 8$ | 0.690 | 0.149 | 0.018 |
| $10 \times 10$ | 5.615 | 0.526 | 0.021 |
| $12 \times 12$ | 39.464 | 53.986 | 0.028 |
| $14 \times 14$ | 244.815 | 129.265 | 0.031 |
| $\vdots$ | - | - | $\vdots$ |
| $36 \times 36$ | - | - | 0.268 |
| $38 \times 38$ | - | - | 0.396 |
| $40 \times 40$ | - | - | 0.448 |

Table 2: Comparison of different solver and different sizes

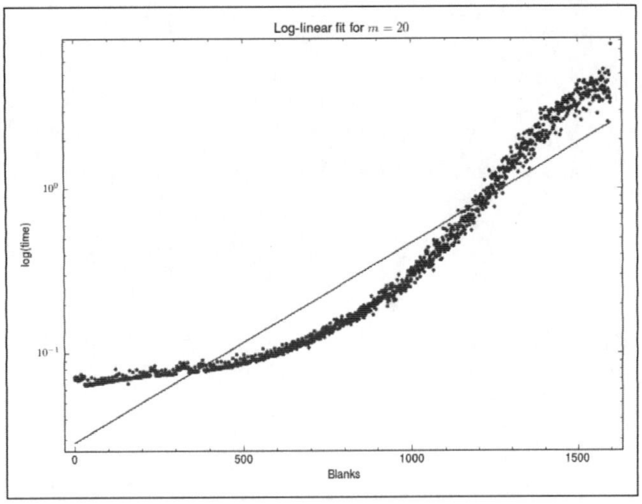


Figure 5: Number of blanks vs. computation time for SMT solver for $40 \times 40$ puzzle (log. scale)

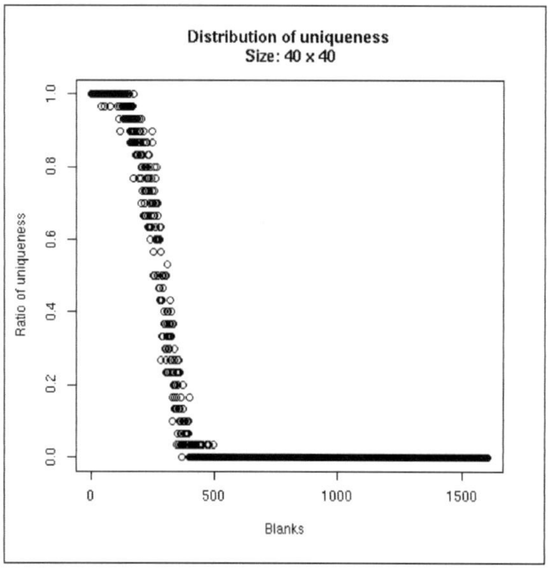


Figure 6: Number of blanks vs. ratio of uniqueness

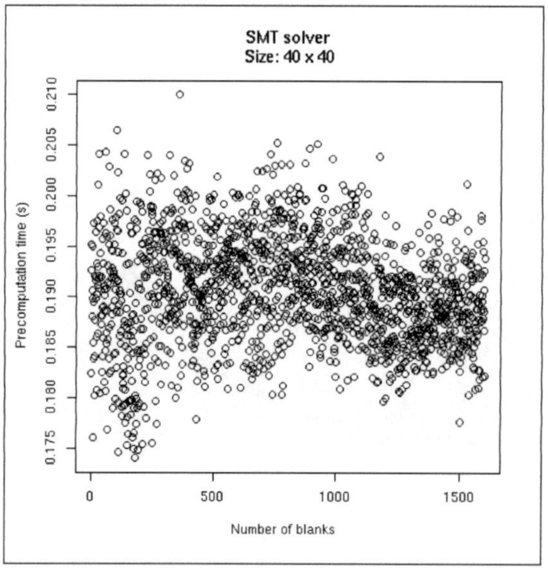


Figure 7: Number of blanks vs. precomputation time for SMT solver

## References

[1] Barrett, Clark and Sebastiani, Roberto and Seshia, Sanjit A. and Tinelli, Cesare. Satisfiability Modulo Theories, Handbook of Satisfiability, chapter 12, pages 737-797, 2008. `https://people.eecs.berkeley.edu/~sseshia/pubdir/SMT-BookChapter.pdf`

[2] Barrett, Clark and Fontaine, Pascal and Tinelli, Cesare. The Satisfiability Modulo Theories Library (SMT-LIB) `www.SMT-LIB.org`

[3] Barrett, Clark and Tinelli, Caesar. Satisfiability Modulo Theories, Handbook on Model Checking. Springer, 2017 (to appear).
Preprint: `http://homepage.cs.uiowa.edu/~tinelli/papers/BarTin-14.pdf`

[4] Blair, Will and Ghalili, Nikka. When Z3 Met Yices,
`http://cs-people.bu.edu/wdblair/CS512/technical-report.pdf`

[5] De Biasi, Marzio. Binary puzzle is NP-complete, `http://nearly42.org`

[6] De Moura, Leonardo and Bjørner, Nikolaj Satisfiability Modulo Theories: An Appetizer. Formal Methods: Foundations and Applications: 12th Brazilian Symposium on Formal Methods, SBMF 2009 Gramado, Brazil, August 19-21, 2009, pages 23-36. `http://dx.doi.org/10.1007/978-3-642-10452-7_3`

[7] De Moura, Leonardo and Bjrner, Nikolaj. Satisfiability modulo theories: introduction and applications. Commun. ACM 54, 9 (September 2011), 69-77. `http://dx.doi.org/10.1145/1995376.1995394`

[8] Dutertre, Bruno and De Moura, Leonardo. Integrating simplex with DPLL($T$),
CSL Technical Report SRI-CSL-06-01. May 23, 2006

[9] Dutertre, Bruno. Yices 2.2, Computer Aided Verification: 26th International Conference, CAV 2014, Held as Part of the Vienna Summer of Logic, VSL 2014, Vienna, Austria, July 18-22, 2014. Proceedings, pages 737-744

[10] Høfler, Andrea. SMT Solver Comparison,
`http://spreadsheets.ist.tugraz.at/wp-content/uploads/2015/06/DS_Hoefler.pdf`

[11] Soos, Mate. CryptoMiniSat, `http://www.msoos.org/cryptominisat2/`

[12] Utomo, Putranto and Pellikaan, Ruud. Binary Puzzles as an Erasure Decoding Problem. *Proc. 36th WIC Symp. on Information Theory in the Benelux*, pp. 129–134 May 2015. `http://www.w-i-c.org/proceedings/proceedings_SITB2015.pdf`.

[13] Utomo, Putranto and Pellikaan, Ruud. On The Rate of Constrained Arrays. *Proc. IndoMS International Conference on Mathematics and Its Applications (IICMA) 2015*, 3 November 2015. `http://www.win.tue.nl/~ruudp/paper/75.pdf`

[14] Utomo, Putranto and Makarim, Rusydi. Solving the Binary Puzzle Proc. 22nd Conference on Application of Computer Algebra, August 2016.
Book of abstract: `http://www.mathematik.uni-kassel.de/ACA2016/docs/ACAproc.pdf`

# On The Partition Dimension of $C_m + P_n$ Graph

**Hidra Vertana, Tri Atmojo Kusmayadi**
Department of Mathematics, Faculty of Mathematics and Natural Sciences, Universitas Sebelas Maret, Surakarta, Indonesia

E-mail: hidravertana@gmail.com, tri.atmojo.kusmayadi@gmail.com

**Abstract.** Let $G$ be a connected graph with $V(G) = \{v_1, v_2, \ldots, v_i\}$ and $E(G) = \{e_1, e_2, \ldots, e_j\}$, where $V(G)$ is vertex set and $E(G)$ is edge set. If $S \subseteq V(G)$ and $v \in V(G)$, then the distance between $v$ and $S$ is defined by $d(v, S) = min\{d(v, x)|x \in S\}$. For an ordered $k$-partition $\Pi = \{S_1, S_2, \ldots, S_k\}$ of $V(G)$, the representation of $v$ with respect to $\Pi$ is $r(v|\Pi)$ with $r(v|\Pi) = (d(v, S_1), d(v, S_2), \ldots, d(v, S_k))$. If the representation of $v \in V(G)$ with respect to $\Pi$ are distinct, so $\Pi$ is called a resolving partition of $V(G)$. The minimum cardinality of resolving partition $\Pi$ is called a partition dimension of $G$, denoted by $pd(G)$. In this paper, we study the partition dimension of a $C_m + P_n$ graph. $C_m + P_n$ graph is a graph formed from join operation of cycle graph $C_m$ with order $m \geq 3$ and path $P_n$ with order $n \geq 2$. $C_m + P_n$ is the union $C_m \cup P_n$ together with all edges $u_a v_b$, for $u_a \in V(C_m)$ and $v_b \in V(P_n)$ with $1 \leq a \leq m$ and $1 \leq b \leq n$. We obtain the partition dimension of $C_m + P_n$ graph is $pd(C_3 + P_n) = g$ where $g$ is the smallest positive integer such that $n \leq 5g - 12$ for $g = 5$ and $n \leq \frac{g^3 - 7g^2 + 20g - 18}{2}$ for $g \geq 6$, and $pd(C_q + P_n) = min\{p + f, r + t, x + y\}$ for $q \geq 4$ and $n \geq 2$ where $p, f, r, t, x$ and $y$ are some positive integers related to the number of partition classes containing vertices of $C_q$ and $P_n$.

## 1. Introduction

Graph theory is a branch of mathematics that developed rapidly. In daily life, the graph theory is used to solve some problems with vertex as discrete objects and edge as the relationship between both of them. One of the interesting concepts in graph theory is so called the partition dimension. The partition dimension of a graph was introduced by Chartrand et al. [1] in 1998. The partition dimension is a variation of the metric dimension introduced by Slater in 1975 (Chartrand et al. [3]).

Let G be a connected graph. For a subset $S$ of $V(G)$ and a vertex $v$ of $G$, the distance between $v$ and $S$ is defined as $d(v, S) = min\{d(v, x)|x \in S\}$. For an ordered $k$-partition $\Pi = \{S_1, S_2, \ldots, S_k\}$ of $V(G)$ and a vertex $v$ of $G$, the representation $v$ with respect to $\Pi$ is $k$-vectors $r(v|\Pi) = (d(v, S_1), d(v, S_2), \ldots, d(v, S_k))$. The partition $\Pi$ is called a resolving partition if the $k$-vectors $r(v|\Pi)$ are distinct, for every $v \in V(G)$. The minimum cardinality of the $k$-partition of $V(G)$ is called the partition dimension of $G$ denoted by $pd(G)$. (Chartrand et al. [2])

Many researches on the partition dimension have applied for specific graph classes. In 2000, Chartrand et al. [2] determined the partition dimension of some graph classes which result $pd(G) = 2$ if and only if $G = P_n$, $pd(G) = n$ if and only if $G = K_n$, and $pd(G) = 3$ if $G = C_n$ for $n \geq 3$. In 2007, Tomescu et al. [5] formed the partition dimension of a wheel graph. Hidayat [4] in 2015 determined the partition dimension of some graph classes, one of them is a

$K_1 + (P_m \odot K_n)$ graph. In this paper, we determine the partition dimension of a $C_m + P_n$ graph for $m \geq 3$ and $n \geq 2$.

## 2. Partition Dimension of $C_m + P_n$ Graph

Let $C_m + P_n$ be a graph formed from join operation of cycle graph with order $m$ ($C_m$) and path with order $n$ ($P_n$). $C_m + P_n$ is a graph with $V(C_m + P_n) = V(C_m) \cup V(P_n)$ and $E(C_m + P_n) = E(C_m) \cup E(P_n) \cup \{u_a v_b | u_a \in V(C_m)$ and $v_b \in V(P_n)\}$ with $1 \leq a \leq m$ and $1 \leq b \leq n$. The vertex set of $C_m + P_n$ graph is $V(C_m + P_n) = \{u_1, u_2, \ldots, u_m, v_1, v_2, \ldots, v_n\}$ and the number of vertices in a $C_m + P_n$ graph is $|V(C_m + P_n)| = m + n$ for $m \geq 3$ and $n \geq 2$. For example of $C_m + P_n$ graph is in Figure 1.

$C_m + P_n$ :

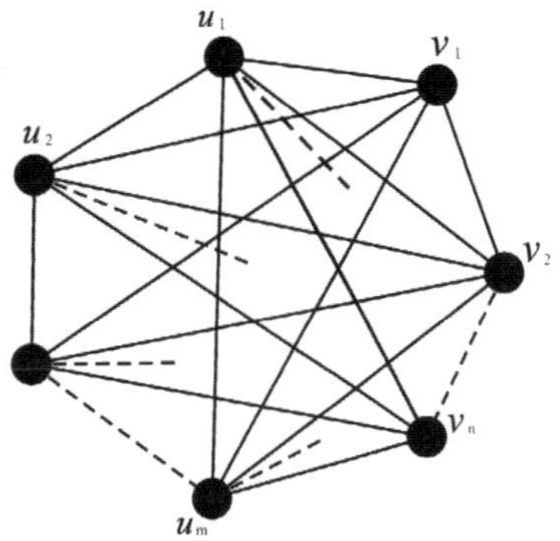


**Figure 1.** $C_m + P_n$ Graph

**Lemma 2.1** *Let $C_3 + P_n$ be a graph with $n \geq 2$, and let $\Pi$ be a resolving partition of $C_3 + P_n$. If vertex $u_1, u_2, u_3 \in V(C_3)$, $v_b \in V(P_n)$ for $b \in [1, n]$, $u_1, v_c \in S_1$, $u_2, v_d \in S_2$, $u_3, v_e \in S_3$ $(1 \leq c \neq d \neq e \leq n)$, and $v_y \in S_j$ for vertex $v_y \in V(P_n) \setminus \{v_c, v_d, v_e\}$ and $j = 4, 5, \ldots, g$ then $|S_i| \leq 1 + \binom{g-3}{0} + \binom{g-3}{1}$ for $i = 1, 2, 3$, $g = 5$ and $|S_i| \leq 1 + \binom{g-3}{0} + \binom{g-3}{1} + \binom{g-3}{2}$ for $i = 1, 2, 3$, $g \geq 6$.*

**Proof.** Let $\Pi = \{S_1, S_2, \ldots, S_g\}$ be a resolving partition of $C_3 + P_n$. If vertex $u_1, u_2, u_3 \in V(C_3)$, $v_b \in V(P_n)$ for $b \in [1, n]$, $u_1, v_c \in S_1$, $u_2, v_d \in S_2$, $u_3, v_e \in S_3$ $(1 \leq c \neq d \neq e \leq n)$, and $v_y \in S_j$ for vertex $v_y \in V(P_n) \setminus \{v_c, v_d, v_e\}$ and $j = 4, 5, \ldots, g$, then we obtain the representation $r(u_1|\Pi) = (0, 1, 1, \ldots, 1)$, $r(u_2|\Pi) = (1, 0, 1, \ldots, 1)$, $r(u_3|\Pi) = (1, 1, 0, \ldots, 1)$, and $r(v_c|\Pi) = (0, 1, 1, \ldots)$. Since the diameter of $C_3 + P_n$ is 2, then other than the three positions of $r(v_c|\Pi)$ can be filled by 1 or 2. If $g = 5$ then there are at most one 1 other than the first three positions, the rest can be filled by 2. If $g \geq 6$ then there are at most two 1's other than the first three positions, then the rest can be filled by 2's. Because vertex $u_1$ has a single representation, so there are at most $1 + \binom{g-3}{0} + \binom{g-3}{1}$ distinct representation for $g = 5$ or at most $1 + \binom{g-3}{0} + \binom{g-3}{1} + \binom{g-3}{2}$ distinct representation for $g \geq 6$. Thus, we obtain $|S_i| \leq 1 + \binom{g-3}{0} + \binom{g-3}{1}$ for $i = 1, 2, 3$, $g = 5$ and $|S_i| \leq 1 + \binom{g-3}{0} + \binom{g-3}{1} + \binom{g-3}{2}$ for $i = 1, 2, 3$, $g \geq 6$. □

**Lemma 2.2** *Let $C_3 + P_n$ be a graph with $n \geq 2$, and let $\Pi$ be a resolving partition of $C_3 + P_n$. If the vertex $u_1, u_2$ or $u_3$ contained in partition classes $S_1, S_2$, or $S_3$, then $|S_j| \leq \binom{g-4}{0} + \binom{g-4}{1}$ for $j = 4, 5$, $g = 5$ and $|S_j| \leq \binom{g-4}{0} + \binom{g-4}{1} + \binom{g-4}{2}$ for $j = 4, 5, \ldots, g$, $g \geq 6$.*

**Proof.** Let $\Pi = \{S_1, S_2, \ldots, S_g\}$ be a resolving partition of $C_3 + P_n$. Consider the partition class other than the $S_1, S_2$, and $S_3$. Suppose the vertex $u_1, u_2$ or $u_3$ are in $S_1, S_2$ or $S_3$ respectively and vertex $v_y \in V(C_3+P_n)\setminus\{u_1,u_2,u_3\}$ for $1 \le y \ne c \ne d \ne e \le n$ is in $S_4$, then the representation $r(v_y|\Pi) = (1,1,1,0,\ldots)$. Therefore $g-4$ positions on the representation of vertex $v_y$ can be filled at most one 1 for $g=5$ and at most two 1's for $g \ge 6$. There are at most $\binom{g-4}{0}+\binom{g-4}{1}$ distinct representation for $g=5$, or at most $\binom{g-4}{0}+\binom{g-4}{1}+\binom{g-4}{2}$ distinct representation for $g \ge 6$. So, $|S_j| \le \binom{g-4}{0}+\binom{g-4}{1}$ for $j=4,5$, $g=5$ and $|S_j| \le \binom{g-4}{0}+\binom{g-4}{1}+\binom{g-4}{2}$ for $j=4,5,\ldots,g$, $g \ge 6$. □

**Theorem 2.3** *Let $C_3+P_n$ be a graph with $n \ge 2$, the partition dimension of $C_3+P_n$ is $pd(C_3+P_n) = g$ with $g$ is the smallest positive integer such that $n \le 5g-12$ for $g=5$ and $n \le \frac{g^3-7g^2+20g-18}{2}$ for $g \ge 6$.*

**Proof.** Let $\Pi = \{S_1,S_2,\ldots,S_g\}$ be resolving partition of $C_3+P_n$ graph for $n \ge 2$. By Lemma 2.1 we have $|S_i| \le 1+\binom{g-3}{0}+\binom{g-3}{1}$ for $i=1,2,3$, $g=5$ and $|S_i| \le 1+\binom{g-3}{0}+\binom{g-3}{1}+\binom{g-3}{2}$ for $i=1,2,3$, $g \ge 6$. By Lemma 2.2 we have $|S_j| \le \binom{g-4}{0}+\binom{g-4}{1}$ for $j=4,5$ and $|S_j| \le \binom{g-4}{0}+\binom{g-4}{1}+\binom{g-4}{2}$ for $j=4,5,\ldots,g$, $g \ge 6$. Since the number of vertex in $C_3+P_n$ graph is $|V(C_3+P_n)| = n+3$, so

$$\begin{aligned}
|V(C_3+P_n)| &= |S_i|+|S_j|,\\
&\le \left(1+\tbinom{g-3}{0}\tbinom{g-3}{1}\right)(3)+\left(\tbinom{g-4}{0}+\tbinom{g-4}{1}\right)(g-3)\\
n &\le 5g-12, \text{for g=5.}
\end{aligned}$$

For $g \ge 6$ by the same way, we obtain $n \le \frac{g^3-7g^2+20g-18}{2}$. Thus, $pd(C_3+P_n) = g$ where $g$ is the smallest positive integer such that $n \le 5g-12$ for $g=5$ and $n \le \frac{g^3-7g^2+20g-18}{2}$ for $g \ge 6$. □

Let $u_a \in V(C_q)$ for $a \in [1,q]$ and $v_b \in V(P_n)$ for $b \in [1,n]$ be vertices of $C_q+P_n$ with $q \ge 4$, $n \ge 2$ and let $\Pi = \{S_1,S_2,\ldots,S_g\}$ be a resolving partition of $C_q+P_n$. We have the following remark.

**Remark 2.4**

*(i) $p$ is a number of partition class containing a vertex $v_b \in V(P_n)$ such that $v_b \in S_i$ for $1 \le i \le p$, and $u_h \in S_1$, $u_w \in S_2$ for $u_h, u_w \in V(C_q)$, $1 \le h \ne w \le q$,*

*(ii) $f = g-p$ is a number of partition class containing a vertex $u_z \in V(C_q)\setminus\{u_h,u_w\}$, $1 \le z \ne h \ne w \le q$,*

*(iii) $r$ is a number of partition class containing a vertex $v_b \in V(P_n)$ such that $v_b \in S_i$ for $1 \le i \le r$, and $u_h \in S_1$ for $1 \le h \le q$,*

*(iv) $t = g-r$ is a number of partition class containing a vertex $u_z \in V(C_q)\setminus\{u_h\}$ for $1 \le z \ne h \le q$,*

*(v) $x$ is a number of partition class containing a vertex $v_b \in V(P_n)$, and*

*(vi) $y = g-x$ is a number of partition class containing a vertex $u_a \in V(C_q)$*

**Lemma 2.5** *Let $C_q+P_n$ be a graph with $q \ge 4$ and $n \ge 2$, let $\Pi = \{S_1,S_2,\ldots,S_p,S_{p+1},\ldots,S_g\}$ be a resolving partition of $C_q+P_n$. If $v_b \in S_i$ for $1 \le i \le p$, $u_h \in S_1$ and $u_w \in S_2$ for $1 \le h \ne w \le q$, then $n \le 2$ for $p=2$, $n \le 2p-1$ for $p=3$, $n \le p^2-p$ for $p=4$ and $n \le \frac{p^3-5p^2+12p-8}{2}$ for $p \ge 5$.*

**Proof.** Let $\Pi = \{S_1,S_2,\ldots,S_p,\ldots,S_g\}$ be a resolving partition of $C_q+P_n$ with $q \ge 4$ and $n \ge 2$. Let $v_b \in S_i$ with $i=1,2,\ldots,p$ and a vertex $u_h \in S_1$ and $u_w \in S_2$ with $1 \le h \ne w \le q$. We obtain $r(u_h|\Pi) = (0,1,\ldots)$, $r(u_w|\Pi) = (1,0,\ldots)$. Let $v_c \in S_1$ and $v_d \in S_2$ with $1 \le c \ne d \le n$, we obtain $r(v_c|\Pi) = (0,1,\ldots,1,\ldots,1)$ and $r(v_d|\Pi) = (1,0,\ldots,1,\ldots,1)$. Since the diameter of $C_q+P_n$ graph is 2, the elements on $r(v_c|\Pi)$ other than the first and second positions can be filled by 1 or 2. There are at most two 1's on representations other than the first and second positions. Therefore, $p-2$ positions on the representation $r(v_c|\Pi)$ can be filled by at most two 1's and the rest can be filled by 2. Let $p=2$, there are at most $\binom{p-2}{0}$ distinct representations of $v_c \in S_1$. Then, for $p=3$ there are at most $\binom{p-2}{0}+\binom{p-2}{1}$ distinct representations of $v_c \in S_1$, and let $p \geq 4$ there are at most $\binom{p-2}{0}+\binom{p-2}{1}+\binom{p-2}{2}$ distinct representation of $v_c \in S_1$. Similarly for $v_d \in S_2 \setminus \{u_w\}$. Furthermore, let $v_y \in S_i \setminus \{S_1, S_2\}$ with $1 \leq y \neq c \neq d \leq n$, we obtain $r(v_y|\Pi) = (1,1,0,\ldots,1,\ldots,1)$. There are at most $p-3$ positions on the representation of $v_y \in S_i \setminus \{S_1,S_2\}$ can be filled with at most two 1's and the rest can be filled by 2. Let $p=3$, there are at most $\binom{p-3}{0}$ distinct representations of $v_y \in S_i \setminus \{S_1,S_2\}$. Then for $p=4$, there are at most $\binom{p-3}{0}+\binom{p-3}{1}$ distinct representations of $v_y \in S_i \setminus \{S_1,S_2\}$, and for $p \geq 5$ there are at most $\binom{p-3}{0}+\binom{p-3}{1}+\binom{p-3}{2}$ distinct representation of $v_y \in S_i \setminus \{S_1,S_2\}$. Since the number of $v_b \in V(P_n)$ is $n$, then

(i) for $p=2$,

$$\begin{aligned} |V(P_n)| &= |v_c \in S_1| + |v_d \in S_2| \\ &\leq \binom{p-2}{0}+\binom{p-2}{0} \\ n &\leq 2, \end{aligned}$$

(ii) for $p=3$,

$$\begin{aligned} |V(P_n)| &= |v_c \in S_1| + |v_d \in S_2| + |v_y \in S_3| \\ &\leq \left(\binom{p-2}{0}+\binom{p-2}{1}\right)(2)+\binom{p-3}{0} \\ n &\leq 2p-1, \end{aligned}$$

(iii) for $p=4$,

$$\begin{aligned} |V(P_n)| &= |v_c \in S_1| + |v_d \in S_2| + |v_y \in S_i|, (i=3,4) \\ &\leq \left(\binom{p-2}{0}+\binom{p-2}{1}+\binom{p-2}{2}\right)(2)+\left(\binom{p-3}{0}+\binom{p-3}{1}\right)(2) \\ n &\leq p^2-p, \end{aligned}$$

(iv) for $p \geq 5$

$$\begin{aligned} |V(P_n)| &= |v_c \in S_1| + |v_d \in S_2| + |v_y \in S_i|, (3 \leq i \leq p) \\ &\leq \left(\binom{p-2}{0}+\binom{p-2}{1}+\binom{p-2}{2}\right)2+\left(\binom{p-3}{0}+\binom{p-3}{1}+\binom{p-3}{2}\right)(p-2) \\ n &\leq \frac{p^3-5p^2+12p-8}{2}. \end{aligned}$$

So, if $v_b \in S_i$ for $1 \leq i \leq p$, $u_h \in S_1$ and $u_w \in S_2$ for $1 \leq h \neq w \leq q$, then $n \leq 2$ for $p=2$, $n \leq 2p-1$ for $p=3$, $n \leq p^2-p$ for $p=4$ and $n \leq \frac{p^3-5p^2+12p-8}{2}$ for $p \geq 5$.

**Lemma 2.6** *Let $C_q+P_n$ be a graph with $q \geq 4$ and $n \geq 2$, let $\Pi = \{S_1, S_2, \ldots, S_p, S_{p+1}, \ldots, S_g\}$ be a resolving partition of $C_q+P_n$. If $v_b \in S_i$ for $1 \leq i \leq p$, $u_z \in V(C_q) \setminus \{u_h, u_w\}$ for $1 \leq z \neq h \neq w \leq q$ contained in the partition class $S_j$ for $j=1,2,\ldots,f$ with $f=g-p$, then $q \leq 2f^2+f$ for $n=2,5,12$ and $f=2$ and $q \leq \frac{f^3+f^2+4f+4}{2}$ for another $p,f$.*

**Proof.** Let $\Pi = \{S_1, S_2, \ldots, S_p, \ldots, S_g\}$ be a resolving partition of $C_q+P_n$ graph with $q \geq 4$ and $n \geq 2$. Let $v_b \in V(P_n)$, $v_b \in S_i$ for $i=1,2,\ldots,p$, $u_h \in S_1$ and $u_w \in S_2$, $u_z \in S_j$ with $u_z \in V(C_q) \setminus \{u_h, u_w\}$, $1 \leq p \neq w \neq z \leq q$ and $j=1,2,\ldots,f$ with $f=g-p$. From Lemma 2.5, we obtain the representations of $u_h$ and $u_w$ with respect to $\Pi$ are $r(u_h|\Pi) = (0,1,1,\ldots,1,\ldots)$

and $r(u_w|\Pi) = (1, 0, 1, \ldots, 1, \ldots)$. Because the diameter of $C_q + P_n$ graph is 2, the element of $r(u_h|\Pi)$ other than the first position and $p-1$ positions can be filled by 1 and 2 so that the $g-p$ position on the representation of $u_h \in S_1$ can be filled at most two 1's and the rest can be filled by 2. There are at most $\binom{g-p}{0} + \binom{g-p}{1} + \binom{g-p}{2}$ distinct representations of $u_h \in S_i$. Since $f = g-p$, so it can be written $\binom{f}{0} + \binom{f}{1} + \binom{f}{2}$ distinct representations of $u_h \in S_1$. Suppose $n = 2, 5, 12$ and $f = 2$, there are two equal representation $r(v_c|\Pi) = r(u_h|\Pi)$ so there at most $\binom{f}{0} + \binom{f}{1} + \binom{f}{2} - 1$ distinct representation for all $u_h \in S_1$. The same for $u_w \in S_2$, there is a $\binom{f}{0} + \binom{f}{1} + \binom{f}{2}$ distinct representations of $u_w \in S_2$, and $\binom{f}{0} + \binom{f}{1} + \binom{f}{2} - 1$ distinct representations for all $u_w \in S_2$ with $n = 3, 8$ and $f = 2$. Furthermore, let $u_z \in S_j$ with $u_z \in V(C_q) \setminus \{u_h, u_w\}$, $1 \le z \ne h \ne w \le q$, $j = 1, 2, \ldots, f$, and $f = g - p$ then the representation of $u_z$ with respect to $\Pi$ is $r(u_z|\Pi) = (1, 1, \ldots, 1, 0, \ldots)$. Since the diameter of $C_q + P_n$ graph is 2, then other than $p$ position and the $(p+1)^{th}$ position of $r(u_z|\Pi)$ can be filled by 1 or 2. There are at most two 1's and the rest filled by 2. If $n = 2, 5, 12$ and $p = 2$, other than the $p$ position and the $(p+1)^{th}$ position can be filled by 1 or 2 so that at most $\binom{f-1}{0} + \binom{f-1}{1}$ distinct representation of $u_z$ for $n = 2, 5, 12$ and $f = 2$. For another $p$ and $f$, the element of $r(u_z|\Pi)$ other than $p$ position and the $(p+1)^{th}$ position can be filled with at most two 1's and the rest filled by 2 so that there $\binom{f-1}{0} + \binom{f-1}{1} + \binom{f-1}{2}$ distinct representations for $u_z$. The number of vertices element $C_q$ is the number of vertex $u_h, u_w$ and $u_z$ in each class partition, then we obtain

$$\begin{aligned} |V(C_q)| &= |u_h, u_w \in S_i| + |u_z \in S_j| \\ &\le \left(\binom{f}{0} + \binom{f}{1} + \binom{f}{2} - 1\right)(2) + \left(\binom{f-1}{0} + \binom{f-1}{1}\right)(f) \\ q &\le 2f^2 + f, \text{ for } n = 2, 5, 12 \text{ and } f = 2, \end{aligned}$$

$$\begin{aligned} |V(C_q)| &= |u_h, u_w \in S_i| + |u_z \in S_j| \\ &\le \left(\binom{f}{0} + \binom{f}{1} + \binom{f}{2}\right)(2) + \left(\binom{f-1}{0} + \binom{f-1}{1} + \binom{f-1}{2}\right)(f) \\ q &\le \frac{f^3 + f^2 + 4f + 4}{2}, \text{ for another } p, f. \end{aligned}$$

So, if $v_b \in S_i$ for $1 \le i \le p$, $u_z \in V(C_q) \setminus \{u_h, u_w\}$ for $1 \le z \ne h \ne w \le q$ contained in the partition class $S_j$ for $j = 1, 2, \ldots, f$ with $f = g - p$, then $q \le 2f^2 + f$ for $n = 2, 5, 12$ and $f = 2$ and $q \le \frac{f^3+f^2+4f+4}{2}$ for another $p, f$.

**Lemma 2.7** *Let $\Pi = \{S_1, S_2, \ldots, S_r, S_{r+1}, \ldots, S_g\}$ be a resolving partition of $C_q + P_n$ graph for $q \ge 4$ and $n \ge 2$. If $v_b \in S_i$ for $i = 1, 2, \ldots, r$ and some $u_h \in V(C_q)$ for $h \in [1, q]$, $u_h \in S_1$, then $n \le r + 1$ for $r = 2$, $n \le \frac{r^2+3r-2}{2}$ for $r = 3$ and $n \le \frac{r^3-3r^2+6r-2}{2}$ for $r \ge 4$.*

**Proof.** Let $\Pi = \{S_1, S_2, \ldots, S_r, S_{r+1}, \ldots, S_g\}$ be a resolving partition of $C_q + P_n$ graph for $q \ge 4$ and $n \ge 2$. Let $v_b \in S_i$ for $i = 1, 2, \ldots, r$, $v_c \in S_1$ with $c \in [1, n]$ and $u_h \in S_1$ with $h \in [1, q]$, we obtain the representation of $v_c \in S_1$ with respect to $\Pi$ is $r(v_c|\Pi) = (0, \ldots)$, and the representation of $u_h \in S_1$ with respect to $\Pi$ is $r(u_h|\Pi) = (0, 1, \ldots, 1, \ldots)$. Since the diameter of $C_q + P_n$ graph is 2, the elements of $r(v_c|\Pi)$ other than the first position can be filled by 1 or 2. There are at most two 1's on representations other than the first position. Therefore, $r - 1$ position on the representation of $v_c \in S_1$ can be filled by two 1's and the rest can be 2. Let $r = 2$, there are at most $\binom{r-1}{0} + \binom{r-1}{1}$ distinct representations of $v_c \in S_1$. Then let $r \ge 3$, there are at most $\binom{r-1}{0} + \binom{r-1}{1} + \binom{r-1}{2}$ distinct representations of $v_c \in S_1$. Let $v_d \in S_i$ with $i = 2, 3, \ldots, r$ and $1 \le d \ne c \le n$, we obtain $r(v_d|\Pi) = (1, 0, \ldots)$. Because the diameter of $C_q + P_n$ graph is 2, the element of $r(v_d|\Pi)$ other than the first and second positions can be filled by 1 or 2. Let $r = 2$, there are at most $\binom{r-2}{0}$ distinct representations of $v_d \in S_i$. Then let $r = 3$, there are at most $\binom{r-2}{0}+\binom{r-2}{1}$ distinct representations of $v_d \in S_i$, and let $r \geq 4$ there are at most $\binom{r-2}{0}+\binom{r-2}{1}+\binom{r-2}{2}$ distinct representations of $v_d \in S_i$. Furthermore, because $|P_n|$ is the sum of the number of vertices $v_b$ in each class partition $i$, we obtain

(i) for $r = 2$,

$$\begin{aligned} |V(P_n)| &= |v_c \in S_1| + |v_d \in S_2| \\ &\leq \binom{r-1}{0}+\binom{r-1}{1}+\binom{r-2}{0} \\ n &\leq r+1, \end{aligned}$$

(ii) for $r = 3$,

$$\begin{aligned} |V(P_n)| &= |v_c \in S_1| + |v_d \in S_i|, (i = 2,3) \\ &\leq \binom{r-1}{0}+\binom{r-1}{1}+\binom{r-1}{2}+\left(\binom{r-2}{0}+\binom{r-2}{1}\right)(2) \\ n &\leq \frac{r^2+3r-2}{2}, \end{aligned}$$

(iii) for $r \geq 4$,

$$\begin{aligned} |V(P_n)| &= |v_c \in S_1| + |S_i|, 2 \leq i \leq r \\ &\leq \binom{r-1}{0}+\binom{r-1}{1}+\binom{r-1}{2}+\left(\binom{r-2}{0}+\binom{r-2}{1}+\binom{r-2}{2}\right)(r-1) \\ n &\leq \frac{r^3-3r^2+6r-2}{2}. \end{aligned}$$

So, if $v_b \in S_i$ for $i = 1,2,\ldots,r$ and some $u_h \in V(C_q)$ for $h \in [1,q]$, $u_h \in S_1$, then $n \leq r+1$ for $r = 2$, $n \leq \frac{r^2+3r-2}{2}$ for $r = 3$ and $n \leq \frac{r^3-3r^2+6r-2}{2}$ for $r \geq 4$.

**Lemma 2.8** *Let* $\Pi = \{S_1, S_2, \ldots, S_r, S_{r+1}, \ldots, S_g\}$ *be a resolving partition of* $C_q + P_n$ *graph for* $q \geq 4$ *and* $n \geq 2$. *If vertex* $u_z \in V(C_q) \setminus \{u_h\}$ *for* $1 \leq z \neq h \leq q$ *contained in the partition class* $S_j$ *for* $j = 1,2,\ldots,t$, $t = g-r$, *then* $q \leq \frac{3t^2+t}{2}$ *for* $n = 3,8$ *and* $t = 2$ *and* $q \leq \frac{t^3+3t+2}{2}$ *for another* $r, t$.

**Proof.** Let $\Pi = \{S_1, S_2, \ldots, S_r, S_{r+1}, \ldots, S_g\}$ be a resolving partition of $C_q + P_n$ graph for $q \geq 4$ and $n \geq 2$. Let $v_b \in S_i$ for $i = 1,2,\ldots,r$, $u_h \subset S_1$, $u_z \in S_j$ for $1 \leq h \neq z \leq q$, $j = 1,2,\ldots,t$ with $t = g-r$. Based on Lemma 2.8, we obtain the representation of $u_h \in S_1$ is $r(u_h|\Pi) = (0, 1\ldots, 1, \ldots)$. Since the diameter of $C_q + P_n$ graph is 2, the elements of $r(u_h|\Pi)$ other than $r$ positions can be filled by 1 or 2. There are at most two 1's on the representation other than $r$ positions. Therefore, $g-r$ position of representation $u_h \in S_1$ can be filled by two 1's and the rest can be 2. There are at most $\binom{g-r}{0}+\binom{g-r}{1}+\binom{g-r}{2}$ distinct representation of $u_h \in S_1$ or at most $\binom{t}{0}+\binom{t}{1}+\binom{t}{2}$. Next, if $v_b \in S_i$ for $i = 1,2,\ldots,r$, we obtain the representation of $u_z \in S_j$ for $1 \leq h \neq z \leq q$ and $j = 1,2,\ldots,t$ with $t = g-r$ is $r(u_z|\Pi) = (1,\ldots,1,0,\ldots)$. If $n = 3,8$ and $t = 2$ then there are two equal representations, so as to $n = 3,8$ and $t = 2$ there are at most $\binom{t}{0}+\binom{t}{1}+\binom{t}{2}-1$ distinct representation of $u_h \in S_1$. Since the diameter of $C_q+P_n$ graph is 2, the elements of representation $r(u_z|\Pi)$ other than $r$ positions and the $r+1$'s position can be filled by 1 or 2. If $n = 3,8$ and $t = 2$, there are at most $\binom{g-r-1}{0}+\binom{g-r-1}{1}$ distinct representation of $u_z \in S_j$ or at most $\binom{t-1}{0}+\binom{t-1}{1}$ since $t = g-r$. Then, for another $r, t$, there are at most $\binom{g-r-1}{0}+\binom{g-r-1}{1}+\binom{g-r-1}{2}$ or can be expressed by $\binom{t-1}{0}+\binom{t-1}{1}+\binom{t-1}{2}$ distinct representation for $u_z \in S_j$. Since $|V(C_q)|$ is the sum of the number of vertices $u_h \in S_1$ and $u_z \in S_j$, then

$$\begin{aligned} |V(C_q)| &= |u_h \in S_1| + |S_j|, 1 \leq j \leq t \\ q &\leq \left(\binom{t}{0}+\binom{t}{1}+\binom{t}{2}-1\right)+\left(\binom{t-1}{0}+\binom{t-1}{1}\right)(t) \\ q &\leq \frac{3t^2+t}{2}, \text{ for } n = 3,8 \text{ and } t = 2 \end{aligned}$$

$$\begin{aligned}|V(C_q)| &= |u_h \in S_1| + |S_j|, 1 \le j \le t \\ q &\le \tbinom{t}{0} + \tbinom{t}{1} + \tbinom{t}{2} + (\tbinom{t-1}{0} + \tbinom{t-1}{1} + \tbinom{t-1}{2})(t) \\ q &\le \frac{t^3+3t+2}{2}, \text{ for another } r, t.\end{aligned}$$

So, If vertex $u_z \in V(C_q) \setminus \{u_h\}$ for $1 \le z \ne h \le q$ contained in the partition class $S_j$ for $j = 1, 2, \ldots, t$, $t = g - r$, then $q \le \frac{3t^2+t}{2}$ for $n = 3, 8$ and $t = 2$ and $q \le \frac{t^3+3t+2}{2}$ for another $r, t$.

**Lemma 2.9** *Let* $\Pi = \{S_1, S_2, \ldots, S_x, S_{x+1}, \ldots, S_g\}$ *be a resolving partition of* $C_q + P_n$ *graph for* $q \ge 4$ *and* $n \ge 2$. *If vertex* $v_b \in S_i$ *for* $i = 1, 2, \ldots, x$, *then* $n \le \frac{x^3-x^2+2x}{2}$ *for* $x \ge 2$.

**Proof.** Let $\Pi = \{S_1, S_2, \ldots, S_x, S_{x+1}, \ldots, S_g\}$ be a resolving partition of $C_q + P_n$ graph for $q \ge 4$ and $n \ge 2$. Let $v_b \in S_i$ for $i = 1, 2, \ldots, x$ and $u_a \in S_j$ for $j = 1, 2, \ldots, y$ with $y = g - x$. Representation of $v_c \in S_1$ for $1 \le c \le n$ is $r(v_c|\Pi) = (0, \ldots)$. Since the diameter of $C_q + P_n$ is 2, the elements of $r(v_c|\Pi)$ other than first position can be filled with 1 or 2. There are $x - 1$ position of the representation $r(v_c|\Pi)$ can be filled at most two 1's and the rest can be filled by 2, so there are at most $\tbinom{x-1}{0} + \tbinom{x-1}{1} + \tbinom{x-1}{2}$. Since the number of $P_n$ is $n$, then

$$\begin{aligned}|V(P_n)| &= |S_i|, 1 \le i \le x \\ &\le (\tbinom{x-1}{0} + \tbinom{x-1}{1} + \tbinom{x-1}{2})(x) \\ n &\le \frac{x^3 - x^2 + 2x}{2}.\end{aligned}$$

So, if *vertex* $v_b \in S_i$ for $i = 1, 2, \ldots, x$ and $u_a \in S_j$ for $j = 1, 2, \ldots, y$ with $y = g - x$, then $q \le \frac{x^3-x^2+2x}{2}$.

**Lemma 2.10** *Let* $\Pi = \{S_1, S_2, \ldots, S_x, S_{x+1}, \ldots, S_g\}$ *be a resolving partition of* $C_q + P_n$ *graph for* $q \ge 4$ *and* $n \ge 2$. *If vertex* $u_a \in V(C_q)$ *contained in partition class* $S_j$ *for* $j = 1, 2, \ldots, y$ *with* $yg - x$, *then* $q \le \frac{y^3-y^2+2y}{2}$ *for* $y \ge 2$.

**Proof.** Let $\Pi = \{S_1, S_2, \ldots, S_x, S_{x+1}, \ldots, S_g\}$ be a resolving partition of $C_q + P_n$ graph for $q \ge 4$ and $n \ge 2$. Let $v_b \in S_i$ for $i = 1, 2, \ldots, x$ and $u_a \in S_j$ for $j = 1, 2, \ldots, y$ with $y = g - x$. Representation of $u_h \in S_y$ for $1 \le h \ne a \le q$ is $r(u_h|\Pi) = (1, \ldots, 1, \ldots, 0)$. Since the diameter of $C_q + P_n$ is 2, the elements of $r(u_h|\Pi)$ other than last position can be filled with 1 or 2. There are at most two's 1 on the representation of $u_h \in S_y$. Therefore, $y - 1$ position of the representation $r(u_h|\Pi)$ can be filled at most two 1's and the rest can be filled by 2, so there are at most $\tbinom{y-1}{0} + \tbinom{y-1}{1} + \tbinom{y-1}{2}$. Since the number of $C_q$ is $q$, then

$$\begin{aligned}|V(C_q)| &= |S_j|, 1 \le j \le y \\ &\le (\tbinom{y-1}{0} + \tbinom{y-1}{1} + \tbinom{y-1}{2})(y) \\ q &\le \frac{y^3 - y^2 + 2y}{2}.\end{aligned}$$

So, if *vertex* $v_b \in S_i$ for $i = 1, 2, \ldots, x$ and $u_a \in S_j$ for $j = 1, 2, \ldots, y$ with $y = g - x$, then $q \le \frac{y^3-y^2+2y}{2}$.

**Theorem 2.11** *Let* $C_q + P_n$ *a graph with* $q \ge 4$ *and* $n \ge 2$, $\Pi = \{S_1, S_2, \ldots, S_g\}$ *be a resolving partition of* $C_q + P_n$. *The partition dimension of* $C_q + P_n$ *is* $pd(C_q + P_n) = g$ *with* $g = min\{(p + f), (t + r), (x + y)\}$, *where* $p, f, r, t, x$ *and* $y$ *as stated in the Remark 2.4.*

**Proof.** Let $C_q + P_n$ be a graph with $q \geq 4$ and $n \geq 2$, and let $\Pi = \{S_1, S_2, \ldots, S_g\}$ be a resolving partition of $C_q + P_n$. From Lemma 2.5 and Lemma 2.6 we have $\Pi = \{S_1, S_2, \ldots, S_p, S_{p+1}, \ldots, S_g\}$ and $g = p + f$ with $p, f$ is the smallest positive integer such that $n \leq 2$ for $p = 2$, $n \leq 2p - 1$ for $p = 3$, $n \leq p^2 - p$ for $p = 4$ and $n \leq \frac{p^3-5p^2+12p-8}{2}$ for $p \geq 5$ and $q \leq 2f^2 + f$ for $p = 2$ and $f = 2$ and $q \leq \frac{f^3+f^2+4f+4}{2}$ for another $p, f$. From Lemma 2.7 and Lemma 2.8 we have $\Pi = \{S_1, S_2, \ldots, S_r, S_{r+1}, \ldots, S_g\}$ and $g = r + t$ where $r, t$ is the smallest positive integer such that $n \leq r + 1$ for $r = 2$, $n \leq \frac{r^2+3r-2}{2}$ for $r = 3$ and $n \leq \frac{r^3-3r^2+6r-2}{2}$ for $r \geq 4$ and $q \leq \frac{3t^2+t}{2}$ for $r = 2$ and $t = 2$ and $q \leq \frac{t^3+3t+2}{2}$ for another $r, t$. From Lemma 2.9 and Lemma 2.10 we have $\Pi = \{S_1, S_2, \ldots, S_x, S_{x+1}, \ldots, S_g\}$ and $g = x + y$ where $x, y$ is the smallest positive integer such that $n \leq \frac{x^3-x^2+2x}{2}$ for $x \geq 2$ and $q \leq \frac{y^3-y^2+2y}{2}$ for $y \geq 2$. Furthermore, because there are three different $g$ values and partition dimension of graph is the minimum cardinality of the resolving partition, then $pd(C_q + P_n) = min\{(p+f), (s+t), (x+y)\}$. □

## 3. Conclusion

Based on the discussion above, we conclude that the partition dimension of $C_m + P_n$ graph for $m \geq 3$ and $n \geq 2$ can be divided into two theorems. The first theorem for $C_3 + P_n$ with $n \geq 2$ as stated in the Theorem 2.3 and the second theorem for $C_q + P_n$ with $q \geq 4$ and $n \geq 2$ as stated in Theorem 2.11.

## 4. References

[1] Chartrand, G., E. Salehi, and P. Zhang, *On the Partition Dimension of a Graph*, Congressus Numerantium, **131** (1998), 55-66.

[2] Chartrand, G., E. Salehi, and P. Zhang, *The Partition Dimension of a Graph*, Aequation Math. **55**(2000), 45-54.

[3] Chartrand, G., L. Eroh, M. Johnson, and O. Oellermann, *Resolvability in Graphs and the Metric Dimension of Graph*, Discrete Appl. Math. **105** (2000), 98-113.

[4] Hidayat, D. W., *Dimensi Partisi pada Beberapa Kelas Graf*, Tugas akhir, FMIPA Universtitas Sebelas Maret, Surakarta, 2015.

[5] Tomescu, I., I. Javaid, and Slamin, *On The Partition Dimension and Connected Partition Dimension of Wheels*, Ars Combinatoria. **84** (2007), 311-318.

# Domination Number of Vertex Amalgamation of Graphs

**Y Wahyuni[1], M I Utoyo[1] and Slamin[2]**

[1]Department of Mathematics, Airlangga University, Surabaya 60115 Indonesia
[2]Information System Study Programme, University of Jember, Jember 68121 Indonesia

E-mail: yayuk-w@fst.unair.ac.id

**Abstract.** For a graph $G = (V, E)$, a subset $S$ of $V$ is called a dominating set if every vertex $x$ in $V$ is either in $S$ or adjacent to a vertex in $S$. The domination number $\boldsymbol{\gamma(G)}$ is the minimum cardinality of the dominating set of $G$. The dominating set of $G$ with a minimum cardinality denoted by $\boldsymbol{\gamma(G)}$-set. Let $G_1, G_2, \ldots, G_t$ be subgraphs of the graph $G$. If the union of all these subgraphs is $G$ and their intersection is $\{v\}$, then we say that $G$ is the vertex-amalgamation of $G_1, G_2, \ldots, G_t$ at vertex $v$. Based on the membership of the common vertex $v$ in the $\boldsymbol{\gamma(G_i)}$-set, there exist three conditions to be considered. First, if $v$ elements of every $\boldsymbol{\gamma(G_i)}$-set, second if there is no $\boldsymbol{\gamma(G_i)}$-set containing $v$, and third if either $v$ is element of $\boldsymbol{\gamma(G_i)}$-set for $\mathbf{1 \le i \le p}$ or there is no $\boldsymbol{\gamma(G_i)}$-set containing $v$ for $\boldsymbol{p < i \le t}$. For these three conditions, the domination number of $G$ as vertex-amalgamation of $G_1, G_2, \ldots, G_t$ at vertex $v$ can be determined.

## 1. Introduction

Several results about domination number $\gamma(G)$ and operation of graphs have been explored by some researchers. Some of these results were obtained by Pavlic and Zerovnik [6], Go and Canoy [2], and Kuziak, Lemanska, and Yero [5]. In this paper, we determined the domination number of vertex amalgamation of graphs.

Suppose $G_1$ and $G_2$ are subgraphs of a graph $G = (V, E)$ and $v \in V$. The graph $G$ is called vertex amalgamation of $G_1$ and $G_2$ at vertex $v$, denoted by $G = G_1 \vee^1_{\{v\}} G_2$, if $G = G_1 \cup G_2$ and $G_1 \cap G_2 = \{v\}$. This vertex amalgamation definition proposed by Yang and Kong [9] and can be generated over more than two subgraphs. Suppose $G_1, G_2, \ldots, G_t$ are subgraphs of $G$ and $v \in V$. If $G = \bigcup_{i=1}^t G_i$ and $\bigcap_{i=1}^t G_i = \{v\}$, then $G$ is a vertex amalgamation of $G_1, G_2, \ldots, G_t$ at vertex $v$, denoted by $G = \vee^1_{\{v\}} \{G_1, G_2, \ldots, G_t\}$. If $G_1, G_2, \ldots, G_t$ has $n_1, n_2, \ldots, n_t$ vertices respectively, then $\vee^1_{\{v\}} \{G_1, G_2, \ldots, G_t\}$ has $\sum_{i=1}^t n_i - t + 1$ vertices.

For a survey of some family of graphs, see [1]. The open neighborhood of $v \in V$ is the set $N(v) = \{w \in V; vw \in E\}$ and the closed neighborhood is $N[v] = N(v) \cup \{v\}$. For a set $S \subseteq V$, the open neighborhood $N(S)$ is defined as $\bigcup_{v \in S} N(v)$ and the closed neighborhood of $S$ is $N[S] = N(S) \cup S$. A set $S$ of vertices of a graph $G$ is called a *dominating set* if each vertex of $V - S$ is adjacent to a member of $S$. It is equivalent to that $N[S] = V$. The domination number of $G$, denoted by $\gamma(G)$, is the minimum cardinality of the dominating set of $G$ [3].

Some known results on domination number of some graphs are $\gamma(P_n) = \left\lceil \frac{n}{3} \right\rceil$ for $n > 1$ and $\gamma(C_n) = \left\lceil \frac{n}{3} \right\rceil$ for $n > 3$ by Klobucar [4], $\gamma(K_n) = 1$, $\gamma(K_{m,n}) = \gamma(K_{n_1, n_2, \ldots, n_k}) = 2$ for $m, n > 1$ and $n_i > 1$ by Snyder [7], $\gamma(F_{n,k}) = n$ for $k > 1$, and $\gamma(B_{n,k}) = n + 1$, for $n, k > 1$ by Wardani [8], $\gamma(L_n) = \left\lfloor \frac{n+1}{2} \right\rfloor$, and $\gamma(P_{n,f}) = 1$ . The bounds of domination number was given by Berge [3], that is, for a graph with order $n$ and the maximum degree $\Delta(G)$ holds $\left\lceil \frac{n}{1+\Delta(G)} \right\rceil \le \gamma(G) \le n - \Delta(G)$.

## 2. Results

We start this section by the observation concerning on the domination number of vertex amalgamation of some complete graphs.

**Observation 1.** If $G = \mathrm{V}_{\{v\}}^{1}\ \{K_{n_1}, K_{n_2}, \dots, K_{n_t}\}$ then $\gamma(G) = 1$.

The domination number of vertex amalgamation of a complete *k*-partite graphs is presented in the following theorem.

**Theorem 1.**

a. For a complete bipartite graph $K_{mn,}$ where $m, n \geq 2$, if $G = \mathrm{V}_{\{v\}}^{1}\ \{K_{m_1,n_1}, K_{m_2,n_2}, \dots, K_{m_t,n_t}\}$ then $\gamma(G) = t + 1$.
b. Let a complete multipartite graph $H_i$ has cardinality for each partite be more than one. If $G = \mathrm{V}_{\{v\}}^{1}\ \{H_1, H_2, \dots, H_t\}$ then $\gamma(G) = t + 1$.

**Proof.**

a. If $G = \mathrm{V}_{\{v\}}^{1}\ \{K_{m_1,n_1}, K_{m_2,n_2}, \dots, K_{m_t,n_t}\}$ where $m_i, n_i \geq 2$ and $m_i \leq n_i$ then the order of $G$ is $|V(G)| = \sum_{i=1}^{t}(m_i + n_i) - t + 1$ and the maximum degree is $\sum_{i=1}^{t} m_i \leq \Delta(G) \leq \sum_{i=1}^{t} n_i$. Using Berge, the upper bound of the domination number of *G* is $(\sum_{i=1}^{t}(m_i + n_i) - t + 1) - \sum_{i=1}^{t} n_i = \sum_{i=1}^{t} m_i - t + 1$. Because $m_i \geq 2$ then $\sum_{i=1}^{t} m_i - t + 1 \geq t + 1$ so that the least upper bound is $\gamma(G) \leq t + 1$. Suppose there is $T \subseteq V(G)$ which $|T| = t$. Let $V(K_{m_i,n_i}) = V_i \cup V_{v_i}$ where $V_{v_i}$ is the vertex partite which consists of *v*. If $v \in T$ then there is $V_j$ such that $x_j \notin T$ for every $x_j \in V_j$. So, for every $y_j \in V_{v_j}$ which $y_j \neq v$ holds $y_j \in V(G) - T$, $y_j \sim x_j$ but $y_j \nsim v$. If $v \notin T$ then there are two conditions. First, there exists a $z_i \in V(K_{m_i,n_i})$ for every *i* such that $z_i \in T$. In this case, every vertex which belongs to the same partite of *z* adjacent with no element of *T*. Second, there exists $K_{m_i,n_i}$ such that no vertex of $K_{m_i,n_i}$ belongs to *T*. In this case, every vertiex of $K_{m_i,n_i}$ except *v* is not adjacent to the element of *T*. It's means that *T* is not dominating set.
b. Let $S = \{v, a_1, a_2, \dots, a_t\}$ where $a_i \in H_i$ and $a_i$ belongs to the diffferent partite with $v$. For every $x \in (V(G) - S)$ holds: (i) if $x$ belongs to the same partite with $v$ then $x \sim a_i$ for some *i*; (ii) if $x$ belongs to the same partite with $a_i$ then $x \sim v$: and (iii) if $x$ belongs to the diferrent partite with neither $v$ and $a_i$ then both $x \sim v$ and $x \sim a_i$ sor some *i*. So *S* is a dominating set. Suppose there is $T \subseteq V(G)$ which $|T| = t$. If $v \in T$ then there is $H_i$ such that no element of this set belongs to $T$ except $v$. It implies that the elements of $H_i$ which belong to the same partite with *v* adjacent just with the vertex in the partite that not consist of *v*. If $v \notin T$ then there are two cases. First, there is exactly a vertex *x* of $H_i$ belongs to *T*. In this case, every vertex belongs to the same partite with *x* has not adjacent with *x*. Second, there exists $H_i$ such that no vertex of $H_i$ belongs to *T*. So, every vertex of this $H_i$ was not adjacent with any element of *T*. ♦

The following theorem presents the domination number of vertex amalgamation of some cycles.

**Theorem 2.** If $G = \mathrm{V}_{\{v\}}^{1}\ \{C_{n_1}, C_{n_2}, \dots, C_{n_t}\}$ then $\gamma(G) = \sum_{i=1}^{t} \left\lceil \frac{n_i}{3} \right\rceil - t + 1$.

**Proof.** Let $V(C_{n_i}) = \{v_{i,j}\ ; i = 1, \dots, t, \text{ and } j = 1, \dots, n_i\}$. Without loss of generality let $v = v_{i,1}$ be the common vertex. Let $S = \{v_{i,1+3j}\ ; i = 1, \dots, t \text{ and } j = 0, 1, \dots, \left\lceil \frac{n_i}{3} \right\rceil - 1\}$. For every $x \in (V(G) - S)$ then $x = v_{i,k}$ where $i = 1, \dots, t$ and $k \in \{2, 3, \dots, n_i\ ;\ k \not\equiv 1 (mod\ 3)\}$ such that there is $y \in S$ with $x \sim y$. So $S$ is a dominating set. It is clear that $|S| = 1 + \sum_{i=1}^{t}\left(\left\lceil \frac{n_i}{3} \right\rceil - 1\right) = \sum_{i=1}^{t}\left\lceil \frac{n_i}{3} \right\rceil - t + 1$. Suppose there exist $T \subseteq V(G)$ which $|T| = \sum_{i=1}^{t}\left(\left\lceil \frac{n_i}{3} \right\rceil - 1\right)$. If $v \in T$ then there is $C_{n_i}$ such that $\gamma(C_{n_i}) = \left\lceil \frac{n_i}{3} \right\rceil - 1$. If $v \notin T$ then $\gamma(C_{n_i}) = \left\lceil \frac{n_i}{3} \right\rceil - 1$ for every $C_{n_i}$. ♦

A specially case of Theorem 2, if the cycles are isomorphic, that is, if $n_1 = n_2 = \cdots = n_t = n$, then $\gamma(G) = t\left\lceil \frac{n}{3} \right\rceil - t + 1$.

The following theorem presents the domination number of vertex amalgamation of some paths.

**Theorem 3.** If $G = \mathrm{V}^1_{\{v\}}\ \{P_{n_1}, P_{n_2}, \dots, P_{n_t}\}$ then

$$\gamma(G) = \begin{cases} \sum_{i=1}^{t} \left\lceil \frac{n_i}{3} \right\rceil - t + 1\,, \text{if } v \text{ belongs to every } \gamma(P_{n_i}) - \text{set} \\ \sum_{i=1}^{t} \left\lceil \frac{n_i}{3} \right\rceil \qquad\quad , \text{ if } v \text{ has not belongs to any } \gamma(P_{n_i}) - \text{set} \end{cases}.$$

**Proof.** Let $U = \{v\} \cup \{w \in V(G); w{\sim}v\}$. For induced subgraph $\langle U \rangle$, it holds $G - \langle U \rangle = \bigcup_{i=1}^{t}(P_{n_i}\text{-}\langle U \rangle)$ such that $\gamma(G - \langle U \rangle) = \gamma\left(\bigcup_{i=1}^{t}(P_{n_i}\text{-}\langle U \rangle)\right) = \sum_{i=1}^{t} \gamma(P_{n_i}\text{-}\langle U \rangle)$. If $v$ belongs to every $\gamma(P_{n_i})$-set then $\gamma(P_{n_i}\text{-}\langle U \rangle) = \gamma(P_{n_i}) - 1 = \left\lceil \frac{n_i}{3} \right\rceil - 1$. It is clear that $\gamma(\langle U \rangle) = 1$. So we have $\gamma(G) = \gamma(G - \langle U \rangle) + \gamma(\langle U \rangle) = \sum_{i=1}^{t}\left(\left\lceil \frac{n_i}{3} \right\rceil - 1\right) + 1 = \sum_{i=1}^{t} \left\lceil \frac{n_i}{3} \right\rceil - t + 1$. If $v$ has not belongs to any $\gamma(P_{n_i})$-set then $v{\sim}z_i$ which $z_i$ belongs to $\gamma(P_{n_i})$-set. In this case $v$ is independent to the $\gamma(P_{n_i})$, so for $\gamma(G)$ too. It implies $\gamma(G) = t\gamma(P_{n_i})$. ♦

The domination number of vertex amalgamation of some ladders is presented in the following theorem.

**Theorem 4.** If $G = \mathrm{V}^1_{\{v\}}\ \{L_{n_1}, L_{n_2}, \dots, L_{n_t}\}$ for ladder graph $L_n$ then

$$\gamma(G) = \begin{cases} \sum_{i=1}^{t} \left\lceil \frac{n_i+1}{2} \right\rceil - t + 1\,, \text{if } v \text{ belongs to every } \gamma(L_{n_i}) - \text{set} \\ \sum_{i=1}^{t} \left\lceil \frac{n_i+1}{2} \right\rceil \qquad\quad , \text{ if } v \text{ has not belongs to any } \gamma(L_{n_i}) - \text{set} \end{cases}. \quad ♦$$

Before proving the dominating number of a vertex amalgamation of some graphs, we present a lemma concerning on the dominating number of the graph $G$ obtained by joining all vertices of a graph $H$ to a vertex $K_1$, that is, $G = K_1 + H$.

**Lemma 1.** For every graph $G$, $\gamma(G) = 1$ if and ony if $G = K_1 + H$ for some graph $H$.

As the diameter of $G = K_1 + H$ is one, it is very easy to see that any one vertex in $G$ can dominate other vertices in $G$. We now have the following theorem.

**Theorem 5.** Let graph $G_i$ which $\gamma(G_i) = 1$ and $G = \mathrm{V}^1_{\{v\}}\ \{G_1, G_2, \dots, G_t\}$.

a. If $v$ belongs to every $\gamma(G_i)$-set then $\gamma(G) = 1$.
b. If $v$ has not belongs to any $\gamma(G_i)$-set then $\gamma(G) = t$.
c. If $v$ has not belongs to any $\gamma(G_i)$-set for $1 \le i \le p$ and $v$ belongs to every $\gamma(G_i)$-set for $p < i \le t$ then $\gamma(G) = p + 1$.

**Proof.** From Lemma 1, we have $G_i = K_1 + H_i$. Let $|V(H_i)| = n_i$ so $|V(G_i)| = n_i + 1$ and $|V(G)| = \sum_{i=1}^{t} n_i + 1$. The degree of every vertex in $\gamma(G_i)$–set is $n_i$ and less than $n_i$ for the others.

a. If $v$ belongs to every $\gamma(G_i)$-set then $v$ has $\sum_{i=1}^{t} n_i$ degree so $\{v\}$ is dominating set.
b. If $v$ has not belong to every $\gamma(G_i)$-set then $v$ adjacent with $t$ vertices which degrees are $n_1, n_2, \dots, n_t$ respectively. These $t$ vertices span the dominating set of $G$. The cardinality of this dominating set is minimum, if there exist a set with $t$-1 cardinality then there exists $G_i$ where the $n_i$ vertices which less than $n_i$ degree is not adjacent to these $t$-1 vertices.
c. We combine (b) condition for $p$ first subgraphs $G_i$ and (a) condition for the others then we have $\gamma(G) = p + 1$. ♦

The next lemma describes the dominating set of vertex amalgamation of some graphs.

**Lemma 2.** Let $G = \mathsf{V}^1_{\{v\}}\ \{G_1, G_2, \dots, G_t\}$. If $S_i$ is a dominating set of $G_i$ for every $i$ = 1,2,..,$t$, then $\bigcup_{i=1}^t S_i$ is a dominating set of $G$.

In the theorems below we notice $S_i$ as a dominating set of $G_i$, then the set $P_i$ is as $\gamma(G_i)$-set, and the set $P$ is as $\gamma(G)$-set.

**Theorem 6.** Let $G = \mathsf{V}^1_{\{v\}}\ \{G_1, G_2, \dots, G_t\}$ and $v$ belongs to every $\gamma(G_i)$-set, then $\gamma(G) = \sum_{i=1}^t \gamma(G_i) - t + 1$.

**Proof.** By Lemma 2, $\bigcup_{i=1}^t S_i$ is a dominating set of $G$ such that $\gamma(G) = \gamma(\bigcup_{i=1}^t G_i) \le |\bigcup_{i=1}^t S_i|$. If $v$ belongs to every $\gamma(G_i)$-set then $v \in S_i$ for every $i$ such that $\bigcap_{i=1}^t S_i = \{v\}$. It implies $|\bigcup_{i=1}^t S_i| = 1 + \sum_{i=1}^t(|S_i| - 1) = \sum_{i=1}^t |S_i| - t + 1$. We have $\gamma(G) = \gamma(\bigcup_{i=1}^t G_i) \le \sum_{i=1}^t |S_i| - t + 1$. The least upper bound reached for $|S_i| = \gamma(G_i)$, so $\gamma(G) \le \sum_{i=1}^t \gamma(G_i) - t + 1$. Suppose there exists $P \subset V(G)$ which $P = \bigcup_{i=1}^t P_i$ for $P_i \subset V(G_i)$ such that $|P| = \sum_{i=1}^t \gamma(G_i) - t$. If $v \in P$ then there exists $P_i$ such that $|P_i| = \gamma(G_i) - 1$. However, there exists $x \in V(G_i) - P_i$ such that for every $y \in P_i$ holds $x \nsim y$. Because $V(G_i) \subset V(G)$ it mean that there is $x \in V(G) - P$ such that for every $y \in P$ holds $x \nsim y$. So, $P$ have not a dominating set. If $v \notin P$ then $|P_i| = \gamma(G_i) - 1$ for every $i$. It means that $P_i$ is not a dominating set with minimum cardinality. So does $P$. ♦

**Theorem 7.** If $G = \mathsf{V}^1_{\{v\}}\ \{G_1, G_2, \dots, G_t\}$ and $v$ has not belong to every $\gamma(G_i)$-set then $\gamma(G) = \sum_{i=1}^t \gamma(G_i)$.

**Proof.** By Lemma 2, we have $\gamma(G) = \gamma(\bigcup_{i=1}^t G_i) \le |\bigcup_{i=1}^t S_i|$. If there is no $\gamma(G_i)$-set consist of $v$ then $v \notin S_i$ for every $i$ such that $\bigcap_{i=1}^t S_i = \emptyset$. It implies $\gamma(G) = \gamma(\bigcup_{i=1}^t G_i) \le |\bigcup_{i=1}^t S_i| = \sum_{i=1}^t |S_i|$. The least upper bound reached for $|S_i| = \gamma(G_i)$, so $\gamma(G) \le \sum_{i=1}^t \gamma(G_i)$. Suppose there is vertices subset $P = \bigcup_{i=1}^t P_i$ which $|P| = \sum_{i=1}^t \gamma(G_i) - 1$. There exist $P_i$ such that $|P_i| = \gamma(G_i) - 1$ for every $i$. It means that $P_i$ is not dominating set with minimum cardinality. So do $P$. ♦

**Corollary 8.** Let $G = \mathsf{V}^1_{\{v\}}\ \{G_1, G_2, \dots, G_t\}$. If $v$ belongs to every $\gamma(G_i)$-set for $1 < i < p$ and $v$ has not belongs to any $\gamma(G_i)$-set for $p + 1 \le i \le t$, then $\gamma(G) = \sum_{i=1}^t \gamma(G_i) - p + 1$.

## 3. Conclusion

We conclude this paper with the domination number of vertex amalgamation of some graphs at a vertex $v$ is like the order of these graphs especially if $v$ belongs to every $\gamma(G_i)$-set, that is $\gamma(\mathsf{V}^1_{\{v\}}\ \{G_1, G_2, \dots, G_t\}) = \sum_{i=1}^t \gamma(G_i) - t + 1$. The following open problems for future work.

**Open problem.** Find the distance domination number of particular classes of graphs and the graphs obtained from graph operations.

**Acknowledgement**

This research is supported by Direktorat Riset dan Pengabdian Masyarakat, Ditjen Penguatan Risbang, Kemenristekdikti.

**References**

[1] Chartrand G and Lesniak L 1996 *Graphs & Digraphs* third edition (Florida Chapman & Hall/CRC)

IOP Conf. Series: Journal of Physics: Conf. Series **855** (2017) 012059 doi:10.1088/1742-6596/855/1/012059

[2] Go C E and Canoy Jr S R 2011 Domination in the corona and join of graphs *International Mathematical Forum* **6 no 16** 763–71
[3] Haynes T W, Hedetniemi S T and Slater P J 1998 *Fundamentals of Domination in Graphs* (New York Marcel Dekker Inc)
[4] Klobucar A 1999 Domination Numbers of Cardinal Product *Mathematica Slovaca* **49 no 4** 387-402
[5] Kuziak D Lemanska M and Yero I G 2014 Domination related parameters in rooted product graphs *arXiv : 1204.0644v2*
[6] Pavlic P and Zerovnik J 2013 A note on the domination number of the Cartesian products of paths and cycles *Kragujevac Journal of Mathematics* **37(2)** 275–85
[7] Snyder K 2011 c-Dominating Sets for Families of Graphs
[8] Wardani DAR *et al* 2014 Domination Number of Special Graphs *Proceedings of Seminar of Mathematics and Mathematics Education* **1 no 1** 78-82
[9] Yang Y and Kong X 2012 The thickness of amalgamations of graphs *arXiv: 1201.6483v1*

# Fisher Scoring Method for Parameter Estimation of Geographically Weighted Ordinal Logistic Regression (GWOLR) Model

Purnami Widyaningsih[1], Dewi Retno Sari Saputro[1], Aulia Nugrahani Putri[1]

[1] Department of Mathematics, Faculty of Mathematics and Natural Sciences, Universitas Sebelas Maret, Indonesia.
Email : purnami_w@staff.uns.ac.id, dewiretnoss@gmail.com ,aulianugrahaniputri@gmail.com

**Abstract.** GWOLR model combines geographically weighted regression (GWR) and (ordinal logistic reression) OLR models. Its parameter estimation employs maximum likelihood estimation. Such parameter estimation, however, yields difficult-to-solve system of nonlinear equations, and therefore numerical approximation approach is required. The iterative approximation approach, in general, uses Newton-Raphson (NR) method. The NR method has a disadvantage—its Hessian matrix is always the second derivatives of each iteration so it does not always produce converging results. With regard to this matter, NR model is modified by substituting its Hessian matrix into Fisher information matrix, which is termed Fisher scoring (FS). The present research seeks to determine GWOLR model parameter estimation using Fisher scoring method and apply the estimation on data of the level of vulnerability to Dengue Hemorrhagic Fever (DHF) in Semarang. The research concludes that health facilities give the greatest contribution to the probability of the number of DHF sufferers in both villages. Based on the number of the sufferers, IR category of DHF in both villages can be determined.

## 1. Introduction

In nondeterministic models, in order to find out the relationship between two or more variables, regression model is used. The variables involve predictor and response variables. The response variable can take the form of either quantitative or qualitative. A qualitative, or so-called categorical, variable is the measurement result of a variable involving two or more possible values (categories). Reference [1] points out that a categorical variable with two categories is termed a binary or dichotomous variable, while that with more than two categories is called polychotomous variable.

A regression model which can accommodate both categorical response variables and categorical (and/or quantitative) predictor variables is defined as a logistic regression model. It represents the probability of an event occurring resulted from predictor variables. There are two types of polychotomous logistic regression model: ordinal logistic regression and nominal logistic regression. The former takes ordered response variable into account, while the latter does not [2].

Meanwhile, Tobler's first law of Geography (Tobler in Anselin [1]), "Everything is related to everything else, but near things are more related than distant things", has laid foundation for regional science, specifically on spatial dimension. Spatial data refer to geographically-oriented data and possess certain coordinate system as a reference framework. This implies that spatial data contain information about geographic location of an area. Spatial analysis, in general, requires location-based data containing the location's characteristics. It includes modeling which indicates a concept of cause and effect relationship by using a method sourced from spatial and non-spatial data sources to predict the presence of a spatial pattern.

One of spatial models associated with regression model is geographically weighted regression (GWR). Logistic regression model has been developed to represent the relationship between response variable and predictor variables regarding a geographical location where data are observed. According to [3], the model is called geographically weighted logistic regression (GWLR). The GWLR model which involves response variables with ordinal scales is geographically weighted ordinal logistic regression (GWOLR). It combines geographically weighted regression (GWR) and (ordinal logistic regression) OLR models.

The GWOLR model includes response variables, predictor variables, and parameters. The parameters, however, are unknown, and therefore a sample-based estimation to obtain parameter values is required. Reference [2] assumes that one method to estimate parameters is maximum likelihood estimation (MLE). Such method, however, results in difficult-to-solve system of nonlinear equations, and therefore a numerical method is required.

Fisher scoring (FS) is a numerical method modified from Newton-Raphson (NR) method using score vectors and Fisher information matrix. The Fisher information plays a key role in statistical inference ([8], [9]). NR iterations employ Hessian matrix of which elements comprise the second derivatives of a likelihood function. The method, however, does not always produce converging results (Farbod [7]). In 2004, [5] compared NR's and FS' strengths in term of the computation of maximum likelihood estimation and concluded that FS was better than NR algorithm since FS algorithm converged when NR algorithm did not. In 2013, [6] applied Fisher scoring algorithm for estimating parameters in binary logistic regression model and found out FS' advantage—its guaranteed convergence. With regard to the gap, the present research seeks to examine parameter estimation using FS method in GWOLR model and provides an example of its application on data of the level of vulnerability to Dengue Hemorrhagic Fever (DHF) in Semarang city of Central Java.

## 2. Research Method

The present research presents a theoretical review examining FS method for GWOLR model parameter estimation. Several deepened theories included the modification of NR and FS, the correlation between Fisher Information and FS, and convergence matter of iterative methods. The results of the review will then be applied on data of the level of vulnerability to DHF in Semarang. The data were obtained from Central Agency on Statistics (*Badan Pusat Statistik—BPS*) and the Department of Health (*Dinas Kesehatan)* of Semarang.

Some research procedures were conducted to achieve the objective of the research. Firstly, we determined the likelihood function of joint probability density function and formulated its ln-likelihood. Afterwards, we determined a solution which enabled its maximization. In this stage, a constraint was found—a solution to system of nonlinear equations was difficult to find, and therefore numerical approximation using Fisher scoring was required. Next, we applied the results of its parameter estimation on data of the level of vulnerability to DHF in Semarang by categorizing DHF sufferers. According to the Ministry of Health (*Kementerian Kesehatan—Kemenkes*), DHF sufferers fall into three categories with regard to incidence rate (IR): mild, moderate, and severe. After that, we determined coordinate of each village (*kelurahan*) in Semarang, distance between villages, weight of each village, and parameter estimation. This way, parameter estimation values and GWOLR model were yielded.

## 3. Results and Discussion

*3.1 GWOLR Model.* GWOLR model is the combination of geographically weighted regression (GWR) and ordinal logistic regression (OLR) models of which each parameter depends on location $(u_i, v_i)$. According to reference [4] GWOLR model with categorical response variable $K$ is expressed as

$$Logit\big(P(Y_i < s|x_i)\big) = \alpha_s(u_i, v_i) + \boldsymbol{x_i}^{\boldsymbol{T}}\boldsymbol{\beta}(u_i, v_i),$$

where $s = 1,2,\dots,K-1$ and $i = 1,2,\dots,n$. Parameters $\alpha_s(u_i, v_i)$ represent intercepts, $\boldsymbol{\beta}(u_i, v_i)$ is the vector of regression coefficient for the $i^{\text{th}}$ location sized $p \times 1$, $\boldsymbol{x_i}^{\boldsymbol{T}}$ is the vector of predictor variables for the $i^{\text{th}}$ location sized $1 \times p$, and $(u_i, v_i)$ represent coordinates (latitude, longitude) of the $i^{\text{th}}$ location.

*3.2 Parameter Estimation.* The research examined samples of observation (response variables) $Y_1, Y_2, \dots, Y_n$ having $K$ category and probability of category towards $x$ $(P_k(x))$, where

$k = 1,2,\dots,K, \sum_{k=1}^{K} P_k(x) = 1$. Since the response variables have $K$ category (multinomial distribution), the likelihood function is expressed as joint probability density function of the multinomial distribution. The joint probability density function is denoted as

$$f(Y_k = y_k) = P_k(x) \qquad k = 1,2,\dots K$$

Therefore, the likelihood function of $n$ samples of observation is formulated by

$$L(u_i, v_i) = \prod_{i=1}^{n} f(y_{i1}) f(y_{i2}) \dots f(y_{iK})$$

$$= \prod_{i=1}^{n} (P_1(x_i)^{y_{i1}} P_2(x_i)^{y_{i2}} \dots P_K(x_i)^{y_{iK}})$$

$$= \prod_{i=1}^{n} \left( \left( \frac{\exp(\alpha_1(u_i, v_i) + \boldsymbol{x_i}^T \boldsymbol{\beta}(u_i, v_i))}{1 + \exp(\alpha_1(u_i, v_i) + \boldsymbol{x_i}^T \boldsymbol{\beta}(u_i, v_i))} \right)^{y_{i1}} \right.$$

$$\left( \frac{\exp(\alpha_2(u_i, v_i) + \boldsymbol{x_i}^T \boldsymbol{\beta}(u_i, v_i))}{1 + \exp(\alpha_2(u_i, v_i) + \boldsymbol{x_i}^T \boldsymbol{\beta}(u_i, v_i))} - \right.$$

$$\left. \frac{\exp(\alpha_1(u_i, v_i) + \boldsymbol{x_i}^T \boldsymbol{\beta}(u_i, v_i))}{1 + \exp(\alpha_1(u_i, v_i) + \boldsymbol{x_i}^T \boldsymbol{\beta}(u_i, v_i))} \right)^{y_{i2}} \dots$$

$$\left. \left( 1 - \frac{\exp(\alpha_{K-1}(u_i, v_i) + \boldsymbol{x_i}^T \boldsymbol{\beta}(u_i, v_i))}{1 + \exp(\alpha_{K-1}(u_i, v_i) + \boldsymbol{x_i}^T \boldsymbol{\beta}(u_i, v_i))} \right)^{y_{iK}} \right).$$

Likelihood function serves as exponential function, and therefore to simply the computation, the likelihood is altered to ln-likelihood function. In spatial analysis, to find out closeness between locations, a weight is required. The weight provided, therefore, is in the form of ln-likelihood. If the weight of each location $(u_i, v_i)$ is denoted $\boldsymbol{w}_{ij}(u_i, v_i)$, then the weighted ln-likelihood function is expressed as

$$ln\, L(u_i, v_i) = \sum_{i=1}^{n} \left( y_{i1} ln \left( \frac{\exp(\alpha_1(u_i, v_i) + \boldsymbol{x_i}^T \boldsymbol{\beta}(u_i, v_i))}{1 + \exp(\alpha_1(u_i, v_i) + \boldsymbol{x_i}^T \boldsymbol{\beta}(u_i, v_i))} \right) + \right.$$

$$\left( y_{i2} ln \frac{\exp(\alpha_2(u_i, v_i) + \boldsymbol{x_i}^T \boldsymbol{\beta}(u_i, v_i))}{1 + \exp(\alpha_2(u_i, v_i) + \boldsymbol{x_i}^T \boldsymbol{\beta}(u_i, v_i))} - \right.$$

$$\left. \frac{\exp(\alpha_1(u_i, v_i) + \boldsymbol{x_i}^T \boldsymbol{\beta}(u_i, v_i))}{1 + \exp(\alpha_1(u_i, v_i) + \boldsymbol{x_i}^T \boldsymbol{\beta}(u_i, v_i))} \right) + \dots + y_{iK} ln$$

$$\left. \left( 1 - \frac{\exp(\alpha_1(u_i, v_i) + \boldsymbol{x_i}^T \boldsymbol{\beta}(u_i, v_i))}{1 + \exp(\alpha_1(u_i, v_i) + \boldsymbol{x_i}^T \boldsymbol{\beta}(u_i, v_i))} \right) \right) \boldsymbol{w}_{ij}(u_i, v_i).$$

Weight $\boldsymbol{w}_{ij}(u_i, v_i)$ is the weight of fixed Gaussian kernel represented as $\boldsymbol{w}_{ij}(u_i, v_i) = exp\left(-\frac{1}{2}\left(\frac{d_{ij}}{h}\right)^2\right)$ where $d_{ij} = \sqrt{(u_i - u_j)^2 + (v_i - v_j)^2}$ is the distance between location $(u_i, v_i)$ and location $(u_j, v_j)$, $h$ is bandwidth (which measures the size of neighborhood). A method used to obtain optimum $h$ is cross validation (CV), represented as $CV(h) = \sum_{i=1}^{n} \sum_{k=1}^{K} (y_{\neq i,k} - P_{\neq i,k}(h))^2$, where $y_{\neq i,k}$ is response variable in which observation at location $(u_i, v_i)$ is negleted from estimation, and $P_{\neq i,k}(h)$ is the value of observation probability estimation at location $(u_i, v_i)$ having $K$ category. Optimum $h$ value is obtained from $h$ which produces minimum CV.

The parameter value which maximizes ln-likelihood function can be determined by computing the first derivatives of In-likelihood function on each parameter:

$$\begin{aligned}
\frac{\partial lnL(u_i,v_i)}{\partial \alpha_1(u_i,v_i)} &= \sum_{i=1}^{n}\left(y_{i1}\frac{1}{1+e_1} - y_{i2}\left(\frac{e_1(1+e_2)}{(1+e_1)(e_2-e_1)}\right)\right)\boldsymbol{w}_{ij}(u_i,v_i) \\
\frac{\partial lnL(u_i,v_i)}{\partial \alpha_2(u_i,v_i)} &= \sum_{i=1}^{n}\left(y_{i2}\left(\frac{e_2(1+e_1)}{(1+e_2)(e_2-e_1)}\right)\right)\boldsymbol{w}_{ij}(u_i,v_i) \\
&\vdots \\
\frac{\partial lnL(u_i,v_i)}{\partial \alpha_{K-1}(u_i,v_i)} &= \sum_{i=1}^{n}\left(y_{iK}\left(\frac{e_{K-1}}{(1+e_{K-1})}\right)\right)\boldsymbol{w}_{ij}(u_i,v_i) \\
\frac{\partial lnL(u_i,v_i)}{\partial \boldsymbol{\beta}(u_i,v_i)} &= \sum_{i=1}^{n}\left(y_{i1}\frac{\boldsymbol{x}_i^T}{1+e_1} - y_{i2}\left(\frac{\boldsymbol{x}_i^T(-1+e_1e_2)}{(1+e_1)(e_2-e_1)}\right)\right. \\
&\quad \left. - y_{iK}\frac{\boldsymbol{x}_i^T(e_{K-1})}{1+e_{K-1}}\right)\boldsymbol{w}_{ij}(u_i,v_i)
\end{aligned} \tag{1}$$

Where $e_1 = \exp(\alpha_1(u_i,v_i) + \boldsymbol{x}_i^T\boldsymbol{\beta}(u_i,v_i))$, $e_2 = \exp(\alpha_2(u_i,v_i) + \boldsymbol{x}_i^T\boldsymbol{\beta}(u_i,v_i))$, and $e_{K-1} = \exp(\alpha_{K-1}(u_i,v_i) + \boldsymbol{x}_i^T\boldsymbol{\beta}(u_i,v_i))$.

Moreover, maximum ln-likelihood function can be determined by its negative second derivatives. System (1) is a system of nonlinear equations. The value of GWOLR model parameter is the (exact) solution of system (1). The solution of system (1) is difficult to be found so that it is later determined numerically using FS method.

In principle, FS method is similar to NR, which is a modification of NR due to convergence matter in NR (Farbod [7]). GWOLR model parameter estimation with FS requires score vectors and Fisher information matrix. A score vector is a vector of which elements are the first derivatives of ln-likelihood function towards each parameter. It is expressed:

$$S = \begin{bmatrix}
\sum_{i=1}^{n}\left(y_{i1}\frac{1}{1+e_1} - y_{i2}\left(\frac{e_1(1+e_2)}{(1+e_1)(e_2-e_1)}\right)\right)\boldsymbol{w}_{ij}(u_i,v_i) \\
\sum_{i=1}^{n}\left(y_{i2}\left(\frac{e_2(1+e_1)}{(1+e_2)(e_2-e_1)}\right)\right)\boldsymbol{w}_{ij}(u_i,v_i) \\
\vdots \\
\sum_{i=1}^{n}\left(y_{iK}\left(\frac{e_{K-1}}{(1+e_{K-1})}\right)\right)\boldsymbol{w}_{ij}(u_i,v_i) \\
\sum_{i=1}^{n}\left(y_{i1}\frac{\boldsymbol{x}_i^T}{1+e_1} - y_{i2}\left(\frac{\boldsymbol{x}_i^T(-1+e_1e_2)}{(1+e_1)(e_2-e_1)}\right) - y_{iK}\frac{\boldsymbol{x}_i^T(e_{K-1})}{1+e_{K-1}}\right)\boldsymbol{w}_{ij}(u_i,v_i)
\end{bmatrix}.$$

Fisher information matrix is a modification of NR algorithm with substitution of its Hessian matrix. Hessian matrix is a matrix whose elements consist of the second derivatives of ln-likelihood function towards each parameter. It is denoted:

$$H = \begin{bmatrix} \frac{\partial^2 lnL(u_i,v_i)}{\partial \alpha_1{}^2(u_i,v_i)} & \frac{\partial^2 lnL(u_i,v_i)}{\partial \alpha_1(u_i,v_i)\partial \alpha_2(u_i,v_i)} & \cdots & \frac{\partial^2 lnL(u_i,v_i)}{\partial \alpha_1(u_i,v_i)\partial \boldsymbol{\beta}(u_i,v_i)} \\ \frac{\partial^2 lnL(u_i,v_i)}{\partial \alpha_2(u_i,v_i)\partial \alpha_1(u_i,v_i)} & \frac{\partial^2 lnL(u_i,v_i)}{\partial \alpha_2{}^2(u_i,v_i)} & \cdots & \frac{\partial^2 lnL(u_i,v_i)}{\partial \alpha_2(u_i,v_i)\partial \boldsymbol{\beta}(u_i,v_i)} \\ \vdots & \vdots & \ddots & \vdots \\ \frac{\partial^2 lnL(u_i,v_i)}{\partial \boldsymbol{\beta}(u_i,v_i)\partial \alpha_1(u_i,v_i)} & \frac{\partial^2 lnL(u_i,v_i)}{\partial \boldsymbol{\beta}(u_i,v_i)\partial \alpha_2(u_i,v_i)} & \cdots & \frac{\partial^2 lnL(u_i,v_i)}{\partial \boldsymbol{\beta}^2(u_i,v_i)} \end{bmatrix}.$$

The use of Hessian matrix in NR algorithm presents a disadvantage: it does not converge, and therefore it has to be modified with Fisher information matrix. The Fisher information matrix is a matrix of which elements consist of the expectation value of the second derivatives of ln-likelihood function towards each parameter. It is expressed as

$$Inf = \begin{bmatrix} E\left(\frac{\partial^2 lnL(u_i,v_i)}{\partial \alpha_1{}^2(u_i,v_i)}\right) & E\left(\frac{\partial^2 lnL(u_i,v_i)}{\partial \alpha_1(u_i,v_i)\partial \alpha_2(u_i,v_i)}\right) & \cdots & E\left(\frac{\partial^2 lnL(u_i,v_i)}{\partial \alpha_1(u_i,v_i)\partial \boldsymbol{\beta}(u_i,v_i)}\right) \\ E\left(\frac{\partial^2 lnL(u_i,v_i)}{\partial \alpha_2(u_i,v_i)\partial \alpha_1(u_i,v_i)}\right) & E\left(\frac{\partial^2 lnL(u_i,v_i)}{\partial \alpha_2{}^2(u_i,v_i)}\right) & \cdots & E\left(\frac{\partial^2 lnL(u_i,v_i)}{\partial \alpha_2(u_i,v_i)\partial \boldsymbol{\beta}(u_i,v_i)}\right) \\ \vdots & \vdots & \ddots & \vdots \\ E\left(\frac{\partial^2 lnL(u_i,v_i)}{\partial \boldsymbol{\beta}(u_i,v_i)\partial \alpha_1(u_i,v_i)}\right) & E\left(\frac{\partial^2 lnL(u_i,v_i)}{\partial \boldsymbol{\beta}(u_i,v_i)\partial \alpha_2(u_i,v_i)}\right) & \cdots & E\left(\frac{\partial^2 lnL(u_i,v_i)}{\partial \boldsymbol{\beta}^2(u_i,v_i)}\right) \end{bmatrix}.$$

The concept of the modification is called Fisher information which is later so-called FS. The followings are FS algorithm.

(1) Choose an initial value ($m$=0) of parameter vector $V_0$ which is obtained from parameter estimation value of ordinal logistic regression model.

(2) Calculate parameters

$$\boldsymbol{V}_{m+1} = \boldsymbol{V}_m + \boldsymbol{Inf}_m{}^{-1}\boldsymbol{S}_m, \text{ where } m = 01,2, \ldots.$$

(3) Calculating the norm $\boldsymbol{V}_{m+1} - \boldsymbol{V}_m = \|\boldsymbol{V}_{m+1} - \boldsymbol{V}_m\|$, where

$$\|\boldsymbol{V}_{m+1} - \boldsymbol{V}_m\| = \sqrt{(\alpha_{1(m+1)}(u_i,v_i) - \alpha_{1(m)}(u_i,v_i))^2 - (\beta_{1(m+1)}(u_i,v_i) - \beta_{p(m)}(u_i,v_i))^2}.$$

$\|\boldsymbol{V}_{m+1} - \boldsymbol{V}_{\boldsymbol{m}}\|$ is used to stop iteration with the criterion of $\|\boldsymbol{V}_{m+1} - \boldsymbol{V}_m\| <$ error tolerance. If the criterion is met, the iteration process stops and the parameter estimation value is $\widehat{\boldsymbol{V}} = \boldsymbol{V}_{m+1}$. On the other hand, if $\|\boldsymbol{V}_{m+1} - \boldsymbol{V}_m\| >$ error tolerance, the iteration process is repeated from step (2) to step (3).

After the parameter estimation value is obtained, GWOLR model is formulated as

$$Logit\big(P(Y_i < s|x_i)\big) = \widehat{\alpha_s}(u_i,v_i) + \boldsymbol{x}_i{}^{\boldsymbol{T}}\widehat{\boldsymbol{\beta}}(u_i,v_i).$$

*3.3 Application.* Data used in the application include data of the level of vulnerability to DHF in Semarang in 2014. The level of vulnerability to DHF comprises three categories (K=3) based on IR, namely mild, moderate, and severe, which are used as response variables, while the predictor variables are population density ($X_1$), the number of population aged 0-14 years ($X_2$), the number of semi-permanent houses ($X_3$), the availability of health facilities ($X_4$), and larvae-free rate ($X_5$), and therefore the p value is 5 ($p$=5). Thus, there are seven parameters, consisting of $V = [\alpha_1(u_i; v_i); \alpha_2(u_i; v_i); \beta_1(u_i; v_i); \beta_2(u_i; v_i); \beta_3(u_i; v_i); \beta_4(u_i; v_i); \beta_5(u_i; v_i)]$

Semarang city covers 16 sub-districts and 177 villages, and therefore $n$=177. Based on equation (3.1), GWOLR has 177 models for intercept parameter 1 and 177 models for intercept parameter 2. Each intercept parameter was provided with 2 samples of villages. The procedures included categorizing DHF sufferers, determining coordinate (latitude, longitude) of each village, calculating the distance between villages, calculating weight of each village and calculating parameter estimation. The initial value used is

$\alpha_1 = -4{:}843250, \alpha_2 = -2{:}804115,\ \beta_1 = 2{:}623848x10^{-6},\ \beta_2 = -4{:}601760x10^{-5},\ \beta_3 = 3{:}561353x10^{-4}, \beta_4 = -7{:}387467x10^{-1}, \hat{a}_5 = -1{:}737042x10^{-2}$ and therefore

$S_0 = [-0.1517446, 1.1334551, \ldots, -74.6099972]^T$ is obtained

The result of parameter estimation with error tolerance of 0.0001 is obtained in 15th iteration. The results of parameter estimation for Kuningan and Tinjomoyo villages are presented in Table 1.

Table 1. The results of GWOLR model parameter estimation

| Village | $\alpha_1(u_i; v_i)$ | $\alpha_2(u_i; v_i)$ | $\beta_1(u_i; v_i)$ | $\beta_2(u_i; v_i)$ | $\beta_3(u_i; v_i)$ | $\beta_4(u_i; v_i)$ | $\beta_5(u_i; v_i)$ |
|---|---|---|---|---|---|---|---|
| Ku | -4.704688 | -2.829467 | 0.000042 | -0.000048 | 0.000533 | -0.638496 | -0.017453 |
| Ti | -4.835065 | -2.821396 | -0.000004 | -0.000047 | 0.000376 | -0.746907 | -0.018490 |

In reference to Table 1, the following GWOLR models are obtained.

(1) GWOLR model of data of DHF for Kuningan village is

$$Logit(P(Y_1 \le 1|x_i)) = -4.704688 + 0.000042x_{i1} - 0.000048x_{i2} + 0.000533x_{i3} - 0.638496x_{i4} - 0.017453x_{i5}$$

$$Logit(P(Y_1 \le 2|x_i)) = -2.829467 + 0.000042x_{i1} - 0.000048x_{i2} + 0.000533x_{i3} - 0.638496x_{i4} - 0.017453x_{i5}$$

(2) GWOLR model of data of DHF for Tinjomoyo village is

$$Logit(P(Y_2 \le 1|x_i)) = -4.835065 - 0.00004x_{i1} - 0.000047x_{i2} + 0.000376x_{i3} - 0.746907x_{i4} - 0.018490x_{i5}$$

$$Logit(P(Y_2 \le 2|x_i)) = -2.821396 + 0.00004x_{i1} - 0.000047x_{i2} + 0.000376x_{i3} - 0.746907x_{i4} - 0.018490x_{i5}$$

On the basis of the models obtained, for Kuningan village, additions of one population ($X_1$) and one unit of semi-permanent house ($X_3$) lead to 0.0042% and 0.0533% increase in the number of DHF sufferers, so the village belongs to moderate or severe DHF IR category. Additions of one population aged 0-14 ($X_2$), one unit of health facilities ($X_4$), and 1% larvae-free rate ($X_5$) cause the probability of the number of DHF sufferers to reduce 0.0048%, 63.8496%, and 1.7453%, so the village is classified mild or moderate DHF IR category.

For Tinjomoyo village, additions of one unit of semi-permanent house ($X_3$) contribute to 0.0376% increase in probability of the number of DHF sufferers, therefore the village is included in moderate or severe DHF IR category. Meanwhile, additions of one population ($X_1$), one population aged 0-14 ($X_2$), one unit of health facilities ($X_4$), and 1% larvae-free rate ($X_5$) cause the probability of the number of DHF sufferers to reduce 0.0004%, 0.0047%, 74.6907%, and 1.849%, so the village belongs to mild or moderate DHF fever IR category.

## 4. Conclusions and Suggestions

(1) The result of GWOLR model parameter estimation using Fisher scoring mehod is: $\widehat{\boldsymbol{V}} = \boldsymbol{V}_{m+1}$ , where $\boldsymbol{V}_{m+1} = \boldsymbol{V}_m + \boldsymbol{Inf}_m{}^{-1}\boldsymbol{S}_m$ . The initial value obtained from the value of ordinal logistic regression model parameter estimation is provided, and therefore the GWOLR model is notated:

$$Logit\big(P(Y_i < s|x_i)\big) = \hat{\alpha}_s(u_i, v_i) + \boldsymbol{x_i}^T\widehat{\boldsymbol{\beta}}(u_i, v_i)$$

, where $s = 1,2, \ldots, K-1$.

(2) The Fisher information is expanded through NR algorithmic modification. It is notated in the form of matrix which is so-called Fisher information matrix. Fisher-information matrix is the modification of NR algorithm which substitutes its Hessian matrix. The fisher information matrix is a matrix of which elements consist of expectation values of the second derivatives of ln-likelihood function towards each parameter, stated as

$$Inf = \begin{bmatrix} E\left(\frac{\partial^2 lnL(u_i,v_i)}{\partial \alpha_1{}^2(u_i,v_i)}\right) & E\left(\frac{\partial^2 lnL(u_i,v_i)}{\partial \alpha_1(u_i,v_i)\partial \alpha_2(u_i,v_i)}\right) & \cdots & E\left(\frac{\partial^2 lnL(u_i,v_i)}{\partial \alpha_1(u_i,v_i)\partial \boldsymbol{\beta}(u_i,v_i)}\right) \\ E\left(\frac{\partial^2 lnL(u_i,v_i)}{\partial \alpha_2(u_i,v_i)\partial \alpha_1(u_i,v_i)}\right) & E\left(\frac{\partial^2 lnL(u_i,v_i)}{\partial \alpha_2{}^2(u_i,v_i)}\right) & \cdots & E\left(\frac{\partial^2 lnL(u_i,v_i)}{\partial \alpha_2(u_i,v_i)\partial \boldsymbol{\beta}(u_i,v_i)}\right) \\ \vdots & \vdots & \ddots & \vdots \\ E\left(\frac{\partial^2 lnL(u_i,v_i)}{\partial \boldsymbol{\beta}(u_i,v_i)\partial \alpha_1(u_i,v_i)}\right) & E\left(\frac{\partial^2 lnL(u_i,v_i)}{\partial \boldsymbol{\beta}(u_i,v_i)\partial \alpha_2(u_i,v_i)}\right) & \cdots & E\left(\frac{\partial^2 lnL(u_i,v_i)}{\partial \boldsymbol{\beta}^2(u_i,v_i)}\right) \end{bmatrix}.$$

The present research employed FS for GWOLR model parameter estimation, while methods for parameter estimation do not only employ FS so that developing other methods is highly possible.

## References

[1] McCullagh, P. and J. A. Nelder, "*Generalized Linear Models*," second ed., Chapman and Hall, 1983.

[2] Hosmer, D. W. and S. Lemeshow, "*Applied Logistic Regression*," John Wiley and Sons, Inc., USA, 2000.

[3] Atkinson, P. M., S. E. German, D. A. sear, and M. J. Clark, "*Exploring the Relations Between Riverbank Erison and Geomorphological Control Using Geographically Weighted Logistic Regression*", Ohio: Ohio State University, vol. 35, pp. 58-82, 2003.

[4] Purhadi, M. Rifada, and P. Wulandari, "*Geographically Weighted Ordinal Logistic Regression Model*," International Journal of Mathematics and Computation, vol. 16, pp. 116-216, 2012.

[5] Schworer, A. and P. Hovey, "*Newton Raphson versus Fisher Scoring Algorithms in Calculating Maximum Likelihood Estimates*," Dayton, 2004.

[6] Marius, O. U. and O. I. C. Anaene, "*Estimating the Fisher's Scoring Matrix Formula from Logistic Model*," American Journal of Theoretical and Applied Statistics," vol. 2, pp. 221-227, 2013.

[7] Farbod, D., Ebrahimpour, M., Ghayournoradi, Z., (2010), *Maximum Likelihood Estimation for Distribution Generated by Cauchy Stable Law*, International Journal of Mathematics and Computation, Vol. 7, No. J10, pp. 23-28.

[8] Lehman, E. L., & Casella, G. (1998). *Theory of Point Estimation 2nd edition*. New York, NY: Springer.

[9] Spanos, A. (1999). *Probability Theory and Statistical Inference: Econometric Modeling with Observational Data*. Cambridge, UK: Cambridge University Press.

[10] Tobler, W.R., 1970. *A Computer Movie Simulating Urban Growth in the Detroit Region*, Economic Geography, 46: 234-240.

[11] Anselin, L., *Spatial econometrics : methods and models,* Kluwer Academic Publishers, Dordrecht, 1998.

# Bayesian Hierarchical Random Intercept Model Based on Three Parameter Gamma Distribution

**Ika Wirawati, Nur Iriawan*, and Irhamah**
Departement of Statistics, Institut Teknologi Sepuluh Nopember Surabaya, Indonesia

E-mail: wirawati_ika15@mhs.statistika.its.ac.id, nur_i@statistika.its.ac.id, irhamah@statistika.its.ac.id

**ABSTRACT.** Hierarchical data structures are common throughout many areas of research. Beforehand, the existence of this type of data was less noticed in the analysis. The appropriate statistical analysis to handle this type of data is the hierarchical linear model (HLM). This article will focus only on random intercept model (RIM), as a subclass of HLM. This model assumes that the intercept of models in the lowest level are varied among those models, and their slopes are fixed. The differences of intercepts were suspected affected by some variables in the upper level. These intercepts, therefore, are regressed against those upper level variables as predictors. The purpose of this paper would demonstrate a proven work of the proposed two level RIM of the modeling on per capita household expenditure in Maluku Utara, which has five characteristics in the first level and three characteristics of districts/cities in the second level. The per capita household expenditure data in the first level were captured by the three parameters Gamma distribution. The model, therefore, would be more complex due to interaction of many parameters for representing the hierarchical structure and distribution pattern of the data. To simplify the estimation processes of parameters, the computational Bayesian method couple with Markov Chain Monte Carlo (MCMC) algorithm and its Gibbs Sampling are employed.

## 1. Introduction

Often encountered in many researches that the data used has a hierarchical structure. For example, in a social research, human populations are structured by hierarchical groupings. Whereby, individuals are grouped within households, households are grouped within community, and at the top level could be a district/city, province or country [9]. Grouping on hierarchical data is built on the principle of the similarities between members of a group, so that the members in the group have similar characteristics [8]. While on the other hand, among the members from a group to the other group members are different, so-called there is variation between groups. Therefore, the hierarchical structure of the data cannot be ignored in the analysis process.

Beforehand, the existence of a hierarchical structure in data was less noticed in the analysis. Conventional statistical analysis will not provide the proper inference and will not be able to capture the real phenomenon if they ignore the existence of this data type [13]. When the observation data has a hierarchical structure, then the appropriate analysis is hierarchical or multilevel models. The hierarchical model has two main advantages [12]. First, it can be used to analyze at several different levels simultaneously in a single statistical analysis. Secondly, it considers the variance at each level of the response variance. In addition, a study had proven that the hierarchical model was better than the classical methods (unilevel model) [15]. This study was performed using hierarchical data simulation.

The hierarchical model was developed for the data analysis involving two or more levels of correlation among variables and parameters. Thus the hierarchical model can be used to examine the relationship between the measured variables at different levels in the hierarchical data structure [12]. Basically, the hierarchical model constructed by two sub-models, micro and macro models. In the two-level hierarchical model, micro models are the regression model between the observed response and predictor in the first/lowest level. Meanwhile macro models are the model that show the relationship between the regression coefficients from micro models to the predictor variables on the second/upper level [13].

This article will only focus on random intercept model (RIM), as a subclass of the hierarchical model. This model assumes that the intercept of models in the lowest level are varied among those models, and their slopes are fixed. The differences of intercepts were suspected affected by some variables in the upper level. These intercepts, therefore, are regressed against those upper level variables as predictors. This paper would demonstrate the work of this two level RIM on the household expenditure in Maluku Utara, which has five characteristics in the first level and three characteristics of districts/cities in the second level.

Distribution of household expenditure has a unique characteristic, which it has a positive value and a long right tail (skewed distribution). Some distribution that represents these properties well are log-normal distribution, log-logistic distribution, gamma distribution, etc. Several previous studies have used these distributions to explain the pattern of household expenditure data, for example [1], [11], [19], and [14]. The data in this study were captured by the three parameters Gamma (Gamma3) distribution. The household expenditure in nine of ten districts/cities in Maluku Utara shown to have the Gamma3 distribution.

The model, therefore, would be more complex due to interaction of many parameters for representing the hierarchical structure and distribution pattern of the data. For the complex hierarchical model, parameter estimation using classical approach become very difficult to be done. Bayesian modeling approach on this issue going to provide a better solution. Bayesian method is very flexible and easy to estimate the parameters of a complex hierarchical models [17]. To simplify the estimation processes of parameters, the computational Bayesian method couple with Markov Chain Monte Carlo (MCMC) algorithm and its Gibbs Sampling are employed.

## 2. Three Parameter Gamma (Gamma3) Distribution

In probability theory and statistics, the Gamma3 distribution is a three parameter family of continuous probability distributions. Consider that $Y_1, Y_2, \dots, Y_n$ , is a random sample from a population with the Gamma3 distribution. The probability distribution function (PDF) can be written as follows [16]:

$$f(y|\alpha,\phi,\lambda) = \frac{\phi^{\alpha}(y-\lambda)^{1-\alpha}}{\Gamma(\alpha)}\exp[-\phi(y-\lambda)], \alpha > 0, \phi > 0, y > \lambda, \tag{1}$$

where $\alpha$ is the shape parameter, $\phi$ is the scale parameter of the distribution. Meanwhile $\lambda$ is the location or threshold parameter. If $\lambda = 0$, then $Y$ would be a gamma distribution with two parameters, are denoted by $\text{Gamma}(\alpha, \phi)$. The mean and variance of the Gamma3 distribution is [2]:

$$E(Y) = \lambda + \frac{\alpha}{\phi},$$

$$V(Y) = \frac{\alpha}{\phi^2}.$$

Gamma distribution has been widely used in many fields, such in the science of Physics, Meteorology, Ecology, and other fields. Even in 1974, Salem and Mount had showed that the gamma distribution is more appropriate than the log-normal distribution to illustrate the income data of people in the United States in 1960 to 1969 [16].

## 3. Random Intercept Model

Basically, hierarchical model is formed by two sub-models, it is micro models (model at the lowest level of the hierarchical structure) and macro models (model at the upper level of the hierarchical structure) [12]. In principle, the depth level of the models hierarchy is not limited, however, for further elaboration is used two-level hierarchical model. The micro models illustrate association between household expenditure as the response and household characteristics as predictors in the first level. While the macro models represent the relationship between coefficients in the micro models with district characteristics as the predictors in the second level.

The two-level hierarchical model can be described as follows [20] and [8]:

1. Micro Model

Let $N$ is number of household in Maluku Utara Province, $n_j$ is number of household in $j^{th}$ district, $m$ is number of districts/cities. The model at the first level of each district/city can be expressed as below:

$$Y_{ij} = \beta_{0j} + \beta_{1j}X_{1ij} + \cdots + \beta_{kj}X_{kij} + e_{ij},\ i = 1,2,\dots,n_j \text{ dan } j = 1,2,\dots,m, \tag{2}$$

or it can also be written in vector as,

$$\mathbf{y}_j = \mathbf{X}_j\boldsymbol{\beta}_j + \mathbf{e}_j, \tag{3}$$

where,

$\mathbf{y}_j = (y_{1j}\quad y_{2j}\quad \cdots\quad y_{n_jj})^T$,

$$\mathbf{X}_j = \begin{pmatrix} 1 & X_{11j} & X_{21j} & \cdots & X_{kij} \\ 1 & X_{12j} & X_{22j} & \cdots & X_{k2j} \\ \vdots & \vdots & \vdots & \vdots & \vdots \\ 1 & X_{1n_jj} & X_{2n_jj} & \cdots & X_{kn_jj} \end{pmatrix},$$

$\boldsymbol{\beta}_j = (\beta_{0j}\quad \beta_{1j}\quad \cdots\quad \beta_{kj})^T$,

$\mathbf{e}_j = (e_{1j}\quad e_{2j}\quad \cdots\quad e_{n_jj})^T$.

$\mathbf{X}$ is predictors matrix indicated a household characteristics, $\boldsymbol{\beta}$ is regression coefficients vector of the micro models and $\mathbf{e}$ is residual vector of the micro models with assumption $\mathbf{e}_j \sim N\left(\mathbf{0}, \sigma_j^2 \boldsymbol{I}_{n_j}\right)$.

2. Macro Model

Regression coefficients of the micro model have varying values between districts/cities. Therefore, these variations will be explained by macro model. Modelling conducted for each regression coefficient in micro models using the predictors at the second level (macro predictors). These predictors represented district characteristic, noted by $\mathbf{W}$. The macro model is described in this following equations,

$$\beta_{rj} = \gamma_{0r} + \sum_{l=1}^{q} \gamma_{lr}W_{lj} + u_{rj},\ r = 0,1,2,\dots,k \tag{4}$$

or it can also be written in vector as,

$$\boldsymbol{\beta}_r = \mathbf{W}\boldsymbol{\gamma}_r + \mathbf{u}_r, \tag{5}$$

where,

$\boldsymbol{\beta}_r = (\beta_{r1} \quad \beta_{r2} \quad \cdots \quad \beta_{rm})^T,$

$$\mathbf{W} = \begin{pmatrix} 1 & w_{11} & w_{21} & \cdots & w_{l1} \\ 1 & w_{12} & w_{22} & \cdots & X_{l2} \\ \vdots & \vdots & \vdots & \vdots & \vdots \\ 1 & w_{1m} & w_{2m} & \cdots & X_{lm} \end{pmatrix},$$

$\boldsymbol{\gamma}_j = (\gamma_{01} \quad \gamma_{1r} \quad \cdots \quad \gamma_{lr})^T,$

$\mathbf{u}_r = (u_{r1} \quad u_{r2} \quad \cdots \quad u_{rm})^T.$

$\boldsymbol{\gamma}$ is regression coefficients vector of the macro models and $\mathbf{u}$ is residual vector of the macro models with assumption $\mathbf{u}_j \sim N(\mathbf{0}, \mathbf{T})$.

Equation (2), (3), (4) and (5) are known as separate equation model of hierarchical model. And the single equation model of hierarchical model is a combination of separate equation model, written as follow [6]:

$$\mathbf{y}_j = \mathbf{X}_j\mathbf{W}_j\boldsymbol{\gamma} + \mathbf{X}_j\mathbf{u}_j + \mathbf{e}_j, \tag{6}$$

Hierarchical model is built by regress all regression coefficients in micro models with predictors in macro model. When the hierarchical model only use intercept in micro models for regressed at the upper level, this model is known as random intercept model (RIM). The micro models of RIM are same with equation (2) and (3), rewritten for more simply as follow:

$$Y_{ij} = \beta_{0j} + \sum_{r=1}^{k} \beta_{rj}X_{rij} + e_{ij}, \tag{7}$$

while the macro model is,

$$\beta_{0j} = \sum_{l=0}^{q} \gamma_{l0}W_{lj} + u_{0j}, \tag{8}$$

and the single equation of RIM is written as follow:

$$Y_{ij} = \sum_{l=0}^{q} \gamma_{l0}W_{lj} + u_{0j}, + \sum_{r=1}^{k} \beta_{rj}X_{rij} + e_{ij}. \tag{9}$$

## 4. Bayesian Methods

Conceptually, Bayesian methods was developed based on Bayes' theorem, which combines formally prior distribution and data information (likelihood function) into the posterior distribution [18]. Consider Bayes' theorem [3]:

$$p(\boldsymbol{\theta}|\mathbf{y}) = \frac{p(\mathbf{y}|\boldsymbol{\theta})p(\boldsymbol{\theta})}{p(\mathbf{y})}, \tag{10}$$

where $\boldsymbol{\theta}$ is parameter vector and $\mathbf{y}$ is vector of observation from the sample. $p(\mathbf{y})$ is defined as normalized constant with respect to $\boldsymbol{\theta}$. Then, the posterior can be explained as a proportional form,

$$p(\boldsymbol{\theta}|\mathbf{y}) \propto p(\mathbf{y}|\boldsymbol{\theta}) \text{ x } p(\boldsymbol{\theta}), \tag{11}$$

Refer to equation (11), the posterior distribution of hierarchical Bayesian models could be specified as follow [20]:

$$p(\boldsymbol{\beta}, \boldsymbol{\gamma}, \boldsymbol{\Omega}, \mathbf{T}|\mathbf{y}) \propto p(\mathbf{y}|\boldsymbol{\beta}, \boldsymbol{\Omega})p_1(\boldsymbol{\beta}|\boldsymbol{\gamma}, \mathbf{T})p_2(\boldsymbol{\gamma}, \boldsymbol{\Omega}, \mathbf{T}), \tag{12}$$

where $\mathbf{\Omega} = Var(\mathbf{y}) = \sigma_j^2 \mathbf{I}_N$, $p_1(\boldsymbol{\beta}|\boldsymbol{\gamma}, \mathbf{T})$ and $p_2(\boldsymbol{\gamma}, \mathbf{\Omega}, \mathbf{T})$ denotes first stage prior and second stage prior.

In parameter estimation of complex model, the process of determining the posterior distribution is usually difficult. This process involves integral equation with large dimensions and needs a long time. A numerical approach as a solution to solve this problem is a Markov Chain Monte Carlo (MCMC) [4]. MCMC approach involves the simulation process by taking a random sample of the posterior distribution. This process is executed following due Markov chain process using Monte Carlo simulation iteratively to obtain convergence condition on posterior [18]. Furthermore, sample parameters of Markov chain are taken after a stationary condition reached. Thus, it can be guaranteed that the drawn sample is a posterior distribution sample of the parameter.

Implementation of MCMC methods in Bayesian analysis requires a proper sampling algorithm to obtain a sample of a distribution. Gibbs sampler is a highly efficient generator, so it is often used as a generator of random variables in data analysis using MCMC [7]. Gibbs sampler can be defined as a simulation technique to generate random variables of the marginal distribution functions without having to calculate the density function [5].

## 5. Data

The data used in this study is sourced from Badan Pusat Statistik (BPS), included data result of National Socioeconomic Survey (Susenas) 2015 and publications of BPS Maluku Utara Province. Data coverage in this study includes 3.749 household samples of Susenas and 10 districts/cities in Maluku Utara Province. The response used in the model is per capita household expenditure (**y**), while the used predictors are household characteristics as predictors at the first level (**X**) and district characteristics as predictors at the second level (**W**). The list of predictors is presented in Table 1.

**Table 1**. List of Predictors at The First and Second Level

| First Level Predictors | | Second Level Predictors | |
|---|---|---|---|
| $\mathbf{X}_1$ : | Head of household age | $\mathbf{W}_1$ : | Per Capita GRDP (million rupiahs) |
| $\mathbf{X}_2$ : | Household dependency ratio | $\mathbf{W}_2$ : | Percentage of villages had roads drivable by four-wheel vehicle all year |
| $\mathbf{X}_3$ : | Head of household employement status | | |
| $\mathbf{X}_4$ : | Head of household educational level | $\mathbf{W}_3$ : | Percentage of original local government revenues to total regional revenue |
| $\mathbf{X}_5$ : | Asset ownership | | |

## 6. Identification of The Model Structure

Goodness of fit test on per capita household expenditure obtained the result that the data in nine districts/cities in Maluku Utara are following the three parameters Gamma distribution. Therefore, this distribution will be based on the hierarchical model analysis in this study. Figure 1 illustrate the three parameters Gamma distribution pattern of response data in Maluku Utara.

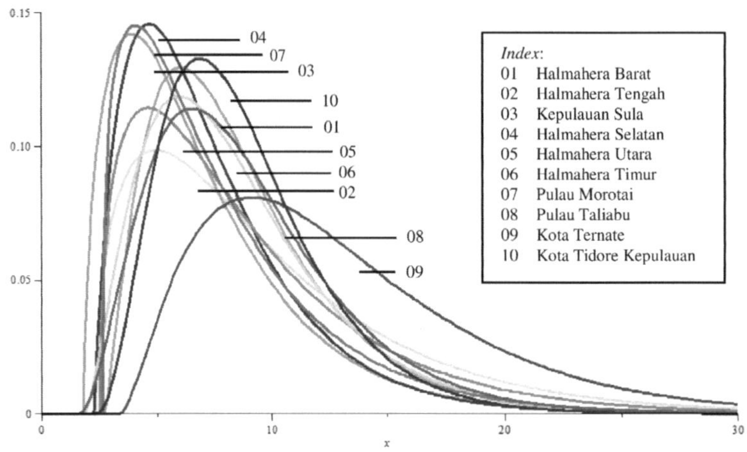


**Figure 1.** The Distribution Plot of per Capita Household Expenditure

Based on the figure, it can be shown that the three parameters Gamma distribution can explained clearly the difference distribution pattern of household expenditure data among the districts/cities. These differences may be caused by the varied parameter values of the three parameters Gamma distribution. In this study, the third parameter (threshold parameter) was assumed varying, while the others are fixed. Variation of the threshold parameter was suspected be affected by household and district characteristics. Thus, the structure of models RIM using Bayesian approach in this study can be formulated as follows:

$$Y_{ij} \sim Gamma3(\alpha_j, \phi_j, \lambda_{ij})$$

$$\lambda_{ij} = \beta_{0j} + \beta_{1j}X_{1ij} + \beta_{2j}X_{2ij} + \beta_{3j}X_{3ij} + \beta_{4j}X_{4ij} + \beta_{5j}X_{5ij} + e_{ij},$$

$$\lambda_{ij} = \beta_{0j} + \sum_{r=1}^{5} \beta_{rj}X_{rij} + e_{ij},$$

where,

$$\beta_{0j} = \gamma_{00} + \gamma_{10}W_{1j} + \gamma_{20}W_{2j} + \gamma_{30}W_{3j} + u_{0j},$$

$$\beta_{0j} = \gamma_{00} + \sum_{l=1}^{3} \gamma_{l0}W_{lj} + u_{0j}$$

Then, the single equation model of RIM can be written as,

$$Y_{ij} \sim Gamma3(\alpha_j, \phi_j, \lambda_{ij})$$

$$\lambda_{ij} = \gamma_{00} + \sum_{l=1}^{3} \gamma_{l0}W_{lj} + \sum_{r=1}^{5} \beta_{rj}X_{rij} + (e_{ij} + u_{0j}). \quad (13)$$

Parameters of the equation (13) was estimated computationally using WinBUGS 1.4. The software was developed using the Gibbs sampler and MCMC. Therefore, it is very suitable to achieve the objectives of this study.

## 7. Main Results

The model structure has been formulated previously are implemented to model the per capita household expenditure data in Maluku Utara. According to the Bayesian hierarchical RIM concept, the parameters of micro and macro models were estimated simultaneously. Modeling was calculated based on the three parameter Gamma using five household predictors and three district predictors. Parameter estimation of model (13) using WinBUGS were conducted by 10,000 iterations, 10 iterations of thin and additional discarded 400 burn in. Thus, the samples used to estimate the parameters of micro and macro models were 9,600 samples. The estimation results are given in Table 2.

The estimation results of the WinBUGS also displayed a credible interval of the estimated posterior. This interval is analogous to the confidence intervals in frequentist. Credible interval is used as the interval of estimated parameter. Thus, the credible interval could also be used as a diagnostic model, which parameter values are zero for the null hypothesis. The estimated parameters are decided significant when credible interval were non zero. Based on Table 2 can be seen that almost all of the coefficients in models of micro significant at 95% credible interval. Only predictor of asset ownership in three districts/cities were not significant, i.e. Halmahera Timur, Pulau Taliabu, and Ternate.

The five household predictors were proven effected on the rate of per capita household expenditure. Four of the five predictors have a positive effect, such head of household age ($X_1$), head of household employment status ($X_3$), head of household educational level ($X_4$), and assets ownership ($X_5$). While the household dependency ratio (X2) has negative effect on per capita household expenditure.

**Table 2. Estimated Coefficients of Bayesian Hierarchical Random Intercept Model**

| Parameters | Estimated Value | Credible Interval (95%) | | Parameters | Estimated Value | Credible Interval (95%) | |
|---|---|---|---|---|---|---|---|
| | | Lower Bounds | Upper Bounds | | | Lower Bounds | Upper Bounds |
| **Estimated Coefficients of Micro Model** | | | | | | | |
| $\beta_{01}$ | 109200 | 91510 | 124900 | $\beta_{31}$ | 4003 | 3039 | 4953 |
| $\beta_{02}$ | 96520 | 73960 | 115300 | $\beta_{32}$ | 4021 | 3026 | 4998 |
| $\beta_{03}$ | 34400 | 18940 | 48750 | $\beta_{33}$ | 4079 | 3112 | 5071 |
| $\beta_{04}$ | -78830 | -85440 | -72590 | $\beta_{34}$ | 3996 | 3000 | 4986 |
| $\beta_{05}$ | 89090 | 75780 | 101100 | $\beta_{35}$ | 3997 | 3022 | 4992 |
| $\beta_{06}$ | -175000 | -189300 | -161200 | $\beta_{36}$ | 4001 | 3023 | 4963 |
| $\beta_{07}$ | 72830 | 54570 | 90300 | $\beta_{37}$ | 4115 | 3135 | 5105 |
| $\beta_{08}$ | -133400 | -148200 | -120700 | $\beta_{38}$ | 3992 | 2999 | 4972 |
| $\beta_{09}$ | -138600 | -150900 | -126500 | $\beta_{39}$ | 3966 | 2967 | 4968 |
| $\beta_{010}$ | 89670 | 65070 | 112900 | $\beta_{310}$ | 4020 | 3030 | 4992 |
| $\beta_{11}$ | 2816 | 2620 | 3016 | $\beta_{41}$ | 8649 | 7911 | 9379 |
| $\beta_{12}$ | 2815 | 2619 | 3012 | $\beta_{42}$ | 8652 | 7928 | 9397 |
| $\beta_{13}$ | 2794 | 2603 | 2996 | $\beta_{43}$ | 8656 | 7924 | 9388 |
| $\beta_{14}$ | 3028 | 2825 | 3229 | $\beta_{44}$ | 8670 | 7946 | 9406 |
| $\beta_{15}$ | 2817 | 2627 | 3015 | $\beta_{45}$ | 8664 | 7897 | 9419 |
| $\beta_{16}$ | 2727 | 2528 | 2928 | $\beta_{46}$ | 8670 | 7919 | 9415 |
| $\beta_{17}$ | 2651 | 2470 | 2837 | $\beta_{47}$ | 8658 | 7915 | 9404 |
| $\beta_{18}$ | 2848 | 2646 | 3048 | $\beta_{48}$ | 8646 | 7903 | 9387 |
| $\beta_{19}$ | 2824 | 2621 | 3025 | $\beta_{49}$ | 8619 | 7883 | 9354 |
| $\beta_{110}$ | 2840 | 2641 | 3040 | $\beta_{410}$ | 8655 | 7900 | 9403 |
| $\beta_{21}$ | -3515 | -4227 | -2792 | $\beta_{51}$ | 1364 | 140.3 | 2594 |
| $\beta_{22}$ | -3448 | -4168 | -2732 | $\beta_{52}$ | 1329 | 81.64 | 2582 |
| $\beta_{23}$ | -3437 | -4148 | -2728 | $\beta_{53}$ | 1337 | 105.2 | 2587 |
| $\beta_{24}$ | -3459 | -4180 | -2750 | $\beta_{54}$ | 1452 | 225.8 | 2697 |
| $\beta_{25}$ | -3420 | -4135 | -2698 | $\beta_{55}$ | 1374 | 114.1 | 2636 |
| $\beta_{26}$ | -3421 | -4152 | -2704 | $\beta_{56}$ | 1192 | -42.74 | 2451 |
| $\beta_{27}$ | -3478 | -4199 | -2770 | $\beta_{57}$ | 1348 | 107.4 | 2589 |
| $\beta_{28}$ | -3401 | -4130 | -2684 | $\beta_{58}$ | 1221 | -55.67 | 2482 |
| $\beta_{29}$ | -3414 | -4118 | -2698 | $\beta_{59}$ | 1217 | -25.06 | 2451 |
| $\beta_{210}$ | -3462 | -4178 | -2746 | $\beta_{510}$ | 1289 | 24.01 | 2557 |
| **Estimated Coefficients of Macro Model** | | | | | | | |
| $\gamma_{00}$ | 23730 | 22250 | 25200 | $\gamma_{20}$ | 325.4 | 103.3 | 551 |
| $\gamma_{10}$ | 1249 | 983.4 | 1517 | $\gamma_{30}$ | 4312 | 4058 | 4570 |

* Not significant coefficient

The results showed that the intercept of micro models have different values among the districts. This condition was significantly affected by the district characteristics. It was shown by all estimated parameter of the district predictors were significant at 95% credible interval. This means that the three districts predictor effected on per capita household expenditure. the third predictors, especially, had positive effects on per capita household expenditure.

## 8. Conclusions

Household and district characteristics had been proven to give significant effects on per capita household expenditure in Maluku Utara. This situation had been explained by the random intercept model based on three parameters Gamma distribution using Bayesian approach. Thus, this model had demonstrated its work to capture the hierarchical structure in the data. Another interesting future perspective is to model all of the coefficient parameters regression at the first level with predictors at level the second level. This will be performed in the author's final project.

## References

[1] Alanko T 1997 Statistical Models for Estimating the Distribution Function of Alcohol Consumption: A Parametric Approach (Vol. 44). *Finnish Foundation for Alcohol Studies; Piscataway, NJ, USA: Rutgers University Center of Alcohol Studies*

[2] Basak I and Balakrishnan, N 2012 Estimation for the three-parameter gamma distribution based on progressively censored data. *Statistical Methodology*, *9*(3), 305-319

[3] Box G E and Tiao G C 1973 *Bayesian Inference in Statistical Analysis* (United State: Addison-Wesley)

[4] Carlin B P and Chib, S 1995 Bayesian Model Choice via Markov Chain Monte Carlo Methods. *Journal of the Royal Statistical Society. Series B (Methodological)*, 473-484

[5] Casella G and George E I 1992 Explaining the Gibbs Sampler *The American Statistician*, **46(3)**, 167-174

[6] De Leeuw J, Meijer E, and Goldstein H 2008 *Handbook of Multilevel Analysis* (New York: Springer)

[7] Gelman A, Carlin J B, Stern H S, and Rubin D B 2014 *Bayesian Data Analysis, Third Edition* (USA: Chapman & Hall/CRC)

[8] Goldstein H 1995 *Multilevel Statistical Models* (London: Edward Arnold)

[9] Goldstein H, Rasbash J, Browne W, Woodhouse G, and Poulain M 2000 Multilevel Models in The Study of Dynamic Household Structures *European Journal of Population/Revue européenne de Démographie*. **16(4)**, 373-387

[11] Ghosh A, Gangopadhyay K and Basu B 2011 Consumer Expenditure Distribution in India, 1983–2007: Evidence of A Long Pareto Tail. *Physica A: Statistical Mechanics and its Applications*, ***390*****(1)**, 83 97

[12] Hox J J, Moerbeek M, and Van de Schoot R 2010 *Multilevel analysis: Techniques and applications*. (New York: Routledge)

[13] Iriawan N 2012 *Modelling and Data-Driven Analysis, Volume 1* (Surabaya: ITS Press)

[14] Ismartini P 2013 *Development in Hierarchical Linear Model with Bayesian Approachon Modelling of per Capita Household Expenditure Data* (Dissertation), Institut Teknologi Sepuluh Nopember, Surabaya

[15] Ismartini P, Iriawan N, and Ulama B S S 2013 Comparison of Unilevel and Multilevel Model for Hierarchical Structure Data Analysis Using Bayesian Approach. *Paper Presented at The Proceeding of the Fourth National Mathematics Conference.*

[16] Johnson N L, Kotz S, and Balakrishnan N 1995 *Continuous Univariate Distribution, Volume 1 Second Edition* (New York: John Wiley and Sons)

[17] Kruschke J K and Vanpaemel W 2015 Bayesian Estimation in Hierarchical Models. *The Oxford Handbook of Computational and Mathematical Psychology*, 279

[18] Ntzoufras I 2009 *Bayesian Modeling in WinBugs* (New Jersey, USA: John Wiley & Sons, Inc)

[19] Pertiwi R 2012 *Modelling per Capita Expenditure of Regencies Kalimantaan Barat Using Hierarchical Bayesian Method* (Surabaya: Institut Teknologi Sepuluh Nopember)

[20] Raudenbush S W and Bryk A S 2002 *Hierarchical Linear Models: Applications and Data Analysis Methods, Second Edition* (Vol. 1). (London: Sage).

# Median Filter Noise Reduction of Image and Backpropagation Neural Network Model for Cervical Cancer Classification

**D. U. Wutsqa, M. Marwah**

Mathematics Department, Yogyakarta State University, Yogyakarta, Indonesia

dhoriva@yahoo.com, marwahmasruroh2@gmail.com

Abstract. In this paper, we consider spatial operation median filter to reduce the noise in the cervical images yielded by colposcopy tool. The backpropagation neural network (BPNN) model is applied to the colposcopy images to classify cervical cancer. The classification process requires an image extraction by using a gray level co -occurrence matrix (GLCM) method to obtain image features that are used as inputs of BPNN model. The advantage of noise reduction is evaluated by comparing the performances of BPNN models with and without spatial operation median filter. The experimental result shows that the spatial operation median filter can improve the accuracy of the BPNN model for cervical cancer classification.
Keywords—spatial operation median filter, backpropagation neural network, cervical cancer classification, colposcopy.

## 1. Introduction

Cervical cancer is one type cancer that occurs in the woman's cervix, it begins with the appearance of abnormal cells (dysplasia). Cervical cancer grows very slowly and the development is very difficult to be detected. Most cases of cervical cancer detected in the hospital have advanced stages, so they are difficult to be treated. Early detection of cervical cancer is very important to prevent death from cervical cancer. Detection of cervical cancer can be performed by pap smear test.

Previous researches have applied soft computing approach to classify the stadium of cervical cancer using the pap smear data. One of which is the three stages method including Adaptive Fuzzy Moving K-means clustering, feature extracting and Fuzzy Min-Max neural network [1]. The other researches involve the radial basis neural network with adaptive fuzzy K-means clustering [2], Support Vector Machine (SVM) classifier, K-nearest neighbour and artificial neural network [3], and multilayer perceptron and bayes network [4].

Detection of cervical cancer can also be analyzed by using a colposcopy test. This test is done using a colposcopy tool by enlarging the image surface of the cervix to clarify the blood vessel image. The result of colposcopy test is in the form of digital image red green blue (rgb). The image quality is often poor due to the salt and pepper noise on the image. Image enhancement to reduce the noise can be done by a spatial operation median filter. This operation can also soften the image so the image has higher pixel density. Median filter has successfully denoised the mammography image of breast cancer [5]. The median fiter has also been applied to remove the noise on the CT scan images of lung regions. Based on that enhanced image, the lung cancer detection is employed by Backpropagation Network [6].

We propose the median fiter to reduce the noise on the colposcopy image and backpropagation neural network (BPNN) to classify the cervical cancer stadium. The BPNN model offers advantages, since it has adaptive learning capability to carry out activities based on the data given earlier. Backpropagation algorithm trains to obtain balance between the ability of the network to recognize input pattern used during training as well as to provide the correct response to the new similar input pattern [7]. The BPNN model has been successfully applied in many fields, such as in medical diagnosis [8-12], the stock price prediction [13], the design optimization [14], and the introduction of instrumental music notation numbers [15].

The cancer detection using digital image based on the soft computing approach requires feature extraction process. Those features are composed as input variables of the model. The values of the features are supposed to characterize the surface structure of the images. The gray level co-occurrence matrix (GLCM) is one method for feature extraction. Many researchers have worked to classify cervical cancer using the feature values obtained by GLCM extraction. They use various methods of soft computing and comparing approaches, such as multilayer perceptron and bayes network [4], SVM technique in machine learning method [16], nearest neighbour classifier, baye's classifier, and ANN classifier [17], and fuzzy neural network [18].

In this study, two BPNN models are generated to classify the cervical cancer stadium. The first model is generated relied on colposcopy image proceed by spatial operation median filter. The second model is generated relied on the original colposcopy image. We compare the performance of both BPNN models.

## 2. Backpropagation Neural Network Model

The BPNN model is a class of multilayer perceptron neural network, learned by using backpropagation algorithm. The other class of multilayer perceptron neural network that used backpropagation algorithm is recurrent neural network. The backpropagation algorithm to train recurrent neural network has been effectively applied in time series forecasting [19].

The architecture of the BPNN model is composed of input layer, one or more hidden layers and output layer. Each layer involves a certain number of elements called neuron. We consider the BPNN model with single hidden layer. Let the input layer consists of $p$ variables $X_1, X_2, \ldots, X_p$ and one bias $X_0$. The hidden layer consists of $m$ neurons $Z_1, Z_2, \ldots, Z_m$ and one bias $Z_0$. The value of bias is ordinarily set as a unit number "one". In this classification problem the output layer Y is set to have single neuron.

In the hidden layer and output layer, each neuron applies activation functions. The general formula of BPNN model is determined by the activation functions. We use a bipolar sigmoid

$$y = f(x) = \frac{1-e^{-x}}{1+e^{-x}}, x \in R \quad (1)$$

as the activation function in the hidden layer, and a linear function

$$y = f(x) = x, \ x \in R \quad (2)$$

as the activation function in the output layer. The BPNN model with the architecture in Figure 1 and the activation function (1) and (2) can be written in the following expression

$$y = \sum_{j=1}^{m} w_j \left( \frac{1-exp(-(v_{0j}+\sum_{i=1}^{p} v_{ij}\, x_i))}{1+exp(-(v_{0j}+\sum_{i=1}^{p} v_{ij}\, x_i))} \right) + w_0 + \varepsilon \quad (3)$$

where $y$ is dependent variable (output), $x_i$ are independent variables (inputs), $i = 0, 1, 2, \ldots, p$, $v_{ij}$ are weights and $v_{oj}$ are bias on hidden layer from input layer, while $w_j$ are weights on output layer from hidden layer, $w_0$ is bias, and $j = 1, 2, 3, \ldots, m$, $\varepsilon$ is the model error. We train the BPNN model (3) to obtain the optimal weights $v_{ij}$ and $w_j$ using *backpropagation* algorithm. The architecture of the BPNN model is presented in Figure 1.

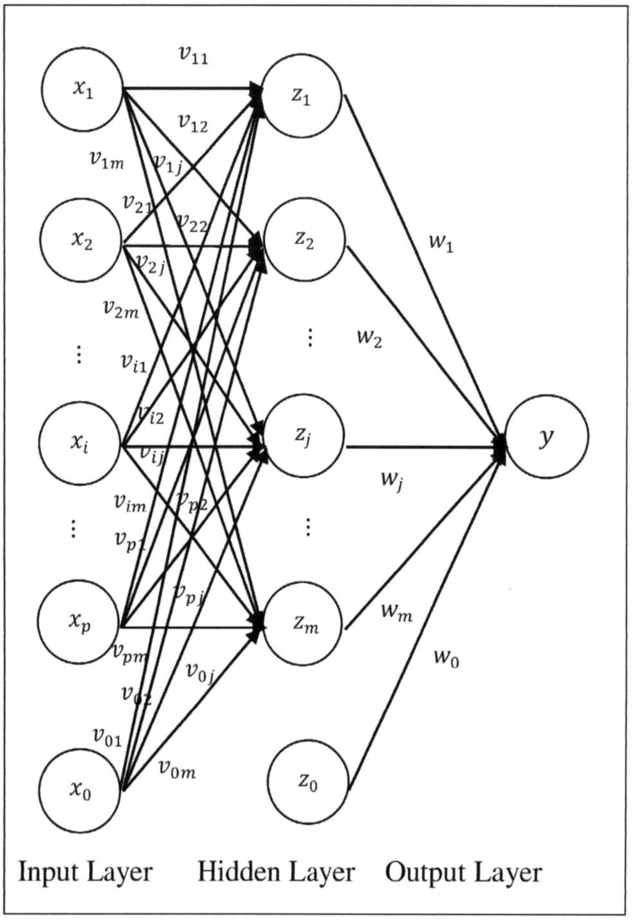


**Figure 1**. The Architecture of the BPNN model with single hidden layer

The optimal weights in equation are (3) are reached by minimizing the function

$$MSE = \frac{1}{n}\sum_{k=1}^{n}(y_k - t_k)^2 \quad (4)$$

where $y_k$ is the $k$th output/predicted value, and $t_k$ is the associated actual observation /target (in this problem is cervical cancer condition), and $n$ is the number of observations.

## 3. Backpropagation Algorithm

Backpropagation algorithm is a supervised learning algorithm that is very popular algorithm in multilayer perceptrons neural network. This algorithm is relied on error correction learning by comparing the output and the target values [20]. Backpropagation algorithm includes three phases, namely feedforward, backpropagation, and weights adjustment.

The algorithm starts with initialize weights by a small random values. Then the algorithm follows the phase described below [21].

**Phase I: Feedforward**

Each input neuron ($X_i$, $i = 1, \ldots, p$) receives input signal $x_i$ and sends this signal to all neurons in the hidden layer. Each hidden neuron ($z_j$, $j = 1, 2, \ldots, m$) sums the weighted input signals $x_i$ and uses a specific activation function to calculate its output signal

$$z_j = f(z\_in_j) = f\left(v_{0j} + \sum_{i=1}^{p} v_{ij}\, x_i\right) \tag{5}$$

and sends this signal to the neuron in the output layer. In this study, we only consider single output neuron. Similarly, the output neuron ($y$) sums the weighted input signals $z_j$ and uses activation function to calculate its output signal

$$y = f(y\_in_j) = f\left(w_0 + \sum_{j=1}^{m} w_j\, z_j\right) \tag{6}$$

**Phase II: Backpropagation**
The error information is computed by comparing the output and the corresponding target as follows

$$\delta = (t - y)f'(y_{\text{in}}) \tag{7}$$

The output utilizes δ to calculate the weights correction, which is used to update weights ($w_j$) later

$$\Delta\, w_j = \alpha.\, \delta.\, z_j \tag{8}$$

and

$$\Delta\, w_0 = \alpha.\, \delta \tag{9}$$

Each hidden neuron ($z_j$, $j = 1, 2, \dots, m$) uses delta input from the output layer, then uses it to calculate its error information term ($\delta_j$)

$$\delta_j = \delta\_in_j\, f'\,(z\_in_j) = \left(\delta\, w_j\right) f'\,(z\_in_j) \tag{10}$$

Then, $\delta_j$ is used to calculate the weights correction term (to update weights ($v_{ij}$) in the layer below)

$$\Delta\, v_{ij} = \alpha.\, \delta_j.\, x_i \tag{11}$$

and bias correction term

$$\Delta\, v_0 = \alpha.\, \delta_j \tag{12}$$

**Phase III: Update weights and bias**
The output neuron ($y$) and each hidden neuron ($z_j$, $j = 1, 2, \dots$, m) update their weights in order to obtain new weights

$$w_{jk}(\text{new}) = w_{jk}(\text{old}) + \Delta\, w_{jk} \tag{13}$$

$$v_{ij}(\text{new}) = v_{ij}(\text{old}) + \Delta v_{ij} \tag{14}$$

The algorithm will stop when the error measure (typically uses MSE) less than error target or the epoch less than maximum epoch (the number of iterations). The error target and the maximum epoch are predetermined by the researcher. The backpropagation algorithm is learned by updating the weights to increase the accuracy.

## 4. Spatial Operation Median filter

Spatial operation also called the local operation is an image enhancement technique through a two-dimensional convolution kernel. This operation delivers an output image whose pixel intensity depends on the pixels around it [22]. The formula of spatial operation can be written as

$$f_B(x, y) = O_{spatial}\,(f_A(x_i, y_i))\ ,\ (x_i, y_i) \in N(x, y) \tag{15}$$

where $f_A(x_i, y_i)$ and $f_B(x, y)$ are input and output images, respectively, and $O_{spatial}$ is spatial operation. The neighborhood $N$ is the set of pixels located around $(x, y)$. The spatial operation intends to reduce the noise rising in the image. The spatial operation can be employed by mean, median, and high-pass filtering methods. In this paper we concern on median filter.

Median filter works by exchanging the pixel in the input image with the pixel median and its neighbourhood according to the size of the specified kernel. The median filter provides a very good ability in noise reduction, especially in the salt-and-pepper form. On the same kernel size, it produces less blurring than the mean filtering. The median filter uses the odd and small kernel-size in order to minimize blurring. We use a median filter with a kernel-size 3x3.

The median filter processes a black and white colours image, so this operation can run only on a grey scale image. To apply this method, the original colour (rgb) image must be converted to the greyscale image.

## 5. Cervical Cancer Classification

Classification addressed in this study is to detect cervical cancer stadium based on colposcopy images drawn from the medical website www.gfmer.ch data set[23]. The images consist of 83 instances that are distributed as normal, cancer stadium I, cancer stadium II, cancer stadium III, and cancer stadium IV. We examine BPNN models for cervical cancer classifications using colposcopy images with and without pre-processing of spatial operation median filter. This operation aims to improve the image quality by softening and reducing noise in the image. The example of normal colposcopy images and the associated matrices before and after spatial operation median filter are presented in Figure 2.

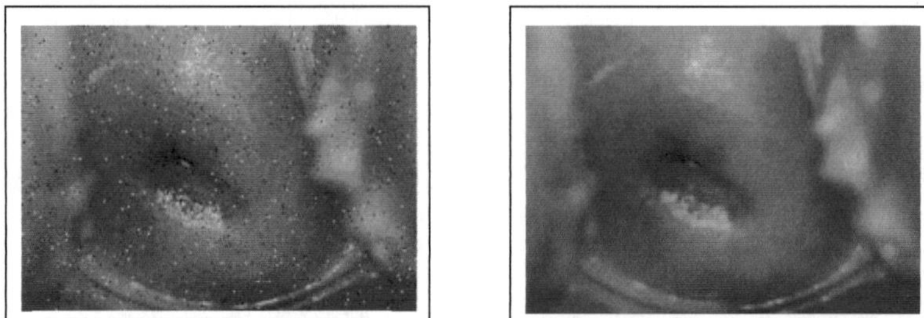

| | | | | |
|---|---|---|---|---|
| 152 | 157 | 161 | 164 | 168 |
| 169 | 97 | 157 | 160 | 164 |
| 217 | 164 | 152 | 155 | 72 |
| 143 | 146 | 149 | 152 | 157 |
| 53 | 144 | 147 | 150 | 155 |

(a)

| | | | | |
|---|---|---|---|---|
| 152 | 157 | 160 | 164 | 164 |
| 152 | 157 | 157 | 160 | 164 |
| 145 | 152 | 152 | 155 | 160 |
| 143 | 147 | 150 | 152 | 155 |
| 142 | 144 | 147 | 150 | 155 |

(b)

**Figure 2**. The grayscale colposcopy images and the associated pixel matrices of (a) the original image (b) the spatial operation median filter image

Figure 2 (a) displays the original colposcopy image that has a lot of salt and pepper noises. They appear as white patches which spread on the image. The noises are represented in the pixel values of the associated matrix that are greatly different from the surrounding values. In that image, the numbers 217, 53, 97, and 72 are pixel values that represent the noise appearances. Figure 2 (b) displays the colposcopy image with spatial operation median filter. We can observe that the associated matrix in figure 2 (b) has nearly the same pixel values. The differences among the pixel values in figure 2 (b) are smaller than those in figure 2 (a). This indicates that the image in figure 2 (b) has the higher pixel density than the image in figure 2 (a). Therefore, the image with median filter spatial operation appears smoother and cleaner compared to the original image.

To build the BPNN model, we begin the process by defining the inputs as the features of colposcopy images extracted by GLCM method. We define 13 features as inputs, namely energy, entropy, contrast, variance, correlation, inverse difference moment (IDM), sum average, sum entropy, sum variance, difference entropy, maximum probability, homogeneity, and dissimilarity [24]. Table 1 provides the maximum, minimum, average, and standard deviation values of each feature of the entire images with and without spatial operation.

**Table 1**. Descriptive of image features

| **Input Variables** | **With Spatial Operation** | | | | **Without Spatial Operation** | | | |
|---|---|---|---|---|---|---|---|---|
| | Max | Min | Average | STDV[a] | Max | Min | Average | STDV[a] |
| Energy | 0.41 | 0.12 | 0.25 | 0.07 | 0.39 | 0.10 | 0.24 | 0.07 |
| Entropy | 2.10 | 1.24 | 1.74 | 0.28 | 2.57 | 1.29 | 1.84 | 0.29 |
| Contrast | 0.19 | 0.04 | 0.08 | 0.03 | 0.29 | 0.05 | 0.12 | 0.05 |
| Variance | 37.25 | 6.33 | 20.96 | 7.53 | 37.24 | 6.39 | 20.87 | 7.37 |
| Correlation | 0.99 | 0.93 | 0.96 | 0.02 | 0.99 | 0.85 | 0.95 | 0.03 |
| IDM | 1.00 | 1.00 | 1.00 | 0.00 | 1.00 | 1.00 | 1.00 | 0.00 |
| Sum Average | 12.10 | 4.79 | 8.74 | 1.67 | 12.11 | 4.81 | 8.72 | 1.64 |
| Sum Entropy | 2.27 | 1.21 | 1.68 | 0.26 | 2.37 | 1.25 | 1.75 | 0.27 |
| Sum Variance | 117.46 | 14.64 | 57.88 | 25.23 | 116.05 | 14.59 | 56.45 | 24.08 |
| Difference Entropy | 0.45 | 0.15 | 0.27 | 0.07 | 0.58 | 0.19 | 0.35 | 0.09 |
| Maximum probability | 0.57 | 0.19 | 0.37 | 0.10 | 0.55 | 0.17 | 0.36 | 0.10 |
| Homogenity | 0.98 | 0.92 | 0.96 | 0.01 | 0.98 | 0.88 | 0.95 | 0.02 |
| Dissimilarity | 0.16 | 0.03 | 0.08 | 0.03 | 0.25 | 0.05 | 0.11 | 0.04 |

[a]STDV = standard deviation

The output variable only involves a single neuron which represents the cervical conditions. It is a categorical variable whose values are one for normal condition, two for cancer stadium I, three for cancer stadium II, four for cancer stadium III, and five for cancer stadium IV.

The BPNN models are evaluated by cross validation. We divide the data of 83 instances become 66 instances as training and 17 instances as testing data sets. The training set aims to build the model, and the testing set aims to evaluate the capability of the obtained model to predict the cervical cancer stadium based on the data that are not used in training process.

The backpropagation algorithm is employed to train the BPNN model for attaining best network architecture. It is a network with the number of hidden neuron that delivers the best performance both in training and testing sets. It is achieved by trial and error starting from a single hidden neuron, and continuing the training process by adding the number of hidden neurons until no significant improvement in the network performance. The performance of the model is addressed to the accuracy value, which is computed as follows

$$\text{Accuracy} = \frac{\text{The number of correct classifications}}{\text{The number of observations}} \times 100\% \qquad (16)$$

In each architecture, the algorithm is exercised to achieve the optimal weights which are attained by minimizing the MSE (4).We do trial and error from a single hidden neuron until 13 hidden neurons. The results of the BPNN models with and without spatial operation in term of accuracy performances are presented in table 2. Table 2 demonstrates that the best architecture of BPNN models with and without median filter spatial operation are the models with respective number of hidden neurons 10 and 12. On the training data, both models reach the perfect performances, which are revealed from percentage of accuracy 100%. All 66 instances that consist of 21 normal, 14 stadium I, 14 stadium II, 10 stadium III, and 7 stadium IV are correctly classified. The experiment proves that reducing the noise by the spatial operation can increase the accuracy of the model, specifically on testing data. That assertion is supported by the result in table 2 that each BPNN model with median filter always performs better than each BPNN model without median filter in the same number of hidden neuron. Specifically, at the optimal number of hidden neuron, the spatial operation median filter can increase the accuracy from 82 % (without spatial operation) to 88.24 % (with spatial operation).

The results of the predicted classifications for each cervical condition on testing data are presented in table 3 and table 4.

IOP Conf. Series: Journal of Physics: Conf. Series **855** (2017) 012062 doi:10.1088/1742-6596/855/1/012062

**Table 2**. The accuracy (%) of BPNN Model

| The number of hidden neuron | With spatial operation | | Without spatial operation | |
|---|---|---|---|---|
| | Training | Testing | Training | Testing |
| 1 | 27.27 | 35.29 | 44 | 29 |
| 2 | 54.55 | 58.82 | 56 | 29 |
| 3 | 54.55 | 52.94 | 77 | 23 |
| 4 | 66.67 | 64.71 | 83 | 59 |
| 5 | 60.61 | 47.06 | 85 | 41 |
| 6 | 81.82 | 64.71 | 92 | 59 |
| 7 | 90.91 | 70.59 | 95 | 41 |
| 8 | 100 | 64.71 | 100 | 59 |
| 9 | 98.48 | 76.47 | 94 | 53 |
| 10 | 100 | 88.24[a] | 100 | 59 |
| 11 | 100 | 64.71 | 100 | 53 |
| 12 | 100 | 88.24 | 100 | 82[a] |
| 13 | 100 | 76.47 | 100 | 47 |
| 14 | 100 | 64.71 | 100 | 29 |

[a] The best BPNN model

**Table 3**. The predicted classifications of BPNN model with spatial operation on testing data

| Actual Diagnosis | The BPNN Classifications | | | | | Total |
|---|---|---|---|---|---|---|
| | Normal | Std.I | Std.II | Std.III | Std.IV | |
| Normal | 4 | | | | | 4 |
| Std.I | | 3 | | 1 | | 4 |
| Std.II | | | 2 | 1 | | 3 |
| Std.III | | | | 3 | | 3 |
| Std.IV | | | | | 3 | 3 |
| Total | 4 | 3 | 2 | 5 | 3 | 17 |

**Table 4**. The predicted classifications of BPNN model without spatial operation on testing data

| Actual Diagnosis | The BPNN Classifications | | | | | Total |
|---|---|---|---|---|---|---|
| | Normal | Std.I | Std.II | Std.III | Std.IV | |
| Normal | 3 | | 1 | | | 4 |
| Std.I | | 3 | | 1 | | 4 |
| Std.II | | | 3 | | | 3 |
| Std.III | | | | 3 | | 3 |
| Std.IV | | | | 1 | 2 | 3 |
| Total | 3 | 3 | 4 | 5 | 2 | 17 |

Referring to table 3 and table 4, it can be perceived that two instances of stadium I and stadium II in BPNN with spatial operation are misclassified as stadium III, and the remaining instances are correctly classified. While three instances contain normal, stadium I and stadium IV in BPNN without spatial operation are misclassified.

If we focus only on the non cancerous and cancerous cervical conditions, we can evaluate the performance of BPNN model in term of specificity and sensitivity. There are four possibilities that can occur in the diagnosis, they are true positive (TP), true negative (TN), false positive (FP), and false negative (FN). The diagnosis refers to true positive if the diagnostic test of cancerous cervical condition leads to cancerous cervical condition. Similarly, the diagnosis refers to true negative if the diagnostic test of non cancerous cervical condition leads to non cancerous cervical condition. The diagnosis refers to false positive if non cancerous cervical condition is detected as cancerous cervical condition. Finally, the diagnosis refers to false negative if cancerous cervical condition is detected as non cancerous cervical condition. The following formulas define the specificity and sensitivity in terms of TP, TN, FN, and FP

$$\text{Sensitivity} = \frac{TP}{TP+FN} \times 100\% \tag{17}$$

$$\text{Specificity} = \frac{TN}{TN+FP} \times 100\% \tag{18}$$

On the training data, all instances are well classified, so the FN and FP values are zero. Thus, the sensitivity and specificity values are 100% for both BPNN models. Table 5 shows the diagnosis results of cervical cancer on testing data which have been grouped in the TP, TN, FP, and FN values.

**Table 5**. The Diagnosis result of the BPNN Model on testing data

| Diagnose | With spatial | | Without spatial | |
|---|---|---|---|---|
| | Cancer | Non Cancer | Present | Non Cancer |
| Cancer | TP=13 | FP=0 | TP=13 | FP=0 |
| Non Cancer | FN=0 | TN=4 | FN=1 | TN=3 |

The sensitivity and specificity values are computed by using (17) and (18). The sensitivity and specificity values are both 100% for the BPNN model with spatial operation, 100% and 75% for the BPNN model without spatial operation, respectively. The BPNN model with spatial operation is as well as the one without spatial operation in terms of sensitivity, both reach 100%. This fact suggests that both types of colposcopy images can be used properly to detect the present of cancer cervical. The specificity value of the BPNN model with spatial operation performs better than that of the BPNN model without spatial operation median filter, means that the operation can increase the capability of the image to detect the absent of cancer cervical.

## 6. Conclusion

In this study, we propose the spatial median filter operation to enhance the quality of cervical colposcopy images, especially to reduce the noise of images in the salt-and-pepper form. The BPNN model is applied to the cervical colposcopy images with and without the spatial operation for classifying stadium of cervical cancer. We extract the colposcopy images by GLCM method obtain 13 parameter values. Those parameters are set as input variables of BPNN model. To evaluate which model is better we split the data as training and testing data set. The backpropagation algorithm is applied to train the model in order to obtain the weights of the model with the best performance. The performance of the model is evaluated by considering the sensitivity, the specificity, and the accuracy values. The result reveals that the BPNN model with the spatial operation median filter is more accurate than the BPNN model without spatial median filter operation.

## Reference

[1] Quteishat A, Al-Batah M, Al-Mofleh A, and Alnabelsi S H 2013 Cervical cancer diagnostic system using adaptive fuzzy moving K-means algorithm and fuzzy min-max neural network *J. of Theoretical and Applied Information Technolog* Vol. 57 No. 1 p 48

[2] Sulaiman S N, Mat-Isa N A, Othman N H, and Ahmad F 2010 Improvement of features extraction process and classification of cervical cancer for the neuralpap system *Proc. Computer Science* **60** (Elsevier ) p 750.

[3] Athinarayanan A and Srinath M V 2016 Classification of cervical cancer cells in pap smear screening test ICTACT *J. on Image and Video Processing* Vol. 6 Issue 4 pp 1234-1238

[4] Lakshmi K and Krishnaveni K 2016 Feature extraction and feature set selection for cervical cancer diagnosis *Indian. J. of Science and Technology* Vol. 9 pp 1-7

[5] Makandar A and Halali B 2015 Breast cancer image enhancement using median filter and CLAHE *Intl. J. on Scientific and Engineering Research* Vol. 6 Issue 4 pp 462-465

[6] Kaur J, Garg N, and Kaur D 2014 An automatic CAD system for early detection of lung tumor using backpropagation network *Int. Conf. on Medical Imaging* IEEE pp 257-261

[7] Siang J J 2005 *Jaringan Syaraf Tiruan: Pemrogramannya Menggunakan Matlab* (Yogyakarta: Andi Yogyakarta)

[8] Nayaz I A, & Chaya P 2014 Segmentation and classification of skin cancer image *Int. J. of Advanced Research in Computer Science and Software Enginering* Vol. 4 Issue 5 pp 1349-1353

[9] Divya, Chhabra R, Kaur S, and Ghosh S 2013 Diabetes detection using artificial neural network and backpropagation algorithm *Int. J. of Scientific & Technology Research* Vol. 2 Issue 1 pp 9-11.

[10] Kumash B S, Verma K, and Thoke A S 2015 Adaptive gradient descent backpropagation for classification of breast tumors in ultrasound imaging *Int. Conf. on Information and Communication Technologies* (Elsevier) pp 1601-1609

[11] Verma M 2014 Medical diagnosis using backpropagation algorithm in ANN *Int. J. of* Science, Enginering and Technology Research Vol 3. Issue 1 pp 94-99

[12] Khan I Y, Zope P H, and Suralkar S R 2013 Importance of artificial neural network in medical diagnosis disease like acute nephritis disease and heart disease *Int. J. of Engineering Science and Innovative Technology* Vol. 2 Issue 2 pp 210-217

[13] Lahmiri S 2014 Wavelet low-and high-frequency components as features for predicting stock prices with backpropagation neural network *J. of King Saud University Computer and Information Sciences* 26 pp 218-227

[14] Li S, Guo Z, Cheng S, and Zhang S 2014 Design optimization of sheet metal stamped parts by CAE simulation and backpropagation neural network *Procedia Engineering* (Elsevier) pp 1023-1028

[15] Christofer A, Kusuma C, Pribadi V and Budiharto W 2015 The notation scanner system using resilient backpropagation method *Int. Conf. on Computer Science and Computational Intelligence* (Elsevier) pp 98-105

[16] Ashok B and Aruna P 2016 Comparison of feature selection methods for diagnosis of cervical cancer using SVM clasifier *J. of Engineering Research and Applications* Vol. 6 Issue 1 pp 94-99.

[17] Praba P S and Priya R Comparison of classification technique for pap smear diagnosis IOSR *J. of Computer Engineering* e-ISSN: 2278-0661 p-ISSN: 2278-8727 pp 11-18

[18] Wutsqa D U and Rahmada P S 2016 Point operation to enhance the performance of fuzzy neural network model for breast cancer classification *Global J. of Pure and Applied Mathematics* ISSN: 0973- 1768 In press

[19] Wutsqa D U, Kusumawati R, and Subekti R 2014 The application of Elman recurrent neural network model for forecasting consumer price index of education, recreation and sports in Yogyakarta The 10th *Proc. Int. Conf. on Natural Computation* (IEEE) pp 192-196

[20] Haykin S 1999 *Neural Network: A Comprehensive Foundation* (New York: Prentice Hall)

[21] Fausett L 1994 *Fundamentals of Neural Networks: Architecture, Algorithms, and Aplications* (New Jersey: Prentice Hall)

[22] Munir R 2004 *Pengolahan Citra Digital dengan Pendekatan Algoritmik* (Bandung: Informatika)

IOP Conf. Series: Journal of Physics: Conf. Series **855** (2017) 012062 doi:10.1088/1742-6596/855/1/012062

[23] http://gfmer.ch
[24] Haralick R M, Shanmugam K, and Dinstein I 1973 Textural features for image classification *IEEE Transactions on Systems, Man and Cybernetics* Vol. SMC-3 No. 6 pp 610-621

# Calculation of Lung Cancer Volume of Target Based on Thorax Computed Tomography Images using Active Contour Segmentation Method for Treatment Planning System

**Fiet Patra Yosandha, Kusworo Adi, and Catur Edi Widodo**

Post Graduate Program of Physics Department of Diponegoro University, Indonesia

E-mail : fiet.patra@st.fisika.undip.ac.id

**Abstract**. In this research, calculation process of the lung cancer volume of target based on computed tomography (CT) thorax images was done. Volume of the target calculation was done in purpose to treatment planning system in radiotherapy. The calculation of the target volume consists of gross tumor volume (GTV), clinical target volume (CTV), planning target volume (PTV) and organs at risk (OAR). The calculation of the target volume was done by adding the target area on each slices and then multiply the result with the slice thickness. Calculations of area using of digital image processing techniques with active contour segmentation method. This segmentation for contouring to obtain the target volume. The calculation of volume produced on each of the targets is 577.2 $cm^3$ for GTV, 769.9 $cm^3$ for CTV, 877.8 $cm^3$ for PTV, 618.7 $cm^3$ for OAR 1, 1,162 $cm^3$ for OAR 2 right, and 1,597 $cm^3$ for OAR 2 left. These values indicate that the image processing techniques developed can be implemented to calculate the lung cancer target volume based on CT thorax images. This research expected to help doctors and medical physicists in determining and contouring the target volume quickly and precisely.

## 1. Introduction

One method to detect lung cancer is an examination of computed tomography (CT). The results of the examination will distinguish normal and abnormal lungs[1]. For the case of not-normal-lung, in this case is lung cancer, one treatment that can be done is radiation therapy or radiotherapy. In radiotherapy, planning is required to determine the amount of radiation dose received by a tissue. Dose planning in radiotherapy is called treatment planning system (TPS).

The initial phase of the TPS is contouring the cancer or target volume. The results of this phase are area and volume of the target. However, the process was done manually, requiring skilled medical personnel, and needing a long time in the process. This process can be done quickly by using segmentation which classifies the object pixels into a regions or parts. Segmentation can be done by several methods, one of them is the active contour segmentation method which is a method using models of closed curves that can move widened or narrowed.

Keshani et al. did a research to segment and identify pulmonary nodul. Image segmentation was performed by using active contour segmentation method and support vector machine (SVM) classification. Image segmentation use masking having the shape of non-isolated nodules and isolated nodules. The classification process was done using various SVM method of variations of the kernel function, named dehmeshki, taghavi, and farag. The results from this research were obtained an average accuracy of 89% [2].

Other researchers, Jin et al. conducted research delineation of lung tumors for therapy planning area in the image GTV PET/CT. Used image segmentation method is optimum contour selection (OCS). Jin et al. propose an contouring automatic algorithm based on optimal selection contour. Lung tumors in PET images are identified with thresholding method. Furthermore, the CT image segmentation is fixed using PET image segmentation results. Then the area GTV on CT images is contoured by using OCS [3].

Sari et al. did a study for the measurement and calculation of the volume of the object of the CT image. Tube-shaped object under investigation is assumed to be a tumor. The images generated from CT modalities, slice per slice. This calculation uses the trapezoid integration method. Before calculating the volume of the object, the object area of the base was calculated. Base area of the object is a slice thickness area in the image converted into a binary image. After obtaining the area of the base object, the volume of the object was calculated. From several variations of the position of the studied object, the object positioned perpendicular to the central ray will get a more accurate volume calculations [4].

In this research was done development of lung cancer detection system of the CT images. This detection to determine lung cancer target volume with image processing segmentation method. Segmentation in this research using active contour method. This segmentation for contouring to obtain the target volume. This research expected to help doctors and medical physicists in determining and contouring the target volume quickly and precisely.

## 2. Literature Study

Target volume is a prerequisite for 3-D treatment plan in obtaining accurate dose. ICRU Report No. 50 and 62 explain the multiple targets and volumes of critical structures that assist in the treatment planning process and provide a basis for comparison of the results of therapy [5, 6]. The following volume is defined as the primary volume associated with 3-D treatment planning that is gross tumor volume (GTV), clinical target volume (CTV) and planning target volume (PTV). Figure 1 shows how the different volume is associated with each other [7]. Part of the target volume Illustration can be seen in Figure 1.

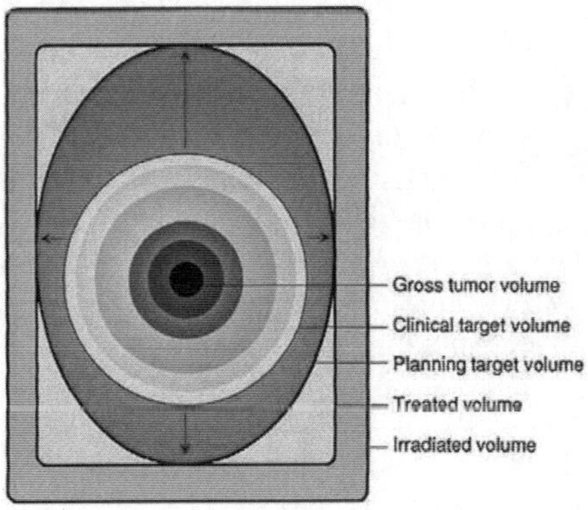


**Figure 1.** Illustration of parts of the target volume [8]

### 2.1. *GTV (Gross Tumor Volume)*

GTV is part of the tumor which is palpable and can be seen visually with the help of tools such as endoscopes, plain to contrast radiography, CT scans, MRI or PET. GTV is part of both primary and metastatic of an organ or whole organ. GTV cannot be identified after surgery treatment [8]. GTV usually performed based on information from the modality diagnostics (pathology and histological report) and clinical examination [7].

### *2.2. CTV (Clinical Target Volume)*
CTV is the tissue volume containing GTV which has proved its sub-clinical microscopic malignant disease that has to be eliminated. CTV usually covers an area that immediately adjacent to the GTV that may be presenting microscopic disease and considered at risk and in need of treatment (e.g. lymph node positive). CTV is a clinical and anatomic volume usually determined by the radiation oncologist, performed after an examination conducted by a specialist such as a pathologist or a radiologist. CTV is usually expressed as a fixed or variable margin around the GTV (e.g. CTV = GTV + 1 cm margin). However, in some cases it is the same as the GTV (e.g. prostate gland increases only). There can be some CTV non-contiguous requiring total dose different to achieve the goal of treatment [7].

### *2.3. PTV (Planning Target Volume)*
PTV are geometric concepts defined to select the right bricks by considering the net effect of all the variations of geometric in order to ensure that the prescribed dose is actually absorbed in the CTV [5]. PTV includes internal target margin [6] and an additional margin for set-up uncertainty, tool tolerances and intra-treatment variations. PTV is described as CTV plus a margin fixed or variable (e.g. PTV = CTV + 1 cm). PTV depends on tools precision such as immobilization and laser devices that require additional margin during the planning [7]. In the process of the formation of this PTV, patient's body movements dynamics factor is noticed, such as the movement of the cardiac, respiratory and other dynamic processes [8].

### *2.4. OAR (Organ at Risk)*
OAR is the organ that is sensitive to radiation. Special attention should be given to the organs at risk, although it is not directly adjacent to the CTV. OAR must have a very low tolerance dose (e.g. eyelens on asopharyngeal or brain tumors). Organ with radiation tolerance dependent on the fractionation scheme should be fully described to minimize the exposure dose of radiation during treatment planning evaluation [7].

### *2.5. Active Contour*
Active contour, which is also called snake, is a segmentation method using a model of closed curves that can move widened or narrowed. This method works by minimizing the energy associated with the contour of the current amount of internal and external energy. Active contour can move widened or narrowed by minimizing energy image using external power, and it is also influenced by the characteristics of the image such as lines or edges.

Active contour is described as a point of control which is followed each other. Determination of the object in the image through active contour is an interactive process. Users must estimate the initial contours determined nearly approximates the shape of the object features. Furthermore, the contour will gravitate toward the features in the image due to the influence of the internal energy [9].

### *2.6. Spatial Resolution*
Spatial resolution reflects the details of the data about the objects that can be tapped from a remote sensing system in a size of the smallest object that can be presented, distinguished and recognized in image called pixel/picture element [10].

The spatial resolution image is the minimum distance between the two objects that are still distinguishable. The image spatial resolution of the imaging system is defined as the ability to distinguish two separate points in space. The higher the spatial resolution of the image, the smaller the distance that can be distinguished [11].

## 3. Method

### *3.1. Instruments and materials*

The used tools are computer and programming method Matlab. Used digital image is the image of the thorax with DICOM format by the acquisition of CT scan modality. This digital image is used as an input image that will be processed through the computational programming. The variables in this study are an area target volume ($mm^2$), the volume of the target ($cm^3$), the spatial resolution (pixels/mm), the number of slice, and the slice thickness (mm).

### *3.2. Procedure*

The process begins by determining image slice conducting cancer and organ that must be protected. The image Slice that there is no cancer and protected organ will not be segmented. On thorax CT image, slice containing cancer and organ that must be protected, initialization masking was done by firstly determining the appropriate iteration. Initialization used in this study is impoly. This process will produce a contour. After that, the target area on each slice is calculated. Then the volume of the target is calculated too by summing the target area multiplied by the slice thickness. The results of these calculations will be obtained the volume of the target of GTV, CTV, PTV and OAR.

## 4. Results

This study, discuss the results of the determination of the target volume and volume of targets in lung cancer by computed tomography images using active contour segmentation method for Treatment Planning System. There are the types of the processed target volume. Those are GTV, CTV, PTV, OAR 1, OAR 2 right, OAR 2 left. Imagery used in this study consists of 75 slice by slice thickness of 3 mm. This image is 512 x 512 pixels and has a spatial resolution of 1.0216 pixels / mm.

### *4.1. Target volume*

Image contouring process have been done on target volume. In this study, the process of image contouring was done using active contour segmentation method. Contouring process begins with segmenting the image of GTV, which the main target volume of the primary tumor in each slice images is contained cancer. Image containing lung cancer is as much as 38 slice occurred in the slice 59 to 96. The GTV contouring results sample performed on slice to 78 are shown in Figure 2.

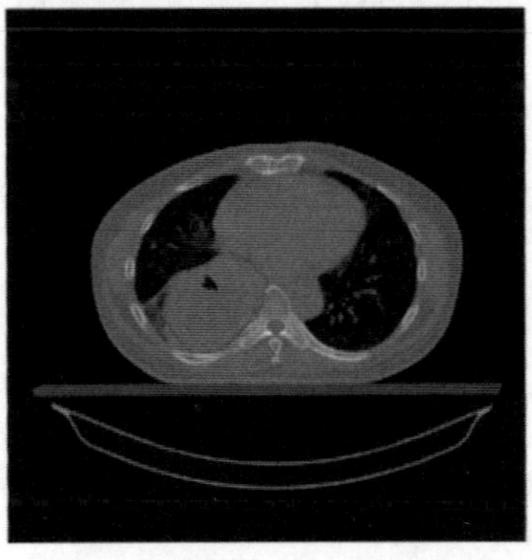

**Figure 2.** Contouring results GTV on slice to 78

Figure 2 shows the GTV contouring process performed on slice to 78. GTV contouring have been done right at the primary tumor that can be seen. After doing GTV contouring, the next process is to segment the CTV. CTV contouring is the margin of 1 cm from the GTV or CTV = GTV + 1 cm. GTV areas is used as masking initialization for CTV contouring. Image used for segmentation CTV is similar to the image used for GTV contouring, as many as 38 slice occurred in the slice 59 to 96. The CTV contouring results sample performed on slice to 78 are shown in Figure 3.

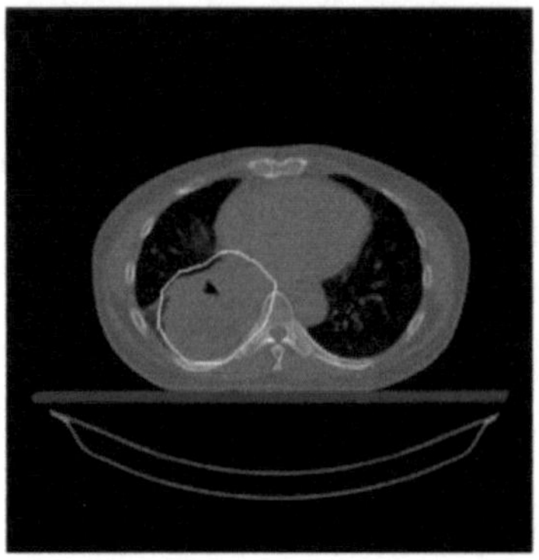

**Figure 3.** Contouring results CTV on a slice to 78

Figure 3 shows the CTV contouring process performed on slice to 78. CTV contouring performed at 1 cm margin of GTV contouring. After doing CTV contouring, the next process is to segment the PTV. PTV contouring is the margin of 1 cm from CTV or PTV = CTV + 1 cm or PTV = GTV + 2 cm. CTV areas is used as masking initialization for PTV contouring. Image used for image PTV segmentation is similar to image used for GTV and CTV contouring, as many as 38 slice occurred in the slice 59 to 96. The PTV contouring results sample performed on slice to 78 are shown in Figure 4.

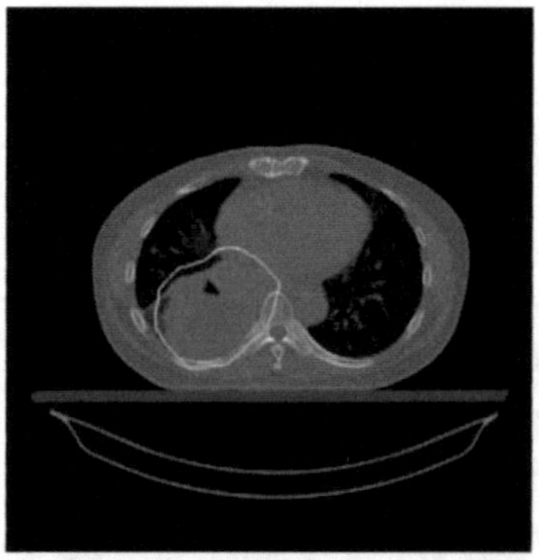

**Figure 4.** Contouring results PTV on slice to 78

Figure 4 shows the PTV contouring process performed on slice to 78. PTV contouring is performed at 1 cm margin of CTV contouring. After doing PTV contouring, the next process is to segment the OAR. In this research, there are two OAR, OAR 1 is an organ that is the cardiac and OAR 2 is the lungs. OAR 2 is divided into two, OAR 2 right represents the right lung and the left lung is represented by OAR 2 left. Contouring process for OAR 1 segmentation, which is the target volume for each the image slice containing of the cardiac organ. Images containing cardiac are as many as 27 slice occurred on slice 69 to 95. The OAR 1 contouring results sample performed on slice to 69 are shown in Figure 5.

IOP Conf. Series: Journal of Physics: Conf. Series **855** (2017) 012063 doi:10.1088/1742-6596/855/1/012063

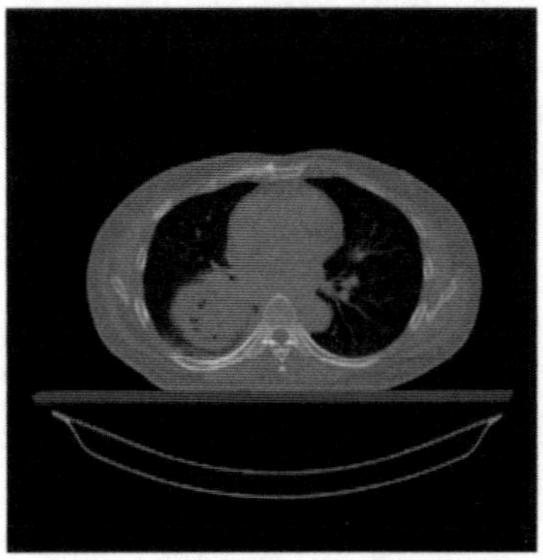

**Figure 5.** Contouring results OAR 1 on slice to 69

Figure 5 shows the image contouring process performed on slice to 69. OAR 1 contouring was done right in the cardiac organ that can be seen. After OAR 1 contouring, OAR 2 right contouring have been done. Contouring process for OAR 2 right segmentation, which is the target volume for each slice images containing right lung organ. The image containing right lung organ is as much as 70 slice occurred on slice 23 to 92. The OAR 2 right contouring results sample performed on slice to 67 are shown in Figure 6.

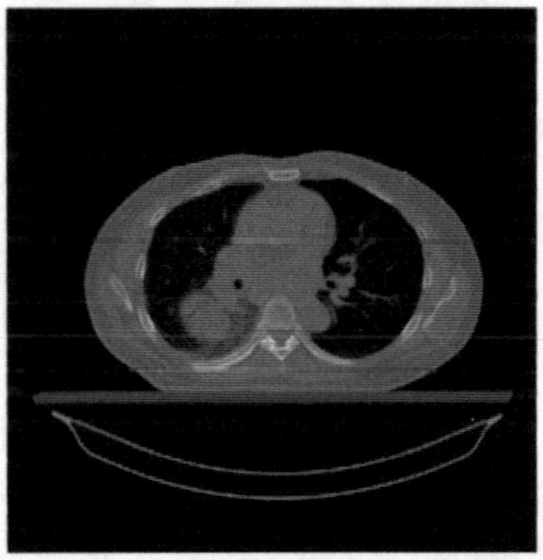

**Figure 6.** Contouring results OAR 2 right on slice to 67

Figure 6 shows image contouring process performed on slice to 67. OAR 2 right contouring was done right on the right side of lung organs that can be seen. After OAR 2 right contouring, OAR 2 left contouring was done. Contouring process for OAR 2 left segmentation, which is the target volume for each slice contained the image of the left lung organ. The image containing the left lung organ as much as 72 slice occurred on slice 24 to 95. The OAR 2 left contouring results sample performed on slice to 70 are shown in Figure 7.

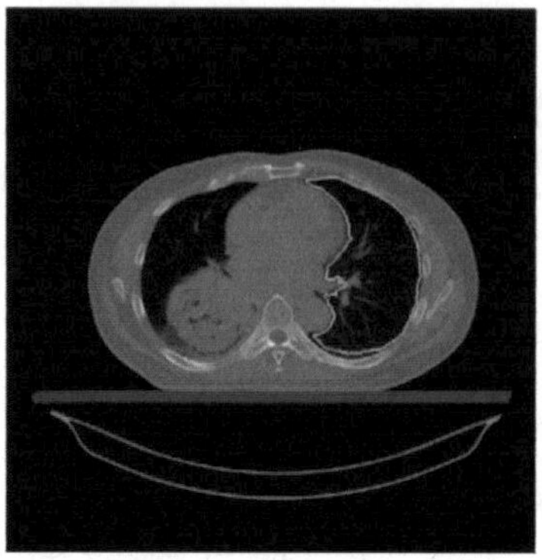

**Figure 7.** Contouring results OAR 2 left on slice to 70

Figure 7 shows image contouring process performed on slice to 70. OAR 2 left contouring was done right on the left side of the lung organ that can be seen. The result of the whole target volume contouring process can be combined into a single image. The whole sample results of target volume contouring process that have been done to slice 69 are shown in Figure 8.

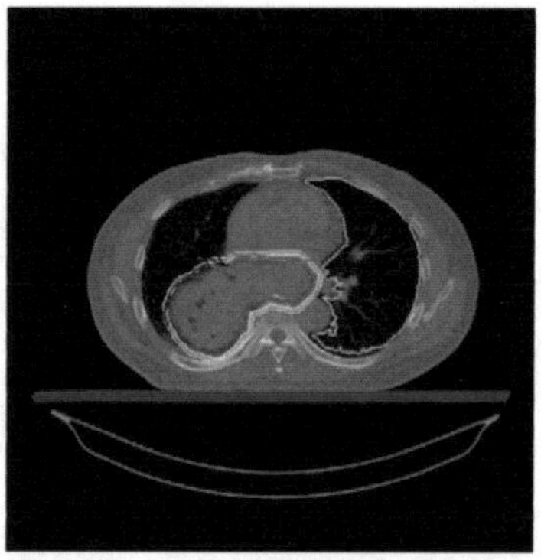

**Figure 8.** The results of the all contouring on slice to 69

Figure 8 shows the entire process of the target volume contouring performed on slice to 69. After the target volume contouring process is complete, then the volume of the target is calculated.

*4.2. Volume Target*

Volume of the target calculation begins by calculating the target area on each slice foremost. Initially the image is segmented by using active contour segmentation method to get the binary image. After obtained a binary image segmentation results for each of the target, then the target area is calculated by counting the number of pixels of 1 in the binary image. The result of this calculation is still in pixels$^2$, so it need to be converted into units cm$^2$ with equation (1).

$$L\ (cm^2) = L\ (pixel)\ /\ resolution^2 \quad (1)$$

Sample of target area calculation results performed on slice 78 are shown in Table 1.

**Table 1.** Results of target area calculation on slice to 78

| Target volume | Calculation of target area ($cm^2$) |
|---|---|
| GTV | 64.26 |
| CTV | 81.65 |
| PTV | 91.67 |
| OAR 1 | 88.47 |
| OAR 2 Right | 52.20 |
| OAR 2 Left | 85.83 |

After calculating the area of the target, then the volume of the target is calculated by summing area of the target multiplied by the slice thickness. The volume of target calculation with developed system can be obtained using equation (2).

$$V = (\sum A) \times S \tag{2}$$

with $V$ is the volume of target ($cm^3$), $A$ is the area of the target ($cm^2$), and $S$ is the slice thickness (mm). The results of calculation of the volume of target using the developed systems are shown in Table 2.

**Table 2.** Results of calculation of the volume of target

| Target volume | Calculation of volume with the system ($cm^3$) |
|---|---|
| GTV | 577.2 |
| CTV | 769.9 |
| PTV | 877.8 |
| OAR 1 | 618.7 |
| OAR 2 Right | 1,162 |
| OAR 2 Left | 1,597 |

Table 2 shows the value of calculation the volume on each of the target.

## 5. Conclusions

The calculation of volume produced on each of the targets is 577.2 $cm^3$ for GTV, 769.9 $cm^3$ for CTV, 877.8 $cm^3$ for PTV, 618.7 $cm^3$ for OAR 1, 1,162 $cm^3$ for OAR 2 right, and 1,597 $cm^3$ for OAR 2 left. These values indicate that the image processing techniques developed can be implemented to calculate the lung cancer target volume based on CT thorax images.

## References

[1] Elizabeth D S, Nehemiah H K, Raj C S R, and Kannan A 2012 Computed-Aided Diagnosis of Lung Cancer Based on Analysis of Significant Slice of Chest Computed Tomography Image *IET Image Process* **6 6** 697-705

[2] Keshani M, Azimifar Z, Tajeripour F, and Boostani R 2012 Lung Nodule Segmentation and Recognition using SVM Classifier and Active Contour Modeling: A Complete Intelligent System *Computers in Biology and Medicine* **43** 287-300

[3] Jin Z, Arimura H, Shioyama Y, Nakamura K, Kuwazuru J, Magome T, Yabu-Uchi H, Honda H, Hirata H, and Sasaki M 2014 Computer-Assisted Delineation of Lung Tumor Regions in Treatment Planning CT Images with PET/CT Image Sets Based on An Optimum Contour Selection Method *Journal of Radiation Research* **55** 1153-1162

[4] Sari R T D, Adi K, and Anam C 2014 Volume Phantom Measurement and Calculation of Computed Tomography (CT) Scan Images *Youngster Physics Journal* **3 4** 221-226

[5] ICRU I993 *Prescribing, Recording, and Reporting Photon Beam Therapy (Report 50) The International Commision on Radiation Units and Measurements* (Maryland: Woodmont Avenue Bethesda)

[6] ICRU 1999 *Prescribing, Recording, and Reporting Photon Beam Therapy (Report 62) The International Commision on Radiation Units and Measurements* (Maryland: Woodmont Avenue Bethesda)

[7] Podgorsak E B 2005 *External Photon Beams: Physical Aspects in Radiation Oncology Physics* (Austria: IAEA Publishing)

[8] Khan F M and Gibbons J P 2014 *Physics of Radiation Therapy Fifth Edition* (Philadelphia: Lippincott Williams & Wilkins)

[9] Caselles V, Kimmel R, and Sapiro G 1997 Geodesic Active Contours *International Journal of Computer Vision* **22** **1** 61-79

[10] Hartono 2010 Remote Sensing and Geographic Information System, and Its Application in Education and Development *Seminar Nasional-PJ dan SIG*

[11] Bushberg J T, Seibert J A, Boone J M, and Leidholdt E M 2002 *The Essential Physics of Medical Imaging* (Philadelphia: Lippincott Williams & Wilkins)

# Parameters Estimation of Geographically Weighted Ordinal Logistic Regression (GWOLR) Model

**Shaifudin Zuhdi, Dewi Retno Sari Saputro, Purnami Widyaningsih**
Mathematics Department, Mathematics and Natural Sciences Faculty, Universitas Sebelas Maret, Indonesia

E-mail: shaifudinzuhdi@gmail.com, dewi.rss@gmail.com

**Abstract.** A regression model is the representation of relationship between independent variable and dependent variable. The dependent variable has categories used in the logistic regression model to calculate odds on. The logistic regression model for dependent variable has levels in the logistics regression model is ordinal. GWOLR model is an ordinal logistic regression model influenced the geographical location of the observation site. Parameters estimation in the model needed to determine the value of a population based on sample. The purpose of this research is to parameters estimation of GWOLR model using R software. Parameter estimation uses the data amount of dengue fever patients in Semarang City. Observation units used are 144 villages in Semarang City. The results of research get GWOLR model locally for each village and to know probability of number dengue fever patient categories.

## 1. Introduction

A regression model is the representation of relationship between independent variable and dependent variable. The dependent variable has categories used in the logistic regression model to calculate odds on.

Logistic regression is extended to handle outcome variables that have more than two ordered categories, there are polytomous logistic regression and ordinal logistic regression. Polytomous logistic regression is used when the categories of the outcome variable are nominal, that is, they do not have any natural order. When the categories of the outcome variable have a natural order, ordinal logistic regression may be appropriate [5].

Statistical methods have been developed to model the relationship between dependent variable and independent variables that depend on the geographic location where the data is observed. For categorical dependent variable have been developed geographically weighted logistic regression (GWLR) model [1]. GWLR model can also be developed for ordinal scale dependent variable, namely geographically weighted ordinal logistic regression (GWOLR) model.

Generally to estimate the parameter of GWOLR model, the likelihood maximum method needs to be employed. This method results in a system of nonlinear equation which is hard to solve. This method includes iteration process which is a repetition process. In the field of programming, iteration is a process consisted of more than one algorithm and is conducted in a loop program which often refers to a repetition program. One of the programming languages which allows the repetition of instruction is the R language (R software).

R software is an integrated software which has the facility to manipulate the data, calculation, and graphical display. R software is supported by a community whose members are actively interacting to each other via internet and guided by a manual or R-help. Parameter estimation on R software can be done by creating an algorithm-based program.

## 2. Ordinal logistic regression model

The OLR model is a model representing the relationship between ordinal-scaled response variable and independent variable which has the characteristic of category and/or continual [3]. Logistic regression model can be called as logit model. If the response variable which has $G$ category employs ordinal scale and $x_i$ refers to the vector of independent variable in $i$-observation, thus the OLR model can be formulated as

$$logit[P(Y_a \leq g|x_a)] = \alpha_g + \boldsymbol{x_i}^T\boldsymbol{\beta}_i, \quad (1)$$

with $g = 1,2,\dots,G-1$, $a = 1,2,\dots,n$, and $i = 1,2,\dots,p$. $P(Y_a \leq g|x_i)$ refers to the cumulative probability which is less or equal to g-category on $x_i$. The parameter $\alpha_g$ refers to the intercept which fulfils the condition of $\alpha_1 \leq \alpha_2 \leq \dots \leq \alpha_{g-1}$ and $\beta_i$ refers to the coefficient vector of the regression which is in mutual accord with $x_i$. Thus, the probability of each $G$ response category is formulated as

$$P_g(x_i) = \frac{exp(\alpha_g + \boldsymbol{x_i}^T\boldsymbol{\beta}_i)}{1 + exp(\alpha_g + \boldsymbol{x_i}^T\boldsymbol{\beta}_i)} - \frac{exp(\alpha_{g-1} + \boldsymbol{x_i}^T\boldsymbol{\beta}_i)}{1 + exp(\alpha_{g-1} + \boldsymbol{x_i}^T\boldsymbol{\beta}_i)}, g = 1,2,\dots,G. \quad (2)$$

The probability value for each response category is used as guidance for classification. An observation will be included to the $g$-response category based on the biggest probability value.

The parameter estimation of OLR model is conducted with maximum likelihood estimation (MLE) method. This method is employed to examine the maximum solution from the likelihood function, but can only derive nonlinear equation system. This nonlinear equation system is conducted using numerical approach, i.e. Newton Raphson method. The Newton Raphson method is employed to conduct the nonlinear equation system until the maximum parameter value is derived. In the OLR model, for instance take $n$ amount of sample on random variable vector which is $Y_1, Y_2, \dots, Y_n$, with $Y_a = [y_{a1} y_{a2} \cdots y_{a,G-1}]^T$ and probability of result from the $g$-category is $p_g(x_i)$, thus the likelihood function is

$$L = \prod_{a=1}^{n}\prod_{g=1}^{G}\left(\frac{exp(\alpha_g + \boldsymbol{x_i}^T\boldsymbol{\beta}_i)}{1 + exp(\alpha_g + \boldsymbol{x_i}^T\boldsymbol{\beta}_i)} - \frac{exp(\alpha_{g-1} + \boldsymbol{x_i}^T\boldsymbol{\beta}_i)}{1 + exp(\alpha_{g-1} + \boldsymbol{x_i}^T\boldsymbol{\beta}_i)}\right)^{y_{ag}}. \quad (3)$$

The MLE principle is employed to estimate the parameter vector $V = [\alpha_1\alpha_2 \dots \alpha_{G-1}\beta_1\beta_2 \cdots \beta_p]^T$ by maximize the likelihood function. To simplify the calculation, the likelihood function is transformed into *ln-likelihood* which is formulated as

$$\ell = \sum_{a=1}^{n}\sum_{g=1}^{G} y_{ag}\, ln\left(\frac{exp(\alpha_g + \boldsymbol{x_i}^T\boldsymbol{\beta}_i)}{1 + exp(\alpha_g + \boldsymbol{x_i}^T\boldsymbol{\beta}_i)} - \frac{exp(\alpha_{g-1} + \boldsymbol{x_i}^T\boldsymbol{\beta}_i)}{1 + exp(\alpha_{g-1} + \boldsymbol{x_i}^T\boldsymbol{\beta}_i)}\right). \quad (4)$$

## 3. Geographically weighted ordinal logistic regression model

The function which is used for GWOLR model is the cumulative logit function. In GWOLR model, each regression parameter depends on geographical location of the data. GWOLR model with response variable in $G$ amount of ordinal-scaled categories is formulated as [7]

$$Logit[P(Y_a \leq g|x_a)] = \boldsymbol{\alpha}_{\boldsymbol{g}}(u_a, v_a) + \boldsymbol{x}_i{}^T\boldsymbol{\beta}_i(u_a, v_a) \quad (5)$$

with $P(Y_a \leq g|x_a)$ refers to the cumulative probability of $g$-category toward $x_a$; $g = 1,2, \ldots, G$ is the amount of response variable category; $\alpha_g(u_a, v_a)$ is the intercept parameter which fulfills the condition of $\alpha_1(u_a, v_a) \leq \alpha_2(u_a, v_a) \leq \cdots \leq \alpha_{G-1}(u_a, v_a)$; $(u_a, v_a)$ is the coordinate point (latitude, longitude) for the location of $a$, and $\beta_i(u_a, v_a) = [\beta_1(u_a, v_a)\beta_2(u_a, v_a) \ldots \beta_p(u_a, v_a)$ is the regression parameter vector for the $a$-location. The probability of each $G$ response category is

$$P_g(x_a) = \frac{exp(\boldsymbol{\alpha}_{\boldsymbol{g}}(u_a,v_a)+\boldsymbol{x}_i{}^T\boldsymbol{\beta}_i(u_a,v_a))}{1+exp(\boldsymbol{\alpha}_{\boldsymbol{g}}(u_a,v_a)+\boldsymbol{x}_i{}^T\boldsymbol{\beta}_i(u_a,v_a))} - \frac{exp(\boldsymbol{\alpha}_{\boldsymbol{g-1}}(u_a,v_a)+\boldsymbol{x}_i{}^T\boldsymbol{\beta}_i(u_a,v_a))}{1+exp(\boldsymbol{\alpha}_{\boldsymbol{g-1}}(u_a,v_a)+\boldsymbol{x}_i{}^T\boldsymbol{\beta}_i(u_a,v_a))}. \quad (6)$$

The parameter estimation of GWOLR model employs the weighted maximum likelihood method. The geographical factor is the weighting factor in GWOLR model. This factor has different value for each location. The weight is given in the form of weighted ln-likelihood for local GWOLR model. For instance if the weight for each location $(u_a, v_a)$ is $w_{ab}(u_a, v_a)$ then the weighted ln-likelihood function is

$$l = \sum_{a=1}^{n}[y_{a1}ln(p_1(x_a)) + y_{a2}ln(p_2(x_a))+\ldots+y_{aG}ln(p_G(x_a))]w_{ab}(u_a, v_a). \quad (7)$$

Further, it uses kernel function as the weight. Based on [6], the weight based on kernel function which is used is adaptive Gaussian and stated as

$$w_{ab}(u_a, v_a) = exp\left(-\frac{1}{2}\left(\frac{d_{ab}}{b}\right)^2\right) \quad (8)$$

with $d_{ab}$ refers to the distance between the location of $(u_a, v_a)$ and the location of $(u_b, v_b)$, $d_{ab} = \sqrt{(u_a - u_b)^2 + (v_a - v_b)^2}$, $b$ refers to the already known nonnegative parameter and called as bandwidth.

The ln-likelihood function and the weighted ln-likelihood will reach maximum point if the first partial derivation toward the estimated parameter equals to zero and the derivation matrix of the both weighted ln-likelihood functions have the characteristic of negative definite. Hence, the derivation of the first and second weighted ln-likelihood functions toward each estimated parameter can be determined.

## 4. R software

R software is a language programming that one of the object-oriented programming with syntax like S language. R can also computing on the language [2]. Instruction that was done by R software is based on code described.

Within computing on the language, the user can be create the goal program that appropriate for statistical modelling or graphic. Parameters estimation in R is used one of the loop instruction, that is instruction was done by R with any condition, if the condition has been correct then R would be stopped the process. Three instruction in R software is: for, while, and repeat. The understanding about statistical method is needed before did implementation in R software [6].

## 5. Data

The data used in this study are secondary data obtained from Badan Pusat Statistik (BPS) and public health office of Semarang city. The observation unit is villages in Semarang city, Central Java, consisting of 144 villages.

In this study, dependent variable (Y) is incidence rate (IR) of dengue fever of village. The incidence rate have three categories: IR $< 20/100.000$ (low), IR $20 - 50/100.000$ (medium), and IR $> 50/100.000$ (high) [4]. Independent variable used are the population density ($X_1$), the ratio of 0-14 years old group ($X_2$), the ratio of semi-permanent houses ($X_3$), the ratio of healthy facilities ($X_4$), and the larva free index ($X_5$).

## 6. Main Results

Based on the variables above, parameters were needed is two intercept parameters and five regression parameters. Parameters estimation has done by R software, create the program consisted loop instruction. For example, parameters estimation of GWOLR model for 10 locations shown in Table 1.

**Table 1.** The Parameters Estimation Result of 10 Villages

| Villages | $\widehat{\alpha_1}$ | $\widehat{\alpha_2}$ | $\widehat{\beta_1}$ | $\widehat{\beta_2}$ | $\widehat{\beta_3}$ | $\widehat{\beta_4}$ | $\widehat{\beta_5}$ |
|---|---|---|---|---|---|---|---|
| Pekunden | 53.3676 | 78.1957 | $-0.0005$ | 0.0084 | 0.0030 | 0.4175 | $-0.9071$ |
| Karang Kidul | $-0.8665$ | 8.2843 | $-0.0001$ | 0.0016 | 0.0011 | 0.1998 | $-0.1132$ |
| Jagalan | 15.5292 | 19.6942 | $-4.8703 \times 10^{-5}$ | $-0.0004$ | $5.0770 \times 10^{-5}$ | 0.0687 | $-0.2192$ |
| Brumbungan | 101.7097 | 123.7707 | $-0.0005$ | 0.0066 | 0.0206 | $-0.1492$ | $-1.4097$ |
| Miroto | 119.6938 | 129.1980 | $-0.0009$ | 0.0057 | 0.0278 | $-0.4242$ | $-1.3881$ |
| Gabahan | 84.1002 | 94.1645 | $-0.0002$ | $-0.0017$ | 0.0102 | $-0.1063$ | $-1.0284$ |
| Kranggan | 61.6866 | 68.4917 | $-7.0066 \times 10^{-5}$ | $-0.0025$ | 0.0033 | 0.0318 | $-0.7502$ |
| Purwodinatan | 52.3171 | 58.8539 | $-9.8767 \times 10^{-5}$ | $-0.0015$ | 0.0019 | 0.0168 | $-0.6436$ |
| Kauman | 59.7941 | 65.9429 | $-8.6251 \times 10^{-5}$ | $-0.0019$ | 0.0021 | 0.0381 | $-0.7235$ |
| Bangunharjo | 25.4967 | 29.0148 | $-7.2769 \times 10^{-5}$ | $-0.0008$ | 0.0004 | 0.0599 | $-0.3156$ |

The parameters estimator could have used to construct the GWOLR model and calculate probability of IR in each villages. From the models could have calculated probability of dependent variable of each villages. Other villages is also estimated the parameters using R software. For the example, GWOLR model of Pekunden village is :

$$P_1(x_{Pekunden}) = \frac{\exp(53.3676 - 0.0005\mathbf{x}_1 + 0.0084\mathbf{x}_2 + 0.0030\mathbf{x}_3 + 0.4175\mathbf{x}_4 - 0.9071\mathbf{x}_5)}{1 + \exp(53.3676 - 0.0005\mathbf{x}_1 + 0.0084\mathbf{x}_2 + 0.0030\mathbf{x}_3 + 0.4175\mathbf{x}_4 - 0.9071\mathbf{x}_5)}$$

$$P_2(x_{Pekunden}) = \frac{\exp(78.1957 - 0.0005x_1 + 0.0084x_2 + 0.0030x_3 + 0.4175x_4 + 0.9071x_5)}{1 + \exp(78.1957 - 0.0005x_1 + 0.0084x_2 + 0.0030x_3 + 0.4175x_4 + 0.9071x_5)} - \frac{\exp(53.3676 - 0.0005x_1 + 0.0084x_2 + 0.0030x_3 + 0.4175x_4 + 0.9071x_5)}{1 + \exp(53.3676 - 0.0005x_1 + 0.0084x_2 + 0.0030x_3 + 0.4175x_4 + 0.9071x_5)}$$

$$P_3(x_{Pekunden}) = 1 - \frac{\exp(78.1957 - 0.0005x_1 + 0.0084x_2 + 0.0030x_3 + 0.4175x_4 + 0.9071x_5)}{1 + \exp(78.1957 - 0.0005x_1 + 0.0084x_2 + 0.0030x_3 + 0.4175x_4 + 0.9071x_5)}.$$

$P_1$ is the probability of low IR, $P_2$ is the probability of medium IR, and $P_3$ is the probability of high IR. The probability of response category are affected by independent variable value. $\beta_1$ is negative so that if the population density increases then low category and medium category have decreased but high category have increased.

## 7. Conclusion

In this paper we have discussed parameters estimation of GWOLR model using R software. Estimation in R software could be easier than manually estimate, because R have package/library for analizing statistics data. User would need to create the syntax program if the package/library not available in R software. Run the program can get the parameters estimator quickly. From the parameters estimator could have construct the model and calculate the probability each categories of dependent variable. GWOLR model is the local model, so one village have one GWOLR model, not the global model. The model parameters obtained by modified ordinal logistic regression model with add coordinate locations of research. From the coordinate locations, get the local model for each village.

## References

[1] Atkinson P M, S E German, D A Sear and M J Clark 2003 *Exploring The Relations Between Riverbank Erison and Geomorphological Control Using Geographically Weighted Logistic Regression* (Ohio: Ohio State University vol 35) pp 58-82

[2] Becker R A, Chambers J M and Wilks A R 1988 *The New S Language: A Programming Environment for Data Analysis and Graphics* (Chapman & Hall)

[3] Hosmer D W and S Lemeshow 2000 *Applied Logistic Regression* (USA: John Willey and Sons Inc.)

[4] Kemenkes RI 2010 *Buletin Jendela Epidemiologi: Demam Berdarah Dengue, Pusat Data dan Surveilans Epidemiologi* (Jakarta)

[5] Kleinbaum D G and Klein M 2010 *Logistic Regression – A Self-Learning Text* 3rd Ed. (New York: Springer)

[6] Micheaux L P, Remy D and Benoit L 2013 *The R Software: Fundamentals of Programming and Statistical Analysis* (New York: Springer)

[7] Purhadi, Rifada M and Wulandari P 2012 Geographically Weighted Ordinal Logistic Regression Model *International Journal of Mathematics and Computation* 16